Numerical Methods

Numerical methods play an important role in solving complex engineering and science problems. This textbook provides essential information on a wide range of numerical techniques, and it is suitable for undergraduate and postgraduate/research students from various engineering and science streams. It covers numerical methods and their analysis to solve nonlinear equations, linear and nonlinear systems of equations, eigenvalue problems, interpolation and curve-fitting problems, splines, numerical differentiation and integration, ordinary and partial differential equations with initial and boundary conditions. C-programs for various numerical methods are presented to enrich problem-solving capabilities. The concepts of error and divergence of numerical methods are described by using unique examples. The introductions to all chapters carry graphical representations of the problems so that readers can visualize and interpret the numerical approximations.

C-Programs are available at www.cambridge.org/9781108716000

Rajesh Kumar Gupta is an associate professor of mathematics at Central University of Haryana and Central University of Punjab (on lien), India. He has more than 13 years of teaching and research experience. He has published 65 research papers in reputed international journals on the applications of Lie symmetry analysis to nonlinear partial differential equations governing important physical phenomena and related fields.

Numerical Methods

Fundamentals and Applications

Rajesh Kumar Gupta

CAMBRIDGE
UNIVERSITY PRESS

Shaftesbury Road, Cambridge CB2 8EA, United Kingdom

One Liberty Plaza, 20th Floor, New York, NY 10006, USA

477 Williamstown Road, Port Melbourne, VIC 3207, Australia

314–321, 3rd Floor, Plot 3, Splendor Forum, Jasola District Centre, New Delhi – 110025, India

103 Penang Road, #05–06/07, Visioncrest Commercial, Singapore 238467

Cambridge University Press is part of Cambridge University Press & Assessment, a department of the University of Cambridge.

We share the University's mission to contribute to society through the pursuit of education, learning and research at the highest international levels of excellence.

www.cambridge.org
Information on this title: www.cambridge.org/9781108716000

First published 2019

A catalogue record for this publication is available from the British Library

Library of Congress Cataloging-in-Publication data
Names: Gupta, Rajesh Kumar, 1979 author.
Title: Numerical methods: fundamentals and applications / Rajesh Kumar Gupta.
Description: Cambridge; New York, NY: Cambridge University Press, 2019. |
Includes bibliographical references and index.
Identifiers: LCCN 2019013359 | ISBN 9781108716000 (alk. paper)
Subjects: LCSH: Numerical analysis—Problems, exercises, etc. | Mathematical notation.
Classification: LCC QA297 .G8725 2019 | DDC 518—dc23 LC record available at
https://lccn.loc.gov/2019013359

ISBN   978-1-108-71600-0   Paperback

Additional resources for this publication at www.cambridge.org/9781108716000

To My Parents

Sh. Murari Lal and Smt. Santosh Devi

To My Teacher

Professor Karanjeet Singh

To My Wife and Children

Dr Usha Rani Gupta and Aaradhya and Reyansh

Contents

Preface

There is no branch of mathematics, however abstract, which may not some day be applied to phenomena of the real world.

Nikolai Ivanovich Lobachevsky
(December 1, 1792–February 24, 1856)
His work is mainly on hyperbolic geometry, also known as Lobachevskian geometry.

The rapid growth of science and technology during the last few decades has made a tremendous change to the nature of various mathematical problems. It is not easy to solve these new problems for analytical solutions by conventional methods. In fact, the study of these mathematical problems for analytical solutions is not only regarded as a difficult endeavor, rather it is almost impossible to get analytical solutions in many cases. The tools for analysis and for obtaining the analytical solutions of complex and nonlinear mathematical systems are limited to very few special categories. Due to this reason, when confronted with such complex problems we usually simplify them by invoking certain restrictions on the problem and then solve it. But these solutions, however, fail to render much needed information about the system. These shortcomings of analytical solutions lead us to seek alternates, and various numerical techniques developed for different types of mathematical problems seem to be excellent options. During the last century, the numerical techniques have witnessed a veritable explosion in research, both in their application to complex mathematical systems and in the very development of these techniques. At many places in this book, we will compare numerical techniques with analytical techniques, and point out various problems which can not be solved through analytical techniques, and to which numerical techniques provide quite good approximate solutions.

Many researchers are using numerical techniques to investigate research problems. Numerical techniques are now widely used in a lot of engineering and science fields. Almost all universities now offer courses on introductory and advanced computer-oriented numerical methods to their engineering and science students, keeping in mind the utilization merits of these techniques. In addition, computer-oriented problems are part of various other courses of engineering/technology.

It gives me immense pleasure in presenting the book to our esteemed readers. This book is written keeping several goals in mind. It provides essential information on various numerical techniques to the students from various engineering and science streams. The aim of the book is to make the subject easy to understand, and to provide in-depth knowledge about various numerical tools in a simple and concise manner.

Students learn best when the course is problem-solution oriented, especially when studying mathematics and computing. This book contains many examples for almost all numerical techniques designed from a problem-solving perspective. In fact, theoretical and practical introductions to numerical techniques and worked examples make this book student-friendly.

While the main emphasis is on problem-solving, sufficient theory and examples are also included in this book to help students understand the basic concepts. The book includes theories related to errors and convergence, limitations of various methods, comparison of various methods for solving a specific type of problem and scope for further improvements, etc.

The practical knowledge of any subject is thought to be an essential part of the curriculum for an engineering student. Numerical methods require tedious and repetitive arithmetic operations, wherein for large-scale problems it is almost impossible to do such cumbersome arithmetic operations manually. Fortunately most numerical techniques are algorithmic in nature, so it is easy to implement them with the aid of a computer. To enrich problem-solving capabilities, we have presented the basic C-programs for a wide range of methods to solve algebraic and transcendental equations, linear and nonlinear systems of equations, eigenvalue problems, interpolation problems, curve fitting and splines, numerical integration, initial and boundary value problems, etc.

The section below provides an overview of the contents of the book. Each chapter contains a brief introduction and it also emphasis the need for numerical techniques for solving specific problems. We have provided exercises in all chapters with the aim of helping students check their capabilities and understanding, and also illustrate how various numerical methods are the better problem solvers.

Chapter-by-chapter Introduction to the Book

The book comprises sixteen chapters.

Chapter 1: Number Systems explains integral and fractional numbers in the binary, octal, decimal and hexadecimal number systems. It also includes the conversion from one number system to another number system.

Chapter 2: Error Analysis primarily presents various types of errors, and some standard remedies to trace and reduce these errors.

Except Chapters 1 and 2, all other chapters of this book have been devoted to numerical techniques which are used to solve some specific type of problems. In each chapter, various numerical methods will be introduced to solve these problems.

Chapter 3: Nonlinear Equations consists of various techniques to solve nonlinear equations in single variable. Primary aim is to determine the value of variable or parameter x, called root of the equation that satisfies the equation

$$f(x) = 0$$

Roots of simple equations like quadratic equation $x^2 - 3x + 2 = 0$ can be obtained easily. But in the case of higher order polynomial equations like $3x^5 + x^4 + 3x^3 - 2x^2 - 3x + 9 = 0$ and transcendental equations viz. $2e^x \cos x - x = 0$, we do not have any general method to compute the roots of these equations. Numerical techniques will be helpful for computing roots of such equations.

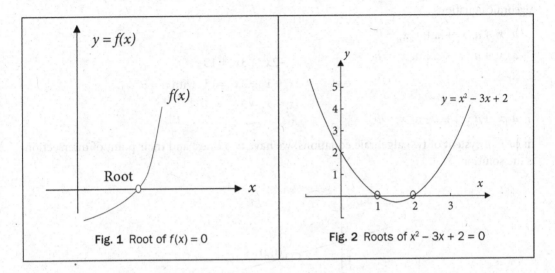

Fig. 1 Root of $f(x) = 0$ **Fig. 2** Roots of $x^2 - 3x + 2 = 0$

These problems are especially valuable in engineering design contexts where due to the complexity of the design equations it is often impossible to solve these equations with analytical methods.

Chapter 4: Nonlinear Systems and Polynomial Equations deals with the numerical techniques to solve the systems of nonlinear equations, say, the system of two equations $\begin{aligned} f(x, y) &= 0 \\ g(x, y) &= 0 \end{aligned}$.

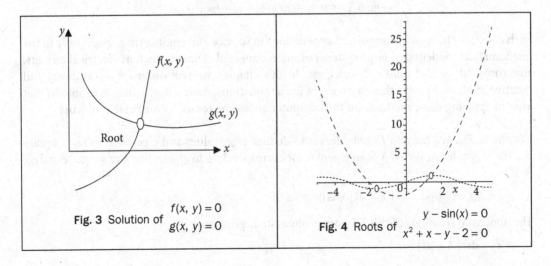

Fig. 3 Solution of $\begin{aligned} f(x, y) &= 0 \\ g(x, y) &= 0 \end{aligned}$ **Fig. 4** Roots of $\begin{aligned} y - \sin(x) &= 0 \\ x^2 + x - y - 2 &= 0 \end{aligned}$

The aim is to find coordinate (x, y), which satisfies these two equations simultaneously. Since there is no general analytical method for the solution of such systems of nonlinear equations, therefore we will apply numerical methods to solve such kind of problems. This chapter also includes some numerical methods for the roots of polynomial equations.

Chapter 5: Systems of Linear Equations is devoted to obtain solution of the system of linear algebraic equations

$$a_{11}x_1 + a_{12}x_2 + \ldots + a_{1n}x_n = b_1$$
$$a_{21}x_1 + a_{22}x_2 + \ldots + a_{2n}x_n = b_2$$
$$\vdots$$
$$\vdots$$
$$a_{n1}x_1 + a_{n2}x_2 + \ldots + a_{nn}x_n = b_n$$

e.g.,

$$x_1 - 2x_2 + 3x_3 = 15$$
$$2x_1 - x_2 + 3x_3 = 15 \text{ with } n = 3.$$
$$x_1 + x_2 - 3x_3 = -9$$

In case of system of two algebraic equations, we have two lines, and their point of intersection is the solution.

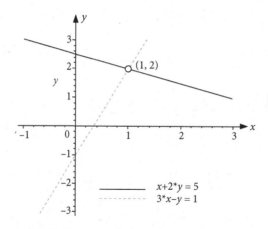

Fig. 5 Linear system in two variables (x, y)

Such equations have many important applications in science and engineering, specifically in the mathematical modeling of large systems of interconnected elements such as electrical circuits, structures, lattice and fluid networks, etc. In this chapter, we will discuss various direct and iterative methods to solve these systems of linear equations. Also, we will discuss problems that arise in applying these methods on the computer and some remedies for these problems.

Chapter 6: Eigenvalues and Eigenvectors is to deduce eigenvalues and eigenvectors for a square matrix A. A column vector X is an eigenvector corresponding to eigenvalue λ of a square matrix A, if

$$AX = \lambda X. \quad \text{(or)} \quad (A - \lambda I)X = 0$$

The nontrivial solutions of this homogeneous system exist, only if

$$p(\lambda) = \det(A - \lambda I) = 0$$

$p(\lambda)$ is the polynomial of degree n for a square matrix of order n. There are only n eigenvalues of matrix A, including repetitions (eigenvalues may be complex). The polynomial $p(\lambda)$ is known as characteristic polynomial, and the equation $p(\lambda) = 0$ is called characteristic equation.

For example, the characteristic equation for the matrix $A = \begin{bmatrix} 1 & 2 \\ 3 & 2 \end{bmatrix}$ is given by

$$p(\lambda) = |A - \lambda I| = \begin{vmatrix} 1-\lambda & 2 \\ 3 & 2-\lambda \end{vmatrix} = (\lambda-4)(\lambda+1) = 0$$

The roots of the characteristic equation give eigenvalues – 1 and 4.

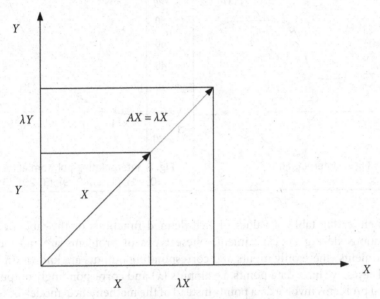

Fig. 6 Eigenvalue λ and eigenvector X of matrix A

These types of problems arise in different streams of science and engineering especially in the case of oscillatory systems like elasticity, vibrations, etc.

Chapter 7: Eigenvalues and Eigenvectors of Real Symmetric Matrices deals with the eigenvalues and eigenvectors of real symmetric matrices. Some methods are applicable only to real symmetric matrices. Since these methods are easy to implement and provide all the eigenvalues and eigenvectors at a time, hence need more exploration.

Chapter 8: Interpolation is most important part of numerical methods, as it deals with the approximation of the data sets with the polynomials. This chapter deals with the task of constructing a polynomial function $P(x)$ of minimum degree, which passes through a given set of discrete data points (x_i, y_i), $i = 0, 1, ..., n$. This polynomial is known as interpolating polynomial. It estimates the value of the dependent variable y for any intermediate value of the independent variable, x.

For example: consider the data set $(0, -1)$, $(1, 1)$, $(2, 9)$, $(3, 29)$, $(5, 129)$. The aim is to construct a polynomial of minimum degree which passes through all these points. We will discuss methods to construct such polynomial. The polynomial $P(x) = x^3 + x - 1$ is the required polynomial and it passes through all these points.

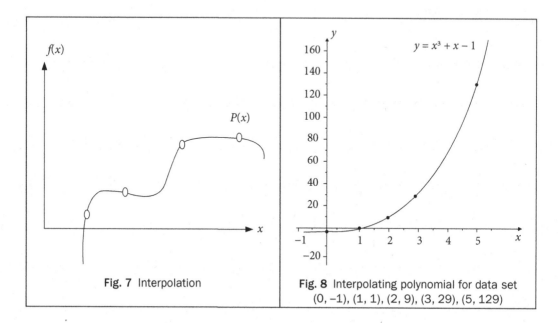

Fig. 7 Interpolation

Fig. 8 Interpolating polynomial for data set $(0, -1)$, $(1, 1)$, $(2, 9)$, $(3, 29)$, $(5, 129)$

A data set is either the table of values of well-defined functions or the table of data points from observations during an experiment. These types of problems are most common in various experiments where only inputs and corresponding outputs are known. In most of the experimental cases, we have data points, i.e., inputs (x) and correspondingly outputs (y). Also, many practical problems involve data points instead of the mathematical model for the problem. For example, Indian government carries out national census after a gap of 10 years to speculate about the development in population of country. Hence, we have populations in these years as follows:

Years	Population (in crores)
1961	43.9235
1971	54.8160
1981	68.3329
1991	84.6421
2001	102.8737
2011	121.0193

This population data is exact up to four decimal digits. But, in intermediate years such as 1977, 2010, etc., we do not have exact population. The numerical techniques can be used to compute approximate populations in these years.

Except for data points, sometimes, we also require approximating different functions with polynomials due to the simple structure of the polynomials. The polynomials are also easy for analysis like differentiation and integration etc.

This chapter is devoted to various techniques for the polynomial approximations of functions and data points. The chapter also includes the piecewise interpolation.

Chapter 9: Finite Operators introduces various finite operators including finite difference operators (forward, backward and central difference operators) and other operators like average or mean operator, shift operator, and differential operator. The chapter contains the relations between these operators. This chapter also presents construction of finite difference tables and the error propagation in these tables.

These finite difference operators are helpful in constructing solutions of difference equations and also used to construct interpolating polynomials for equally spaced points, as discussed in Chapter 10.

Chapter 10: Interpolation for Equal Intervals and Bivariate Interpolation contains some interpolation methods for equally spaced points. The methods discussed in Chapter 8 are applicable for both unequally as well as equally spaced points. Rather, the interpolating polynomial obtained from any formula is unique, but for equally spaced points, the calculations for interpolation become simpler and hence need more exploration.

We will also discuss the extension of interpolation from one independent variable to two independent variables known as bivariate interpolation.

Chapter 11: Splines, Curve Fitting, and Other Approximating Curves discusses approximations of data set other than interpolation. In interpolation, we fit a polynomial of the degree $\leq n$ to $(n+1)$ data points. But if the data set is large, say 50 data points, then it is impractical to fit a polynomial of degree 49 to the data set. In this case, other approximation techniques like least squares curve fitting, spline fitting, etc., can be used. In this chapter, we will discuss different approximation techniques which have certain advantages over interpolation in some real time problems.

Curve fitting is to construct an approximate function $f(x)$ (like exponential, polynomial, logistic curve, etc.) for a table of data points.

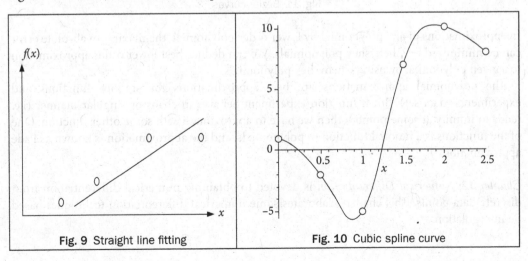

Fig. 9 Straight line fitting   **Fig. 10** Cubic spline curve

Interpolating polynomials have global effect, i.e., if we change a point in the data set, then complete polynomial will change. Also if we change the order of data points, the interpolating polynomial remain same, which is not recommended for certain applications like computer graphics and designing, etc. In these cases, we can apply Bězier and B-Spline curves.

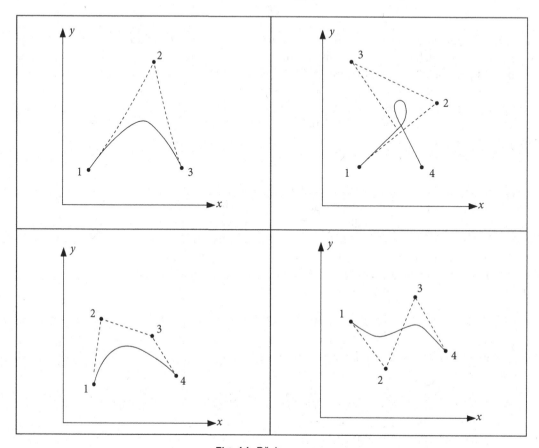

Fig. 11 Bězier curves

In approximations of any polynomial by lower order polynomial, the maximum absolute error can be minimized by Chebyshev polynomials. We can deduce best lower order approximation to a given polynomial by using Chebyshev polynomials.

The polynomial approximations are best approximations for smooth functions and experiments (data set). But if function/experiment behaves in chaos or singular manner (i.e. tends to infinity at some points), then we have to approximate with some other function. One of the functions is a rational function of polynomials, and the approximation is known as Padé approximation.

Chapter 12: Numerical Differentiation is devoted to obtaining numerical differentiation from discrete data points. This chapter elaborates some numerical differentiation techniques based on interpolation.

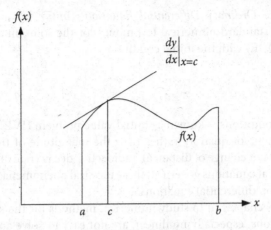

Fig. 12 Differentiation

Chapter 13: Numerical Integration deals with approximating the finite integral of the functions, which are complicated enough to integrate analytically. For example, we don't have exact closed form solutions of integrations like $\int_0^\pi \sqrt{1+\cos^2 x}\ dx$, $\int_1^2 \frac{\sin x}{x} dx$, $\int_0^2 e^{-x^2} dx$ etc. In these cases, we can simply apply numerical methods for the approximate solutions. Sometimes we have to find the integration from a set of discrete data points $\{(x_i, y_i),\ i = 0, 1, ..., n\}$. It is not possible to integrate data points analytically, so it is imperative to approximate these integrations by numerical methods. For example, the value of integral $\int_0^5 y(x)dx$ for the given data set $(0, -1)$, $(1, 1)$, $(2, 9)$, $(3, 29)$, $(5, 129)$ can be obtained only through numerical methods.

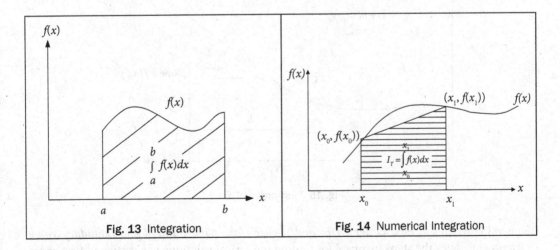

Fig. 13 Integration   **Fig. 14** Numerical Integration

Chapter 14: First Order Ordinary Differential Equations: Initial Value Problems provides a detailed description of standard numerical techniques for the solution of first order ordinary differential equation (ODE) with the initial condition

$$\frac{dy}{dx} = f(x, y), \ y(x_0) = y_0$$

The ODE with initial conditions is known as initial value problem (IVP). Most of the physical laws have a rate of change of quantity rather than the magnitude of the quantity itself; e.g., velocity of any fluid (rate of change of distance), radioactive decay (rate of change of radioactive material), etc. Differential equations govern all these physical phenomena. This chapter contains some basic definitions on differential equations.

The main aim of this chapter in to study numerical methods for the solutions of first order IVP. Differential equations, especially nonlinear, are not easy to solve analytically, as very few analytical methods exist in the literature for a limited class of differential equations. Hence, numerical methods play an important role in the theories of the differential equations.

Consider the following examples

i) $\quad \frac{dy}{dx} = x + y^2, \ y(1) = 2$

ii) $\quad \frac{d^2 y}{dx^2} = x \frac{dy}{dx} + \sin y; \quad y(0) = 1, \ y'(0) = 1,$ etc.

These examples are difficult to solve analytically, but we can use numerical techniques for approximate solutions of such ODEs.

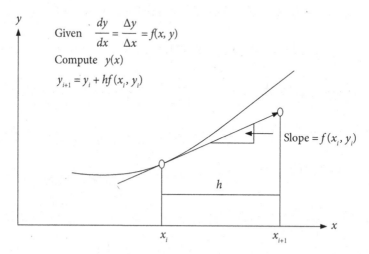

Fig. 15 First order ODE

Chapter 15: Systems of First Order ODEs and Higher Order ODEs: Initial and Boundary Value Problems elucidates the steps involved for finding numerical solutions of a system of first order ODEs and higher order ODEs with initial and boundary conditions, for examples

Systems of First Order ODEs:	
i) $\dfrac{dy}{dx}=x+y-z^2$ $\dfrac{dz}{dx}=z-\sin(xy)$ $y(0)=1,\ z(0)=-1$	ii) $\dfrac{dy}{dx}=w+\sin(x)y-z^2$ $\dfrac{dz}{dx}=z^2-\sin(xy)$ $\dfrac{dw}{dx}=x+w-2y$ $y(1)=1,\ z(1)=-1,\ w(1)=1.3$

Second and Higher Order Initial Value Problems

i) $\dfrac{d^2y}{dx^2}+x\dfrac{dy}{dx}+y=3;\qquad\qquad y(0)=1,\ y'(0)=2$

ii) $\dfrac{d^3y}{dx^3}+\sin x\dfrac{d^2y}{dx^2}+xy=\cos x;\quad y(0)=1,\ y'(0)=2,\ y''(0)=2$

Second and Higher Order Boundary Value Problems

i) $x^2\dfrac{d^2y}{dx^2}+(x-1)\dfrac{dy}{dx}+y=3;\quad y(0)+2y'(0)=1,\ y(1)=3$

ii) $\dfrac{d^3y}{dx^3}+\sin x\dfrac{d^2y}{dx^2}+xy=\cos x;\quad y(0)=1,\ y'(1)=2,\ y(3)+y''(3)=-4$

In last chapter, we have described various numerical methods for the solutions of the first order ODE $\dfrac{dy}{dx}=f(x,y);\ y(x_0)=y_0$. In this chapter, we will generalize these methods to find the numerical solutions of system of first order ODEs.

The chapter deals with the conversion of higher order ODEs to the systems of first order ODEs. This chapter also includes the finite difference approximations of derivatives and further solutions of boundary value problems using these finite differences.

Chapter 16: Partial Differential Equations: Finite Difference Methods presents various finite difference methods for the solutions of some standard linear partial differential equations (PDEs). The finite difference method is a simple and most commonly used method to solve PDEs. In this method, we select some node points in the domain of the PDE. Various derivative terms in the PDE and the derivate boundary conditions are replaced by their finite difference approximations at these node points. The PDE is converted to a set of linear algebraic equations at node points. This system of linear algebraic equations can be solved by any direct/iterative procedure discussed in Chapter 5. The solution of this system of linear equations leads to the solution of PDE at node points. An important advantage of this method is that the procedure is algorithmic, and the calculations can be carried out on the computer. So, the solutions can be obtained in a systematic and easy way.

PDEs are of great significance in describing the systems in which the behavior of any physical quantity depends on two or more independent variables. Laplace and Poisson equations (steady-state flow, fluid mechanics, electromagnetic theory and torsion problems), heat conduction equation (temperature distribution) and wave equation (vibrations, fluid dynamics, etc.) are some important examples of second order linear PDEs. Numerical techniques for the solution of PDEs include finite difference methods (FDMs), finite volume methods (FVMs) and finite element methods (FEMs). This chapter contains only a few finite difference techniques for the solutions of following PDEs governing some important physical phenomena.

Parabolic Equation (Heat Conduction or Diffusion Equation)

$$\frac{\partial u}{\partial t} = c\frac{\partial^2 u}{\partial x^2}$$ (1-Dimensional heat conduction equation)

$$\frac{\partial u}{\partial t} = c\left(\frac{\partial^2 u}{\partial x^2} + \frac{\partial^2 u}{\partial y^2}\right) = c\nabla^2 u$$ (2-Dimensional heat conduction equation)

Elliptic Equation (Laplace and Poisson Equations)

$$\nabla^2 u \equiv \frac{\partial^2 u}{\partial x^2} + \frac{\partial^2 u}{\partial y^2} = 0$$ (Laplace equation in 2-dimensions)

$$\nabla^2 u \equiv \frac{\partial^2 u}{\partial x^2} + \frac{\partial^2 u}{\partial y^2} = f(x, y)$$ (Poisson equation in 2-dimensions)

Hyperbolic Equation (Wave Equation)

$$\frac{\partial^2 u}{\partial t^2} = c^2 \frac{\partial^2 u}{\partial x^2}$$ (1-Dimensional wave equation)

The primary focus is on the preliminary material and the basic concepts of the finite difference techniques used in the book along with their application procedures to derive the numerical solutions of the PDEs.

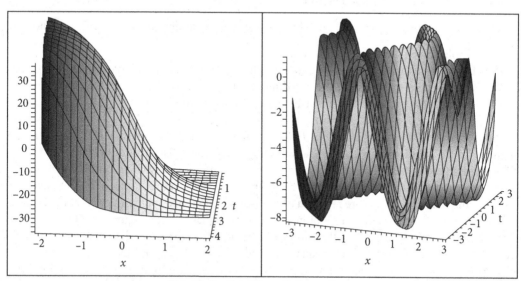

Fig. 16 Partial differential equations

Any Information concerning corrections/errors in this book will be gratefully received.

Rajesh Kumar Gupta
rajeshateli@gmail.com

Acknowledgments

I owe this work to the grace of Almighty, whose divine light provided me strength to complete this book.

It is a great pleasure to thank my mathematics teachers for their expert guidance, support, and encouragement.

I take this opportunity to thank the authorities, my colleagues and students at Thapar University, Patiala, and at the Central University of Punjab, Bathinda, for their support, suggestions and constructive criticism.

I want to thank reviewers and staff at Cambridge University Press who worked to ensure the quality publication of this book.

I am also grateful to my parents, Sh. Murari Lal and Smt. Santosh Devi, brother, Shiv Shanker, sisters, Hemlata and Poonam, brother's wife, Suman, my wife, Usha, and children, Aastha, Akshit, Yashvi, Aadhya, Aaradhya and Reyansh, for providing me a lovely environment in our home.

My friends Dr Harsh and Himani, Dr Amit Kumar, Dr Anoop and Kamal, Dr Amit Bhardwaj, Dr Sunil Singla, Dr Khusneet Jindal, Dr Aklank, Dr Rajendra and Geeta, Dr Phool Singh, Yashpal, Nardeep, Gandhi, Sanjay and Aaditya, PhD scholars and all my well-wishers deserve my heartfelt gratitude for their love and constant support.

Number Systems

> All the mathematical sciences are founded on relations between physical laws and laws of numbers, so that the aim of exact science is to reduce the problems of nature to the determination of quantities by operations with numbers.
>
> In a few years, all great physical constants will have been approximately estimated, and that the only occupation which will be left to men of science will be to carry these measurements to another place of decimals.
>
> **James Clerk Maxwell**
> (June 13, 1831–November 5, 1879)
> He pioneered the classical theory of "Electromagnetism".

1.1 Introduction

In everyday life, we are habituated to doing arithmetic using numbers based on the decimal system. Any number in the decimal number system, say for example, 349.15, can be expressed as a polynomial in the base or radix 10 with integral coefficients 0 to 9.

$$(349.15)_{10} = 3 \times 10^2 + 4 \times 10^1 + 9 \times 10^0 + 1 \times 10^{-1} + 5 \times 10^{-2}$$

In number 349.15, 349 is an integral part and .15 is a fractional part. Note that the subscript (10) in the number $(349.15)_{10}$ denotes the base of the number system.

There is no intrinsic reason to use the decimal number system. Computers read electrical signals, and the state of an electrical impulse is either on or off. Hence, binary system, with base 2 and with integer coefficients 0 and 1, is convenient for computers. However, most computer users prefer to work with the familiar decimal system. It is cumbersome to work with the binary number system, as a large number of binary digits are required to represent even a moderate-sized decimal number. Hence the octal and hexadecimal number systems are also used for this purpose. If the base is two, eight or sixteen, the number is called as the binary, octal or hexadecimal number, respectively. Any number $x = (a_n a_{n-1} \ldots a_1 a_0 . b_1 b_2 \ldots)_\beta$ with base β can be represented as follows

$$x = a_n\beta^n + a_{n-1}\beta^{n-1} + \ldots + a_1\beta + a_0\beta^0 + b_1\beta^{-1} + b_2\beta^{-2} \ldots \tag{1.1}$$

The number system with base β contains numbers from 0 to $\beta-1$. For examples, decimal number system, with base 10, contains numbers from 0 to 9. Similarly, binary system, with base 2, contains numbers 0 and 1.

Table 1.1 Binary, Octal, Decimal and Hexadecimal Numbers

Binary Base: 2 Digits: 0, 1	Octal Base: 8 Digits: 0, 1, 2, 3, 4, 5, 6, 7	Decimal Base: 10 Digits: 0, 1, 2, 3, 4, 5, 6, 7, 8, 9	Hexadecimal Base: 16 Digits: 0, 1, 2, 3, 4, 5, 6, 7, 8, 9, A, B, C, D, E, F
0000	00	00	0
0001	01	01	1
0010	02	02	2
0011	03	03	3
0100	04	04	4
0101	05	05	5
0110	06	06	6
0111	07	07	7
1000	10	08	8
1001	11	09	9
1010	12	10	A
1011	13	11	B
1100	14	12	C
1101	15	13	D
1110	16	14	E
1111	17	15	F

To work with the computer-preferred binary and the people-preferred decimal, and also with the octal and hexadecimal number systems, it is imperative to have algorithms for conversion from one number system to another number system. In the next two sections, some algorithms are discussed to convert the integral and fractional parts of a number from one number system to another number system.

1.2 Representation of Integers

The arithmetic for various number systems with some examples has been discussed in this section. We will use this for conversion of integers in different number systems.

Example ——————————————————————————————————— **1.1**

Explore the addition and multiplication in the decimal, binary, octal and hexadecimal number systems with some examples.

Decimal Arithmetic (For base 10, digits are 0 ... 9)
$(1295)_{10} + (357)_{10} = (1652)_{10}$
$(734)_{10} \times (46)_{10} = (33764)_{10}$

Binary Arithmetic (For base 2, digits are 0 and 1)
$(101011)_2 + (11011)_2 = (1000110)_2$
$(11101)_2 \times (1001)_2 = (100000101)_2$

Octal Arithmetic (For base 8, digits are 0 ... 7)
$(1635)_8 + (274)_8 = (2131)_8$
$(752)_8 \times (23)_8 = (22136)_8$

Hexadecimal Arithmetic (For base 16, digits are 0 ... 9, A, B, C, D, E, F)
$(5AB7)_{16} + (F63)_{16} = (6A1A)_{16}$
$(A4B)_{16} \times (7A)_{16} = (4E7BE)_{16}$

Note: Arithmetic for numbers with base β:
Consider the addition of two numbers $(1635)_8$ and $(274)_8$ in the octal number system with the base $\beta = 8$. Note that, the addition of numbers 5 and 4 will produce number 9. For $\beta = 8$, we have 1 carry, and the remaining number is 1. Similarly, other calculations give the following result

$$\begin{array}{r} 1\ 1\ 1 \quad \text{Carry} \\ (1\ 6\ 3\ 5)_8 \\ +\ (2\ 7\ 4)_8 \\ \hline (2\ 1\ 3\ 1)_8 \end{array}$$
$$\Rightarrow (1635)_8 + (274)_8 = (2131)_8$$

Similarly, consider the multiplication of two numbers. For example, multiplication of numbers 7 and 5 will produce number 35. In octal system (base $\beta = 8$), for number 32, we have 4 carry; and remaining is 3. So, final result is $(7)_8 \times (5)_8 = (43)_8$.

1.2.1 Conversion from Any Number System to the Decimal Number System

Conversion from any number system to the decimal form may be obtained directly from the definition (1.1)

$$x = a_n\beta^n + a_{n-1}\beta^{n-1} + \ldots + a_1\beta + a_0\beta^0 + b_1\beta^{-1} + b_2\beta^{-2}\ldots$$

Some of the examples are as follows

Example ——————————————————————————————— **1.2**

$(1101.101)2 = 1 \times 2^3 + 1 \times 2^2 + 0 \times 2^1 + 1 \times 2^0 + 1 \times 2^{-1} + 0 \times 2^{-2} + 1 \times 2^{-3} = (13.625)_{10}$

$(347.623)8 = 3 \times 8^2 + 4 \times 8^1 + 7 \times 8^0 + 6 \times 8^{-1} + 2 \times 8^{-2} + 3 \times 8^{-3} = (231.787109375)_{10}$

$(A5F.B42)16 = 10 \times 16^2 + 5 \times 16^1 + 15 \times 16^0 + 11 \times 16^{-1} + 4 \times 16^{-2} + 2 \times 16^{-3}$
$\qquad\qquad = (2655.70361328125)_{10}$

1.2.2 Conversion between Binary, Octal and Hexadecimal Number Systems

Conversion in the binary, octal and hexadecimal can be accomplished easily since four/ three binary digits make one hexadecimal/octal digit, respectively. To convert from the binary to the octal/hexadecimal, we have to partition the binary digits in groups of three/ four (starting from right in an integral part and from left in fractional part) and then replace each group by its equivalent octal/hexadecimal digit. To convert from octal and hexadecimal, we have to replace all digits by their binary equivalents.

Example ——————————————————————————————— **1.3**

$(1101.101)_2 = (001\ 101.\ 101) = (\underbrace{001}_{1}\ \underbrace{101}_{5}.\ \underbrace{101}_{5}) = (15.5)_8$

$(1101.101)_2 = (1101.\ 1010) = (\underbrace{1101}_{D}.\ \underbrace{1010}_{A}) = (D.A)_{16}$

$(347.623)_8 = (\underbrace{011}_{3}\ \underbrace{100}_{4}\ \underbrace{111}_{7}.\ \underbrace{110}_{6}\ \underbrace{010}_{2}\ \underbrace{011}_{3}) = (11100111.110010011)_2$

$(A5F.B42)_{16} = (\underbrace{1010}_{A}\ \underbrace{0101}_{5}\ \underbrace{1111}_{F}.\ \underbrace{1011}_{B}\ \underbrace{0100}_{4}\ \underbrace{0010}_{2}) = (101001011111.101101000010)_2$

1.2.3 Conversion from Decimal Number System to Any Other Number System

The conversion of the integer N in decimal number system to another number system can be easily obtained in a systematic manner described as follows. Let there be a number N with base β

$$N = a_n\beta^n + a_{n-1}\beta^{n-1} + \ldots + a_1\beta + a_0$$

Division by the base β will give

$$\frac{N}{\beta} = a_n\beta^{n-1} + a_{n-1}\beta^{n-2} + \ldots + a_1 + \frac{a_0}{\beta}$$

The digit a_0 is the remainder after the base β divides the number N. Let us consider the above equation in the form

$$\frac{N}{\beta} = N_0 + \frac{a_0}{\beta}, \text{ where } N_0 = a_n\beta^{n-1} + a_{n-1}\beta^{n-2} + \ldots + a_1.$$

On dividing N_0 by base β, we get

$$\frac{N_0}{\beta} = a_n\beta^{n-2} + a_{n-1}\beta^{n-3} + \ldots + \frac{a_1}{\beta}$$

The number a_1 is the remainder. We can continue the process till the quotient is 0. Apparently, the conversion from decimal number system to a number system with base β can be achieved by the following algorithm.

$$N = \beta N_0 + a_0$$

$$N_0 = \beta N_1 + a_1$$

$$N_1 = \beta N_2 + a_2$$

$$\vdots$$

till the quotient is 0.

Example   1.4

Convert the decimal number $(231)_{10}$ into its binary equivalent.

Ans.

$231 = 115 \times 2 + 1$	$N_0 = 115$	$a_0 = 1$
$115 = 57 \times 2 + 1$	$N_1 = 57$	$a_1 = 1$
$57 = 28 \times 2 + 1$	$N_2 = 28$	$a_2 = 1$
$28 = 14 \times 2 + 0$	$N_3 = 14$	$a_3 = 0$
$14 = 7 \times 2 + 0$	$N_4 = 7$	$a_4 = 0$
$7 = 3 \times 2 + 1$	$N_5 = 3$	$a_5 = 1$
$3 = 1 \times 2 + 1$	$N_6 = 1$	$a_6 = 1$
$1 = 0 \times 2 + 1$	$N_7 = 0$	$a_7 = 1$

Thus the binary representation of the decimal number $(231)_{10}$ is $(11100111)_2$. It can be computed from the expression $\left(a_n a_{n-1} \ldots a_1 a_0\right)_2$.

Example ────────────────────────────────────── 1.5 ──

Compute the hexadecimal equivalent of the decimal number $(2655)_{10}$.

Ans.

$2655 = 165 \times 16 + 15$	$N_0 = 165$	$a_0 = (15)_{10} = (F)_{16}$
$165 = 10 \times 16 + 5$	$N_1 = 10$	$a_1 = (5)_{10} = (5)_{16}$
$10 = 0 \times 16 + 10$	$N_2 = 0$	$a_2 = (10)_{10} = (A)_{16}$

So, $(A5F)_{16}$ is hexadecimal equivalent of $(2655)_{10}$.

1.2.4 Conversion from One Number System to Any Other Number System

So far, we have discussed the algorithms for conversion of integers in some number systems. The following recursive algorithm can be utilized for conversion of integers in any general number systems.

Algorithm 1.1

Consider a number N with the coefficients $a_n, a_{n-1}, ..., a_0$

$$N = a_n \beta^n + a_{n-1} \beta^{n-1} + ... + a_1 \beta + a_0 \beta^0$$

Calculate the following numbers $b_n, b_{n-1}, ..., b_0$ recursively using

$$b_n = a_n$$

$$b_i = a_i + b_{i+1} \beta, \quad i = n-1, \ n-2, ..., 0$$

Then $b_0 = N$.

Example ────────────────────────────────────── 1.6 ──

Convert the binary number $(110111)_2$ into its decimal equivalent.

Ans.

$$(110111)_2 = 1 \times 2^5 + 1 \times 2^4 + 0 \times 2^3 + 1 \times 2^2 + 1 \times 2^1 + 1 \times 2^0$$

Since the conversion is from binary to decimal, we will use decimal arithmetic for this conversion. Note that each digit in the following calculation is in decimal number system.

$$b_5 = a_5 = 1$$
$$b_4 = a_4 + b_5\beta = 1 + 1 \times 2 = 3$$
$$b_3 = 0 + 3 \times 2 = 6$$
$$b_2 = 1 + 6 \times 2 = 13$$
$$b_1 = 1 + 13 \times 2 = 27$$
$$b_0 = 1 + 27 \times 2 = 55$$

Example —————————————————————————————— 1.7

Compute the binary equivalent of the decimal number $(231)_{10}$ using recursive algorithm 1.1.

Ans.

$$(231)_{10} = 2 \times 10^2 + 3 \times 10^1 + 1 \times 10^0 = (10)_2 \times (1010)_2^2 + (11)_2 \times (1010)_2 + (1)_2$$

This conversion uses binary arithmetic as follows

$$b_2 = a_2 = (10)_2$$
$$b_1 = a_1 + b_2\beta = (11)_2 + (10)_2 \times (1010)_2 = (10111)_2$$
$$b_0 = a_0 + b_1\beta = (1)_2 + (10111)_2 \times (1010)_2 = (11100111)_2$$

Example —————————————————————————————— 1.8

Compute the octal equivalent of the decimal number $(231)_{10}$.

Ans.

$$(231)_{10} = 2 \times 10^2 + 3 \times 10^1 + 1 \times 10^0 = (2)_8 \times (12)_8^2 + (3)_8 \times (12)_8 + (1)_8$$

On using octal arithmetic in the Algorithm 1.1, we have

$$b_2 = a_2 = (2)_8$$
$$b_1 = a_1 + b_2\beta = (3)_8 + (2)_8 \times (12)_8 = (27)_8$$
$$b_0 = a_0 + b_1\beta = (1)_8 + (27)_8 \times (12)_8 = (1)_8 + (346)_8 = (347)_8$$

Example ── 1.9 ──

Convert the decimal number $(2655)_{10}$ into hexadecimal number.

Ans.

$(2655)_{16} = 2 \times 10^3 + 6 \times 10^2 + 5 \times 10^1 + 5 \times 10^0$

$$= (2)_{16} \times (A)_{16}^3 + (6)_{16} \times (A)_{16}^2 + (5)_{16} \times (A)_{16} + (5)$$
$$b_3 = a_3 = (2)_{16}$$

$$b_2 = a_2 + b_3\beta = (6)_{16} + (2)_{16} \times (A)_{16} = (6)_{16} + (14)_{16} = (1A)_{16}$$

$$b_1 = a_1 + b_2\beta = (5)_{16} + (1A)_{16} \times (A)_{16} = (5)_{16} + (104)_{16} = (109)_{16}$$

$$b_0 = a_0 + b_1\beta = (5)_{16} + (109)_{16} \times (A)_{16} = (5)_{16} + (A5A)_{16} = (A5F)_{16}$$

1.3 Representation of Fractions

In a number system with base β, the fractional part can always be written as follows

$$x_F = \sum_{k=1}^{\infty} b_k \beta^{-k}$$

where b_k is a non-negative integer less than the number β. If $b_k = 0$ for all k greater than a positive integer, then the fractional part is said to be terminating otherwise non-terminating. For example $\frac{1}{4} = 0.25$ is terminating, while $\frac{1}{6} = 0.166666...$ is non-terminating. Conversion of the fractional part from one number system to another number system can be achieved with the help of the following algorithm.

───────────────────────── **Algorithm 1.2** ─────────────────────────

On multiplying the fraction $x_F = \sum_{k=1}^{\infty} b_k \beta^{-k} = .b_1 b_2 b_3 \dots$ with base β, we get

$$\beta x_F = \sum_{k=1}^{\infty} b_k \beta^{-k+1} = b_1 + \sum_{k=1}^{\infty} b_{k+1} \beta^{-k}$$

Thus the number b_1 is an integral part of the product βx_F. On repeating the process, we find that b_2 is an integral part of $\beta(\beta x_F)_F$, b_3 is an integral part of $\beta(\beta(\beta x_F)_F)_F$ and so on. One can easily conclude the following algorithm for a general base β from the procedure above.

Let $\quad c_0 = x_F$

$$b_1 = (\beta c_0)_I, \qquad c_1 = (\beta c_0)_F$$
$$b_2 = (\beta c_1)_I, \qquad c_2 = (\beta c_1)_F$$
$$\vdots$$

where subscript I denotes an integral part, while subscript F denotes the fractional part.

Example 1.10

Calculate the binary equivalent of the decimal number $(.3125)_{10}$ using the recursive algorithm 1.2.

Ans.

Let $\quad c_0 = (.3125)_{10}$

$$2(.3125)_{10} = (.6250)_{10} \qquad b_1 = 0 \qquad c_1 = (.6250)_{10}$$
$$2(.6250)_{10} = (1.250)_{10} \qquad b_2 = 1 \qquad c_2 = (.250)_{10}$$
$$2(.250)_{10} = (.50)_{10} \qquad b_3 = 0 \qquad c_3 = (.50)_{10}$$
$$2(.50)_{10} = (1.00)_{10} \qquad b_4 = 1 \qquad c_4 = (0)_{10}$$

The binary equivalent of $(.3125)_{10}$ is $(.b_1 b_2 b_3 b_4)_2 = (.0101)_2$. This example has a terminating binary fraction, but not each terminating decimal fraction will give a terminating binary fraction, and this is true for other number systems also.

Example 1.11

Find the binary equivalent of the decimal number $(0.3)_{10}$.

Ans.

Let $\quad c_0 = (.3)_{10}$

$$2(.3)_{10} = (.6)_{10} \qquad b_1 = 0 \qquad c_1 = (.6)_{10}$$
$$2(.6)_{10} = (1.2)_{10} \qquad b_2 = 1 \qquad c_2 = (.2)_{10}$$
$$2(.2)_{10} = (.4)_{10} \qquad b_3 = 0 \qquad c_3 = (.4)_{10}$$
$$2(.4)_{10} = (.8)_{10} \qquad b_4 = 0 \qquad c_4 = (.8)_{10}$$
$$2(.8)_{10} = (1.6)_{10} \qquad b_5 = 1 \qquad c_5 = (.6)_{10}$$
$$\vdots$$

Since the digits are repeating, we can conclude that the binary equivalent of $(.3)_{10}$ is a non-terminating fraction $(.0\ 1001\ 1001\ 1001\ \ldots)_2$ (or) $(.01001)$

Example ——————————————————————————————————————— **1.12** ——

Find the decimal representation of the binary number $(.0101)_2$.

Ans.

Using the algorithm 1.2 and binary arithmetic, we get

$$c_0 = (.0101)_2$$

$$(1010)_2 (.0101)_2 = (11.0010)_2 \qquad b_1 = (11)_2 = (3)_{10} \qquad c_1 = (.0010)_2$$

$$(1010)_2 (.0010)_2 = (1.010)_2 \qquad b_2 = (1)_2 = (1)_{10} \qquad c_2 = (.010)_2$$

$$(1010)_2 (.010)_2 = (10.10)_2 \qquad b_3 = (10)_2 = (2)_{10} \qquad c_3 = (.10)_2$$

$$(1010)_2 (.10)_2 = (101.0)_2 \qquad b_4 = (101)_2 = (5)_{10} \qquad c_4 = (0)_2$$

Hence $(.3125)_{10}$ is decimal equivalent of the binary fraction $(.0101)_2$.

Example ——————————————————————————————————————— **1.13** ——

Convert the octal fraction $(.71)_8$ to its equivalent decimal representation.

Ans.

Let $\qquad c_0 = (.71)_8$

$$(12)_8 (.71)_8 = (10.72)_8 \qquad b_1 = (10)_8 = (8)_{10} \qquad c_1 = (.72)_8$$

$$(12)_8 (.72)_8 = (11.04)_8 \qquad b_2 = (11)_8 = (9)_{10} \qquad c_2 = (.04)_8$$

$$(12)_8 (.04)_8 = (0.5)_8 \qquad b_3 = (0)_8 = (0)_{10} \qquad c_3 = (.5)_8$$

$$(12)_8 (.5)_8 = (6.2)_8 \qquad b_4 = (6)_8 = (6)_{10} \qquad c_4 = (.2)_8$$

$$(12)_8 (.2)_8 = (2.4)_8 \qquad b_5 = (2)_8 = (2)_{10} \qquad c_5 = (.4)_8$$

$$(12)_8 (.4)_8 = (5.0)_8 \qquad b_6 = (5)_8 = (5)_{10} \qquad c_6 = (0)_8$$

The decimal representation is $(.890625)_{10}$.

Example ——————————————————————————————————————— **1.14** ——

Convert the hexadecimal fraction $(.B4)_{16}$ to its equivalent decimal representation.

Ans.

Let $\qquad c_0 = (.B4)_{16}$

$$(A)_{16} (.B4)_{16} = (7.08)_{16} \qquad b_1 = (7)_{16} = (7)_{10} \qquad c_1 = (.08)_{16}$$

$$(A)_{16} (.08)_{16} = (0.5)_{16} \qquad b_2 = (0)_{10} \qquad c_2 = (.5)_{16}$$

$$(A)_{16}(.5)_{16} = (3.2)_{16} \qquad b_3 = (3)_{10} \qquad c_3 = (.2)_{16}$$
$$(A)_{16}(.2)_{16} = (1.4)_{16} \qquad b_4 = (1)_{10} \qquad c_4 = (.4)_{16}$$
$$(A)_{16}(.4)_{16} = (2.8)_{16} \qquad b_5 = (2)_{10} \qquad c_5 = (.8)_{16}$$
$$(A)_{16}(.8)_{16} = (5.0)_{16} \qquad b_6 = (5)_{10} \qquad c_6 = (0)_{16}$$

The decimal representation is $(.703125)_{10}$.

For conversion from one number system to another number system, one can separately convert the integral and fractional part and then combine them. For example, the decimal equivalent of the number $(.B4)_{16}$ is $(.703125)_{10}$ and decimal equivalent of the number $(A5F)_{16}$ is $(2655)_{10}$. Therefore decimal equivalent of the number $(A5F.B4)_{16}$ is $(2655.703125)_{10}$.

Exercise 1

1. Perform the given arithmetic in the following examples, where the subscript in the number represents the base of the number system:

   a) $(583)_{10} + (3057)_{10}$
   c) $(10110111)_2 + (101011)_2$
   e) $(6047)_8 + (165)_8$
   g) $(3A73)_{16} + (E84)_{16}$

   b) $(312)_{10} \times (281)_{10}$
   d) $(10101)_2 \times (1101)_2$
   f) $(536)_8 \times (37)_8$
   h) $(85D)_{16} \times (23)_{16}$

   Ans. a) $(3640)_{10}$ b) $(87672)_{10}$ c) $(11100010)_2$ d) $(100010001)_2$
   e) $(6234)_8$ f) $(25142)_8$ g) $(48F7)_{16}$ h) $(124B7)_{16}$

2. Convert the following numbers into their decimal equivalents:

   a) $(11011.110)_2$ b) $(67.243)_8$ c) $(2A7.3F)_{16}$

   Ans. a) $(27.75)_{10}$ b) $(55.31835938)_{10}$ c) $(679.2460938)_{10}$

3. Find the binary, octal and hexadecimal forms of the following numbers:

   a) $(101101.110)_2$ b) $(573.42)_8$ c) $(A05.9A)_{16}$

   Ans. a) $[(55.6)_8, (2D.C)_{16}]$ b) $[(101111011.10001)_2, (17B.88)_{16}]$
   c) $[(101000000101.10011010)_2, (5005.464)_8]$

4. Compute the binary, octal and hexadecimal equivalents of the decimal number $(5680)_{10}$.

   Ans. $(1011000110000)_2, (13060)_8, (1630)_{16}$

5. Use the algorithm 1.1 for the following conversions:

   a) $(1101101)_2$ in decimal
   c) $(237)_8$ in decimal
   e) $(2AD3)_{16}$ in decimal
   g) $(438)_{10}$ in binary

   b) $(5691)_{10}$ in octal
   d) $(110111)_2$ in hexadecimal
   f) $(4529)_{10}$ in hexadecimal
   h) $(110111)_2$ in octal

   Ans. a) $(109)_{10}$ b) $(13070)_8$ c) $(159)_{10}$ d) $(37)_{16}$
   e) $(10963)_{10}$ f) $(11B1)_{16}$ g) $(110110110)_2$ h) $(67)_8$

6. Obtain the following conversions for the fractional numbers with the aid of recursive algorithm 1.2

      a) $(.1101101)_2$ in decimal   b) $(.50)_{10}$ in octal

      c) $(.237)_8$ in decimal   d) $(.A3)_{16}$ in decimal

      e) $(.45)_{10}$ in hexadecimal   f) $(.325)_{10}$ in binary

    Ans. a) $(.8515625000)_{10}$ b) $(.40)_8$ c) $(.3105468750)_{10}$

           d) $(.1367187500)_{10}$ e) $(.7\overline{3})_{16}$ f) $(.0101001)_2$

7. Obtain the decimal equivalents of the numbers $(A23.4D)_{16}$, $(126.54)_8$, $(10101.11)_2$.

    Ans. $(2595.300781)_{10}$, $(86.6875)_{10}$, $(21.750000)_{10}$

8. Compute the binary, octal and hexadecimal equivalents of the decimal number $(238.40)_{10}$.

    Ans. $(11101110.01\overline{100})_2$, $(356.3\overline{146})_8$, $(EE.\overline{6})_{16}$

9. Calculate the decimal equivalent of the octal number $(.647)_8$ with the aid of the recursive algorithm.

    Ans. $(.8261718750)_{10}$

Error Analysis

I claim to be a simple individual liable to err like any other fellow mortal. I own, however, that I have humility enough to confess my errors and to retrace my steps.

Mohandas Karamchand Gandhi (Mahatma Gandhi)
(October 2, 1869–January 30, 1948)
He embraced non-violent civil disobedience and led India to independence
from British rule.

Numerical methods use arithmetic operations to solve complex mathematical problems. The numerical processes are algorithmic, so these methods can be applied easily with the advent of high-speed computers. In fact, the development of more efficient computers has played a vital role in a veritable explosion in the usage of numerical techniques for engineering and scientific problems. The common characteristic of numerical techniques is that all these involve cumbersome arithmetic operations. During the implementation of the numerical techniques on a computer, we often come across various types of errors. The precisions (number of digits in the representation of a number) of a numerical solution can be diminished by these several possible errors. This chapter deals with various types of errors, and some standard remedies to trace and reduce these errors.

In Section 2.1, measurement of the error will be discussed. Section 2.2 presents the various sources of errors in mathematical modeling of a real world problem. The study of errors during the implementation of numerical methods for the solution of a mathematical model is the primary objective of Section 2.3. The last section is about some interesting discussion on error.

2.1 Absolute, Relative and Percentage Errors

The difference between the exact value and an approximate value of a quantity is called error in the measurement. Its absolute value is called absolute error. Let x be the exact value and $\tilde{x}$ be an approximate value of a given quantity; then the absolute error is given by

Absolute error = $E_a = |x - \tilde{x}|$

Absolute error is not a complete measurement of the error. For example, let absolute error in any quantity be 0.1 m. This information is not complete until we define the quantity for the 0.1 m error. If the 0.1 m error is in 10000 m, then it is small enough to be ignored. But, we cannot neglect 0.1 m error if it is in 1 m. In fact, the error in any quantity depends on the size of that quantity, so relative error and percentage error are the best measurements of the error.

The relative and percentage errors are defined as follows

$$\text{Relative error} = E_r = \left| \frac{x - \tilde{x}}{x} \right|$$

$$\text{Percentage error} = E_p = 100 E_r = 100 \left| \frac{x - \tilde{x}}{x} \right|$$

Let there exist a number $\varepsilon > 0$, such that $|x - \tilde{x}| \leq \varepsilon$. Then ε is an upper limit of the absolute error and measures the absolute accuracy.

The relative and percentage errors are independent of the units of the quantities used while the absolute error is expressed in terms of these units.

Example ──────────────────────────────── **2.1**

An approximation to the value of π is given by $\dfrac{22}{7}$, while its true value in 8 decimal digits

is 3.1415926. Calculate the absolute, relative and percentage errors in the approximation.

Ans. Exact value = $x = 3.1415926$

$$\text{Approximate value} = \tilde{x} = \frac{22}{7} = 3.1428571$$

$$E_a = |x - \tilde{x}| = |-0.0012645| = 0.0012645$$

$$E_r = \left| \frac{x - \tilde{x}}{x} \right| = \frac{0.0012645}{3.1415926} = 0.000402502$$

$$E_p = 100 E_r = 100 \left| \frac{x - \tilde{x}}{x} \right| = .0402502\%$$

To recognize the major sources of errors and then how to quantify or minimize these errors in the numerical computations are the primary objectives of this chapter.

Analysing any real world problem involves the following three major steps: the first step is to convert the real world problem into a mathematical model; the second step is to solve that model analytically or numerically; and the last step is to analyze the obtained solution for its physical and real-time significance.

After the analysis part is complete, we implement the model for its application.

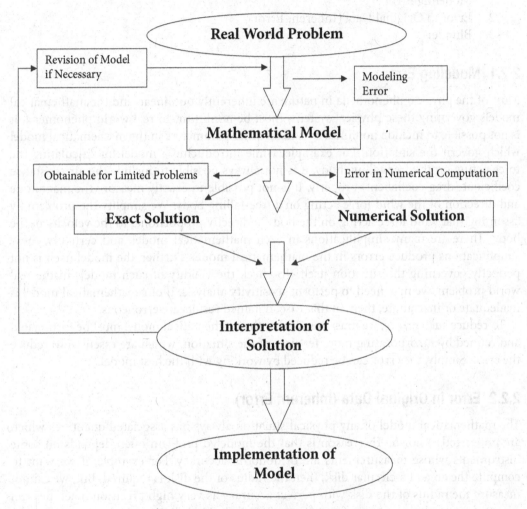

Fig. 2.1 Steps in solving a real world problem

In fact, the error is a multifaceted problem, but it mainly arises during two stages: error occurred during the mathematical modeling of the real world problem, and error when we solve the mathematical model numerically.

In this chapter, we will discuss different types of errors: those encountered during the first step (modeling) and the second step (mathematical model to the solution).

2.2 Errors in Modeling of Real World Problems

The errors in modeling are not directly connected with numerical techniques, but they have a profound impact on the success of a model. Thus, before implementation of a numerical method to the mathematical model, we must have knowledge of the following errors

1. Modeling Error
2. Error in Original Data (Inherent Error)
3. Blunder

2.2.1 Modeling Error

Most of the physical phenomena in nature are inherently nonlinear, and the mathematical models governing these physical systems must be nonlinear. In real world phenomena, it is not possible to include nonlinearity and all other parameters in the mathematical model which govern the situation. For example, while introducing a model for calculating the force acting on the free falling body, it is not always possible to include the air resistance coefficient (drag coefficient) properly. It is not possible to exactly measure the magnitude and direction of the wind force acting on a free-falling body. We simplify the problem by assuming that wind force acting on the body is directly proportional to the velocity of the body. There are many simplifications in each mathematical model, and certainly, these simplifications produce errors in the mathematical model. Further, the model itself is not perfectly governing the situation itself. To check the validity of such models in the real world problem, we may need to perform sensitivity analysis. If our mathematical model is inadequate or inaccurate, then, in that case, obtained results are erroneous.

To reduce such errors, we must ensure that the mathematical model must be formulated and refined by incorporating more features of the situation, which are essential to reduce the error. Simply, the error can be reduced by working with the best model.

2.2.2 Error in Original Data (Inherent Error)

The mathematical model of any physical situation always has associated quantities which are imperfectly known. The reason is that the modeled problem often depends on some instruments whose measurements are of doubtful accuracy. For example, if we want to compute the area of a circular disk, then the radius of the disk is required. But, we cannot measure the radius of the disk with perfect accuracy as very high precision machines can measure up to the accuracy of maximum 5 to 6 decimal digits. Inherent errors can be minimized using high precision computing systems and by taking better data.

2.2.3 Blunder

There is extensive use of the computer in applications of various numerical techniques; chances that the computers make mistakes are very less. But, during the implementation of algorithms on computers, we can make mistakes at various steps, like problem formulations, selection of numerical procedures, programming, and result interpretations, etc. These lead to blunders or gross errors. Some frequent and common types of errors are as follows

i)   Inaccurate or inadequate knowledge of the nature of the problem.
ii)   Avoiding certain important features of the problem during formulation of the problem.
iii) Some wrong assumptions during the formulation of the problem.
iv)   Error in selecting the mathematical equation, which describes a part of the problem.
v)   Errors in input data.
vi)   Selection of an inappropriate numerical process to determine a solution of the mathematical model.
vii) Implementing a wrong algorithm or avoiding certain important features of a mathematical model in the algorithm.
viii) Starting with a wrong initial guess.
ix)   Other simple mistakes like misprints, wrong subscripts in variables, forgetting unit conversion, forgetting negative sign, etc.
x)   Implementing infinite series without having knowledge of the convergence.

These errors can be reduced to a large extent by acquiring a hold over various intricacies of the real world phenomena, mathematical modeling of the phenomena and the numerical methods for the solutions of these mathematical models. We must carefully examine the results to avoid such blunders, and a test run with known results is worthwhile in this regard. Test problems more often reveal the mistake and permit its correction.

2.3 Errors in Implementation of Numerical Methods

In this section, we will discuss those errors, which are due to the way that computers store numbers and do arithmetic. In any numerical computation, we come across following types of errors

i)   Round-off Error
ii)   Overflow and Underflow
iii) Floating Point Arithmetic and Propagated Error
iv)   Truncation Error
v)   Machine eps (Epsilon)
vi)   Epilogue
vii) Loss of Significance: Condition and Stability

There are several potential sources of errors in numerical computation. But, *round-off* and *truncation* errors can occur in any numerical computation.

2.3.1 Round-off Error

During the implementation of a numerical algorithm with computing devices mainly calculator and computer, we have to work with a finite number of digits in representing a number. The number of digits depends on the word length of the computing device and software. The scientific calculations are carried out in floating point arithmetic. It is necessary to have knowledge of floating point representations of numbers and the basic arithmetic operations performed by the computer (+, -, *, /) in these representations.

Floating Point Representation of Numbers

To understand the major sources of error during the implementation of numerical algorithms, it is necessary to discuss how the computer stores the numbers.

An m-digits floating point number in the base β is of the following form

$$x = \pm \left(.d_1 d_2 d_3 \cdots d_m \right)_\beta \beta^n$$

where $\left(.d_1 d_2 d_3 \cdots d_m \right)_\beta$ is called as a mantissa and the integer n is called the exponent. A non-zero number is said to be normalized if $d_1 \neq 0$.

All the real numbers are stored in normalized form in the computer to avoid wastage of computer memory on storing useless non-significant zeroes. For example, 0.002345 can be represented in a wasteful manner as $(0.002345)10^0$ which is wasting two important decimal points. However, the normalized form is $(0.2345)10^{-2}$, which eliminates these useless zeroes; also known as spurious zeroes.

If we want to enter the number 234.1205, then this number stored in the computer in normalized form, i.e., $(0.2341205)10^3$. Similarly, the number 0.00008671213 stored in the computer in normalized form $(0.8671213)10^{-4}$.

The digits used in mantissa to express a number are called as significant digits or significant figures. More precisely, *digits in the normalized form mantissa of a number are significant digits.*

a) All non-zero digits are significant. For examples, the numbers 3.1416, 4.7894 and 34.211 have five significant digits each.

b) All zeroes between non-zero digits are significant. For examples, the numbers 3.0156 and 7.5608 have five significant digits each.

c) Trailing zeroes following a decimal point are significant. So, the numbers 3.5070 and 76.500 have five significant digits each.
(Why the number 5.1 has two significant digits, and number 5.10 has three significant digits? To explain this, let us assume we are reading Chapter 5 of a book, and it contains 12 sections. The number 5.1 represents first section of Chapter 5, while the number 5.10 represents tenth section of Chapter 5.)

d) Zeroes between the decimal point and preceding a non-zero digit are not significant. i.e., the numbers 0.0023401 and 0.00023401 have five significant digits each.

e) Trailing zeroes are significant if the decimal point is not present, i.e., the numbers 45067000 and 45000 have eight and five significant digits, respectively.

To compute the significant digits in a number, simply convert the number in the normalized form and then compute the significant digits.

There is a limit on the mantissa (m) and exponent (n) as the storage capacity of any machine is finite. The precision or length m of the floating point numbers usually depends on the word length of the computer and software, and it may vary widely. For example, in single precision (float variable, 32 bits), the C-programming allows 23 bits for mantissa (m), 8 bits for exponent (n), and 1 bit for sign ($\pm$). Similarly, double variable gives 52 bits for mantissa, 11 bits for exponent, and 1 bit for sign. Note that the calculations in double precision require more storage and more running time as compared to single precision.

To understand the limit on storing capacity of the computer, consider the number 10/3. Since the computer can enter the number only in normalized form, hence the computer first solves $10/3 = 3.333333...$, and then stores the number. There are infinite numbers of 3's in the expression, but computer will store the number up to its capacity. Let the capacity of the computer be ten digits (i.e., mantissa limit $m <= 10$), then the number will store as $(0.3333333333)10^1$. All computing devices represent such numbers with some imprecision. For examples, $5/3 = 1.6666666...$, $\sqrt{2} = 1.414213...$ and $\pi = 3.141592...$ cannot be expressed by a finite number of digits, since the computer cannot store $50/3$, $\sqrt{2}$, etc. These numbers may be approximated by rounding off the last precision to m-digits floating point number. For example, let $m = 6$, then we can approximate $50/3$, $\sqrt{2}$ and π by numbers 16.6667, 1.41421 and 3.14159, respectively.

This process of rounding off the numbers during the computation will give rise to round off errors.

Rounding and Chopping

Rounding and chopping are two commonly used ways of converting a given real number x into its m-digits floating point representation $fl(x)$. In the case of chopping, the number x is retained up to m-digits, and remaining digits are simply chopped off. For example, consider 6-digits floating point representation, then

$$x_1 = \frac{2}{3} \qquad\qquad fl(x_1) = 0.666666$$

$$x_2 = 3456789 \qquad\qquad fl(x_2) = (.345678)10^7$$

$$x_3 = -0.0011223344 \qquad\qquad fl(x_3) = -(.112233)10^{-2}$$

In rounding, the normalized floating point number $fl(x)$ is chosen such that it is nearest to the number x. In the case of a tie, some special rules such as symmetric rounding can be used. Rules to round off a number to m significant figures are as follows

i)   Discard all digits to the right of m-th digit.
ii) If the last discarded number is

 a) less than half of base β in the $(m + 1)$th place, leave the m-th digit unchanged;
 b) greater than half of base β in the $(m + 1)$th place, increase the m-th digit by unity;
 c) exactly half of base β in the $(m + 1)$th place, increase the m-th digit by unity if it is odd, otherwise leave the m-th digit unchanged. It is known as symmetric rounding around even number. Similarly, we can have symmetric rounding about odd number.

Consider the following numbers with 6-digits floating point representation

$$x_1 = \frac{2}{3} \qquad\qquad fl(x_1) = 0.666667$$

$$x_2 = 3456789 \qquad\qquad fl(x_2) = (.345679)10^7$$

$$x_3 = -0.0011223344 \qquad fl(x_3) = -(.112233)10^{-2}$$

The difference between x and $fl(x)$ is called the round-off error. If the number is correct up to p decimal points, then the maximum absolute error in chopping and rounding is given by

$$\text{Absolute error} = E_a = |x - fl(x)| \leq \begin{cases} \dfrac{1}{2}\beta^{-p} & \text{in rounding} \\ \beta^{-p} & \text{in chopping} \end{cases}$$

For example, if the number 12.345 ($\beta = 10$) is correct up to digits mentioned, then the maximum absolute error in this number is .001 in the case of chopping, and it is .0005 in the case of rounding.

The relative error in the floating point representation of x is as follows

$$\text{Relative Error} = |\delta| = \left| \frac{x - fl(x)}{x} \right|$$

Let the number be correct up to m significant digits in normalized form. Then the maximum relative error is given by the following expression

$$|\delta| \leq \begin{cases} \dfrac{1}{2}\beta^{1-m} & \text{in rounding} \\ \beta^{1-m} & \text{in chopping} \end{cases}$$

For example, let us assume that the number 123.45 ($\beta = 10$) is correct up to digits mentioned. It contains five significant digits, so the maximum relative error in this number is .0001 in the case of chopping, and it is .00005 in the case of rounding.

Note: *It is worth mentioning here that generally we use rounding. Until it is not mentioned to use chopping specifically, we will use rounding for computational work throughout the book.*

Example   **2.2**

Consider the irrational number $\pi = 3.14159265358979...$ It has an infinite number of digits. So, computer representation of the number π will produce the round-off error depending on the number of significant digits in arithmetic. In Table 2.1, we are presenting the absolute and percentage errors for 1,2,…,6 significant digits, while considering the exact value of $\pi = 3.141593$.

Table 2.1

Number of digits	Approximation for π (Rounding)	Absolute error	Percentage error
1	3	0.141593	4.507045%
2	3.1	0.041593	1.323946%
3	3.14	0.001593	0.050707%
4	3.142	0.000407	0.012955%
5	3.1416	0.000007	0.000234%
6	3.14159	0.000003	0.000095%

Example ———————————————————————— **2.3**

Compute the absolute and relative errors in the four significant digits approximations of the numbers 124678 and 345.635.

Ans. Four significant digits approximations of the numbers 124678 and 345.635 are as follows

$$fl(x) = \begin{cases} (.1247)10^6 & x = 124678 \\ (.3456)10^3 & x = 345.635 \end{cases} \qquad \text{rounding}$$

$$fl(x) = \begin{cases} (.1246)10^6 & x = 124678 \\ (.3456)10^3 & x = 345.635 \end{cases} \qquad \text{chopping}$$

$$\text{Absolute error} = E_a = |x - fl(x)| = \begin{cases} |124678 - 124700| = 22 \\ |345.635 - 345.6| = .035 \end{cases} \qquad \text{rounding}$$

$$E_a = |x - fl(x)| = \begin{cases} |124678 - 124600| = 78 \\ |345.635 - 345.6| = .035 \end{cases} \qquad \text{chopping}$$

$$\text{Relative error} = E_r = \left| \frac{x - fl(x)}{x} \right| = \begin{cases} (.1764545)10^{-3} \\ (.1012628)10^{-3} \end{cases} \qquad \text{rounding}$$

$$E_r = \left| \frac{x - fl(x)}{x} \right| = \begin{cases} (.6256116)10^{-3} \\ (.1012628)10^{-3} \end{cases} \qquad \text{chopping}$$

Example ——————————————————————————— **2.4**

The true value of π correct to 8-significant digits is 3.1415926. Calculate the absolute and relative error.

Ans. The value is correct up to 7-decimal digits, so the maximum absolute errors in case of rounding and chopping are $\frac{1}{2}10^{-7}$ and 10^{-7}, respectively.

$$\text{Relative error in rounding} = E_r = \left|\frac{X - X'}{X}\right| = \frac{0.5 \times 10^{-7}}{3.1415926} = 0.15915494 \times 10^{-7}$$

$$\text{Relative error in chopping} = E_r = \frac{10^{-7}}{3.1415926} = 0.31830989 \times 10^{-7}$$

It is easy to see that, in between every two numbers, there are infinitely many numbers, which we cannot represent exactly using the computer. Let us consider the machine with 6–digits floating point arithmetic. Consider any two numbers, say 2.12346 and 2.12347. Then, it is easy to see that we cannot represent the in-between numbers like 2.1234652, 2.12346521, 2.1234603112, etc. and these are infinitely many numbers.

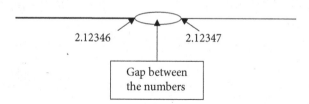

Fig. 2.2 Gaps between floating point numbers

In fact, using computer, we can only represent finite numbers of real numbers and in between every two such numbers, we have infinite numbers, which cannot be represented by the computer.

2.3.2 Overflow and Underflow

The normalized form for an m-digits non-zero floating point number in the base β is given by

$$x = \pm \left(.d_1 d_2 d_3 \cdots d_m\right)_\beta \beta^n, \qquad d_1 \neq 0$$

where $\left(.d_1 d_2 d_3 \cdots d_m\right)_\beta$ is called as mantissa and the integer n is called as exponent.

The exponent n is restricted to a range $l < n < L$, for integers l and L; generally $l = -L$. This limit varies widely and depends on the computational device used. If in the floating point representation of a number x, the exponent n exceeds the limit, i.e., either $|x| \geq \beta^L$ (overflow) or $0 \leq |x| \leq \beta^{l-1}$ (underflow), it results either in a stop or else $fl(x)$ is represented by a special number (either 0 or infinity). These special numbers are not subject to the usual rules of arithmetic when combined with ordinary floating point numbers.

Let a hypothetical computer with maximum ten digits mantissa and exponent range $(-20, 20)$ in the decimal number system, then the overflow and underflow can be structured in the following figure

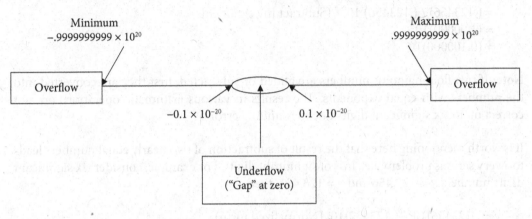

Fig. 2.3 Overflow and underflow

Rather the limit is quite awesome, but it is not able to represent physical quantities like Avogadro's number (6.022×10^{23}) and Plank's constant $(6.626 \times 10^{-34} \, J.s.)$, etc.

2.3.3 Floating Point Arithmetic and Error Propagation

In the last section, we have discussed the errors in number representations. These errors further propagate while performing basic arithmetic operations using a computer. The result of an arithmetic operation is usually not accurate to the same length as the numbers used for the operations. The floating point numbers are first converted into the normalized forms as soon as they enter in the computer.

Here we will explain the arithmetic operations with 6-significant digits numbers. For example, let us take numbers $x = 123.456$ and $y = 12.3456$ with six significant digits. The various arithmetic operations $(+, -, *, /)$ on these two numbers are as follows

$x + y = (.123456)10^3 + (.123456)10^2$ (Normalized form)

$\quad = (.123456)10^3 + (.012346)10^3$ (Equal exponent using symmetric rounding)

$\quad = (.135802)10^3$

$$x - y = (.123456)10^3 - (.123456)10^2$$
$$= (.123456)10^3 - (.012346)10^3 \text{ (Equal exponent using symmetric rounding)}$$
$$= (.111110)10^3$$

$$x * y = (.123456)10^3 * (.123456)10^2$$
$$= (.123456) * (.123456) \ 10^{3+2} \text{ (Add the exponents)}$$
$$= (.015241)10^5$$
$$= (.152410)10^4$$

$$x / y = (.123456)10^3 / (.123456)10^2$$
$$= (.123456) / (.123456) \ 10^{3-2} \text{ (Subtract the exponents)}$$
$$= (1.00000)10^1$$
$$= (0.100000)10^2$$

Note: If two floating point numbers are added or subtracted, first they are converted into the numbers with equal exponents. The results in various arithmetic operations are not correct up to six significant digits due to rounding errors.

It is worth mentioning here that the result of subtraction of two nearly equal numbers leads to a very serious problem, i.e., loss of significant digits. For example, consider six significant digits numbers $x = 123.456$ and $y = 123.432$, then

$$x - y = (.123456)10^3 - (.123432)10^3 \text{ (Normalized form)}$$
$$= (.000024)10^3 \text{ (Result containing only two significant digits, four non-significant}$$
zeroes are appended)

This subtraction of two nearly equal numbers is called as subtractive cancellation or loss of significance. It is a classical example of computer handling mathematics can create a numerical problem. We will discuss it, in detail, in Section 2.3.7.

2.3.3.1 Propagated Error in Arithmetic Operations

Propagated errors are the errors in the succeeding steps of a process due to an earlier error in the input. For example, error in the division of two numbers due to local errors in the numbers. In this section, we will see how errors in numbers may propagate through basic mathematical operations viz. addition, subtraction, multiplication, and division of two numbers.

Consider any two numbers x_1 and x_2. Let the errors in the numbers x_1 and x_2 be δx_1 and δx_2, respectively. Then errors in the addition, subtraction, multiplication, and division of these two numbers are as follows

     i)   **Addition**
Let $X = x_1 + x_2$ and the error in X is δX.

$$X + \delta X = x_1 + \delta x_1 + x_2 + \delta x_2$$

$$\delta X = \delta x_1 + \delta x_2$$

Absolute Error $= |\delta X| \le |\delta x_1| + |\delta x_2|$

Relative Error $= \dfrac{|\delta X|}{|X|} \le \dfrac{|\delta x_1|}{|X|} + \dfrac{|\delta x_2|}{|X|}$ $\qquad$ (2.1)

ii) Subtraction

Similarly, the error in subtraction $X = x_1 - x_2$ is given by

$$\delta X = \delta x_1 - \delta x_2$$

Absolute Error $= |\delta X| \le |\delta x_1| + |\delta x_2|$

Relative Error $= \dfrac{|\delta X|}{|X|} \le \dfrac{|\delta x_1|}{|X|} + \dfrac{|\delta x_2|}{|X|}$ $\qquad$ (2.2)

iii) Multiplication

Let $X = x_1 x_2$, then

$$X + \delta X = (x_1 + \delta x_1)(x_2 + \delta x_2)$$
$$= x_1 x_2 + x_2 \delta x_1 + x_1 \delta x_2 + \delta x_1 \delta x_2$$

Neglecting second order term $(\delta x_1 \delta x_2)$, the error in the multiplication of two numbers is as follows

$$\delta X = x_2 \delta x_1 + x_1 \delta x_2$$

Absolute Error $= |\delta X| \le |x_2 \delta x_1| + |x_1 \delta x_2|$

Relative Error $= \dfrac{|\delta X|}{|X|} \le \dfrac{|\delta x_1|}{|x_1|} + \dfrac{|\delta x_2|}{|x_2|}$ $\qquad$ (2.3)

iv) Division

Let $X = \dfrac{x_1}{x_2}$, then

$$X + \delta X = \frac{x_1 + \delta x_1}{x_2 + \delta x_2} = \left(\frac{x_1 + \delta x_1}{x_2 + \delta x_2} \right) \left(\frac{x_2 - \delta x_2}{x_2 - \delta x_2} \right)$$

$$= \frac{x_1 x_2 + x_2 \delta x_1 - x_1 \delta x_2 - \delta x_1 \delta x_2}{x_2^2 - \delta x_2^2}$$

On neglecting the second order terms ($\delta x_1 \delta x_2$ and δx_2^2), the error is given by

$$\delta X = \frac{x_2 \delta x_1 - x_1 \delta x_2}{x_2^2}$$

Absolute Error $= |\delta X| \le \frac{|\delta x_1|}{|x_2|} + \frac{|x_1 \delta x_2|}{x_2^2}$

Relative error $= \frac{|\delta X|}{|X|} \le \frac{|\delta x_1|}{|x_1|} + \frac{|\delta x_2|}{|x_2|}$ \hfill (2.4)

Example \hfill **2.5**

The numbers $x_1 = 0.123$ and $x_2 = 12.37$ are correct up to the significant digits in the numbers. Compute the relative errors in the addition, subtraction, multiplication and division of these two numbers. Consider symmetric rounding.

Ans. Absolute errors in the numbers $x_1 = 0.123$ and $x_2 = 12.37$ are $\delta x_1 = .0005$ and $\delta x_2 = .005$, respectively.

Using the formulae (2.1 – 2.4) for various error terms, we have

Relative error in the addition $= \left| \frac{\delta X}{X} \right| \le \left| \frac{\delta x_1}{X} \right| + \left| \frac{\delta x_2}{X} \right|$, where $X = x_1 + x_2$

$$= \frac{.0005}{12.493} + \frac{.005}{12.493} = .000440246538$$

Relative error in subtraction $= \left| \frac{\delta X}{X} \right| \le \left| \frac{\delta x_1}{X} \right| + \left| \frac{\delta x_2}{X} \right|$, where $X = x_1 - x_2$

$$= \frac{.0005}{12.247} + \frac{.005}{12.247} = .000449089573$$

Relative error in multiplication and division $= \left| \frac{\delta x_1}{x_1} \right| + \left| \frac{\delta x_2}{x_2} \right|$

$$= \frac{.0005}{.123} + \frac{.005}{12.37} = .004469244369$$

Example ── 2.6

Calculate the absolute and relative errors in the expression $a + \dfrac{5b}{c} - 3bc$, if the measurements

of $a = 3.5435$, $b = .2588$ and $c = 1.0150$ are possibly correct up to four decimal points.

Ans. Let $x = a + \dfrac{5b}{c} - 3bc = A + 5B - 3C$, where $A = a$, $B = \dfrac{b}{c}$ and $C = bc$.

Value of $x = a + \dfrac{5b}{c} - 3bc = 4.03033$

Error in a, b and c is $\delta a = \delta b = \delta c = .00005$

Absolute error in $A = |\delta A| = .00005$

Absolute error in $B = |\delta B| = \dfrac{|c\delta b| + |b\delta c|}{c^2} = \dfrac{(1.015 + 0.2588) \times .00005}{(1.015)^2} = .00006182$

Absolute error in $C = |\delta C| = |c\delta b| + |b\delta c| = (1.015 + 0.2588) \times .00005 = .00006369$

Absolute error in $x = |\delta x| \le |\delta A| + 5|\delta B| + 3|\delta C|$

$$= .00005 + 5(.00006182) + 3(.00006369)$$

$$= .0005502$$

Relative error in $x = \left|\dfrac{\delta x}{x}\right| = \dfrac{.0005502}{4.03033} = .0001365$

Percentage error in $x = 0.01365\%$

2.3.3.2 Error Propagation in Function of Single Variable

Let us consider a function $f(x)$ of a single variable, x. Assume that the variable x has some error and its approximate value is $\tilde{x}$. The effect of error in the value of x on the value of function $f(x)$ is given by

$$\Delta f(x) = |f(x) - f(\tilde{x})|$$

Evaluating $\Delta f(x)$ is difficult as the exact value of x is unknown and hence exact $f(x)$ is unknown. But if $\tilde{x}$ is close to x and the function $f(x)$ is infinitely differentiable in some interval containing the points $\tilde{x}$ and x, then Taylor series can be employed as follows

$$f(x) = f(\tilde{x}) + (x - \tilde{x})f'(\tilde{x}) + \dfrac{(x - \tilde{x})^2}{2!} f''(\tilde{x}) + \dots$$

Since the difference $(x - \tilde{x})$ is very small, hence neglecting the second and higher order terms of $(x - \tilde{x})$ will give following relation

$$f(x) - f(\tilde{x}) \simeq (x - \tilde{x}) f'(\tilde{x})$$

or $\quad \begin{aligned} |\Delta f(x)| &\simeq |x - \tilde{x}| |f'(\tilde{x})| \\ &\simeq \Delta x |f'(\tilde{x})| \end{aligned}$ (2.5)

where $\Delta f(x) = |f(x) - f(\tilde{x})|$ represents the estimated error in the function value and $\Delta x = |x - \tilde{x}|$ is the estimated error of x.

Example ─── 2.7 ──────

Let $\tilde{x} = 3.42$ be an approximate value of the variable x with an error bound $\Delta x = 0.003$. Compute the resulting error bound in the function value $f(x) = x^3$.

Ans. From Eq. (2.5), the resulting error in the function $f(x)$ is given by

$$\Delta f(x) = (0.003)3(3.42)^2 = 0.1052676$$

Note that the approximate function value is $f(3.42) = 40.001688$. Therefore, the predicted value of $f(x)$ must be in the range

$$f(3.42) = 40.001688 \pm 0.1052676$$

Equivalent Statement for Example 2.7: Let us assume that we want to compute the volume of a cube. We measure its length with a machine and find out that it is 3.42m. Let us also assume that the machine can measure with maximum error 0.003m. Find the volume of the cube.

2.3.3.3 Error Propagation in Function of More than One Variable

<u>General Error Formula</u>

The approach above can be generalized to the function of more than one independent variable. Let $y = f(x_1, x_2, ..., x_n)$ be a function of n-independent variables $x_1, x_2, ..., x_n$. Let $\delta x_1, \delta x_2, ..., \delta x_n$ be the errors in calculating the variables $x_1, x_2, ..., x_n$, respectively. Let error in y be δy, i.e.,

$$y + \delta y = f(x_1 + \delta x_1, x_2 + \delta x_2, ..., x_n + \delta x_n)$$

When the required partial derivatives exist, then Taylor's series expansion is given by

$$y + \delta y = f(x_1, x_2, ..., x_n) + \left(\frac{\partial f}{\partial x_1} \delta x_1 + \frac{\partial f}{\partial x_2} \delta x_2 + ... + \frac{\partial f}{\partial x_n} \delta x_n \right) +$$

$$\text{+ terms involving second and higher powers of } \delta x_1, \delta x_2, ..., \delta x_n \qquad (2.6)$$

The errors in the numbers $x_1, x_2, ..., x_n$ are small enough to neglect the second and higher degree terms of the numbers $\delta x_1, \delta x_2, ..., \delta x_n$. We can obtain the following result from Eq. (2.6)

$$\delta y \approx \frac{\partial f}{\partial x_1} \delta x_1 + \frac{\partial f}{\partial x_2} \delta x_2 + ... + \frac{\partial f}{\partial x_n} \delta x_n \qquad (2.7)$$

Equation (2.7) is known as the general error formula. Since the error term may be of any sign, (+)ve or (−)ve, we can take absolute values of the terms in the expression.

$$|\delta y| \approx \left| \frac{\partial f}{\partial x_1} \right| |\delta x_1| + \left| \frac{\partial f}{\partial x_2} \right| |\delta x_2| + ... + \left| \frac{\partial f}{\partial x_n} \right| |\delta x_n|$$

Example   **2.8**

Compute the absolute and relative errors in the function $f(x, y, z) = \dfrac{y^2 \sin(x)}{4z^3}$ at $x = 1$ and $y = z = 5$, if the errors in the values of x, y and z are 0.05.

Ans. On using general error formula (2.7), the error $\delta f(x, y, z)$ in $f(x, y, z)$ is given by

$$\delta f(x, y, z) = \frac{\delta f}{\delta x} \delta x + \frac{\delta f}{\delta y} \delta y + \frac{\delta f}{\delta z} \delta z$$

$$= \frac{y^2 \cos(x)}{4z^3} \delta x + \frac{y \sin(x)}{2z^3} \delta y - \frac{3y^2 \sin(x)}{4z^4} \delta z$$

Absolute error $= \left| \dfrac{y^2 \cos(x)}{4z^3} \delta x \right| + \left| \dfrac{y \sin(x)}{2z^3} \delta y \right| + \left| \dfrac{3y^2 \sin(x)}{4z^4} \delta z \right|$

$$= .001350756 + .000841471 + .001262206$$

$$= .003454433$$

Relative error $= \dfrac{\delta f(x, y, z)}{f(x, y, z)} = \dfrac{.003454433}{.04207355} = .082104624$

Example ————————————————————————————————— 2.9 —

The radius r and height h of a right circular cylinder are measured as .25 m and 2.4 m, respectively, with a maximum error of 5%. Compute the resulting percentage error in the volume of the cylinder. Assume the value of π is exact for calculation.

Ans. The value of π is exact for calculation, so the volume $V = \pi r^2 h$ is dependent only on radius r and height h of the cylinder i.e., $V = V(r,h)$. Therefore, the error $\delta V(r,h)$ in the volume is given by

$$\delta V(r,h) = \frac{\partial V}{\partial r}\delta r + \frac{\partial V}{\partial h}\delta h = (2\pi rh)\delta r + (\pi r^2)\delta h$$

The radius r and height h of the cylinder are measured with a maximum error of 5% i.e.

$$\frac{\delta r}{r} = \frac{\delta h}{h} = 0.05$$

The relative error in volume $V(r,h)$ is given by

$$R.E. = \frac{\delta V(r,h)}{V}$$

$$= \frac{1}{\pi r^2 h}\left((2\pi rh)\delta r + (\pi r^2)\delta h\right)$$

$$= 2\frac{\delta r}{r} + \frac{\delta h}{h}$$

$$= 2(0.05) + 0.05 = 0.15$$

Percentage error in the volume of cylinder = $R.E. \times 100 = 15\%$

2.3.4 Truncation Error

An infinite power series (generally Taylor series) represents the local behavior of a given function $f(x)$ near a given point $x = a$. Approximation of an infinite power series with its finite number of terms, while neglecting remaining terms, leads to the *truncation error*. If we approximate the power series by the n-th order polynomial, then truncation error is of order $n + 1$.

Taylor series for the function $f(x)$ at the point $x = a$ is given by

$$f(x) = f(a) + (x-a)f'(a) + \frac{(x-a)^2}{2!}f''(a) + \cdots + \frac{(x-a)^n}{(n)!}f^{(n)}(a) + \cdots$$

(Or) $f(x) = f(a) + (x-a)f'(a) + \frac{(x-a)^2}{2!}f''(a) + \cdots + \frac{(x-a)^n}{(n)!}f^{(n)}(a) + R_n(x)$

where $R_n(x) = \frac{(x-a)^{n+1}}{(n+1)!}f^{(n+1)}(\xi)$ for some ξ between a and x.

On replacing $x = a + h$, we get following form of the Taylor series

$$f(a+h) = f(a) + (h)f'(a) + \frac{(h)^2}{2!}f''(a) + \cdots + \frac{(h)^n}{(n)!}f^{(n)}(a) + R_n(x)$$

where $R_n(x) = \frac{(h)^{n+1}}{(n+1)!}f^{(n+1)}(\xi); \ a < \xi < a+h$.

For a convergent series, $R_n(x) \to 0$ as $n \to \infty$. Since it is not possible to compute an infinite number of terms, we approximate the function $f(x)$ by first n-terms, and neglecting higher order terms. Then the error is given by remainder term $R_n(x)$. The exact value of ξ is not known, so the value of ξ is such that the error term considered is maximum.

Example   **2.10**

Use the following Taylor series expansion to compute the value of irrational number e.

$$e^x = 1 + x + \frac{x^2}{2!} + \frac{x^3}{3!} + \cdots$$

Create a table for absolute and percentage errors with numbers of terms $n = 1, 2, \ldots 6$ of Taylor series approximations. For the exact value of e, use $e = 2.718282$.

Ans. Computation of exact value of e^x requires an infinitely long series. Approximating e^x with the Taylor series to n terms gives an inexact answer. Table 2.2 contains Taylor series approximations of e of order $n = 1, 2, \ldots 6$. It also contains absolute and percentage errors in these approximations

Table 2.2

Number of terms	Taylor Series of e^x	Approximation for the function e	Absolute error	Percentage error
1	$e^x = 1$	1	1.718282	63.212058%
2	$e^x = 1 + x$	2	0.718282	26.424116%
3	$e^x = 1 + x + \dfrac{x^2}{2!}$	2.500000	0.218282	8.030146%
4	$e^x = 1 + x + \dfrac{x^2}{2!} + \dfrac{x^3}{3!}$	2.666667	0.051615	1.898810%
5	$e^x = 1 + x + \dfrac{x^2}{2!} + \dfrac{x^3}{3!} + \dfrac{x^4}{4!}$	2.708333	0.009948	0.365966%
6	$e^x = 1 + x + \dfrac{x^2}{2!} + \dfrac{x^3}{3!} + \dfrac{x^4}{4!} + \dfrac{x^5}{5!}$	2.716666	0.001616	0.059449%

Example ─────────────────────────────── **2.11**

Calculate the number of terms required in Taylor series approximation of $\sin(x)$ to compute the value of $\sin\left(\dfrac{\pi}{12}\right)$ correct up to 4-decimal places.

Ans. Using Taylor series of $\sin(x)$ at point $x = 0$, we have

$$\sin(x) = x - \frac{x^3}{3!} + \frac{x^5}{5!} + \cdots + \frac{(x)^{2n-1}}{(2n-1)!}(-1)^{n-1} + R_{2n-1}(x)$$

If we retain only first $2n-1$ terms in this expression, then the error term is given by

$$R_{2n-1}(x) = \frac{(x)^{2n}}{(2n)!} f^{(2n)}(\xi); \quad 0 < \xi < x. \text{ at } x = \frac{\pi}{12} = 0.2618$$

The maximum value of $f^{(2n)}(\xi)$ is 1. The error term must be less than .00005 for 4-decimal points accuracy

$$R_{2n-1}(x) = \frac{(0.2618)^{2n}}{(2n)!} \le .00005$$

$$\Rightarrow 2n \ge 5$$

Hence, 4-decimal points accuracy can be achieved by computing more than five terms of Taylor series.

Example ────────────────────────────────── **2.12**

The Gauss error function $erf(x) = \dfrac{2}{\sqrt{\pi}} \displaystyle\int_0^x e^{-t^2} \, dt$ is used widely in probability theory (e.g.,

normal distribution), statistics and partial differential equations. But the exact integral is not available for a finite value of x, so we use approximations. For example, one way is to use Taylor polynomial for the function e^{-t^2} and compute the resulting integration.

Compute the approximate value of the error function $erf(0.1) = \dfrac{2}{\sqrt{\pi}} \displaystyle\int_0^{0.1} e^{-t^2} \, dt$ by using first four terms of the Taylor series.

Ans. Taylor series of e^{-t^2} at $t = 0$ is given by

$$e^{-t^2} = 1 - t^2 + \frac{t^4}{2!} - \frac{t^6}{3!} + \cdots$$

Using the Taylor polynomial of first four terms, we have

$$erf(x) = \frac{2}{\sqrt{\pi}} \int_0^x e^{-t^2} \, dt$$

$$= \frac{2}{\sqrt{\pi}} \int_0^x \left(1 - t^2 + \frac{t^4}{2!} - \frac{t^6}{3!} \right) dt$$

$$= \frac{2}{\sqrt{\pi}} \left(x - \frac{x^3}{3} + \frac{x^5}{10} - \frac{x^7}{42} \right)$$

At $x = 0.1$, we have

$$erf(0.1) = \frac{2}{\sqrt{\pi}} \int_0^{0.1} e^{-t^2} \, dt = 0.112463$$

2.3.5 Machine eps (Epsilon)

Machine epsilon for a given machine, for example a computer, is defined as the smallest positive number which, when added to 1, gives a number different from 1. In fact, the machine epsilon defines the lowest floating point number, which can take part in the arithmetic for a given machine. Machine epsilon depends on round-off of the floating point numbers. Since rounding is machine dependent, so machine epsilon also varies with the machine. Machine epsilon characterizes computer arithmetic in numerical analysis. The quantity is also called as macheps or unit round-off, and it has the symbol epsilon ε.

2.3.6 Epilogue

In following chapters, we will see that several alternative numerical methods are available for the solution of any problem. In the selection of any method, we have to keep in mind all aspects of the problems and the method itself. Only from experience can we develop the skill for right selection and this skill has a prominent role in effective implementation of the method. Following are the deciding factors for selection of a numerical method and its implementation to the problem.

1. Type of mathematical problem
2. Computer available
3. Development cost
4. Characteristics of the numerical method
5. Mathematical behavior of the problem
6. Ease of application
7. Maintenance

2.3.7 Loss of Significance: Condition and Stability

In this section, we will study the two related concepts of condition and stability for function and process, respectively. The condition is used to describe the sensitivity of the function and stability is used to describe the sensitivity of the process.

Condition:

The sensitivity of the function $f(x)$ with the change in the argument x is described by the condition number (CN). It is a relative change in the function $f(x)$ for per unit relative change in x. CN of the function $f(x)$ at any point x is given by

$$CN = \frac{\left|\frac{f(x)-f(\tilde{x})}{f(x)}\right|}{\left|\frac{x-\tilde{x}}{x}\right|} = \left|\frac{f(x)-f(\tilde{x})}{x-\tilde{x}}\right| \left|\frac{x}{f(x)}\right|$$

For small change in x, Lagrange mean value theorem gives

$$\frac{f(x)-f(\tilde{x})}{x-\tilde{x}} \approx f'(x)$$

So, CN is given by

$$CN = \left|\frac{xf'(x)}{f(x)}\right| \tag{2.8}$$

If CN ≤ 1, then the function $f(x)$ is said to be well-conditioned. Otherwise, it is said to be ill-conditioned. The function with large CN is more ill-conditioned as compared to the function with small CN.

Note: *Let us consider a mathematical model of any system, in which variable x gives input, and output is the function f(x). If a small relative change in x (input) produces a large relative change in output f(x), then the system is said to be a sensitive system as fluctuation in input may break the system. Mathematically, if CN is large, then the function is more sensitive to changes and function is ill-conditioned.*

Example 2.13

Find the CNs of the functions $f(x) = \sqrt{x}$ and x^3.

Ans. Using Eq. (2.8), we have

$$\text{CN of the function } \sqrt{x} = \left|\frac{xf'(x)}{f(x)}\right| = \left|\frac{x\left(\frac{1}{2}x^{\frac{-1}{2}}\right)}{\sqrt{x}}\right| = \frac{1}{2}$$

$$\text{CN of the function } x^3 = \left|\frac{xf'(x)}{f(x)}\right| = \left|\frac{x\left(3x^2\right)}{x^3}\right| = 3$$

CN of the function $\sqrt{x}$ is less than 1, so the function $\sqrt{x}$ is well conditioned. The function x^3 is an ill-conditioned function as CN > 1.

Example 2.14

Check the condition of the function $f(x) = \dfrac{1}{1-2x+x^2}$ at $x = 1.01$.

Ans.

$$f(x) = \frac{1}{1-2x+x^2} = \frac{1}{(1-x)^2}$$

$$\text{CN} = \left|\frac{xf'(x)}{f(x)}\right|_{x=1.01} = \left|\frac{x\left(\dfrac{-2}{(1-x)^3}\right)}{\left(\dfrac{1}{(1-x)^2}\right)}\right|_{x=1.01} = 202$$

The function $f(x) = \dfrac{1}{1-2x+x^2}$ at $x = 1.01$ is highly ill-conditioned function. The function has a singular point $x = 1$, so near this point, there are sharp changes in the function value, which make the function highly ill-conditioned.

Example ── **2.15**

Find the CN of the function $f(x) = \sqrt{x+1} - \sqrt{x}$ at point $x = 11111$.

Ans.

$$CN = \left| \frac{xf'(x)}{f(x)} \right| = \left| \frac{x\left(\dfrac{1}{2\sqrt{x+1}} - \dfrac{1}{2\sqrt{x}} \right)}{\sqrt{x+1} - \sqrt{x}} \right|_{x=11111} \approx \frac{1}{2}$$

Example ── **2.16**

Compute the function $f(x) = \sqrt{x+1} - \sqrt{x} = \dfrac{1}{\sqrt{x+1} + \sqrt{x}}$ by using both the formulae at point $x = 11111$. Use six significant digits floating point rounding arithmetic.

Ans. We have two formulas $f(x) = \sqrt{x+1} - \sqrt{x}$ and $f(x) = \dfrac{1}{\sqrt{x+1} + \sqrt{x}}$ to compute the function $f(x)$ at point $x = 11111$. We will use both the formulas with six significant digits arithmetic, and see that both the processes will produce different results for the same function.

Process-I: $f(x) = \sqrt{x+1} - \sqrt{x}$

$$f(x) = \sqrt{11112} - \sqrt{11111}$$
$$= 105.413 - 105.409$$
$$= .004$$

Process-II: $f(x) = \dfrac{1}{\sqrt{x+1} + \sqrt{x}}$

$$f(x) = \frac{1}{\sqrt{11112} + \sqrt{11111}}$$
$$= \frac{1}{105.413 + 105.409}$$
$$= \frac{1}{210.822}$$
$$= 0.00474334$$

Note that, the exact result up to 6 significant digits is .00474330.

Note: Here, it is candidly seen that if we compute the function $f(x) = \sqrt{x+1} - \sqrt{x}$ directly, then it is error-prone. This is due to the fact that if we subtract two approximately equal numbers, then there is a loss of significant digits. For example in Process-I, when we subtract 105.413 and 105.409, then these two numbers are correct up to six significant digits, but the result .004 contains only one significant digit. Since there is a loss of five significant digits, so the result obtained is highly erroneous. This step can be avoided by rationalizing the function $f(x)$. The result obtained in Process-II after rationalization is correct up to five significant digits.

Stability of the Process:
It is clear from Example 2.16 that computation of the same function from two different processes can produce different results. There are following two major phases for computation of the function value $f(x)$:

   i)   First phase is to check the condition of the function by computing the CN of the function.

   ii)   Second phase is to check the stability of the process involved in the computation of the function. The stability of process can be checked by calculating the condition of each step in the process.

The function $f(x) = 1/(1-x^2)$ is ill-conditioned (CN $\gg 1$) near $x = \pm 1$. If the function is ill-conditioned then whatever process we will use, it tends to error. So every process will produce an error in computation of the function value $f(x) = 1/(1-x^2)$ near $x = \pm 1$.

The function $f(x) = \sqrt{x+1} - \sqrt{x}$ at $x = 11111$ is well conditioned (CN $\approx 1/2$, Example 2.15). If the function is well conditioned, then we have to compute the function value by the stable process. If even a single step of the process is ill-conditioned, then the whole process is an unstable process, and we have to switch over to any other alternate stable process.

Example —————————————————————————— 2.17 —————

Discuss the stability of the Processes-I and II in Example 2.16. Hence, validate the results that the Processes-I yields erroneous result and Process-II produces a more accurate result for the same function $f(x)$.

Ans.
We will calculate the CN of each step involved in both the Processes-I and II.

Process-I: $f(x) = \sqrt{x+1} - \sqrt{x}$

$$f(x) = \sqrt{11112} - \sqrt{11111}$$
$$= 105.413 - 105.409$$
$$= .004$$

Various computational steps in the process are as follows

$x_1 = 11111$ $\qquad\qquad$ ($f(x)$ = Constant, CN = 0)

$x_2 = x_1 + 1 = 11112$ $\qquad\qquad$ ($f(x) = x + 1$, CN = 1)

$x_3 = \sqrt{x_2} = 105.413$ $\qquad\qquad$ ($f(x) = \sqrt{x}$, CN = ½)

$x_4 = \sqrt{x_1} = 105.409$ $\qquad\qquad$ ($f(x) = \sqrt{x}$, CN = ½)

$x_5 = x_4 - x_3 = .004$ $\qquad\qquad$ ($f(x) = x - x_3$ and $f(x) = x_4 - x$, CN ≈ 26352)

In the last step $x_5 = x_4 - x_3$, we can assume the function $f(x)$ of variable x_3 or x_4. Let $f(x) = x_4 - x$, so condition for this step is given by

$$CN = \left|\frac{xf'(x)}{f(x)}\right| = \left|\frac{x(-1)}{x_4 - x}\right| = \left|\frac{105.409}{.004}\right| \approx 26352$$

This step is not a stable step as CN is very large. So the whole process is an unstable process due to this step. That's why the result obtained from this process is highly erroneous.

Process-II: $f(x) = \dfrac{1}{\sqrt{x+1} + \sqrt{x}}$

We will check the conditions of each step in Process-II, and conclude that each step in this process is well conditioned.

$$f(x) = \frac{1}{\sqrt{11112} + \sqrt{11111}}$$

$$= \frac{1}{105.413 + 105.409}$$

$$= \frac{1}{210.822}$$

$$= 0.00474334$$

Various steps involved in this process are as follows

$x_1 = 11111$

$x_2 = x_1 + 1 = 11112$

$x_3 = \sqrt{x_2} = 105.413$

$x_4 = \sqrt{x_1} = 105.409$

$x_5 = x_4 + x_3 = 210.822$

$x_6 = \dfrac{1}{x_5} = 0.00474334$

The first four steps in the process are well conditioned as discussed in Process-I. For the fifth step, let $f(x) = x_4 + x$. The condition for this step is given by

$$CN = \left| \frac{xf'(x)}{f(x)} \right| = \left| \frac{x(1)}{x_4 + x} \right| = \left| \frac{105.409}{222.822} \right| \approx \frac{1}{2}$$

The last step is $f(x) = \dfrac{1}{x}$, and the condition for this step is given by

$$CN = \left| \frac{xf'(x)}{f(x)} \right| = \left| \frac{x \left(\dfrac{-1}{x^2} \right)}{\dfrac{1}{x}} \right| = 1$$

From above discussion, it is clear that all the steps in Process-II are well conditioned, and hence this process is a stable process. Since the process is stable, so the result obtained is accurate to five significant digits.

Note: *Even a single step in the process can make the whole process unstable. So we have to be extra careful during a large process, and must avoid the steps (if possible) with the loss of significant digits. We can use any alternate approach like rationalization, Taylor series expansion, etc. to avoid loss of significant digits.*

Example ── **2.18**

Discuss the stability of the function $f(x) = 1 - \cos(x)$, when x is nearly equal to zero. Find a stable way to compute the function $f(x)$.

Ans. If we directly compute the function $f(x) = 1 - \cos(x)$ at $x \approx 0$, then it will lead to subtraction of two nearly equal numbers and produce loss of significance. To avoid this loss, we can use any of the following three alternates

i) $\quad f(x) = 1 - \cos(x)$

$$= 1 - \left(1 - \frac{x^2}{2!} + \frac{x^4}{4!} - \frac{x^6}{6!} + \cdots \right)$$

$$= \frac{x^2}{2!} - \frac{x^4}{4!} + \frac{x^6}{6!} - \cdots$$

ii) $f(x) = 1 - \cos(x)$

$$= \frac{1 - \cos^2(x)}{1 + \cos(x)}$$

$$= \frac{2\sin^2(x)}{1 + \cos(x)}$$

iii) $f(x) = 1 - \cos(x)$

$$= 2\sin^2 \frac{x}{2}$$

Example ────────────────────────────── **2.19**

Calculate the roots of the equation $x^2 + 123x + 0.5 = 0$ using five digits floating point chopping arithmetic.

Ans. The roots of the quadratic equation $ax^2 + bx + c = 0$ are given by

$$x = \frac{-b \pm \sqrt{b^2 - 4ac}}{2a}$$

The roots of the equation $x^2 + 123x + 0.5 = 0$ using five digits floating point chopping arithmetic are given by

Root 1. $x_1 = \dfrac{-b + \sqrt{b^2 - 4ac}}{2a}$

$b^2 = 15129$

$b^2 - 4ac = 15127$

$\sqrt{b^2 - 4ac} = 122.99$

$x_1 = \dfrac{-123 + 122.99}{2} = -0.0005$

Root 2. $x_2 = \dfrac{-b - \sqrt{b^2 - 4ac}}{2a} = \dfrac{-123 - 122.99}{2} = -122.995 = -122.99$

The roots of the equation correct up to some significant digits are $x_1 = -0.004065175$ and $x_2 = -122.995934825$. The root $x_2 = -122.99$ is correctly calculated up to five significant digits. But the root $x_1 = -0.0005$ is not correct even up to one significant digit. This error is due to loss of significant digits which occurs due to subtraction of two nearly equal numbers (123 and 122.99).

To avoid the loss of significant digits, we will rationalize the formula for x_1, and then compute the root.

$$x_1 = \frac{-b+\sqrt{b^2-4ac}}{2a} \times \frac{b+\sqrt{b^2-4ac}}{b+\sqrt{b^2-4ac}}$$

$$= \frac{-2c}{b+\sqrt{b^2-4ac}}$$

$$= \frac{-1}{123+122.99} = -.0040652$$

This value is correct up to five significant digits.

Note: There are two ways to produce the results with desired accuracy. One way is to use stable processes for the computation and another way is to use the computing device with very high precisions. For example, we want to compute the roots of the equation $x^2 + 123x + 0.5 = 0$ correct up to five significant digits. In that case, we can use the computing device with more than ten digits floating point arithmetic, such that the results can be obtained up to desired accuracy even after the loss of significance.

2.4 Some Interesting Facts about Error

a) Let us assume we are doing six significant digits arithmetic on a hypothetical computer. If we want to add a small number $x = 0.000123$ to a large number $y = 123.456$ using this computer, then

$x + y = (.123456)10^3 + (.123000)10^{-3}$ (Normalized form)

$= (.123456)10^3 + (.000000)10^3$ (Equal exponent using symmetric rounding)

$= (.123456)10^3$ (Result, we missed the addition!)

This type of situations occurred commonly during the computations of infinite series. In these series, the initial terms are comparatively large. So, usually after adding some terms of the series, we are in a situation of adding a small term to a very large term. It may produce high rounding error in the computation. To avoid this kind of error, we can use backward sum of the series instead of forward sum, such that the each new term is compatible with the magnitude of accumulated sum.

b) In the case of series with mixed signs (like Taylor series of sin(x)), sometimes individual terms are larger than the summation itself. For example, in Taylor series of sin(2.13), the first term is 2.13. It is called as smearing, and we should use these kinds of series with extra care.

c) While performing arithmetic computations in a numerical method, the steps involving large number of arithmetic operations must be computed in double precisions. Such operations are error-prone to round-off error. For example, in Gauss-Seidel method for the solution of system of linear equations, the inner product

$$\sum_{i=1}^{n} x_i y_i = x_1 y_1 + x_2 y_2 + \cdots + x_n y_n \text{ is a common operation, and such computations}$$

must be made in double precisions.

The accumulated rounding error can create the disastrous results; the following two examples of rounding errors are picked from the internet.

(*http://mathworld.wolfram.com/RoundoffError.html*)

1. An index was started with initial value 1000.000 for Vancouver stock exchange (McCullough and Vinod 1999) in 1982. Three decimal digits chopping arithmetic has been used to compute the index for each change in market value for next 22 months. The computed value was 524.881, while its correct value up to three decimal points is 1009.811.

2. The Ariane rocket was launched on June 4, 1996 (European Space Agency 1996). In the 37th second of flight, a 64-bits floating point number was converted to a 16-bits number by the inertial reference system of the rocket. It was an overflow error, but the guidance system interpreted it as flight data, which led the rocket to getting destroyed.

Exercise 2

1. Define normalized form and hence the number of significant digits for floating point numbers with examples.

2. Find out the number of significant digits in the numbers 788500, 0.4785, .003523, 0.2300, and 7.880.

   Ans. 6, 4, 4, 4, 4

3. Compute the absolute errors (A.E.) and relative errors (R.E.) in the four significant digits chopping approximations of the numbers 234168 and 64.2685.

   Ans. A.E. = 68, 0.0085;   R.E. = 0.000290389, 0.000132257

4. If β based real number $x = (0.d_1 d_2 d_3 .... d_n d_{n+1} ....)_\beta \times \beta^e$ is chopped to n digits and $fl(x)$ is its representation, then show that

$$0 \le \frac{x - fl(x)}{x} \le \beta^{1-n}$$

5. If x is any number in decimal number system and $fl(x)$ is its machine representation up to n digits, then for rounding, show that

$$\left| \frac{x - fl(x)}{x} \right| \le \frac{1}{2} \times 10^{1-n}$$

6. The true value of e(exponential) correct to 10-significant digits is 2.718281828. Calculate absolute and relative errors, if we approximate this value by 2.718.

Ans. A.E. = .000281828, R.E. = 0.000103678

7. Find the relative errors for the following cases. Also, determine the number of significant digits in the approximations:

$x = 2.71828182$ and $\bar{x} = 2.7182$

Ans. R.E. = 0.0000301, 5 Significant Digits

$y = 28350$ and $\bar{y} = 28000$

Ans. R.E. = 0.0123457, At least 2 Significant Digits

$z = 0.000067$ and $\bar{z} = 0.00006$.

Ans. R.E. = 0.104478, 1 Significant Digits

8. Define the terms error, absolute error, relative error and significant digits. The numbers $x = 1.28$ and $y = 0.786$ are correct to the digits specified. Find estimates to the relative errors in $x + y, x - y, xy$, and x/y.

Ans. R.E. in $x + y = 0.00266215$, $x - y = 0.0111336$, xy and $x/y = 0.00454238$

9. Consider the following decimal numbers with a four digits normalized mantissas, $a = 0.2473*10^4$, $b = 0.8125*10^3$, $c = 0.1523*10^1$

Perform the following operations in four significant digits symmetric rounding and indicate the errors in the results.

i) $a + b - c$ **Ans.** $0.3283*10^4$, Error = 0.0000977

ii) b/c **Ans.** $0.5335*10^3$, Error = -0.0001346

iii) $a - b$ **Ans.** $0.1661*10^4$, Error = -0.00005

iv) $b/(a + c)$ **Ans.** 0.3283, Error = 0.00004611

10. The numbers $x_1 = 0.643$ and $x_2 = 1.631$ are correct to the significant digits in the numbers. Compute the relative errors in the addition, subtraction, multiplication and division of these two numbers.

Ans. R.E. in $x_1 + x_2 = 0.00043975$, R.E. in $x_1 - x_2 = 0.001012145$,

R.E. in $x_1 x_2$ and $x_1 / x_2 = 0.001084164$

11. Calculate the absolute and relative errors in the expression $3a - 2bc + \dfrac{b}{a}$, if the measurement of $a = 3.5435$, $b = .2588$ and $c = 1.0150$ are possible only to correct up to four decimal points.

    Ans. Absolute Error $= 0.0002925$, Relative Error $= 0.00002874$

12. Estimate the error in evaluating the function $f(x) = e^{2x^2} \sin(x)$ near the point $x = 1$, if the absolute error in value of x is 10^{-4}.

    Ans. Absolute Error $= 0.0028863$

13. The maximum error tolerance in the measurement of the area of a given circle is 0.1%. What is maximum relative error allowed in the measurement of the diameter?

    Ans. 0.05

14. Compute the resulting error in the function $f(x) = x^3$ for value of $\bar{x} = 2.38$ with an error $\Delta \bar{x} = 0.005$

    Ans. Absolute Error $= 0.084966$

15. Find the maximum possible error in the computed value of the hyperbolic sine function $\sinh(x) = \dfrac{e^x - e^{-x}}{2}$ at the point $x = 1$, if the maximum possible error in the value of x is $|dx| = 0.01$.

    Ans. 0.01543

16. Let the function $u = \dfrac{4x^2 y^3}{z^4}$ and errors in the values of variables x, y, z are 0.001. Find the relative error in the function u at $x = y = z = 1$.

    Ans. 0.009

17. The radius r and height h of a right circular cylinder are measured as 2.5 m and 1.6 m, respectively, with a maximum error of 2%. Compute the resulting percentage error measured in the volume of the cylinder by the formula $V = \pi r^2 h$. Assume the value of π is exact for calculation.

    Ans. 0.06

18. Consider a function $u = e^x \sin(y) + x \ln(z)$. Let the variables x, y and z be measured with maximum possible errors of ± 0.01, $\pm \left(2^\circ = \dfrac{\pi}{90} \right)$ and ± 0.5, respectively. Estimate the maximum possible error in computing the function u for $x = 0.1$, $y = \dfrac{\pi}{4}$ and $z = 50$.

    Ans. 0.4976

19. The voltage V in an electrical circuit satisfies the law $V = IR$, where I is the current and R is the resistance and their starting values are $I = 5$ amp, $R = 600$ ohms, respectively. Let us assume that after a certain time, resistance is changed 0.15% due to heating, and we changed the current I with 5%. Compute the percentage change in the voltage V.

    Ans. 5.15%

20. The length of a simple pendulum measured is $l = 0.362$ m, while the constants $\pi = 3.1416$ and $g = 9.8$ m/sec^2 are correct to the specified digits. Compute the relative error in the time-period $T = 2\pi \sqrt{\dfrac{l}{g}}$.

    Ans. 0.0032575

21. Compute the absolute and relative errors in the function $f(x,y,z)=y^2e^z\cos(x)$ at $x=1.5$ $y=2.3$ and $z=5$, if the error in the values of x, y and z are 0.05.

Ans. Absolute Error = 44.3483, Relative Error = 0.7985

22. Calculate the number of terms required in Taylor series approximation of the function $\cos(x)$ to compute the value of $\cos\left(\dfrac{\pi}{12}\right)$ correct up to 4-decimal places.

Ans. 5

23. Find the number of terms of the Taylor series expansion of the function e^x required to compute the value of e correct to six decimal places.

Ans. 10

24. Discuss CN and stability of the function $y=\sec(x)$ in the interval $\left[0,\dfrac{\pi}{2}\right]$.

Ans. CN $=x\tan(x)$; as we move from 0 to $\dfrac{\pi}{2}$ in the interval $\left[0,\dfrac{\pi}{2}\right]$, the CN increase and hence function $y=\sec(x)$ become ill-conditioned.

25. Calculate the function $f(x)=\cos(x)-\sin(x)$ at the point $x=0.785398$ using 6-decimal digits floating-point round-off arithmetic. Discuss the condition and stability of process involved.

Ans. $f(x)=\cos(x)-\sin(x)=0$ at $x=0.785398$ with 6-decimal digits floating-point round-off arithmetic.
CN of $f(x)=\cos(x)-\sin(x)$ at $x=0.785398$ is approximately 0, hence function is well conditioned. But the process is not a stable process.
We can use any of the following stable processes for computation purpose

i) $\quad f(x)=\dfrac{\cos(2x)}{\cos(x)+\sin(x)}$

ii) $\quad f(x)=1-x-\dfrac{x^2}{2!}+\dfrac{x^3}{3!}+\dfrac{x^4}{4!}-\dfrac{x^5}{5!}-\dfrac{x^6}{6!}+\cdots$

26. Discuss the condition and stability of the function $f(x)=x-\sqrt{x^2-1}$ at $x=11111$, using six significant digits floating point rounding arithmetic. Find a stable way to compute the function.

Ans. $f(x)=\dfrac{1}{x+\sqrt{x^2-1}}$

27. Evaluate roots of the quadratic equation $x^2+234.56x+1.2345=0$, with the minimum loss of significant digits. Use five significant digits chopping arithmetic.

Ans. -234.55, -0.0052632

28. Avoiding loss of significance, find the smallest root of the quadratic equation $x^2-500x+2=0$ by using five significant digits rounding arithmetic.

Ans. 0.004, 500

29. Discuss the condition and stability of the function $f(x)=x-\sin(x)$, when x is nearly equal to zero. Find a stable way to compute the function $f(x)$.

Ans. $\dfrac{x^3}{3!}-\dfrac{x^5}{5!}+\dfrac{x^7}{7!}-\dfrac{x^9}{9!}+\cdots$

30. Subtraction of nearly equal numbers leads to loss of significant digits. Obtain equivalent formulas for the following functions to avoid loss of significance.

a) $\cos^2(x) - \sin^2(x)$ for $x \approx \dfrac{\pi}{4}$

b) $x - \sin(x)$ for $x \approx 0$

c) $x - \sqrt{x^2 - 1}$ for large x

d) $1 - \cos^2(x)$ for $x \approx 0$

e) $\sqrt{1 + \cos(x)}$ for $x \approx \pi$

Ans. a) $\cos(2x)$, b) $\dfrac{x^3}{3!} - \dfrac{x^5}{5!} + \dfrac{x^7}{7!} - \dfrac{x^9}{9!} + \cdots$, c) $\dfrac{1}{x + \sqrt{x^2 - 1}}$, d) $\sin^2(x)$, e) $\sqrt{\dfrac{\sin^2(x)}{1 - \cos(x)}}$

Nonlinear Equations

Truth is ever to be found in simplicity, and not in the multiplicity and confusion of things.

Sir Isaac Newton
(December 25, 1642–March 20, 1726)
He was a great mathematician and physicist. He pioneered 'classical mechanics'.

3.1 Introduction

Mathematical models for many problems in different branches of science and engineering are formulated as

$$f(x) = 0 \tag{3.1}$$

where the variables x and $f(x)$ may be real or complex, and scalar or vector quantities. In this chapter, the variables x and $f(x)$ are real and scalar quantities. The value of x, which satisfies the Eq. (3.1), is called the root of the equation. It is also known as the zero of the function $f(x)$. For example, the quadratic equation

$$x^2 - 3x + 2 = 0$$

has roots 1 and 2.

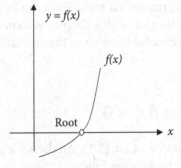

Fig. 3.1 Root of $f(x) = 0$

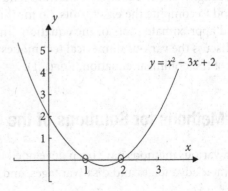

Fig. 3.2 Root of $x^2 - 3x + 2 = 0$

Root finding is also essential in many branches of mathematics. For example, the critical points of the function $f(x)$ are the roots of the equation $f'(x) = 0$. Eigenvalues of a square matrix A are the roots of the characteristic equation.

$$p(\lambda) = \det(A - \lambda I) = 0$$

where $p(\lambda)$ is a polynomial of degree n (order of matrix A).

The nonlinear equations can be categorized broadly as polynomial equations and transcendental equations as follows

3.1.1 Polynomial Equations

The polynomial equations are given by

$$y = f(x) = a_n x^n + a_{n-1} x^{n-1} + \ldots + a_1 x + a_0 = 0$$

This equation is an nth degree polynomial equation, and has exactly n roots. These roots may be real or complex. Some examples of polynomial equations are

i) $3x^3 + x^2 - 9 = 0$
ii) $x^2 - 4x - 5 = 0$

3.1.2 Transcendental Equations

An equation which is not a polynomial equation is a transcendental equation. These equations involve trigonometric, exponential and logarithmic functions, etc. A few examples of the transcendental equations are as follows

i) $3\sin(x) - e^{-x} = 0$
ii) $3x^2 - 2\cos x = 0$
iii) $2e^x \sin x - \ln(x) = 0$

Transcendental equations may have finite or infinite numbers of real roots or may not have real roots at all.

The roots of simple equations are easy to compute by the direct methods. But in the case of higher order equations and transcendental equations, there is no general analytical method to compute the exact roots. So for this purpose, numerical techniques can be used to find approximate roots of the equation. The main objective of this chapter is to present and discuss the various numerical techniques which are useful for finding the approximate roots of the nonlinear equation, Eq. (3.1).

3.2 Methods for Solutions of the Equation $f(x) = 0$

So far, various methods have been developed for the solution of Eq. (3.1). All these methods have their advantages and disadvantages, and broadly categories as follows

i) Direct analytical methods
ii) Graphical methods
iii) Trial and error methods
iv) Iterative methods

In this section, we shall have a brief idea of these methods and conclude that iterative methods for finding numerical approximations for the roots of Eq. (3.1) are the best methods for the complex and complicated equations.

3.2.1 Direct Analytical Methods

We can solve the nonlinear equation by direct analytical methods in certain simple cases. For example, the roots of the quadratic equation $ax^2 + bx + c = 0$ are given by $x = \dfrac{-b \pm \sqrt{b^2 - 4ac}}{2a}$. Similarly, roots of cubic and quartic equations can be obtained using Cardano and Ferrari methods, respectively. The roots obtained by direct methods are exact roots of the equations. But these methods can be applied to some very special categories of the equations. The roots of higher order polynomial equations/ transcendental equations (like $x^5 + 2x^3 + 3x^2 + 5x + 6 = 0$ and $2e^{-x} + x^3 \sin x = 0$) cannot be obtained from direct analytical methods. We don't have direct methods even for the solutions of simple transcendental equations.

3.2.2 Graphical Methods

Plotting of the function $f(x)$ with x-axis gives the root of the equation $f(x) = 0$. The points where the curve $f(x)$ crosses the x-axis, are the roots of the equation.

Solutions obtained using graphical methods are not accurate. But graphs of some standard curves are helpful in tracing the interval in which the root of the equation lies and are also important for an initial guess about the roots, etc. Let us discuss a few examples.

Case 1. Equations $x^2 + x + 1 = 0$, $x - e^x = 0$, $x - \ln(x) = 0$ with no real roots;

The graph (Fig. 3.3) of $y = x^2 + x + 1$ has no point of intersection with x – axis, so the equation $x^2 + x + 1 = 0$ has no real root.

Any equation $f(x) = 0$ can be rewritten as $f_1(x) = f_2(x)$, and points of intersections of the curves $y = f_1(x)$ and $y = f_2(x)$ provide the roots of the equation $f(x) = 0$. For example, consider the graphs of $y = e^x$ and $y = x$, then the points of intersection of these two curves are the roots of equation $x - e^x = 0$ ($x = e^x$). It is easy to see that there is no point of intersection (Fig. 3.4), so the equation $x - e^x = 0$ has no real root. Similarly, we can easily find that equation $x - \ln(x) = 0$ also has no real root (Fig. 3.5).

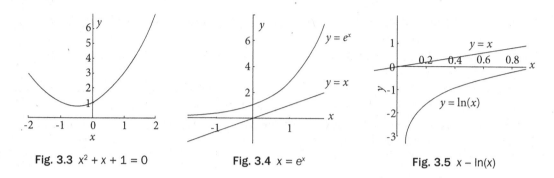

Fig. 3.3 $x^2 + x + 1 = 0$ **Fig. 3.4** $x = e^x$ **Fig. 3.5** $x - \ln(x)$

Case 2. Equations with finite numbers of real roots like $x^3 - 3x^2 + 2x = 0$, $x - \cos x = 0$, etc.

The points of intersection of two curves $y = \cos(x)$ and $y = x$ are the roots of equation $x - \cos x = 0$ (Fig. 3.6). There is only one point of intersection, so the equation $x - \cos x = 0$ has only one real root. Similarly, the graph of the function $y = x^3 - 3x^2 + 2x$ provides that the equation $x^3 - 3x^2 + 2x = 0$ has three real roots (Fig. 3.7).

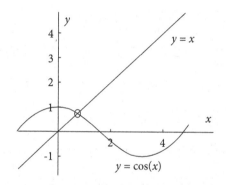

Fig. 3.6 $x = \cos x$, Root ≈ 0.7390851322

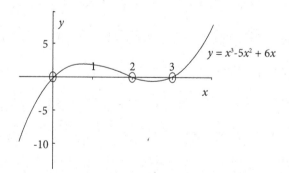

Fig. 3.7 $x^3 - 5x^2 + 6x = 0$, Roots = 0, 2, 3

Case 3. Equations with infinite numbers of real roots like $e^x - \cos x = 0$, $x - \tan x = 0$, $e^{-x} - \sin x = 0$, etc. The following graphs show that these equations have infinitely many real roots.

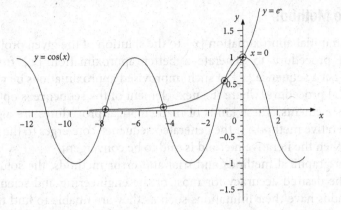

Fig. 3.8 $e^x = \cos(x)$, Roots $\approx$ 0, −1.2927, −4.7213, −7.8536, ...

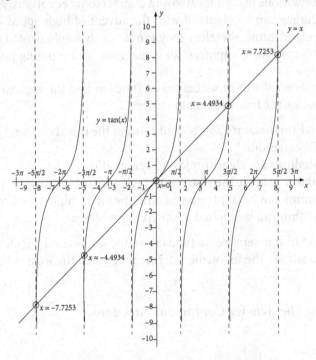

Fig. 3.9 $x = \tan x$, Roots $\approx$ 0, ±4.4934, ±7.7253, ...

3.2.3 Trial and Error Methods

Other approaches to obtain the approximate solutions are the trial and error techniques. These methods involve a series of guesses for the root of the equation (3.1) and check whether the value of the function is close to zero. The value of x, where function is close to zero, is the approximate root of the equation.

The trial and error methods are cumbersome and time-consuming. These methods are not algorithmic, so programming is not possible. Also, approaches in these methods vary from problem to problem. So, these methods are no longer in use.

3.2.4 Iterative Methods

Let us assume an initial approximation $\{x_0\}$ to the solution of the given problem. Consider a mathematical procedure to generate a better approximation $\{x_1\}$ from this initial approximation. Let a sequence $\{x_n\}$ of such improvised approximations be generated using this mathematical procedure, where the next element of the sequence is obtained with the help of previous elements. These kinds of mathematical procedures are algorithmic, and known as the iterative methods. If the generated sequence converges to the solution of the given problem, then the iterative method is said to be convergent.

In the case of graphical methods and trial and error methods, the solutions obtained are not up to the desired accuracy for most of the engineering and scientific problems. The direct methods have their limitations such as they are unable to find the roots of the equations when nature of the equations becomes complicated. The iterative methods are able to provide the solutions of complicated and cumbersome equations. Also the solutions with very high accuracy can be obtained with the advent of high-speed computers. Since these methods are algorithmic, therefore they can be easily implemented on the computer. Due to easy availability of the computers, these methods are becoming popular.

The iterative methods are the most successful methods to find the approximate roots of the nonlinear equations for the following main reasons.

1. The iterative methods are able to find roots of the transcendental and higher order polynomial equations.
2. These procedures are algorithmic and systematic.
3. These methods provide the result with high accuracy.
4. The algorithms can be implemented with the aid of high-speed computers.
5. Same algorithm can be applied to different problems.

An important aspect of the iterative methods is to find an interval which contains the root of the nonlinear equation. The following intermediate value theorem is helpful in locating the roots:

Intermediate Value Theorem for Continuous Functions

Theorem 3.1

Let $f(x)$ be a continuous function on the interval I. If a and b are two points in I and if $f(a) \le c \le f(b)$; then there exists a number ξ in $[a, b]$, such that $f(\xi) = c$.

Since we are interested in finding the zeroes of the function $f(x)$ i.e., $f(\xi) = 0$, so we can modify this theorem as follows.

Let $f(x)$ be a continuous function on the interval I. Let a and b be two points in the interval I, such that $f(a) \le 0 \le f(b)$; then there exists a number ξ in the interval I, such that $f(\xi) = 0$.

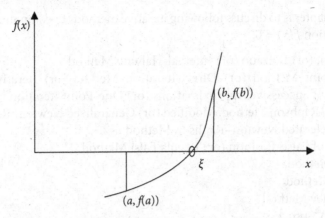

Fig. 3.10 Intermediate value theorem for continuous functions

Note: Consider the equation $f(x) = 0$, where the function $f(x)$ is continuous in an interval I. The prime objective is to find the two points a and b in the interval I, such that $f(a)$ and $f(b)$ are of opposite signs $f(a).f(b) \leq 0$. Then according to the Intermediate value theorem, there exists at least one root of the equation $f(x) = 0$ in the interval I.

Example   **3.1**

Consider the function $f(x) = x^3 + x - 1$, we have

   a)   The function $f(x)$ is continuous in interval $[0, 1]$.

   b) $f(0) = -1, f(1) = 1$ $\Rightarrow f(0).f(1) < 0$

Using intermediate value theorem (3.1), there exists a number ξ in $(0, 1)$ such that $f(\xi) = 0$. So at least one root of the equation $f(x) = x^3 + x - 1 = 0$ lies in the interval $(0, 1)$.

Note: It is easy to see that continuity of the function $f(x)$ is important for the intermediate value theorem. For example, the function $f(x) = x - \tan(x)$ satisfies the condition $f(1).f(2) < 0$. But, there is no real root of the equation $x - \tan(x) = 0$ in the interval $(1, 2)$, because the function $f(x) = x - \tan(x)$ has a discontinuity at the point $\dfrac{\pi}{2} \approx 1.5708 \in (1, 2)$.

In fact, for every small interval (a, b) that contains $\pm\dfrac{n\pi}{2}$, $f(x)$ satisfies the condition $f(a).f(b) < 0$. But there is no real root in that interval.

The aim of this chapter is to discuss following iterative methods to compute the approximate roots of the equation $f(x) = 0$.

1. Bisection, (or) Bolzano (or) Interval-Halving Method
2. Fixed-Point Method (or) Direct-Iteration Method (or) Iterative Method (or) Method of Successive Approximations (or) One-Point Iteration
3. Newton–Raphson Method, Modified (or) Generalized Newton–Raphson Method, and Accelerated Newton–Raphson Method
4. Method of False-Position (or) Regula Falsi Method
5. Secant Method
6. Muller Method
7. Chebyshev Method
8. Aitken Δ^2-process

3.3 Bisection (or) Bolzano (or) Interval-Halving Method

Bisection method is the simplest iterative method for finding the real roots of an equation, $f(x) = 0$. It is based on repeated applications of intermediate value theorem. We need two initial guesses, a and b, which satisfy the intermediate value theorem, i.e., $f(a)f(b) < 0$ and function $f(x)$ must be continuous in (a, b).

Let a and b be two initial guesses, which bracket the root. The first approximation to root is the midpoint of the interval (a, b), i.e.,

$$x_0 = \frac{a+b}{2}$$

If $f(a)f(x_0) < 0$ then according to intermediate value theorem, the root will lie in the interval (a, x_0). So the next approximation is given by

$$x_1 = \frac{a+x_0}{2}$$

Else, the root will lie in the interval (x_0, b) and the next approximation is given by

$$x_1 = \frac{b+x_0}{2}$$

This process can be continued to generate a sequence $x_1, x_2, x_3 \cdots$ of approximations to the root. Let x_{n-3}, x_{n-2} and x_{n-1} be three latest approximations to the root, then the nth approximation x_n is given by

if $f(x_{n-1})f(x_{n-2}) \le 0$ $\qquad\qquad x_n = \dfrac{x_{n-1}+x_{n-2}}{2}$

else $\qquad\qquad\qquad\qquad\qquad x_n = \dfrac{x_{n-1}+x_{n-3}}{2}$

The sequence is generated till the last approximation x_n is correct up to the desired accuracy as discussed after Example 3.2. The graphical representation of the method is given in Fig. 3.11.

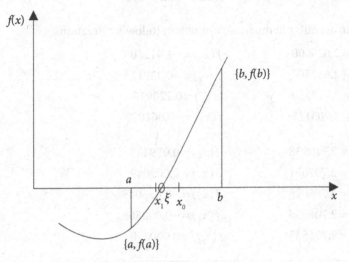

Fig. 3.11 Bisection method.

Example   **3.2**

Use intermediate value theorem to prove that the equation $x^3 - 4x - 9 = 0$ has a root in the interval (2, 3). Obtain the ten iterations of the Bisection method to compute the approximate root of the equation.

Ans. The function $f(x) = x^3 - 4x - 9$ is continuous in the interval (2, 3) and

$$f(2) = -9, \ f(3) = 6$$

According to intermediate value theorem, at least one root of the equation $x^3 - 4x - 9 = 0$ lies in the interval (2, 3). To obtain iterations of Bisection method, let us start with initial approximations

$$a = 2, \ b = 3$$

The first approximation to root is the midpoint of the interval (a, b).

$$x_0 = \frac{a+b}{2} = \frac{2+3}{2} = 2.5$$

$f(2.5) = -3.375000$, while $f(3) = 6$

So, root lies in the interval (2.5, 3)

$$x_1 = \frac{2.5+3}{2} = 2.75, \qquad f(2.75) = 0.796875$$

Proceeding in a similar manner, we can obtain following iterations

$x_2 = 2.625000$	$f(x_2) = -1.412109$
$x_3 = 2.687500$	$f(x_3) = -0.339111$
$x_4 = 2.718750$	$f(x_4) = 0.220917$
$x_5 = 2.703125$	$f(x_5) = -0.061077$
$x_6 = 2.710938$	$f(x_6) = 0.079423$
$x_7 = 2.707031$	$f(x_7) = 0.009049$
$x_8 = 2.705078$	$f(x_8) = -0.026045$
$x_9 = 2.706055$	$f(x_9) = -0.008506$
$x_{10} = 2.706543$	$f(x_{10}) = 0.000270$

In this example, ten iterations were computed, but we can also compute more iterations. In general, numerical methods will not find the exact root; rather they find some suitably accurate approximations to the root. The numerical algorithms refine these approximations with iterations. Now the question is where to stop these iterations? It is necessary to impose certain conditions under which the iteration process is terminated.

An error tolerance needs to be specified, i.e., either the value of the function is close to zero $\left(|f(x_n)| < \delta\right)$ or the absolute difference between last two consecutive iterations is below the tolerance limit $\left(|(x_n - x_{n-1})| < \varepsilon\right)$ or both.

Let us define error tolerance $\delta = 0.0005$ in the Example 3.2. We will stop iterations if the absolute value of the function is less than $\delta = 0.0005$. At 10th iteration, we have $x_{10} = 2.706543$, and function value is $f(x_{10}) = 0.000270$, which is less than error tolerance $\delta = 0.0005$. So, we can stop here and say that approximate root is $x_{10} = 2.706543$.

Similarly, say we define $\varepsilon = 0.002$. The difference between 8th and 9th approximations is less than $\varepsilon = 0.002$. Therefore, we can stop iterations and say that approximate root is $x_9 = 2.706055$.

Note: *It is worth mentioning here that the tolerance limit defines the accuracy of the root. Say, we want the approximate root to be exact up to five decimal places, then the tolerance limit is $\varepsilon = 0.000005$. But in that case, we need to perform more iterations.*

In this chapter, we will perform the iterations without specifying the error tolerance just to avoid the repetitions of the processes.

Example ———————————————— 3.3

Use Bisection method to compute the root of the equation $\cos x - 5x + 1 = 0$ in the interval $(0, 1)$. Note that the function $\cos(x)$ is to be evaluated in radians.

Ans. The function $f(x) = \cos x - 5x + 1$ is continuous and

$$f(0) = +ve, \ f(1) = -ve$$

Intermediate value theorem implies that there exists a root of the equation $\cos x - 5x + 1 = 0$ in the interval $(0, 1)$.

Let $a = 0$, $b = 1$ be the initial approximations to the root. 1st approximation to root is the midpoint of the interval (a, b).

$$x_0 = \frac{a+b}{2} = \frac{0+1}{2} = 0.5$$

$$f(0.5) = -ve, \text{ while } f(0) = +ve$$

$\Rightarrow$ Root lies in the interval $(0, 0.5)$, continuing in this way, we can easily find the following approximations to the root

$$x_1 = 0.250000$$
$$x_2 = 0.375000$$
$$x_3 = 0.437500$$
$$x_4 = 0.406250$$
$$x_5 = 0.390625$$
$$x_6 = 0.382812$$
$$\vdots$$
$$x_{12} = 0.385742$$

Note: *Any trigonometric functions like $\cos(x)$, $\sin(x)$, etc. must be computed in radians not in degree. So, we must keep it in mind that the calculator/computer must be in the radian mode. For example, here in this question, we have to compute the value of $\cos(x)$ at Radian x, not in degree x.*

Example ———————————————— 3.4

Calculate the first positive root of the equation $x - \tan(x) = 0$ with the help of Bisection method.

Ans.
It is easy to see from the Fig. 3.9 that the equation $x - \tan x = 0$ possesses an infinite number of roots. Also, the function $f(x) = x - \tan(x)$ has discontinuities at the points

$\pm \left(\dfrac{\pi}{2}, \dfrac{3\pi}{2}, \dfrac{5\pi}{2} \dots \right)$. So, to compute the first positive root of the equation $x - \tan(x) = 0$,

we cannot continue with the interval $(4, 5)$ due to discontinuity at the point $\dfrac{3\pi}{2} = 4.7124$.

The root lies in the interval $\left(\pi, \dfrac{3\pi}{2} \right)$ and function is also continuous in this interval

so that we can continue with an interval $\left(\pi, \dfrac{3\pi}{2} \right)$. For easy computations, we can also

proceed with smaller interval $(4, 4.5)$ as $f(4) = +ve$, $f(4.5) = -ve$.

Using Bisection method, 1st approximation to root is given by

$$x_0 = \frac{a+b}{2} = \frac{4+4.5}{2} = 4.25$$

The function, $f(4.25) = +ve$, while $f(4.5) = -ve$, so

$$x_1 = \frac{4.25 + 4.5}{2} = 4.375$$

Proceeding in a similar manner, following iterations can be computed easily

$$x_2 = 4.437500$$
$$x_3 = 4.468750$$
$$x_4 = 4.484375$$
$$x_5 = 4.492188$$
$$x_6 = 4.496094$$
$$\vdots$$
$$x_{12} = 4.493164$$

Number of Iterations in Bisection Method

Let ξ be the exact root of the equation $f(x) = 0$, and initial approximations for the root in Bisection method are a and b, then

$$\left| x_0 - \xi \right| \le \frac{b-a}{2}$$

$$\left| x_1 - \xi \right| \le \frac{b-a}{2^2}$$

$$\vdots$$

$$\left| x_n - \xi \right| \le \frac{b-a}{2^{n-1}}$$

Since ξ is the exact root and x_n is the nth approximation to the root, so $\varepsilon_n = \xi - x_n$ is the error in the nth approximation x_n. Let the permissible error be ε, then the number of iterations (n) required to get this accuracy is given by the following formula

$$\frac{b-a}{2^{n+1}} \le \varepsilon$$

$$\Rightarrow n \ge \frac{\ln(b-a) - \ln \varepsilon}{\ln 2} - 1 \qquad (3.2)$$

For example: Let error tolerance be $\varepsilon = .005$. If we start with an interval of unit length $(b - a = 1)$, then the required number of iterations are given by

$$n \ge \frac{\ln(1) - \ln(.005)}{\ln 2} - 1 = 6.6438 \approx 7$$

So, minimum seven iterations of Bisection method are required to get the root correct up to two decimal places $(\varepsilon = .005)$, and it excludes the initial iteration x_0.

3.4 Fixed-Point Method (or) Direct-Iteration Method (or) Method of Successive-Approximations (or) Iterative Method (or) One-Point-Iteration Method

Let us define the fixed point (or point of attraction) of a function $\phi(x)$ to start with Fixed-Point method.

Definition 3.1

A real number c is a fixed point of the function $\phi(x)$, if and only if we have $\phi(c) = c$, i.e., point c is fixed under the mapping function ϕ.

For example $x = 1, 2$ are the fixed points of the function $\phi(x) = \dfrac{x^2 + 2}{3}$ as $\phi(1) = 1$, $\phi(2) = 2$.

Geometrically, the fixed points of a function $\phi(x)$ are the points of intersection of the curve $y = \phi(x)$ and line $y = x$. Let

$$f(x) = \phi(x) - x$$

Then the following statements are equivalent
- i) C is the zero of the function $f(x)$
- ii) C is the root of the equation $f(x) = 0$
- iii) C is the fixed point of function $\phi(x)$

For example, consider the following equation

$$f(x) = x^2 - 3x + 2 = 0$$

(or) $$\phi(x) - x = \frac{x^2 + 2}{3} - x = 0$$

(or) $$x = \frac{x^2 + 2}{3} = \phi(x)$$

Now, $x = 1, 2$ are the roots of the equation $x^2 - 3x + 2 = 0$, and also fixed points of the function $\phi(x) = \dfrac{x^2 + 2}{3}$.

Fixed-Point Method

Consider the equation $f(x) = 0$ and $f(x) = \phi(x) - x$. Let x_0 be any initial approximation to the root of the equation $f(x) = 0$, then the next approximation can be calculated by the value of the function $\phi(x)$ at this point.

$$x_1 = \phi(x_0)$$

Proceeding in a similar manner, we can easily find the various approximations to the fixed point.

$$x_2 = \phi(x_1)$$

$$x_3 = \phi(x_2)$$

$$\vdots$$

$$\vdots$$

$$x_{n+1} = \phi(x_n) \tag{3.3}$$

Example   **3.5**

Find the root of the equation $\cos x - 5x + 1 = 0$ with the aid of Fixed-Point method. Perform only five iterations and start with an initial approximation $x_0 = 0.5$.

Ans. Rewriting the given equation as follows

$$f(x) = \cos x - 5x + 1 = 0$$

$$x = \frac{\cos x + 1}{5} = \phi(x)$$

The iteration function is $\phi(x)=\dfrac{(\cos x+1)}{.\ 5}$. Starting with an initial approximation $x_0=0.5$, we have following next approximation

$$x_1=\phi(x_0)=\frac{(\cos(0.5)+1)}{5}=0.375517$$

Similarly, the Fixed-Point method (3.3) generates the following successive approximations for the root of the given equation.

$$x_2=\phi(x_1)=\frac{(\cos(0.375517)+1)}{5}=0.386064$$

$$x_3=\phi(x_2)=0.385280$$

$$x_4=\phi(x_3)=0.385339$$

$$x_5=\phi(x_4)=0.385334$$

Note: We can have infinite number of iteration functions for any given equation. For example

$$f(x)=\cos x-5x+1=0$$

$$x=\frac{\cos x+1}{5}=\phi(x)$$

$$x=\cos x-4x+1=\phi(x)$$

$$x=\frac{\cos x-3x+1}{2}=\phi(x)$$

$$\vdots$$

Which iteration function is good enough? We will discuss various criteria for the selection of iteration function in coming sections and examples.

Example ───────────────────────────────── **3.6**

Prove that the function $\phi(x)=\tan^{-1}(x)+\pi$ can be an iteration function for the equation $x-\tan(x)=0$. Use this iteration function to compute the approximate root corrects up to five decimal places. Take initial approximation $x_0=4.5$.

Ans. First, we rewrite the given equation as follows

$$x-\tan(x)=0$$

$$x=\tan(x)$$

$$x=\tan(x-\pi)$$

$$\tan^{-1}(x) = x - \pi$$
$$x = \tan^{-1}(x) + \pi = \phi(x)$$

So, we can have iteration function $\phi(x) = \tan^{-1}(x) + \pi$ (we have some specific reason for selection of this iteration function, which will be discussed in Section 3.8.2 of this chapter).

Starting with initial approximation $x_0 = 4.5$, the iterations for the Fixed-Point method (3.3) are as follows

$$x_1 = \phi(x_0) = \tan^{-1}(4.5) + 3.14159265 = 4.493720$$

$$x_2 = \phi(x_1) = \tan^{-1}(4.493720) + 3.14159265 = 4.493424$$

Similarly, other approximations are given by

$$x_3 = 4.493410, \quad x_4 = 4.493410$$

The difference between last two iterations is less than 0.000005, so the root $x = 4.493410$ is correct to five decimal places.

Let us introduce the idea of convergence and divergence for the sequence generated by iterative methods with an example.

Example ─── **3.7** ───

The equation $x^3 - 4x - 9 = 0$ has a root in the interval (2, 3). Use Fixed-Point method to show that the iteration function $\phi(x) = \dfrac{(x^3 - 9)}{4}$ generates a divergent sequence, while the

iteration function $\phi(x) = -\dfrac{x^3 - 19x - 9}{15}$ generates a convergent sequence. Start with any

initial approximation in the interval (2, 3).

Ans. Using the given equation, we can have following iteration functions

$$x^3 - 4x - 9 = 0$$

$$x = \frac{(x^3 - 9)}{4} = \phi(x)$$

$$x = -\frac{x^3 - 19x - 9}{15} = \phi(x)$$

The root of the equation $x^3 - 4x - 9 = 0$ lies in the interval (2, 3). So, we can continue with any initial approximation in this interval say $x_0 = 2.7$. Using Fixed-Point method with iteration function $\phi(x) = \dfrac{(x^3 - 9)}{4}$, we have following successive approximations

$$x_1 = \phi(x_0) = \frac{((2.7)^3 - 9)}{4} = 2.670750$$

$$x_2 = \phi(x_1) = \frac{((2.670750)^3 - 9)}{4} = 2.512554$$

$$x_3 = 1.715393$$

$$x_4 = -0.988084$$

$$x_5 = -2.491169$$

$$x_6 = -6.114999$$

$$x_7 = -59.414871$$

$$x_8 = -52437.757812$$

$$x_9 = -36047266250752.000000$$

$$\vdots$$

It is easy to see that sequence generated by the Fixed-Point method is diverging towards $-\infty$. So, the root cannot be obtained with this iteration function $\phi(x)$.

Let us change the iteration function $\phi(x) = -\dfrac{x^3 - 19x - 9}{15}$. Using Fixed-Point method with any initial approximation in (2, 3), say $x_0 = 2.5$, we have following successive approximation

$$x_1 = \phi(x_0) = -\frac{x_0^3 - 19x_0 - 9}{15} = 2.722500$$

Similarly, other iterations are as follows

$$x_2 = 2.702678$$

$$x_3 = 2.707284$$

$$x_4 = 2.706378$$

$$x_5 = 2.706558$$

$$x_6 = 2.706522$$

$$x_7 = 2.706529$$

$$x_8 = 2.706528$$

$$\vdots$$

Now, the iterations are converging to the root. This iteration function satisfies certain convergence conditions, which we will discuss in Section 3.8.2 of this chapter.

Example ── **3.8**

Use the Fixed-Point method to compute the root of the equation $x - e^{-x} = 0$ in the interval (0, 1).

Ans. Let us rewrite the given equation as follows

$$x = e^{-x}$$

We will use the iteration function $\phi(x) = e^{-x}$ for the Fixed-Point method. Since $f(0) = -1$ and $f(1) = 0.631$, so the root of the equation lies in (0, 1). Let us start with the initial approximation $x_0 = 0.5$.

$$x_1 = \phi(x_0) = e^{-(0.5)} = 0.606531$$
$$x_2 = \phi(x_1) = e^{-(0.606531)} = 0.545239$$

Similarly, we can compute following iterations

$x_3 = 0.579703$	$x_7 = 0.568438$	$x_{11} = 0.567277$	$x_{15} = 0.567157$
$x_4 = 0.560065$	$x_8 = 0.566409$	$x_{12} = 0.567067$	$x_{16} = 0.567135$
$x_5 = 0.571172$	$x_9 = 0.567560$	$x_{13} = 0.567186$	$x_{17} = 0.567148$
$x_6 = 0.564863$	$x_{10} = 0.566907$	$x_{14} = 0.567119$	$x_{18} = 0.567141$

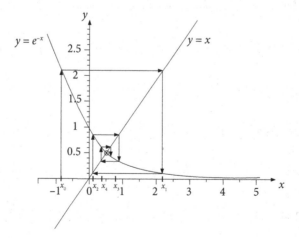

Fig. 3.12 Fixed Point oscillations around the root of the equation $x = e^{-x}$, Root ≈ 0.567141

It is easy to see from the graph that, iterations are oscillating around the root (0.56714). Initial iteration is 0.5 (left to the root), then iteration is 0.606531 (right to the root), next iteration is 0.545239 (again left to the root) and so on.

3.5 Newton–Raphson (NR) Method

Let x_0 be an initial approximation to the root of the equation $f(x) = 0$, and h be the error in this approximation. Then $x_0 + h$ is the exact root of the equation $f(x) = 0$, hence

$$f(x_0 + h) = 0$$

The Taylor Series expansion of $f(x_0 + h)$ is given by

$$f(x_0 + h) = f(x_0) + (h)f'(x_0) + \frac{(h)^2}{2!} f''(x_0) + \cdots$$

where the prime (') denotes derivative on x. If our initial approximation is close to root, then h is a very small quantity. Setting the quadratic and higher terms of h to zero, we get

$$f(x_0) + (h)f'(x_0) = 0$$

$$\Rightarrow h = \frac{-f(x_0)}{f'(x_0)}$$

Let $x_1 = x_0 + h$ be the next approximation to the root, so

$$x_1 = x_0 - \frac{f(x_0)}{f'(x_0)}$$

Similarly, the approximation x_1 can be refined further for a better approximation and so on. In general, the subsequent iterations can be defined as follows

$$x_{n+1} = x_n - \frac{f(x_n)}{f'(x_n)} \quad n = 0, 1, 2 \ldots \tag{3.4}$$

Example   **3.9**

Use Newton–Raphson method with initial approximation, $x_0 = 2.5$, to compute the root of the equation $x^3 - 4x - 9 = 0$ correct to five decimal places.

Ans. Initial approximation is $x_0 = 2.5$. On using Newton–Raphson method (3.4), we have following successive approximations

$$x_1 = x_0 - \frac{f(x_0)}{f'(x_0)} = x_0 - \frac{x_0^3 - 4x_0 - 9}{3x_0^2 - 4} = 2.5 - \frac{(2.5)^3 - 4(2.5) - 9}{3(2.5)^2 - 4} = 2.728814$$

$$x_2 = x_1 - \frac{f(x_1)}{f'(x_1)} = 2.728814 - \frac{(2.728814)^3 - 4(2.728814) - 9}{3(2.728814)^2 - 4} = 2.706749$$

$$x_3 = 2.706528$$

$$x_4 = 2.706528$$

Example ─── **3.10** ───────

Compute the root of the equation $\cos x - 5x + 1 = 0$ with the help of Newton–Raphson method. Use an initial approximation $x_0 = 0.5$ for computation.

Ans. The root of the equation lies in the interval $(0, 1)$. Let initial approximation be $x_0 = 0.5$. Now, using Newton–Raphson method (3.4), we have following successive approximations

$$x_1 = x_0 - \frac{f(x_0)}{f'(x_0)} = x_0 - \frac{\cos(x_0) - 5x_0 + 1}{-\sin(x_0) - 5} = 0.5 - \frac{\cos(0.5) - 5(0.5) + 1}{-\sin(0.5) - 5} = 0.386408$$

$$x_2 = x_1 - \frac{f(x_1)}{f'(x_1)} = 0.386408 - \frac{\cos(0.386408) - 5(0.386408) + 1}{-\sin(0.386408) - 5} = 0.385335$$

$$x_3 = 0.385335$$

The root of the equation $x - \tan(x) = 0$ with Newton–Raphson method will be discussed in Section 3.8.3.

Example ─── **3.11** ───────

Derive Newton–Raphson iterative formula for computing the value of $\sqrt{M}$, where M is a positive real number. Hence compute $\sqrt{40}$.

Ans. Let $x = \sqrt{M}$, then

$$f(x) = x^2 - M = 0$$

The aim is to compute the root of this equation using Newton–Raphson method. The Newton–Raphson method (3.4) for this equation is given by

$$x_{n+1} = x_n - \frac{f(x_n)}{f'(x_n)}$$

$$x_{n+1} = x_n - \frac{x_n^2 - M}{2x_n} = \frac{1}{2}\left(x_n + \frac{M}{x_n}\right)$$

To compute $\sqrt{40}$, the corresponding equation is $x^2 - 40 = 0$. The root of this equation lies in the interval $(6, 7)$. Let initial approximation be $x_0 = 6.5$. Using afore-mentioned iterative formula, we have

$$x_1 = \frac{1}{2}\left(x_0 + \frac{40}{x_0}\right) = \frac{1}{2}\left(6.5 + \frac{40}{6.5}\right) = 6.326923$$

$$x_2 = \frac{1}{2}\left(x_1 + \frac{40}{x_1}\right) = \frac{1}{2}\left(6.326923 + \frac{40}{6.326923}\right) = 6.324556$$

$$x_3 = 6.324555$$

Note that, similar kind of formula can be obtained for the kth root of a number M, i.e., $\sqrt[k]{M}$. Let $x = \sqrt[k]{M}$, then $f(x) = x^k - M = 0$.

Geometrical Interpretation of Newton–Raphson Method

Let $(x_0, f(x_0))$ be any point on the curve $y = f(x)$, then equation of tangent at this point is

$$y - f(x_0) = f'(x_0)(x - x_0)$$

Let this tangent cut the x-axis at the point $x = x_1$. Value of y is zero at this point, so the equation of tangent is given by

$$-f(x_0) = f'(x_0)(x_1 - x_0)$$

i.e., $\qquad x_1 = x_0 - \dfrac{f(x_0)}{f'(x_0)}$

In Newton–Raphson method, the function $f(x)$ is approximated with the tangent of $f(x)$ at point $x = x_0$. Next approximation is the point of intersection of the tangent (at the point $x = x_0$) and the x-axis. In general, the iteration x_{n+1} is the value of x at which tangent at the point $(x_n, f(x_n))$ cuts the x-axis.

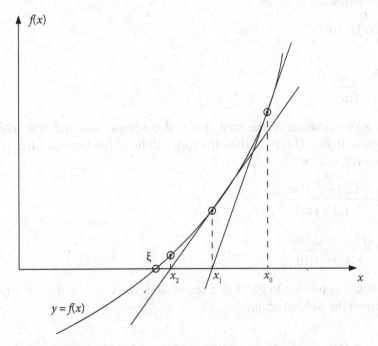

Fig. 3.13 Graphical interpretation of the Newton–Raphson method

Note: *We can easily observe from the iterations of Newton–Raphson method in Examples 3.9–11 that root with very high accuracy is obtained in very small number of iterations compared to other methods. It is due to the high order of convergence of Newton–Raphson method. But the method converges to the root only for closer initial approximations with certain conditions on the function. So, this method can also be used to refine the approximations of the root that are obtained by some other methods in very less number of iterations. All these aspects, we will discuss in sections 3.8 and 3.9 of this chapter.*

3.6 Regula Falsi Method (or) Method of False Position

Like Bisection method, the False-Position method also requires two initial guesses, a and b, which bracket the root. Let the root of the equation $f(x) = 0$ lie in the interval $[a, b]$, and the function $f(x)$ be a continuous function. Then, the next approximation is the point of intersection of the chord joining these two points $\{(a, f(a)), (b, f(b))\}$ and the x-axis. This point of intersection gives a "false position" of the root, and it is the reason for the name, method of false position or in Latin, *regula falsi*. The equation of the chord joining two points $(a, f(a))$ and $(b, f(b))$ is as follows

$$y - f(b) = \frac{f(b) - f(a)}{b - a}(x - b)$$

Let x_0 be the point of intersection of this chord and x-axis. At point $x = x_0$, this equation is reduced to the following equation

$$-f(b) = \frac{f(b) - f(a)}{b - a}(x_0 - b)$$

$$x_0 = \frac{af(b) - bf(a)}{f(b) - f(a)} \tag{3.5}$$

So, x_0 is the approximation to the root. We will continue with the new interval which brackets the root. If $f(a)f(x_0) < 0$, then the root will lie in the interval (a, x_0), so the next approximation is given by

$$x_1 = \frac{af(x_0) - x_0 f(a)}{f(x_0) - f(a)}$$

else $\quad x_1 = \frac{bf(x_0) - x_0 f(b)}{f(x_0) - f(b)} \tag{3.6}$

We will continue this process to generate a sequence $x_1, x_2, x_3 \cdots$, till the nth approximation x_n is correct up to the desired accuracy.

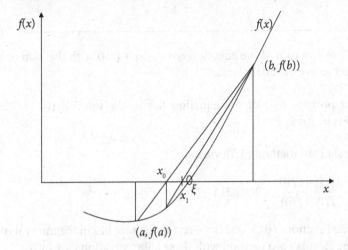

Fig. 3.14 Graphical representation of Regula Falsi method

Example ────────────────────────────── 3.12

Use Regula Falsi method to compute the root of the equation $x^3 - 4x - 9 = 0$ in the interval (2, 3). Perform only six iterations and start with the initial approximations 2 and 3.

Ans. The initial approximations are $a = 2$ and $b = 3$. Using Regula Falsi method, we have following next approximation from Eq.(3.5)

$$x_0 = \frac{2f(3) - 3f(2)}{f(3) - f(2)} = 2.6$$

The function $f(2.6) = -ve$, so the root lies in the interval (2.6, 3).

On applying the Regula Falsi method (3.6) with approximations 2.6 and 3, we have

$$x_1 = \frac{2.6 f(3) - 3f(2.6)}{f(3) - f(2.6)} = 2.693252$$

As $f(2.693252) = -ve$, so the root lies in between 2.693252 and 3. Continuing with these approximations, we have

$$x_2 = \frac{2.693252 f(3) - 3f(2.693252)}{f(3) - f(2.693252)} = 2.704918$$

Similarly, we have

$$x_3 = 2.706333$$
$$x_4 = 2.706504$$
$$x_5 = 2.706525$$
$$x_6 = 2.706528$$

Example ── 3.13 ──────

Find the first positive root of the equation $\cos x - 5x + 1 = 0$ with the help of Regula Falsi method correct to five decimal places.

Ans. The first positive root of the equation lies in the interval $(0, 1)$. Let the initial approximations be $a = 0$, $b = 1$.

On using Regula Falsi method (3.5), we have

$$x_0 = \frac{0f(1) - 1f(0)}{f(1) - f(0)} = 0.366321$$

The value of the function $f(0.366321) = +ve$. So, the root lies in the interval $(0.366321, 1)$. Again applying Regula Falsi method with these approximations, we have

$$x_1 = \frac{0.366321f(1) - 1f(0.366321)}{f(1) - f(0.366321)} = 0.384476$$

Similarly, we have

$$x_2 = 0.385296$$
$$x_3 = 0.385333$$
$$x_4 = 0.385334$$

Example ── 3.14 ──────

Apply Regula Falsi method with initial approximations 4 and 4.5 to compute the root of the equation $x - \tan(x) = 0$.

Ans. Applying Regula Falsi method with initial approximations 4 and 4.5, we have following successive approximations to the root

$$x_0 = \frac{4f(4.5) - 4.5f(4)}{f(4.5) - f(4)} = 4.476954$$

$$x_1 = \frac{4.476954f(4.5) - 4.5f(4.476954)}{f(4.5) - f(4.476954)} = 4.492898$$

$$x_2 = 4.493393$$
$$x_3 = 4.493409$$

3.7 Secant Method

In this method, we proceed like the Regula Falsi method, i.e., we approximate our curve $f(x)$ with the chord joining two points on the curve. The only difference is that we don't check the intermediate value property for the next two approximation points.

Let x_0 and x_1 be the two initial approximations, then the next approximation, as in Regula Falsi method, is given by

$$x_2 = \frac{x_0 f(x_1) - x_1 f(x_0)}{f(x_1) - f(x_0)}$$

Now, we will use the approximations x_1 and x_2 to compute the next approximation x_3.

$$x_3 = \frac{x_1 f(x_2) - x_2 f(x_1)}{f(x_2) - f(x_1)}$$

Similarly, the approximation x_4 is given by

$$x_4 = \frac{x_2 f(x_3) - x_3 f(x_2)}{f(x_3) - f(x_2)}$$

Note that the next approximation is computed with latest available approximations. We do not check the intermediate value property in Secant method. So, the approximations need not to bracket the root.

In general, let x_{n-1} and x_n be two approximations at any step. The next approximation is given by the equation of the chord joining these two points $\left(x_{n-1}, f(x_{n-1})\right)$ and $\left(x_n, f(x_n)\right)$

$$x_{n+1} = \frac{x_{n-1} f(x_n) - x_n f(x_{n-1})}{f(x_n) - f(x_{n-1})}$$

$$\text{(or)} \quad x_{n+1} = x_n - \frac{x_n - x_{n-1}}{f(x_n) - f(x_{n-1})} f(x_n) \qquad n = 1, 2, 3, \ldots \qquad (3.7)$$

Note: *In both the methods (Regula Falsi and Secant), we approximate our curves with the chord joining two points. The difference is that, in Regula Falsi method, we replace one of the two old approximations, so that the root is always between these two new approximations. But in the case of Secant method, the oldest point is always replaced with a new approximation. It is not necessary for Secant method that next two guesses will bracket the root.*

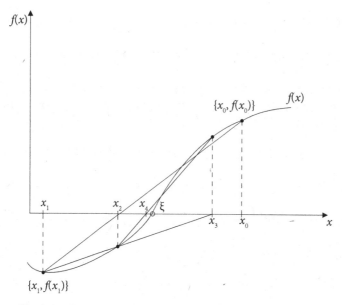

Fig. 3.15 Geometrical representation of the Secant method

Example — 3.15

Find the approximate zero of the function $x^3 - 4x - 9$ using Secant method correct to five decimal places.

Ans. The Secant method (3.7) is given by

$$x_{n+1} = \frac{x_{n-1} f(x_n) - x_n f(x_{n-1})}{f(x_n) - f(x_{n-1})}; \qquad n = 1, 2, \cdots$$

Let initial approximations be $x_0 = 2$ and $x_1 = 3$. Using Secant method, we have following next approximation

$$x_2 = \frac{x_0 f(x_1) - x_1 f(x_0)}{f(x_1) - f(x_0)} = \frac{2 f(3) - 3 f(2)}{f(3) - f(2)} = 2.6$$

Similarly, we can obtain the following approximations by using the Secant method

$$x_3 = \frac{x_1 f(x_2) - x_2 f(x_1)}{f(x_2) - f(x_1)} = \frac{3 f(2.6) - 2.6 f(3)}{f(2.6) - f(3)} = 2.693252$$

$$x_4 = 2.707193$$

$$x_5 = 2.706524$$

$$x_6 = 2.706528$$

Example 3.16

Obtain the root of the equation $\cos x - 5x + 1 = 0$ with the help of Secant method.

Ans. On using Secant method (3.7) with initial approximations $x_0 = 0$ and $x_1 = 1$, we have

$$x_2 = \frac{x_0 f(x_1) - x_1 f(x_0)}{f(x_1) - f(x_0)} = \frac{0 f(1) - 1 f(0)}{f(1) - f(0)} = 0.366321$$

Similarly, the approximations are as follows

$$x_3 = \frac{x_1 f(x_2) - x_2 f(x_1)}{f(x_2) - f(x_1)} = 0.384476$$

$$x_4 = 0.385336$$

$$x_5 = 0.385335$$

$$x_6 = 0.385335$$

Example 3.17

Apply Secant method to compute the first positive root of equation $x - \tan(x) = 0$, which is close to 4.5. Procced with initial approximations $x_0 = 4$ and $x_1 = 5$, and conclude that we cannot obtain this root with these initial approximations. Also, show that the Secant method with initial approximations 4.4 and 4.6 converges to this root.

Ans. Let initial approximations be $x_0 = 4$ and $x_1 = 5$. Next approximation for Secant method (3.7) is given by

$$x_2 = \frac{x_0 f(x_1) - x_1 f(x_0)}{f(x_1) - f(x_0)} = \frac{4 f(5) - 5 f(4)}{f(5) - f(4)} = 3.486817$$

Secant method provides the following approximations

$$x_3 = \frac{x_1 f(x_2) - x_2 f(x_1)}{f(x_2) - f(x_1)} = 2.586052$$

$$x_4 = 38.867615$$

$$x_5 = -0.906766$$

$$x_6 = -1.315063$$

$$x_7 = -0.835925$$

$$x_8 = -0.778018$$

$$x_9 = -0.588269$$

$$x_{10} = -0.471978$$

$$\vdots$$

$$x_{29} = -0.001788$$

$$x_{30} = -0.001349$$

It is easy to see that Secant method is converging to the root $\xi = 0$. So, we cannot obtain the first positive root with the initial approximations $x_0 = 4$ and $x_1 = 5$. The convergence to the root $\xi = 0$ is also very slow, which is due to the fact that the equation has double root $\xi = 0$ ($f(0) = 0$, $f'(0) = 0$). We will discuss about multiple roots in Section 3.9.

But if we continue with initial approximations $x_0 = 4.4$ and $x_1 = 4.6$, then the next approximation using Secant method is given by

$$x_2 = \frac{x_0 f(x_1) - x_1 f(x_0)}{f(x_1) - f(x_0)} = \frac{4.4 f(4.6) - 4.6 f(4.4)}{f(4.6) - f(4.4)} = 4.446862$$

On using $x_1 = 4.6$ and $x_2 = 4.446862$, we have

$$x_3 = \frac{x_1 f(x_2) - x_2 f(x_1)}{f(x_2) - f(x_1)} = 4.470296$$

Similarly, other approximations are as follows

$$x_4 = 4.498540$$
$$x_5 = 4.492849$$
$$x_6 = 4.493396$$
$$x_7 = 4.493410$$
$$x_8 = 4.493410$$

Now, the method is converging to the root for these initial approximations.

Note: *In the case of closed ends method like Bisection method, if f(x) is continuous in (a, b) and root lies in (a, b), our method must converge to the root. But in the case of open end methods like Fixed-Point method, we have to satisfy certain conditions for convergence. We will discuss the convergence criteria for various methods in next section.*

If the convergence criteria are not satisfied for any method, then it is recommended to start with closer approximations to the root. Rather, starting with close approximation is not the guarantee for convergence.

3.8 Convergence Criteria

So far, we have generated the sequences $\{x_n\}$ of approximations to the root of the equation $f(x) = 0$ by various iterative methods. In this section, we will discuss the criteria for the convergence of these sequences. Once we ensure that sequence is converging to the root, then in Section 3.9 we will discuss how rapidly the sequence will converge to the root (order of convergence).

Definition 3.2

A sequence $\{a_n\}$, $n = 1, 2, 3, 4 \ldots$ is a convergent sequence if it approaches to a unique finite limit.

For example, the sequence $\left\{1, \dfrac{1}{2}, \dfrac{1}{3}, \dfrac{1}{4}, \cdots\right\}$ or $\left\{\dfrac{1}{n}\right\}$ is a convergent sequence and converging to number zero. But the sequence $\{n\}$ is divergent as it tends to infinity. Also the sequence $\left\{(-1)^n\right\}$ is divergent as limit is oscillating between -1 and 1, i.e., not unique.

3.8.1 Convergence of Bisection Method

In Bisection method, we need two initial guesses, a and b, which bracket the root, i.e., $f(a)f(b) < 0$ and also $f(x)$ must be a continuous function in (a, b). If these two conditions hold, then it is easy to see that Bisection method is converging to the root.

Let ξ be the exact root of the equation $f(x) = 0$, and $\{x_n\}$ is a sequence generated by the Bisection method. Then, we have

$$|x_0 - \xi| \le \frac{b-a}{2}$$

$$|x_1 - \xi| \le \frac{b-a}{2^2}$$

$$\vdots$$

$$|x_n - \xi| \le \frac{b-a}{2^{n+1}} \tag{3.8}$$

Since the quantities b and a are finite, so

$$\lim_{n \to \infty} \frac{b-a}{2^{n+1}} \to 0$$

This implies

$$\lim_{n \to \infty} |x_n - \xi| \to 0$$

$$\lim_{n \to \infty} x_n \to \xi$$

It proves that sequence $\{x_n\}$ is a convergent sequence.

3.8.2 Convergence of Fixed-Point Method

In Fixed-Point method, first, we rewrite the equation $f(x) = 0$ as $x = \phi(x)$. For each equation $f(x) = 0$, there are infinite numbers of iteration function $\phi(x)$. For example, in the case of the equation $x - e^{-x} = 0$, we can take iteration functions e^{-x}, $-\ln(x)$, $2x - e^{-x}$, $\dfrac{3x - e^{-x}}{2}$, and so on. So which one is to select? Since the method is an iterative procedure, so the convergence of the Fixed-Point method will provide the answer. The function which satisfies the convergence criterion is the iteration function.

For convergence of Fixed-Point method in any interval $[a, b]$, the following three requirements should be met

i)   For any initial approximation x_0 in $[a, b]$, we can compute successively x_1, x_2, $\cdots$ for given iteration function $\phi(x)$. Consider the iteration function $\phi(x) = \ln(x)$, let $x_0 = 20$, then 2.995732, 1.0971887, 0.09275118, −2.37783484 are successive approximations x_1, x_2, x_3. But, it is not possible to compute x_4 and further iterations. Hence, we cannot take $\ln(x)$ as iteration function. Similarly, other examples are $-\sqrt{x}$, $\tan(x)$ (at $\pi/2$), etc.

ii) There exists a fixed point ξ of $\phi(x)$ in $[a, b]$.

iii) The sequence x_1, x_2, $\cdots$ will converge to fixed point ξ.

The following three theorems (3.2–4) establish the conditions on the iteration function $\phi(x)$, which need to be satisfied for the existence of fixed point and convergence of the Fixed-Point method to a unique fixed point.

Theorem 3.2

Let the iteration function $\phi(x) \in [a, b]$, $\forall x \in [a, b]$ and $\phi(x)$ be defined on the interval $[a, b]$. Then, we can compute x_1, x_2, $\cdots$ using Fixed-Point iteration for any initial approximation x_0 in $[a, b]$.

Proof: Let x_0 be any point in $[a, b]$, then

$$x_1 = \phi(x_0)$$

Since the iteration function $\phi(x)$ is defined on the interval $[a, b]$, so we can easily compute x_1.

Also, $\phi(x) \in [a, b]$ $\quad$ $\forall x \in [a, b]$ thus $x_1 \in [a, b]$.

Similarly, we can calculate $x_2 = \phi(x_1)$ and other iterations of Fixed-Point method.

Example ── **3.18** ──

Consider the equation $x - \cos(x) = 0$. Let the iteration function be $\phi(x) = \cos(x)$ This iteration function is defined on $[0, 1]$ and also

$$\phi(x) \in [0, 1] \qquad \forall x \in [0, 1]$$

Therefore, we can compute fixed point iterations $x_1, x_2, x_3 \ldots$ for any initial approximation $x_0 \in [0, 1]$.

Theorem 3.3

Let iteration function $\phi(x) \in [a, b]$, $\forall x \in [a, b]$, and function $\phi(x)$ be continuous on $[a, b]$, then there exists a fixed point of $\phi(x)$ in $[a, b]$.

Proof: Since $\phi(x) \in [a, b]$, $\forall x \in [a, b]$, we have

$$\phi(a) \geq a \text{ and } \phi(b) \leq b$$

If $a = \phi(a)$ then a is the fixed point of $\phi(x)$ and if $b = \phi(b)$ then b is the fixed point of the function $\phi(x)$. If both are not true, then

$$a < \phi(a) \text{ and } b > \phi(b) \tag{3.9}$$

Let us consider the following function

$$f(x) = x - \phi(x)$$

The iteration function $\phi(x)$ is a continuous function, so $f(x)$ is also a continuous function. Using inequalities (3.9), we have

$$f(a) = a - \phi(a) < 0$$

$$f(b) = b - \phi(b) > 0$$

These inequalities imply $f(a)f(b) < 0$, and also the function $f(x)$ is a continuous function. On using intermediate value theorem, there exists a number $\xi \in (a, b)$ such that $f(\xi) = 0$.

Hence, there exists a fixed point ξ of $\phi(x)$ in $[a, b]$.

Example ── **3.19** ──

Consider the equation $x - \cos(x) = 0$. The iteration function $\phi(x) = \cos(x)$ is continuous on $[0, 1]$ and

$$\phi(x) \in [0, 1] \qquad \forall x \in [0, 1]$$

Hence, there exists a fixed point of $\phi(x) = \cos(x)$ in $[0, 1]$.

Theorem 3.4

Let $\phi(x) \in [a, b]\ \forall x \in [a, b]$, and $\phi(x)$ be differentiable on the interval $[a, b]$. Let there be a non-negative constant $k < 1$, such that

$$|\phi'(x)| \le k \quad \forall x \in [a, b]$$

Then the sequence $\{x_n\}$ generated using Fixed-Point method with any initial approximation x_0 in the interval $[a, b]$, will converge to the fixed point of the function $\phi(x)$ in the interval $[a, b]$

Proof: The function $\phi(x)$ is differentiable on $[a, b]$, so it is also a continuous function. Also, we have

$$\phi(x) \in [a, b]\ \forall x \in [a, b]$$

So, according to Theorem 3.3, there exists a fixed point $\xi \in (a, b)$ of the iteration function, $\phi(x)$. ξ is the fixed point of iteration function, $\phi(x)$.

$$\Rightarrow \xi = \phi(\xi) \tag{3.10}$$

The fixed point iterations (3.3) are given by

$$x_{n+1} = \phi(x_n) \tag{3.11}$$

On using equations (3.10, 3.11), we have

$$\xi - x_{n+1} = \phi(\xi) - \phi(x_n)$$

Let $\varepsilon_n = \xi - x_n$ be the error term at nth iteration. Then, we have

$$\varepsilon_{n+1} = \xi - x_{n+1}$$
$$= \phi(\xi) - \phi(x_n) \tag{3.12}$$

Using Lagrange mean value theorem, we have

$$\phi(\xi) - \phi(x_n) = \phi'(\lambda_n)(\xi - x_n) \text{ for some } \lambda_n \in (x_n, \xi) \text{ or } (\xi, x_n)$$
$$= \phi'(\lambda_n)\varepsilon_n$$
$$\le k\varepsilon_n$$

On using equation (3.12), we have

$$\Rightarrow \varepsilon_{n+1} \le k\varepsilon_n \tag{3.13}$$

Similarly, we can prove that $\varepsilon_n \le k\varepsilon_{n-1}$. On using this in the inequality (3.13), we have

$$\varepsilon_{n+1} \le k^2 \varepsilon_{n-1}$$

Repeatedly applying the same procedure, we get

$$\varepsilon_{n+1} \le k^{(n+1)} \varepsilon_0$$

Since the ε_0 is the error in the initial approximation, it is a finite quantity and $k < 1$, this implies

$$\lim_{n \to \infty} \varepsilon_{n+1} \to 0 \text{ (or) } \lim_{n \to \infty} x_n \to \xi$$

The sequence generated with Fixed-Point method converges to ξ.

Now, we have to prove that the fixed point is unique. Let there be another fixed point ζ of the iteration function $\phi(x)$, i.e., $\zeta = \phi(\zeta)$. If we generate the sequence with initial approximation, ζ, then it converges to the point ζ. But we have already proved that it converges to ξ, this shows that $\xi = \zeta$. So, fixed point is unique.

Note: *The condition $|\phi'(x)| < 1$ provides that method will converge to the root of the equation. But if the equation has more than one root, in that case, it is necessary to satisfy the following condition*

$$\phi(x) \in (a, b) \qquad \forall x \in (a, b)$$

If both the conditions are satisfied, then the method will converge to the unique root that lies between a and b.

To illustrate this fact, let us consider the equation $x - \tan(x) = 0$. We can take iteration function, $\phi(x) = \tan^{-1}(x)$, which satisfies the condition $|\phi'(x)| < 1$. But if we continue to this iteration function, then we will get the root $x = 0$ for any initial approximation. Since it possesses an infinite number of roots, so to calculate another root that is in fixed interval, we have to proceed as follows

First, we rewrite the given equation as follows

$$x - \tan(x) = 0$$
$$x - \tan(x - m\pi) = 0, \qquad m \text{ is any integer}$$
$$x = \tan(x - m\pi)$$
$$x = \tan^{-1}(x) + m\pi$$

then, we select m such that the iteration function satisfies the condition $\phi(x) \in (a, b) \qquad \forall x \in (a, b)$.

For example, if we want to compute the root in the interval, $(\pi, 2\pi)$, then m must be 1. So, the iteration function must be $x = \tan^{-1}(x) + \pi$. Similarly, for the root in the interval (198, 200), m must be 63, etc. Similar expressions for negative roots e.g. first negative root is given by the iteration function, $x = \tan^{-1}(x) - \pi$.

Example ———————————————————————————— **3.20**

Compute the first positive root, root near 30 and root in the interval (199, 200) for the equation, $x - \tan(x) = 0$.

Ans.

The iterative functions are as follows

$$\phi(x) = \tan^{-1}(x) + \pi \text{ (first positive root)},$$
$$\phi(x) = \tan^{-1}(x) + 9\pi \text{ (root near 30)}$$
$$\phi(x) = \tan^{-1}(x) + 63\pi \text{ (root in the interval (199, 200))}.$$

Table 3.1 has computations of Fixed-Point method for the first positive root, root near 30 and root in the interval (199, 200).

Table 3.1

1. First positive root:	2. Root near 30:
$\phi(x) = \tan^{-1}(x) + \pi$	$\phi(x) = \tan^{-1}(x) + 9\pi$.
$\phi(x) = \tan^{-1}(x) + 3.14159265$	$\phi(x) = \tan^{-1}(x) + 9(3.14159265)$
Let initial approximation be $x_0 = 4.5$, then	Let initial approximation be $x_0 = 30$, then
$x_1 = \phi(x_0) = \tan^{-1}(4.5) + 3.14159265 = 4.493720$	$x_1 = \phi(x_0) = \tan^{-1}(30) + 9(3.14159265)$
$x_2 = \phi(x_1) = \tan^{-1}(4.493720) + 3.14159265$	$\qquad = 29.811810$
$\qquad = 4.493424$	$x_2 = 29.811600$
$x_3 = 4.493410$	$x_3 = 29.811599$
$x_4 = 4.493410$	$x_4 = 29.811599$
Let us start with initial approximation $x_0 = 4$, then	**3. Root in the interval (199, 200):**
$x_1 = \phi(x_0) = \tan^{-1}(4) + 3.14159265 = 4.467410$	$\phi(x) = \tan^{-1}(x) + 63\pi$
$x_2 = \phi(x_1) = \tan^{-1}(4.467410) + 3.14159265$	
$\qquad = 4.492176$	Let initial approximation be $x_0 = 200$, then
Similarly, we have other approximations as follows	$x_1 = \phi(x_0) = \tan^{-1}(200) + 63(3.14159265)$
$\qquad x_3 = 4.493351$	$\qquad = 199.486130$
$\qquad x_4 = 4.493407$	$x_2 = 199.486125$
$\qquad x_5 = 4.493409$	$x_3 = 199.486121$
$\qquad x_6 = 4.493410$	$x_4 = 199.486121$

Geometrical Representations of Convergence Conditions of Fixed-Point Method

Now, we will describe the following four different cases geometrically, when tangent to iteration function $\phi(x)$ has a slope in the intervals $(-\infty, -1]$, $(-1, 0]$, $(0,1)$ and $[1, \infty)$.

Case 1. $0 \leq \theta < 45°$ or $0 \leq \phi'(x) < 1$

Monotonously Convergent Sequence

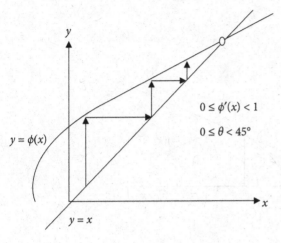

$0 \leq \phi'(x) < 1$

$0 \leq \theta < 45°$

$y = \phi(x)$

$y = x$

Case 2. $45° \leq \theta < 90°$ or $\phi'(x) \geq 1$

Monotonously Divergent Sequence

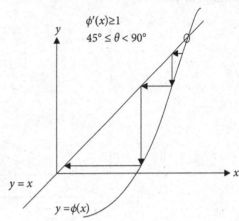

$\phi'(x) \geq 1$

$45° \leq \theta < 90°$

$y = x$

$y = \phi(x)$

Case 3. $90° \leq \theta \leq 135°$ or $\phi'(x) \leq -1$

Oscillatory Divergent Sequence

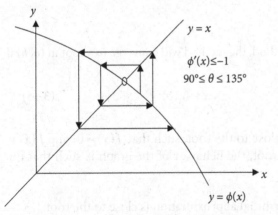

$y = x$

$\phi'(x) \leq -1$

$90° \leq \theta \leq 135°$

$y = \phi(x)$

Case 4. $135° < \theta \leq 180°$ or $-1 < \phi'(x) \leq 0$

Oscillatory Convergent Sequence

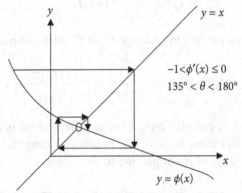

$y = x$

$-1 < \phi'(x) \leq 0$

$135° < \theta < 180°$

$y = \phi(x)$

Fig. 3.16 Geometrical representations of convergence conditions of Fixed-Point method

3.8.3 Convergence of Newton–Raphson Method

The iterations of Newton–Raphson method (3.4) are given by

$$x_{n+1} = x_n - \frac{f(x_n)}{f'(x_n)}$$

If function $f'(x)$ vanishes at any intermediate iteration point, then the method fails to converge. For example, consider the equation $x^3 + x^2 - 1 = 0$, root of this equation lies in the interval $(0, 1)$. But we cannot continue with the initial approximation $x_0 = 0$, as at this point $f'(x)$ vanishes.

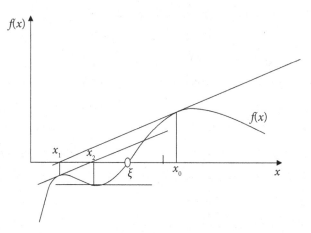

Fig. 3.17 Divergence of the Newton–Raphson algorithm

The Newton–Raphson method is equivalent to Fixed-Point method.

$$x_{n+1} = x_n - \frac{f(x_n)}{f'(x_n)} = \phi(x_n)$$

According to convergence of Fixed-Point method, the method will converge to a root in (a, b) if

$$|\phi'(x)| = \left[x - \frac{f(x)}{f'(x)}\right]' = \left|\frac{f(x)f''(x)}{(f'(x))^2}\right| < 1, \qquad \forall x \in (a,b) \qquad (3.14)$$

It is possible only if the approximation is close to the root, such that $f(x) \approx 0$; and $f'(x)$ is not close to zero. Geometrically, near the root, the behavior of the graph is such that the tangent is not parallel to the y-axis.

Newton–Raphson method converges if the initial approximation is close to the root.

Example ── **3.21**

Compute the first positive root of the equation $x - \tan(x) = 0$ with the help of Newton–Raphson method. Start with initial approximations 4, 4.25. 4.4 and 4.5.

Ans. Let initial approximation be $x_0 = 4$. Using Newton–Raphson method (3.4), we have following successive approximations

$$x_1 = x_0 - \frac{f(x_0)}{f'(x_0)}$$

$$= x_0 - \frac{x_0 - \tan(x_0)}{1 - \sec^2(x_0)}$$

$$= 4 - \frac{4 - \tan(4)}{1 - \sec^2(4)}$$

$$= 6.120158$$

$$x_2 = x_1 - \frac{f(x_1)}{f'(x_1)}$$

$$= x_1 - \frac{x_1 - \tan(x_1)}{1 - \sec^2(x_1)}$$

$$= 6.120158 - \frac{6.120158 - \tan(6.120158)}{1 - \sec^2(6.120158)}$$

$$= 238.403442$$

$x_3 = 1948.533203$ $\qquad$ $x_4 = 4231.167480$

$x_5 = 15102.013672$ $\qquad$ $x_6 = 110992.460938$

$x_7 = 1965980672.000000$ $\qquad$ $x_8 = 63844655104.000000$

$x_9 = 70245744640.000000$ $\qquad$ $x_{10} = 137027641344.000000$

$x_{11} = 288688865280.000000$ $\qquad$ $x_{12} = 869090721792.000000$

$x_{13} = 9176400618641208380000.000000$

$\vdots$

The following table contains Newton–Raphson iterations with initial approximations 4.25. 4.4 and 4.5.

Table 3.2

$x_0 = 4.25,$	$x_0 = 4.4$	It is easy to see that, the method is divergent for initial approximations 4 and 4.25. While it converges for initial approximations 4.4 and 4.5.
$x_1 = 4.807401$	$x_1 = 4.535981$	
$x_2 = 4.946359$	$x_2 = 4.501860$	
$x_3 = 5.465665$	$x_3 = 4.493745$	
$x_4 = 11.209617$	$x_4 = 4.493410$	
$x_5 = 11.956651$	$x_0 = 4.5$	
$x_6 = 37.894390$	$x_1 = 4.493614$	
$x_7 = 1001.399841$	$x_2 = 4.493410$	
$\vdots$	$x_3 = 4.493410$	
$x_{20} = 4145085440.000000$		

This example shows that the initial approximation must be close to the root for convergence of Newton–Raphson method. But this is only qualitative idea about the convergence. We never know that how close we should start for convergence of the Newton–Raphson method. If we want the method to converge to the root for any initial approximation in the given interval, then the function $f(x)$ needs to satisfy some conditions.

The following theorem provides the sufficient condition for the convergence of Newton–Raphson method. Due to complexity, we are not presenting the proof of the theorem.

Theorem 3.5

Let the function $f(x)$ be a twice continuously differentiable function on the interval $[a, b]$, and satisfies the following conditions

   i) $f(a)f(b)<0$.

   ii) $f'(x)\neq 0, \forall x \in [a, b]$.

   iii) $f''(x)<0$ or $f''(x)>0$ for each point in the interval $[a, b]$.

   iv) $\left|\dfrac{f(a)}{f'(a)}\right|<b-a$ and $\left|\dfrac{f(b)}{f'(b)}\right|<b-a$.

Then Newton–Raphson method will converge to the unique root of the equation $f(x) = 0$ in the interval $[a, b]$ for any initial approximation in the interval $[a, b]$.

Example ———————————————————————— **3.22**

Prove that the function $f(x) = \cos x - 5x + 1$ satisfies all the four convergence conditions of Newton–Raphson method (Theorem 3.5) in the interval $(0, 1)$.

   i) $f(0)f(1)<0$.

   ii) $f'(x)=-\sin x - 5 \neq 0, \forall x \in [0, 1]$.

   iii) $f''(x)=-\cos x < 0$ for each point in the interval $[0, 1]$.

   iv) $\left|\dfrac{f(0)}{f'(0)}\right|=0.4<1$ and $\left|\dfrac{f(1)}{f'(1)}\right|=\left|\dfrac{-3.459697694131}{-5.841470984808}\right|=0.5922648<1$.

Hence, Newton–Raphson method will converge to the root for any choice of initial approximation in the interval $(0, 1)$.

Note: These conditions are very strict, and in general, the function $f(x)$ does not satisfy all these conditions. So, we can first compute the approximate root with convergent methods like Bisection, etc. to obtain a close approximation to the root and then continue with Newton–Raphson method using that close approximation.

3.8.4 Convergence of Regula Falsi Method

In Regula Falsi method, if the function $f(x)$ is a convex and continuous function in (a, b) and two initial guesses bracket the root $(f(a)f(b) < 0)$, then the method is always converging to the root. The proof of this result is not provided due to complexity.

3.8.5 Convergence of Secant Method

The convergence criterion of Secant method is the same as Newton–Raphson method (Theorem 3.5). We will discuss that Secant method and Newton–Raphson method have a similar structure. Newton–Raphson method (3.4) is given by

$$x_{n+1} = x_n - \frac{f(x_n)}{f'(x_n)}$$

For large n, the iterations x_n and x_{n+1} are close enough that we can replace the derivative $f'(x)$ by a finite difference based on starting two points $\{x_{n-1}, f(x_{n-1})\}$ and $\{x_n, f(x_n)\}$.

$$f'(x_n) = \frac{f(x_n) - f(x_{n-1})}{x_n - x_{n-1}}$$

We get

$$x_{n+1} = x_n - \frac{f(x_n)}{\left(\dfrac{f(x_n) - f(x_{n-1})}{x_n - x_{n-1}}\right)} = \frac{x_{n-1}f(x_n) - x_n f(x_{n-1})}{f(x_n) - f(x_{n-1})}$$

It is formula for Secant method (3.7).

Due to similar structure, the convergence criterion of Secant method is same as Newton–Raphson method (Theorem 3.5).

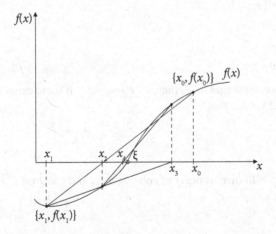

Fig. 3.18 Convergence of the Secant method

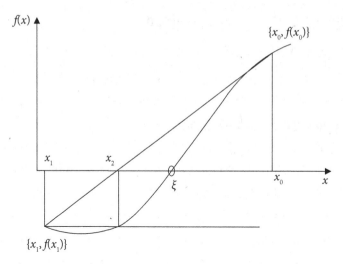

Fig. 3.19 Divergence of the Secant method

3.9 Order of Convergence

Let us consider a sequence $\{x_n\}$ converges to a limit point, ξ. Then the order of convergence mainly describes the speed of convergence. For example, both the sequences $\left\{\dfrac{1}{n}\right\}$ and $\left\{\dfrac{1}{n^2}\right\}$ converge to zero. The order of convergence of the sequence $\left\{\dfrac{1}{n^2}\right\}$ is more than the sequence $\left\{\dfrac{1}{n}\right\}$. Because, after a few terms say 50 terms, the term $\left\{\dfrac{1}{n^2}\right\} = \dfrac{1}{2500}$ is more close to zero as compared to the term $\left\{\dfrac{1}{n}\right\} = \dfrac{1}{50}$.

Definition 3.3

If $\{x_n\}$ is any sequence converges to ξ, then let $\varepsilon_n = \xi - x_n$. If there exists a positive constant C and a constant $p \geq 1$, such that

$$\lim_{n \to \infty} \frac{|\varepsilon_{n+1}|}{|\varepsilon_n|^p} = C$$

Then, the constant p is known as order of convergence, and constant C as asymptotic error constant.

Consider a convergent sequence $\{x_n\}$, then $\lim_{n\to\infty}|\varepsilon_n|\to 0$. So, the error term ε_n is close to zero for large n. It is easy to see that, if p (order of convergence) is large, then sequence converges more rapidly as $\lim_{n\to\infty}|\varepsilon_{n+1}|=C|\varepsilon_n|^p$. For example, let us presume that the error at the tenth step (ε_{10}) is 0.002. If $p = 1$, then error at eleventh step (ε_{11}) is $0.002C$. But if $p = 2$, then the error at eleventh step is $0.00004C$.

The order of convergence plays an important role in the theory of any iterative procedure that is producing a convergent sequence to the exact solution. The method converges faster to the solution for high order of convergence. Therefore, it requires a lesser number of iterations for a given accuracy.

Let ξ be the exact root of the equation $f(x) = 0$, and $\{x_n\}$ is a sequence generated by an iterative method. Let the sequence $\{x_n\}$ converge to limit point ξ. So, $\varepsilon_n = \xi - x_n$ is the error in the nth approximation x_n. Then the iterative method is said to be of order p, if

$$\lim_{n\to\infty}\frac{|\varepsilon_{n+1}|}{|\varepsilon_n|^p}=C \tag{3.15}$$

Note: The asymptotic error constant (C) plays an important role in the convergence of the method with linear rate of convergence (i.e., $p = 1$). We must have $|C| < 1$ for the convergence of the first order method for order of convergence $p > 1$, the asymptotic error constant is less important.

3.9.1 Order of Convergence for Bisection Method

Let ξ be the exact root of the equation $f(x) = 0$, and x_n is the nth approximation to the root from Bisection method starting with initial interval (a, b). So, $\varepsilon_n = \xi - x_n$ is the error in nth approximation x_n. Then the upper bound to the error in the nth and $(n + 1)$th approximations are $\varepsilon_n = \dfrac{b-a}{2^{n+1}}$ and $\varepsilon_{n+1} = \dfrac{b-a}{2^{n+2}}$ respectively. Hence, for Bisection method, we have

$$\lim_{n\to\infty}\frac{|\varepsilon_{n+1}|}{|\varepsilon_n|}=\frac{1}{2} \tag{3.16}$$

On comparing Eqs. (3.15) and (3.16), we have

$$\Rightarrow p = 1,\ C = \frac{1}{2}$$

So the order of convergence for Bisection method is one (linear convergence). The Bisection method is very slow due to a linear rate of convergence. Therefore, a large number of iterations are required to obtain a root correct up to the desired accuracy as compared to other methods. But its simple structure and always converging nature are two important reasons for its popularity.

3.9.2 Order of Convergence for Fixed-Point Method

Consider the equation $f(x) = 0$ and $f(x) = \phi(x) - x$. Let ξ be the fixed point of the iteration function, $\phi(x)$.

$$\Rightarrow \xi = \phi(\xi) \tag{3.17}$$

Let x_n and x_{n+1} be two consecutive iterations of the Fixed-Point method, then we have

$$x_{n+1} = \phi(x_n) \tag{3.18}$$

On subtracting Eq. (3.18) from Eq. (3.17), we get

$$\xi - x_{n+1} = \phi(\xi) - \phi(x_n) \tag{3.19}$$

$$\varepsilon_{n+1} = \phi(\xi) - \phi(\xi - \varepsilon_n)$$

On using Taylor series expansion, we have

$$\varepsilon_{n+1} = \phi(\xi) - \left(\phi(\xi) - \varepsilon_n \phi'(\xi) + \frac{\varepsilon_n^2}{2!} \phi''(\xi) - \cdots \right) \tag{3.20}$$

$$\varepsilon_{n+1} = \varepsilon_n \phi'(\xi) - \frac{\varepsilon_n^2}{2!} \phi''(\xi) + \cdots$$

$$\Rightarrow \lim_{n \to \infty} \frac{|\varepsilon_{n+1}|}{|\varepsilon_n|} = |\phi'(\xi)|$$

On comparing with Eq. (3.15), the order of convergence of Fixed-Point method is one. The method will converge to the exact root, if the following condition holds

$$|\phi'| < 1 \tag{3.21}$$

Note: Consider that an iteration function $\phi(x)$ satisfies the following equations

$$\phi'(\xi) = \phi''(\xi) = \cdots = \phi^{(m-1)}(\xi) = 0, \text{ and } \phi^{(m)}(\xi) \neq 0$$

Then, the Eq. (3.20) reduces to

$$\varepsilon_{n+1} = (-1)^{m-1} \frac{\varepsilon_n^m}{m!} \phi^{(m)}(\xi) + (-1)^m \frac{\varepsilon_n^{m+1}}{(m+1)!} \phi^{(m+1)}(\xi) + \cdots$$

$$\Rightarrow \lim_{n \to \infty} \frac{|\varepsilon_{n+1}|}{|\varepsilon_n|^m} = \frac{1}{m!} |\phi^{(m)}(\xi)| \tag{3.22}$$

So, the order of convergence increases to m for this iteration function, $\phi(x)$.

Example ——————————————————————————— **3.23**

Find the values of constants A and B, such that the following iteration formula will converge to the point $\xi = 1$.

$$x_{n+1} = A + Bx_n + x_n^3$$

Also, find the constants A and B for which order of convergence is highest.

Ans. The iteration function is as follows

$$\phi(x) = A + Bx + x^3$$

The iteration function will converge to $\xi = 1$, if

$$\phi(1) = 1 = A + B + 1$$

and $\quad |\phi'(1)| = |B + 3| < 1.$

These two equations imply the following conditions on constants A and B

$$A = -B \text{ and } -4 \le B \le -2$$

For higher convergence, we have

$$|\phi'(1)| = |B + 3| = 0 \quad \Rightarrow B = -3 \text{ and } A = 3$$

If these two conditions are true, then the sequence converges to $\xi = 1$, and order of convergence is quadratic (2).

Example ——————————————————————————— **3.24**

Find the values of constants A, B, and C, such that the following iteration function gives a sequence with the highest order of convergence and has a limit point $\xi = \sqrt{a}$.

$$x_{n+1} = Ax_n^{-1} + Bx_n + Cx_n^3$$

Use the obtained formula to compute the value of $\sqrt{40}$. Starting with the initial approximation, $x_0 = 6$.

Ans. The iteration function is as follows

$$x_{n+1} = Ax_n^{-1} + Bx_n + Cx_n^3 = \phi(x_n) \tag{3.23}$$

We have to find the constants A, B, and C, such that this iteration function gives a sequence with limit point $\xi = a^{1/2}$ and has the highest order of convergence. Therefore, we must have

$$\phi\left(a^{1/2}\right) = Aa^{-1/2} + Ba^{1/2} + Ca^{3/2} = a^{1/2}$$

$$\phi'\left(a^{1/2}\right) = -Aa^{-1} + B + 3Ca = 0$$

$$\phi''\left(a^{1/2}\right) = 2Aa^{-3/2} + 6Ca^{1/2} = 0$$

On solving these three equations for A, B, and C, we have

$$A = \frac{3a}{8}, \ B = \frac{3}{4} \ \text{and} \ C = \frac{-1}{8a}$$

On using these values in Eq. (3.23), we get following iteration function

$$x_{n+1} = Ax_n^{-1} + Bx_n + Cx_n^3 = \frac{3a}{8} x_n^{-1} + \frac{3}{4} x_n - \frac{1}{8a} x_n^3 \tag{3.24}$$

This iteration function converges to $\xi = \sqrt{a}$. Since $\phi'\left(a^{1/2}\right) = \phi''\left(a^{1/2}\right)'' = 0$, so according to Eq. (3.22), the sequence (3.24) has cubic convergence.

The iteration formula (3.24) for the value of $\sqrt{40}$ is given by

$$x_{n+1} = \frac{120}{8} x_n^{-1} + \frac{3}{4} x_n - \frac{1}{320} x_n^3$$

On using the initial approximation, $x_0 = 6$, we get following iterations

$$x_1 = \frac{120}{8} x_0^{-1} + \frac{3}{4} x_0 - \frac{1}{320} x_0^3 = 6.325000$$

$$x_2 = 6.324555$$

$$x_3 = 6.324555$$

It provides the value of $\sqrt{40}$ correct up to five decimal places only in two iterations.

3.9.3 Order of Convergence for Newton–Raphson Method

Consider the Newton–Raphson method (3.4) converges to a root ξ of the equation, $f(x) = 0$. Let $\varepsilon_n = \xi - x_n$ be the error in nth approximation, x_n.

$$x_{n+1} = x_n - \frac{f(x_n)}{f'(x_n)}$$

$$\xi - \varepsilon_{n+1} = \xi - \varepsilon_n - \frac{f(\xi - \varepsilon_n)}{f'(\xi - \varepsilon_n)}$$

On using Taylor Series expansion of the function $f(x)$ about the point $x = \xi$, we have

$$\varepsilon_{n+1} = \varepsilon_n + \frac{f(\xi) - \varepsilon_n f'(\xi) + \dfrac{\varepsilon_n^2}{2!} f''(\xi) - \dfrac{\varepsilon_n^3}{3!} f'''(\xi) + \cdots}{f'(\xi) - \varepsilon_n f''(\xi) + \dfrac{\varepsilon_n^2}{2!} f'''(\xi) - \dfrac{\varepsilon_n^3}{3!} f^{iv}(\xi) + \cdots} \tag{3.25}$$

Since ξ is the exact root $\Rightarrow f(\xi) = 0$

Let ξ be simple root (i.e., multiplicity one), then $f'(\xi) \neq 0$. On dividing the numerator and denominator in Eq. (3.25) with $f'(\xi)$, we get

$$\varepsilon_{n+1} = \varepsilon_n + \frac{-\varepsilon_n + \dfrac{\varepsilon_n^2}{2!} \dfrac{f''(\xi)}{f'(\xi)} - \dfrac{\varepsilon_n^3}{3!} \dfrac{f'''(\xi)}{f'(\xi)} + \cdots}{1 - \left(\varepsilon_n \dfrac{f''(\xi)}{f'(\xi)} - \dfrac{\varepsilon_n^2}{2!} \dfrac{f'''(\xi)}{f'(\xi)} + \dfrac{\varepsilon_n^3}{3!} \dfrac{f^{iv}(\xi)}{f'(\xi)} - \cdots \right)}$$

$$\varepsilon_{n+1} = \varepsilon_n + \left[-\varepsilon_n + \frac{\varepsilon_n^2}{2!} \frac{f''(\xi)}{f'(\xi)} - \frac{\varepsilon_n^3}{3!} \frac{f'''(\xi)}{f'(\xi)} + \cdots \right] \left[1 - \underbrace{\left(\varepsilon_n \frac{f''(\xi)}{f'(\xi)} - \frac{\varepsilon_n^2}{2!} \frac{f'''(\xi)}{f'(\xi)} + \frac{\varepsilon_n^3}{3!} \frac{f^{iv}(\xi)}{f'(\xi)} - \cdots \right)}_{z} \right]^{-1} \tag{3.26}$$

Let $\quad z = \varepsilon_n \dfrac{f''(\xi)}{f'(\xi)} - \dfrac{\varepsilon_n^2}{2!} \dfrac{f'''(\xi)}{f'(\xi)} + \dfrac{\varepsilon_n^3}{3!} \dfrac{f^{iv}(\xi)}{f'(\xi)} - \cdots$

Since ε_n is the error term and as $\lim\limits_{n \to \infty} \varepsilon_n \to 0$, so we have $z \ll 1$

On using the expansion $(1-z)^{-1} = 1 + z + z^2 + \cdots$, for $|z| < 1$ in the Eq. (3.26), we obtain

$$\varepsilon_{n+1} = \varepsilon_n + \left[-\varepsilon_n + \frac{\varepsilon_n^2}{2!} \frac{f''(\xi)}{f'(\xi)} - \frac{\varepsilon_n^3}{3!} \frac{f'''(\xi)}{f'(\xi)} + \cdots \right] \left[\begin{array}{l} 1 + \left(\varepsilon_n \dfrac{f''(\xi)}{f'(\xi)} - \dfrac{\varepsilon_n^2}{2!} \dfrac{f'''(\xi)}{f'(\xi)} + \dfrac{\varepsilon_n^3}{3!} \dfrac{f^{iv}(\xi)}{f'(\xi)} - \cdots \right) \\ + \left(\varepsilon_n \dfrac{f''(\xi)}{f'(\xi)} - \dfrac{\varepsilon_n^2}{2!} \dfrac{f'''(\xi)}{f'(\xi)} + \dfrac{\varepsilon_n^3}{3!} \dfrac{f^{iv}(\xi)}{f'(\xi)} - \cdots \right)^2 \\ + \cdots \end{array} \right]$$

$$\varepsilon_{n+1} = \varepsilon_n + \left[-\varepsilon_n + \frac{\varepsilon_n^2}{2!} \frac{f''(\xi)}{f'(\xi)} + O(\varepsilon_n^3) \right] \left[1 + \varepsilon_n \frac{f''(\xi)}{f'(\xi)} + O(\varepsilon_n^2) \right]$$

$$\varepsilon_{n+1} = -\frac{\varepsilon_n^2}{2}\frac{f''(\xi)}{f'(\xi)} + O(\varepsilon_n^3) \tag{3.27}$$

$$\Rightarrow \lim_{n\to\infty}\frac{|\varepsilon_{n+1}|}{|\varepsilon_n|^2} = \left|\frac{1}{2}\frac{f''(\xi)}{f'(\xi)}\right| \tag{3.28}$$

The equations (3.28) and (3.15) imply that, the order of convergence of Newton–Raphson method is 2 (quadratic convergence).

Note: We assume that ξ is a simple root (i.e., multiplicity one, then $f'(\xi)\neq 0$) of the equation $f(x) = 0$. For example, consider the equation $f(x) = x^2 - 3x + 2 = (x-1)(x-2) = 0$. So, $x=1$ is a simple root of this equation, as $f(1) = 0$, but $f'(1) \neq 0$. In case of multiple root ξ of order (or multiplicity) m, we have

$$f'(\xi) = f''(\xi) = \cdots f^{m-1}(\xi) = 0 \text{ and } f^m(\xi) \neq 0$$

For example, $x = 1$ is root of order two for the equation

$$f(x) = x^3 - x^2 - x + 1 = (x-1)^2(x+1) = 0$$

where $f(1) = f'(1) = 0$ and $f''(1) \neq 0$

In next section, we will prove that the order of convergence of Newton–Raphson method for the multiple roots is one. So, the modified Newton–Raphson method will be introduced in next section, which has quadratic convergence for the multiple roots.

Order of Convergence for Newton–Raphson Method (Multiple Root)

In the case of multiple roots of order m, the Newton–Raphson method has convergence as follows. Continuing with equation (3.25), we have

$$\varepsilon_{n+1} = \varepsilon_n + \frac{f(\xi) - \varepsilon_n f'(\xi) + \dfrac{\varepsilon_n^2}{2!}f''(\xi) - \dfrac{\varepsilon_n^3}{3!}f'''(\xi) + \cdots}{f'(\xi) - \varepsilon_n f''(\xi) + \dfrac{\varepsilon_n^2}{2!}f'''(\xi) - \dfrac{\varepsilon_n^3}{3!}f^{iv}(\xi) + \cdots}$$

Consider the equation $f(x) = 0$ has multiple root ξ of order m, then

$$f'(\xi) = f''(\xi) = \cdots f^{m-1}(\xi) = 0 \text{ and } f^m(\xi) \neq 0$$

So, Eq. (3.25) reduces to the following equation

$$
\varepsilon_{n+1} = \varepsilon_n + \cfrac{\dfrac{(-1)^m \varepsilon_n^m}{m!} f^m(\xi) + \dfrac{(-1)^{m+1} \varepsilon_n^{m+1}}{(m+1)!} f^{(m+1)}(\xi) + \dfrac{(-1)^{m+2} \varepsilon_n^{m+2}}{(m+2)!} f^{(m+2)}(\xi) + \cdots}{\dfrac{(-1)^{m-1} \varepsilon_n^{m-1}}{(m-1)!} f^{(m)}(\xi) + \dfrac{(-1)^m \varepsilon_n^m}{m!} f^{m+1}(\xi) + \dfrac{(-1)^{m+1} \varepsilon_n^{m+1}}{(m+1)!} f^{(m+2)}(\xi) + \cdots}
$$

On dividing the numerator and denominator by $\dfrac{(-1)^{m-1} \varepsilon_n^{m-1}}{(m-1)!} f^{(m)}(\xi)$, we have

$$
\varepsilon_{n+1} = \varepsilon_n + \cfrac{-\dfrac{\varepsilon_n}{m} + \dfrac{\varepsilon_n^2}{m(m+1)} \dfrac{f^{(m+1)}(\xi)}{f^{(m)}(\xi)} - \dfrac{\varepsilon_n^3}{m(m+1)(m+2)} \dfrac{f^{(m+2)}(\xi)}{f^{(m)}(\xi)} + \cdots}{1 - \dfrac{\varepsilon_n}{m} \dfrac{f^{m+1}(\xi)}{f^{(m)}(\xi)} + \dfrac{\varepsilon_n^2}{m(m+1)} \dfrac{f^{(m+2)}(\xi)}{f^{(m)}(\xi)} - \dfrac{\varepsilon_n^3}{m(m+1)(m+2)} \dfrac{f^{(m+3)}(\xi)}{f^{(m)}(\xi)} + \cdots}
$$

$$
\varepsilon_{n+1} = \varepsilon_n + \cfrac{-\dfrac{\varepsilon_n}{m} + \dfrac{\varepsilon_n^2}{m(m+1)} \dfrac{f^{(m+1)}(\xi)}{f^{(m)}(\xi)} - \dfrac{\varepsilon_n^3}{m(m+1)(m+2)} \dfrac{f^{(m+2)}(\xi)}{f^{(m)}(\xi)} + \cdots}{1 - \underbrace{\left(\dfrac{\varepsilon_n}{m} \dfrac{f^{m+1}(\xi)}{f^{(m)}(\xi)} - \dfrac{\varepsilon_n^2}{m(m+1)} \dfrac{f^{(m+2)}(\xi)}{f^{(m)}(\xi)} + \dfrac{\varepsilon_n^3}{m(m+1)(m+2)} \dfrac{f^{(m+3)}(\xi)}{f^{(m)}(\xi)} - \cdots \right)}_{z}}
\tag{3.29}
$$

On using the expansion, $(1-z)^{-1} = 1 + z + z^2 + \cdots$, for $|z| < 1$, the expression (3.29) can be rewritten as

$$
\varepsilon_{n+1} = \varepsilon_n + \left[-\dfrac{\varepsilon_n}{m} + \dfrac{\varepsilon_n^2}{m(m+1)} \dfrac{f^{(m+1)}(\xi)}{f^{(m)}(\xi)} - \dfrac{\varepsilon_n^3}{m(m+1)(m+2)} \dfrac{f^{(m+2)}(\xi)}{f^{(m)}(\xi)} + \cdots \right]
$$
$$
\left[1 + \left(\dfrac{\varepsilon_n}{m} \dfrac{f^{m+1}(\xi)}{f^{(m)}(\xi)} - \dfrac{\varepsilon_n^2}{m(m+1)} \dfrac{f^{(m+2)}(\xi)}{f^{(m)}(\xi)} + \dfrac{\varepsilon_n^3}{m(m+1)(m+2)} \dfrac{f^{(m+3)}(\xi)}{f^{(m)}(\xi)} - \cdots \right) \right.
$$
$$
\left. + \left(\dfrac{\varepsilon_n}{m} \dfrac{f^{m+1}(\xi)}{f^{(m)}(\xi)} - \dfrac{\varepsilon_n^2}{m(m+1)} \dfrac{f^{(m+2)}(\xi)}{f^{(m)}(\xi)} + \dfrac{\varepsilon_n^3}{m(m+1)(m+2)} \dfrac{f^{(m+3)}(\xi)}{f^{(m)}(\xi)} - \cdots \right)^2 + \cdots \right]
$$

$$
\varepsilon_{n+1} = \varepsilon_n \left(1 - \dfrac{1}{m} \right) - \dfrac{\varepsilon_n^2}{m^2(m+1)} \dfrac{f^{(m+1)}(\xi)}{f^{(m)}(\xi)} + O(\varepsilon_n^3)
\tag{3.30}
$$

So the order of convergence of Newton–Raphson method for multiple roots is linear (1).

Modified (or) Generalized Newton–Raphson Method

It is clear from the previous expressions that the term $\varepsilon_n \left(1 - \dfrac{1}{m}\right)$ in Eq. (3.30) vanishes if we multiply the term $\dfrac{f(x)}{f'(x)}$ in Newton–Raphson formula with multiplicity m. It will increase the order of convergence from linear (1) to quadratic (2). Therefore, the Newton–Raphson method for multiple roots has been modified as follows

$$x_{n+1} = x_n - m \frac{f(x_n)}{f'(x_n)} \qquad\qquad (3.31)$$

It is known as modified (or) generalized Newton–Raphson method. Proceeding in a similar manner, it is easy to see that, the modified Newton–Raphson method (3.31) has second order convergence for multiple roots of order m.

In case of modified Newton–Raphson method, the Eq. (3.30) reduced to the following equation

$$\varepsilon_{n+1} = -\frac{\varepsilon_n^2}{m(m+1)} \frac{f^{(m+1)}(\xi)}{f^{(m)}(\xi)} + O(\varepsilon_n^3)$$

Note that the modified Newton–Raphson method is simply Newton–Raphson method for simple root ($m = 1$).

Example ———————————————————————————— **3.25** —————

Solve the equation $x^3 - .642x^2 - 3.538959x + 3.490082 = 0$ with simple and modified Newton–Raphson methods to calculate the double root of the equation. Start with an initial approximation $x_0 = 1$.

Ans. We have following functions for given equation

$$f(x) = x^3 - .642x^2 - 3.538959x + 3.490082$$

and $f'(x) = 3x^2 - 1.284x - 3.538959$

Newton–Raphson method (3.4) will produce the following successive approximations to the root with an initial approximation, $x_0 = 1$.

$$x_1 = x_0 - \frac{f(x_0)}{f'(x_0)} = 1.169572 \qquad\qquad x_2 = 1.247139$$

$$x_3 = 1.284494 \qquad\qquad x_4 = 1.302849$$

$$x_5 = 1.311950 \qquad\qquad x_6 = 1.316481$$

$$x_7 = 1.318742 \qquad\qquad x_8 = 1.319871$$

$$x_9 = 1.320436 \qquad\qquad x_{10} = 1.320718$$

$$x_{11} = 1.320859 \qquad\qquad x_{12} = 1.320930$$

$$x_{13} = 1.320965 \qquad\qquad x_{14} = 1.320997$$

Since the root is a double root ($m = 2$), therefore the modified Newton–Raphson method (3.31) is as follows

$$x_{n+1} = x_n - 2\frac{f(x_n)}{f'(x_n)}$$

The modified Newton–Raphson method will produce the following successive approximations with an initial approximation $x_0 = 1$.

$$x_1 = x_0 - 2\frac{f(x_0)}{f'(x_0)} = 1.339144$$

$$x_2 = 1.321049$$
$$x_3 = 1.321000$$

$$x_4 = 1.321000$$

Note that the Newton–Raphson method requires around 15 iterations for the accuracy of five decimal places, while only three iterations are sufficient for the same accuracy in the case of modified Newton–Raphson method.

We can use the modified Newton–Raphson method if the multiplicity m of the root is known in advance. But, in general, the modified Newton–Raphson method is not applicable as we do not have multiplicity m. So, the accelerated Newton–Raphson method will be helpful in this case.

Accelerated Newton–Raphson Method

Let the function $f(x)$ have a zero ξ of multiplicity m, then the function $f'(x)$ has zero ξ of multiplicity $m - 1$. So, the function $g(x) = \dfrac{f(x)}{f'(x)}$ has a zero ξ of multiplicity 1, i.e., simple zero. So, the Newton–Raphson method will be applicable for the function $g(x)$ to compute this zero, i.e.,

$$x_{n+1} = x_n - \frac{g(x_n)}{g'(x_n)}$$

where $g(x) = \dfrac{f(x)}{f'(x)}$ \hfill (3.32)

This method is known as the accelerated Newton–Raphson method.

Example ──────────────────────────────── **3.26** ──────

Solve the equation $x^3 - .642x^2 - 3.538959x + 3.490082 = 0$ with accelerated Newton–Raphson method. Start with initial approximation 1.

$$f(x) = x^3 - .642x^2 - 3.538959x + 3.490082$$

$$f'(x) = 3x^2 - 1.284x - 3.538959$$

$$g(x) = \frac{f(x)}{f'(x)} = \frac{x^3 - .642x^2 - 3.538959x + 3.490082}{3x^2 - 1.284x - 3.538959}$$

$$g'(x) = 1 - \frac{\left(x^3 - .642x^2 - 3.538959x + 3.490082\right)\left(6x - 1.284\right)}{\left(3x^2 - 1.284x - 3.538959\right)^2}$$

The accelerated Newton–Raphson method (3.32) is given by

$$x_{n+1} = x_n - \frac{g(x_n)}{g'(x_n)}$$

Let initial approximation be $x_0 = 1$, then we get

$$x_1 = x_0 - \frac{g(x_0)}{g'(x_0)}$$
$$= 1.302097145$$

$$x_2 = x_1 - \frac{g(x_1)}{g'(x_1)}$$

$$= 1.320945602$$

$$x_3 = x_2 - \frac{g(x_2)}{g'(x_2)}$$

$$= 1.320991553$$

$$x_4 = 1.321000$$

3.9.4 Order of Convergence for Secant Method

Consider the Secant method converges to a root ξ of the equation, $f(x) = 0$. Let $\varepsilon_n = \xi - x_n$ be the error in nth approximation, x_n. Let x_{n-1} and x_n be two successive approximations, then the next approximation x_{n+1} with Secant method (3.7) is given by

$$x_{n+1} = \frac{x_{n-1} f(x_n) - x_n f(x_{n-1})}{f(x_n) - f(x_{n-1})}$$

(or) $\quad x_{n+1} = x_n - \dfrac{x_n - x_{n-1}}{f(x_n) - f(x_{n-1})} f(x_n)$

$$\xi - \varepsilon_{n+1} = (\xi - \varepsilon_n) - \frac{(\xi - \varepsilon_n) - (\xi - \varepsilon_{n-1})}{f(\xi - \varepsilon_n) - f(\xi - \varepsilon_{n-1})} f(\xi - \varepsilon_n)$$

$$\varepsilon_{n+1} = \varepsilon_n - \frac{\varepsilon_n - \varepsilon_{n-1}}{f(\xi - \varepsilon_n) - f(\xi - \varepsilon_{n-1})} f(\xi - \varepsilon_n)$$

On expanding $f(x)$ around the root $x = \xi$ with the help of Taylor series, we get

$$\varepsilon_{n+1} = \varepsilon_n - \frac{(\varepsilon_n - \varepsilon_{n-1})\left(f(\xi) - \varepsilon_n f'(\xi) + \frac{\varepsilon_n^2}{2!} f''(\xi) - \cdots\right)}{\left(f(\xi) - \varepsilon_n f'(\xi) + \frac{\varepsilon_n^2}{2!} f''(\xi) - \cdots\right) - \left(f(\xi) - \varepsilon_{n-1} f'(\xi) + \frac{\varepsilon_{n-1}^2}{2!} f''(\xi) - \cdots\right)}$$

Since ξ is the exact root $\Rightarrow f(\xi) = 0$, hence we have

$$\varepsilon_{n+1} = \varepsilon_n - \frac{(\varepsilon_n - \varepsilon_{n-1})\left(-\varepsilon_n f'(\xi) + \frac{\varepsilon_n^2}{2!} f''(\xi) - \cdots\right)}{\left(-\varepsilon_n f'(\xi) + \frac{\varepsilon_n^2}{2!} f''(\xi) - \cdots\right) - \left(-\varepsilon_{n-1} f'(\xi) + \frac{\varepsilon_{n-1}^2}{2!} f''(\xi) - \cdots\right)}$$

Assuming ξ is simple root, then $f'(\xi) \neq 0$. On dividing the numerator and denominator with $f'(\xi)$, we get

$$\varepsilon_{n+1} = \varepsilon_n - \frac{(\varepsilon_n - \varepsilon_{n-1})\left(-\varepsilon_n + \dfrac{\varepsilon_n^2}{2!}\dfrac{f''(\xi)}{f'(\xi)} - \cdots\right)}{\left(-\varepsilon_n + \dfrac{\varepsilon_n^2}{2!}\dfrac{f''(\xi)}{f'(\xi)} - \cdots\right) - \left(-\varepsilon_{n-1} + \dfrac{\varepsilon_{n-1}^2}{2!}\dfrac{f''(\xi)}{f'(\xi)} - \cdots\right)}$$

On simplifying the formula in denominator, we have

$$\varepsilon_{n+1} = \varepsilon_n - \frac{(\varepsilon_n - \varepsilon_{n-1})\left(-\varepsilon_n + \dfrac{\varepsilon_n^2}{2!}\dfrac{f''(\xi)}{f'(\xi)} - \cdots\right)}{-(\varepsilon_n - \varepsilon_{n-1})\left(1 - \dfrac{(\varepsilon_n + \varepsilon_{n-1})}{2!}\dfrac{f''(\xi)}{f'(\xi)} + \cdots\right)} = \varepsilon_n + \frac{\left(-\varepsilon_n + \dfrac{\varepsilon_n^2}{2!}\dfrac{f''(\xi)}{f'(\xi)} - \cdots\right)}{\left(1 - \dfrac{(\varepsilon_n + \varepsilon_{n-1})}{2!}\dfrac{f''(\xi)}{f'(\xi)} + \cdots\right)}$$

$$\varepsilon_{n+1} = \varepsilon_n - \varepsilon_n\left(1 - \dfrac{\varepsilon_n}{2!}\dfrac{f''(\xi)}{f'(\xi)} - \cdots\right)\left(1 - \underbrace{\dfrac{(\varepsilon_n + \varepsilon_{n-1})}{2!}\dfrac{f''(\xi)}{f'(\xi)} + \cdots}_{z}\right)^{-1}$$

on using the expansion $(1-z)^{-1} = 1 + z + z^2 + \cdots$, we get

$$\varepsilon_{n+1} = \varepsilon_n - \varepsilon_n\left(1 - \dfrac{\varepsilon_n}{2!}\dfrac{f''(\xi)}{f'(\xi)} - \cdots\right)\left(1 + \left(\dfrac{(\varepsilon_n + \varepsilon_{n-1})}{2!}\dfrac{f''(\xi)}{f'(\xi)} - \cdots\right) + \left(\dfrac{(\varepsilon_n + \varepsilon_{n-1})}{2!}\dfrac{f''(\xi)}{f'(\xi)} - \cdots\right)^2 + \cdots\right)$$

$$= \varepsilon_n - \varepsilon_n\left(1 - \dfrac{\varepsilon_n}{2!}\dfrac{f''(\xi)}{f'(\xi)} + \dfrac{(\varepsilon_n + \varepsilon_{n-1})}{2!}\dfrac{f''(\xi)}{f'(\xi)} + \cdots\right)$$

$$= \frac{-1}{2}(\varepsilon_n \varepsilon_{n-1})\dfrac{f''(\xi)}{f'(\xi)} + O(\varepsilon_n^2) \tag{3.33}$$

This expression involves both ε_{n-1} and ε_n. To simplify the expression in terms of ε_n only, let us assume that the order of convergence for Secant method is p, and asymptotic error constant is a. We have

$$\varepsilon_{n+1} = a(\varepsilon_n)^p \tag{3.34}$$

and $\quad \varepsilon_n = a\left(\varepsilon_{n-1}\right)^p$

$$\Rightarrow \varepsilon_{n-1} = \left(\frac{1}{a}\right)^{1/p}\left(\varepsilon_n\right)^{1/p}$$

Using, this value of ε_{n-1} in the Eq. (3.33), we have

$$\varepsilon_{n+1} = \frac{-1}{2}\left(\varepsilon_n\left(\varepsilon_n\right)^{1/p}\right)\left(\frac{1}{a}\right)^{1/p}\frac{f''(\xi)}{f'(\xi)} + O(\varepsilon_n^2)$$

$$= \frac{-1}{2}\left(\left(\varepsilon_n\right)^{1+(1/p)}\right)\left(\frac{1}{a}\right)^{1/p}\frac{f''(\xi)}{f'(\xi)} + O(\varepsilon_n^2) \tag{3.35}$$

On comparing the above expression with Eq. (3.34), we have

$$a = \frac{-1}{2}\left(\frac{1}{a}\right)^{1/p}\frac{f''(\xi)}{f'(\xi)} \quad \Rightarrow |a| = \left|\frac{f''(\xi)}{2f'(\xi)}\right|^{\frac{p}{p+1}}$$

$$p = 1 + (1/p) \quad \Rightarrow p = \frac{1}{2}\left(1 \pm \sqrt{5}\right)$$

On neglecting the −ve value of p, we have

$$p = 1.618$$

Using these two relations, we have

$$|\varepsilon_{n+1}| = \left|\frac{1}{2}\frac{f''(\xi)}{f'(\xi)}\right|^{0.618}|\varepsilon_n|^{1.618} + O(\varepsilon_n^2) \tag{3.36}$$

Therefore, the order of convergence for Secant method is 1.618 (super-linear).

3.9.5 Order of Convergence for Regula Falsi Method

In Regula Falsi method, there is no general formula for next iteration in terms of previous iterations. So, we cannot obtain the order of convergence in general. But if the function $f(x)$ is convex in the interval (x_0, x_1) containing a root, in that case, one iteration point either x_0 or x_1 is always fixed, and other varies with the approximations (see Fig. 3.20 for graphical interpretation). A function $f(x)$ is said to be convex if the chord, joining any two points of the curve, always lies above the curve.

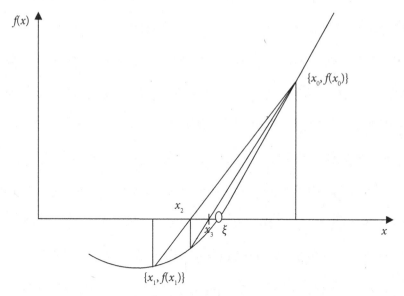

Fig. 3.20 Order of convergence of Regula Falsi method if function f(x) is convex

Let us assume that point x_0 is fixed; then the function $f(x)$ is approximated by the straight line joining the points $\left(x_0,\ f(x_0)\right)$ and $\left(x_n,\ f(x_n)\right), n = 1,2,3,\cdots$. Proceeding in a similar way as in Secant method, our equation for the order of convergence is as follows

$$\varepsilon_{n+1} = \frac{-1}{2}\left(\varepsilon_0\ \frac{f''(\xi)}{f'(\xi)}\right)\varepsilon_n$$

Since the error in first approximation i.e., ε_0 is fixed, so we have

$$\varepsilon_{n+1} = C\varepsilon_n, \text{ where } C = \frac{-1}{2}\left(\varepsilon_0\ \frac{f''(\xi)}{f'(\xi)}\right)$$

This implies that the Regula Falsi method has linear convergence.

If the function $f(x)$ is not a convex function, then the Regula Falsi method may have a higher order of convergence, but at most 1.618 as that of Secant method.

So far, we have discussed the methods in which the approximating curves are straight lines (tangent in Newton–Raphson method and chord joining two points in Regula Falsi and Secant methods). These methods are also known as linear interpolation methods. Now, we will discuss the methods in which the approximating curves are second order polynomials instead of straight line. The Muller method and Chebyshev method are two such methods, we will discuss these methods in sections 3.10 and 3.11 respectively. The Aitken method will be discussed in Section 3.12. It is an extension of the Fixed-Point method with higher rate of convergence

3.10 Muller Method

In this method, we will approximate the function $y = f(x)$ by a second-degree curve $p_2(x)$ in the neighborhood of the root.

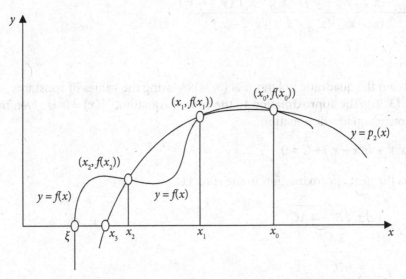

Fig. 3.21 Graphical representation of Muller method

Let x_{i-2}, x_{i-1} and x_i be the approximations to a root of the equation, $f(x) = 0$, then

$$y_{i-2} = f(x_{i-2}), \ y_{i-1} = f(x_{i-1}) \text{ and } y_i = f(x_i)$$

Let the approximating curve be a quadratic polynomial of the following form

$$y = A(x - x_i)^2 + B(x - x_i) + C \tag{3.34}$$

This parabola passes through the points (x_{i-2}, y_{i-2}), (x_{i-1}, y_{i-1}) and (x_i, y_i). So, we must have

$$y_{i-2} = A(x_{i-2} - x_i)^2 + B(x_{i-2} - x_i) + C$$

$$y_{i-1} = A(x_{i-1} - x_i)^2 + B(x_{i-1} - x_i) + C$$

$$y_i = C \tag{3.35}$$

Solving Eqs. (3.35) for the constants A, B and C, we have

$$A = \frac{(x_{i-1} - x_i)(y_{i-2} - y_i) - (x_{i-2} - x_i)(y_{i-1} - y_i)}{(x_i - x_{i-1})(x_{i-1} - x_{i-2})(x_{i-2} - x_i)}$$

$$B = \frac{(x_{i-2} - x_i)^2(y_{i-1} - y_i) - (x_{i-1} - x_i)^2(y_{i-2} - y_i)}{(x_i - x_{i-1})(x_{i-1} - x_{i-2})(x_{i-2} - x_i)}$$

$$C = y_i \qquad\qquad\qquad (3.36)$$

We can obtain the quadratic polynomial (3.34) by using the values of constants A, B and C from Eqs. (3.36). The approximation to the root of equation, $f(x) = 0$ is given by the root of the following quadratic equation

$$A(x - x_i)^2 + B(x - x_i) + C = 0$$

Let x_{i+1} be the next approximation to the root, i.e.

$$x_{i+1} - x_i = \frac{-B \pm \sqrt{B^2 - 4AC}}{2A}$$

$$x_{i+1} = x_i + \frac{-2C}{B \pm \sqrt{B^2 - 4AC}} \qquad\qquad (3.37)$$

Note that the sign in the denominator of the Eq. (3.37) is chosen, so that the denominator becomes largest in the magnitude. It is to reduce the loss of significance in the approximation x_{i+1} (as discussed in Chapter 2). The method can be used to obtain the complex root, when $\sqrt{B^2 - 4AC} < 0$

Example ── **3.27** ──

Compute the approximate root of the equation $x^3 - 4x - 9 = 0$ correct to six decimal places. Use Muller method with initial approximations 2, 3 and 4.

Ans. Let $x_0 = 2$, $x_1 = 3$ and $x_2 = 4$ be the three initial approximations for the root of the equation

$$y = f(x) = x^3 - 4x - 9 = 0$$

This implies

$$\begin{array}{ccc} x_0 = 2 & x_1 = 3 & x_2 = 4 \\ y_0 = -9 & y_1 = 6 & y_2 = 39 \end{array}$$

Let $y = A(x - x_2)^2 + B(x - x_2) + C$ be the parabola passing through the points $(x_0, y_0), (x_1, y_1)$ and (x_2, y_2). We have

$$y_0 = A(x_0 - x_2)^2 + B(x_0 - x_2) + C$$

$$y_1 = A(x_1 - x_2)^2 + B(x_1 - x_2) + C$$

$$y_2 = C$$

Using the values of $(x_0, y_0), (x_1, y_1)$ and (x_2, y_2), and solving these equations for different constants, we have

$$C = 39$$

$$-9 = 4A - 2B + 39$$

$$6 = A - B + 39 \qquad \Rightarrow A = 9,\ B = 42,\ C = 39$$

Let x_3 be the next approximation. From Eq. (3.37), we have

$$x_3 = x_2 + \frac{-2C}{B + \sqrt{B^2 - 4AC}} = 4 + \frac{-2(39)}{42 + \sqrt{(42)^2 - 4(9)(39)}} = 2.720759$$

Note that the +ve sign in the denominator as the value of B is +ve.

On using the following values of $(x_1, y_1), (x_2, y_2)$ and (x_3, y_3)

$$x_1 = 3 \qquad x_2 = 4 \qquad x_3 = 2.720759$$

$$y_1 = 6 \qquad y_2 = 39 \qquad y_3 = 0.257463$$

the next approximation of the Muller method (3.37) is as follows

$$x_4 = 2.706220$$

Similarly, we can obtain following approximations

$$x_5 = 2.706528$$

$$x_6 = 2.706528$$

Example ─── 3.28 ──

Perform three iterations of the Muller method to compute the approximate root of the equation, $\cos x - 5x + 1 = 0$. Assume the first three initial approximations for the root are 0, 1 and 2.

Ans. The values of function $y = f(x) = \cos x - 5x + 1 = 0$ at initial approximations are given by

$$x_0 = 0 \qquad\qquad x_1 = 1 \qquad\qquad x_2 = 2$$
$$y_0 = 2 \qquad\qquad y_1 = -3.45970 \qquad\qquad y_2 = -9.41615$$

Let $y = A(x - x_2)^2 + B(x - x_2) + C$ be the approximating curve, so we get

$$y_0 = A(x_0 - x_2)^2 + B(x_0 - x_2) + C$$

$$y_1 = A(x_1 - x_2)^2 + B(x_1 - x_2) + C$$

$$y_2 = C$$

On using the values of (x_0, y_0), (x_1, y_1) and (x_2, y_2) in above equations, and further solving these equations, we obtain

$$A = -0.248376, \quad B = -6.204825, \quad C = -9.416147$$

Using the $-$ve sign in the denominator (as the value of B is $-$ve), we get following iteration of Muller method (3.37)

$$x_3 = x_2 + \frac{-2C}{B - \sqrt{B^2 - 4AC}} = 0.377006$$

The new values of A, B, and C with points (x_1, y_1), (x_2, y_2) and (x_3, y_3) are as follows

$$A = -0.204124, \quad B = -5.497988, \quad C = 0.044743$$

On using these values, we can easily compute the value of $x_4 = 0.385141$

Similarly, the results for the next iteration are as follows

$$A = -0.283223, \quad B = -5.374220, \quad C = 0.001039$$

$$x_5 = 0.385335$$

Example ———————————————————————————— **3.29**

Obtain the root of the equation $x - \tan(x) = 0$ corrects up to five decimal places. Use Muller method and start with initial approximations 4, 4.5 and 5.

Ans. The first three initial approximations for the root are $x_0 = 4$, $x_1 = 4.5$ and $x_2 = 5$. Using these initial approximations, we have the following iterations of Muller method (3.37)

$$A = 22.994715, \quad B = 28.533052, \quad C = 8.380515$$

$$x_3 = 4.522664$$

$$A = 86.288124, \quad B = -22.197088, \quad C = -0.684731$$

$$x_4 = 4.494828$$

$$A = 84.237091, \quad B = -25.907677, \quad C = -0.028839$$

$$x_5 = 4.493719$$

$$A = -110.777313, \quad B = -20.233555, \quad C = -0.006261$$

$$x_6 = 4.493409$$

$$A = -95.975304, \quad B = -20.190479, \quad C = 0.000006$$

$$x_7 = 4.493410$$

Example ———————————————————————————— **3.30**

Perform five iterations of the Muller method to compute the approximate root of the equation, $x^4 - 5x^3 + 6x^2 - 3x + 2 = 0$. Use initial approximations 0, 1 and 2.

Ans. The first three initial approximations for the root are $x_0 = 0$, $x_1 = 1$ and $x_2 = 2$. Using these initial approximations, we have following iterations (3.37)

$$A = -2.000000, \quad B = -7.000000, \quad C = -4.000000$$

$$x_3 = 1.280776$$

$A = -2.921165, \quad B = -3.719223, \quad C = 0.186021$

$x_4 = 1.328968$

$A = -2.720578, \quad B = -4.125710, \quad C = 0.006487$

$x_5 = 1.327394$

$A = -3.350903, \quad B = -4.145539, \quad C = 0.000046$

$x_6 = 1.327406$

$A = -3.338567, \quad B = -4.145632, \quad C = 0.000000$

$x_7 = 1.327406$

It is easy to see that this method extracts a quadratic factor. So we can compute two roots simultaneously. This method can also be used to compute the complex roots of the nonlinear equation.

3.11　Chebyshev Method

In this method, we approximate the function with the following second degree polynomial

$$f(x) = Ax^2 + Bx + C \tag{3.38}$$

On differentiating this equation twice, we get

$$f'(x) = 2Ax + B$$

$$f''(x) = 2A$$

Let x_n be the nth approximation to root, then we have

$$f(x_n) = Ax_n^2 + Bx_n + C \tag{3.39}$$

$$f'(x_n) = 2Ax_n + B \tag{3.40}$$

$$f''(x_n) = 2A \tag{3.41}$$

Solving these three Eqs. (3.39–3.41) for the constants A, B and C, we get

$$A = \frac{f''(x_n)}{2}$$

$$B = f'(x_n) - x_n f''(x_n)$$

$$C = f(x_n) + \frac{x_n^2 f''(x_n)}{2} - x_n f'(x_n)$$

Using these values of the constants A, B and C in Eq. (3.38), we get

$$f(x) = \frac{f''(x_n)}{2} x^2 + \left(f'(x_n) - x_n f''(x_n)\right)x + f(x_n) + \frac{x_n^2 f''(x_n)}{2} - x_n f'(x_n)$$

Let the next approximation $x = x_{n+1}$ be the root, then $f(x_{n+1}) = 0$. So, we have

$$f(x_{n+1}) = \frac{f''(x_n)}{2} x_{n+1}^2 + \left(f'(x_n) - x_n f''(x_n)\right)x_{n+1} + f(x_n) + \frac{x_n^2 f''(x_n)}{2} - x_n f'(x_n) = 0$$

$$\Rightarrow f''(x_n)\left(x_{n+1} - x_n\right)^2 + 2\left(x_{n+1} - x_n\right)f'(x_n) + 2f(x_n) = 0$$

This equation can be rearranged to the following relation

$$x_{n+1} - x_n = \frac{-f''(x_n)}{2f'(x_n)}\left(x_{n+1} - x_n\right)^2 - \frac{f(x_n)}{f'(x_n)}$$

On using Newton–Raphson method $\left(x_{n+1} - x_n = -\dfrac{f(x_n)}{f'(x_n)}\right)$ in R.H.S. of the above equation,

we get

$$x_{n+1} - x_n = \frac{-f''(x_n)}{2f'(x_n)}\left(\frac{f(x_n)}{f'(x_n)}\right)^2 - \frac{f(x_n)}{f'(x_n)}$$

Finally, we have following Chebyshev formula

$$x_{n+1} = x_n - \frac{\left(f(x_n)\right)^2 f''(x_n)}{2\left(f'(x_n)\right)^3} - \frac{f(x_n)}{f'(x_n)} \tag{3.42}$$

Example ───────────────────────────────────── 3.31 ───────

Compute the root of the equation $x - e^{-x} = 0$ correct to six decimal places. Use Chebyshev method with initial approximation 0.5.

Ans. We have

$$f(x) = x - e^{-x} \qquad\qquad f'(x) = 1 + e^{-x} \qquad\qquad f''(x) = -e^{-x} \qquad\qquad x_0 = 0.5$$

Starting with initial approximation, $x_0 = 0.5$, the Chebyshev method (3.42) gives

$$x_{n+1} = x_n - \frac{\left(f(x_n)\right)^2 f''(x_n)}{2\left(f'(x_n)\right)^3} - \frac{f(x_n)}{f'(x_n)} = x_n - \frac{\left(x_n - e^{-x_n}\right)^2 \left(-e^{-x_n}\right)}{2\left(1 + e^{-x_n}\right)^3} - \frac{x_n - e^{-x_n}}{1 + e^{-x_n}}$$

$$\Rightarrow x_1 = x_0 - \frac{\left(x_0 - e^{-x_0}\right)^2 \left(-e^{-x_0}\right)}{2\left(1 + e^{-x_0}\right)^3} - \frac{x_0 - e^{-x_0}}{1 + e^{-x_0}} = 0.5 - \frac{\left(0.5 - e^{-0.5}\right)^2 \left(-e^{-0.5}\right)}{2\left(1 + e^{-0.5}\right)^3} - \frac{0.5 - e^{-0.5}}{1 + e^{-0.5}}$$

$$= 0.567141$$

Similarly, next two iterations can be computed and given by

$$x_2 = 0.567143$$

$$x_3 = 0.567143$$

Example ───────────────────────────────────── 3.32 ───────

Use Chebyshev method to calculate the root of the equation, $x^3 - 4x - 9 = 0$. Perform five iterations with an initial approximation, $x_0 = 2$.

Ans. We have

$$f(x) = x^3 - 4x - 9 \qquad\qquad f'(x) = 3x^2 - 4 \qquad\qquad f''(x) = 6x$$

Chebyshev method (3.42) with initial approximation $x_0 = 2$ produces the following result

$$x_1 = x_0 - \frac{\left(f(x_0)\right)^2 f''(x_0)}{2\left(f'(x_0)\right)^3} - \frac{f(x_0)}{f'(x_0)} = 2.175781$$

Similarly, other approximations are as follows

$$x_2 = 2.564529$$

$$x_3 = 2.705266$$

$$x_4 = 2.706528$$

$$x_5 = 2.706528$$

Example ──────────────────────────────── **3.33**

Obtain the root of the equation, $\cos x - 5x + 1 = 0$, correct to six decimal places with the aid of Chebyshev method.

Ans. Starting with initial approximation, $x_0 = 0$, we have following iterations for Chebyshev method (3.42)

$$x_1 = 0.384000$$

$$x_2 = 0.385335$$

$$x_3 = 0.385335$$

Example ──────────────────────────────── **3.34**

Perform four iterations of Chebyshev method with initial approximations 4 and 4.5 for the equation, $x - \tan(x) = 0$. Conclude the results.

Ans. The four iterations of Chebyshev method (3.42) with initial approximation $x_0 = 4$ are given by

$$x_1 = -2.966685$$

$$x_2 = -59233.894531$$

$$x_3 = -4741133824.000000$$

$$x_4 = -6709247677974358000000.000000$$

It is easy to see that method diverges for this initial approximation.

Starting with initial approximation, $x_0 = 4.5$, we have

$$x_1 = x_0 - \frac{\left(f(x_0)\right)^2 f''(x_0)}{2\left(f'(x_0)\right)^3} - \frac{f(x_0)}{f'(x_0)} = -4.493416$$

$$x_2 = 4.493410$$

$$x_3 = 4.493410$$

For initial approximation, $x_0 = 4.5$, the method is converging towards the root 4.493410.

Note that the Chebyshev method may diverge like Newton–Raphson method. So it is recommended to start with an approximation that is close to the root.

3.12 Aitken Δ^2 Process: Acceleration of Convergence of Fixed-Point Method

The Fixed-Point method has linear convergence. It can be improved with the Aitken method. Assume the function $f(x) = \phi(x) - x$. Let ξ be the exact root of the equation $f(x) = 0$ (or) ξ be the fixed point of the function $\phi(x)$.

$$\Rightarrow \xi = \phi(\xi) \tag{3.43}$$

Let x_n, x_{n+1} and x_{n+2} be three consecutive iterations of the Fixed-Point method, i.e.,

$$x_{n+1} = \phi(x_n) \tag{3.44}$$

$$x_{n+2} = \phi(x_{n+1}) \tag{3.45}$$

With the help of Lagrange mean value theorem, we have

$$\phi(x_{n+1}) - \phi(\xi) = \phi'(c_1)(x_{n+1} - \xi) \qquad \text{for some } c_1 \in (x_{n+1}, \xi) \text{ or } (\xi, x_{n+1}) \tag{3.46}$$

On using Eqs. (3.43), (3.45) and (3.46), we have

$$x_{n+2} - \xi = \phi(x_{n+1}) - \phi(\xi) = \phi'(c_1)(x_{n+1} - \xi) \tag{3.47}$$

Proceeding in a similar manner, we have

$$x_{n+1} - \xi = \phi(x_n) - \phi(\xi) = \phi'(c_2)(x_n - \xi) \quad \text{for some } c_2 \in (x_n, \xi) \text{ or } (\xi, x_n) \quad (3.48)$$

As $\lim n \to \infty$, we have $x_n \to \xi$. So, with some error, we can assume that $c_1 = c_2 = c$. On eliminating $\phi'(c)$ from Eqs. (3.47) and (3.48), we have

$$(x_{n+2} - \xi)(x_n - \xi) = (x_{n+1} - \xi)^2$$

On solving this equation for ξ, we get

$$\xi = \frac{x_{n+2} x_n - x_{n+1}^2}{x_{n+2} - 2x_{n+1} + x_n} = x_n - \frac{(x_{n+1} - x_n)^2}{x_{n+2} - 2x_{n+1} + x_n}$$

Since the error is introduced with assumption, hence we can take ξ as the next approximation x for the Aitken method

$$x = x_n - \frac{(\Delta x_n)^2}{\Delta^2 x_n} \qquad (3.49)$$

where $\Delta x_n = x_{n+1} - x_n$ and $\Delta^2 x_n = x_{n+2} - 2x_{n+1} + x_n$ are first and second forward differences respectively.

Example ———————————————————————————— **3.35**

Compute the root of the equation $x - e^{-x} = 0$ correct to 5 decimal places using Aitken process.

Ans. Since $f(0) = -1$ and $f(1) = 0.631$, so the root of the equation lies in the interval $(0, 1)$. The iteration function $\phi(x) = e^{-x}$ satisfies both the convergence conditions, therefore, we can continue with any initial approximation in the interval $(0, 1)$.

First Iteration of Aitken process

Let $x_0 = 0.5$. We compute x_1 and x_2 from the Fixed-Point method as follows

$$x_1 = \phi(x_0) = e^{-(0.5)} = 0.606531$$

$$x_2 = \phi(x_1) = e^{-(0.606531)} = 0.545239$$

The Aitken process (3.49) gives the following approximation to the root

$$x = x_0 - \frac{(x_1 - x_0)^2}{x_2 - 2x_1 + x_0}$$

$$x = 0.5 - \frac{(0.606531 - 0.5)^2}{0.545239 - 2(0.606531) + 0.5} = 0.567624$$

Second Iteration of Aitken process

Let $x_0 = 0.567624$,

$$x_1 = \phi(x_0) = e^{-(0.567624)} = 0.566871$$

$$x_2 = \phi(x_1) = e^{-(0.566871)} = 0.567298$$

$$x = x_0 - \frac{(x_1 - x_0)^2}{x_2 - 2x_1 + x_0}$$

$$x = 0.567624 - \frac{(0.566871 - 0.567624)^2}{0.567298 - 2(0.566871) + 0.567624} = 0.567143$$

Third Iteration of Aitken process

Proceeding in a similar way with $x_0 = 0.567143$, we have

$$x = 0.567143$$

The root corrects up to 5 decimal places.

Example ── **3.36**

Perform two iterations of Aitken process for the equation $x^3 - 4x - 9 = 0$ with iteration

function $\phi(x) = \frac{(x^3 - 9)}{4}$ and initial approximation 2.

Ans. First Iteration of Aitken process

$$x_0 = 2$$

$$x_1 = \phi(x_0) = \frac{((2)^3 - 9)}{4} = -0.25$$

$$x_2 = \phi(x_1) = \frac{((-0.25)^3 - 9)}{4} = -2.253906$$

Using Aitken process (3.49), we get

$$x = x_0 - \frac{(x_1 - x_0)^2}{x_2 - 2x_1 + x_0}$$

$$x = 2 - \frac{(-0.25 - 2)^2}{-2.253906 - 2(-0.25) + 2} = -18.571428$$

Second Iteration of Aitken process

$$x_0 = -18.571428$$

$$x_1 = -1603.561890$$

$$x_2 = -1030854080.000000$$

It shows the divergence of the method.

Note: We can use the Aitken process with the iterative function $\phi(x) = -\frac{x^3 - 20x - 9}{16}$ to get the convergent result.

Example 3.37

Compute the root of the equation $\cos x - 5x + 1 = 0$ correct to 5 decimal places with Aitken process.

Ans. Let us consider the iterative function $\phi(x) = \frac{(\cos x + 1)}{5}$. The root of this equation lies in the interval $(0, 1)$.

First Iteration of Aitken process:

Let the initial approximation be $x_0 = 0.5$.

$$x_1 = \phi(x_0) = \frac{(\cos(0.5) + 1)}{5} = 0.375517$$

$$x_2 = \phi(x_1) = \frac{(\cos(0.375517) + 1)}{5} = 0.386064$$

$$x = x_0 - \frac{(x_1 - x_0)^2}{x_2 - 2x_1 + x_0} = 0.385240$$

Second Iteration of Aitken process

$$x_0 = 0.385240$$

$$x_1 = 0.385342$$

$$x_2 = 0.385334$$

$$x = 0.385335$$

Example ──────────────────────────────────── 3.38

Use Aitken process to compute the first positive root of the equation $x - \tan(x) = 0$. Start with initial approximation 4.5.

Ans. Let the iteration function be $\phi(x) = \tan^{-1}(x) + \pi$. Starting with initial approximation $x_0 = 4.5$, we have

$$x_1 = \phi(x_0) = \tan^{-1}(4.5) + 3.14159265 = 4.493720$$

$$x_2 = \phi(x_1) = \tan^{-1}(4.493720) + 3.14159265 = 4.493424$$

On using Aitken process (3.49), we have following approximation

$$x = 4.493410$$

Consider the sequences $\{x_n\}$ and $\{\hat{x}_n\}$ generated by the Fixed-Point method and Aitken process, respectively. Then, the following theorem provides the idea about the order of convergence of Aitken process.

Theorem 3.6

Consider a sequence $\{x_n\}$ converges linearly to the limit point ξ and that

$$\lim_{n \to \infty} \frac{|x_{n+1} - \xi|}{|x_n - \xi|} < 1$$

Then the Aitken sequence $\{\hat{x}_n\}$ converges more rapidly than $\{x_n\}$, such that

$$\lim_{n \to \infty} \frac{|\hat{x}_n - \xi|}{|x_n - \xi|} = 0$$

Table 3.3 Formulation of Methods

Method	Formulation (Next Iteration x_{n+1}) for the roots of the equation $f(x) = 0$	Next Approximation (Graphical Interpretation)
Bisection	$x_{n+1} = \dfrac{a_n + b_n}{2}$ (interval (a_n, b_n) contains the root)	Mid-point of last two approximations (such that root lies in the interval)
Regula Falsi	$x_{n+1} = \dfrac{f(b_n)a_n - f(a_n)b_n}{f(b_n) - f(a_n)}$ (interval (a_n, b_n) contains the root)	Intersection of the chord joining last two approximations and the x-axis (such that root lies between these two approximations)
Secant	$x_{n+1} = \dfrac{f(x_n)x_{n-1} - f(x_{n-1})x_n}{f(x_n) - f(x_{n-1})}$	Intersection of the chord joining last two approximations and the x-axis
Newton–Raphson	$x_{n+1} = x_n - \dfrac{f(x_n)}{f'(x_n)}$	Intersection of the tangent at last approximation and the x-axis
Fixed-Point	$x_{n+1} = \phi(x_n)$ ($f(x) = 0$ is rewritten in the form $x = \phi(x)$)	Value of the function $\phi(x)$ at the point x_n
Muller	$x_{n+1} = x_n + \dfrac{-2C}{B \pm \sqrt{B^2 - 4AC}}$, where $A = \dfrac{(x_{n-1} - x_n)(y_{n-2} - y_n) - (x_{n-2} - x_n)(y_{n-1} - y_n)}{(x_n - x_{n-1})(x_{n-1} - x_{n-2})(x_{n-2} - x_n)}$ $B = \dfrac{(x_{n-2} - x_n)^2(y_{n-1} - y_n) - (x_{n-1} - x_n)^2(y_{n-2} - y_n)}{(x_n - x_{n-1})(x_{n-1} - x_{n-2})(x_{n-2} - x_n)}$ $C = y_n$	Intersection of the interpolating quadratic polynomial and the x-axis
Chebyshev	$x_{n+1} = x_n - \dfrac{\left(f(x_n)\right)^2 f''(x_n)}{2\left(f'(x_n)\right)^3} - \dfrac{f(x_n)}{f'(x_n)}$	Intersection of the quadratic interpolation of the inverse function of $f(x)$ and the x-axis
Aitken Δ^2 Process	$x_{n+1} = \phi(x_n)$ $x_{n+2} = \phi(x_{n+1})$ $x_{n+3} = x_n - \dfrac{(x_{n+1} - x_n)^2}{x_{n+2} - 2x_{n+1} + x_n}$	Intersection of the extrapolating polynomial for the last three iterations (x_n, x_{n+1} and x_{n+2}) with the x-axis

Table 3.4 Properties and Convergence of Methods

Type	Bisection	Regula Falsi	Fixed-Point	Secant	N. R.	Muller	Chebyshev	Aitken Δ^2
	Bracketing	Bracketing	Open End	Open End	Open End	Open End	Open End	Open End
Number of Initial Approximations	2	2	1	2	1	3	1	3
Convergence	Always	Always	–	–	–	–	–	–
Order of Convergence	Linear (1)	Linear (1)	Linear (1)	Super-linear (1.618)	Quadratic (2)	1.84	Cubic (3)	Linear (1) and higher than Fixed-Point method
Programming	Easy	Easy*	Easy	Easy*	Easy**	Moderate	Moderate**	Moderate
Stopping Criterion	In all the iterative methods, stopping criterion is either, last two iterations matches up to desired decimal points, (or) when the value of function f(x) is near to zero (whatever we required, say 0.000001, etc.)							
Accuracy Check	We can always check our root for correctness by simply putting the value in the equation. For example, we obtain root ($x = 0.385335$) of the equation $f(x) = \cos x - 5x + 1 = 0$. Now $$f(0.385335) = \cos(0.385335) - 5(0.385335) + 1$$ $$= 0.926673 - 1.926675 + 1$$ $$= -0.000002$$ Which is nearly equal to zero, hence root obtained is correct.							

* Refer point 9 of next section
** Refer point 10 of next section

In this chapter, we have discussed various numerical techniques to compute the roots of nonlinear equations. Each method has its advantages and disadvantages. Sometimes the nature of the problem plays a prominent role in the selection of the numerical techniques. Here, we are enlisting a few important characteristics of all these methods.

1. Bisection method has simplest structure and easy to understand. It is surely convergent if the function $f(x)$ is continuous in the interval (a, b) and $f(a)f(b) < 0$. To compute the root of a given accuracy, the number of iterations (3.2) are known in advance, which is not possible in any other method. The major disadvantage of this method is its slow convergence.

2. Regula Falsi and Secant methods are linear interpolation methods, and generate iterations in same manner, but Regula Falsi method tests that the successive iterations always bracket the root. Regula Falsi method is slow to converge as compared to Secant method, as it converges from one side. Bisection and Regula Falsi methods have slow convergence, but always converging nature is an important criterion for their selection to solve any problem. A major disadvantage for both the methods is to compute the interval that contains the root. The open end methods, like Secant, Newton–Raphson etc., do not require this condition.

3. Regula Falsi method has fast convergence as compared to Bisection method for most of the problems. But, in certain cases, the Bisection method gives better results as compared Regula Falsi method. Also, the computational effort for one iteration of Bisection method is much less than the Regula Falsi method.

4. The flexibility to select iteration function $\phi(x)$ is a major advantage for the fixed point method. We can select the iteration function $\phi(x)$, which satisfies the convergence conditions and also has a higher order of convergence. We need to compute only one function value for the iteration of Fixed-Point method, while the Secant and Regula Falsi methods require two function values. The convergence conditions for the Fixed-Point method are much relaxed as compared to Secant and Newton–Raphson methods.

5. The orders of convergence for Secant and Newton–Raphson methods are super-linear (1.618) and quadratic (2), respectively. The Fixed-Point method has linear rate of convergence.

6. The orders of convergence are very high for Newton–Raphson and Chebyshev methods, and only few iterations are required for a very high accuracy. But, the computation of the derivative terms is not possible without symbolic software like Maple, Mathematica, etc.

7. The divergence occurs frequently in open end methods, such as Secant and Newton–Raphson methods, so the initial approximation must be sufficiently close to the exact root for these methods. But, these methods have high rate of convergence. So, a sufficiently accurate initial approximation can be obtained from Bisection method, and it can be refined with the help of these methods.

8. Secant and Newton–Raphson methods have problems in obtaining multiple roots. The modified and accelerated Newton–Raphson methods are recommended for faster convergence in such cases.

9. The formulae for Secant and Regula Falsi methods have the term $f(x_n) - f(x_{n-1})$ in denominator. After some iterations, the difference between two successive iterations

3.13 Summary and Observations

$(x_{n-1}$ and $x_n)$ is very small. So, we have subtraction of two nearly equal numbers $(f(x_n) - f(x_{n-1}))$ in the denominator. It may lead to overflow in the programming.

10. The programming of Newton–Raphson and Chebyshev methods require the computation of derivatives of the function $f(x)$, which is not possible in general programming software like C-programming, Fortran, etc. For this, we have to find the derivatives manually and use these as user-defined functions (Refer the C-programs). Consider the function $f(x)$ is not known explicitly, but it is produced within the other problem. The computation of derivatives is not possible in such cases, so we have to use other methods. For example, the characteristics equation is produced in the eigenvalue problem, and we have to compute the roots of this equation within the program to compute the eigenvalues.

11. Muller method has high order of convergence (1.84), and generally gives the root with any initial approximation. The method can also be used to obtain complex roots.

12. Aitken process is used to accelerate the convergence of a linearly convergent sequence. In this chapter, the convergence of Fixed-Point method has been speed up with the help of this method. But, Aitken process can also be used with other methods.

13. The computational works, convergence criteria, implementation of algorithm on computer, and order of convergence are some important factors in selection of an iterative procedure to solve a particular problem.

Exercise 3

1. Compute the point of intersection of the curve $\sin x$ and $x^3 - 1$ with the help of Bisection method.

   Ans. 1.24905

2. a) Use Bisection method to obtain first positive root of the following equation

   $$x^3 + 2x^2 - 16x + 5 = 0$$

   Ans. (0.328171)

   b) Use the Bisection method to find zeroes of the following functions

        i) $f(x) = \sin x - x^3 - 5.3$ in the interval (-2, -1)   **Ans.** -1.84330

        ii) $f(x) = x \ln(x) - 2$ in the interval (2, 3)   **Ans.** 2.34575

3. Find the positive root of the equation $xe^x = 2$, which lies in the interval (0, 1) and correct to five decimal places. Use the following methods.

        (a)   Secant method

        (b)   Newton–Raphson method

   Ans. 0.8526055

4. Prove that the iteration function $x = \frac{1}{2} e^{x/2}$ is not suitable to compute the root of the equation $e^x - 4x^2 = 0$ in the internal (4, 5). Find appropriate iteration function and use it to compute the root.

   Ans. iteration function $x = \ln(4x^2)$, Root = 4.306585

5. Write down the iteration formula to calculate a real root of the equation $\cos x = 3x - 1$ by the Fixed-Point method, such that it satisfies the convergence conditions. Also, use it to compute the root correct to four decimal places.

   Ans. 0.6071

6. Use Newton–Raphson method to compute the root of the equation $e^{2x} - 2e^x - x + \sin(x) = 0$ in the interval $(-2, -1)$, correct to four decimal places. (-1.4148)

7. Find the smallest root of the equation $e^{-x} = 2\cos x$ using Newton–Raphson method.

   Ans. 1.453674

8. Use the Newton–Raphson method to derive the formula $x_{n+1} = \dfrac{1}{2}\left(x_n + \dfrac{a}{x_n}\right),\ n = 0,1,2,\cdots$ for finding the square root of the positive number a. Use it to find $\sqrt{3}$ correct to 5 decimal places.

   Ans. 1.73205

9. Use the Newton–Raphson method to derive the iterative formulae to determine
   i) Square and cube root of a positive number.
   ii) Reciprocal of a number.
   Use them to obtain square root, cube root and reciprocal of 26 correct to four decimal places.

   Ans. 5.0990, 2.9625, 0.03846

10. The bacteria concentration in reservoir varies as $C = 3e^{-1.4t} + e^{-0.2t}$. Using Newton–Raphson method, calculate the time required for the bacteria concentration to be 0.7.

    Ans. 2.488459

11. Compute the root of the equation $x^3 - 23x - 85 = 0$ with the aid of Regula Falsi method correct to two decimal places.

    Ans. 6.08

12. Find the positive root of the equation $x^2 - 6e^{-x} = 0$, with initial approximations 2.5 and 2 using Secant method.

    Ans. 1.28705

13. Show that Newton–Raphson method oscillates for any initial approximation to the following function

$$f(x) = \begin{cases} \sqrt{x} & x \geq 0 \\ \sqrt{-x} & x < 0 \end{cases}$$

   Ans. Since $x_{n+1} = x_n - \dfrac{f(x_n)}{f'(x_n)} = x_n - 2x_n = -x_n$, hence for any initial approximation x_0, the Newton–Raphson method will oscillate between x_0 and $-x_0$.

14. Find the cube root of 15 correct to four significant digits by Fixed-Point and Secant methods.

    Ans. 2.4662

15. Find the roots of the equation $f(x) = (x^2 - 3x + 1)^2 = 0$ using accelerated Newton–Raphson method, correct to four significant figures. Assuming multiplicity $m = 2$ and starting with the initial approximation $x_0 = 0, 2$.

    Ans. 0.381966, 2.61803

16. Find the multiple roots of the equation

$$27x^5 + 27x^4 + 36x^3 + 28x^2 + 9x + 1 = 0$$

with the aid of Newton–Raphson, accelerated Newton–Raphson and Secant methods. Start with an initial approximation $x_0 = 0$.

Ans. 0.3333

17. Show that the equation $(1-x)\sin(1-x) = 0$ has a double root at the point $x = 1$. Compute the root by using the Newton–Raphson method and modified Newton–Raphson method with $m = 2$. Take initial approximation $x_0 = 0$ for both the methods.

18. Perform the Fixed-Point method to find all the roots of the equation $5x^3 - 20x + 3 = 0$, starting with initial approximation 0.5. (0.150169, 4.39516, –4.54533)

19. Determine the intervals of unit length which contain the roots of following equations and then compute the roots of these equations by Bisection, Regula Falsi, Secant, Newton–Raphson, Fixed-Point, Muller, Chebyshev and Aitken Δ^2 methods

    a) $x^3 - 3x^2 - 2x + 2 = 0$ **Ans.** 0.585786, -1, 3.41421

    b) $3x^3 - 5x + 2 = 0$ **Ans.** 0.457427, 1, -1.45743

    c) $2x^4 + 3x^3 - 5x^2 - 2x + 1 = 0$ **Ans.** 0.311836, –0.602027, 1.13557, -2.34538

20. Compute all the three roots of the cubic equation $x^3 - 3x^2 - 5x + 1 = 0$ with the aid of Muller method, which are in the intervals (-2, -1), (0, 1) and (4, 5).

Ans. -1.32887, 0.181442. 4.14743

21. Apply the Muller method with $x_0 = 0$, $x_1 = 0.1$, $x_2 = 0.2$, to find the roots of the following equations

    i) $\cos x - 5x + x^2 = 0$ **Ans.** 0.204184

    ii) $x^4 - 3x^2 - 5x + 1 = 0$ **Ans.** 0.180635

22. All the following equations have one root in the interval (0, 1). Compute the roots of these equations by Bisection, Regula Falsi, Secant, Newton–Raphson, Fixed-Point, Muller, Chebyshev and Aitken Δ^2 methods

    a) $2x - \cos(x) = 0$ **Ans.** 0.450184

    b) $x + 3\sin(x) = 2e^{-x}$ **Ans.** 0.355878

    c) $x^3 + 3x^2 - 5\cos(x) = 0$ **Ans.** 0.895651

23. In process of computing extreme values of the function, $f(x)$, first we find the critical points (values of x where $f'(x) = 0$). These critical points are possible points of maxima and minima (extreme values). Compute these critical points for the following functions and hence find the extreme values for the following functions:

    a) $x^2 - 2e^x \cos(x)$ b) $x^4 - 3x^3 + x + 1$ c) $x^3 + 3x^2 - 5\cos(x)$

(Hint: Find the zeroes of $f'(x)$)

Ans.   a) 0.557664, 3.97962, 7.064312 ..., (Infinite number of extreme points

      b) 0.364091, 2.19826, -0.312356 (Only three)

      c) 0, -2.44051 (only two)

24. Find the x-coordinate of the point on the curve $y = e^x$, which is closest to the origin. Use Newton–Raphson method.

    Ans. The distance between the point (x, e^x) from the origin $(0, 0)$ is $f(x) = (x-0)^2 + (e^x - 0)^2$. To compute point of minima, we have $f'(x) = x + e^{2x} = 0$, which implies $x = -0.5671433$

25. The van der Waals equation for a fluid (*n moles*) is an equation relating the density of gases and liquids (fluids) to the pressure (p), volume (V), and temperature (T) conditions

    $$\left(p + \frac{n^2 a}{V^2}\right)(V - nb) = nRT$$

    On using $v = V/n$, we have

    $$\left(p + \frac{a}{v^2}\right)(v - b) = RT$$

    where p is the pressure, $v = V/n$ is the partial molar volume, R is the universal gas constant, and T is the absolute temperature. The factors a and b are constants dependent on the nature of the fluid. The constant "a" measures the attractive forces between the molecules. The constant "b" is the actual volume of a mole of molecules.

    a) Consider the Carbon dioxide gas (5.00 moles) at 40.0 atmosphere pressure and 323 K temperature. Van der Waals constants (a and b) for the carbon dioxide gas are given by: $a = 3.610 \text{ L}^2 \text{ atm mol}^{-2}$; $b = 0.0429 \text{ L mol}^{-1}$. Compute the partial molar volume of the given gas for these values. Given gas constant $R = 0.08205746 \text{ L atm K}^{-1} \text{ mol}^{-1}$.

    Ans. The van der Waals equation is

    $$\left(p + \frac{a}{v^2}\right)(v - b) = RT$$

    $$\left(40 + \frac{3.61}{v^2}\right)(v - 0.0429) = (0.08205746)(323)$$

    $$40v^3 - 28.22056v^2 + 3.61v - 0.154869 = 0$$

    The equation has only one real root in the interval (0, 1). Compute this root by any method, and the root is given by 0.55562574. So the partial molar volume for the Carbon dioxide is $v = 0.55562574 \text{ L mol}^{-1}$.

    b) Consider the Helium gas (5.00 moles) at 50.0 atmosphere pressure and 373 K temperature. Van der Waals constants (a and b) for the Helium gas are given by: $a = 0.0341 \text{ L}^2 \text{ atm mol}^{-2}$; $b = 0.0238 \text{ L mol}^{-1}$. Compute the partial molar volume. Given gas constant $R = 0.08205746 \text{ L atm K}^{-1} \text{ mol}^{-1}$.

    Ans. The van der Waals equation is

    $$\left(p + \frac{a}{v^2}\right)(v - b) = RT$$

    $$\left(50 + \frac{0.0341}{v^2}\right)(v - 0.0238) = (0.08205746)(373)$$

    $$50v^3 - 31.79743258v^2 + 0.0341v - 0.00081158 = 0$$

The equation has only one real root in the interval (0, 1). Compute this root by any method, and the root is given by 0.634914757. So the partial molar volume for the Carbon dioxide is $v = 0.634914757$ L mol^{-1}.

26. In general, the iteration method ($x_{n+1} = \phi(x_n)$) for computing the roots of an equation $f(x) = 0$ is of linear convergence (order of convergence is 1). To increase the order of convergence, we select the iteration function $\phi(x) = x + \lambda_1 f(x) + \lambda_2 f^2(x) + \cdots$, where λ_i's are arbitrary constants to be determined. Compute the λ_1 such that the order of convergence is quadratic (2); λ_1 and λ_2, such that the order of convergence is cubic (3).

Use these derived methods to compute the roots of the following equations correct to six decimal places

   a) $2x - \cos(x) = 0$ **Ans.** 0.450184
   b) $x + 3\sin(x) = 2e^{-x}$ **Ans.** 0.355878
   c) $x^3 + 3x^2 - 5\cos(x) = 0$ **Ans.** 0.895651

27. In the Bisection algorithm, let M denote the length of the interval $[a, b]$. Let $\{x_0, x_1, ...\}$ represent successive midpoints generated by the Bisection method. Show that, $|x_{i+1} - x_i| = M / 2^{i+2}$ and find the number of iterations required to guarantee an approximation to a root to accuracy ε.

28. The function $f(x) = \begin{cases} 0.1 & x \leq 1.2 \\ -0.1 & x > 1.2 \end{cases}$ changes its sign in the interval (1, 2); that is $f(a)f(b) < 0$.

   What does the Bisection algorithm locate? Is there any zero of the function $f(x)$?

   Ans. $x = 1.2$ is a point of discontinuity, so, the Bisection method is not applicable to this function. In fact, the function $f(x)$ has no zero.

29. Use convergence conditions of Fixed-Point method to determine an iterative function for the equation $x = \tan(x)$ in the interval (199, 200). Also, find the solution correct to four decimal places.

   Ans. 199.48612

30. Let the function $f(x): [a, b] \to [a, b]$ be a continuous function, then prove that there exists a fixed point of the function $f(x)$ in the interval $[a, b]$.

31. Describe the False-Position method geometrically.

32. Find the value of constant C, such that the sequence generated by the formula $x_{n+1} = \dfrac{x_n}{2}\left(C - \dfrac{x_n^2}{a^2}\right)$ has second order convergence with the limit point a.

   Ans. $C = 3$

33. For what value of K, the iteration function $x_{n+1} = 2 - Kx_n + \left(\dfrac{K}{2} - 1\right)x_n^2$ will have quadratic convergence to the fixed point $\xi = 2$.

   Ans. $K = 4$

34. Assume that the error of a Fixed-Point iteration satisfies the recurrence relation $e_{n+1} = ke_n$ for some constant k, mod (k) <1. Find an expression for the number of iterations N required to reduce the initial error e_0 by a factor $10^{-m} (m > 0)$.

35. For each of the following function locate an interval containing the smallest positive zero and show that four conditions of Theorem 3.5 for the convergence of Newton–Raphson and Secant methods are satisfied.

(a) $x - e^{-x} = 0$

(b) $x^3 - x - 1 = 0$

(c) $e^{-x^2} - \cos x = 0$

36. For Newton–Raphson method, show that if $f(\xi) = 0, f'(\xi) \neq 0$ and if the function $f(x)$ is twice continuously differentiable then $\phi'(\xi) = 0$ and $\phi''(\xi) = f''(\xi) / f'(\xi)$.

37. The cubic equation $2x^3 + 3x^2 - 3x - 10 = 0$ has a root near $x = 1.5$. Find at least two iteration functions for iterative methods which satisfy the convergence conditions, and use these iterative functions to compute the root.

Ans. 1.54975

38. The Newton–Raphson method has been modified for multiple roots. Similarly, the Chebyshev method can also be modified for multiple roots of multiplicity m.Obtain the values of constants A and B in following modified Chebyshev iteration method, such that the method has cubic convergence.

$$x_{n+1} = x_n - A\frac{(f(x_n))^2 f''(x_n)}{2(f'(x_n))^3} - B\frac{f(x_n)}{f'(x_n)} \quad n = 0, 1, 2 \cdots$$

Ans. $A = m^2$, $B = \dfrac{m(3-m)}{2}$

39. Find the rate of convergence of the following two sequences (both converges to $\sqrt{a}$)

$$x_{n+1} = \frac{x_n}{2}\left(1 + \frac{a}{x^2_n}\right), \qquad x_{n+1} = \frac{x_n}{2}\left(3 - \frac{x^2_n}{a}\right)$$

Ans. Rate of convergence is quadratic for both the sequences

Chapter 4

Nonlinear Systems and Polynomial Equations

We have discussed various methods for the numerical solutions of a single nonlinear equation $f(x) = 0$ in the last chapter. This chapter presents numerical methods for the solution of a system of nonlinear equations. A system of m nonlinear equations in m variables is given by

$$
\begin{aligned}
f_1(x_1, x_2, \cdots, x_m) &= 0 \\
f_2(x_1, x_2, \cdots, x_m) &= 0 \\
&\vdots \\
f_m(x_1, x_2, \cdots, x_m) &= 0
\end{aligned}
\tag{4.1}
$$

For example, the following system of two equations, are in two variables (x, y)

$$
\sin(x\,y) - x^2 - 5x - y = 0
$$
$$
\cos(x + y) + y^2 - x - 6y + 2 = 0
$$

In this chapter, following methods have been discussed for the solutions of the system (4.1)

    i)   Fixed-Point method
    ii)   Seidel Iteration method
    iii)   Newton–Raphson method.

These methods are also useful in finding complex roots of the equation, $F(z) = 0$, where z is a complex variable.

The methods presented in the last chapter can solve nonlinear equations including polynomial equations. This chapter also includes the methods which are exclusively for solutions of polynomial equations. These methods are better from computational point of view, to extract quadratic factor (which is helpful for complex roots), to obtain all the roots in single application, and to avoid divergence of the methods, etc. This chapter deals with the following methods for the solutions of polynomial equations.

i)   Birge–Vieta method
ii) Lin–Bairstow method
iii) Graeffe root squaring method.

4.1 Fixed-Point Method

Consider the following system of two nonlinear equations

$$f(x, y) = 0$$
$$g(x, y) = 0 \qquad (4.2)$$

On a similar pattern as in Fixed-Point method for single equation, first, we rewrite the system as follows

$$x = \phi(x, y)$$
$$y = \psi(x, y)$$

The zeroes of the functions, $f(x, y)$ and $g(x, y)$ are the fixed points of the functions, $\phi(x, y)$ and $\psi(x, y)$, respectively. Let (x_0, y_0) be any initial approximation to the root of the system (4.2). Then the next approximation can be calculated by the values of the functions $\phi(x, y)$ and $\psi(x, y)$ at this point.

$$x_1 = \phi(x_0, y_0) \qquad\qquad y_1 = \psi(x_0, y_0)$$

Proceeding in a similar manner, we can compute the following approximations

$$x_2 = \phi(x_1, y_1) \qquad\qquad y_2 = \psi(x_1, y_1)$$

$$x_3 = \phi(x_2, y_2) \qquad\qquad y_3 = \psi(x_2, y_2)$$

$$\vdots$$

In general, the Fixed-Point method for the system (4.2) is given by

$$x_{k+1} = \phi(x_k, y_k) \qquad y_{k+1} = \psi(x_k, y_k) \qquad k = 0, 1, 2, \dots \qquad (4.3)$$

Note that the subscript denotes the iteration in Eqs. (4.3).

Example   **4.1**

Compute the solution of nonlinear system $\begin{array}{l} \sin(x\,y)-x^2-5x-y=0 \\ \cos(x+y)+y^2-x-6y+2=0 \end{array}$ with the help of Fixed-Point method. Use an initial approximation $(0, 0)$.

Ans. The given system can be written as follows

$$x = (\sin(x\,y)-x^2-y)/5 = \phi(x,\ y)$$
$$y = (\cos(x+y)+y^2-x+2)/6 = \psi(x,\ y)$$

Let $(0, 0)$ be the initial approximation, (x_0, y_0), then the first approximation of the Fixed Point iteration (4.3) is given by

$$x_1 = \phi(x_0, y_0) \qquad\qquad y_1 = \psi(x_0, y_0)$$
$$x_1 = (\sin(x_0\,y_0)-x_0^2-y_0)/5 = \phi(x_0, y_0) = 0$$
$$y_1 = (\cos(x_0+y_0)+y_0^2-x_0+2)/6 = \psi(x_0, y_0) = 0.5$$

Similarly, the second approximation is as follows

$$x_2 = (\sin(x_1\,y_1)-x_1^2-y_1)/5 = \phi(x_1, y_1) = -0.1$$
$$y_2 = (\cos(x_1+y_1)+y_1^2-x_1+2)/6 = \psi(x_1, y_1) = 0.521264$$

Proceeding in a similar manner, we can obtain the following iterations of Fixed-Point method

$$x_3 = -0.116673 \quad y_3 = 0.547381$$

$$x_4 = -0.124963 \quad y_4 = 0.554162$$

$$x_5 = -0.127794 \quad y_5 = 0.556893$$

$$x_6 = -0.128866 \quad y_6 = 0.557878$$

$$x_7 = -0.129263 \quad y_7 = 0.558245$$

$$x_8 = -0.129410 \quad y_8 = 0.558382$$

$$x_9 = -0.129465 \quad y_9 = 0.558432$$

$$x_{10} = -0.129486 \quad y_{10} = 0.558451$$

So, the approximate solution is given by

$$x = -0.129486 \quad y = 0.558451$$

Convergence Conditions for Fixed-Point Method

Let (ξ, η) be the exact solution of the system (4.2), then

$$\xi = \phi(\xi, \eta)$$
$$\eta = \psi(\xi, \eta) \tag{4.4}$$

The Fixed-Point iterations (4.3) are given by

$$\begin{aligned} x_{k+1} &= \phi(x_k, y_k) \\ y_{k+1} &= \psi(x_k, y_k) \end{aligned} \quad k = 0, 1, 2, \dots \tag{4.5}$$

Let the iterations converge to the exact solution, (ξ, η), then

$$\lim_{k \to \infty} x_{k+1} \to \xi \quad \text{and} \quad \lim_{k \to \infty} y_{k+1} \to \eta$$

On subtracting Eq. (4.5) from Eq. (4.4), we have

$$\begin{aligned} \xi - x_{k+1} &= \phi(\xi, \eta) - \phi(x_k, y_k) \\ \eta - y_{k+1} &= \psi(\xi, \eta) - \psi(x_k, y_k) \end{aligned}$$

Let us consider the error at k^{th} step $\varepsilon_k = \xi - x_k$ and $\delta_k = \eta - y_k$, then we have

$$\begin{aligned} \varepsilon_{k+1} &= \phi(x_k + \varepsilon_k, y_k + \delta_k) - \phi(x_k, y_k) \\ \delta_{k+1} &= \psi(x_k + \varepsilon_k, y_k + \delta_k) - \psi(x_k, y_k) \end{aligned}$$

Using Taylor series expansion and neglecting the second and higher order terms of ε_k and δ_k, we get

$$\begin{aligned} \varepsilon_{k+1} &= \left[\phi(x_k, y_k) + \varepsilon_k \phi_x(x_k, y_k) + \delta_k \phi_y(x_k, y_k) \right] - \phi(x_k, y_k) \\ \delta_{k+1} &= \left[\psi(x_k, y_k) + \varepsilon_k \psi_x(x_k, y_k) + \delta_k \psi_y(x_k, y_k) \right] - \psi(x_k, y_k) \end{aligned}$$

In Matrix form, we have

$$\begin{bmatrix} \varepsilon_{k+1} \\ \delta_{k+1} \end{bmatrix} = \begin{bmatrix} \phi_x & \phi_y \\ \psi_x & \psi_y \end{bmatrix}_{(x_k, y_k)} \begin{bmatrix} \varepsilon_k \\ \delta_k \end{bmatrix}$$

$$E_{k+1} = A_k E_k$$

where $E_{k+1} = \begin{bmatrix} \varepsilon_{k+1} \\ \delta_{k+1} \end{bmatrix}$, $A_k = \begin{bmatrix} \phi_x & \phi_y \\ \psi_x & \psi_y \end{bmatrix}_{(x_k, y_k)}$ and $E_k = \begin{bmatrix} \varepsilon_k \\ \delta_k \end{bmatrix}$ are the error vector at $(k+1)^{th}$

step, the Jacobian of iteration matrix at k^{th} step and error vector at k^{th} step, respectively.

So, the sufficient condition for the convergence of the Fixed-Point method is that for each k, we have following condition on iteration matrix (A_k)

$$\|A_k\| < 1$$

where $\|\ \|$ is a suitable norm. For example, let us assume the maximum absolute row sum as a norm. Then, Fixed-Point method will converge to the solution, if the iteration functions ϕ and ψ satisfy the following conditions

$$\left|\phi_x(x_k, y_k)\right| + \left|\phi_y(x_k, y_k)\right| < 1$$

$$\left|\psi_x(x_k, y_k)\right| + \left|\psi_y(x_k, y_k)\right| < 1 \tag{4.6}$$

A sufficient and necessary condition for the convergence of Fixed-Point method is that for each k, we have

$$\rho(A_k) < 1 \tag{4.7}$$

where $\rho(A_k)$ is spectral radius (largest eigenvalue in magnitude) of matrix A_k.

Note: *Convergence condition (4.6) is similar to Fixed-Point method for a single equation in Chapter 3.*

So far, we have discussed Fixed-Point method for a system of two equations. Now, we will generalize it to the Fixed-Point method for the solution of system (4.1) of m nonlinear equations.

First, rewrite the system (4.1) as follows

$$x_1 = \phi_1(x_1, x_2, \cdots, x_m)$$
$$x_2 = \phi_2(x_1, x_2, \cdots, x_m)$$
$$\vdots$$
$$x_m = \phi_m(x_1, x_2, \cdots, x_m)$$

Then, use the initial approximation $x_1^{(0)}, x_2^{(0)}, \cdots, x_m^{(0)}$ to compute next approximation as follows

$$x_1^{(1)} = \phi_1(x_1^{(0)}, x_2^{(0)}, \cdots, x_m^{(0)})$$
$$x_2^{(1)} = \phi_2(x_1^{(0)}, x_2^{(0)}, \cdots, x_m^{(0)})$$
$$\vdots$$
$$x_m^{(1)} = \phi_m(x_1^{(0)}, x_2^{(0)}, \cdots, x_m^{(0)})$$

It is worth mentioning here that the subscript i in $x_i^{(j)}$ denotes the variable and the superscript j in $x_i^{(j)}$ denotes the iteration.

In general, the Fixed-Point iterations are given by

$$x_1^{(k+1)} = \phi_1(x_1^{(k)}, x_2^{(k)}, \cdots, x_m^{(k)})$$
$$x_2^{(k+1)} = \phi_2(x_1^{(k)}, x_2^{(k)}, \cdots, x_m^{(k)}) \qquad k = 0,1,2\cdots \qquad (4.8)$$
$$\vdots$$
$$x_m^{(k+1)} = \phi_m(x_1^{(k)}, x_2^{(k)}, \cdots, x_m^{(k)})$$

Note: The necessary and sufficient convergence condition for Fixed-Point method for the system of nonlinear equations, Eq. (4.1) can be obtained easily by extending conditions (4.6) and (4.7) as follows

Consider the following iteration matrix

$$A_k = \begin{bmatrix} \dfrac{\partial \phi_1}{\partial x_1} & \dfrac{\partial \phi_1}{\partial x_2} & \cdots & \dfrac{\partial \phi_1}{\partial x_m} \\[2ex] \dfrac{\partial \phi_2}{\partial x_1} & \dfrac{\partial \phi_2}{\partial x_2} & \cdots & \dfrac{\partial \phi_2}{\partial x_m} \\[2ex] \vdots & & & \\[1ex] \dfrac{\partial \phi_m}{\partial x_1} & \dfrac{\partial \phi_m}{\partial x_2} & \cdots & \dfrac{\partial \phi_m}{\partial x_m} \end{bmatrix}_{(x_1^k, x_2^k, \cdots, x_m^k)} \qquad k = 0,1,2\cdots$$

If $\rho(A_k)$ (spectral radius of the matrix A_k) is less than 1 for each k, then the Fixed-Point method will converge to the solution.

While sufficient condition is as follows

$$\left|\dfrac{\partial \phi_i}{\partial x_1}\right| + \left|\dfrac{\partial \phi_i}{\partial x_2}\right| + \cdots + \left|\dfrac{\partial \phi_i}{\partial x_m}\right| < 1 \qquad \forall i = 1,2,\cdots,m \qquad (4.9)$$

Example   4.2

Solve the following system of nonlinear equations

$$e^{-(x_1+x_3)} + 8x_3 + 2x_2 - 1 = 0$$
$$\cos(x_2 x_3) + 3\sin x_1 + 10x_1 - 2x_3 = 0$$
$$2x_2^2 + 10x_2 - x_3 + 2x_1 + 2 = 0$$

with the help of Fixed-Point method. Consider the initial approximation, $x_1^{(0)} = x_2^{(0)} = x_3^{(0)} = 0$.

Ans. First, we rewrite the system as follows, keeping in mind the convergence conditions

$$x_1 = \frac{-1}{10}\left(\cos(x_2 x_3) + 3\sin x_1 - 2x_3\right)$$

$$x_2 = \frac{-1}{10}\left(2x_2^2 - x_3 + 2x_1 + 2\right)$$

$$x_3 = \frac{-1}{8}\left(e^{-(x_1+x_3)} + 2x_2 - 1\right)$$

Here, we write x_3 from the first equation, and x_1, x_2 from the second and third equations, respectively. The iteration functions are as follows

$$\phi_1(x_1, x_2, x_3) = \frac{-1}{10}\left(\cos(x_2 x_3) + 3\sin x_1 - 2x_3\right)$$

$$\phi_2(x_1, x_2, x_3) = \frac{-1}{10}\left(2x_2^2 - x_3 + 2x_1 + 2\right)$$

$$\phi_3(x_1, x_2, x_3) = \frac{-1}{8}\left(e^{-(x_1+x_3)} + 2x_2 - 1\right)$$

The initial approximation is $x_1^{(0)} = x_2^{(0)} = x_3^{(0)} = 0$, where the subscript and superscript denote the variables and iterations, respectively. For example, here 1, 2 and 3 in subscripts represent the variables while superscript 0 represents the initial approximation. Starting with initial approximation, $x_1^{(0)} = x_2^{(0)} = x_3^{(0)} = 0$, we get our first approximation as follows

$$x_1^{(1)} = \phi_1(x_1^{(0)}, x_2^{(0)}, x_3^{(0)}) = \frac{-1}{10}\left(\cos(0) + 3\sin(0) - 2(0)\right) = \frac{-1}{10}(1) = -0.1$$

$$x_2^{(1)} = \phi_2(x_1^{(0)}, x_2^{(0)}, x_3^{(0)}) = \frac{-1}{10}\left(2(0) - 0 + 2(0) + 2\right) = \frac{-1}{10}(2) = -0.2$$

$$x_3^{(1)} = \phi_3(x_1^{(0)}, x_2^{(0)}, x_3^{(0)}) = \frac{-1}{8}\left(e^{-(0)} + 2(0) - 1\right) = \frac{-1}{8}(0) = 0$$

On using the values of $x_1^{(1)}$, $x_2^{(1)}, x_3^{(1)}$, the second approximation is given by

$$x_1^{(2)} = \phi_1(x_1^{(1)}, x_2^{(1)}, x_3^{(1)}) = \frac{-1}{10}\left(\cos(0) + 3\sin(-0.1) - 2(0)\right) = -0.070050$$

$$x_2^{(2)} = \phi_2(x_1^{(1)}, x_2^{(1)}, x_3^{(1)}) = \frac{-1}{10}\left(2(-0.5)^2 - 0 + 2(-0.1) + 2\right) = -0.188000$$

$$x_3^{(2)} = \phi_3(x_1^{(1)}, x_2^{(1)}, x_3^{(1)}) = \frac{-1}{8}\left(e^{-(-0.1)} + 2(-0.5) - 1\right) = 0.036854$$

Similarly, other iterations are given by

$$x_1^{(3)} = -0.071629 \quad x_2^{(3)} = -0.189373 \quad x_3^{(3)} = 0.042781$$

$$x_1^{(4)} = -0.069970 \quad x_2^{(4)} = -0.188569 \quad x_3^{(4)} = 0.043685$$

$$x_1^{(5)} = -0.070286 \quad x_2^{(5)} = -0.188749 \quad x_3^{(5)} = 0.043813$$

$$x_1^{(6)} = -0.070166 \quad x_2^{(6)} = -0.188687 \quad x_3^{(6)} = 0.043834$$

$$x_1^{(7)} = -0.070197 \quad x_2^{(7)} = -0.188704 \quad x_3^{(7)} = 0.043837$$

$$x_1^{(8)} = -0.070187 \quad x_2^{(8)} = -0.188699 \quad x_3^{(8)} = 0.043837$$

$$x_1^{(9)} = -0.070190 \quad x_2^{(9)} = -0.188700 \quad x_3^{(9)} = 0.043837$$

4.2 Seidel Iteration Method

The Fixed-Point method can be modified by using the most recent approximations of the variables in calculating the next iterations. Let (x_0, y_0) be any initial approximation to the root of the system (4.2). The next approximate value x_1 can be calculated by the value of the function $\phi(x, y)$ at the point (x_0, y_0). In general, x_1 is better approximation of x as compared to x_0. Therefore, the value of y_1 can be calculated by the value of the function $\psi(x, y)$ at the point (x_1, y_0). Proceeding in a similar manner, we can easily find the various approximations to the Seidel method as follows

$$x_1 = \phi(x_0, y_0) \qquad\qquad y_1 = \psi(x_1, y_0)$$

$$x_2 = \phi(x_1, y_1) \qquad\qquad y_2 = \psi(x_2, y_1)$$

$$x_3 = \phi(x_2, y_2) \qquad\qquad y_3 = \psi(x_3, y_2)$$

$$\vdots \qquad\qquad\qquad \vdots$$

In general, the Seidel iterations for the solution of the system (4.2) are given by

$$x_{k+1} = \phi(x_k, y_k) \qquad y_{k+1} = \psi(x_{k+1}, y_k) \qquad k = 0, 1, 2, \ldots \qquad (4.10)$$

Note: Convergence conditions for Seidel method are same as of Fixed-Point method. But the order of convergence of Seidel method is higher than Fixed-Point method.

Example 4.3

Use Seidel method to compute the root of nonlinear system in Example 4.1

Ans. Rewrite the given system as follows

$$x = (\sin(x\,y) - x^2 - y)/5 = \phi(x,\,y)$$
$$y = (\cos(x + y) + y^2 - x + 2)/6 = \psi(x,\,y)$$

On using the initial approximation $(x_0,\,y_0) = (0,\,0)$, the first approximation of Seidel iterations (4.10) is given by

$$x_1 = \phi(x_0,\,y_0) = (\sin(x_0\,y_0) - x_0^{\,2} - y_0)/5 = 0$$
$$y_1 = \psi(x_1,\,y_0) = (\cos(x_1 + y_0) + y_0^{\,2} - x_1 + 2)/6 = 0.5$$

Using these values, we can easily compute the second iteration as follows

$$x_2 = (\sin(x_1\,y_1) - x_1^{\,2} - y_1)/5 = \phi(x_1,\,y_1) = -0.1$$
$$y_2 = (\cos(x_2 + y_1) + y_1^{\,2} - x_2 + 2)/6 = \psi(x_2,\,y_1) = 0.545177$$

Similarly, other iterations are given by

$$x_3 = -0.121933 \quad y_3 = 0.555152$$

$$x_4 = -0.127532 \quad y_4 = 0.557614$$

$$x_5 = -0.128986 \quad y_5 = 0.558243$$

$$x_6 = -0.129365 \quad y_6 = 0.558405$$

$$x_7 = -0.129463 \quad y_7 = 0.558448$$

Note: *It is easy to point out here that the Seidel method converges with a faster rate than the Fixed-Point method. Therefore, the root of an accuracy of four decimal points can be obtained in seven iterations with Seidel method, and Fixed-Point method requires twelve iterations for a similar accuracy.*

The Seidel method (4.10) can be extended for the solution of the system (4.1) as follows

Rewrite the system (4.1) as follows

$$x_1 = \phi_1(x_1, x_2, \cdots, x_m)$$
$$x_2 = \phi_2(x_1, x_2, \cdots, x_m)$$
$$\vdots$$
$$x_m = \phi_m(x_1, x_2, \cdots, x_m)$$

Then, use the initial approximation $x_1^{(0)}, x_2^{(0)}, \cdots, x_m^{(0)}$ to compute next approximation as follows

$$x_1^{(1)} = \phi_1(x_1^{(0)}, x_2^{(0)}, x_3^{(0)}, \cdots, x_m^{(0)})$$
$$x_2^{(1)} = \phi_2(x_1^{(1)}, x_2^{(0)}, x_3^{(0)}, \cdots, x_m^{(0)})$$
$$x_3^{(1)} = \phi_3(x_1^{(1)}, x_2^{(1)}, x_3^{(0)} \cdots, x_m^{(0)})$$
$$\vdots$$
$$x_m^{(1)} = \phi_m(x_1^{(1)}, x_2^{(1)}, x_3^{(1)} \cdots, x_m^{(0)})$$

$$(4.11)$$

Note that the latest available approximations are used here. Similarly, other approximations for Seidel method are as follows

$$x_1^{(k+1)} = \phi_1(x_1^{(k)}, x_2^{(k)}, x_3^{(k)}, \cdots, x_m^{(k)})$$
$$x_2^{(k+1)} = \phi_2(x_1^{(k+1)}, x_2^{(k)}, x_3^{(k)}, \cdots, x_m^{(k)})$$
$$x_3^{(k+1)} = \phi_3(x_1^{(k+1)}, x_2^{(k+1)}, x_3^{(k)} \cdots, x_m^{(k)})$$
$$\vdots$$
$$x_m^{(k+1)} = \phi_m(x_1^{(k+1)}, x_2^{(k+1)}, x_3^{(k+1)} \cdots, x_{n-1}^{(k+1)}, x_m^{(k)})$$

$$k = 0,1,2 \cdots$$

$$(4.12)$$

The subscript i and superscript j in the variable $x_i^{(j)}$ denote variable and iteration, respectively.

Note: Convergence conditions for Seidel method are same as that of Fixed-Point method, and the rate of convergence of Seidel method is higher than Fixed-Point method.

Example 4.4

Use Seidel iterative method to solve the following system

$$e^{-(x_1+x_3)} + 8x_3 + 2x_2 - 1 = 0$$
$$\cos(x_2 x_3) + 3\sin x_1 + 10x_1 - 2x_3 = 0$$
$$2x_2^2 + 10x_2 - x_3 + 2x_1 + 2 = 0$$

Take initial approximation, $x_1^{(0)} = x_2^{(0)} = x_3^{(0)} = 0$.

Ans. First, rewrite the system as follows

$$x_1 = \phi_1(x_1, x_2, x_3) = \frac{-1}{10}\left(\cos(x_2 x_3) + 3\sin x_1 - 2x_3\right)$$

$$x_2 = \phi_2(x_1, x_2, x_3) = \frac{-1}{10}\left(2x_2^2 - x_3 + 2x_1 + 2\right)$$

$$x_3 = \phi_3(x_1, x_2, x_3) = \frac{-1}{8}\left(e^{-(x_1 + x_3)} + 2x_2 - 1\right)$$

Starting with initial approximation, $x_1^{(0)} = x_2^{(0)} = x_3^{(0)} = 0$, we will get our first approximation from Eq. (4.11) as follows

$$x_1^{(1)} = \phi_1(x_1^{(0)}, x_2^{(0)}, x_3^{(0)}) = \frac{-1}{10}\left(\cos(0) + 3\sin(0) - 2(0)\right) = \frac{-1}{10}(1) = -0.1$$

$$x_2^{(1)} = \phi_2(x_1^{(1)}, x_2^{(0)}, x_3^{(0)}) = \frac{-1}{10}\left(2(0) - 0 + 2(-0.1) + 2\right) = \frac{-1}{10}(1.8) = -0.18$$

$$x_3^{(1)} = \phi_3(x_1^{(1)}, x_2^{(1)}, x_3^{(0)}) = \frac{-1}{8}\left(e^{-(-0.1)} + 2(-0.18) - 1\right) = \frac{-1}{8}(-0.254829) = 0.031854$$

Similarly, second approximation to the root is given by

$$x_1^{(2)} = \phi_1(x_1^{(1)}, x_2^{(1)}, x_3^{(1)}) = \frac{-1}{10}\left(\cos((-0.18)(0.031854)) + 3\sin(-0.1) - 2(0.031854)\right) = -0.063678$$

$$x_2^{(2)} = \phi_2(x_1^{(2)}, x_2^{(1)}, x_3^{(1)}) = \frac{-1}{10}\left(2(-0.18)^2 - 0.031854 + 2(-0.063678) + 2\right) = -0.190559$$

$$x_3^{(2)} = \phi_3(x_1^{(2)}, x_2^{(2)}, x_3^{(1)}) = \frac{-1}{8}\left(e^{-(-0.063678 + 0.031854)} + 2(-0.190559) - 1\right) = 0.043598$$

On computing other iterations from Eq. (4.12) in a similar manner, we have

$$x_1^{(3)} = -0.072187 \quad x_2^{(3)} = -0.188465 \quad x_3^{(3)} = 0.043491$$

$$x_1^{(4)} = -0.069661 \quad x_2^{(4)} = -0.188822 \quad x_3^{(4)} = 0.043891$$

$$x_1^{(5)} = -0.070337 \quad x_2^{(5)} = -0.188674 \quad x_3^{(5)} = 0.043819$$

$$x_1^{(6)} = -0.070149 \quad x_2^{(6)} = -0.188708 \quad x_3^{(6)} = 0.043842$$

$$x_1^{(7)} = -0.070201 \quad x_2^{(7)} = -0.188698 \quad x_3^{(7)} = 0.043836$$

$$x_1^{(8)} = -0.070187 \quad x_2^{(8)} = -0.188700 \quad x_3^{(8)} = 0.043838$$

4.3 Newton–Raphson (NR) Method

Let (x_0, y_0) be initial approximation to the exact root (ξ, η) of the system (4.2) of two nonlinear equations

$$f(x, y) = 0$$
$$g(x, y) = 0$$

Let (h, k) be the error in the approximation (x_0, y_0). Then, $(x_0 + h, y_0 + k)$ is the exact root of the system, so

$$f(x_0 + h, y_0 + k) = 0$$
$$g(x_0 + h, y_0 + k) = 0$$

Let the functions $f(x, y)$ and $g(x, y)$ be differentiable functions. On using the Taylor Series expansion, we obtain

$$f(x_0, y_0) + \left(h\left(\frac{\partial f}{\partial x}\right)_{(x_0, y_0)} + k\left(\frac{\partial f}{\partial y}\right)_{(x_0, y_0)} \right) + \text{second and higher order terms} = 0$$

$$g(x_0, y_0) + \left(h\left(\frac{\partial g}{\partial x}\right)_{(x_0, y_0)} + k\left(\frac{\partial g}{\partial y}\right)_{(x_0, y_0)} \right) + \text{second and higher order terms} = 0$$

Let the initial approximation be close to the root, then h and k are very small quantities. On setting the quadratic and higher terms to zero, we get

$$f(x_0, y_0) + \left(h\left(\frac{\partial f}{\partial x}\right)_{(x_0, y_0)} + k\left(\frac{\partial f}{\partial y}\right)_{(x_0, y_0)} \right) = 0$$

$$g(x_0, y_0) + \left(h\left(\frac{\partial g}{\partial x}\right)_{(x_0, y_0)} + k\left(\frac{\partial g}{\partial y}\right)_{(x_0, y_0)} \right) = 0$$

On solving these two equations for h and k, we have

$$h = \frac{g f_y - f g_y}{f_x g_y - g_x f_y}\bigg|_{(x_0, y_0)}$$

$$k = \frac{f g_x - g f_x}{f_x g_y - g_x f_y}\bigg|_{(x_0, y_0)} \tag{4.13}$$

Since we neglected the second and higher order terms in Taylor series expansion, we will get approximate values of h and k. The next approximation to the root is given by

$$x_1 = x_0 + h$$
$$y_1 = y_0 + k$$

(4.14)

Proceeding in a similar manner, we can further compute other approximations to the root. In general, the Newton–Raphson method for the solution of the system (4.2) is given by

$$x_{i+1} = x_i + \frac{g f_y - f g_y}{f_x g_y - g_x f_y}\bigg|_{(x_i, y_i)}$$

$$y_{i+1} = y_i + \frac{f g_x - g f_x}{f_x g_y - g_x f_y}\bigg|_{(x_i, y_i)} \qquad i = 0, 1, 2, \ldots$$

(4.15)

Example ———————————————————————————————————— **4.5**

Solve the following system of nonlinear equations using Newton–Raphson method

$$\sin(x y) - x^2 - 5x - y = 0$$
$$\cos(x + y) + y^2 - x - 6y + 2 = 0$$

Use initial approximations, $x_0 = y_0 = 0$.

Ans. The functions corresponding to the given system are given by

$$f(x, y) = \sin(x y) - x^2 - 5x - y$$
$$g(x, y) = \cos(x + y) + y^2 - x - 6y + 2$$

$$f_x = y \cos(x y) - 2x - 5 \qquad\qquad f_y = x \cos(x y) - 1$$
$$g_x = -\sin(x + y) - 1 \qquad\qquad g_y = -\sin(x + y) + 2y - 6$$

First iteration

Starting with initial approximations $(x_0 = y_0 = 0)$ and using the Newton–Raphson (NR) method (4.13), we have

$$h = \frac{g f_y - f g_y}{f_x g_y - g_x f_y}\bigg|_{(x_0, y_0)} = \frac{g f_y - f g_y}{f_x g_y - g_x f_y}\bigg|_{(0,0)}$$

$$= \frac{\left(\cos(x+y)+y^2-x-6y+2\right)\left(x\cos(x\,y)-1\right)-\left(\sin(x\,y)-x^2-5x-y\right)\left(-\sin(x+y)+2y-6\right)}{\left(y\cos(x\,y)-2x-5\right)\left(-\sin(x+y)+2y-6\right)-\left(-\sin(x+y)-1\right)\left(x\cos(x\,y)-1\right)}\Bigg|_{(0,0)}$$

$$= \frac{(3)(-1)-(0)(-6)}{(-6)(-5)-(-1)(-1)} = \frac{-3}{29} = -0.103448$$

$$k = \frac{f\,g_x - g\,f_x}{f_x g_y - g_x f_y}\Bigg|_{(x_0,y_0)} = \frac{f\,g_x - g\,f_x}{f_x g_y - g_x f_y}\Bigg|_{(0,0)}$$

$$= \frac{\left(\sin(x\,y)-x^2-5x-y\right)\left(-\sin(x+y)-1\right)-\left(\cos(x+y)+y^2-x-6y+2\right)\left(y\cos(x\,y)-2x-5\right)}{\left(y\cos(x\,y)-2x-5\right)\left(-\sin(x+y)+2y-6\right)-\left(-\sin(x+y)-1\right)\left(x\cos(x\,y)-1\right)}\Bigg|_{(0,0)}$$

$$= \frac{(0)(-1)-(3)(-5)}{(-5)(-6)-(-1)(-1)} = \frac{15}{29} = 0.517241$$

On using these values of h and k in Eqs. (4.14), we get the following next approximation

$$x_1 = x_0 + h = 0 + (-0.103448) = -0.103448$$

$$y_1 = y_0 + k = 0 + 0.517241 = 0.517241$$

Second iteration

Proceeding with the approximation, $x_1 = -0.103448$, $y_1 = 0.517241$, we have

$$h = \frac{g\,f_y - f\,g_y}{f_x g_y - g_x f_y}\Bigg|_{(x_1,y_1)} = \frac{g\,f_y - f\,g_y}{f_x g_y - g_x f_y}\Bigg|_{(-0.103448,\,0.517241)} = -0.025531$$

$$k = \frac{f\,g_x - g\,f_x}{f_x g_y - g_x f_y}\Bigg|_{(x_1,y_1)} = \frac{f\,g_x - g\,f_x}{f_x g_y - g_x f_y}\Bigg|_{(-0.103448,\,0.517241)} = 0.040789$$

On using these values of h and k, the next iteration from Eq. (4.15) is as follows

$$x_2 = x_1 + h = -0.103448 - 0.025531 = -0.128979$$

$$y_2 = y_1 + k = 0.517241 + 0.040789 = 0.558030$$

Similarly, we can obtain the following iterations

$$x_3 = -0.129498 \quad y_3 = 0.558462$$

$$x_4 = -0.129498 \quad y_4 = 0.558463$$

Accelerated Newton–Raphson (NR) Method

Convergence of Newton–Raphson method (4.15) can be accelerated, if we use the latest approximation. For this, the value of x_{i+1} is computed as in Eqs. (4.15), but the value of y_{i+1} is computed using the values at the point (x_{i+1}, y_i) instead of at the point, (x_i, y_i).

$$x_{i+1} = x_i + \left. \frac{g f_y - f g_y}{f_x g_y - g_x f_y} \right|_{(x_i, y_i)}$$

$$y_{i+1} = y_i + \left. \frac{f g_x - g f_x}{f_x g_y - g_x f_y} \right|_{(x_{i+1}, y_i)} \qquad i = 0, 1, 2, \ldots \qquad (4.16)$$

Example 4.6

Use accelerated NR method to solve the following nonlinear system

$$\sin(x\,y) - x^2 - 5x - y = 0$$
$$\cos(x + y) + y^2 - x - 6y + 2 = 0$$

Assume initial approximation $x_0 = y_0 = 0$.

Ans. We have

$$f(x,\ y) = \sin(x\,y) - x^2 - 5x - y$$
$$g(x,\ y) = \cos(x + y) + y^2 - x - 6y + 2$$

$$f_x = y\cos(x\,y) - 2x - 5 \qquad\qquad f_y = x\cos(x\,y) - 1$$
$$g_x = -\sin(x + y) - 1 \qquad\qquad g_y = -\sin(x + y) + 2y - 6$$

First iteration

Starting with initial approximations $x_0 = y_0 = 0$ and using the accelerated Newton–Raphson formula (4.16), we get

$$h = \frac{g f_y - f g_y}{f_x g_y - g_x f_y}\bigg|_{(x_0, y_0)} = \frac{g f_y - f g_y}{f_x g_y - g_x f_y}\bigg|_{(0,0)} = -0.103448$$

$$x_1 = x_0 + h = 0 + (-0.103448) = -0.103448$$

The value of k is computed at the point $x = -0.103448$, $y = 0$

$$k = \frac{f g_x - g f_x}{f_x g_y - g_x f_y}\bigg|_{(x_1, y_0)} = \frac{f g_x - g f_x}{f_x g_y - g_x f_y}\bigg|_{(-0.103448, 0)} = 0.527800$$

$$y_1 = y_0 + k = 0 + 0.527800 = 0.527800$$

Second iteration

Proceeding with the approximation, $x_1 = -0.103448$, $y_1 = 0.527800$, we get following Newton–Raphson iteration (4.16)

$$h = \frac{g f_y - f g_y}{f_x g_y - g_x f_y}\bigg|_{(x_1, y_1)} = \frac{g f_y - f g_y}{f_x g_y - g_x f_y}\bigg|_{(-0.103448,\, 0.527800)} = -0.025633$$

$$x_2 = x_1 + h = -0.103448 - 0.025633 = -0.129081$$

$$k = \frac{f g_x - g f_x}{f_x g_y - g_x f_y}\bigg|_{(x_2, y_1)} = \frac{f g_x - g f_x}{f_x g_y - g_x f_y}\bigg|_{(-0.129081,\, 0.527800)} = 0.030557$$

$$y_2 = y_1 + k = 0.527800 + 0.030557 = 0.558357$$

Similarly, further approximations for the solution are given by

$$x_3 = -0.129498 \quad y_3 = 0.558463$$

$$x_4 = -0.129498 \quad y_4 = 0.558463$$

So far, we have discussed the Newton–Raphson method for system of two nonlinear equations. Now, we will extend the Newton–Raphson method (4.15) for the solution of the system (4.1).

Consider the system (4.1) in vector form

$$f_1(x_1, x_2, \cdots, x_m) = 0$$
$$f_2(x_1, x_2, \cdots, x_m) = 0 \quad \text{or} \quad F(X) = 0 \tag{4.17}$$
$$\vdots$$
$$f_m(x_1, x_2, \cdots, x_m) = 0$$

where $X = \left[x_1, x_2, \cdots, x_m \right]^T$ and $F = \left[f_1, f_2, \cdots, f_m \right]^T$.

Let $X^k = \left[x_1^{\ k}, x_2^{\ k}, \cdots, x_m^{\ k} \right]^T$ be an approximation to the exact root, $\left[\xi_1, \xi_2, \cdots, \xi_m \right]^T$. Also, assume that $\delta X^k = \left[\delta x_1^{\ k}, \delta x_2^{\ k}, \cdots, \delta x_m^{\ k} \right]^T$ be the change in $X^k = \left[x_1^{\ k}, x_2^{\ k}, \cdots, x_m^{\ k} \right]^T$ such that $X^k + \delta X^k = \left[x_1^{\ k} + \delta x_1^{\ k}, x_2^{\ k} + \delta x_2^{\ k}, \cdots, x_m^{\ k} + \delta x_m^{\ k} \right]^T$ is the exact root of the system (4.17)

$$f_1(x_1^{\ k} + \delta x_1^{\ k}, x_2^{\ k} + \delta x_2^{\ k}, \cdots, x_m^{\ k} + \delta x_m^{\ k}) = 0$$
$$f_2(x_1^{\ k} + \delta x_1^{\ k}, x_2^{\ k} + \delta x_2^{\ k}, \cdots, x_m^{\ k} + \delta x_m^{\ k}) = 0$$
$$\vdots$$
$$f_m(x_1^{\ k} + \delta x_1^{\ k}, x_2^{\ k} + \delta x_2^{\ k}, \cdots, x_m^{\ k} + \delta x_m^{\ k}) = 0 \tag{4.18}$$

The Taylor series expansion for any function $g(x_1, x_2, ..., x_m)$ is given by

$$g(x_1 + \delta x_1, x_2 + \delta x_2, ..., x_m + \delta x_m) = g(x_1, x_2, ..., x_m) + \left(\delta x_1 \frac{\partial}{\partial x_1} + \delta x_2 \frac{\partial}{\partial x_2} + ... + \delta x_m \frac{\partial}{\partial x_m} \right) g$$

$$+ \frac{1}{2!} \left(\delta x_1 \frac{\partial}{\partial x_1} + \delta x_2 \frac{\partial}{\partial x_2} + ... + \delta x_m \frac{\partial}{\partial x_m} \right)^2 g + \text{terms involving third and higher powers}$$

With the help of above expression, the Taylor series expansion of the system (4.18) about the point $X^k = \left[x_1^{\ k}, x_2^{\ k}, \cdots, x_m^{\ k} \right]^T$ is given by

$$f_1(x_1^{\ k}, x_2^{\ k}, \cdots, x_m^{\ k}) + \left(\delta x_1^k \frac{\partial}{\partial x_1} + \delta x_2^k \frac{\partial}{\partial x_2} + \cdots + \delta x_m^k \frac{\partial}{\partial x_m} \right) f_1(x_1^{\ k}, x_2^{\ k}, \cdots, x_m^{\ k}) + \cdots = 0$$

$$f_2(x_1^{\ k}, x_2^{\ k}, \cdots, x_m^{\ k}) + \left(\delta x_1^k \frac{\partial}{\partial x_1} + \delta x_2^k \frac{\partial}{\partial x_2} + \cdots + \delta x_m^k \frac{\partial}{\partial x_m} \right) f_2(x_1^{\ k}, x_2^{\ k}, \cdots, x_m^{\ k}) + \cdots = 0$$

$$\vdots$$

$$f_m(x_1^{\ k}, x_2^{\ k}, \cdots, x_m^{\ k}) + \left(\delta x_1^k \frac{\partial}{\partial x_1} + \delta x_2^k \frac{\partial}{\partial x_2} + \cdots + \delta x_m^k \frac{\partial}{\partial x_m} \right) f_m(x_1^{\ k}, x_2^{\ k}, \cdots, x_m^{\ k}) + \cdots = 0$$

On neglecting the second and higher power of the error vector, $\delta X^k = \left[\delta x_1^k, \delta x_2^k, \cdots, \delta x_m^k \right]^T$, we get

$$f_1(x_1^k, x_2^k, \cdots, x_m^k) + \left(\delta x_1^k \frac{\partial}{\partial x_1} + \delta x_2^k \frac{\partial}{\partial x_2} + \cdots + \delta x_m^k \frac{\partial}{\partial x_m} \right) f_1(x_1^k, x_2^k, \cdots, x_m^k) = 0$$

$$f_2(x_1^k, x_2^k, \cdots, x_m^k) + \left(\delta x_1^k \frac{\partial}{\partial x_1} + \delta x_2^k \frac{\partial}{\partial x_2} + \cdots + \delta x_m^k \frac{\partial}{\partial x_m} \right) f_2(x_1^k, x_2^k, \cdots, x_m^k) = 0$$

$$\vdots$$

$$f_m(x_1^k, x_2^k, \cdots, x_m^k) + \left(\delta x_1^k \frac{\partial}{\partial x_1} + \delta x_2^k \frac{\partial}{\partial x_2} + \cdots + \delta x_m^k \frac{\partial}{\partial x_m} \right) f_m(x_1^k, x_2^k, \cdots, x_m^k) = 0$$

In vector form, the above system is given by

$$F\left(X^K\right) + \delta X^k . J_k[F] = \overline{0} \tag{4.19}$$

where the Jacobian matrix $J_k[F]$ is as follows

$$J_k[F] = \begin{bmatrix} \dfrac{\partial f_1}{\partial x_1} & \dfrac{\partial f_1}{\partial x_2} & \cdots & \dfrac{\partial f_1}{\partial x_m} \\[2mm] \dfrac{\partial f_2}{\partial x_1} & \dfrac{\partial f_2}{\partial x_2} & \cdots & \dfrac{\partial f_2}{\partial x_m} \\[2mm] \vdots & & & \\[1mm] \dfrac{\partial f_m}{\partial x_1} & \dfrac{\partial f_m}{\partial x_2} & \cdots & \dfrac{\partial f_m}{\partial x_m} \end{bmatrix}_{(x_1^k, x_2^k, \cdots, x_m^k)} \tag{4.20}$$

On solving the system (4.19) for $\delta X^k = \left[\delta x_1^k, \delta x_2^k, \cdots, \delta x_m^k \right]^T$, we get

$$\delta X^k = -J_k^{-1}[F] F(X^k) \tag{4.21}$$

Therefore, the next approximation is given by

$$X^{k+1} = X^k + \delta X^k = \left[x_1^k + \delta x_1^k, x_2^k + \delta x_2^k, \cdots, x_m^k + \delta x_m^k \right]^T \tag{4.22}$$

Example — 4.7

Solve the following system of nonlinear equation using Newton–Raphson method.

$$\cos(x_2 x_3) + 3 \sin x_1 + 10 x_1 - 2 x_3 = 0$$
$$2 x_2^2 + 10 x_2 - x_3 + 2 x_1 + 2 = 0$$
$$e^{-(x_1 + x_3)} + 8 x_3 + 2 x_2 - 1 = 0$$

Assume initial approximation is $x_1^{(0)} = x_2^{(0)} = x_3^{(0)} = 0$.

Ans. We have

$$f_1(x_1, x_2, x_3) = \cos(x_2 x_3) + 3\sin x_1 + 10x_1 - 2x_3 = 0$$
$$f_2(x_1, x_2, x_3) = 2x_2^2 + 10x_2 - x_3 + 2x_1 + 2 = 0$$
$$f_3(x_1, x_2, x_3) = e^{-(x_1+x_3)} + 8x_3 + 2x_2 - 1 = 0$$

Initial approximation is $x_1^{(0)} = x_2^{(0)} = x_3^{(0)} = 0$.

First Iteration

Compute the Jacobian matrix from Eq. (4.20) as follows

$$J_k[F] = \begin{bmatrix} \dfrac{\partial f_1}{\partial x_1} & \dfrac{\partial f_1}{\partial x_2} & \dfrac{\partial f_1}{\partial x_3} \\ \dfrac{\partial f_2}{\partial x_1} & \dfrac{\partial f_2}{\partial x_2} & \dfrac{\partial f_2}{\partial x_3} \\ \dfrac{\partial f_3}{\partial x_1} & \dfrac{\partial f_3}{\partial x_2} & \dfrac{\partial f_3}{\partial x_3} \end{bmatrix}_{(x_1^{(0)}, x_2^{(0)}, x_3^{(0)})}$$

$$= \begin{bmatrix} 3\cos x_1 + 10 & -x_3\sin(x_2 x_3) & -x_2\sin(x_2 x_3) - 2 \\ 2 & 4x_2 + 10 - 1 & \\ -e^{-(x_1+x_3)} & 2 & -e^{-(x_1+x_3)} + 8 \end{bmatrix}_{(0,0,0)}$$

$$= \begin{bmatrix} 13 & 0 & -2 \\ 2 & 10 & -1 \\ -1 & 2 & 7 \end{bmatrix}$$

On using Eq. (4.21), we get

$$\delta X^{(0)} = -J_0^{-1}[F]F(X^0)$$

$$\begin{bmatrix} \delta x_1^{(0)} \\ \delta x_2^{(0)} \\ \delta x_3^{(0)} \end{bmatrix} = -\begin{bmatrix} 13 & 0 & -2 \\ 2 & 10 & -1 \\ -1 & 2 & 7 \end{bmatrix}^{-1}\begin{bmatrix} f_1 \\ f_2 \\ f_3 \end{bmatrix}_{(0,0,0)}$$

$$= -\begin{bmatrix} 13 & 0 & -2 \\ 2 & 10 & -1 \\ -1 & 2 & 7 \end{bmatrix}^{-1}\begin{bmatrix} 1 \\ 2 \\ 0 \end{bmatrix}$$

$$= -\begin{bmatrix} \dfrac{18}{227} & \dfrac{-1}{227} & \dfrac{5}{227} \\ \dfrac{-13}{908} & \dfrac{89}{908} & \dfrac{9}{908} \\ \dfrac{7}{454} & \dfrac{-13}{454} & \dfrac{65}{454} \end{bmatrix} \begin{bmatrix} 1 \\ 2 \\ 0 \end{bmatrix}$$

$$= \begin{bmatrix} \dfrac{-16}{227} \\ \dfrac{-165}{908} \\ \dfrac{19}{454} \end{bmatrix} = \begin{bmatrix} -0.070485 \\ -0.181718 \\ 0.041850 \end{bmatrix}$$

$$\Rightarrow \delta x_1^{(0)} = -0.070485, \; \delta x_2^{(0)} = -0.181718, \; \delta x_3^{(0)} = 0.041850$$

From Eq. (4.22), the first approximation to the solution is given by

$$x_1^{(1)} = x_1^{(0)} + \delta x_1^{(0)} = -0.070485$$
$$x_2^{(1)} = x_2^{(0)} + \delta x_2^{(2)} = -0.181718$$
$$x_3^{(1)} = x_3^{(0)} + \delta x_3^{(0)} = 0.041850$$

Second Iteration

Proceeding in a similar manner, the second iteration of Newton–Raphson formula (4.22) is computed as follows

$$\delta X^1 = -J_1^{-1}[F]F(X^1)$$

$$= \begin{bmatrix} 3\cos x_1 + 10 & -x_3 \sin(x_2 x_3) & -x_2 \sin(x_2 x_3) - 2 \\ 2 & 4x_2 + 10 & -1 \\ -e^{-(x_1 + x_3)} & 2 & -e^{-(x_1 + x_3)} + 8 \end{bmatrix}_{(-0.070485, -0.181718, 0.041850)}^{-1} \begin{bmatrix} f_1 \\ f_2 \\ f_3 \end{bmatrix}_{(-0.070485, -0.181718, 0.041850)}$$

$$\begin{bmatrix} \delta x_1^{(1)} \\ \delta x_2^{(1)} \\ \delta x_3^{(1)} \end{bmatrix} = \begin{bmatrix} 0.000295 \\ -0.006972 \\ 0.001984 \end{bmatrix}$$

So, the second approximation to the root is given by

$$x_1^{(2)} = x_1^{(1)} + \delta x_1^{(1)} = -0.070485 + 0.000295 = -0.070190$$
$$x_2^{(2)} = x_2^{(1)} + \delta x_2^{(1)} = -0.181718 - 0.006972 = -0.188690$$
$$x_3^{(2)} = x_3^{(1)} + \delta x_3^{(1)} = 0.041850 + 0.001984 = 0.043834$$

Third Iteration

$x_1^{(3)} = -0.070190$

$x_2^{(3)} = -0.188700$

$x_3^{(3)} = 0.043837$

Fourth Iteration

$x_1^{(4)} = -0.070190$

$x_2^{(4)} = -0.188700$

$x_3^{(4)} = 0.043837$

Convergence Conditions for NR Method

The NR method is given by

$$X^{k+1} - X^k = \delta X^k = -J_k^{-1}[F]F(X^k) \qquad\qquad k = 0, 1, 2 \ldots$$

where $\delta X^k = X^{k+1} - X^k$ is the error vector.

Hence, a sufficient condition for convergence is that for each k, we have

$$\left\| J_k^{-1}[F] \right\| < 1, \text{ for a suitable norm.}$$

While the sufficient and necessary condition for convergence of NR method is that the spectral radius of Jacobian matrix is less than 1.

$$\rho\left(J_k^{-1}[F]\right) < 1 \tag{4.23}$$

4.4 Complex Roots

Let us consider a complex equation in complex variable z

$$F(z) = 0$$

We separate the real and imaginary parts to solve this equation

$$F(z) = F(x + iy) = f(x, y) + i\,g(x, y)$$

The function $F(z)$ will vanish if and only if the real and imaginary parts both vanish separately. It produces the system (4.2), for which we have already discussed the solution.

Example — 4.8

Use NR method to compute all three roots of the following complex equation

$$F(z) = z^3 - 4iz^2 - 3e^z = 0$$

where the variable z is complex.

Ans. Let $z = x + iy$, the equation is given by

$$F(z) = (x + iy)^3 - 4i(x + iy)^2 - 3e^{(x+iy)} = 0$$

$$F(z) = x^3 + 3ix^2 y - 3xy^2 - iy^3 - 4i(x^2 - y^2 + 2ixy) - 3e^x(\cos y + i\sin y) = 0$$

On separating real and imaginary parts, we get

$$x^3 - 3xy^2 + 8xy - 3e^x \cos y = 0$$

$$3x^2 y - y^3 - 4x^2 + 4y^2 - 3e^x \sin y = 0$$

Let, the functions be as follows

$$f(x, y) = x^3 - 3xy^2 + 8xy - 3e^x \cos y$$

$$g(x, y) = 3x^2 y - y^3 - 4x^2 + 4y^2 - 3e^x \sin y$$

On differentiating the functions $f(x, y)$ and $g(x, y)$, we have

$$f_x = 3x^2 - 3y^2 + 8y - 3e^x \cos y \qquad f_y = -6xy + 8x + 3e^x \sin y$$

$$g_x = 6xy - 8x - 3e^x \sin y \qquad g_y = 3x^2 - 3y^2 + 8y - 3e^x \cos y$$

We will use Newton–Raphson method (4.15) to compute all the three roots of the system by starting with different initial approximations.

First Root

On using the Newton–Raphson method with initial approximations $x_0 = y_0 = -0.5$, we have following iterations

First iteration

$$h = \left. \frac{g f_y - f g_y}{f_x g_y - g_x f_y} \right|_{(x_0, y_0)} = \left. \frac{g f_y - f g_y}{f_x g_y - g_x f_y} \right|_{(-0.5, -0.5)} = -0.004313$$

$$k = \left. \frac{f g_x - g f_x}{f_x g_y - g_x f_y} \right|_{(x_0, y_0)} = \left. \frac{f g_x - g f_x}{f_x g_y - g_x f_y} \right|_{(-0.5, -0.5)} = 0.106287$$

$$x_1 = x_0 + h = -0.5 + (-0.004313) = -0.504313$$

$$y_1 = y_0 + k = -0.5 + 0.106287 = -0.393713$$

Second iteration

$$h = \frac{g\,f_y - f\,g_y}{f_x g_y - g_x f_y}\bigg|_{(x_1,\,y_1)} = \frac{g\,f_y - f\,g_y}{f_x g_y - g_x f_y}\bigg|_{(-0.504313,\,-0.393713)} = -0.004468$$

$$k = \frac{f\,g_x - g\,f_x}{f_x g_y - g_x f_y}\bigg|_{(x_1,\,y_1)} = \frac{f\,g_x - g\,f_x}{f_x g_y - g_x f_y}\bigg|_{(-0.504313,\,-0.393713)} = 0.007054$$

$$x_2 = x_1 + h = -0.504313 - 0.004468 = -0.508781$$

$$y_2 = y_1 + k = -0.393713 + 0.007054 = -0.386659$$

Third Iteration

$$x_3 = -0.508831 \quad y_3 = -0.386661$$

So, the approximate root to the given complex equation is given by

$$z = x + iy = -0.508831 - 0.386661\,i$$

Note that we will get the same root with initial approximation $x_0 = y_0 = 0$, but the number of iterations increases in this case. Iterations with initial approximation $x_0 = y_0 = 0$ are as follows

$$x_1 = -1.000000 \quad y_1 = 0.000000$$
$$x_2 = -0.467584 \quad y_2 = -0.136748$$
$$x_3 = -0.491771 \quad y_3 = -0.449375$$
$$x_4 = -0.506468 \quad y_4 = -0.388324$$
$$x_5 = -0.508827 \quad y_5 = -0.386656$$

Second Root

Proceeding in a similar manner, we can obtain the second complex root of the equation. The following iterations of Newton–Raphson methods are obtained with the initial approximation, $x_0 = y_0 = 0.5$.

$$x_1 = 0.029637 \quad y_1 = 0.998927$$
$$x_2 = 0.367184 \quad y_2 = 1.149769$$
$$x_3 = 0.311058 \quad y_3 = 1.153530$$
$$x_4 = 0.310100 \quad y_4 = 1.154557$$
$$x_5 = 0.310101 \quad y_5 = 1.154558$$

The second root of given complex equation is as follows

$$z = x + iy = 0.310101 + 1.154558\,i.$$

Third Root

Proceeding with the initial approximation, $x_0 = y_0 = 5$, we have

$$x_1 = 3.719573 \quad y_1 = 4.877457$$
$$x_2 = 2.464507 \quad y_2 = 4.503765$$
$$x_3 = 1.460492 \quad y_3 = 4.017134$$
$$x_4 = 0.642656 \quad y_4 = 3.686140$$
$$x_5 = -0.132443 \quad y_5 = 3.843745$$
$$x_6 = 0.206784 \quad y_6 = 4.194346$$
$$x_7 = 0.096327 \quad y_7 = 4.160002$$
$$x_8 = 0.091169 \quad y_8 = 4.165862$$
$$x_9 = 0.091203 \quad y_9 = 4.165862$$

So, the third root of the given complex equation is given by

$$z = x + iy = 0.091203 + 4.165862\,i$$

4.5 Polynomial Equations

Before we discuss the methods to compute the roots of polynomial equation of degree n

$$P_n(x) = a_0 x^n + a_1 x^{n-1} + \cdots\cdots + a_{n-1}x + a_n = 0, \quad a_0 \neq 0 \tag{4.24}$$

where $a_0, a_1, a_2, \cdots a_{n-1}, a_n$ are all real numbers, it is advantageous to have the following information

i) The exact number of real (+ve and –ve) and complex roots
ii) The intervals containing roots

Descartes rule and Strum theorem will be helpful in this regard.

4.5.1 Descartes Rule of Signs

Arrange all the terms of a polynomial in order. If a + sign follows a –sign (or) vice versa, then a change of signs is said to occur. Otherwise, it is a permanence or continuation of signs.

The number of positive real roots of the equation $P_n(x) = 0$ cannot exceed the number of sign changes in the polynomial, $P_n(x)$. The number of negative real roots cannot exceed the number of sign changes in the polynomial, $P_n(-x)$.

Example ── **4.9** ──

The polynomial equation

$$P_4(x) = x^4 - 3x^3 + 2x^2 + x - 4 = 0$$

has maximum three real +ve roots, as there are three sign changes in $P_4(x)$. Also, the polynomial $P_4(-x) = x^4 + 3x^3 + 2x^2 - x - 4$ has only one sign change, so there is maximum one negative real root of the equation.

This rule provides the maximum number of positive and negative real roots, but not the exact number of real roots of polynomial equations. The exact number of real roots can be obtained from Strum theorem.

4.5.2 Strum Sequence

A Sturm sequence or Strum chain for a polynomial function $f(x)$ is a finite sequence of polynomials

$$f_0(x),\ f_1(x),\ f_2(x),\ f_3(x),\ \cdots,\ f_m(x)$$

of decreasing degrees.

This sequence is constructed by the following algorithm

$$f_0(x) = f(x)$$

$$f_1(x) = f'(x)$$

$$f_2(x) = -\text{remainder}(f_0, f_1)$$

$$f_3(x) = -\text{remainder}(f_1, f_2)$$

$$\vdots$$

$$f_m(x) = -\text{remainder}(f_{m-2}, f_{m-1}) \tag{4.25}$$

The last term $f_m(x)$ must be a constant term.

The sequence $f_0(x),\ f_1(x),\ f_2(x),\ f_3(x),\ \cdots,\ f_m(x)$ is known as Strum chain or Strum sequence. To avoid rational coefficients, we can multiply or divide the elements of a Strum sequence by any positive constants.

Note the following points about Strum sequence

1. The function $f_0(x) = f(x)$ is squarefree, i.e. no repeated root
2. If $f_i(x) = 0$, $0 < i < m$, then sign$(f_{i-1}(x)) = -$ sign$(f_{i+1}(x))$

Strum Theorem: Let $f(x)$, $f_1(x)$, $f_2(x)$, $f_3(x)$, $\cdots$ be a Sturm sequence. Let g(a) and g(b) be numbers of sign changes (ignoring zeroes) in Strum sequence at $x = a$ and $x = b$, respectively for two real numbers $a < b$. Then the number of distinct real roots of the equation $f(x) = 0$ in the interval [a, b] is given by

g(a) – g(b), provided f(a) ≠ 0 and f(b) ≠ 0

A polynomial equation of degree n has exactly n number of roots. So, if we have s number of real roots (the root of multiplicity r must be counted r times), then the number of complex roots is $n - s$.

Example — 4.10 —

Obtain the Strum sequence and compute the numbers of real and complex roots of the following polynomial equation

$$x^3 - 3x + 1 = 0$$

Ans. The first two terms of Strum sequence (4.25) are as follows

$$f_0(x) = f(x) = x^3 - 3x + 1$$

$$f_1(x) = f'(x) = 3x^2 - 3 = 3(x^2 - 1) \text{ (or) } x^2 - 1$$

(To avoid rational coefficients, we can multiply or divide the elements of a Strum sequence by any positive constant).

The other terms of Strum sequence (4.25) can be computed as follows

$$f_2(x) = -\text{remainder}(f_0, f_1) = 2x - 1 \text{ (or) } x - \frac{1}{2}$$

$$f_3(x) = -\text{remainder}(f_1, f_2) = 1$$

Finally, the Strum sequence is given by

$$\left[x^3 - 3x + 1, \ x^2 - 1, x - \frac{1}{2}, 1 \right]$$

Let g(a) be number of sign changes (ignoring zeroes) in the Strum sequence at $x = a$. To obtain number of real and complex roots, we have

x	$f_0(x)$	$f_1(x)$	$f_2(x)$	$f_3(x)$	$g(x)$
$-\infty$	$-$	$+$	$-$	$+$	3
∞	$+$	$+$	$+$	$+$	0

According to Strum theorem, the number of distinct real roots of the equation $f(x)=0$ in the interval $[a, b]$ is g(a) – g(b). Since $g(-\infty)=3$ and $g(\infty)=0$, hence the equation has three distinct real roots in the interval $(-\infty, \infty)$.

Note: We can also construct the following table to obtain the intervals containing roots of the equation in Example 4.10.

Table 4.1

x	$f_0(x) = f(x)$	$f_1(x)$	$f_2(x)$	$f_3(x)$	$g(x)$
-2	$-$	$+$	$-$	$+$	3
-1	$+$	0	$-$	$+$	2
0	$+$	$-$	$-$	$+$	2
1	$-$	0	$+$	$+$	1
2	$+$	$+$	$+$	$+$	0

All three roots are real, and lie in the intervals $(-2, -1)$, $(0, 1)$ and $(1, 2)$. Since the function $f(x)$ is continuous, hence we can also use Intermediate value theorem to compute the intervals containing roots.

Note that the approximate roots of the equation are 0.347296355, 1.532088886 and −1.87938524.

Example ── **4.11**

Obtain the Strum sequence and compute the numbers of real and complex roots of the following polynomial equation

$$x^4 - 4x^3 + 8x^2 - 4x + 3 = 0$$

Ans. The Strum sequence (4.25) can be computed as follows

$$f_0(x) = f(x) = x^4 - 4x^3 + 8x^2 - 4x + 3$$

$$f_1(x) = f'(x) = 4x^3 - 12x^2 + 16x - 4 \text{ (or) } f_1(x) = x^3 - 3x^2 + 4x - 1$$

$$f_2(x) = -\text{remainder}(f_0, f_1) = -x^2 - x - 2$$

$$f_3(x) = -\text{remainder}(f_1, f_2) = -x - \frac{7}{6}$$

$$f_4(x) = -\text{remainder}(f_2, f_3) = 1$$

The Strum sequence is given by

$$\left[x^4 - 4x^3 + 8x^2 - 4x + 3, x^3 - 3x^2 + 4x - 1, -x^2 - x - 2, -x - \frac{7}{6}, 1 \right]$$

Let $g(a)$ be number of sign changes (ignoring zeroes) in Strum sequence at $x = a$. To obtain number of real and complex roots, we have

x	$f_0(x)$	$f_1(x)$	$f_2(x)$	$f_3(x)$	$f_4(x)$	$g(x)$
$-\infty$	$+$	$-$	$-$	$+$	$+$	2
∞	$+$	$+$	$-$	$-$	$+$	2

So, all the roots of the equation are complex.

Example 4.12

Obtain the Strum sequence and compute the numbers of real and complex roots of the following polynomial equation

$$x^4 + 2x^3 - 2x^2 - 4x + 1 = 0$$

Ans. The Strum sequence is as follows

$$\left[x^4 + 2x^3 - 2x^2 - 4x + 1, x^3 + \frac{3}{2}x^2 - x - 1, x^2 + \frac{10}{7}x - \frac{6}{7}, x + \frac{23}{6}, -1 \right]$$

x	$f_0(x)$	$f_1(x)$	$f_2(x)$	$f_3(x)$	$f_4(x)$	$g(x)$
$-\infty$	$+$	$-$	$+$	$-$	$-$	3
∞	$+$	$+$	$+$	$+$	$-$	1

The numbers of sign changes are given by

$$g(-\infty) = 3 \text{ and } g(\infty) = 1.$$

So, the equation has two distinct real roots in $(-\infty, \infty)$ according to Strum theorem. The remaining two roots are complex.

$$\left[x^4 + 2x^3 - 2x^2 - 4x + 1, x^3 + \frac{3}{2}x^2 - x - 1, x^2 + \frac{10}{7}x - \frac{6}{7}, x + \frac{23}{6}, -1 \right]$$

To obtain the intervals of real roots, we have used the Intermediate value theorem as follows

x	$-\infty$	-3	-2	-1	0	1	2	∞
$f(x)$	$+$	$+$	$+$	$+$	$+$	$-$	$+$	$+$

We have

$$f(0)f(1) < 0 \quad \text{and} \quad f(1)f(2) < 0.$$

According to intermediate value theorem, the intervals (0, 1) and (1, 2) contain roots of the given equation. Note that two real roots are 0.2302926 and 1.3321898.

So far, we have discussed the methods which are applicable for both transcendental equations as well as for the polynomial Eq. (4.24).

Polynomial equations have a simple structure and are easy to examine. So, many methods have been developed especially for the solutions of polynomial equations with the following advantages

i)   Less computational efforts are required
ii)   Quadratic factors can be extracted (which is helpful for complex roots)
iii) All the roots can be obtained at a time
iv)   Divergence can be avoided, etc.

In this section, we will discuss following important methods to compute the roots of the polynomial equations.

i)   Birge–Vieta (or) Horner Method (extract linear factor from polynomial)
ii)   Lin–Bairstow Method (extract quadratic factor from polynomial)
iii) Graeffe Root Squaring Method

Note that algorithm used in Birge–Vieta method is Horner algorithm, so some authors called this method as Horner method.

4.6 Birge–Vieta (or) Horner Method

In this method, we will extract a linear factor $(x - p)$ from the polynomial

$$P_n(x) = a_0 x^n + a_1 x^{n-1} + \cdots + a_{n-1}x + a_n$$

For this, let p_0 be an initial approximation for p. If we divide $P_n(x)$ by $(x - p_0)$, then let constant R be the remainder and $Q_{n-1}(x)$ be the quotient.

$$P_n(x) = (x - p_0)Q_{n-1}(x) + R \tag{4.26}$$

where $Q_{n-1}(x) = b_0 x^{n-1} + b_1 x^{n-2} + \cdots + b_{n-2}x + b_{n-1}$.

Now, we can use any iterative method to improve the value P_0. Let p be the exact root of the equation $P_n(x) = 0$, so $(x - p)$ is a factor of $P_n(x)$. Let us apply Newton–Raphson method for this purpose

$$p_1 = p_0 - \frac{P_n(p_0)}{P_n'(p_0)} \tag{4.27}$$

Now, we will compute $P_n(p_0)$ and $P_n'(p_0)$ in terms of coefficients of the polynomial equation. On comparing the coefficients of like powers of x on both sides of the equation (4.26), we get

$$
\begin{aligned}
a_0 &= b_0 & b_0 &= a_0 \\
a_1 &= b_1 - b_0 p_0 & b_1 &= a_1 + b_0 p_0 \\
a_2 &= b_2 - b_1 p_0 & b_2 &= a_2 + b_1 p_0 \\
&\ \ \vdots & & \\
a_k &= b_k - b_{k-1} p_0 & b_k &= a_k + b_{k-1} p_0 \\
&\ \ \vdots & & \\
a_{n-1} &= b_{n-1} - b_{n-2} p_0 & b_{n-1} &= a_{n-1} + b_{n-2} p_0 \\
a_n &= R - b_{n-1} p_0 & R &= a_n + b_{n-1} p_0
\end{aligned}
\tag{4.28}
$$

Note that these calculations with the help of synthetic division are as follows

$$
\begin{array}{c|cccccc}
 & a_0 & a_1 & a_2 & \cdots & a_{n-1} & a_n \\
p_0 & & p_0 b_0 & p_0 b_1 & \cdots & p_0 b_{n-2} & p_0 b_{n-1} \\
\hline
 & b_0 & b_1 & b_2 & \cdots & b_{n-1} & b_n = R
\end{array}
\tag{4.29}
$$

It is easy to see from Eq. (4.26) that

$$P_n(p_0) = R \tag{4.30}$$

On differentiating Eq. (4.26), we have

$$P_n'(x) = (x - p_0)Q_{n-1}'(x) + Q_{n-1}(x)$$

$$P_n'(p_0) = Q_{n-1}(p_0)$$

As we compute the $P_n(p_0)$ by Eqs. (4.28), on a similar pattern the $Q_{n-1}(p_0)$ can be computed as follows

$$
p_0 \begin{array}{cccccc}
b_0 & b_1 & b_2 & \cdots & b_{n-1} & \\
 & p_0 c_0 & p_0 c_1 & \cdots & p_0 c_{n-2} & \\
\hline
c_0 & c_1 & c_2 & \cdots & c_{n-1} = Q_{n-1}(p_0) = P_n'(p_0) &
\end{array}
\tag{4.31}
$$

The algorithms (4.29) and (4.31) can be clubbed together to produce the following algorithm

$$
p_0 \begin{array}{cccccc}
a_0 & a_1 & a_2 & \cdots & a_{n-1} & a_n \\
 & p_0 b_0 & p_0 b_1 & \cdots & p_0 b_{n-2} & p_0 b_{n-1} \\
\hline
b_0 & b_1 & b_2 & \cdots & b_{n-1} & b_n = P_n(p_0) \\
 & p_0 c_0 & p_0 c_1 & \cdots & p_0 c_{n-2} & \\
\hline
c_0 & c_1 & c_2 & \cdots & c_{n-1} = P_n'(p_0) &
\end{array}
\tag{4.32}
$$

The Newton–Raphson method (4.27) is given by

$$
P_1 = p_0 - \frac{P_n(p_0)}{P_n'(p_0)} = p_0 - \frac{b_n}{c_{n-1}}
\tag{4.33}
$$

Similarly, we can compute other iterations as follows

$$
p_{k+1} = p_k - \frac{b_n}{c_{n-1}}, \quad k = 0,1,2,\dots
\tag{4.34}
$$

It is known as Birge–Vieta method.

Example   4.13

Compute the root of equation $x^3 - 4x - 9 = 0$ using Birge–Vieta method, starting with an initial approximation, $p_0 = 2$.

Ans. The iterations of the Birge–Vieta method (4.34) are computed as follows

First Approximation
The synthetic division table (4.32) for the given equation with initial approximation $p_0 = 2$ is given by

$$2|1 \qquad 0 \qquad -4 \qquad -9$$
$$\qquad\qquad 2 \qquad 4 \qquad 0$$
$$2|1 \qquad 2 \qquad 0 \qquad -9 = b_3$$
$$\qquad\qquad 2 \qquad 8$$
$$\qquad 1 \qquad 4 \qquad 8 = c_2$$

On using Birge–Vieta method (4.34) with $k = 0$ and $n = 3$, we have

$$p_1 = p_0 - \frac{b_3}{c_2}$$

$$\Rightarrow p_1 = 2 - \frac{(-9)}{8} = 3.125$$

Second Approximation

$$3.125|1 \qquad 0 \qquad -4 \qquad -9$$
$$\qquad\qquad 3.125 \qquad 9.765625 \qquad 18.017578$$
$$3.125|1 \qquad 3.125 \qquad 5.765625 \qquad 9.017578 = b_3$$
$$\qquad\qquad 3.125 \qquad 19.53125 \qquad 42.328$$
$$\qquad 1 \qquad 6.25 \qquad 25.296875 = c_2$$

$$p_2 = p_1 - \frac{b_3}{c_2}$$

$$\Rightarrow p_2 = 3.125 - \frac{(9.017578)}{25.296875} = 2.768530$$

Third Approximation

$$2.76853|1 \qquad 0 \qquad -4 \qquad -9$$
$$\qquad\qquad 2.76853 \qquad 7.664758 \qquad 10.145993$$
$$2.76853|1 \qquad 2.76853 \qquad 3.664758 \qquad 1.145993 = b_3$$
$$\qquad\qquad 2.76853 \qquad 15.329517$$
$$\qquad 1 \qquad 5.53706 \qquad 18.994275 = c_2$$

$$p_3 = p_2 - \frac{b_3}{c_2}$$

$$\Rightarrow p_3 = 2.76853 - \frac{(1.145993)}{18.994275} = 2.708196$$

Proceeding in a similar manner, the fourth and fifth iterations are 2.706528 and 2.706528, respectively. So, approximate root of the equation is 2.706528.

Note: The Birge–Vieta method uses the Newton–Raphson algorithm to compute the root of the polynomial equation. The question is why to use Birge–Vieta method (4.34) for polynomial equations? In the Newton–Raphson method (4.27), we have to compute the differentiation of the function, which is not possible through programming. Also, we have to specify the function in the program. While Birge–Vieta method uses only numerical computation, so the programming is possible for any general polynomial equations. But, Birge–Vieta method can solve polynomial equations only.

4.7 Lin–Bairstow Method

In this method, we extract a quadratic factor $(x^2 + px + q)$ from the polynomial

$$P_n(x) = a_0 x^n + a_1 x^{n-1} + \cdots + a_{n-1} x + a_n \tag{4.35}$$

Let us divide the polynomial (4.35) by a quadratic factor, $(x^2 + px + q)$. Then, we obtain a quotient polynomial $Q_{n-2}(x)$ of degree $n - 2$, and a remainder term $Rx + S$ which is a polynomial of degree one. So, we have

$$P_n(x) = (x^2 + px + q)Q_{n-2}(x) + Rx + S \tag{4.36}$$

where $Q_{n-2}(x) = b_0 x^{n-2} + b_1 x^{n-3} + \cdots + b_{n-3} x + b_{n-2}$.

If we want to extract a quadratic factor $(x^2 + px + q)$ from the polynomial $P_n(x)$, then the remainder term $Rx + S$ must vanish. So, the aim is to find p and q, such that

$$R(p, q) = S(p, q) = 0$$

Let us solve this system of two simultaneous equations with the help of Newton–Raphson method. On starting with an initial approximation (p_0, q_0) for (p, q), the next approximation is given by

$$p_1 = p_0 + \Delta p$$
$$q_1 = q_0 + \Delta q$$

where

$$\Delta p = \frac{SR_q - RS_q}{R_p S_q - S_p R_q}\bigg|_{(p_0, q_0)}$$

$$\Delta q = \frac{RS_p - SR_p}{R_p S_q - S_p R_q}\bigg|_{(p_0, q_0)} \tag{4.37}$$

The aim is to evaluate functions R, S, R_p, S_p, R_q, S_q in terms of the coefficients of the polynomial (4.35).

On comparing the L.H.S. and R.H.S. in Eq. (4.36), we have

$$a_0 = b_0 \qquad\qquad\qquad b_0 = a_0$$
$$a_1 = b_1 + b_0 p \qquad\qquad b_1 = a_1 - b_0 p$$
$$a_2 = b_2 + b_1 p + b_0 q \qquad\quad b_2 = a_2 - b_1 p - b_0 q$$
$$\vdots \qquad\qquad\qquad\qquad \vdots$$
$$a_k = b_k + b_{k-1} p + b_{k-2} q \qquad b_k = a_k - b_{k-1} p - b_{k-2} q$$
$$\vdots \qquad\qquad\qquad\qquad \vdots$$
$$a_{n-1} = R + b_{n-2} p + b_{n-3} q \qquad R = a_{n-1} - b_{n-2} p - b_{n-3} q$$
$$a_n = S + b_{n-2} q \qquad\qquad\quad S = a_n - b_{n-2} q \tag{4.38}$$

Consider the constants b_{n-1} and b_n as follows

$$b_{n-1} = R$$
$$b_n = S - p b_{n-1} \tag{4.39}$$

The following recursion formula can be obtained from Eqs. (4.38) and (4.39).

$$b_{-1} = 0$$
$$b_0 = a_0 \tag{4.40}$$
$$b_k = a_k - p b_{k-1} - q b_{k-2}, \qquad k = 1, 2, \cdots, n$$

Note that all a_k's are constants for a given polynomial (4.35). But, all b_k's are dependent on p and q (as if we divide the $P_n(x)$ with $x^2 + px + q$, then $Q_{n-2}(x)$ will change according to the value of p and q). On differentiating the Eq. (4.40) with respect to the variables p and q, we obtain

$$\frac{-\partial b_k}{\partial p} = b_{k-1} + p \frac{\partial b_{k-1}}{\partial p} + q \frac{\partial b_{k-2}}{\partial p}; \quad \frac{\partial b_0}{\partial p} = \frac{\partial b_{-1}}{\partial p} = 0$$

$$\frac{-\partial b_k}{\partial q} = b_{k-2} + p\frac{\partial b_{k-1}}{\partial q} + q\frac{\partial b_{k-2}}{\partial q}; \quad \frac{\partial b_0}{\partial q} = \frac{\partial b_{-1}}{\partial q} = 0 \tag{4.41}$$

Let us introduce two new constants c_{k-1} and c_{k-2}, such that

$$c_{k-1} = \frac{-\partial b_k}{\partial p}, \ c_{k-2} = \frac{-\partial b_k}{\partial q} \tag{4.42}$$

Using the expressions (4.42) in Eqs. (4.41), we have .

$$c_{k-1} = b_{k-1} - pc_{k-2} - qc_{k-3}$$

$$c_{k-2} = b_{k-2} - pc_{k-3} - qc_{k-4}$$

The following recursion operators define the expressions above

$$c_k = b_k - pc_{k-1} - qc_{k-2}, \quad k = 1, 2, \cdots, n-1$$

$$c_{-1} = 0, \ c_0 = \frac{-\partial b_1}{\partial p} = \frac{-\partial}{\partial p}(a_1 - pb_0) = b_0 \tag{4.43}$$

Differentiating the Eqs. (4.39), and using the expressions (4.42), we obtain various partial derivatives of R and S as given below

$$R_p = \frac{\partial b_{n-1}}{\partial p} = -c_{n-2}$$

$$R_q = \frac{\partial b_{n-1}}{\partial q} = -c_{n-3}$$

$$S_p = \frac{\partial}{\partial p}(b_n + pb_{n-1}) = -c_{n-1} + b_{n-1} - pc_{n-2}$$

$$S_q = \frac{\partial}{\partial q}(b_n + pb_{n-1}) = -c_{n-2} - pc_{n-3}$$

From Eqs. (4.37 and 4.39) and the above expressions, we have

$$\Delta p = \frac{SR_q - RS_q}{R_p S_q - S_p R_q}\Bigg|_{(p_0,q_0)} = \frac{b_{n-1}c_{n-2} - b_n c_{n-3}}{c_{n-2}^2 - c_{n-3}(c_{n-1} - b_{n-1})}$$

$$\Delta q = \frac{RS_p - SR_p}{R_p S_q - S_p R_q}\Bigg|_{(p_0,q_0)} = \frac{b_n c_{n-2} - b_{n-1}(c_{n-1} - b_{n-1})}{c_{n-2}^2 - c_{n-3}(c_{n-1} - b_{n-1})} \tag{4.44}$$

On using these values, we can find the next approximations for p and q as follows

$$p_1 = p_0 + \Delta p$$
$$q_1 = q_0 + \Delta q \tag{4.45}$$

The constants b_i's and c_i's can be computed from Eqs.(4.40) and Eqs.(4.43) respectively. These constants are used to compute the Δp and Δq from Eqs.(4.44). Then, we can use Eqs. (4.45) to get the improved values of p and q. The computations in Eqs.(4.40) and Eqs.(4.43) can be summarized in following table, starting with initial approximation (p_0, q_0).

	a_0	a_1	a_2	...	a_n
$-p_0$		$-p_0 b_0$	$-p_0 b_1$		$-p_0 b_{n-1}$
$-q_0$			$-q_0 b_0$		$-q_0 b_{n-2}$
	b_0	b_1	b_2	...	b_n
$-p_0$		$-p_0 c_0$	$-p_0 c_1$		$-p_0 c_{n-1}$
$-q_0$			$-q_0 c_0$		$-q_0 c_{n-2}$
	c_0	c_1	c_2		c_n

Example ——————————————————————— 4.14 —

Extract a quadratic factor with the aid of Lin–Bairstow method from the quartic equation $x^4 - 15x^2 - 10x + 24 = 0$ and hence determine all the roots. Start with an initial approximation, $p = 1, q = 2$.

Ans. The coefficients for the equation $x^4 - 15x^2 - 10x + 24 = 0$ are as follows

$$a_0 = 1, \ a_1 = 0, \ a_2 = -15, \ a_3 = -10, \ a_4 = 24$$

First Iteration

To apply the Lin–Bairstow method, we will compute the constants b_i's and c_i's from the recursion formulas (4.40) and (4.43).

$$b_{-1} = c_{-1} = 0$$
$$b_0 = c_0 = a_0$$
$$b_k = a_k - pb_{k-1} - qb_{k-2}, \qquad k = 1, 2, ..., n$$
$$c_k = b_k - pc_{k-1} - qc_{k-2}, \qquad k = 1, 2, ..., n-1$$

These recursion formulas are implemented in the following table.

	1	0	-15	-10	24
-1		-1	1	16	-8
-2			-2	2	32
-1	$1 = b_0$	$-1 = b_1$	$-16 = b_2$	$8 = b_3$	$48 = b_4$
-2		-1	2	16	
			-2	4	
	$1 = c_0$	$-2 = c_1$	$-16 = c_2$	$28 = c_3$	

From the Eqs. (4.44), we get

$$\Delta p = \frac{b_{n-1}c_{n-2} - b_n c_{n-3}}{c_{n-2}^2 - c_{n-3}(c_{n-1} - b_{n-1})} = \frac{8(-16) - 48(-2)}{(-16)^2 + 2(28-8)} = \frac{-32}{296} = -0.108108$$

$$\Delta q = \frac{b_n c_{n-2} - b_{n-1}(c_{n-1} - b_{n-1})}{c_{n-2}^2 - c_{n-3}(c_{n-1} - b_{n-1})} = \frac{48(-16) - 8(28-8)}{296} = -3.135135$$

The next approximations (4.45) for p and q are given by

$$p_1 = p_0 + \Delta p = 1 - 0.108108 = 0.891892$$
$$q_1 = q_0 + \Delta q = 2 - 3.135135 = -1.135135$$

Second Iteration

The following table provides the values of constants b_i's and c_i's.

	1	0	-15	-10	24
-0.891892		-0.891892	0.795471	11.656488	-0.574441
1.135135			1.135135	-1.012418	-14.835526
-0.891892	$1 = b_0$	$-0.891892 = b_1$	$-13.069394 = b_2$	$0.644070 = b_3$	$8.590033 = b_4$
1.135135		-0.891892	1.590943	9.225121	
			1.135135	-2.024836	
	$1 = c_0$	$-1.783784 = c_1$	$-10.343317 = c_2$	$7.844355 = c_3$	

The values of Δp and Δq are given by

$$\Delta p = \frac{b_{n-1}c_{n-2} - b_n c_{n-3}}{c_{n-2}^2 - c_{n-3}(c_{n-1} - b_{n-1})}$$

$$= \frac{(0.644070)(-10.343317) - (8.590033)(-1.783784)}{(-10.343317)^2 + (1.783784)(7.844355 - 0.644070)}$$

$$= \frac{8.660943}{119.827960} = 0.072278$$

$$\Delta q = \frac{b_n c_{n-2} - b_{n-1}(c_{n-1} - b_{n-1})}{c_{n-2}^2 - c_{n-3}(c_{n-1} - b_{n-1})}$$

$$= \frac{(8.590033)(-10.343317) - (0.644070)(7.844355 - 0.644070)}{119.827960}$$

$$= \frac{-93.486922}{119.827960} = -0.780176$$

On using these values, we can find the next approximations for p and q as follows

$$p_1 = p_0 + \Delta p = 0.891892 + .072278 = 0.964170$$
$$q_1 = q_0 + \Delta q = -1.135135 - 0.780176 = -1.915311$$

Similarly, the third iteration of Lin–Bairstow method provides the following results.

$$b_1 = -0.964170, \quad b_2 = -12.155065, \quad b_3 = -0.127136, \quad b_4 = 0.841849$$
$$c_1 = -1.928340, \quad c_2 = -8.380506, \quad c_3 = 4.259725$$

$$\Delta p = 0.034169, \quad \Delta q = -0.082567$$
$$p = 0.998339 \approx 1, \quad q = -1.997878 \approx -2$$

So, the polynomial $x^2 + 0.998339x - 1.997878$ is the required quadratic factor. Note that the exact factor is the polynomial $x^2 + x - 2$. On dividing the polynomial $x^4 - 15x^2 - 10x + 24$ by this factor, we get

$$x^4 - 15x^2 - 10x + 24 = \left(x^2 + x - 2\right)\left(x^2 - x - 12\right) = (x-1)(x+2)(x+3)(x-4)$$

Hence, roots are 1, –2, –3 and 4.

4.8 Graeffe Root Squaring Method

Graeffe root squaring method computes all the roots (real or complex) of the real coefficients polynomial equation. In this method, we construct the new equation, whose roots are square of the roots of the given equation. The process of root squaring is repeated till certain number. Consider a polynomial equation.

$$a_0 x^n + a_1 x^{n-1} + \ldots + a_{n-1} x + a_n = 0 \qquad (4.46)$$

with roots $\alpha_1, \alpha_2, \alpha_3, \ldots, \alpha_n$. Now, we will construct the equation with the roots $-\alpha_1^2, -\alpha_2^2, -\alpha_3^2, \ldots, -\alpha_n^2$. To separate the even and odd powers of x, rewrite the equation (4.46) as follows

$$a_0 x^n + a_2 x^{n-2} + \ldots = -\left(a_1 x^{n-1} + a_3 x^{n-3} + \ldots \right)$$

On squaring both sides and simplifying the resulting equation, we get

$$\left(a_0 x^n + a_2 x^{n-2} + \ldots \right)^2 = \left(a_1 x^{n-1} + a_3 x^{n-3} + \ldots \right)^2$$

$$a_0^2 x^{2n} - \left(a_1^2 - 2a_0 a_2 \right) x^{2n-2} + \left(a_2^2 - 2a_1 a_3 + 2a_0 a_4 \right) x^{2n-4} + \ldots$$

$$\ldots + (-1)^{n-1} \left(a_{n-1}^2 - 2a_{n-2} a_n \right)^2 x^2 + (-1)^n a_n^2 = 0$$

By using the expression $y = -x^2$ in the above equation, we have

$$b_0 y^n + b_1 y^{n-1} + \ldots + b_{n-1} y + b_n = 0 \qquad (4.47)$$

where

$$b_0 = a_0^2$$

$$b_1 = a_1^2 - 2a_0 a_2$$

$$b_2 = a_2^2 - 2a_1 a_3 + 2a_0 a_4$$

$$\vdots$$

$$b_k = a_k^2 - 2a_{k-1} a_{k+1} + 2a_{k-2} a_{k+2} - \ldots, \text{ till coefficients are available for cross product}$$

$$\vdots$$

$$b_{n-1} = a_{n-1}^2 - 2a_{n-2} a_n$$

$$b_n = a_n^2 \qquad (4.48)$$

This new equation in y has roots $-\alpha_1^2, -\alpha_2^2, -\alpha_3^2, \ldots, -\alpha_n^2$. Equation (4.46) is converted into Eq. (4.47), such that roots of Eq. (4.47) are squares (in magnitude) of the roots of Eq. (4.46). Similarly, we can convert the Eq. (4.47) into a new equation with roots $-\alpha_1^{2^2}, -\alpha_2^{2^2}, -\alpha_3^{2^2}, \ldots, -\alpha_n^{2^2}$. This root squaring process is repeated sufficient number of times say m-times. Let the final equation be

$$A_0 z^n + A_1 z^{n-1} + \ldots + A_{n-1} z + A_n = 0 \qquad (4.49)$$

So, the roots of this equation are $-\alpha_1^{2^m}, -\alpha_2^{2^m}, -\alpha_3^{2^m}, \ldots, -\alpha_n^{2^m}$.

Let $\beta_i = -\alpha_i^{2^m}$; $\quad i = 1, 2, \ldots, n$.

If the roots $\alpha_1, \alpha_2, \alpha_3, \ldots, \alpha_n$ of Eq. (4.46) differ in magnitude, then there is a large difference in the magnitude of the roots $\beta_1, \beta_2, \beta_3, \ldots, \beta_n$. From Eq. (4.49), we have

$$\frac{-A_1}{A_0} = \sum_{i=1}^{n} \beta_i \approx \beta_1$$

$$\frac{A_2}{A_0} = \sum_{i=1}^{n} \sum_{j=1}^{n} \beta_i \beta_j \approx \beta_1 \beta_2$$

$$\frac{-A_3}{A_0} = \sum_{i=1}^{n} \sum_{j=1}^{n} \sum_{k=1}^{n} \beta_i \beta_j \beta_k \approx \beta_1 \beta_2 \beta_3$$

$$\vdots$$

$$(-1)^n \frac{A_n}{A_0} = \beta_1 \beta_2 \cdots \beta_n$$

From these equations, we can easily obtain the following results

$$\beta_i = \frac{-A_i}{A_{i-1}}; \quad i = 1, 2, \ldots, n$$

We can use these equations to compute the roots of Eq. (4.46) as follows

$$-\alpha_i^{2^m} = \beta_i = \frac{-A_i}{A_{i-1}}; \quad i = 1, 2, \ldots, n$$

$$\Rightarrow |\alpha_i|^{2^m} = \frac{|A_i|}{|A_{i-1}|}; \quad i = 1, 2, \ldots, n$$

$$\ln|\alpha_i| = 2^{-m} \left(\ln|A_i| - \ln|A_{i-1}| \right); \quad i = 1, 2, \ldots, n \tag{4.50}$$

Convergence of this method is fast if the roots of given equation are sufficiently different in magnitude. We can also apply this method to obtain the double roots as well as if the roots are complex conjugates.

Equal Roots

If after a few iterations the magnitude of the coefficient A_k is nearly half of the square of the coefficient A_k in the previous step, then we can identify that α_k is double root. The following procedure can be used to compute this root.

$$\beta_k \approx \frac{-A_k}{A_{k-1}} \text{ and } \beta_{k+1} \approx \frac{-A_{k+1}}{A_k}$$

Since the root is double root, so $\beta_k = \beta_{k+1}$

$$\Rightarrow \beta_k \beta_{k+1} \simeq \beta_k^2 \simeq \left| \frac{A_{k+1}}{A_{k-1}} \right|$$

Finally, the root is given by the following equation

$$\left| \beta_k^2 \right| = \left| \alpha_k \right|^{2(2^m)} \simeq \left| \frac{A_{k+1}}{A_{k-1}} \right| \tag{4.51}$$

Complex Roots

In the case of complex roots, the coefficients fluctuate both in magnitude and sign in successive steps. Let the complex roots be $\alpha_k, \alpha_{k+1} = \mu_k e^{\pm i v_k}$. The following relations determine these roots.

$$\mu_k^{2(2^m)} \simeq \left| \frac{A_{k+1}}{A_{k-1}} \right| \text{ and } 2\mu_k^m \cos(m v_k) \simeq \frac{A_{k+1}}{A_{k-1}}$$

If the equation has only one complex pair $\alpha_k, \alpha_{k+1} = u \pm iv$, then the sum of the roots is given by

$$\alpha_1 + \alpha_2 + \cdots + \alpha_{k-1} + 2u + \alpha_{k+2} + \cdots + \alpha_n = \frac{-a_1}{a_0} \tag{4.52}$$

This equation is used to compute the value of variable u. Then, we will compute the value of v from the following equation

$$u^2 + v^2 = \mu_k^2, \text{ where } \mu_k^{2(2^m)} \simeq \left| \frac{A_{k+1}}{A_{k-1}} \right| \tag{4.53}$$

Example ──────────────────────────────── **4.15** ─

Compute the roots of the quartic equation, $x^4 - 15x^2 - 10x + 24 = 0$ with the help of Graeffe root squaring method.

Ans. The coefficients for the equation $x^4 - 15x^2 - 10x + 24 = 0$ are as follows

$a_0 = 1, a_1 = 0, a_2 = -15, a_3 = -10, a_4 = 24$.

Equations (4.48) provide the values of the coefficients b_i's as follows

$$b_0 = a_0^2$$
$$b_1 = a_1^2 - 2a_0 a_2$$
$$b_2 = a_2^2 - 2a_1 a_3 + 2a_0 a_4$$
$$\vdots$$

$$b_k = a_k^2 - 2a_{k-1}a_{k+1} + 2a_{k-2}a_{k+2} - \dots, \text{ till coefficients are available for cross product}$$

$$\vdots$$

$$b_{n-1} = a_{n-1}^2 - 2a_{n-2}a_n$$

$$b_n = a_n^2.$$

The following table presents the calculations of all the coefficients till the fourth step (i.e., $m = 4$).

m					
0	1	0	−15	−10	24
	1	0	225	100	576
		30	0	720	
			48		
1	1	30	273	820	576
	1	900	74529	672400	331776
		−546	−49200	−314496	
			1152		
2	1	354	26481	357904	331776
	1	125316	701243361	128095273216	110075314176
		−52962	−253396032	−17571520512	
			663552		
3	1	72354	448510881	110523752704	110075314176
	1	5235101316	201162010375396161	12215499911774947311616	1211657479094510655 8976
		−897021762	−15993671206290432	−98739952274859098112	
			220150628352		
4	1	4338079554	185168559319734081	12116759959500088213504	1211657479094510655 8976

Now, we will use the coefficients at last step to compute the roots by the Eqs. (4.50)

$$\ln|\alpha_i| = 2^{-m}\left(\ln|A_i| - \ln|A_{i-1}|\right); \qquad i = 1, 2, \dots, n$$

The magnitudes of roots are given by

$$\ln \alpha_1 = 2^{-4}\left(\ln(4338079554) - \ln(1)\right) = 2^{-4}(22.190697588 - 0) = 1.3869186$$

$$\Rightarrow \alpha_1 = e^{1.3869186} = 4.0025 \approx 4$$

$$\ln \alpha_2 = 2^{-4}\left(\ln(185168559319734081) - \ln(4338079554)\right)$$

$$= 2^{-4}(39.760043 - 22.190700) = 1.098084$$

$$\Rightarrow \alpha_2 = e^{1.098084} = 2.9984 \approx 3$$

$$\ln \alpha_3 = 2^{-4} \left(\ln(12116759959500088213504) - \ln(185168559319734081) \right)$$
$$= 2^{-4}(50.848876 - 39.760043) = 0.693052$$
$$\Rightarrow \alpha_3 = e^{0.693052} = 1.9998 \approx 2$$

$$\log \alpha_4 = 2^{-4} \left(\log(12116574790945106558976) - \log(12116759959500088213504) \right)$$
$$= 2^{-4}(50.848861 - 50.848876) = -0.000001$$
$$\Rightarrow \alpha_4 = e^{0.000001} = 0.999999 \approx 1$$

The magnitudes of the roots are 1, 2, 3 and 4. After substituting these values with +ve and −ve signs in the original equation, we can easily obtain that the exact roots are 4, −3, −2, 1

Example ── **4.16**

Use Graeffe root squaring method to compute the roots of the equation, $x^4 - 4x^3 - 5x^2 + 18x - 10 = 0$.

Ans. The coefficients of the given equation are $a_0 = 1$, $a_1 = -4$, $a_2 = -5$, $a_3 = 18$, $a_4 = -10$.

Proceeding in a similar manner as in the previous question, the calculations of all the coefficients till the fourth step are presented in the following table.

m					
0	1	−4	−5	18	−10
	1	16	25	324	100
		10	144	−100	
			−20		
1	1	26	149	224	100
	1	676	22201	50176	10000
		−298	−11648	−29800	
			200		
2	1	378	10753	20376	10000
	1	142884	115627009	415181376	100000000
		−21506	−15404256	−215060000	
			20000		
3	1	121378	100242753	200121376	100000000
	1	14732618884	10048609529019009	40048565132133376	10000000000000000
		−200485506	−48580664752256	−20048550600000000	
			200000000		
4	1	14532133378	10000029064266753	20000014532133376	10000000000000000

The following Eqs. (4.50) produce the roots

$$\ln |\alpha_i| = 2^{-m} \left(\ln |A_i| - \ln |A_{i-1}| \right); \qquad i = 1, 2, \ldots, n$$

The magnitudes of the roots are given by

$$\ln \alpha_1 = 2^{-4}\left(\ln(14532133378) - \ln(1)\right) = 2^{-4}(23.399628 - 0) = 1.462477$$

$$\Rightarrow \alpha_1 = e^{1.462477} = 4.3166386, \text{ exact root up to seven decimal places is } 4.31662479$$

$$\ln \alpha_2 = 2^{-4}\left(\ln(10000029064266753) - \ln(14532133378)\right)$$

$$= 2^{-4}(36.841364 - 23.399628) = 0.8401085$$

$$\Rightarrow \alpha_2 = e^{0.8401085} = 2.3166183, \text{ exact root up to seven decimal places is } -2.31662479$$

It is easy to see that the magnitude of the coefficient A_3 is nearly half of the square of the coefficient A_3 in the previous step, so we can identify that α_3 is double root. To compute this root, we can use Eq. (4.51) as follows

$$|\alpha_3|^{2(2^4)} \approx \left|\frac{A_4}{A_2}\right|$$

$$|\alpha_3|^{2(2^4)} \approx \left|\frac{A_4}{A_2}\right| = \left|\frac{10000000000000000}{10000029064266753}\right| = 0.9999971$$

$$|\alpha_3| = 1$$

The absolute values of the roots are 1, 1, 2.3166183 and 4.3166386, while the exact roots are 1, 1, −2.3166183 and 4.3166386

Example   **4.17**

Perform four iterations of Graeffe root squaring method to compute all the roots of the following equation $x^4 + 3x^3 - 5x^2 + 2x + 31 = 0$

The following table presents all the coefficients.

m					
0	1	3	−5	2	31
	1	9	25	4	961
		10	−12	310	
			62		
1	1	19	75	314	961
	1	361	5625	98596	923521
		−150	−11932	−144150	
			1922		
2	1	211	−4385	−45554	923521
	1	44521	19228225	2075166916	852891037441
		8770	19223788	8099279170	
			1847042		
3	1	53291	40299055	10174446086	852891037441
	1	2839930681	1624013833893025	103519353156920719396	727423121747185263828481
		−80598110	−1084412812738052	−68741405653683836510	
			1705782074882		
4	1	2759332571	541306803229855	34777947503236882886	727423121747185263828481

In successive steps, the coefficient A_2 fluctuates both in magnitude and sign. So, we can easily identify complex roots. First, we will determine the real roots A_1 and A_4 as follows

$$\ln \alpha_1 = 2^{-4} \left(\ln(2759332571) - \ln(1) \right) = 2^{-4}(21.738255 - 0) = 1.358641$$

$$\Rightarrow \alpha_1 = e^{1.358641} = 3.890902, \text{ exact root up to seven decimal places is } -3.8908844$$

$$\ln \alpha_4 = 2^{-4} \left(\ln(72742312174718526382848 1) - \ln(34777947503236882886) \right)$$

$$= 2^{-4}(54.943795 - 44.995515) = 0.6217675$$

$$\Rightarrow \alpha_2 = e^{0.6217675} = 1.862217.$$

The exact roots are given by

$$\alpha_1 = -3.890902 \text{ and } \alpha_4 = -1.862217.$$

The equation has only one complex pair $\alpha_k, \alpha_{k+1} = u \pm iv$. The sum of the roots is given by

$$\alpha_1 + 2u + \alpha_4 = \frac{-3}{1} = -3$$

$$\Rightarrow u = 1.376560$$

Now, we will compute the value of v from Eq. (4.53) as follows

$$\mu_2^{2(2^4)} \approx \left| \frac{A_3}{A_1} \right| = \left| \frac{34777947503236882886}{2759332571} \right|$$

$$\Rightarrow \ln \mu_2 = 2^{-5} \left(\ln(34777947503236882886) - \ln(2759332571) \right)$$

$$= 2^{-5} \left(44.995515 - 21.738255 \right) = 0.726789$$

$$\mu_2 = 2.0684282$$

By using $u^2 + v^2 = \mu_2^2$, we get

$$v = \sqrt{2.3834778} = 1.5438516$$

Hence, complex pair of roots is $\alpha_k, \alpha_{k+1} = u \pm iv = 1.376560 \pm i1.5438516$

Table 4.2 Methods for Solutions of the Systems of Nonlinear Equations

Method	Formulation	Convergence	Rate of convergence	Initial approximations required	Breadth of application	Programming
Fixed Point	$x_1^{(k+1)} = \phi_1(x_1^{(k)}, x_2^{(k)}, \cdots, x_n^{(k)})$ $x_2^{(k+1)} = \phi_2(x_1^{(k)}, x_2^{(k)}, \cdots, x_n^{(k)})$ $\cdots$ $x_n^{(k+1)} = \phi_n(x_1^{(k)}, x_2^{(k)}, \cdots, x_n^{(k)})$.	May diverge	1	1	General	Easy
Seidel	$x_1^{(k+1)} = \phi_1(x_1^{(k)}, x_2^{(k)}, x_3^{(k)}, \cdots, x_n^{(k)})$ $x_2^{(k+1)} = \phi_2(x_1^{(k+1)}, x_2^{(k)}, x_3^{(k)}, \cdots, x_n^{(k)})$ $x_3^{(k+1)} = \phi_3(x_1^{(k+1)}, x_2^{(k+1)}, x_3^{(k)}, \cdots, x_n^{(k)})$ $\cdots$ $x_n^{(k+1)} = \phi_n(x_1^{(k+1)}, x_2^{(k+1)}, x_3^{(k+1)}, \cdots, x_{n-1}^{(k+1)}, x_n^{(k)})$.	May diverge	1	1	General	Easy
Newton–Raphson	$X^{k+1} = X^k + \delta X^k = \left[x_1^k + \delta x_1^k, x_2^k + \delta x_2^k, \cdots x_n^k + \delta x_n^k\right]^T$ where $\delta X^k = -J_k^{-1}[F]F(X^k)$ and $$J_k[F] = \begin{bmatrix} \dfrac{\partial f_1}{\partial x_1} & \dfrac{\partial f_1}{\partial x_2} & \cdots & \dfrac{\partial f_1}{\partial x_n} \\[2mm] \dfrac{\partial f_2}{\partial x_1} & \dfrac{\partial f_2}{\partial x_2} & \cdots & \dfrac{\partial f_2}{\partial x_n} \\[2mm] \cdots & & & \\[2mm] \dfrac{\partial f_n}{\partial x_1} & \dfrac{\partial f_n}{\partial x_2} & \cdots & \dfrac{\partial f_n}{\partial x_n} \end{bmatrix}_{(x_1^k, x_2^k, \cdots x_n^k)}$$	May diverge	2	1	General	Difficult

Table 4.3 Methods for the Solutions of the Polynomial Equations

Method	Formulation	Convergence	Rate of convergence	Initial approximations required	Breadth of application	Programming
Birge–Vieta	$p_{k+1} = p_k - \dfrac{b_n}{c_{n-1}}$, where b_n and c_{n-1} are given by $$\begin{array}{c|cccccc} & a_0 & a_1 & a_2 & \cdots & a_{n-1} & a_n \\ p_k & & p_k b_0 & p_k b_1 & \cdots & p_k b_{n-2} & p_k b_{n-1} \\ \hline & b_0 & b_1 & b_2 & \cdots & b_{n-1} & b_n \\ p_k & & p_k c_0 & p_k c_1 & \cdots & p_k c_{n-2} & \\ \hline & c_0 & c_1 & c_2 & \cdots & c_{n-1} & \end{array}$$	May Diverge	2	1	Polynomial equations only	Moderate
Lin–Bairstow	$p_{k+1} = p_k + \Delta p$ with $q_{k+1} = q_k + \Delta q$ $$\Delta p = \frac{b_{n-1}c_{n-2} - b_n c_{n-3}}{c_{n-2}^2 - c_{n-3}(c_{n-1} - b_{n-1})}$$ $$\Delta q = \frac{b_n c_{n-2} - b_{n-1}(c_{n-1} - b_{n-1})}{c_{n-2}^2 - c_{n-3}(c_{n-1} - b_{n-1})}$$ where b_i's and c_i's are as given by $$\begin{array}{c|ccccc} & a_0 & a_1 & a_2 & \cdots & a_n \\ -p_0 & & -p_0 b_0 & -p_0 b_1 & \cdots & -p_0 b_{n-1} \\ -q_0 & & & -q_0 b_0 & \cdots & -q_0 b_{n-2} \\ \hline & b_0 & b_1 & b_2 & \cdots & b_n \end{array}$$ $$\begin{array}{c|ccccc} & b_0 & b_1 & b_2 & \cdots & b_n \\ -p_0 & & -p_0 c_0 & -p_0 c_1 & \cdots & -p_0 c_{n-1} \\ -q_0 & & & -q_0 c_0 & \cdots & -q_0 c_{n-2} \\ \hline & c_0 & c_1 & c_2 & \cdots & c_n \end{array}$$	May Diverge	2	1	Extract quadratic factor from Polynomial equations, so it can compute complex roots also	Moderate
Graeffe Root Squaring	Create the new equations whose roots are squares of the roots of original equations and then use these to compute roots of original equations	May Diverge	2	No initial guess required	Polynomial equations only	Moderate

Exercise 4

1. Solve the following systems of simultaneous equations with the aid of Fixed-Point method and Seidel iteration method.

   a)
   $$4x + y - \sin(x) = 0$$
   $$\cos(xy) - e^{-x} + 6y = -2$$
   Start with initial approximation $(x = 0, y = 0)$

   Ans. $(x = 0.117115, y = -0.351612)$

   b)
   $$3x + y^2 + \cos(x) = 3$$
   $$e^{-x} + 4y - \sin(x + y) = 1$$
   Start with initial approximation $(x = 0, y = 0)$

   Ans. $(x = -2.70674, y = -3.46802)$

   c)
   $$x^2 - 2y^3 - xy + 8y = 3,$$
   $$2x^3 + y^2 - 7x = 0.$$
   Start with initial approximation $(x = 1, y = -1)$

   Ans. $(x = 1.55244, y = -1.83960)$

2. Solve the following systems of equations with the help of Newton–Raphson method

   a)
   $$\cos(x) + y - xe^y = 0 \; ;$$
   $$y^2 - \sin(x) = 2$$
   Start with initial approximation $(x = -5, y = -1)$

   Ans. $(x = -5.40668, y = -1.66388)$

   b)
   $$5y - 2xy - \ln(x) = 0 \; ;$$
   $$x^2 \tan(y) - y^2 = 0$$
   Start with initial approximation $(x = 1, y = -5)$

   Ans. $(x = 2.55904, y = -7.95705)$

3. Solve the following system of nonlinear equations using Newton–Raphson method starting with the initial approximation $(0, 0)$. Perform two iterations only.

   $$3x^3 + 2y^2 - xy + 16y + 3 = 0,$$
   $$2x^3 - 3y + 11x = 0.$$

   Ans.

First Iteration	$x = -0.051136$	$y = -0.187500$
Second Iteration	$x = -0.052182$	$y = -0.191428$
Third Iteration	$x = -0.052182$	$y = -0.191430$

4. Compute the roots of the following equations, where z is the complex variable

   a) $z^3 + 2z + 3 = 0$

   Ans. $(-1, 0.5 \pm 1.65831\,i)$

   b) $iz^3 + 2z + 3 = 0$

   Ans. $(-1.00638 + 0.337588\,i, 1.48412 + 1.11040\,i, -0.477743 - 1.44799\,i)$

   c) $z^4 - 3z^3 + 20z^2 - 4z + 3 = 0$

   Ans. $(0.0921810 \pm 0.383062\,i, 1.40782 \pm 4.16458\,i)$

5. Use Fixed–Point, Seidel iteration and Newton–Raphson methods for the solution of the system
    $$\sin(x_1 + x_2) + 4x_2 = 1$$
    of simultaneous equations $3e^{-x_2} + 7x_1 - x_3 = 4$
    $$x_2 - 6x_3 + x_1 + 2x_1^2 = -2.$$

    Ans. ($x_1 = 0.263187$, $x_2 = 0.149689$, $x_3 = 0.425235$)

6. Find the intervals containing roots of the following equation
    $$x^3 - 11x^2 + 30x - 22 = 0.$$
    Also, compute all the roots of the equation with the aid of Birge–Vieta method.

    Ans. (1.21433, 2.47977, 7.30590)

7. Find the Strum sequences and obtain the numbers of real and complex roots for the following algebraic equations. If the root is multiple, then find its multiplicity also.

    a) $x^3 + 3x^2 + 2 = 0$

        Ans. Strum Sequence: $[x^3 + 3x^2 + 2, x^2 + 2x, -1 + x, -1]$
        (Real $=1$, Complex $=2$)

    b) $x^3 - 2x^2 - 4x + 1 = 0$

        Ans. Strum Sequence $\left[x^3 - 2x^2 - 4x + 1, x^2 - \dfrac{4}{3}x - \dfrac{4}{3}, x - \dfrac{1}{32}, 1 \right]$

        (All 3 roots are real)

    c) $4x^4 + 4x^3 + 4x^2 + 4x + 1 = 0$

        Ans. Strum Sequence: $\left[x^4 + x^3 + x^2 + x + \dfrac{1}{4}, x^3 + \dfrac{3}{4}x^2 + \dfrac{1}{2}x + \dfrac{1}{4}, -x^2 - 2x - \dfrac{3}{5}, -x - \dfrac{5}{12}, -1 \right]$

        (Real $=2$, Complex $=2$)

8. Extract a quadratic factor from following polynomial equation with the help of Lin–Bairstow method (perform only two iterations)
    $$x^4 - 5x^3 + 18x^2 + 38x + 51 = 0, \qquad\qquad (p,q) = (2,2)$$

    Ans. First Iteration $p = 1.78966$ $q = 1.81379$
    Second Iteration $p = 1.77840$ $q = 1.80538$
    Third Iteration $p = 1.77837$ $q = 1.80536$

9. Determine all the real roots of the following equations by Birge–Vieta, Lin–Bairstow and Graeffe root squaring methods

    a) $x^3 - 3x^2 - 2x + 2 = 0$ **Ans.** (0.585786, −1, 3.41421)

    b) $3x^3 - 5x + 2 = 0$ **Ans.** (0.457427, 1, −1.45743)

    c) $2x^4 + 3x^3 - 5x^2 - 2x + 1 = 0$ **Ans.** (0.311836, −0.602027, 1.13557, −2.34538)

10. Use Graeffe root squaring method to compute the roots of the following polynomial equations.

    a) $x^3 + 2x + 3 = 0$ **Ans.** (−1, 0.5 ± 1.6583i)

    b) $x^4 - 2x^3 - 13x^2 + 38x - 24 = 0$ **Ans.** (1, 2, 3, −4)

    c) $x^4 - 17x^2 + 36x - 20 = 0$ **Ans.** (1, 2, 2, −5)

    d) $4x^4 + 4x^3 + 5x^2 + 4x + 1 = 0$ **Ans.** (−0.5, −0.5, ±i)

    e) $x^4 - 3x^2 - 5x + 1 = 0$ **Ans.** (0.1806, 2.2429, −1.2118 ± 0.9999i)

Systems of Linear Equations

Chapter 5

Mathematics is the queen of the sciences.
I have had my results for a long time: but I do not yet know how I am to arrive at them.

Johann Carl Friedrich Gauss
(April 30, 1777–February 23, 1855)
He was a great mathematician having impact on many fields such as algebra, differential geometry, number theory, statistics, mechanics, astronomy, and optics.

5.1 Introduction

The systems of linear equations arise in the modeling of many physical and engineering problems. The linear system of equations with m equations in n variables $x_1, x_2, ..., x_n$ has the following form.

$$a_{11}x_1 + a_{12}x_2 + \cdots + a_{1n}x_n = b_1$$
$$a_{21}x_1 + a_{22}x_2 + \cdots + a_{2n}x_n = b_2$$
$$\vdots$$
$$a_{m1}x_1 + a_{m2}x_2 + \cdots + a_{mn}x_n = b_m$$

(Or), equivalently

$$\sum_{j=1}^{n} a_{ij}x_j = b_i; \quad 1 \le i \le m$$

The matrix form of the system is given by

$$AX = B$$

where

$$A = \begin{bmatrix} a_{11} & a_{12}\dots & a_{1n} \\ a_{21} & a_{22}\dots & a_{2n} \\ \vdots & & \\ a_{m1} & a_{m2}\dots & a_{mn} \end{bmatrix}, \qquad X = \begin{bmatrix} x_1 \\ x_2 \\ \vdots \\ x_n \end{bmatrix}, \qquad B = \begin{bmatrix} b_1 \\ b_2 \\ \vdots \\ b_m \end{bmatrix}$$

The matrix A is a coefficient matrix, and vector X is a solution vector.

If each element of vector B is zero, then the system is called homogeneous system. Otherwise, it is a nonhomogeneous system. For any homogeneous system, zero solution is always a solution, and it is also known as trivial solution.

The system of linear equations may have a unique solution, an infinite number of solutions, or no solution. Here, we are presenting all these cases with some simple examples.

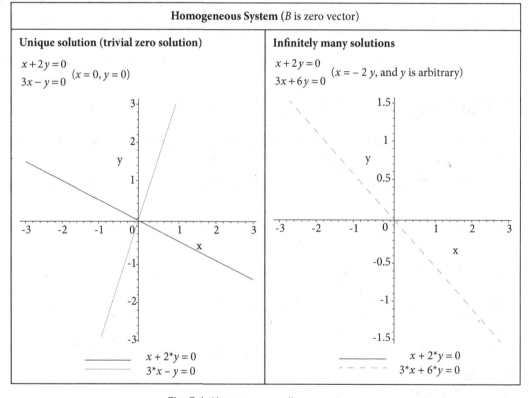

Homogeneous System (B is zero vector)	
Unique solution (trivial zero solution)	**Infinitely many solutions**
$x+2y=0$ $3x-y=0$ $(x=0, y=0)$	$x+2y=0$ $3x+6y=0$ $(x=-2y,$ and y is arbitrary$)$

Fig. 5.1 Homogeneous linear systems

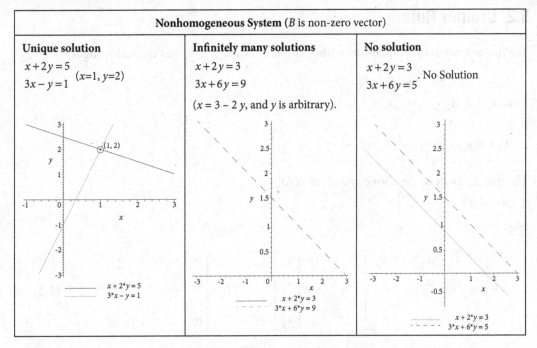

Nonhomogeneous System (B is non-zero vector)		
Unique solution $x + 2y = 5$ $3x - y = 1$ $(x=1, y=2)$	**Infinitely many solutions** $x + 2y = 3$ $3x + 6y = 9$ $(x = 3 - 2\,y$, and y is arbitrary$)$.	**No solution** $x + 2y = 3$ $3x + 6y = 5$. No Solution

Fig. 5.2 Nonhomogeneous linear systems

In this chapter, linear systems with unique solutions have been discussed. There are many direct and iterative methods for the solutions of such systems. Both types of methods have some advantages and disadvantages. It depends on the size and structure of the coefficient matrix A, available computer resources, and solution strategies adopted. This chapter deals with following direct and iterative methods.

Direct Methods

1. Cramer Rule
2. Matrix Inversion Method
3. LU – Decomposition Method, Factorization Method, Triangularization Method
   - i) Crout Method
   - ii) Doolittle Method
   - iii) Cholesky Method
4. Gauss Elimination Method
5. Gauss–Jordan Method

Iterative Methods

1. Jacobi Method (or) Method of Simultaneous Displacement
2. Gauss–Seidel Method (or) Liebmann Method (or) Method of Successive Displacement
3. Relaxation Methods

5.2 Cramer Rule

The linear system of n equations with n variables x_1, x_2, ..., x_n has the following form

$$a_{11}x_1 + a_{12}x_2 + \cdots + a_{1n}x_n = b_1$$
$$a_{21}x_1 + a_{22}x_2 + \cdots + a_{2n}x_n = b_2$$
$$\vdots \qquad\qquad\qquad\qquad\qquad\qquad (5.1)$$
$$a_{n1}x_1 + a_{n2}x_2 + \cdots + a_{nn}x_n = b_n$$

The matrix form of the above system is as follows

$$AX = B$$

where

$$A = \begin{bmatrix} a_{11} & a_{12}\cdots & a_{1n} \\ a_{21} & a_{22}\cdots & a_{2n} \\ \vdots & & \\ a_{n1} & a_{n2}\cdots & a_{nn} \end{bmatrix}, \qquad X = \begin{bmatrix} x_1 \\ x_2 \\ \vdots \\ x_n \end{bmatrix}, \qquad B = \begin{bmatrix} b_1 \\ b_2 \\ \vdots \\ b_n \end{bmatrix} \qquad (5.2)$$

Let C_{ij} be the cofactor associated with the element a_{ij} of the matrix A. Now, multiply the first equation with C_{1j}, the second equation with C_{2j},..., the last equation with C_{nj}; and add all these equations to get the following expression

$$C_{1j}\left(a_{11}x_1 + a_{12}x_2 + \cdots + a_{1n}x_n\right) + C_{2j}\left(a_{21}x_1 + a_{22}x_2 + \cdots + a_{2n}x_n\right)$$
$$\cdots + C_{nj}\left(a_{n1}x_1 + a_{n2}x_2 + \cdots + a_{nn}x_n\right) = C_{1j}b_1 + C_{2j}b_2 + \cdots + C_{nj}b_n$$

On collecting the coefficients of the variables x_1, x_2, ..., x_n in the above expression, we have

$$x_1\sum_{i=1}^{n}a_{i1}C_{ij} + x_2\sum_{i=1}^{n}a_{i2}C_{ij} + \cdots + x_j\sum_{i=1}^{n}a_{ij}C_{ij} + \cdots + x_n\sum_{i=1}^{n}a_{in}C_{ij} = \sum_{i=1}^{n}b_iC_{ij} \qquad (5.3)$$

The coefficient $\displaystyle\sum_{\substack{i=1 \\ k\neq j}}^{n}a_{ik}C_{ij}$ is the expansion of the determinant obtained from matrix A by

replacing the kth row by the jth row. This determinant is zero as it contains two identical

rows. Therefore, all the coefficients $\displaystyle\sum_{\substack{i=1 \\ k\neq j}}^{n}a_{ik}C_{ij}$ in the L.H.S. vanish except the coefficients of

x_j, i.e., $\det A = \displaystyle\sum_{i=1}^{n}a_{ij}C_{ij}$. The system (5.3) reduces to the following expression

$$x_j \det A = \sum_{i=1}^{n} b_i C_{ij}$$

Let the matrix A be non-singular matrix ($\det A \neq 0$) and $\det A = D$.

$$\Rightarrow x_j = \frac{\sum_{i=1}^{n} b_i C_{ij}}{\det A} = \frac{D_j}{D} \tag{5.4}$$

where $D_j = \det A_j$, $\qquad j = 1, 2, \ldots, n$.

Matrix A_j is derived from the matrix A by replacing the jth column with the vector $B = \left[b_1, b_2, \ldots, b_n \right]^T$.

Example 5.1

Solve the following system of linear equations with the help of Cramer rule

$$x_1 - 2x_2 + 3x_3 = 15$$
$$2x_1 - x_2 + 3x_3 = 15$$
$$x_1 + x_2 - 3x_3 = -9$$

Ans.

First, we will compute the following determinants

$$D = \det A$$

and $\qquad D_j = \det A_j, \qquad\qquad j = 1, 2, 3$

where A_j is the matrix derived from the matrix A by replacing the jth column with the vector B.

$$D = \begin{vmatrix} 1 & -2 & 3 \\ 2 & -1 & 3 \\ 1 & 1 & -3 \end{vmatrix} = -9 \qquad\qquad D_1 = \begin{vmatrix} 15 & -2 & 3 \\ 15 & -1 & 3 \\ -9 & 1 & -3 \end{vmatrix} = -18$$

$$D_2 = \begin{vmatrix} 1 & 15 & 3 \\ 2 & 15 & 3 \\ 1 & -9 & -3 \end{vmatrix} = 18 \qquad\qquad D_3 = \begin{vmatrix} 1 & -2 & 15 \\ 2 & -1 & 15 \\ 1 & 1 & -9 \end{vmatrix} = -27$$

Using Eq. (5.4), the solution of system of equations is given by

$$x_1 = \frac{D_1}{D} = 2; \quad x_2 = \frac{D_2}{D} = -2; \quad x_3 = \frac{D_3}{D} = 3$$

Example ─── 5.2 ──

Use Cramer method to compute the solution of the following system of linear equations

$$x_1 + x_2 + 2x_3 - x_4 = 5$$
$$3x_1 + 2x_2 + x_3 + 4x_4 = 2$$
$$x_1 - 2x_2 + 3x_3 + x_4 = 1$$
$$x_1 + 3x_2 - 2x_3 - 3x_4 = 5$$

Ans.

The determinants $D = \det A$ and $D_j = \det A_j, j = 1, 2, 3, 4$ are given by

$$D = \begin{vmatrix} 1 & 1 & 2 & -1 \\ 3 & 2 & 1 & 4 \\ 1 & -2 & 3 & 1 \\ 1 & 3 & -2 & -3 \end{vmatrix} = 74$$

$$D_1 = \begin{vmatrix} 5 & 1 & 2 & -1 \\ 2 & 2 & 1 & 4 \\ 1 & -2 & 3 & 1 \\ 5 & 3 & -2 & -3 \end{vmatrix} = 74 \qquad D_2 = \begin{vmatrix} 1 & 5 & 2 & -1 \\ 3 & 2 & 1 & 4 \\ 1 & 1 & 3 & 1 \\ 1 & 5 & -2 & -3 \end{vmatrix} = 74$$

$$D_3 = \begin{vmatrix} 1 & 1 & 5 & -1 \\ 3 & 2 & 2 & 4 \\ 1 & -2 & 1 & 1 \\ 1 & 3 & 5 & -3 \end{vmatrix} = 74 \qquad D_4 = \begin{vmatrix} 1 & 1 & 2 & 5 \\ 3 & 2 & 1 & 2 \\ 1 & -2 & 3 & 1 \\ 1 & 3 & -2 & 5 \end{vmatrix} = -74$$

On using these determinant, we get the following solution

$$x_1 = \frac{D_1}{D} = 1; \; x_2 = \frac{D_2}{D} = 1; \; x_3 = \frac{D_3}{D} = 1; \; x_4 = \frac{D_4}{D} = -1$$

It is worth mentioning here that the calculation of higher order determinants requires a large number of arithmetic operations. In the case of floating point arithmetic, each arithmetic operation tends to round off error. So, the method produces large round-off error for higher order systems. But, this method has some theoretical importance.

5.3 Matrix Inversion Method

The system (5.1) can be written in the matrix form as follows

$$AX = B \tag{5.5}$$

where

$$A = \begin{bmatrix} a_{11} & a_{12}\cdots & a_{1n} \\ a_{21} & a_{22}\cdots & a_{2n} \\ \vdots & & \\ a_{n1} & a_{n2}\cdots & a_{nn} \end{bmatrix}, \quad X = \begin{bmatrix} x_1 \\ x_2 \\ \vdots \\ x_n \end{bmatrix}, \quad B = \begin{bmatrix} b_1 \\ b_2 \\ \vdots \\ b_n \end{bmatrix}$$

Consider that matrix A is a nonsingular matrix (invertible), then A^{-1} exists. Pre-multiplying Eq. (5.5) with A^{-1}, we get

$$A^{-1}AX = A^{-1}B$$
$$\left(A^{-1}A\right)X = A^{-1}B$$
$$(I)X = A^{-1}B$$
$$IX = A^{-1}B$$
$$X = A^{-1}B \tag{5.6}$$

Equation (5.6) provides the solution of the given system (5.1).

Example 5.3

Solve the following system of linear equations with matrix inversion method

$$3x_1 - x_2 + x_3 = 1$$
$$2x_1 + 3x_2 + x_3 = 4$$
$$3x_1 + x_2 - 2x_3 = 6$$

Ans. The system of equations in matrix form is as follows

$$AX = B$$

where $A = \begin{bmatrix} 3 & -1 & 1 \\ 2 & 3 & 1 \\ 3 & 1 & -2 \end{bmatrix}$, $X = \begin{bmatrix} x_1 \\ x_2 \\ x_3 \end{bmatrix}$, $B = \begin{bmatrix} 1 \\ 4 \\ 6 \end{bmatrix}$

The matrix inverse (A^{-1}) of the coefficient matrix A is given by

$$A^{-1} = \frac{1}{35}\begin{bmatrix} 7 & 1 & 4 \\ -7 & 9 & 1 \\ 7 & 6 & -11 \end{bmatrix}$$

On using equation (5.6), we have

$$X = A^{-1}B = \frac{1}{35}\begin{bmatrix} 7 & 1 & 4 \\ -7 & 9 & 1 \\ 7 & 6 & -11 \end{bmatrix}\begin{bmatrix} 1 \\ 4 \\ 6 \end{bmatrix} = \begin{bmatrix} 1 \\ 1 \\ -1 \end{bmatrix}.$$

Therefore, the solution is given by

$$x_1 = 1, \; x_2 = 1, \; x_3 = -1$$

Example   5.4

Use the matrix inversion method to solve the following system of linear equations

$$x_1 - 2x_2 + x_3 = 0$$
$$2x_1 + 3x_2 - 3x_3 = -1$$
$$3x_1 + x_2 - x_3 = 2$$

Ans. Matrix inverse is given by

$$A^{-1} = \frac{1}{7}\begin{bmatrix} 0 & -1 & 3 \\ -7 & -4 & 5 \\ -7 & -7 & 7 \end{bmatrix}$$

The solution of the given system is as follows

$$X = A^{-1}B = \frac{1}{7}\begin{bmatrix} 0 & -1 & 3 \\ -7 & -4 & 5 \\ -7 & -7 & 7 \end{bmatrix}\begin{bmatrix} 0 \\ -1 \\ 2 \end{bmatrix} = \begin{bmatrix} 1 \\ 2 \\ 3 \end{bmatrix}$$

Example   5.5

Use Matrix inversion method to compute the solution of the linear system in Example 5.2.

Ans.
The system of equations in matrix form is given by

$$AX = B$$

where $A = \begin{bmatrix} 1 & 1 & 2 & -1 \\ 3 & 2 & 1 & 4 \\ 1 & -2 & 3 & 1 \\ 1 & 3 & -2 & -3 \end{bmatrix}$, $X = \begin{bmatrix} x_1 \\ x_2 \\ x_3 \\ x_4 \end{bmatrix}$, $B = \begin{bmatrix} 5 \\ 2 \\ 1 \\ 5 \end{bmatrix}$

Matrix inverse is given by

$$A^{-1} = \frac{1}{74} \begin{bmatrix} -37 & 8 & 48 & 39 \\ 37 & 6 & -38 & -17 \\ 37 & -2 & -12 & -19 \\ 0 & 10 & -14 & -16 \end{bmatrix}$$

The solution of the given system is as follows

$$X = A^{-1}B = \frac{1}{74} \begin{bmatrix} -37 & 8 & 48 & 39 \\ 37 & 6 & -38 & -17 \\ 37 & -2 & -12 & -19 \\ 0 & 10 & -14 & -16 \end{bmatrix} \begin{bmatrix} 5 \\ 2 \\ 1 \\ 5 \end{bmatrix} = \begin{bmatrix} 1 \\ 1 \\ 1 \\ -1 \end{bmatrix}$$

Example   **5.6**

Solve the following system of linear equations with matrix inversion method

$$x_1 - 2x_2 + x_3 = 0$$
$$2x_1 + 3x_2 - 3x_3 = -1$$
$$3x_1 + x_2 - 2x_3 = -1$$

Ans. The determinant of matrix A vanishes

$$|A| = 0$$

Matrix inverse (A^{-1}) is not possible for singular matrix A. So, we cannot apply matrix inversion method for the solution of the given system.

Note: It is easy to see that the third equation in the system is the addition of the first and second equations. So this system has only two independent equations, and one equation in the system is redundant. This system consists of two equations in three variables and has infinitely many solutions.

Similarly, if we assume the following system of equations

$$x_1 - 2x_2 + x_3 = 0$$
$$2x_1 + 3x_2 - 3x_3 = -1$$
$$3x_1 + x_2 - 2x_3 = 1$$

The matrix A is a singular matrix. Adding the first and second equations, and comparing with the third equation, we obtain $-1 = 1$. The system is inconsistent, so no solution exists for this system.

5.4 LU Decomposition (or) Factorization (or) Triangularization Method

In this method, the coefficient matrix A is factorized into the product of two triangular matrices such that one matrix is lower triangular L and the other matrix is upper triangular U, i.e.,

$A = LU$

where $L = \begin{bmatrix} l_{11} & 0 & \cdots & 0 \\ l_{21} & l_{22} & \cdots & 0 \\ \vdots & & & \\ l_{n1} & l_{n2} & \cdots & l_{nn} \end{bmatrix}$ and $U = \begin{bmatrix} u_{11} & u_{12} & \cdots & u_{1n} \\ 0 & u_{22} & \cdots & u_{2n} \\ \vdots & & & \\ 0 & 0 & \cdots & u_{nn} \end{bmatrix}$ are lower

and upper triangular matrices, respectively. The matrices L and U have to be computed, such that

$$A = LU = \begin{bmatrix} l_{11} & 0 & \cdots & 0 \\ l_{21} & l_{22} & \cdots & 0 \\ \vdots & & & \\ l_{n1} & l_{n2} & \cdots & l_{nn} \end{bmatrix} \begin{bmatrix} u_{11} & u_{12} & \cdots & u_{1n} \\ 0 & u_{22} & \cdots & u_{2n} \\ \vdots & & & \\ 0 & 0 & \cdots & u_{nn} \end{bmatrix}$$

(5.7)

$$= \begin{bmatrix} l_{11}u_{11} & l_{11}u_{12} & \cdots & l_{11}u_{1n} \\ l_{21}u_{11} & l_{21}u_{12}+l_{22}u_{22} & \cdots & l_{21}u_{1n}+l_{22}u_{2n} \\ \vdots & & & \\ l_{n1}u_{11} & l_{n1}u_{12}+l_{n2}u_{22} & \cdots & l_{n1}u_{1n}+l_{n2}u_{2n}+\cdots+l_{nn}u_{nn} \end{bmatrix}$$

After comparing the elements of both the matrices, we get the following relations

$$l_{i1}u_{1j} + l_{i2}u_{2j} + \cdots + l_{in}u_{nj} = a_{ij} \qquad 1 \le i, j \le n$$

where $l_{ij} = 0, \quad j > i$ and $u_{ij} = 0, \quad i > j$

This set contains n^2 equations. But, the total number of variables is $(n^2 + n)$ in lower and upper triangular matrices. So, we have to predefine n variables for a unique solution. For convenience, let us consider

either $l_{ii} = 1$ (or) $u_{ii} = 1; 1 \le i \le n$

Accordingly, we have following two methods

5.4.1 Doolittle Method

$$l_{ii} = 1; \, 1 \leq i \leq n \tag{5.8}$$

5.4.2 Crout Method

$$u_{ii} = 1; \, 1 \leq i \leq n \tag{5.9}$$

Here, we will discuss the computation of lower and upper triangular matrices in Crout method i.e., we have $u_{ii} = 1$, $1 \leq i \leq n$. The similar procedure can be used in Doolittle method. We have

$$l_{i1}u_{1j} + l_{i2}u_{2j} + \cdots + l_{in}u_{nj} = a_{ij} \qquad 1 \leq i, j \leq n$$

where $u_{ii} = 1$, $1 \leq i \leq n$; $l_{ij} = 0$, $j > i$; and $u_{ij} = 0$, $i > j$.

From (5.7) and $u_{11} = 1$; it is clear that the first columns of matrix L and A are identical. So, we have

$$l_{i1} = a_{i1}, 1 \leq i \leq n$$

The first rows of both the matrices in (5.7) produce the first row of matrix U as follows

$$u_{11} = 1 \text{ and } u_{1j} = \frac{a_{1j}}{l_{11}}, 2 \leq j \leq n$$

Now, we will compute second column of matrix L and second row of matrix U as follows

$$l_{i2} = a_{i2} - l_{i1}u_{12}, 2 \leq i \leq n$$

$$u_{22} = 1 \text{ and } u_{2j} = \frac{a_{2j} - l_{21}u_{1j}}{l_{22}}, 3 \leq j \leq n$$

In general, we can compute kth column and kth row of matrices L and U, respectively by using following equations

$$l_{ik} = a_{ik} - \sum_{j=1}^{k-1} l_{ij}u_{jk}, k \leq i \leq n$$

$$u_{kk} = 1 \text{ and } u_{kj} = \frac{a_{kj} - \sum_{m=1}^{k-1} l_{km}u_{mj}}{l_{kk}}, k+1 \leq j \leq n$$

After computing the matrices L and U, the system of equations (5.5) is given by

$$AX = B$$

$$LUX = B$$

Let $UX = Y$, then the above system reduces to

$$LY = B \tag{5.10}$$

The system $LY = B$ is the lower triangular system. So, the vector Y can be easily determined by using forward substitution. The vector X can be easily computed by using back substitution from the following upper triangular system

$$UX = Y \tag{5.11}$$

Example ── **5.7**

Use Crout and Doolittle methods to calculate the solution of the following system of linear equations

$$3x_1 - x_2 + x_3 = 1$$
$$2x_1 + 3x_2 + x_3 = 4$$
$$3x_1 + x_2 - 2x_3 = 6$$

Ans.

Crout method (5.9): First, we decompose the coefficient matrix A into the product of lower and upper triangular matrices with diagonal elements in upper triangular matrix as unity, i.e.

$$\begin{bmatrix} 3 & -1 & 1 \\ 2 & 3 & 1 \\ 3 & 1 & -2 \end{bmatrix} = \underbrace{\begin{bmatrix} l_{11} & 0 & 0 \\ l_{21} & l_{22} & 0 \\ l_{31} & l_{32} & l_{33} \end{bmatrix}}_{L} \underbrace{\begin{bmatrix} 1 & u_{12} & u_{13} \\ 0 & 1 & u_{23} \\ 0 & 0 & 1 \end{bmatrix}}_{U}$$

$$= \begin{bmatrix} l_{11} & l_{11}u_{12} & l_{11}u_{13} \\ l_{21} & l_{21}u_{12} + l_{22} & l_{21}u_{13} + l_{22}u_{23} \\ l_{31} & l_{31}u_{12} + l_{32} & l_{31}u_{13} + l_{32}u_{23} + l_{33} \end{bmatrix}$$

After equating the terms on both sides, we obtain following set of equations

$$l_{11} = 3, \ l_{11}u_{12} = -1, \ l_{11}u_{13} = 1$$
$$l_{21} = 2, \ l_{21}u_{12} + l_{22} = 3, \ l_{21}u_{13} + l_{22}u_{23} = 1$$
$$l_{31} = 3, l_{31}u_{12} + l_{32} = 1, \ l_{31}u_{13} + l_{32}u_{23} + l_{33} = -2$$

The solution of this system produces the values of l_{ij} and u_{ij} as follows

First Column: $l_{11} = 3, l_{21} = 2, l_{31} = 3$

First Row: $u_{12} = -1/l_{11} = -1/3$ and $u_{13} = 1/l_{11} = 1/3$

Second Column: $l_{22} = 3 - l_{21}u_{12} = \dfrac{11}{3}$ and $l_{32} = 1 - l_{31}u_{12} = 2$

Second Row: $u_{23} = (1 - l_{21}u_{13})/l_{22} = \dfrac{1}{11}$

Third Column: $l_{33} = -2 - l_{31}u_{13} + l_{32}u_{23} = \dfrac{-35}{11}$

So, we can easily write the coefficient matrix A in terms of the matrices L and U as follows

$$\begin{bmatrix} 3 & -1 & 1 \\ 2 & 3 & 1 \\ 3 & 1 & -2 \end{bmatrix} = \begin{bmatrix} 3 & 0 & 0 \\ 2 & \dfrac{11}{3} & 0 \\ 3 & 2 & \dfrac{-35}{11} \end{bmatrix} \begin{bmatrix} 1 & \dfrac{-1}{3} & \dfrac{1}{3} \\ 0 & 1 & \dfrac{1}{11} \\ 0 & 0 & 1 \end{bmatrix}$$

The system (5.10) $LY = B$ is given by

$$\begin{bmatrix} 3 & 0 & 0 \\ 2 & \dfrac{11}{3} & 0 \\ 3 & 2 & \dfrac{-35}{11} \end{bmatrix} \begin{bmatrix} y_1 \\ y_2 \\ y_3 \end{bmatrix} = \begin{bmatrix} 1 \\ 4 \\ 6 \end{bmatrix}$$

This system of equations can be rewritten as follows

$$3y_1 + 0y_2 + 0y_3 = 1$$

$$2y_1 + \dfrac{11}{3}y_2 + 0y_3 = 4$$

$$3y_1 + 2y_2 - \dfrac{35}{11}y_3 = 6$$

From the first equation, we get

$$y_1 = \dfrac{1}{3}$$

On substituting this value in the second equation, we have $y_2 = \dfrac{10}{11}$, and from the last

equation $y_3 = -1$.

On using these values of y_1, y_2 and y_3 in the system (5.11)

$$UX = Y$$

we have

$$\begin{bmatrix} 1 & \dfrac{-1}{3} & \dfrac{1}{3} \\ 0 & 1 & \dfrac{1}{11} \\ 0 & 0 & 1 \end{bmatrix} \begin{bmatrix} x_1 \\ x_2 \\ x_3 \end{bmatrix} = \begin{bmatrix} \dfrac{1}{3} \\ \dfrac{10}{11} \\ -1 \end{bmatrix}$$

From the last equation $x_3 = -1$, using this value in the second equation $x_2 = 1$, and the first equation gives $x_1 = 1$. So, the solution is given by

$$x_1 = 1, x_2 = 1, x_3 = -1$$

Doolittle method (5.8): First, we decompose the coefficient matrix A in the product of lower and upper triangular matrices with diagonal elements in the lower triangular matrix as unity.

$$A = LU$$

$$\begin{bmatrix} 3 & -1 & 1 \\ 2 & 3 & 1 \\ 3 & 1 & -2 \end{bmatrix} = \underbrace{\begin{bmatrix} 1 & 0 & 0 \\ l_{21} & 1 & 0 \\ l_{31} & l_{32} & 1 \end{bmatrix}}_{L} \underbrace{\begin{bmatrix} u_{11} & u_{12} & u_{13} \\ 0 & u_{22} & u_{23} \\ 0 & 0 & u_{33} \end{bmatrix}}_{U}$$

Proceeding in a similar manner as in Crout method, we obtain

$$\begin{bmatrix} 3 & -1 & 1 \\ 2 & 3 & 1 \\ 3 & 1 & -2 \end{bmatrix} = \begin{bmatrix} 1 & 0 & 0 \\ \dfrac{2}{3} & 1 & 0 \\ 1 & \dfrac{6}{11} & 1 \end{bmatrix} \begin{bmatrix} 3 & -1 & 1 \\ 0 & \dfrac{11}{3} & \dfrac{1}{3} \\ 0 & 0 & \dfrac{-35}{11} \end{bmatrix}$$

First, we solve $LY = B$ by using forward substitution

$$\begin{bmatrix} 1 & 0 & 0 \\ \dfrac{2}{3} & 1 & 0 \\ 1 & \dfrac{6}{11} & 1 \end{bmatrix} \begin{bmatrix} y_1 \\ y_2 \\ y_3 \end{bmatrix} = \begin{bmatrix} 1 \\ 4 \\ 6 \end{bmatrix}$$

The solution is, $y_1 = 1$, $y_2 = \dfrac{10}{3}$, $y_3 = \dfrac{35}{11}$. Now, we solve $UX = Y$ by using backward substitutions

$$
\begin{bmatrix}
3 & -1 & 1 \\
0 & \dfrac{11}{3} & \dfrac{1}{3} \\
0 & 0 & \dfrac{-35}{11}
\end{bmatrix}
\begin{bmatrix}
x_1 \\ x_2 \\ x_3
\end{bmatrix}
=
\begin{bmatrix}
1 \\ \dfrac{10}{3} \\ \dfrac{35}{11}
\end{bmatrix}
$$

On solving this system of equations, final solution is given by
$$x_1 = x_2 = 1 \text{ and } x_3 = -1$$

Example 5.8

Solve the linear system of Example 5.2 with the help of Doolittle method.

Ans. The LU factorization of matrix A is given by

$$
A =
\begin{bmatrix}
1 & 1 & 2 & -1 \\
3 & 2 & 1 & 4 \\
1 & -2 & 3 & 1 \\
1 & 3 & -2 & -3
\end{bmatrix}
=
\begin{bmatrix}
1 & 0 & 0 & 0 \\
3 & 1 & 0 & 0 \\
1 & 3 & 1 & 0 \\
1 & -2 & -\dfrac{7}{8} & 1
\end{bmatrix}
\begin{bmatrix}
1 & 1 & 2 & -1 \\
0 & -1 & -5 & 7 \\
0 & 0 & 16 & -19 \\
0 & 0 & 0 & -\dfrac{37}{8}
\end{bmatrix}
$$

The solution of the system $LY = B$ using forward substitution is given by

$$
\begin{bmatrix}
1 & 0 & 0 & 0 \\
3 & 1 & 0 & 0 \\
1 & 3 & 1 & 0 \\
1 & -2 & -\dfrac{7}{8} & 1
\end{bmatrix}
\begin{bmatrix}
y_1 \\ y_2 \\ y_3 \\ y_4
\end{bmatrix}
=
\begin{bmatrix}
5 \\ 2 \\ 1 \\ 5
\end{bmatrix}
$$

The solution is, $y_1 = 5$, $y_2 = -13$, $y_3 = 35$, $y_4 = \dfrac{37}{8}$. Now, we will use backward substitution to solve the system $UX = Y$ as follows

$$
\begin{bmatrix}
1 & 1 & 2 & -1 \\
0 & -1 & -5 & 7 \\
0 & 0 & 16 & -19 \\
0 & 0 & 0 & -\dfrac{37}{8}
\end{bmatrix}
\begin{bmatrix}
x_1 \\
x_2 \\
x_3 \\
x_4
\end{bmatrix}
=
\begin{bmatrix}
5 \\
-13 \\
35 \\
\dfrac{37}{8}
\end{bmatrix}
$$

The final solution is given by

$$x_1 = x_2 = x_3 = 1 \text{ and } x_4 = -1$$

Example ———————————————————————————————————— **5.9**

Solve the following system of linear equations with the help of LU-decomposition method

$$3x_1 - 3x_2 + x_3 = 4$$
$$-2x_1 + 2x_2 + x_3 = -1$$
$$x_1 + x_2 + 2x_3 = 3$$

Ans. Crout method: The coefficient matrix A can be written as the product of lower and upper triangular matrices as follows

$$
\begin{bmatrix}
3 & -3 & 1 \\
-2 & 2 & 1 \\
1 & 1 & 2
\end{bmatrix}
=
\begin{bmatrix}
l_{11} & 0 & 0 \\
l_{21} & l_{22} & 0 \\
l_{31} & l_{32} & l_{33}
\end{bmatrix}
\begin{bmatrix}
1 & u_{12} & u_{13} \\
0 & 1 & u_{23} \\
0 & 0 & 1
\end{bmatrix}
$$

After equating the terms on both sides, we obtain

$$l_{11} = 3, \; l_{11}u_{12} = -3, \; l_{11}u_{13} = 1$$
$$l_{21} = -2, \; l_{21}u_{12} + l_{22} = 2, \; l_{21}u_{13} + l_{22}u_{23} = 1$$
$$l_{31} = 1, \; l_{31}u_{12} + l_{32} = 1, \; l_{31}u_{13} + l_{32}u_{23} + l_{33} = 2$$

The solution of these equations is as follows

$$l_{11} = 3, \; u_{12} = -1, \; u_{13} = \frac{1}{3}$$

$$l_{21} = -2, \; l_{22} = 0$$

Since the element $l_{22} = 0$, so we cannot solve the equation $l_{21}u_{13} + l_{22}u_{23} = 1$ for the variable u_{23}.

The method fails as the element $l_{22} = 0$ is zero.

Similarly, in Doolittle method, the element $u_{22} = 0$ is zero, so method fails again.

Note: Rather, the system has a unique solution $x_1 = 1$, $x_2 = 0$, $x_3 = 1$, but LU-decomposition method does not work here. The first two rows of the system are a linear multiple of each other till the first two terms. Hence, the pivot element (l_{22} and u_{22}) becomes zero and the method fails. The solution can be obtained by interchanging any of first two rows with the third row.

We will see that the other direct methods like Gauss elimination and Gauss–Jordan also fail to render solution without pivoting (interchange of rows), as the pivot element becomes zero.

So far, we have discussed the LU decomposition method for a general coefficient matrix, but if the coefficient matrix A is a positive definite symmetric matrix, then the method becomes simpler and is known as a Cholesky method.

A square matrix A is a positive definite symmetric matrix if it is symmetric and $X^T A X > 0$ for each nonzero column vector X.

Example — 5.10

Prove that matrix $A = \begin{bmatrix} 3 & -1 & 1 \\ -1 & 3 & 1 \\ 1 & 1 & 2 \end{bmatrix}$ is positive definite symmetric matrix.

Ans: Let nonzero vector be $X = \begin{bmatrix} a \\ b \\ c \end{bmatrix}$. Then, we have

$$X^T AX = \begin{bmatrix} a & b & c \end{bmatrix} \begin{bmatrix} 3 & -1 & 1 \\ -1 & 3 & 1 \\ 1 & 1 & 2 \end{bmatrix} \begin{bmatrix} a \\ b \\ c \end{bmatrix}$$

$$= \begin{bmatrix} a & b & c \end{bmatrix} \begin{bmatrix} 3a - b + c \\ -a + 3b + c \\ a + b + 2c \end{bmatrix}$$

$$= 3a^2 - ab + ac - ab + 3b^2 + bc + ac + bc + 2c^2$$

$$= (a^2 + b^2 - 2ab) + (a^2 + c^2 + 2ac) + (b^2 + c^2 + 2bc) + a^2 + b^2$$

$$= (a-b)^2 + (a+c)^2 + (b+c)^2 + a^2 + b^2$$

which is always positive for each nonzero vector X. The matrix A is symmetric matrix and $X^T A X > 0$ for each nonzero X.

So, the matrix A is positive definite symmetric matrix.

Example ── **5.11**

Show that the matrix $A = \begin{bmatrix} 2 & -1 & 1 \\ -1 & 2 & 1 \\ 1 & 1 & 2 \end{bmatrix}$ is not positive definite symmetric matrix.

Ans:

For nonzero vector $X = \begin{bmatrix} a \\ b \\ c \end{bmatrix}$, we have

$$X^T AX = (a-b)^2 + (a+c)^2 + (b+c)^2$$

The scalar X^TAX can be zero for a, b, c such that $a = b = -c$. For example $X = \begin{bmatrix} 1 \\ 1 \\ -1 \end{bmatrix}$.

The matrix A is symmetric, but it does not satisfy $X^TAX > 0$ for each nonzero X. Hence, the matrix A is not positive definite symmetric matrix.

A symmetric matrix A is positive definite symmetric matrix, if any one of the following properties holds

1. All its eigen values are positive
2. All its leading principal minors are positive
3. $a_{ii} > 0$ and $a_{ii} > \sum_{j \neq i} |a_{ij}|$ for each i
4. All pivots are positive

5.4.3 Cholesky Method

In case of positive definite symmetric matrix A, there exists a unique decomposition of matrix A, known as Cholesky decomposition

$A = LL^T$ $\qquad\qquad$ (5.12)

where L is a lower triangular matrix and L^T is its transpose.

Therefore, the system $AX = B$ can be written as follows

$LL^TX = B$

Let $L^TX = Y$, then

$LY = B$ $\qquad\qquad$ (5.13)

First we compute vector Y using forward substitution from Eq. (5.13) and then compute vector X from the equation

$L^T X = Y$ $\qquad\qquad$ (5.14)

The matrix A can also be decomposed as $A = UU^T$, where U is an upper triangular matrix.

Example ── 5.12 ────

Solve the following system of linear equations with the aid of Cholesky method

$$3x_1 - x_2 + x_3 = 2$$
$$-x_1 + 3x_2 + x_3 = 6$$
$$x_1 + x_2 + 2x_3 = 5$$

Ans.

The matrix $A = \begin{bmatrix} 3 & -1 & 1 \\ -1 & 3 & 1 \\ 1 & 1 & 2 \end{bmatrix}$ is a positive definite matrix (refer to Example

(5.10)). Using Eq. (5.12) to decompose the matrix, we get

$$A = LL^T$$

where L is a lower triangular matrix and L^T is the transpose of L, i.e.

$$\begin{bmatrix} 3 & -1 & 1 \\ -1 & 3 & 1 \\ 1 & 1 & 2 \end{bmatrix} = \begin{bmatrix} l_{11} & 0 & 0 \\ l_{21} & l_{22} & 0 \\ l_{31} & l_{32} & l_{33} \end{bmatrix} \begin{bmatrix} l_{11} & l_{21} & l_{31} \\ 0 & l_{22} & l_{32} \\ 0 & 0 & l_{33} \end{bmatrix}$$

On comparing both sides and solving the resulting equations, we get matrix L as follows

$$L = \begin{bmatrix} \sqrt{3} & 0 & 0 \\ \dfrac{-\sqrt{3}}{3} & \dfrac{2\sqrt{6}}{3} & 0 \\ \dfrac{\sqrt{3}}{3} & \dfrac{\sqrt{6}}{3} & 1 \end{bmatrix}$$

The system $AX = B$ can be written as follows

$$LL^T X = B$$

Let $L^T X = Y$, then $LY = B$. Compute vector Y from the equation $LY = B$.

$$\begin{bmatrix} \sqrt{3} & 0 & 0 \\ \dfrac{-\sqrt{3}}{3} & \dfrac{2\sqrt{6}}{3} & 0 \\ \dfrac{\sqrt{3}}{3} & \dfrac{\sqrt{6}}{3} & 1 \end{bmatrix} \begin{bmatrix} y_1 \\ y_2 \\ y_3 \end{bmatrix} = \begin{bmatrix} 2 \\ 6 \\ 5 \end{bmatrix}$$

The solution of this system of equations is given by

$$y_1 = \frac{2\sqrt{3}}{3}, \; y_2 = \frac{5\sqrt{6}}{3}, \; y_3 = 1$$

On computing the vector X from the equation $L^T X = Y$, we have

$$\begin{bmatrix} \sqrt{3} & \dfrac{-\sqrt{3}}{3} & \dfrac{\sqrt{3}}{3} \\ 0 & \dfrac{2\sqrt{6}}{3} & \dfrac{\sqrt{6}}{3} \\ 0 & 0 & 1 \end{bmatrix} \begin{bmatrix} x_1 \\ x_2 \\ x_3 \end{bmatrix} = \begin{bmatrix} \dfrac{2\sqrt{3}}{3} \\ \dfrac{5\sqrt{6}}{3} \\ 1 \end{bmatrix}$$

On solving this system of equations by back substitution, we get the following solution

$$x_1 = 1, \; x_2 = 2, \; x_3 = 1$$

5.5 Gauss Elimination Method

Gauss elimination method is one of the simplest algorithmic procedures in all known direct methods with a minimum number of arithmetic operations and hence most widely used direct method. Before we discuss this method; it is necessary to have knowledge of elementary row operations described as follows

Elementary Row Operations

Let us denote the ith equation of the system (5.1) with the symbol, R_i; $1 \le i \le n$. The solution of a system of Eqs. (5.1) remains unaltered, if we perform the following row operations with the equations.

i)   The interchange of any two equations, $R_i \leftrightarrow R_j$.
ii) The multiplication of any equation by a nonzero constant, $R_i \to kR_i$; $k \ne 0$.
iii) The addition of an equation with the scalar multiple of another equation $R_i \to R_i + kR_j$.

These operations are known as elementary row operations. These elementary row operations are used in Gauss Elimination method. An algorithmic overview of the method is as follows

Let us rewrite the system (5.1) into the augmented matrix form $[A:B]$ as follows

$$[A:B] = \begin{bmatrix} a_{11} & a_{12} & a_{13} \cdots & a_{1n} & : b_1 \\ a_{21} & a_{22} & a_{23} \cdots & a_{2n} & : b_2 \\ a_{31} & a_{32} & a_{33} \cdots & a_{3n} & : b_3 \\ \vdots & & & & \vdots \\ a_{n1} & a_{n2} & a_{n3} \cdots & a_{nn} & : b_n \end{bmatrix} \qquad (5.15)$$

In Gauss Elimination method, the solution of system (5.1) is obtained in two phases: In the first phase, the system of linear equations is converted into an equivalent upper triangular system with the aid of elementary row operations. In the second phase, the solution is obtained by back substitutions.

First Phase (Conversion to upper triangular form)
In the first phase, we will convert the augmented matrix $[A{:}B]$ into the upper triangular matrix by applying elementary row operations.

We use the first equation to vanish the coefficients of x_1 from the remaining equations by applying the elementary row operations, $R_2 \to R_2 - \dfrac{a_{21}}{a_{11}} R_1, R_3 \to R_3 - \dfrac{a_{31}}{a_{11}} R_1, \cdots,$

$R_n \to R_n - \dfrac{a_{n1}}{a_{11}} R_1$ on the augmented matrix to get following row-equivalent matrix

$$[A{:}B] \sim \begin{bmatrix} a_{11} & a_{12} & a_{13} \cdots & a_{1n} & : b_1 \\ 0 & a_{22} & a_{23} \cdots & a_{2n} & : b_2 \\ 0 & a_{32} & a_{33} \cdots & a_{3n} & : b_3 \\ \vdots & & \vdots & & \\ 0 & a_{n2} & a_{n3} \cdots & a_{nn} & : b_n \end{bmatrix}$$

At this step, the element a_{11} is pivot element and $a_{11} \neq 0$.

Note that, the elements in last $(n - 1)$ rows of augmented matrix are changed due to operations applied on these elements. But we can continue with the same notions without any loss of generality.

Similarly, we use the second equation (a_{22} is pivot element and $a_{22} \neq 0$) to vanish the coefficients of x_2 from the equations below the second equation by applying the elementary row operations $R_3 \to R_3 - \dfrac{a_{32}}{a_{22}} R_2, R_4 \to R_4 - \dfrac{a_{42}}{a_{22}} R_2, \cdots, R_n \to R_n - \dfrac{a_{n2}}{a_{22}} R_2$ on the augmented

matrix to get following row-equivalent matrix

$$[A{:}B] \sim \begin{bmatrix} a_{11} & a_{12} & a_{13} \cdots & a_{1n} & : b_1 \\ 0 & a_{22} & a_{23} \cdots & a_{2n} & : b_2 \\ 0 & 0 & a_{33} \cdots & a_{3n} & : b_3 \\ \vdots & & \vdots & & \\ 0 & 0 & a_{n3} \cdots & a_{nn} & : b_n \end{bmatrix}$$

On continuing in a similar manner by using 3rd, 4th,..., $(n-1)^{th}$ equations, we will vanish all the lower triangular elements to obtain the following upper triangular matrix

$$[A:B] \sim \begin{bmatrix} a_{11} & a_{12} & a_{13}\cdots & a_{1n} & : b_1 \\ 0 & a_{22} & a_{23}\cdots & a_{2n} & : b_2 \\ 0 & 0 & a_{33}\cdots & a_{3n} & : b_3 \\ \vdots & & & & \\ 0 & 0 & 0 \cdots & a_{nn} & : b_n \end{bmatrix}$$

In algebraic form, this system can be written as follows

$$\begin{aligned} a_{11}x_1 + a_{12}x_2 + a_{13}x_2 + \cdots & \quad +a_{1n-1}x_{n-1} + a_{1n}x_n = b_1 \\ a_{22}x_2 + a_{23}x_3 + \cdots & \quad +a_{2n-1}x_{n-1} + a_{2n}x_n = b_2 \\ \vdots \quad \vdots & \quad \vdots \quad \vdots \\ & a_{n-1\,n-1}x_{n-1} + a_{n-1n}x_n = b_{n-1} \\ & a_{nn}x_n = b_n \end{aligned} \qquad (5.16)$$

We have converted the augmented matrix in the upper triangular matrix by applying elementary row operations. The solution of system (5.1) remains unaltered on performing the elementary row operations. Therefore, the solution of the system (5.16) is a solution of the system (5.1).

Second Phase (Obtain the solution from upper triangular form using back substitutions)
In the second phase, we will compute the solution from the upper triangular system (5.16) as follows: From last equation of system (5.16), we can easily compute the variable x_n as follows

$$x_n = \frac{b_n}{a_{nn}}$$

On using the value of x_n in second last equation, we have

$$x_{n-1} = \frac{1}{a_{n-1n-1}}\left(b_{n-1} - a_{n-1n}x_n\right)$$

Similarly, the remaining equations of the system (5.16) produce all other x_i's as follows

$$x_i = \frac{1}{a_{ii}}\left(b_i - \sum_{j=i+1}^{n} a_{ij}x_j\right) \qquad i = n-1, n-2,...,1 \qquad (5.17)$$

Example _____ 5.13

Use Gauss elimination method to compute the solution of the following system of linear equations

$$x_1 + x_2 + 2x_3 = 5$$
$$3x_1 + 2x_2 + x_3 = 8$$
$$x_1 - 2x_2 + 3x_3 = 0$$

Ans. The associated augmented matrix is as follows

$$[A:B] = \begin{bmatrix} 1 & 1 & 2 & 5 \\ 3 & 2 & 1 & 8 \\ 1 & -2 & 3 & 0 \end{bmatrix}$$

We use the first equation to eliminate x_1 from the remaining equations by applying the elementary row operations $R_2 \to R_2 - 3R_1$ and $R_3 \to R_3 - R_1$ on the augmented matrix. We get following row-equivalent matrix

$$[A:B] \sim \begin{bmatrix} 1 & 1 & 2 & 5 \\ 0 & -1 & -5 & -7 \\ 0 & -3 & 1 & -5 \end{bmatrix}$$

Similarly, on applying $R_3 \to R_3 - 3R_2$ to eliminate x_2 from 3rd equation, we obtain an equivalent upper triangular system

$$[A:B] \sim \begin{bmatrix} 1 & 1 & 2 & 5 \\ 0 & -1 & -5 & -7 \\ 0 & 0 & 16 & 16 \end{bmatrix}$$

Equivalently, the system of linear equations in algebraic form is given by

$$x_1 + x_2 + 2x_3 = 5$$
$$-x_2 - 5x_3 = -7$$
$$16x_3 = 16$$

Now, we will make use of back substitutions to find the solution. From the last equation, we get

$$x_3 = 1$$

On using the value of x_3 in the second last equation, we deduce

$$x_2 = 2$$

Similarly, the value of $x_1 = 1$ is obtained from back substitution in the first equation. The final solution is as follows

$$x_1 = 1, x_2 = 2, x_3 = 1$$

Example ─── **5.14**

Use Gauss elimination method to compute the solution of the following system of linear equations

$$x_1 + x_2 + 2x_3 - x_4 = 5$$
$$3x_1 + 2x_2 + x_3 + 4x_4 = 2$$
$$x_1 - 2x_2 + 3x_3 + x_4 = 1$$
$$x_1 + 3x_2 - 2x_3 - 3x_4 = 5$$

Ans. The associated augmented matrix is as follows

$$[A:B] = \begin{bmatrix} 1 & 1 & 2 & -1 & 5 \\ 3 & 2 & 1 & 4 & 2 \\ 1 & -2 & 3 & 1 & 1 \\ 1 & 3 & -2 & -3 & 5 \end{bmatrix} \qquad (5.18)$$

We use the first equation to eliminate x_1 from the remaining equations by applying the elementary row operations $R_2 \to R_2 - 3R_1$, $R_3 \to R_3 - R_1$ and $R_4 \to R_4 - R_1$ on the augmented matrix to get following row-equivalent matrix

$$[A:B] \sim \begin{bmatrix} 1 & 1 & 2 & -1 & 5 \\ 0 & -1 & -5 & 7 & -13 \\ 0 & -3 & 1 & 2 & -4 \\ 0 & 2 & -4 & -2 & 0 \end{bmatrix} \qquad (5.19)$$

Similarly, applying $R_3 \to R_3 - 3R_2$ and $R_4 \to R_4 + 2R_2$ to eliminate x_2 from the third and fourth equations, we obtain

$$[A:B] \sim \begin{bmatrix} 1 & 1 & 2 & -1 & 5 \\ 0 & -1 & -5 & 7 & -13 \\ 0 & 0 & 16 & -19 & 35 \\ 0 & 0 & -14 & 12 & -26 \end{bmatrix} \qquad (5.20)$$

The elementary row operation $R_4 \to R_4 + \dfrac{7}{8} R_3$ reduces the system (5.20) into an equivalent upper triangular system

$$[A:B] \sim \begin{bmatrix} 1 & 1 & 2 & -1 & 5 \\ 0 & -1 & -5 & 7 & -13 \\ 0 & 0 & 16 & -19 & 35 \\ 0 & 0 & 0 & \dfrac{-37}{8} & \dfrac{37}{8} \end{bmatrix} \qquad (5.21)$$

This system is an upper triangular system. Equivalently, the system of linear equations in algebraic form is given by

$$x_1 + x_2 + 2x_3 - x_4 = 5$$
$$-x_2 - 5x_3 + 7x_4 = -13$$
$$16x_3 - 19x_4 = 35$$
$$\frac{-37}{8} x_4 = \frac{37}{8}$$

Now, we will make use of back substitutions to derive the solution. From the last equation, we get

$$x_4 = -1$$

On using the value of x_4 in second last equation, we deduce

$$x_3 = -1$$

Similarly, the values of x_2 and x_1 are obtained from back substitution in the second and the first equation, respectively. The final solution is as follows

$$x_1 = x_2 = x_3 = 1, x_4 = -1$$

Note: It is easy to see that system (5.18) is a 4x4 system, (4 equations in 4 variables). After one step, the system (5.19) has last three equations in three variables (x_2, x_3 and x_4). Similarly, the system (5.20) has last two equations in two variables (x_3 and x_4). In the system (5.21), we simply have last equation in one variable. Means, in Gauss elimination method, we are reducing the order of the system one by one.

Also, it is worth to notice that the upper triangular system obtained in above example is same as obtained in LU decomposition method in Example 5.8. It shows the similarity of LU decomposition method with the Gauss Elimination method.

5.5.1 Operational Counts for Gauss Elimination Method

Direct methods have fixed numbers of arithmetic operations for the solution of a system of linear algebraic equations. The numbers of multiplications and divisions are called as operational counts for the method. We have already discussed the rounding and chopping in arithmetic operations (+, -, *, /), so the number of arithmetic operations is an important criterion in the selection of method. The direct method with a lesser number of arithmetic operations is less error-prone as compared to the method with more number of arithmetic operations.

In general, multiplication/ division requires more time on a computational device as compared to addition/subtraction. Also, the time required for multiplication and division on a computer is approximately equal. So, the operational count (number of multiplications and divisions) is an important criterion for the selection of method. In this section, we will obtain the operational counts (divisions and multiplications) for Gauss elimination method. This problem is dealt in following two phases: first operational counts for reduction to an upper-triangular system, and then operational counts for back substitutions.

First Phase (Conversion to upper triangular form)

In the first phase, we convert the given augmented matrix in the upper triangular matrix by applying elementary row operations.

In the first step, we use the first equation to vanish the coefficients of x_1 from the remaining equations by applying the elementary row operations

$$R_2 \to R_2 - \frac{a_{21}}{a_{11}} R_1, \; R_3 \to R_3 - \frac{a_{31}}{a_{11}} R_1 \cdots R_n \to R_n - \frac{a_{n1}}{a_{11}} R_1 \quad \text{on the augmented matrix.}$$

In second step, we use the second equation to vanish the coefficients of x_2 from the equations below second equation by applying the elementary row operations

$$R_3 \to R_3 - \frac{a_{32}}{a_{22}} R_2, \; R_4 \to R_4 - \frac{a_{42}}{a_{22}} R_2 \cdots R_n \to R_n - \frac{a_{n2}}{a_{22}} R_2 \quad \text{on the resulting augmented matrix}$$

and so on.

Number of Divisions:

First step (division of $a_{21}, a_{31} ..., a_{n1}$ by first pivot element a_{11}): $n-1$

Second step (division of $a_{32}, a_{42} ..., a_{n2}$ by second pivot element a_{22}): $n-2$

$\vdots$

$(n-1)^{\text{th}}$ step (division of $(n-1)^{\text{th}}$ equation by $(n-1)^{\text{th}}$ pivot element $a_{n-1\,n-1}$):   1

Hence, total number of division $= \sum (n-1) = \dfrac{(n-1)n}{2}$

Number of Multiplications:

First step $\left(R_2 \to R_2 - \dfrac{a_{21}}{a_{11}} R_1, \; R_3 \to R_3 - \dfrac{a_{31}}{a_{11}} R_1 \cdots R_n \to R_n - \dfrac{a_{n1}}{a_{11}} R_1 \right)$

Second equation: n

Third equation: n

$\vdots$

n^{th} equation: n

The total number of multiplication in the first step: $n(n-1)$

Similarly in the second step, the number of multiplications: $(n-1)(n-2)$

$\vdots$

Hence, total numbers of multiplications are $\sum n(n-1) = \sum (n^2 - n) = \dfrac{n}{3}(n+1)(n-1)$

Second Phase (Solution from upper triangular system using back substitutions)

In the first phase, we have converted the matrix in the upper triangular matrix by applying elementary row operations. In the second phase, we will compute the solution from this upper triangular system.

Number of Divisions:

From the last equation, we compute $x_n = \dfrac{b_n}{a_{nn}}$. Similarly, in the computation of x_i form i^{th} equation, we have a division with a constant a_{ii}.

$$x_i = \frac{1}{a_{ii}}\left(b_i - \sum_{j=i+1}^{n} a_{ij} x_j\right) \qquad i = n-1, n-2, \ldots, 1$$

Hence, total number of divisions: $\qquad n$

Number of Multiplications:

In the computation of x_i form i^{th} equation, we have a term $\sum_{j=i+1}^{n} a_{ij} x_j$ with the numbers of multiplications is $n-i$. Hence total number of multiplications is given by

$$\sum (n-1) = \frac{(n-1)n}{2}$$

Total number of operational counts in both the phases:

The total number of divisions in both phases:

$$\frac{(n-1)n}{2} + n = \frac{n(n+1)}{2}$$

The total number of multiplications in both phases:

$$\frac{n}{3}(n+1)(n-1) + \frac{(n-1)n}{2} = \frac{n(n-1)(2n+5)}{6}$$

Total number of operational counts:

$$\frac{n(n+1)}{2} + \frac{n(n-1)(2n+5)}{6} = \frac{n(n^2+3n-1)}{3}$$

5.5.1 Thomas Algorithm (Tridiagonal Matrix Algorithm)

Tridiagonal system of linear equations contains non-zero elements only at diagonal, lower diagonal and upper diagonal of the matrix A. We will come across tridiagonal systems in Chapters 15 and 16 during the solutions of the differential equations with boundary conditions. These systems have simple structures, and therefore require less computational efforts. In this section, the Thomas algorithm also known as Tridiagonal Matrix Algorithm (TDMA) will be discussed.

Consider the following tridiagonal system

$$
\begin{bmatrix}
b_1 & c_1 & 0 & 0 & \cdots & 0 & 0 & 0 \\
a_2 & b_2 & c_2 & 0 & \cdots & 0 & 0 & 0 \\
0 & a_3 & b_3 & c_3 & \cdots & 0 & 0 & 0 \\
 & \ddots & \ddots & \ddots & & & & \\
 & & & & \ddots & \ddots & \ddots & \\
0 & 0 & 0 & 0 & \cdots & a_{n-1} & b_{n-1} & c_{n-1} \\
0 & 0 & 0 & 0 & \cdots & 0 & a_n & b_n
\end{bmatrix}
\begin{bmatrix}
x_1 \\ x_2 \\ \vdots \\ \\ x_{n-1} \\ x_n
\end{bmatrix}
=
\begin{bmatrix}
d_1 \\ d_2 \\ \vdots \\ \\ d_{n-1} \\ d_n
\end{bmatrix}
$$

$$(5.22)$$

This system of equations can be written as follows

$$
a_i x_{i-1} + b_i x_i + c_i x_{i+1} = d_i \quad i = 1, 2, \ldots, n \tag{5.23}
$$

where $a_1 = c_n = 0$.

Now, we will apply Thomas algorithm for the solution of system (5.22). It is worthy of mentioning here that the Thomas algorithm is simplified form of Gauss elimination method for tridiagonal systems.

It is easy to see from the system (5.22) that the variable x_i can be expressed in terms of x_{i+1} as follows

$$
x_i = P_i x_{i+1} + Q_i \tag{5.24}
$$

$$
\text{and} \quad x_{i-1} = P_{i-1} x_i + Q_{i-1} \tag{5.25}
$$

On substituting the equation (5.25) in (5.23), we get

$$
a_i \left(P_{i-1} x_i + Q_{i-1} \right) + b_i x_i + c_i x_{i+1} = d_i
$$

$$
x_i = \frac{-c_i}{b_i + a_i P_{i-1}} x_{i+1} + \frac{d_i - a_i Q_{i-1}}{b_i + a_i P_{i-1}} \tag{5.26}
$$

On comparing equations (5.24) and (5.26), we get

$$
P_i = \frac{-c_i}{b_i + a_i P_{i-1}} \qquad Q_i = \frac{d_i - a_i Q_{i-1}}{b_i + a_i P_{i-1}} \tag{5.27}
$$

These recurrence relations can be used to compute the values of the constants P and Q. To start recurrence relation, we require initial values P_0 and Q_0. These values can be computed from equation (5.23) for $i = 1$

$$x_1 = \frac{-c_1}{b_1} x_2 + \frac{d_1}{b_1}$$

On comparing this equation with (5.26) with $i = 1$, we have

$$P_0 = Q_0 = 0 \qquad (5.28)$$

So, we can compute all the constants P and Q using equations (5.28) and (5.27). The system (5.24) implies the upper triangular system of equations. Now, we will compute the values of variables x_i using back substitutions. Using the constant $c_n = 0$ in the equation (5.27), we get $P_n = 0$. Using equation (5.24), we have

$$x_n = Q_n \qquad (5.29)$$

Now, we can use equation (5.24) to compute $x_{n-1}, x_{n-2}, ..., x_1$.

Note that the Thomas algorithm requires $O(n)$ operations as compared to $O(n^3)$ operations of Gauss elimination method.

Example 5.15

Use Thomas algorithm to compute the solution of the following system of linear equations

$$x_1 + x_2 = 1$$
$$3x_1 + 2x_2 + x_3 = 5$$
$$2x_2 + 3x_3 + x_4 = 2$$
$$-2x_3 - 3x_4 = -5$$

Ans. The associated matrix form for the tridiagonal system is as follows

$$
\begin{bmatrix}
1 & 1 & 0 & 0 \\
3 & 2 & 1 & 0 \\
0 & 2 & 3 & 1 \\
0 & 0 & -2 & -3
\end{bmatrix}
\begin{bmatrix}
x_1 \\
x_2 \\
x_3 \\
x_4
\end{bmatrix}
=
\begin{bmatrix}
1 \\
5 \\
2 \\
-5
\end{bmatrix}
$$

On comparing with system (5.22), we have

$$a_2 = 3,\ a_3 = 2,\ a_4 = -2$$
$$b_1 = 1,\ b_2 = 2,\ b_3 = 3,\ b_4 = -3$$
$$c_1 = 1,\ c_2 = 1,\ c_3 = 1$$
$$d_1 = 1,\ d_2 = 5,\ d_3 = 2,\ d_4 = -5$$

We can easily compute the constants P and Q from equations (5.27) with $P_0 = Q_0 = 0$.

$$P_i = \frac{-c_i}{b_i + a_i P_{i-1}} \qquad Q_i = \frac{d_i - a_i Q_{i-1}}{b_i + a_i P_{i-1}}$$

These recurrence relations provide following results

$$
\begin{aligned}
P_1 &= -1 & Q_1 &= 1 \\
P_2 &= 1 & Q_2 &= -2 \\
P_3 &= -0.2 & Q_3 &= 1.2 \\
P_4 &= 0 & Q_4 &= 1
\end{aligned}
$$

On using equation (5.29), we get

$$x_4 = Q_4 = 1$$

Now, the equation (5.24) gives following values

$$
\begin{aligned}
x_3 &= P_3 x_4 + Q_3 = 1 \\
x_2 &= P_2 x_3 + Q_2 = -1 \\
x_1 &= P_1 x_2 + Q_1 = 2
\end{aligned}
$$

The final solution is as follows

$$x_1 = 2,\ x_2 = -1,\ x_3 = 1,\ x_4 = 1$$

Example ── **5.16** ──

Use Thomas algorithm to compute the solution of the following system of linear equations

$$
\begin{aligned}
1.5625 y_2 - 3.125 y_1 &= -1.3671875 \\
1.125 y_3 - 2.125 y_2 + 0.875 y_1 &= 0.03125 \\
-1.125 y_3 + 0.3125 y_2 &= -2.0546875
\end{aligned}
$$

Ans. Comparing with system (5.22), we have

$$
\begin{aligned}
a_2 &= 0.875,\ a_3 = 0.3125 \\
b_1 &= -0.3125,\ b_2 = -2.125,\ b_3 = -1.125 \\
c_1 &= 1.5625,\ c_2 = 1.125 \\
d_1 &= -1.3671875,\ d_2 = 0.03125,\ d_3 = -2.0546875
\end{aligned}
$$

The recurrence relations (5.27) with $P_0 = Q_0 = 0$ provide following results

$$
\begin{aligned}
P_1 &= 0.500000 & Q_1 &= 0.437500 \\
P_2 &= 0.666667 & Q_2 &= 0.208333 \\
P_3 &= 0.000000 & Q_3 &= 2.312500
\end{aligned}
$$

On using equation (5.29) and (5.24), we get

$$y_1 = 1.3125,\ y_2 = 1.75,\ y_3 = 2.3125$$

Note: *The tridiagonal system of Example 5.16 is the resultant system of linear equations arises during solutions of boundary value problem (BVP) in Example 15.10 of Chapter 15. It is worth mentioning here that during solutions of BVPs, we come across tridiagonal systems with large number of equations. Therefore, the Thomas algorithm reduces the computational efforts and hence the rounding error to great extent.*

5.6 Gauss–Jordan Method

In this method, the given system of linear equations (5.1) is converted into an equivalent diagonal system with the aid of elementary row operations. First, we convert the given augmented matrix in the upper triangular matrix by applying elementary row operations as in Gauss elimination method

$$[A:B] \sim \begin{bmatrix} a_{11} & a_{12} & a_{13}\cdots & a_{1n} & : b_1 \\ 0 & a_{22} & a_{23}\cdots & a_{2n} & : b_2 \\ 0 & 0 & a_{33}\cdots & a_{3n} & : b_3 \\ \vdots & & & & \vdots \\ 0 & 0 & 0\cdots & a_{nn} & : b_n \end{bmatrix}$$

Then, we use the last equation to vanish the coefficients of x_n from the remaining equations by applying the elementary row operations. Similarly, the elementary row operations are used in reverse order to eliminate the coefficients of x_{n-1}, x_{n-2} and x_1 to get the equivalent diagonal matrix as follows

$$[A:B] \sim \begin{bmatrix} a_{11} & 0 & 0\cdots & 0 & : b_1 \\ 0 & a_{22} & 0\cdots & 0 & : b_2 \\ 0 & 0 & a_{33}\cdots & 0 & : b_3 \\ \vdots & & & & \vdots \\ 0 & 0 & 0\cdots & a_{nn} & : b_n \end{bmatrix}$$

We can directly solve this system of get the solutions of system 5.1.

Rather, the Gauss elimination and Gauss–Jordan Method are similar, but the number of operational counts is less in Gauss elimination method. Due to this reason, the Gauss elimination method is more popular than Gauss–Jordan and in fact from all other direct methods. But Gauss–Jordan method is useful from other points, for example, we use this method to compute the inverse of a matrix etc.

Example ── 5.17 ─

Use Gauss–Jordan method to compute the solution of the linear system of Example 5.13.

$$x_1 + x_2 + 2x_3 = 5$$
$$3x_1 + 2x_2 + x_3 = 8$$
$$x_1 - 2x_2 + 3x_3 = 0$$

Ans. We have already obtained following upper triangular matrix in Example 5.13

$$\begin{bmatrix} A:B \end{bmatrix} \sim \begin{bmatrix} 1 & 1 & 2 & 5 \\ 0 & -1 & -5 & -7 \\ 0 & 0 & 16 & 16 \end{bmatrix}$$

Till these steps, the Gauss Elimination and Gauss–Jordan Methods are same. In the Gauss–Jordan Method, this system is reduced further to vanish all other non-diagonal entries in the matrix A. We use the third equation to eliminate x_3 from the remaining equations by applying the following elementary row operations

$$R_1 \to R_1 - \frac{1}{8} R_3, \text{ and } R_2 \to R_2 + \frac{5}{16} R_3$$

We get following row-equivalent matrix after these elementary row operations.

$$\begin{bmatrix} A:B \end{bmatrix} \sim \begin{bmatrix} 1 & 1 & 0 & 3 \\ 0 & -1 & 0 & -2 \\ 0 & 0 & 16 & 16 \end{bmatrix}$$

On applying $R_1 \to R_1 + R_2$ to eliminate x_2 from the first equation, we get

$$\begin{bmatrix} A:B \end{bmatrix} \sim \begin{bmatrix} 1 & 0 & 0 & 1 \\ 0 & -1 & 0 & -2 \\ 0 & 0 & 16 & 16 \end{bmatrix}$$

We can easily solve these equations to get the following result.

$$x_1 = \frac{1}{1} = 1, \ x_2 = \frac{-2}{-1} = 2, \ x_3 = \frac{16}{16} = 1$$

Example ── 5.18 ─

Use Gauss–Jordan Method to compute the solution of following system of linear equations

$$x_1 - 3x_2 + 2x_3 + x_4 = 5$$
$$2x_1 - 5x_2 + 3x_3 + 2x_4 = 9$$
$$3x_1 + x_2 + 2x_3 + 4x_4 = 6$$
$$x_1 + 3x_2 - 2x_3 + 3x_4 = 1$$

Ans. The associated augmented matrix is given by

$$[A:B] = \begin{bmatrix} 1 & -3 & 2 & 1 & 5 \\ 2 & -5 & 3 & 2 & 9 \\ 3 & 1 & 2 & 4 & 6 \\ 1 & 3 & -2 & 3 & 1 \end{bmatrix}$$

On applying the elementary row operations $R_2 \to R_2 - 2R_1$, $R_3 \to R_3 - 3R_1$ and $R_4 \to R_4 - R_1$ on the augmented matrix, we get

$$[A:B] = \begin{bmatrix} 1 & -3 & 2 & 1 & 5 \\ 0 & 1 & -1 & 0 & -1 \\ 0 & 10 & -4 & 1 & -9 \\ 0 & 6 & -4 & 2 & -4 \end{bmatrix}$$

We implemented $R_3 \to R_3 - 10R_2$, $R_4 \to R_4 - 6R_2$ to obtain the following reduced system

$$[A:B] = \begin{bmatrix} 1 & -3 & 2 & 1 & 5 \\ 0 & 1 & -1 & 0 & -1 \\ 0 & 0 & 6 & 1 & 1 \\ 0 & 0 & 2 & 2 & 2 \end{bmatrix}$$

The elementary row operation $R_4 \to R_4 - \dfrac{1}{3}R_3$ reduces the system into following equivalent upper triangular system

$$[A:B] = \begin{bmatrix} 1 & -3 & 2 & 1 & 5 \\ 0 & 1 & -1 & 0 & -1 \\ 0 & 0 & 6 & 1 & 1 \\ 0 & 0 & 0 & \dfrac{5}{3} & \dfrac{5}{3} \end{bmatrix}$$

This matrix is an upper triangular matrix. Now, this system is reduced further to vanish all other non-diagonal entries in the matrix A. We use the fourth equation to eliminate x_4 from the remaining equations by applying the following elementary row operations

$$R_1 \to R_1 - \frac{3}{5}R_4, \; R_2 \to R_2 - 0R_4, \; \text{and} \; R_3 \to R_3 - \frac{3}{5}R_4$$

We get the following row-equivalent matrix after these elementary row operations.

$$[A:B] = \begin{bmatrix} 1 & -3 & 2 & 0 & 4 \\ 0 & 1 & -1 & 0 & -1 \\ 0 & 0 & 6 & 0 & 0 \\ 0 & 0 & 0 & \dfrac{5}{3} & \dfrac{5}{3} \end{bmatrix}$$

By applying the elementary row operations $R_1 \to R_1 - \dfrac{1}{3}R_3$ and $R_2 \to R_2 + \dfrac{1}{6}R_3$, we will

eliminate x_3 from the remaining first and second equations as follows

$$[A:B] = \begin{bmatrix} 1 & -3 & 0 & 0 & 4 \\ 0 & 1 & 0 & 0 & -1 \\ 0 & 0 & 6 & 0 & 0 \\ 0 & 0 & 0 & \dfrac{5}{3} & \dfrac{5}{3} \end{bmatrix}$$

On applying $R_1 \to R_1 + 3R_2$ to eliminate x_2 from the first equation, we get

$$[A:B] = \begin{bmatrix} 1 & 0 & 0 & 0 & 1 \\ 0 & 1 & 0 & 0 & -1 \\ 0 & 0 & 6 & 0 & 0 \\ 0 & 0 & 0 & \dfrac{5}{3} & \dfrac{5}{3} \end{bmatrix}$$

The solution from this reduced system is as follows

$$x_1 = 1, \; x_2 = -1, \; x_3 = 0, \; x_4 = 1$$

5.7 Comparison of Direct Methods

We have discussed various direct methods for the solution of a system of linear equations (5.1) without considering round-off error. For each arithmetic operation, there is a chance for loss of significant digits. Therefore, these methods may produce erroneous results beyond desired accuracy. In the case of small systems (with lesser number of equations, say systems of three or four equations), few digits higher arithmetic (than required number of accurate digits) can produce good results. But in the case of large systems, say system with 20 equations, we have very large numbers of operational counts and loss of significant digits may produce an erroneous result.

In direct methods, once the error is committed, it cannot be recovered by iterations, etc. Therefore, the number of operational counts is a major tool for comparison of direct methods. We have presented the operational counts for Gauss elimination method only. Table 5.1 gives the operational counts for large n.

Table 5.1

Sr. No.	Method	No. of operational counts for large n
1	Gauss Elimination	$\approx n^3 / 3$
2	Gauss Jordan	$\approx n^3 / 2$
3	Matrix Inversion	$\approx n^3$
4	Cramer Rule	$\approx n^3$
5	LU-Decomposition (Crout and Doolittle methods)	$\approx n^3 / 3$
6	Cholesky (only positive definite matrix)	$\approx n^3 / 6$

The number of operational counts is minimum in the Gauss elimination and LU decomposition methods. In Gauss Elimination method, pivoting can be used to reduce the error. Therefore, the Gauss elimination method produces better results as compared to other direct methods for large systems.

5.8 Pivoting Strategies for Gauss Elimination Method

So far, we have applied direct methods without discussing the round-off error caused due to finite digits floating point arithmetic. The round-off error mainly depends on the total number of operational counts, and this number is minimum in case of Gauss elimination method. Therefore, at this moment, we are presenting Gauss elimination method for explanations. In this section, we will discuss partial, scaled and complete pivoting strategies to reduce the round-off error.

Partial Pivoting: At each stage (i^{th}), we make the pivot element (a_{ii}) largest in magnitude by the elements below pivot element in that column, by simple row conversion. It is to ensure that magnitudes of the multipliers are less than 1 for each row. That is, at i^{th} stage, we simply want

$$|a_{ii}| \geq |a_{ji}|; \qquad j = i+1, i+2, ..., n \tag{5.30}$$

Example ———————————————————————————— 5.19

Compute the solution of the following system of linear equations with the help of Gauss elimination method without pivoting and with partial pivoting. Use only seven significant digits rounding arithmetic for computations.

$$5.42x_1 + 16.78x_2 + 0.78x_3 = 38.0014$$
$$0.423x_1 + 2.3x_2 + 23.46x_3 = 53.97833$$
$$26.73x_1 + 1.274x_2 + 2.45x_3 = 68.74938$$

Compare the results with the exact solution, $x_1 = 2.310000$, $x_2 = 1.420000$, $x_3 = 2.120000$.

Ans.

Without Pivoting

Proceeding in a similar manner as in Example 5.13, we eliminate x_1 from the second and third equations with the help of the first equation by using seven significant digits arithmetic.

$$5.42x_1 + 16.78x_2 + 0.78x_3 = 38.0014$$
$$0.990417x_2 + 23.39912x_3 = 51.01254$$
$$-81.4805x_2 - 1.396753x_3 = -118.6634$$

On eliminating x_2 from the third equation, we get

$$5.42x_1 + 16.78x_2 + 0.78x_3 = 38.0014$$
$$0.990417x_2 + 23.39912x_3 = 51.01254$$
$$1923.622x_3 = 4078.082$$

The solution is obtained using back substitutions, and it is given by

$$x_3 = 2.120002$$
$$x_2 = 1.419968$$
$$x_1 = 2.3101$$

On comparison with exact result, one can easily see that the error is at seventh, sixth and fifth significant digits in the values of variables x_3, x_2 and x_1, respectively. So, we can use pivoting to reduce the error. It is only for a system of order three, and we are using 7-digits floating points arithmetic. The error increases significantly in the case of a large system.

With Partial Pivoting

In the first column [5.44, 0.423, 26.73] of the given system, the largest element in magnitude is 26.73. Therefore, we change the position of the last row to the first row, to make pivot element (a_{11}) largest.

$$26.73x_1 + 1.274x_2 + 2.45x_3 = 68.74938$$
$$5.42x_1 + 16.78x_2 + 0.78x_3 = 38.0014$$
$$0.423x_1 + 2.3x_2 + 23.46x_3 = 53.97833$$

On applying elementary row operations to eliminate x_1 from second and third rows, we obtain

$$26.73x_1 + 1.274x_2 + 2.45x_3 = 68.74938$$
$$16.52167x_2 + 0.2832174x_3 = 24.0612$$
$$2.279839x_2 + 23.42123x_3 = 52.89038$$

In pivot column [16.52167, 2.279839], the largest element in magnitude is 16.52167. There is no need to interchange rows. Now, eliminating x_2 from the third row, we get the following upper triangular system.

$$26.73x_1 + 1.274x_2 + 2.45x_3 = 68.74938$$
$$16.52167x_2 + 0.2832174x_3 = 24.0612$$
$$23.38215x_3 = 49.57016$$

We can easily obtain the following solution by using back substitutions.

$$x_3 = 2.120000$$
$$x_2 = 1.420000$$
$$x_1 = 2.310000$$

The result is correct up to seven significant digits.

Note: It is not always possible to get the correct result with partial pivoting, especially if the variation in the magnitude of elements of matrix A is large. In that case, we can use scaled partial pivoting and complete pivoting strategies. We will first discuss these two pivoting strategies, and then take examples to show the effectiveness of these strategies.

Scaled Partial Pivoting: In this pivoting strategy, the pivot element is scaled largest element in magnitude in its row.

Consider the largest elements in magnitude for each row in the matrix A are as follows

$$S_i = \max_{1 \le j \le n} |a_{ij}| \quad 1 \le i \le n$$

A vector $S = \begin{bmatrix} S_1, S_2, ..., S_n \end{bmatrix}$ is defined using these values. We define following scaled pivot vector for first pivot element.

$$R = \left[\left| \frac{a_{11}}{S_1} \right|, \left| \frac{a_{21}}{S_2} \right|, ..., \left| \frac{a_{n1}}{S_n} \right| \right] \qquad (5.31)$$

Let the largest scaled pivot element be $\left|\dfrac{a_{i1}}{S_i}\right|$, then i^{th} row is the pivot row. We repeat the process for all pivot elements with the vector S.

Complete Pivoting: At each stage (i^{th}), we select the pivot element that is absolutely largest in the pivot row and rows below pivot element in the matrix A by using row as well as column conversion. Mathematically, we have

$$|a_{ii}| \geq |a_{jk}|; \qquad j,k=i,\ i+1,\ i+2,...,n \tag{5.32}$$

Example ── **5.20** ──

Solve the following system of linear equations by Gauss elimination method with partial, scaled partial and complete pivoting using three significant digits floating points rounding arithmetic.

$$x_1 + x_2 + 2x_3 = 1$$
$$2x_1 - x_2 + 100x_3 = 53$$
$$3x_1 + x_2 + 200x_3 = 102$$

Compare the obtained results for each case with the exact solution, $x_1 = 1$, $x_2 = -1$, $x_3 = 0.5$.

Ans.

The augmented matrix for the given system of linear equations is as follows

$$[A:B] = \begin{bmatrix} 1 & 1 & 2 & 1 \\ 2 & -1 & 100 & 53 \\ 3 & 1 & 200 & 102 \end{bmatrix}$$

Now, we will use Gauss Elimination method to obtain the solution of the given system with all three pivoting strategies as follows.

With Partial Pivoting

Since the element 3 is the largest in the first column of augmented matrix $[A:B]$, hence interchanging the last and first rows $R_1 \leftrightarrow R_3$, we have

$$\begin{bmatrix} 3 & 1 & 200 & 102 \\ 2 & -1 & 100 & 53 \\ 1 & 1 & 2 & 1 \end{bmatrix}$$

On applying the elementary row operations $R_2 \to R_2 - \left(\frac{2}{3} = .667\right) R_1$ and

$R_3 \to R_3 - \left(\frac{1}{3} = .333\right) R_1$ with three significant digits rounding arithmetic, we have

$$\begin{bmatrix} 3 & 1 & 200 & 102 \\ 0 & -1.67 & -33.0 & -15.0 \\ 0 & 0.667 & -64.6 & -33.0 \end{bmatrix}$$

The elementary row operation $R_3 \to R_3 + \left(\dfrac{0.667}{1.67} = .399\right) R_2$ produces the following upper triangular matrix.

$$\begin{bmatrix} 3 & 1 & 200 & 102 \\ 0 & -1.67 & -33.0 & -15.0 \\ 0 & 0 & -77.8 & -39.0 \end{bmatrix}$$

On using back substitutions, we get

$$x_3 = 0.501$$
$$x_2 = -0.898$$
$$x_1 = 0.967$$

The result obtained is highly erroneous. The value of the variable x_1 is correct up to two significant digits; the variable x_3 is correct up to one significant digit; while the variable x_2 is incorrect even for the first significant digit.

With Scaled Partial Pivoting

In the augmented matrix, the attached largest elements for each row in the matrix A are as follows

$$[A:B] = \begin{bmatrix} 1 & 1 & 2 & 1 \\ 2 & -1 & 100 & 53 \\ 3 & 1 & 200 & 102 \end{bmatrix} \begin{matrix} \to 2 \\ \to 100 \\ \to 200 \end{matrix}$$

The vector $S = \begin{bmatrix} S_1, S_2, S_3 \end{bmatrix}$ of the largest elements in magnitude for each row of the matrix A is as follows

$$S = \begin{bmatrix} S_1, S_2, S_3 \end{bmatrix} = [2, 100, 200]$$

The scaled pivot vector is given by

$$R = \left[\frac{|a_{11}|}{S_1}, \frac{|a_{21}|}{S_2}, \frac{|a_{31}|}{S_3} \right] = \left[\frac{1}{2}, \frac{2}{100}, \frac{3}{200} \right]$$

Since the first element in this vector is the largest in magnitude, therefore the first row is the pivot row, and interchange of rows is not required. Apply $R_2 \rightarrow R_2 - (2)R_1$ and $R_3 \rightarrow R_3 - (3)R_1$, by using three significant digits rounding arithmetic to get the following system.

$$[A:B] = \begin{bmatrix} 1 & 1 & 2 & 1 \\ 0 & -3.00 & 96.0 & 51.0 \\ 0 & -2.00 & 194 & 99.0 \end{bmatrix} \begin{matrix} \rightarrow 2 \\ \rightarrow 100 \\ \rightarrow 200 \end{matrix}$$

On computing scaled pivot vector, we have

$$R = \left[\frac{|a_{22}|}{S_2}, \frac{|a_{32}|}{S_3} \right] = \left[\frac{3}{100}, \frac{2}{200} \right]$$

Again, there is no need of interchanging the rows. Applying, $R_3 \rightarrow R_3 - \left(\frac{2}{3} = .667 \right) R_2$, we have

$$[A:B] = \begin{bmatrix} 1 & 1 & 2 & 1 \\ 0 & -3.00 & 96.0 & 51.0 \\ 0 & 0 & 130 & 65.0 \end{bmatrix}$$

On solving this system, we obtained the following solution by using three significant digits arithmetic.

$$x_3 = 0.500$$
$$x_2 = -1.00$$
$$x_1 = 1.00$$

The result obtained is the exact result.

With Complete Pivoting

The variables x_1, x_2, x_3 are attached to the first, second and third columns of the augmented matrix, respectively.

$$\begin{matrix} & x_1 & x_2 & x_3 & \\ [A:B] = & \begin{bmatrix} 1 & 1 & 2 & 1 \\ 2 & -1 & 100 & 53 \\ 3 & 1 & 200 & 102 \end{bmatrix} \end{matrix}$$

The absolutely largest element in the matrix A is 200. So, this element must be the pivot element. Interchanging the first and last rows, and then the first and third columns ($R_1 \leftrightarrow R_3$ and $C_1 \leftrightarrow C_3$), we have

$$= \begin{bmatrix} \overset{x_3}{200} & \overset{x_2}{1} & \overset{x_1}{3} & 102 \\ 100 & -1 & 2 & 53 \\ 2 & 1 & 1 & 1 \end{bmatrix}$$

On applying the row operations $R_2 \rightarrow R_2 - (.5)R_1$ and $R_3 \rightarrow R_3 - (.01)R_1$, we get

$$= \begin{bmatrix} \overset{x_3}{200} & \overset{x_2}{1} & \overset{x_1}{3} & 102 \\ 0 & -1.5 & .5 & 2 \\ 0 & .99 & .97 & -.02 \end{bmatrix}$$

Since the element -1.5 is absolutely largest element in second and third rows of matrix A, so there is no need to interchange the rows. Applying $R_3 \rightarrow R_3 + (.66)R_1$, we obtained the following upper triangular matrix

$$= \begin{bmatrix} \overset{x_3}{200} & \overset{x_2}{1} & \overset{x_1}{3} & 102 \\ 0 & -1.5 & .5 & 2 \\ 0 & 0 & 1.30 & 1.30 \end{bmatrix}$$

Solving this system of equations, we have

$$x_1 = 1.00$$
$$x_2 = -1.00$$
$$x_3 = 0.500$$

The result obtained is the exact result.

Note: *In this example, scaled and complete pivoting strategies are providing exact results, while the partial pivoting is producing erroneous results. In general, complete pivoting is better than partial and scaled pivoting, and it becomes clear from the following example. In case of complete pivoting, we are taking care of complete matrix A for largest element and hence the rounding error is minimized.*

Example   5.21

Solve the following system of linear equations by Gauss elimination method. Use partial, scaled partial and complete pivoting strategies with three significant digits floating points rounding arithmetic.

$$3x_1 + x_2 + 200x_3 = 102$$
$$2x_1 - x_2 + 100x_3 = 53$$
$$x_1 + 5x_2 + 2x_3 = -3$$

Compare the obtained results with exact solutions, $x_1 = 1$, $x_2 = -1$, $x_3 = 0.5$.

Ans.

The augmented matrix corresponding to the given system is given by

$$[A:B] = \begin{bmatrix} 3 & 1 & 200 & 102 \\ 2 & -1 & 100 & 53 \\ 1 & 5 & 2 & -3 \end{bmatrix}$$

The results obtained with various pivoting strategies are as follows.

With Partial Pivoting

The absolutely largest entry in the first column is 3, so no interchange of rows is required.

Applying $R_2 \rightarrow R_2 - \left(\dfrac{2}{3} = .667 \right) R_1$ and $R_3 \rightarrow R_3 - \left(\dfrac{1}{3} = .333 \right) R_1$ with three significant

digits rounding arithmetic, we get

$$\begin{bmatrix} 3 & 1 & 200 & 102 \\ 0 & -1.67 & -33.0 & -15.0 \\ 0 & 4.67 & -64.6 & -37.0 \end{bmatrix}$$

Now, the element 4.67 is largest of magnitude, therefore on interchanging the second and third rows, we have

$$\begin{bmatrix} 3 & 1 & 200 & 102 \\ 0 & 4.67 & -64.6 & -37.0 \\ 0 & -1.67 & -33.0 & -15.0 \end{bmatrix}$$

Applying $R_3 \rightarrow R_3 + \left(\dfrac{1.67}{4.67} = .358 \right) R_2$, we have

$$\begin{bmatrix} 3 & 1 & 200 & 102 \\ 0 & 4.67 & -64.6 & -37.0 \\ 0 & 0 & -56.1 & -28.2 \end{bmatrix}$$

The solution is given by

$$x_3 = 0.503$$
$$x_2 = -0.964$$
$$x_1 = 0.833$$

With Scaled Partial Pivoting

Consider the augmented matrix with attached absolutely largest elements for each rows.

$$[A:B] = \begin{bmatrix} 3 & 1 & 200 & 102 \\ 2 & -1 & 100 & 53 \\ 1 & 5 & 2 & -3 \end{bmatrix} \begin{matrix} \to 200 \\ \to 100 \\ \to 5 \end{matrix}$$

The scaled pivot vector $S = \begin{bmatrix} S_1, S_2, S_3 \end{bmatrix}$ is as follows

$$S = \begin{bmatrix} S_1, S_2, S_3 \end{bmatrix} = [200, 100, 5]$$

We define a scaled pivot vector as follows

$$R = \left[\left| \frac{a_{11}}{S_1} \right|, \left| \frac{a_{21}}{S_2} \right|, \left| \frac{a_{31}}{S_3} \right| \right] = \left[\frac{3}{200}, \frac{2}{100}, \frac{1}{5} \right]$$

Since the third element in this vector is largest, so after interchanging the first and third rows, we have

$$[A:B] = \begin{bmatrix} 1 & 5 & 2 & -3 \\ 2 & -1 & 100 & 53 \\ 3 & 1 & 200 & 102 \end{bmatrix} \begin{matrix} \to 5 \\ \to 100 \\ \to 200 \end{matrix}$$

Applying $R_2 \to R_2 - (2)R_1$ and $R_3 \to R_3 - (3)R_1$, we have

$$[A:B] = \begin{bmatrix} 1 & 5 & 2 & -3 \\ 0 & -11 & 96 & 59 \\ 0 & -14 & 194 & 111 \end{bmatrix} \begin{matrix} \to 5 \\ \to 100 \\ \to 200 \end{matrix}$$

The scaled pivot vector is given by

$$R = \left[\left| \frac{a_{22}}{S_2} \right|, \left| \frac{a_{32}}{S_3} \right| \right] = \left[\frac{11}{100}, \frac{14}{200} \right]$$

So, interchange of rows is not required. Applying $R_3 \to R_3 - \left(\dfrac{14}{11} = 1.27 \right) R_2$, we have

$$[A:B] = \begin{bmatrix} 1 & 5 & 2 & -3 \\ 0 & -11 & 96 & 59 \\ 0 & 0 & 72 & 36.1 \end{bmatrix} \begin{matrix} \rightarrow 5 \\ \rightarrow 100 \\ \rightarrow 200 \end{matrix}$$

On solving this system, we get

$$x_3 = 0.501$$
$$x_2 = -0.991$$
$$x_1 = 0.960$$

With Complete Pivoting

The columns of augmented matrix are attached with the variables x_1, x_2 and x_3 as follows

$$\begin{matrix} x_1 & x_2 & x_3 \end{matrix}$$
$$[A:B] = \begin{bmatrix} 3 & 1 & 200 & 102 \\ 2 & -1 & 100 & 53 \\ 1 & 5 & 2 & -3 \end{bmatrix}$$

On interchanging the first and third columns, $C_1 \leftrightarrow C_3$, we have

$$\begin{matrix} x_3 & x_2 & x_1 \end{matrix}$$
$$[A:B] = \begin{bmatrix} 200 & 1 & 3 & 102 \\ 100 & -1 & 2 & 53 \\ 2 & 5 & 1 & -3 \end{bmatrix}$$

The elementary row operations $R_2 \rightarrow R_2 - (.5)R_1$ and $R_3 \rightarrow R_3 - (.01)R_1$ produce the following equivalent system

$$\begin{matrix} x_3 & x_2 & x_1 \end{matrix}$$
$$= \begin{bmatrix} 200 & 1 & 3 & 102 \\ 0 & -1.5 & .5 & 2 \\ 0 & 4.99 & .97 & -4.02 \end{bmatrix}$$

Since the element 4.99 is largest in second and third rows of matrix A, so we have to interchange second and third rows.

$$\begin{matrix} x_3 & x_2 & x_1 \end{matrix}$$
$$= \begin{bmatrix} 200 & 1 & 3 & 102 \\ 0 & 4.99 & .97 & -4.02 \\ 0 & -1.5 & .5 & 2 \end{bmatrix}$$

On applying $R_3 \rightarrow R_3 + (.301)R_1$, we get

$$
\begin{array}{ccc}
x_3 & x_2 & x_1
\end{array}
$$

$$
= \begin{bmatrix}
200 & 1 & 3 & 102 \\
0 & 4.99 & .97 & -4.02 \\
0 & 0 & .792 & .790
\end{bmatrix}.
$$

The solution of this system of equations is given by

$$x_1 = .997$$
$$x_2 = -1.00$$
$$x_3 = 0.500$$

Note: The complete pivoting is the best strategy for the direct methods, but its programming is very difficult as it involves both rows and columns interchange. The columns interchange also involves the changes of attached variables, which is difficult to manage in programming.

5.9 Iterative Methods

In direct methods, as a reference example of Gauss elimination method, we have discussed the round-off error and some remedies to reduce it. Certain scientific problems produce millions of equations, and large numbers of arithmetic computations are involved. The direct methods are easy to implement, but round-off error is significant in case of large systems. The iterative procedures can be used for solution of such systems. The iterative methods may require large numbers of iterations to produce the result with higher accuracy. But, once the algorithms for these methods are implemented, these iterations can be easily computed with the advent of high-speed computers.

For more accuracy and lesser computational work, direct and iterative methods can be mixed up. First, we can apply the direct method to compute the solution and then further improve this solution for more accuracy with an iterative procedure.

However, the iterative procedures are not always converging to the solutions and rate of convergence is second important criteria in the applications of these methods. In Section 5.13, we will discuss the convergence criteria and see that these methods are not convergent for any arbitrary linear system of equations. The coming sections deal with the following three iterative methods.

1. Jacobi Method (or) Method of Simultaneous Displacement
2. Gauss–Seidel Method (or) Liebmann Method (or) Method of Successive Displacement
3. Successive Over Relaxation Method

5.10 Jacobi Method (or) Method of Simultaneous Displacement

The linear system of equations (5.1) can be rewritten as follows

$$x_1 = \frac{1}{a_{11}}\left[b_1 - (a_{12}x_2 + a_{13}x_3 + \cdots + a_{1n}x_n)\right]$$

$$x_2 = \frac{1}{a_{22}}\left[b_2 - (a_{21}x_1 + a_{23}x_3 + \cdots + a_{2n}x_n)\right]$$

$$\vdots$$

$$x_n = \frac{1}{a_{nn}}\left[b_n - (a_{n1}x_1 + a_{n2}x_2 + \cdots + a_{n\,n-1}x_{n-1})\right]$$

Some initial approximation is required to compute the vector, $\left[x_1, x_2, ..., x_n\right]$, and let that approximation be $\left[x_1^{(0)}, x_2^{(0)}, ..., x_n^{(0)}\right]$. We use these values in the above expressions to get the next approximation $\left[x_1^{(1)}, x_2^{(1)}, ..., x_n^{(1)}\right]$ of the Jacobi method.

$$x_1^{(1)} = \frac{1}{a_{11}}\left[b_1 - (a_{12}x_2^{(0)} + a_{13}x_3^{(0)} + \cdots + a_{1n}x_n^{(0)})\right]$$

$$x_2^{(1)} = \frac{1}{a_{22}}\left[b_2 - (a_{21}x_1^{(0)} + a_{23}x_3^{(0)} + \cdots + a_{2n}x_n^{(0)})\right]$$

$$\vdots$$

$$x_n^{(1)} = \frac{1}{a_{nn}}\left[b_n - (a_{n1}x_1^{(0)} + a_{n2}x_2^{(0)} + \cdots + a_{n\,n-1}x_{n-1}^{(0)})\right] \tag{5.33}$$

The subscripts and superscripts denote variables and iterations, respectively. For example, x_3^1 denotes the first iteration of the variable x_3.

Similarly, the first approximation $\left[x_1^{(1)}, x_2^{(1)}, ..., x_n^{(1)}\right]$ is used to compute the second iteration of Jacobi method. The process is repeated till the desired accuracy is obtained. The $(k+1)$th iteration can be obtained from kth iteration by the following Jacobi iteration formula

$$x_1^{(k+1)} = \frac{1}{a_{11}}\left[b_1 - (a_{12}x_2^{(k)} + a_{13}x_3^{(k)} + \cdots + a_{1n}x_n^{(k)})\right]$$

$$x_2^{(k+1)} = \frac{1}{a_{22}}\left[b_2 - (a_{21}x_1^{(k)} + a_{23}x_3^{(k)} + \cdots + a_{2n}x_n^{(k)})\right]$$

$$\vdots$$

$$x_n^{(k+1)} = \frac{1}{a_{nn}}\left[b_n - (a_{n1}x_1^{(k)} + a_{n2}x_2^{(k)} + \cdots + a_{n\,n-1}x_{n-1}^{(k)})\right] \qquad k = 0,1,2,... \tag{5.34}$$

The above Jacobi iteration formula (5.34) can be written as follows

$$x_i^{(k+1)} = \frac{1}{a_{ii}}\left(b_i - \sum_{\substack{j=1 \\ j \neq i}}^{n} a_{ij}x_j^{(k)}\right); \qquad 1 \leq i \leq n \quad k = 0,1,2,... \tag{5.35}$$

Example 5.22

Solve the following system of linear equations using Jacobi iterative procedure

$$3x_1 + x_2 - x_3 = 3$$
$$2x_1 + 4x_2 + x_3 = 7$$
$$x_1 - x_2 + 4x_3 = 4$$

Consider the initial approximation, $x_1^{(0)} = 0$, $x_2^{(0)} = 0$, $x_3^{(0)} = 0$. Apply Jacobi method till the last two consecutive iterations have difference less than 0.0005.

Ans. To solve the system, first we rewrite the system as follows

$$x_1 = \frac{1}{3}(3 - x_2 + x_3)$$

$$x_2 = \frac{1}{4}(7 - 2x_1 - x_3)$$

$$x_3 = \frac{1}{4}(4 - x_1 + x_2)$$

On using the initial approximation, $x_1^{(0)} = 0$, $x_2^{(0)} = 0$, $x_3^{(0)} = 0$, we get the first approximation of Jacobi method (5.33) as follows

$$x_1^{(1)} = \frac{1}{3}(3 - x_2^{(0)} + x_3^{(0)}) = 1$$

$$x_2^{(1)} = \frac{1}{4}(7 - 2x_1^{(0)} - x_3^{(0)}) = 1.75$$

$$x_3^{(1)} = \frac{1}{4}(4 - x_1^{(0)} + x_2^{(0)}) = 1$$

The first approximation $x_1^{(1)} = 1$, $x_2^{(1)} = 1.75$, $x_3^{(1)} = 1$ provides the next approximation as follows

$$x_1^{(2)} = \frac{1}{3}(3 - x_2^{(1)} + x_3^{(1)}) = 0.75$$

$$x_2^{(2)} = \frac{1}{4}(7 - 2x_1^{(1)} - x_3^{(1)}) = 1$$

$$x_3^{(2)} = \frac{1}{4}(4 - x_1^{(1)} + x_2^{(1)}) = 1.1875$$

Similarly, we can compute other iterations. The values of the variables x_1, x_2 and x_3 at the first ten iterations are as follows

Iteration 1
1.0000 1.7500 1.0000

Iteration 2
0.7500 1.0000 1.1875

Iteration 3
1.0625 1.0781 1.0625

Iteration 4
0.9948 0.9531 1.0039

Iteration 5
1.0169 1.0016 0.9896

Iteration 6
0.9960 0.9941 0.9962

Iteration 7
1.0007 1.0030 0.9995

Iteration 8
0.9989 0.9998 1.0006

Iteration 9
1.0003 1.0004 1.0002

Iteration 10
0.9999 0.9998 1.0000

The differences in the values of x_i's at the ninth and tenth iterations are less than 0.0005. So, after ten iterations, the solution is $x_1 = 0.9999$, $x_2 = 0.9998$, $x_3 = 1.0000$.

The exact solution is $x_1 = 1$, $x_2 = 1$, $x_3 = 1$.

Note: We can continue with any initial approximation, for example, $x_1^{(0)} = 5$, $x_2^{(0)} = -6$, $x_3^{(0)} = 8$. But if our initial approximation is far away from the solution, then we need to do more number of iterations for the desired accuracy. In the absence of any better approximation, it is advisable to continue with zero initial approximation.

Example   **5.23**

Compute seven iterations of the Jacobi method for the following system of linear equations

$$7x_1 - 3x_2 + 2x_3 + x_4 = 12$$
$$2x_1 - 6x_2 + x_3 + 2x_4 = 6$$
$$x_1 + x_2 + 5x_3 + 2x_4 = 12$$
$$x_1 + 3x_2 - 2x_3 + 8x_4 = 5$$

Consider the initial approximation, $x_1^{(0)} = 0$, $x_2^{(0)} = 0$, $x_3^{(0)} = 0$, $x_4^{(0)} = 0$.

Ans. Rewrite the given system as follows

$$x_1 = \frac{1}{7}\left(12 + 3x_2 - 2x_3 - x_4\right)$$

$$x_2 = \frac{-1}{6}\left(6 - 2x_1 - x_3 - 2x_4\right)$$

$$x_3 = \frac{1}{5}\left(12 - x_1 - x_2 - 2x_4\right)$$

$$x_4 = \frac{1}{8}\left(5 - x_1 - 3x_2 + 2x_3\right)$$

Using the initial approximation, $x_1^{(0)} = 0$, $x_2^{(0)} = 0$, $x_3^{(0)} = 0$, $x_4^{(0)} = 0$, we get the first approximation of Jacobi method (5.33) as follows

$$x_1^{(1)} = \frac{1}{7}\left(12 + 3x_2^{(0)} - 2x_3^{(0)} - x_4^{(0)}\right) = \frac{12}{7} = 1.714286$$

$$x_2^{(1)} = \frac{-1}{6}\left(6 - 2x_1^{(0)} - x_3^{(0)} - 2x_4^{(0)}\right) = -1.000000$$

$$x_3^{(1)} = \frac{1}{5}\left(12 - x_1^{(0)} - x_2^{(0)} - 2x_4^{(0)}\right) = 2.400000$$

$$x_4^{(1)} = \frac{1}{8}\left(5 - x_1^{(0)} - 3x_2^{(0)} + 2x_3^{(0)}\right) = 0.625000$$

The next approximation can be computed as follows

$$x_1^{(2)} = \frac{1}{7}\left(12 + 3x_2^{(1)} - 2x_3^{(1)} - x_4^{(1)}\right) = 0.510714$$

$$x_2^{(2)} = \frac{-1}{6}\left(6 - 2x_1^{(1)} - x_3^{(1)} - 2x_4^{(1)}\right) = 0.179762$$

$$x_3^{(2)} = \frac{1}{5}\left(12 - x_1^{(1)} - x_2^{(1)} - 2x_4^{(1)}\right) = 2.007143$$

$$x_4^{(2)} = \frac{1}{8}\left(5 - x_1^{(1)} - 3x_2^{(1)} + 2x_3^{(1)}\right) = 1.385714$$

Similarly, we can compute following iterations

Iteration 3
    1.019898   −0.033333   1.707619   0.995536

Iteration 4
    1.069889   −0.043586   1.804473   0.936918

Iteration 5

        1.046197   −0.030319   1.819972   0.958727

Iteration 6

        1.044339   −0.028363   1.813334   0.960588

Iteration 7

        1.046808   −0.029469   1.812570   0.958427

Note that the exact solution is, $x_1 = 1$, $x_2 = 0$, $x_3 = 2$, $x_4 = 1$.

It is worth to note that we are using values at last approximation to compute the next approximation. For example, we are using $x_1^{(0)}$ in computation of $x_2^{(1)}$. But, we have already computed $x_1^{(1)}$, and it can be used in computation of $x_2^{(1)}$. Since the approximation $x_1^{(1)}$ is better as compared to $x_1^{(0)}$, hence it will produce better results. Similarly, we can use other latest available approximations (if available) instead of old approximations. We will discuss Gauss–Seidel method in next section. The latest available values of variables are used in this method.

5.11 Gauss–Seidel Method (or) Method of Successive Displacement (or) Liebmann Method

On a similar pattern as in Jacobi method, consider the system of equations (5.1) in the following form

$$x_1 = \frac{1}{a_{11}}\left[b_1 - (a_{12}x_2 + a_{13}x_3 + \cdots + a_{1n}x_n)\right]$$

$$x_2 = \frac{1}{a_{22}}\left[b_2 - (a_{21}x_1 + a_{23}x_3 + \cdots + a_{2n}x_n)\right]$$

$$\vdots$$

$$x_n = \frac{1}{a_{nn}}\left[b_n - (a_{n1}x_1 + a_{n2}x_2 + \cdots + a_{n\,n-1}x_{n-1})\right] \tag{5.36}$$

Let the initial approximation be $\left[x_1^{(0)}, x_2^{(0)}, \ldots, x_n^{(0)}\right]$. In Gauss–Seidel method, the latest available values of the variables are used; while in Jacobi method, values in last approximation are used to get a new approximation. Let the next approximation of Gauss–Seidel iteration be $\left[x_1^{(1)}, x_2^{(1)}, \ldots, x_n^{(1)}\right]$.

First, we calculate the approximation of variable x_1 as follows

$$x_1^{(1)} = \frac{1}{a_{11}}\left[b_1 - (a_{12}x_2^{(0)} + a_{13}x_3^{(0)} + \cdots + a_{1n}x_n^{(0)})\right]$$

To calculate $x_2^{(1)}$, we use $x_1^{(1)}$ instead of $x_1^{(0)}$ (as used in Jacobi method).

$$x_2^{(1)} = \frac{1}{a_{22}} \left[b_2 - (a_{21}x_1^{(1)} + a_{23}x_3^{(0)} + \cdots + a_{2n}x_n^{(0)}) \right]$$

Proceeding in a similar manner, the first approximation to the ith variable is computed as follows

$$x_i^{(1)} = \frac{1}{a_{ii}} \left[b_i - (a_{i1}x_1^{(1)} + a_{i2}x_2^{(1)} + \cdots + a_{ii-1}x_{i-1}^{(1)} + a_{ii+1}x_{i+1}^{(0)} + a_{ii+2}x_{i+2}^{(0)} + \cdots + a_{in}x_n^{(0)}) \right]$$

Similarly, the last variable is calculated as follows

$$x_n^{(1)} = \frac{1}{a_{nn}} \left[b_n - (a_{n1}x_1^{(1)} + a_{n2}x_2^{(1)} + \cdots + a_{n\,n-1}x_{n-1}^{(1)}) \right]$$

On a similar pattern, we use this first approximation $\left[x_1^{(1)}, x_2^{(1)}, \ldots, x_n^{(1)} \right]$ to compute the second iteration. In general, the $(k+1)$th iteration of Gauss–Seidel method can be obtained from kth iteration by the following formula.

$$x_1^{(k+1)} = \frac{1}{a_{11}} \left[b_1 - (a_{12}x_2^{(k)} + a_{13}x_3^{(k)} + \cdots + a_{1n}x_n^{(k)}) \right]$$

$$x_2^{(k+1)} = \frac{1}{a_{22}} \left[b_2 - (a_{21}x_1^{(k+1)} + a_{23}x_3^{(k)} + \cdots + a_{2n}x_n^{(k)}) \right]$$

$$\vdots$$

$$x_n^{(k+1)} = \frac{1}{a_{nn}} \left[b_n - (a_{n1}x_1^{(k+1)} + a_{n2}x_2^{(k+1)} + \cdots + a_{n\,n-1}x_{n-1}^{(k+1)}) \right] \qquad k = 0,1,2,\ldots \qquad (5.37)$$

The above system can be written as follows

$$x_i^{(k+1)} = \frac{1}{a_{ii}} \left(b_i - \sum_{j=1}^{i-1} a_{ij}x_j^{(k+1)} - \sum_{j=i+1}^{n} a_{ij}x_j^{(k)} \right); \qquad 1 \le i \le n \quad k = 0,1,2,\ldots \qquad (5.38)$$

Example   5.24

Solve the following system of linear equations corrects up to three decimal places using the Gauss–Seidel iterative procedure. Take zero vector as the initial solution vector.

$$3x_1 + x_2 - x_3 = 3$$
$$2x_1 + 4x_2 + x_3 = 7$$
$$x_1 - x_2 + 4x_3 = 4$$

Show that at the fifth iteration; the solution is correct to 3 decimal places. The exact solution is for the system is $x_1 = 1$, $x_2 = 1$, $x_3 = 1$.

Ans. To solve the system, we rewrite the system as follows

$$x_1 = \frac{1}{3}(3 - x_2 + x_3)$$

$$x_2 = \frac{1}{4}(7 - 2x_1 - x_3)$$

$$x_3 = \frac{1}{4}(4 - x_1 + x_2)$$

The initial approximation is $x_1^{(0)} = 0$, $x_2^{(0)} = 0$, $x_3^{(0)} = 0$. First approximation is given by

$$x_1^{(1)} = \frac{1}{3}(3 - x_2^{(0)} + x_3^{(0)}) = 1$$

To compute $x_2^{(1)}$, we use latest values i.e. $x_1^{(1)} = 1$, $x_3^{(0)} = 0$

$$x_2^{(1)} = \frac{1}{4}(7 - 2x_1^{(1)} - x_3^{(0)}) = 1.25$$

Similarly the value of $x_3^{(1)}$ is computed by using $x_1^{(1)} = 1$, $x_2^{(1)} = 1.25$

$$x_3^{(1)} = \frac{1}{4}(4 - x_1^{(1)} + x_2^{(1)}) = 1.0625$$

From this first approximation, $x_1^{(1)} = 1$, $x_2^{(1)} = 1.25$, $x_3^{(1)} = 1.0625$, we can compute the second approximation as follows

$$x_1^{(2)} = \frac{1}{3}(3 - x_2^{(1)} + x_3^{(1)}) = 0.9375$$

$$x_2^{(2)} = \frac{1}{4}(7 - 2x_1^{(2)} - x_3^{(1)}) = 1.0156$$

$$x_3^{(2)} = \frac{1}{4}(4 - x_1^{(2)} + x_2^{(2)}) = 1.0195$$

The approximate values of the variables x_1, x_2 and x_3 at first seven iterations are as follows

Iteration 1
1.0000 1.2500 1.0625

Iteration 2
0.9375 1.0156 1.0195

Iteration 3
1.0013 0.9945 0.9983

Iteration 4
1.0013 0.9998 0.9996

Iteration 5
0.9999 1.0001 1.0000

Iteration 6
1.0000 1.0000 1.0000

Iteration 7
1.0000 1.0000 1.0000

After seven iterations, we have $x_1 = 1$, $x_2 = 1$, $x_3 = 1$.
The exact solution is $x_1 = 1$, $x_2 = 1$, $x_3 = 1$.

Note: In Gauss–Seidel method, the result correct up to three decimal places can be obtained in the 5th iteration as the error in values of x is less than .0005. Hence, it is easy to see that for the same accuracy, the Jacobi method requires nine iterations, while Gauss–Seidel requires five iterations only. In Section 5.13, we will see that the Gauss–Seidel method is faster than Jacobi method. Hence, we prefer Gauss–Seidel method over Jacobi method to solve the system of linear equations.

Example ——————————————————————————————————— **5.25**

Solve the following system of linear equations with the aid of Gauss–Seidel method.

$$5.13x_1 - 1.70x_2 + 2.83x_3 = 11.3569$$
$$-1.20x_1 - 5.03x_2 + 2.91x_3 = 9.63028$$
$$0.23x_1 + 1.78x_2 - 8.32x_3 = 15.7821$$

Ans. Rewrite the given system as follows

$$x_1 = \frac{1}{5.13}(11.3569 + 1.70x_2 - 2.83x_3)$$

$$x_2 = \frac{-1}{5.03}(9.63028 + 1.20x_1 - 2.91x_3)$$

$$x_3 = \frac{-1}{8.32}(15.7821 - 0.23x_1 - 1.78x_2)$$

Using the initial approximation $x_1^{(0)} = 0$, $x_2^{(0)} = 0$, $x_3^{(0)} = 0$ in the above system, we get the first approximation as follows

$$x_1^{(1)} = 2.213821$$
$$x_2^{(1)} = -2.442717$$
$$x_3^{(1)} = -2.358288$$

Similarly, we can easily compute the following approximations

Iteration 2
2.705309   -3.924308   -2.661676

Iteration 3
2.381700   -4.022624   -2.691656

Iteration 4

2.365658   –4.036140   –2.694992

Iteration 5

2.363019   –4.037441   –2.695343

Iteration 6

2.362782   –4.037587   –2.695380

Iteration 7

2.362754   –4.037602   –2.695385

Iteration 8

2.362751   –4.037604   –2.695385

Iteration 9

2.362751   –4.037604   –2.695385

After 9 iterations, we have

$$x_1 = 2.362751, x_2 = -4.037604, x_3 = -2.695385.$$

Example 5.26

Compute seven iterations of the Gauss-Seidel method for the linear system of Example 5.23.

Ans. Rewrite the given system as follows

$$x_1 = \frac{1}{7}\left(12 + 3x_2 - 2x_3 - x_4\right)$$

$$x_2 = \frac{-1}{6}\left(6 - 2x_1 - x_3 - 2x_4\right)$$

$$x_3 = \frac{1}{5}\left(12 - x_1 - x_2 - 2x_4\right)$$

$$x_4 = \frac{1}{8}\left(5 - x_1 - 3x_2 + 2x_3\right)$$

On using the initial approximation, $x_1^{(0)} = 0$, $x_2^{(0)} = 0$, $x_3^{(0)} = 0$, $x_4^{(0)} = 0$, we get the first approximation of Gauss–Seidel method (5.29) as follows

$$x_1^{(1)} = \frac{1}{7}\left(12 + 3x_2^{(0)} - 2x_3^{(0)} - x_4^{(0)}\right) = 1.714286$$

$$x_2^{(1)} = \frac{-1}{6}\left(6 - 2x_1^{(1)} - x_3^{(0)} - 2x_4^{(0)}\right) = -0.428571$$

$$x_3^{(1)} = \frac{1}{5}\left(12 - x_1^{(1)} - x_2^{(1)} - 2x_4^{(0)}\right) = 2.142857$$

$$x_4^{(1)} = \frac{1}{8}\left(5 - x_1^{(1)} - 3x_2^{(1)} + 2x_3^{(1)}\right) = 1.107143$$

Similarly the next approximation can be computed as follows

$$x_1^{(2)} = \frac{1}{7}\left(12 + 3x_2^{(1)} - 2x_3^{(1)} - x_4^{(1)}\right) = 0.760204$$

$$x_2^{(2)} = \frac{-1}{6}\left(6 - 2x_1^{(2)} - x_3^{(1)} - 2x_4^{(1)}\right) = -0.020408$$

$$x_3^{(2)} = \frac{1}{5}\left(12 - x_1^{(2)} - x_2^{(2)} - 2x_4^{(1)}\right) = 1.809184$$

$$x_4^{(2)} = \frac{1}{8}\left(5 - x_1^{(2)} - 3x_2^{(2)} + 2x_3^{(2)}\right) = 0.989923$$

Similarly, we have following iterations

Iteration 3
1.047212   −0.019424   1.798473   0.951001

Iteration 4
1.056254   −0.031169   1.814583   0.958302

Iteration 5
1.045575   −0.029610   1.813486   0.958779

Iteration 6
1.046488   −0.029330   1.813057   0.958452

Iteration 7
1.046778   −0.029414   1.813146   0.958470

5.12 Relaxation Method

The system (5.36) can be rewritten as follows

$$x_1 = x_1 - \frac{1}{a_{11}}a_{11}x_1 + \frac{1}{a_{11}}\left[b_1 - (a_{12}x_2 + a_{13}x_3 + \cdots + a_{1n}x_n)\right]$$

$$x_2 = x_2 - \frac{1}{a_{22}}a_{22}x_2 + \frac{1}{a_{22}}\left[b_2 - (a_{21}x_1 + a_{23}x_3 + \cdots + a_{2n}x_n)\right]$$

$$\vdots$$

$$x_n = x_n - \frac{1}{a_{nn}}a_{nn}x_n + \frac{1}{a_{nn}}\left[b_n - (a_{n1}x_1 + a_{n2}x_2 + \cdots + a_{n\,n-1}x_{n-1})\right]$$

Or

$$x_1 = x_1 + \frac{1}{a_{11}}\left[b_1 - (a_{11}x_1 + a_{12}x_2 + a_{13}x_3 + \cdots + a_{1n}x_n)\right]$$

$$x_2 = x_2 + \frac{1}{a_{22}}\left[b_2 - (a_{21}x_1 + a_{22}x_2 + a_{23}x_3 + \cdots + a_{2n}x_n)\right]$$

$$\vdots$$

$$x_n = x_n + \frac{1}{a_{nn}}\left[b_n - (a_{n1}x_1 + a_{n2}x_2 + \cdots + a_{n\,n-1}x_{n-1} + a_{nn}x_n)\right] \tag{5.39}$$

Let us define r_i's as follows

$$r_1 = b_1 - (a_{11}x_1 + a_{12}x_2 + \cdots + a_{1n}x_n)$$
$$r_2 = b_2 - (a_{21}x_1 + a_{22}x_2 + \cdots + a_{2n}x_n)$$
$$\vdots \qquad\qquad \text{(or)} \qquad r_i = b_i - \sum_{j=1}^{n} a_{ij}x_j ; 1 \le i \le n$$
$$r_n = b_n - (a_{n1}x_1 + a_{n2}x_2 + \cdots + a_{nn}x_n)$$

where r_i's are known as the residuals. For an exact solution $\left(x_1^e, x_2^e, ..., x_n^e\right)$, these residuals must vanish.

$$r_1 = b_1 - (a_{11}x_1^e + a_{12}x_2^e + \cdots + a_{1n}x_n^e) = 0$$
$$r_2 = b_2 - (a_{21}x_1^e + a_{22}x_2^e + \cdots + a_{2n}x_n^e) = 0$$
$$\vdots \qquad\qquad \text{(or)} \qquad r_i = b_i - \sum_{j=1}^{n} a_{ij}x_j^{(e)} = 0; \quad 1 \le i \le n$$
$$r_n = b_n - (a_{n1}x_1^e + a_{n2}x_2^e + \cdots + a_{nn}x_n^e) = 0$$

Our aim is to find those values of x_1, x_2, ..., x_n, for which values of these residuals are zeroes. So, the iteration process can be speed up by multiplying these residual terms with a factor ω in the system (5.39),

i.e.,

$$x_1 = x_1 + \frac{\omega}{a_{11}}\left[b_1 - (a_{11}x_1 + a_{12}x_2 + a_{13}x_3 + \cdots + a_{1n}x_n)\right]$$

$$x_2 = x_2 + \frac{\omega}{a_{22}}\left[b_2 - (a_{21}x_1 + a_{22}x_2 + a_{23}x_3 + \cdots + a_{2n}x_n)\right]$$

$$\vdots$$

$$x_n = x_n + \frac{\omega}{a_{nn}}\left[b_n - (a_{n1}x_1 + a_{n2}x_2 + \cdots + a_{n\,n-1}x_{n-1} + a_{nn}x_n)\right]$$

Or

$$x_1 = (1-\omega)x_1 + \frac{\omega}{a_{11}}\left[b_1 - (a_{12}x_2 + a_{13}x_3 + \cdots + a_{1n}x_n)\right]$$

$$x_2 = (1-\omega)x_2 + \frac{\omega}{a_{22}}\left[b_2 - (a_{21}x_1 + a_{23}x_3 + \cdots + a_{2n}x_n)\right]$$

$$\vdots$$

$$x_n = (1-\omega)x_n + \frac{\omega}{a_{nn}}\left[b_n - (a_{n1}x_1 + a_{n2}x_2 + \cdots + a_{n\,n-1}x_{n-1})\right] \qquad (5.40)$$

If we use simultaneous displacement (as in Jacobi iterations) for the system (5.40), then the relaxation method is as follows

$$x_i^{(k+1)} = (1-\omega)x_i^{(k)} + \frac{\omega}{a_{ii}}\left(b_i - \sum_{\substack{j=1 \\ j\neq i}}^{n} a_{ij}x_j^{(k)}\right); \quad 1\leq i \leq n \;\; k = 0,1,2,\dots \qquad (5.41)$$

Similarly, the relaxation method for successive displacement (Gauss–Seidel iterations) is as follows

$$x_i^{(k+1)} = (1-\omega)x_i^{(k)} + \frac{\omega}{a_{ii}}\left(b_i - \sum_{j=1}^{i-1} a_{ij}x_j^{(k+1)} - \sum_{j=i+1}^{n} a_{ij}x_j^{(k)}\right); \quad 1\leq i \leq n \;\; k = 0,1,2,\dots \qquad (5.42)$$

The methods involving systems (5.41) or (5.42) are called as relaxation methods. In case $0 < \omega < 1$, the methods are called as under-relaxation methods, while for $\omega > 1$, the methods are called as over-relaxation methods. These relaxation methods are used to accelerate the convergence of Jacobi and Gauss–Seidel methods.

Since successive displacement (Gauss–Seidel) has faster convergence than simultaneous displacement (Jacobi), hence formula (5.42) is suggested. The method (5.42) (successive displacements) with over-relaxation is known as a successive over-relaxation (SOR) method.

Example _____ 5.27

Solve the following system of linear equations corrects up to three decimal places using Gauss–Seidel relaxation scheme with relaxation parameter 0.9.

$$3x_1 - x_2 + x_3 = 5$$
$$x_1 - 3x_2 + x_3 = 5$$
$$x_1 + 2x_2 + 4x_3 = 7$$

Consider the initial approximation, $x_1^{(0)} = 0$, $x_2^{(0)} = 0$, $x_3^{(0)} = 0$.

Ans. Rewrite the system of equations in the form (5.40) with $\omega = 0.9$

$$x_1 = 0.1x_1 + \frac{0.9}{3}(5 + x_2 - x_3)$$

$$x_2 = 0.1x_2 - \frac{0.9}{3}(5 - x_1 - x_3)$$

$$x_3 = 0.1x_3 + \frac{0.9}{4}(7 - x_1 - 2x_2)$$

Using the initial approximation $x_1^{(0)} = 0$, $x_2^{(0)} = 0$, $x_3^{(0)} = 0$ in the above system, we get following first approximation (use formula 5.42)

$$x_1^{(1)} = 0.1x_1^{(0)} + \frac{0.9}{3}(5 + x_2^{(0)} - x_3^{(0)}) = 1.5$$

$$x_2^{(1)} = 0.1x_2^{(0)} - \frac{0.9}{3}(5 - x_1^{(1)} - x_3^{(0)}) = -1.05$$

$$x_3^{(1)} = 0.1x_3^{(0)} + \frac{0.9}{4}(7 - x_1^{(1)} - 2x_2^{(1)}) = 1.71$$

Similarly, other approximations are given by

Iteration 2
0.822000   −0.845400   1.941480

Iteration 3
0.746136   −0.778255   1.951482

Iteration 4
0.755692   −0.765673   1.944670

Iteration 5
0.762466   −0.764426   1.941904

Iteration 6
0.764348   −0.764567   1.941267

Iteration 7
0.764684   −0.764671   1.941175

Iteration 8
0.764715   −0.764700   1.941172

After eight iterations, the approximate solution is as follows

$$x_1 = 0.764715, x_2 = -0.7647, x_3 = 1.941172$$

Example ─── 5.28 ──

Solve the system in Example 5.24 correct up to 0.000001 using Gauss–Seidel relaxation scheme with relaxation parameters 0.9, 1 and 1.1.

Ans. The system of equations in Example 5.24 is as follows

$$3x_1 + x_2 - x_3 = 3$$
$$2x_1 + 4x_2 + x_3 = 7$$
$$x_1 - x_2 + 4x_3 = 4$$

Using Gauss–Seidel relaxation scheme (5.42) for the given system, we have

$$x_1^{(k+1)} = (1-\omega)x_1^{(k)} + \frac{\omega}{3}\left(3 - x_2^{(k)} + x_3^{(k)}\right)$$

$$x_2^{(k+1)} = (1-\omega)x_2^{(k)} + \frac{\omega}{4}\left(7 - 2x_1^{(k+1)} - x_3^{(k)}\right)$$

$$x_3^{(k+1)} = (1-\omega)x_3^{(k)} + \frac{\omega}{4}\left(4 - x_1^{(k+1)} + x_2^{(k+1)}\right)$$

Using the initial approximation $x_1^{(0)} = 0$, $x_2^{(0)} = 0$, $x_3^{(0)} = 0$ in the above system, we get following approximations for different values of relaxation parameter

Relaxation parameter $\omega = 0.9$			
Iteration (k)	$x_1^{(k)}$	$x_2^{(k)}$	$x_3^{(k)}$
1	0.900000	1.170000	0.960750
2	0.927225	1.058580	1.025630
3	0.982837	1.007814	1.008183
4	0.998394	0.999663	1.001104
5	1.000272	0.999596	0.999958
6	1.000136	0.999908	0.999945
7	1.000025	0.999992	0.999987
8	1.000001	1.000002	0.999999
9	0.999999	1.000001	1.000000
10	1.000000	1.000000	1.000000

It is easy to see that the difference between the ninth iteration and the tenth iteration is less than 0.000001. So, total ten iterations are required for the accuracy of 0.000001.

Relaxation parameter $\omega = 1.0$			
Iteration (k)	$x_1^{(k)}$	$x_2^{(k)}$	$x_3^{(k)}$
1	1.000000	1.250000	1.062500
2	0.937500	1.015625	1.019531
3	1.001302	0.994466	0.998291
4	1.001275	0.999790	0.999629
5	0.999946	1.000120	1.000043
6	0.999975	1.000002	1.000007
7	1.000002	0.999997	0.999999
8	1.000000	1.000000	1.000000
9	1.000000	1.000000	1.000000

Relaxation parameter $\omega = 1.1$			
Iteration (k)	$x_1^{(k)}$	$x_2^{(k)}$	$x_3^{(k)}$
1	1.100000	1.320000	1.160500
2	0.931517	0.961528	0.992203
3	1.018096	0.996039	0.994714
4	0.997705	1.003112	1.002016
5	0.999828	0.999229	0.999634
6	1.000166	1.000087	1.000015
7	0.999957	1.000011	1.000013
8	1.000005	0.999992	0.999995
9	1.000000	1.000002	1.000001
10	1.000000	1.000000	1.000000
11	1.000000	1.000000	1.000000

Note: Here, we observe that Gauss–Seidel method ($\omega = 1.0$) requires a minimum number of iterations (9 iterations) for the accuracy of 0.000001 as compared to relaxation method with relaxation parameters $\omega = 0.9$ and 1.1.

But, this is not the case always, which will be clear from the following example.

Example   5.29

Solve the following tridiagonal system of linear equations with an accuracy of 0.000001 using Gauss–Seidel method

$$3x_1 - 2x_2 = 18$$
$$-2x_1 + 3x_2 - x_3 = -19$$
$$-x_2 + 3x_3 = 9$$

Also, solve the system using SOR method with relaxation parameter, $\omega = 1.2$. Consider the initial approximation, $x_1^{(0)} = 0$, $x_2^{(0)} = 0$, $x_3^{(0)} = 0$. Compare the results obtained from both the methods. Also, compare the results with the exact solution, $x_1 = 4$, $x_2 = -3$, $x_3 = 2$.

Ans.
The Gauss–Seidel iterations are given by

Gauss–Seidel Iteration (k)	$x_1^{(k)}$	$x_2^{(k)}$	$x_3^{(k)}$
1	6.000000	-2.333333	2.222222
2	4.444445	-2.629629	2.123457
3	4.246914	-2.794238	2.068587
4	4.137175	-2.885688	2.038104
5	4.076208	-2.936493	2.021169
6	4.042338	-2.964719	2.011760
7	4.023521	-2.980399	2.006534
8	4.013067	-2.989111	2.003630
9	4.007259	-2.993951	2.002017
10	4.004033	-2.996639	2.001120
11	4.002241	-2.998133	2.000623
12	4.001245	-2.998962	2.000346

13	4.000692	−2.999423	2.000192
14	4.000384	−2.999680	2.000107
15	4.000214	−2.999822	2.000059
16	4.000119	−2.999901	2.000033
17	4.000066	−2.999945	2.000018
18	4.000036	−2.999970	2.000010
19	4.000020	−2.999984	2.000005
20	4.000011	−2.999991	2.000003
21	4.000006	−2.999995	2.000002
22	4.000003	−2.999997	2.000001
23	4.000002	−2.999998	2.000000
24	4.000001	−2.999999	2.000000

It is easy to see that the difference between 23rd iteration and 24th iteration is less than 0.000001. So, total 24 iterations of the Gauss–Seidel method are required for the accuracy of 0.000001.

Let us solve this system using SOR method (5.42) with relaxation parameter, $\omega = 1.2$. The SOR iterations are given by

SOR Iteration (k)	$x_1^{(k)}$	$x_2^{(k)}$	$x_3^{(k)}$
1	7.200000	−1.840000	2.864000
2	4.288000	−2.656000	1.964800
3	4.217600	−2.908800	2.043520
4	4.029440	−2.977280	2.000384
5	4.012288	−2.994560	2.002099
6	4.001894	−2.998733	2.000087
7	4.000635	−2.999711	2.000098
8	4.000104	−2.999935	2.000006

9	4.000031	−2.999985	2.000005
10	4.000005	−2.999997	2.000000
11	4.000001	−3.000000	2.000000
12	4.000000	−3.000000	2.000000
13	4.000000	−3.000000	2.000000

It is easy to see that the difference between 12th iteration and 13th iteration is less than 0.000001. So, total 13 iterations of the SOR method are required for the accuracy of 0.000001. It is easy to conclude that the number of iterations required in SOR method (13) is pretty less than the number of iterations of the Gauss–Seidel method (24).

Note: Consider the relaxation method (5.42) with different relaxation parameter ω for the solution of the system in Example 5.29. The numbers of iterations required for the accuracy of 0.000001 with different values of relaxation parameter ω are given in the following table.

Sr. No.	Relaxation parameter ω	Number of iterations
1	0.60	44
2	0.75	37
3	0.90	29
4	0.95	26
5	1.00	24
6	1.05	21
7	1.10	19
8	1.15	16
9	1.20	13
10	1.25	13
11	1.30	14
12	1.40	19
13	1.50	24

It is easy to see that for relaxation parameter $\omega = 1.2$ and 1.25, the minimum number of iterations are required for a given accuracy 0.000001. So, these values of relaxation parameter are appropriate for this system. It is further worth mentioning here that the best relaxation parameter varies for different systems. So, the obvious question is which relaxation parameter is the best relaxation parameter for a given system? For a general system (5.1), we don't have a complete answer to this question. But some theorems are useful in this regard, which we will state in the next sections.

Example —— **5.30**

Solve the following set of linear equations with the help SOR method (5.42) with relaxation parameter $\omega = 1.1$.

$$4x_1 - x_2 - x_3 = 99.875$$
$$2x_1 - 4x_2 + x_4 = -99.6875$$
$$2x_1 - 4x_3 + x_4 = 0.3125$$
$$x_2 + x_3 - 2x_4 = 0.25$$

Ans. Gauss–Seidel relaxation scheme (5.42) is given by

$$x_i^{(k+1)} = (1-\omega)x_i^{(k)} + \frac{\omega}{a_{ii}}\left(b_i - \sum_{j=1}^{i-1} a_{ij}x_j^{(k+1)} - \sum_{j=i+1}^{n} a_{ij}x_j^{(k)} \right); \qquad 1 \le i \le n$$

We have following equations for the given system with relaxation parameter $\omega = 1.1$

$$x_1^{(k+1)} = (-0.1)x_1^{(k)} + \frac{1.1}{4}\left(99.875 + x_2^{(k)} + x_3^{(k)}\right)$$

$$x_2^{(k+1)} = (-0.1)x_2^{(k)} + \frac{1.1}{4}\left(99.6875 + 2x_1^{(k+1)} + x_4^{(k)}\right)$$

$$x_3^{(k+1)} = (-0.1)x_3^{(k)} - \frac{1.1}{4}\left(0.3125 - 2x_1^{(k+1)} - x_4^{(k+1)}\right)$$

$$x_4^{(k+1)} = (-0.1)x_4^{(k)} - \frac{1.1}{2}\left(0.25 - x_2^{(k+1)} - x_3^{(k+1)}\right)$$

On using the initial approximation $x_1^{(0)} = 0$, $x_2^{(0)} = 0$, $x_3^{(0)} = 0$, $x_4^{(0)} = 0$ in the above system, we get the following first 10 iterations

Iteration 1
27.465626   42.520157   15.020157   31.509674

Iteration 2
40.542648   54.125664   29.375664   42.637264

Iteration 3
46.374222   59.232567   34.207565   46.990845

Iteration 4
48.524235   61.101612   36.104115   48.626564

Iteration 5
49.344776   61.815834   36.815582   49.247124

Iteration 6
49.654785   62.085575   37.085594   49.481934

Iteration 7
49.772217   62.187759   37.187752   49.570839

Iteration 8
49.816673   62.226437   37.226440   49.604500

Iteration 9
49.833504   62.241085   37.241085   49.617241

Iteration 10
49.839874   62.246632   37.246628   49.622070

So, the final solution after 10 iterations is given by

$x_1^{(10)} = 49.839874,\ x_2^{(10)} = 62.246632,\ x_3^{(10)} = 37.246628,\ x_4^{(10)} = 49.622070$

Note: In Chapters 15 and 16, we have many examples of linear systems arising during the solutions of differential equations. These examples also contain systems of four and six simultaneous equations. We will use methods discussed in this chapter for the solutions of these systems. So, readers interested in more solved examples may refer to Chapters 15 and 16.

5.13 Convergence Criteria for Iterative Methods

The iterative methods (Jacobi, Gauss–Seidel, and Relaxation methods) may converge or diverge depending on the convergence conditions that the linear system (5.1) needs to satisfy. Now, we will discuss the convergence criteria for these iterative methods.

Let $\{x_n\}$ be any sequence that converges to ξ and

$$\varepsilon_n = \xi - x_n$$

Consider there exists a positive constant C and a constant $p \geq 1$, such that

$$\lim_{n \to \infty} \frac{|\varepsilon_{n+1}|}{|\varepsilon_n|^p} = C$$

Then p is called as the order of convergence and C is the asymptotic error constant. For convergence, we must have $|\varepsilon_{n+1}| < |\varepsilon_n|$ after sufficiently large n, i.e., error at $(n+1)$th step is less than the error at nth step after some iterations.

Convergence of Jacobi Method
The kth iteration of Jacobi method (5.35) is as follows

$$x_i^{(k+1)} = \frac{1}{a_{ii}} \left(b_i - \sum_{\substack{j=1 \\ j \neq i}}^{n} a_{ij} x_j^{(k)} \right); \quad 1 \leq i \leq n,\ k = 0,1,2,\dots$$

Let vector $\xi = \begin{bmatrix} \xi_1 & \xi_2 & \cdots & \xi_n \end{bmatrix}^T$ be the exact solution, then

$$\xi_i = \frac{1}{a_{ii}} \left(b_i - \sum_{\substack{j=1 \\ j \neq i}}^{n} a_{ij} \xi_j \right); \quad 1 \leq i \leq n$$

From above two equations, we can easily conclude that

$$\xi_i - x_i^{(k+1)} = \frac{-1}{a_{ii}} \sum_{\substack{j=1 \\ j \neq i}}^{n} a_{ij} \left(\xi_j - x_j^{(k)} \right); \quad 1 \leq i \leq n$$

Let the term $\varepsilon_j^{(k)} = \xi_j - x_j^{(k)}$ be the error at k^{th} iteration of the jth variable. The above expression takes the form

$$\varepsilon_i^{(k+1)} = \frac{-1}{a_{ii}} \sum_{\substack{j=1 \\ j \neq i}}^{n} a_{ij} \varepsilon_j^{(k)}; \qquad 1 \leq i \leq n$$

$$\left| \varepsilon_i^{(k+1)} \right| \leq \frac{1}{|a_{ii}|} \sum_{\substack{j=1 \\ j \neq i}}^{n} \left| a_{ij} \right| \left| \varepsilon_j^{(k)} \right|; \qquad 1 \leq i \leq n$$

i.e., $\qquad \left\| \varepsilon_i^{(k+1)} \right\| \leq \frac{1}{|a_{ii}|} \sum_{\substack{j=1 \\ j \neq i}}^{n} \left| a_{ij} \right| \left\| \varepsilon_j^{(k)} \right\|; \qquad 1 \leq i \leq n$

$$\left\| \varepsilon^{(k+1)} \right\| \leq \frac{1}{|a_{ii}|} \sum_{\substack{j=1 \\ j \neq i}}^{n} \left| a_{ij} \right| \left\| \varepsilon^{(k)} \right\|; \qquad 1 \leq i \leq n \qquad (5.43)$$

Let $\qquad \lambda = \max_i \left\{ \frac{1}{|a_{ii}|} \sum_{\substack{j=1 \\ j \neq i}}^{n} \left| a_{ij} \right| \right\}; \quad 1 \leq i \leq n$

Therefore, the system (5.43) is given by

$$\left\| \varepsilon^{(k+1)} \right\| \leq \lambda \left\| \varepsilon^{(k)} \right\|; \qquad 1 \leq i \leq n \qquad (5.44)$$

Hence Jacobi method is linearly convergent as the order of convergence is one. Also

$$\left\| \varepsilon^{(k+1)} \right\| \leq \lambda \left\| \varepsilon^{(k)} \right\| \leq \lambda^2 \left\| \varepsilon^{(k-1)} \right\| \leq \ldots \leq \lambda^k \left\| \varepsilon^{(0)} \right\|$$

If we have

$$\lambda = \max_i \left\{ \frac{1}{|a_{ii}|} \sum_{\substack{j=1 \\ j \neq i}}^{n} \left| a_{ij} \right| \right\} < 1; \qquad 1 \leq i \leq n \qquad (5.45)$$

then $\lambda^k \to 0$ as $k \to \infty$ and hence $\left\| \varepsilon^{(k+1)} \right\| \to 0$.

If the system of linear equations (5.1) satisfies the conditions (5.45), then the Jacobi method will converge for any initial approximation.

The matrix A is said to be diagonally dominant matrix if it satisfies the following conditions

$$\sum_{\substack{j=1 \\ j \neq i}}^{n} \left| a_{ij} \right| \leq \left| a_{ii} \right|; \qquad 1 \leq i \leq n$$

The convergence conditions (5.45) are as follows

$$\sum_{\substack{j=1 \\ j \neq i}}^{n} \left| a_{ij} \right| < \left| a_{ii} \right|; \qquad 1 \leq i \leq n \qquad (5.46)$$

The condition (5.46) is known as strictly diagonally dominance condition. So if our matrix A in the system (5.1) is a strictly diagonally dominant matrix, then Jacobi method converges to the solution for any initial approximation. The condition represents that absolute value of each diagonal element in the coefficient matrix A must be greater than the sum of the absolute values of other elements in the row containing that diagonal element.

Note that the condition (5.46) is much broader condition than required for the convergence of the method. So, this condition is *sufficient* condition, but not a *necessary* condition. If the condition (5.46) is true, the Jacobi method must converge. But there may be systems which are not strictly diagonally dominant, and method converges to the solutions for these systems. This statement may be well described by the examples in the next section.

The following theorem will be helpful for computing the *upper bound of the error* at k^{th} iteration of Jacobi method.

Theorem 5.1

Let us reconsider the relation

$$\left\| \varepsilon^{(k+1)} \right\| \leq \lambda \left\| \varepsilon^{(k)} \right\| = \lambda \left\| e^{(k)} + \varepsilon^{(k+1)} \right\|$$
$$\leq \lambda \left\| e^{(k)} \right\| + \lambda \left\| \varepsilon^{(k+1)} \right\|$$
$$\leq \frac{\lambda}{1-\lambda} \left\| e^{(k)} \right\|$$
$$\Rightarrow \left\| \varepsilon^{(k+1)} \right\| \leq \frac{\lambda}{1-\lambda} \left\| e^{(k)} \right\| \qquad (5.47)$$

where $e^{(k)}$ is the difference between kth iteration and $(k+1)$th iteration i.e. $e^{(k)} = \left| \varepsilon^{(k+1)} - \varepsilon^{(k)} \right|$. It will be helpful in computing the upper bound of the error in some particular cases.

Convergence of Gauss–Seidel Method:

The kth iteration of Gauss–Seidel method (5.38) is given by

$$x_i^{(k+1)} = \frac{1}{a_{ii}}\left(b_i - \sum_{j=1}^{i-1} a_{ij}x_j^{(k+1)} - \sum_{j=i+1}^{n} a_{ij}x_j^{(k)} \right); \qquad 1 \le i \le n, \ k = 0,1,2,\ldots$$

Proceeding in a similar manner as in Jacobi method, we have

$$\left| \varepsilon_i^{(k+1)} \right| \le \frac{1}{|a_{ii}|}\left(\sum_{j=1}^{i-1} |a_{ij}|\left|\varepsilon_j^{(k+1)}\right| + \sum_{j=i+1}^{n} |a_{ij}|\left|\varepsilon_j^{(k)}\right| \right); \qquad 1 \le i \le n \tag{5.48}$$

where ξ_j's; $1 \le j \le n$ are exact values of the variables and $\varepsilon_j^{(k)} = \xi_j - x_j^{(k)}$ is the error at kth iteration of the jth variable.

Let $\displaystyle \lambda = \max_i \left\{ \frac{1}{|a_{ii}|}\sum_{\substack{j=1 \\ j\ne i}}^{n} |a_{ij}| \right\}, \ \lambda_i = \max_i \left\{ \frac{1}{|a_{ii}|}\sum_{j=1}^{i-1} |a_{ij}| \right\}; \quad 1 \le i \le n$

The system (5.48) takes the form

$$\left| \varepsilon^{(k+1)} \right| \le \lambda_i \left| \varepsilon^{(k+1)} \right| + (\lambda - \lambda_i)\left| \varepsilon^{(k)} \right|$$

i.e., $\qquad \left\| \varepsilon^{(k+1)} \right\| \le \lambda_i \left\| \varepsilon^{(k+1)} \right\| + (\lambda - \lambda_i)\left\| \varepsilon^{(k)} \right\|$

$$\left\| \varepsilon^{(k+1)} \right\| \le \frac{(\lambda - \lambda_i)}{1 - \lambda_i}\left\| \varepsilon^{(k)} \right\| \tag{5.49}$$

Therefore, order of convergence of Gauss Seidel method is one.

From system (5.49), we have following convergence conditions

$$\frac{(\lambda - \lambda_i)}{1 - \lambda_i} < 1 \ \text{(or)} \ 0 \le \lambda_i \le \lambda < 1$$

$$\Rightarrow \frac{(\lambda - \lambda_i)}{1 - \lambda_i} \le \lambda$$

The convergence condition (5.49) is similar to Jacobi method, $\left(\left\| \varepsilon^{(k+1)} \right\| \le \lambda \left\| \varepsilon^{(k)} \right\| \right)$. However,

$\dfrac{(\lambda - \lambda_i)}{1 - \lambda_i} \le \lambda$, so Gauss–Seidel method converges faster than Jacobi method.

Example ———————————————————————————————— **5.31**

Which of the following systems is a strictly diagonally dominant system?

a)
$$6x_1 + 3x_2 + x_3 = 10$$
$$x_1 + 6x_2 + 4x_3 = 11$$
$$3x_1 - 2x_2 + 8x_3 = 9$$

b)
$$x_1 - 3x_2 + x_3 = 5$$
$$x_1 + 4x_2 + 2x_3 = 7$$
$$3x_1 - x_2 + 5x_3 = 5$$

c)
$$3x_1 - 2x_2 + x_3 = 2$$
$$2x_1 - 4x_2 + x_3 = -1$$
$$x_1 + 2x_2 + 5x_3 = 8$$

Ans. For strictly diagonally dominant system, our coefficient matrix must satisfy the condition (5.46) given by

$$\sum_{\substack{j=1 \\ j \neq i}}^{n} |a_{ij}| < |a_{ii}|; \qquad 1 \leq i \leq n$$

Let us check the condition for each of the given systems.

a) The coefficient matrix $A = \begin{bmatrix} 6 & 3 & 1 \\ 1 & 6 & 4 \\ 3 & -2 & 8 \end{bmatrix}$ is a strictly diagonally dominant

matrix as the absolute values of all the diagonal elements are greater than the sum of the absolute values of other elements in the row containing that diagonal element. We have

$$|6| > |3| + |1|, \qquad |6| > |1| + |4|, \qquad |8| > |3| + |-2|$$

b) The coefficient matrix $A = \begin{bmatrix} 1 & -3 & 1 \\ 1 & 4 & 2 \\ 3 & -1 & 5 \end{bmatrix}$ is not strictly diagonally

dominant matrix as for the first row, we have $|1| < |-3| + |1|$.

c) The coefficient matrix $A = \begin{bmatrix} 3 & -2 & 1 \\ 2 & -4 & 1 \\ 1 & 2 & 5 \end{bmatrix}$ is not strictly diagonally

dominant but simply diagonally dominant matrix. The second and third rows are satisfying the strict diagonally dominance criterion, but the first row is not satisfying the criterion as $|3| = |-2| + |1|$.

Example ——————————————————————————————— **5.32** —————

Consider the following system of linear equations

$$x_1 - 3x_2 + x_3 = 5$$
$$x_1 + 2x_2 + 4x_3 = 7$$
$$3x_1 - x_2 + x_3 = 5$$

i. Compute the ten iterations of the Jacobi and Gauss–Seidel methods with initial approximation $[0\ 0\ 0]^T$, and show that the methods are diverging for the given system.

ii. State the sufficient convergence condition (diagonally dominance) for the Jacobi and Gauss–Seidel methods. Rearrange the equations such that the system satisfies the convergence condition.

iii. Solve the given system correct up to three decimal places. Start with the initial approximation $[0\ 0\ 0]^T$.

Ans.

i. On using the initial approximation $[0\ 0\ 0]^T$, the Jacobi iterations (5.35) are as follows

Jacobi Iteration (k)	$x_1^{(k)}$	$x_2^{(k)}$	$x_3^{(k)}$
1	5.000000	3.500000	5.000000
2	10.500000	−9.000000	−6.500000
3	−15.500000	11.250000	−35.500000
4	74.250000	82.250000	62.750000
5	189.000000	−159.125000	−135.500000
6	−336.875000	180.000000	−721.125000
7	1266.125000	1614.187500	1195.625000
8	3651.937500	−3020.812500	−2179.187500
9	−6878.250000	2535.906250	−13971.625000
10	21584.343750	31385.875000	23175.656250

The Gauss–Seidel iterations (5.38) are given by

Gauss–Seidel Iteration (k)	$x_1^{(k)}$	$x_2^{(k)}$	$x_3^{(k)}$
1	5.000000	1.000000	−9.000000
2	17.000000	13.000000	−33.000000
3	77.000000	31.000000	−195.000000
4	293.000000	247.000000	−627.000000
5	1373.000000	571.000000	−3543.000000
6	5261.000000	4459.000000	−11319.000000
7	24701.000000	10291.000000	−63807.000000
8	94685.000000	80275.000000	−203775.000000
9	444605.000000	185251.000000	−1148559.000000
10	1704317.000000	1444963.000000	−3667983.000000

It is easy to see that both the Jacobi and Gauss–Seidel methods diverge for the given system.

    ii.   The sufficient condition for convergence of Jacobi and Gauss–Seidel methods is that the coefficient matrix is strictly diagonally dominant (5.46).

$$\sum_{\substack{j=1 \\ j \neq i}}^{n} \left| a_{ij} \right| < \left| a_{ii} \right|; \qquad 1 \leq i \leq n$$

The system is not satisfying the convergence conditions, so Jacobi and Gauss–Seidel methods are diverging for the given system in the present form.

On rearranging the equations in given system, we have

$$3x_1 - x_2 + x_3 = 5$$
$$x_1 - 3x_2 + x_3 = 5$$
$$x_1 + 2x_2 + 4x_3 = 7$$

This system is strictly diagonally dominant system, as we have

$$|3| > |-1| + |1|, \qquad |-3| > |1| + |1|, \qquad |4| > |1| + |2|$$

So the Jacobi and Gauss–Seidel methods will converge to the solution for any initial approximation.

iii. The Jacobi iterations (5.35) with initial approximation $[0\ 0\ 0]^T$ are given by

Jacobi Iteration (k)	$x_1^{(k)}$	$x_2^{(k)}$	$x_3^{(k)}$
1	1.6667	−1.6667	1.7500
2	0.5278	−0.5278	2.1667
3	0.7685	−0.7685	1.8819
4	0.7832	−0.7832	1.9421
5	0.7582	−0.7582	1.9458
6	0.7653	−0.7653	1.9396
7	0.7650	−0.7650	1.9413
8	0.7645	−0.7645	1.9413
9	0.7647	−0.7647	1.9411
10	0.7647	−0.7647	1.9412

After ten iterations of Jacobi method, the approximate solution correct up to three decimal places is as follows

$$x_1 = 0.7647, x_2 = -0.7647, \ x_3 = 1.9412$$

The Gauss–Seidel method provides following iterations for the given system.

Gauss–Seidel Iteration (k)	$x_1^{(k)}$	$x_2^{(k)}$	$x_3^{(k)}$
1	1.6667	−1.1111	1.8889
2	0.6667	−0.8148	1.9907
3	0.7315	−0.7593	1.9468
4	0.7647	−0.7629	1.9403
5	0.7656	−0.7647	1.9409
6	0.7648	−0.7648	1.9412
7	0.7647	−0.7647	1.9412

After seven iterations of the Gauss–Seidel method, the approximate solution correct up to three decimal places is as follows

$$x_1 = 0.7647, x_2 = -0.7647, \ x_3 = 1.9412$$

Note: For convergence of Jacobi and Gauss–Seidel iterative procedures, the strictly diagonally dominance condition of coefficient matrix A is only *sufficient condition*. The necessary and sufficient conditions for convergence of various iterative formulas are obtained in next section.

5.14 Matrix Forms and Convergence of Iterative Methods

Consider the system (5.1) of linear equations

$$AX = B$$

Let us write the coefficient matrix A as follows

$$A = L + D + U$$

where L is a lower triangular matrix with zero diagonal elements, D is a diagonal matrix, and U is an upper triangular matrix with zero diagonal elements. Then, the system (5.1) is given by

$$(L + D + U)X = B$$

Now, we will construct the matrix forms for various iterative methods.

Matrix Form of Jacobi Method
Consider the Jacobi iteration formula (5.35)

$$x_i^{(k+1)} = \frac{1}{a_{ii}} \left(b_i - \sum_{\substack{j=1 \\ j \neq i}}^{n} a_{ij} x_j^{(k)} \right); \qquad 1 \leq i \leq n$$

To determine the matrix form of Jacobi method, rewrite it as follows

$$a_{ii} x_i^{(k+1)} = -\sum_{\substack{j=1 \\ j \neq i}}^{n} a_{ij} x_j^{(k)} + b_i; \qquad 1 \leq i \leq n$$

In matrix form, we can write the above equations as follows

$$DX^{(k+1)} = -(L + U)X^{(k)} + B$$

$$X^{(k+1)} = -D^{-1}(L + U)X^{(k)} + D^{-1}B; \text{ for each } k = 0, 1, 2,\ldots \qquad (5.50)$$

We can write the system (5.50) as follows

$$X^{(k+1)} = T_j X^{(k)} + C_j \qquad\qquad (5.51)$$

where $T_j = -D^{-1}(L+U)$ and $C_j = D^{-1}B$

The iteration formula (5.50) is matrix form of the Jacobi method (5.35). We can use either formula (5.35) or (5.50) to obtain the Jacobi iterations.

Note: We can also obtain the matrix form of the Jacobi iteration formula directly, by using following procedure. Consider the linear system (5.1)

$$AX = B$$
$$(L+D+U)X = B$$
$$DX = -(L+U)X + B$$
$$X = -D^{-1}(L+U)X + D^{-1}B$$

Using the kth iteration $X^{(k)}$, we can obtain the Jacobi iterative formula (5.50)

$$X^{(k+1)} = -D^{-1}(L+U)X^{(k)} + D^{-1}B; \text{ for each } k = 0, 1, 2,\dots$$

Matrix Form of Gauss–Seidel Method
The Gauss–Seidel iteration formula (5.38) is given by

$$x_i^{(k+1)} = \frac{1}{a_{ii}}\left(b_i - \sum_{j=1}^{i-1} a_{ij}x_j^{(k+1)} - \sum_{j=i+1}^{n} a_{ij}x_j^{(k)} \right); \qquad 1 \le i \le n$$

$$a_{ii}x_i^{(k+1)} = b_i - \sum_{j=1}^{i-1} a_{ij}x_j^{(k+1)} - \sum_{j=i+1}^{n} a_{ij}x_j^{(k)}; \qquad 1 \le i \le n$$

$$a_{ii}x_i^{(k+1)} + \sum_{j=1}^{i-1} a_{ij}x_j^{(k+1)} = -\sum_{j=i+1}^{n} a_{ij}x_j^{(k)} + b_i; \qquad 1 \le i \le n$$

$$\left(a_{ii} + \sum_{j=1}^{i-1} a_{ij} \right)x_i^{(k+1)} = -\sum_{j=i+1}^{n} a_{ij}x_j^{(k)} + b_i; \qquad 1 \le i \le n$$

In matrix form, we have

$$(D+L)X^{(k+1)} = -UX^{(k)} + B$$

It gives the following Gauss–Seidel formula

$$X^{(k+1)} = -(D+L)^{-1}UX^{(k)} + (D+L)^{-1}B \text{ for each } k = 0, 1, 2, \ldots \tag{5.52}$$

We can write this system as follows

$$X^{(k+1)} = T_{GS}X^{(k)} + C_{GS} \tag{5.53}$$

where $T_{GS} = -(D+L)^{-1}U$ and $C_{GS} = (D+L)^{-1}B$.

Note that the Gauss–Seidel formulas (5.38) and (5.52) are equivalent formulas and iterations obtained from both the formulas are equal.

Matrix Form of Relaxation Method

a) Relaxation Method with Jacobi Iteration

The relaxation method with Jacobi iteration (5.41) is given by

$$x_i^{(k+1)} = (1-\omega)x_i^{(k)} + \frac{\omega}{a_{ii}}\left(b_i - \sum_{\substack{j=1 \\ j \neq i}}^{n} a_{ij}x_j^{(k)}\right); \qquad 1 \leq i \leq n$$

$$a_{ii}x_i^{(k+1)} = (1-\omega)a_{ii}x_i^{(k)} + \omega\left(b_i - \sum_{\substack{j=1 \\ j \neq i}}^{n} a_{ij}x_j^{(k)}\right); \qquad 1 \leq i \leq n$$

$$a_{ii}x_i^{(k+1)} = a_{ii}x_i^{(k)} + \omega\left(b_i - \sum_{j=1}^{n} a_{ij}x_j^{(k)}\right); \qquad 1 \leq i \leq n$$

$$a_{ii}x_i^{(k+1)} = a_{ii}x_i^{(k)} - \omega\sum_{j=1}^{n} a_{ij}x_j^{(k)} + \omega b_i; \qquad 1 \leq i \leq n$$

In matrix form, we can write the above equations as follows

$$DX^{(k+1)} = DX^{(k)} - \omega AX^{(k)} + \omega B$$

$$X^{(k+1)} = X^{(k)} - \omega D^{-1}AX^{(k)} + \omega D^{-1}B$$

$$X^{(k+1)} = \left(I - \omega D^{-1}A\right)X^{(k)} + \omega D^{-1}B \text{ ; for each } k = 0, 1, 2, \ldots \tag{5.54}$$

The system (5.54) can also be written as follows

$$X^{(k+1)} = T_{RJ}X^{(k)} + C_{RJ} \tag{5.55}$$

where $T_{RJ} = I - \omega D^{-1}A$ and $C_{RJ} = \omega D^{-1}B$.

b) Relaxation Method with Gauss–Seidel Iteration

The relaxation method with Gauss–Seidel iteration (5.42) is as follows

$$x_i^{(k+1)} = (1-\omega)x_i^{(k)} + \frac{\omega}{a_{ii}}\left(b_i - \sum_{j=1}^{i-1}a_{ij}x_j^{(k+1)} - \sum_{j=i+1}^{n}a_{ij}x_j^{(k)}\right); \qquad 1 \le i \le n$$

Or $$a_{ii}x_i^{(k+1)} = (1-\omega)a_{ii}x_i^{(k)} + \omega\left(b_i - \sum_{j=1}^{i-1}a_{ij}x_j^{(k+1)} - \sum_{j=i+1}^{n}a_{ij}x_j^{(k)}\right); \qquad 1 \le i \le n$$

Or $$\left(a_{ii} + \omega\sum_{j=1}^{i-1}a_{ij}\right)x_j^{(k+1)} = (1-\omega)a_{ii}x_i^{(k)} + \omega\left(b_i - \sum_{j=i+1}^{n}a_{ij}x_j^{(k)}\right); \qquad 1 \le i \le n$$

In vector form, we have

$$(D+\omega L)X^{(k+1)} = (1-\omega)DX^{(k)} - \omega UX^{(k)} + \omega B$$

$$(D+\omega L)X^{(k+1)} = ((1-\omega)D - \omega U)X^{(k)} + \omega B$$

$$X^{(k+1)} = (D+\omega L)^{-1}((1-\omega)D - \omega U)X^{(k)} + \omega(D+\omega L)^{-1}B \text{ for each } k = 0, 1, 2,....$$

$$(5.56)$$

The system (5.56) can also be written as follows

$$X^{(k+1)} = T_{RGS}X^{(k)} + C_{RGS} \qquad (5.57)$$

where $T_{RGS} = (D+\omega L)^{-1}((1-\omega)D - \omega U)$ and $C_{RGS} = \omega(D+\omega L)^{-1}B$

Example — 5.33

Solve the following system of linear equations with the help of matrix form of Gauss–Seidel method

$$4x_1 - 2x_2 + x_3 = 4$$
$$x_1 - 4x_2 + x_3 = 0$$
$$x_1 + 2x_2 + 5x_3 = 7$$

Use initial approximation vector, $X^{(0)} = [0 \quad 0 \quad 0]^T$.

Ans. Matrix $A = L + D + U$ can be written as follows

$$\begin{bmatrix} 4 & -2 & 1 \\ 1 & -4 & 1 \\ 1 & 2 & 5 \end{bmatrix} = \begin{bmatrix} 0 & 0 & 0 \\ 1 & 0 & 0 \\ 1 & 2 & 0 \end{bmatrix} + \begin{bmatrix} 4 & 0 & 0 \\ 0 & -4 & 0 \\ 0 & 0 & 5 \end{bmatrix} + \begin{bmatrix} 0 & -2 & 1 \\ 0 & 0 & 1 \\ 0 & 0 & 0 \end{bmatrix}$$
$$\quad\quad A \quad\quad\quad\quad\quad\quad L \quad\quad\quad\quad\quad\quad D \quad\quad\quad\quad\quad\quad U$$

Using the Gauss–Seidel method (5.52), we have

$$X^{(k+1)} = -(D+L)^{-1} U X^{(k)} + (D+L)^{-1} B$$

$$\begin{bmatrix} x_1^{(k+1)} \\ x_2^{(k+1)} \\ x_3^{(k+1)} \end{bmatrix} = -\left(\begin{bmatrix} 4 & 0 & 0 \\ 0 & -4 & 0 \\ 0 & 0 & 5 \end{bmatrix} + \begin{bmatrix} 0 & 0 & 0 \\ 1 & 0 & 0 \\ 1 & 2 & 0 \end{bmatrix} \right)^{-1} \begin{bmatrix} 0 & -2 & 1 \\ 0 & 0 & 1 \\ 0 & 0 & 0 \end{bmatrix} \begin{bmatrix} x_1^{(k)} \\ x_2^{(k)} \\ x_3^{(k)} \end{bmatrix}$$

$$+ \left(\begin{bmatrix} 4 & 0 & 0 \\ 0 & -4 & 0 \\ 0 & 0 & 5 \end{bmatrix} + \begin{bmatrix} 0 & 0 & 0 \\ 1 & 0 & 0 \\ 1 & 2 & 0 \end{bmatrix} \right)^{-1} \begin{bmatrix} 4 \\ 0 \\ 7 \end{bmatrix}$$

Using $k = 0$, and initial vector, $X^{(0)} = [0 \quad 0 \quad 0]^T$ in the above system, the first iteration $X^{(1)}$ is given by

$$X^{(1)} = \begin{bmatrix} 1 & 0.25 & 1.1 \end{bmatrix}^T$$

Similarly, the other iterations are given by

Iteration 2
0.850000   0.487500   1.035000

Iteration 3
0.985000   0.505000   1.001000

Iteration 4
1.002250   0.500813   0.999225

Iteration 5
1.000600   0.499956   0.999897

Iteration 6
1.000004   0.499975   1.000009

After six iterations, the solution is as follows

$$x_1 = 1.000004,\ x_2 = 0.499975,\ x_3 = 1.000009$$

Example ── **5.34** ──────

Solve the following system of linear equations with the help of matrix form of relaxation method with successive displacement

$$3x_1 - x_2 + x_3 = 5$$
$$x_1 - 3x_2 + x_3 = 5$$
$$x_1 + 2x_2 + 4x_3 = 7$$

Use initial approximation vector $X^{(0)} = [0 \quad 0 \quad 0]^T$ and relaxation parameter, $\omega = 1.02943725$.

Ans. Matrix $A = L + D + U$ can be written as follows

$$
\underbrace{\begin{bmatrix} 3 & -1 & 1 \\ 1 & -3 & 1 \\ 1 & 2 & 4 \end{bmatrix}}_{A} = \underbrace{\begin{bmatrix} 0 & 0 & 0 \\ 1 & 0 & 0 \\ 1 & 2 & 0 \end{bmatrix}}_{L} + \underbrace{\begin{bmatrix} 3 & 0 & 0 \\ 0 & -3 & 0 \\ 0 & 0 & 4 \end{bmatrix}}_{D} + \underbrace{\begin{bmatrix} 0 & -1 & 1 \\ 0 & 0 & 1 \\ 0 & 0 & 0 \end{bmatrix}}_{U}
$$

Using the matrix form (5.54), we have

$$X^{(k+1)} = (D + \omega L)^{-1} \left((1-\omega)D - \omega U \right) X^{(k)} + \omega (D + \omega L)^{-1} B$$

$$
\begin{bmatrix} x_1^{(k+1)} \\ x_2^{(k+1)} \\ x_3^{(k+1)} \end{bmatrix} = \left(\begin{bmatrix} 3 & 0 & 0 \\ 0 & -3 & 0 \\ 0 & 0 & 4 \end{bmatrix} + 1.02943725 \begin{bmatrix} 0 & 0 & 0 \\ 1 & 0 & 0 \\ 1 & 2 & 0 \end{bmatrix} \right)^{-1}
$$

$$
\left(-0.02943725 \begin{bmatrix} 3 & 0 & 0 \\ 0 & -3 & 0 \\ 0 & 0 & 4 \end{bmatrix} - 1.02943725 \begin{bmatrix} 0 & -1 & 1 \\ 0 & 0 & 1 \\ 0 & 0 & 0 \end{bmatrix} \right) \begin{bmatrix} x_1^{(k)} \\ x_2^{(k)} \\ x_3^{(k)} \end{bmatrix}
$$

$$
+ 1.02943725 \left(\begin{bmatrix} 3 & 0 & 0 \\ 0 & -3 & 0 \\ 0 & 0 & 4 \end{bmatrix} + 1.02943725 \begin{bmatrix} 0 & 0 & 0 \\ 1 & 0 & 0 \\ 1 & 2 & 0 \end{bmatrix} \right)^{-1} \begin{bmatrix} 5 \\ 5 \\ 7 \end{bmatrix}
$$

Using the initial vector $X^{(0)} = [0 \quad 0 \quad 0]^T$ in the above system, the first iteration $X^{(1)}$ is given by

$$X^{(1)} = \begin{bmatrix} 1.715729 & -1.126984 & 1.940036 \end{bmatrix}^T$$

Similarly, the other iterations are given by

Iteration 2
0.612787   −0.806563   2.001852

Iteration 3
0.733994   −0.753192   1.941368

Iteration 4
0.769495   −0.763336   1.939233

Iteration 5
0.765702   −0.765071   1.941165

Iteration 6
0.764555   −0.764751   1.941239

Iteration 7
0.764674   −0.764694   1.941177

Iteration 8
0.764711   −0.764704   1.941175

After eight iterations, the solution is as follows

$$x_1 = 0.764711, \ x_2 = -0.764704, \ x_3 = 1.941175$$

Convergence Results for Iterative Methods

It is clear from the matrix forms [(5.51), (5.53), (5.55) and (5.57)] of the iterative formulas that the iterative methods have the following form

$$X^{(k+1)} = T X^{(k)} + C, \quad \text{for each } k = 0, 1, 2, \ldots \tag{5.58}$$

where T and C are fixed matrices for a given system and $X^{(0)}$ is arbitrary. The matrix T is known as iteration matrix. Note that the iteration matrices (T) for Jacobi (5.51), Gauss–Seidel (5.53), and relaxation methods (5.55, 5.57) are $T_J = -D^{-1}(L+U)$, $T_{GS} = -(D+L)^{-1}U$ and $T_{RJ} = I - \omega D^{-1}A$, $T_{RGS} = (D+\omega L)^{-1}\big((1-\omega)D - \omega U\big)$, respectively. To study the convergence of iterative methods (Jacobi, Gauss–Seidel, and Relaxation), we need to analyze the formula (5.58).

It is not possible to conclude on the convergence properties of the iterative methods for the general system (5.1) of linear equations. However, there exist some special classes of the coefficient matrix A, for which we have convergence results for the iterative methods. Here, we are stating some useful results regarding the convergence of the iterative methods.

Definition 5.1

Consider any matrix $A \in C^{n \times n}$ and let $\lambda_1, \lambda_2, ..., \lambda_n$ be the eigenvalues (real or complex) of this matrix, then the spectral radius $\rho(A)$ is defined as follows

$$\rho(A) = \max\{|\lambda_1|, |\lambda_2|, ..., |\lambda_n|\}$$

Theorem 5.2

For any initial approximation, $X^{(0)}$, the sequence $\{X^{(k)}\}_{k=1}^{\infty}$ generated by the formula (5.58)

$$X^{(k+1)} = T X^{(k)} + C, \qquad \text{for each } k = 0, 1, 2, ...$$

converges to the unique solution of the system $X = T X + C$ if and only if the spectral radius $\rho(T) < 1$.

The theorem (5.2) provides the necessary and sufficient condition for the convergence of general iterative formula (5.58). Using Theorem 5.2, we can easily prove the following result for the convergence of iterative methods.

Corollary: For the convergence of iterative methods (Jacobi, Gauss–Seidel, and Relaxation), the necessary and sufficient condition is that the absolute values of all the eigenvalues of iteration matrix T are less than unity.

$$|\lambda_i(T)| < 1 \qquad i = 1, 2, ..., n$$

where the iteration matrices (T) for Jacobi, Gauss–Seidel and relaxation methods are given by following expressions

$$T_J = -D^{-1}(L+U)$$
$$T_{GS} = -(D+L)^{-1}U$$
$$T_{RJ} = I - \omega D^{-1} A$$
$$T_{RGS} = (D+\omega L)^{-1}((1-\omega)D - \omega U)$$

Since it is very difficult to compute the eigenvalues of iteration matrix T, therefore we use this condition only when it requires.

Theorem 5.3

Let the coefficient matrix A be a strictly diagonally dominant matrix, then the relaxation methods converge if $0 < \omega \leq 1$.

Theorem 5.4

The relaxation method with Jacobi iterations is convergent if the corresponding Jacobi method is convergent and $0 < \omega \leq 1$.

Theorem 5.5 (Kahan)

The relaxation methods fail to converge if the relaxation parameter $\omega \leq 0$ or $\omega \geq 2$.

Theorem 5.6

If the coefficient matrix A is a positive definite symmetric matrix, then the Gauss–Seidel method is convergent.

Theorem 5.7 (Ostrowski–Reich)

If the coefficient matrix A is a positive definite matrix, then the relaxation methods are convergent for any choice of initial approximate vector, if and only if $\omega \in (0,2)$.

We have already discussed that the relaxation methods have more or less rapid convergence depending on the choice of relaxation parameter. For the optimal relaxation parameter with the highest convergence, the following theorem is useful.

Theorem 5.8

If the coefficient matrix A is positive definite and tridiagonal, then

$$\rho(T_{GS}) = \left[\rho(T_J)\right]^2 < 1$$

The optimal relaxation parameter for the relaxation methods is given by

$$\omega_{opt} = \frac{2}{1 + \sqrt{1 - \left[\rho(T_J)\right]^2}}$$

Example ──────────────────────────────────── 5.35

Compute the optimal relaxation parameter for the system of linear equations in Example 5.29.

Ans. The system of linear equations in Example 5.29 is given by

$$3x_1 - 2x_2 = 18$$
$$-2x_1 + 3x_2 - x_3 = -19$$
$$-x_2 + 3x_3 = 9$$

Consider the matrix $A = L + D + U$ as follows

$$
\begin{bmatrix} 3 & -2 & 0 \\ -2 & 3 & -1 \\ 0 & -1 & 3 \end{bmatrix} = \begin{bmatrix} 0 & 0 & 0 \\ -2 & 0 & 0 \\ 0 & -1 & 0 \end{bmatrix} + \begin{bmatrix} 3 & 0 & 0 \\ 0 & 3 & 0 \\ 0 & 0 & 3 \end{bmatrix} + \begin{bmatrix} 0 & -2 & 0 \\ 0 & 0 & -1 \\ 0 & 0 & 0 \end{bmatrix}
$$
$$\quad A \qquad\qquad\qquad L \qquad\qquad\qquad D \qquad\qquad\qquad U$$

The iteration matrix for Jacobi method is given by

$$T_J = -D^{-1}(L+U)$$

$$
T_J = -D^{-1}(L+U) = -\begin{bmatrix} 1/3 & 0 & 0 \\ 0 & 1/3 & 0 \\ 0 & 0 & 1/3 \end{bmatrix}\begin{bmatrix} 0 & -2 & 0 \\ -2 & 0 & -1 \\ 0 & -1 & 0 \end{bmatrix} = \begin{bmatrix} 0 & 2/3 & 0 \\ 2/3 & 0 & 1/3 \\ 0 & 1/3 & 0 \end{bmatrix}
$$

The eigenvalues of iteration matrix T_J are the elements of the vector $\begin{bmatrix} 0 \\ \dfrac{\sqrt{5}}{3} \\ \dfrac{-\sqrt{5}}{3} \end{bmatrix}$

On using the eigenvalue $\rho(T_J) = \dfrac{\sqrt{5}}{3}$, the optimal relaxation parameter is given by

$$\omega_{opt} = \frac{2}{1+\sqrt{1-\left[\rho(T_J)\right]^2}} = \frac{2}{1+\sqrt{1-\left[\dfrac{\sqrt{5}}{3}\right]^2}} = 1.2$$

The optimal relaxation parameter is given by $\omega = 1.2$. With this value of relaxation parameter, the relaxation method requires a minimum number of iterations as given in Example 5.29.

Table 5.2 Formulae for Iterative Methods

Method	Formulation (Next Iteration $x_i^{(k+1)}$, $1 \le i \le n$)	Matrix Form ($X^{(k+1)} = T X^{(k)} + C$)
Jacobi	$x_i^{(k+1)} = \dfrac{1}{a_{ii}}\left(b_i - \displaystyle\sum_{\substack{j=1 \\ j \ne i}}^{n} a_{ij} x_j^{(k)} \right);\ 1 \le i \le n$	$X^{(k+1)} = -D^{-1}(L+U)X^{(k)} + D^{-1}B$
Gauss–Seidel	$x_i^{(k+1)} = \dfrac{1}{a_{ii}}\left(b_i - \displaystyle\sum_{j=1}^{i-1} a_{ij} x_j^{(k+1)} - \sum_{j=i+1}^{n} a_{ij} x_j^{(k)} \right);\ 1 \le i \le n$	$X^{(k+1)} = -(D+L)^{-1}UX^{(k)} + (D+L)^{-1}B$
Relaxation (Jacobi)	$x_i^{(k+1)} = (1-\omega)x_i^{(k)} + \dfrac{\omega}{a_{ii}}\left(b_i - \displaystyle\sum_{\substack{j=1 \\ j \ne i}}^{n} a_{ij} x_j^{(k)} \right);\ 1 \le i \le n$	$X^{(k+1)} = \left(I - \omega D^{-1}A\right)X^{(k)} + \omega D^{-1}B$
Relaxation (Gauss–Seidel)	$x_i^{(k+1)} = (1-\omega)x_i^{(k)} + \dfrac{\omega}{a_{ii}}\left(b_i - \displaystyle\sum_{j=1}^{i-1} a_{ij} x_j^{(k+1)} - \sum_{j=i+1}^{n} a_{ij} x_j^{(k)} \right);\ 1 \le i \le n$	$X^{(k+1)} = (D+\omega L)^{-1}\left((1-\omega)D - \omega U\right)X^{(k)} + \omega(D+\omega L)^{-1}B$
Stopping Criterion	In all the iterative methods, stopping criterion is last two iterations matches up to desired decimal points (or whatever accuracy we required, say 0.000001, etc.)	

The necessary and sufficient condition for the convergence of iterative methods (Jacobi, Gauss–Seidel, and Relaxation), is that the absolute values of all the eigenvalues of iteration matrix T are less than unity.

$$|\lambda_i(T)| < 1;\ \text{for each } i = 1, 2, \ldots, n$$

The iteration matrices (T) for Jacobi, Gauss–Seidel, and relaxation methods are $T_J = -D^{-1}(L+U)$, $T_{GS} = -(D+L)^{-1}U$ and $T_{RJ} = I - \omega D^{-1}A$,

$T_{RGS} = (D+\omega L)^{-1}\left((1-\omega)D - \omega U\right)$, respectively.

Order of convergence is linear (1) for all the methods. But the Gauss–Seidel method converges faster than Jacobi method. The relaxation method is also faster for appropriate choice of relaxation parameter ω.

5.15 Discussion

This section deals with some observations about the direct and iterative methods for the solution of a system of linear equations.

i) Direct methods are applicable to any general system (5.1) of linear equations, while the iterative methods require very cumbersome calculations and converge for only a limited type of systems which satisfy the convergence conditions. The question is why to use iterative methods? In direct methods, as a reference example of Gauss elimination method, we have discussed the round-off errors and some remedies to reduce these errors. The direct methods require large number of arithmetic computations for the higher-order systems, so the round-off error is significant. Iterative procedures can be used for such systems. The iterative methods may require large numbers of iterations to produce the result with higher accuracy. But, once the algorithms for these methods are implemented, these iterations can be computed with the advent of high-speed computers. In Chapter 16, we will see that large systems of linear equations are produced during the solutions of linear PDEs. In fact, we need to solve large systems of linear equations to reduce the truncation error arises during the finite difference approximations of differential coefficients in PDEs. In this case, it is imperative to use iterative methods for the solutions to avoid the large round-off error in case of direct methods.

ii) In iterative procedures, the main question is when to stop the iterations? In general, we do not have the exact solution, so the accuracy of iteration is not known. Many books follow that when last two iterations are equal up to significant digits required, then we stop the iterations. The last iteration is our final solution corrects up to the number of significant digits mentioned. But this is not true in general, consider the following example with Gauss–Seidel method

$$6x_1 + 7x_2 + 8x_3 = 0$$
$$4x_1 + 9x_2 + 7x_3 = 12$$
$$6x_1 + 8x_2 + 9x_3 = 4$$

Use C – Program (Turbo C) and declare the elements of augmented matrix [A: B] and the variables (x_1, x_2, x_3) as float variables and start with the initial approximation (100, 10, 10). Then at 37th iteration, the result is (−5.4356, 1.7629, 2.4878) and at 38th iteration, the result is (−5.3738, 1.7867, 2.4388). Rather, the difference between these two consecutive iterations is less than 0.05, but the result is not correct even up to one decimal place as the exact solution is (−5, 2, 2).

iii) If the system satisfies the convergence condition (spectral radius of iteration matrix is less than unity or diagonally dominant system), then both Jacobi and Gauss–Seidel methods must converge to a unique solution of the system. Otherwise, no conclusion can be drawn about the convergence of the methods. In that case, both the methods may converge or diverge simultaneously (Example 5.24); (or) the Jacobi method may fail to converge, while the Gauss–Seidel scheme may be convergent; (or) vice-versa. For example, consider the following system

$$-4x_1 + 4x_2 - 8x_3 = -8$$
$$-5x_1 + 8x_2 - 7x_3 = -4$$
$$3x_1 + 4x_2 - 6x_3 = 1$$

Note that the solution for this system is a unit vector. Let the initial approximation vector be $X^{(0)} = [0\ 0\ 0]^T$. The Jacobi method is convergent for this system, while the Gauss–Seidel method fails to converge.

iv) Many books claim that the iterative methods may converge to the solution, if the initial approximation is close to the solution, but this is not true. If the method is converging for any initial approximation, then it will converge for each initial approximation. If the initial approximation is far from the solution, then the number of iterations increases for the desired accuracy. For example, consider the following system

$$5x_1 + 7x_2 + 8x_3 = 5$$
$$4x_1 + 9x_2 + 7x_3 = 12$$
$$6x_1 + 8x_2 + 9x_3 = 4$$

The iterative methods diverge even for approximations that are very close to the solutions. For example, start with the initial approximation (–4, 1.5, 1.5) and even for a closer initial solution (try yourself; except when the x_2, x_3 is 2, 2, as these are the exact values and in next approximation, we will get the exact result). Similar is the case for following two systems with the solution (1, –1, 1).

The iterative methods converge for the linear system, $\begin{aligned}3x_1 + 3x_2 + 4x_3 &= 4\\ 4x_1 + 7x_2 + 6x_3 &= 3,\\ 5x_1 + 7x_2 + 8x_3 &= 6\end{aligned}$

even if the initial approximation is very far from the exact solution. While the

iterative methods diverge for the system, $\begin{aligned}2x_1 + 3x_2 + 4x_3 &= 4\\ 4x_1 + 7x_2 + 6x_3 &= 3,\\ 5x_1 + 7x_2 + 8x_3 &= 6\end{aligned}$ even the initial

approximation is very close to the exact solution.

Comparison between Direct Methods and Iterative Methods:

In direct methods, if the arithmetic is exact, and the matrix **A** is non-singular, then the solution obtained will be exact. But computer arithmetic is not exact in general; there will be some round-off and truncation error in the result. The numbers of operational counts are high for a large system. So, there is a chance of high round-off error. Pivoting is the only alternate to reduce this error.

In direct methods, once the error is committed, it cannot be recovered. But in iterative procedures, we can increase the number of iterations to enhance the accuracy of the solutions. Especially, in the case of large systems, we can use iterative procedures, provided the availability of a reasonably good computer.

It is also worth to mention here that in SOR method, each time we have to compute optimal relaxation parameter for different systems, which is a very cumbersome job.

5.16 Applications

Electrical Network:
In Electrical Network, determining currents and voltages in resistor circuits is a common problem. Following two conservation laws are used for this purpose.

Kirchhoff's First Law (Conservation of Charge)
Algebraic sum of all the currents flowing through a junction must be zero

$$\sum I = 0$$

where the current flowing into the junction is considered to be positive and flowing out is considered negative.

Kirchhoff's Second Law (Conservation of Energy)
The algebraic sum of the products $I R$ in any closed loop is equal to the algebraic sum of voltage drops.

$$\sum I R = \sum V$$

Example   **5.36**

Use Kirchhoff's laws to determine the currents I_1, I_2 and I_3 for the electrical network shown in the following figure. Note that units for current, resistance and voltage drop in an electrical network are ampere (amp), ohm(Ω) and volt(V), respectively.

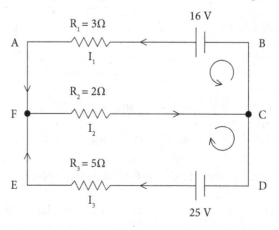

Fig. 5.3 Electrical network 1

Ans.

Kirchhoff's first law at Junction F produces the following equation

$$\sum I = I_1 - I_2 + I_3 = 0 \tag{5.59}$$

The currents I_1 and I_3 are positive as these are flowing towards the junction, and the current I_2 is negative as it is flowing out the junction. Kirchhoff's law at junction C produces the same equation.

Kirchhoff's second law along closed loops AFCB and CDEF produces following two equations

$$I_1 R_1 + I_2 R_2 = 3I_1 + 2I_2 = 16 \tag{5.60}$$
$$I_2 R_2 + I_3 R_3 = 2I_2 + 5I_3 = 25 \tag{5.61}$$

We have the following system of three linear equations (5.59–5.61) in the variables I_1, I_2 and I_3.

$$I_1 - I_2 + I_3 = 0$$
$$3I_1 + 2I_2 = 16$$
$$2I_2 + 5I_3 = 25$$

The solution of system by using Gauss elimination method is given by

$$I_1 = 2, I_2 = 5 \text{ and } I_3 = 3 \text{ amps.}$$

Example ———————————————————————————————— **5.37**

Use Kirchhoff's laws to determine the currents for the electrical network shown in the following figure.

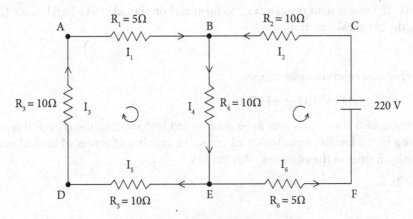

Fig. 5.4 Electrical network 2

Ans.

Kirchhoff's first law at Junctions A, B, D, E produces the following equations

$$-I_1 + I_3 = 0$$
$$I_1 + I_2 - I_4 = 0$$
$$I_4 - I_5 - I_6 = 0$$
$$-I_3 + I_5 = 0 \tag{5.62}$$

Kirchhoff's second law along closed loops ABED and CBEF produces following two equations

$$I_1 R_1 + I_3 R_3 + I_4 R_4 + I_5 R_5 = 5I_1 + 10I_3 + 10I_4 + 10I_5 = 0$$
$$I_2 R_2 + I_4 R_4 + I_6 R_6 = 10I_2 + 10I_4 + 5I_6 = 220 \tag{5.63}$$

The systems (5.62) and (5.63) have six linear equations, and solution of these equations is as follows

$$I_1 = I_3 = I_5 = -2.8387, \ I_2 = I_6 = 9.9355 \text{ and } I_4 = 7.0968 \text{ amps.}$$

Chemical Equations:

In chemical engineering, the law of conservation of mass is used for balancing a chemical equation. It states that "*mass is neither created nor destroyed in any chemical reaction.*" Therefore, balancing of a chemical equation requires the equal number of atoms on both sides of a chemical reaction.

Example ———————————————————————————— **5.38**

Ethane (C_2H_6) reacts with oxygen (O_2) to form carbon dioxide (CO_2) and water (H_2O). Balance the chemical equation.

Ans.

Let the chemical reaction be as follows

$$xC_2H_6 + yO_2 \rightarrow zCO_2 + wH_2O$$

where x, y, z, and w are constants to be determined by balancing the chemical reaction. According to the law of conservation of mass, the number of atoms of each element is equal on both sides of the equation. This implies

$$2x = z$$
$$6x = 2w$$
$$2y = 2z + w$$

The solution of this system is given by

$$z = 2x$$
$$w = 3x$$
$$y = \frac{7}{2}x$$

Since we want integral values of the variables x, y, z and w. Let $x = 2$, we get $y = 7, z = 4$ and $w = 6$. The balanced equation is given by

$$2C_2H_6 + 7O_2 \rightarrow 4CO_2 + 6H_2O$$

Beside many other applications in engineering and sciences, the system of linear equations frequently occurs in solutions of differential equations by finite difference methods, and other numerical methods. Kindly refer to Chapters 15 and 16 for these applications.

Exercise 5

1. Use Cramer, matrix inversion, LU-decomposition (both Crout and Doolittle), Gauss elimination and Gauss–Jordan Methods for the solutions of the following systems of equations

   a) $\begin{bmatrix} 3 & 2 & 0 \\ 2 & 3 & -1 \\ 0 & -1 & 2 \end{bmatrix} X = \begin{bmatrix} 5 \\ 4 \\ 1 \end{bmatrix}$
   b) $\begin{bmatrix} 4 & -1 & 0 \\ -1 & 4 & -1 \\ 0 & -1 & 4 \end{bmatrix} X = \begin{bmatrix} 3 \\ 2 \\ 3 \end{bmatrix}$

   Ans. a), b): $[1 \quad 1 \quad 1]^T$

2. Solve the following systems of linear equations with the help of Cramer, matrix inversion, triangularization (or) LU-decomposition (both Crout and Doolittle), Gauss elimination and Gauss–Jordan Methods

   a) $\begin{aligned} 2x_1 - x_2 + 3x_3 &= 6 \\ x_1 + 3x_2 - x_3 &= -3 \\ -2x_1 - x_2 - 3x_3 &= -4 \end{aligned}$
   b) $\begin{aligned} x_1 + x_2 + x_3 &= 3 \\ 2x_1 + 3x_2 + x_3 &= 5 \\ x_1 - x_2 - x_3 &= -3 \end{aligned}$
   c) $\begin{aligned} x_1 - 3x_2 + x_3 &= 5 \\ x_1 + 2x_2 + 4x_3 &= 3 \\ 3x_1 - x_2 + x_3 &= 5 \end{aligned}$

   Ans. a) $[1 \quad -1 \quad 1]^T$ b) $[0 \quad 1 \quad 2]^T$ c) $[1 \quad -1 \quad 1]^T$

3. Show that the following system cannot be solved with the help of LU-decomposition method

   $$3x_1 - 3x_2 + x_3 = 4$$
   $$-2x_1 + 2x_2 + x_3 = -1$$
   $$x_1 + x_2 + 2x_3 = 3$$

4. Solve the following system of linear equations by Cholesky method

   $$2x_1 + x_2 - x_3 = 6$$
   $$x_1 - 3x_2 + 5x_3 = 11$$
   $$-x_1 + 5x_2 + 4x_3 = 13$$

   Ans. $\begin{bmatrix} \dfrac{59}{17} & \dfrac{24}{17} & \dfrac{40}{17} \end{bmatrix}^T$

5. Find a parabola, $y = a + bx + cx^2$, which passes through the points $(1, 4)$, $(2, 8)$ and $(3, 14)$.

    Ans. $2 + x + x^2$

6. Solve the following system of linear equations by Gauss elimination method with partial and complete pivoting using three significant digits floating points rounding arithmetic.

    $$2x_1 + x_2 + 2x_3 = 4$$
    $$2x_1 - x_2 + 100x_3 = 100$$
    $$3x_1 + 50x_2 + x_3 = 52.5$$

    Compare the obtained results for each case with the exact solution $x_1 = 0.5$, $x_2 = 1$, $x_3 = 1$.

    Ans.

    1. Partial Pivoting:

        Upper triangular matrix: $\begin{bmatrix} 3 & 1 & 200 & 102 \\ 0 & -1.67 & -33.0 & -15.0 \\ 0 & 0 & -77.8 & -39.0 \end{bmatrix}$

        Solution: $x_1 = 0.567$, $\quad x_2 = 0.997$, $\quad x_3 = 1.00$

    2. Complete Pivoting:

        $$\quad\quad\quad x_3 \quad\quad x_2 \quad\quad x_1$$

        Upper triangular matrix: $\begin{bmatrix} 100 & -1 & 2 & 100 \\ 0 & 50.0 & 2.98 & 51.5 \\ 0 & 0 & 1.90 & 0.95 \end{bmatrix}$

        Solution: $x_1 = 0.500$, $\quad x_2 = 1.00$, $\quad x_3 = 1.00$

7. Solve the given system of linear equations by Gauss elimination method with and without partial pivoting. Use five significant digits floating points rounding arithmetic.

    $$5.32x_1 + 2.630x_2 - 11.31x_3 = 27.52$$
    $$2.3x_1 + 18.21x_2 + 5.16x_3 = 43.173$$
    $$-20.7x_1 + 13.51x_2 - 7.4x_3 = -29.683$$

    Compare the solution with the exact solution up to five significant digits
    $[3.0141 \quad 2.1371 \quad -0.51852]^T$.

8. Carry out four significant digits chopping arithmetic to solve the following linear system using Gauss elimination without pivoting and with partial, scaled and complete pivoting strategies and compare it with the exact solution.

    $$-3.1x_1 + 1.2x_2 + 0.7 = 5.673$$
    $$2.3x_1 + 5.3x_2 + 1.6x_3 = 7.203$$
    $$-0.3x_1 + 2.4x_2 + 6.2x_3 = 4.123$$

    Ans. Exact solution up to four significant digits $[-1.127 \quad 1.884 \quad -0.1189]^T$

9. Solve the following system of linear equations with the help of Gauss elimination method. Use seven significant digits rounding arithmetic.

$$27.534x_1 - 8.432x_2 + 2.783x_3 = 51.8932$$
$$13.098x_1 - 45.210x_2 + 7.231x_3 = 43.7638$$
$$-2.134x_1 + 3.564x_2 - 17.230x_3 = -76.4368$$

Use this solution as initial approximation for the Gauss–Seidel method and solve the system to obtain solution correct up to six decimal places

Ans. Exact solution up to seven significant digits $[1.498103 \quad 0.150872 \quad 4.281924]^T$

10. Perform five iterations of Jacobi and Gauss–Seidel iterative methods to solve the following system of linear equations. Consider the initial approximation is $x_1^{(0)} = 0$, $x_2^{(0)} = 0$, $x_3^{(0)} = 0$.

$$16x_1 + 2x_2 - 3x_3 = 34$$
$$23x_1 + 42x_2 + 4x_3 = 88$$
$$5x_1 - 9x_2 + 32x_3 = 1$$

Compare the obtained results with the exact solution $[2 \quad 1 \quad 0]^T$.

Ans.

Jacobi iterations

Iteration 1	2.1250	2.0952	0.0312
Iteration 2	1.8690	0.9286	0.2885
Iteration 3	2.0630	1.0443	0.0004
Iteration 4	1.9945	0.9655	0.0026
Iteration 5	2.0048	1.0027	-0.0089

Gauss–Seidel iterations

Iteration 1	2.125000	0.931548	-0.038783
Iteration 2	2.001285	1.002990	0.000640
Iteration 3	1.999746	1.000078	0.000062
Iteration 4	2.000002	0.999993	-0.000002
Iteration 5	2.000000	1.000000	0.000000

11. Perform 5 iterations of Gauss Seidel method in matrix form to solve the following systems of linear equations.

a)
$$5x_1 + x_2 - 2x_3 = 8$$
$$2x_1 + 3x_2 + x_3 = 4$$
$$x_1 - 3x_2 + 6x_3 = -8$$

b)
$$7x_1 + 3x_2 - x_3 = 3$$
$$2x_1 - 4x_2 + x_3 = 7$$
$$x_1 - 2x_2 + 4x_3 = 7$$

Start with the initial approximation, $x_1^{(0)} = 0$, $x_2^{(0)} = 0$, $x_3^{(0)} = 0$.

Ans.

a) Gauss–Seidel iterations

Iteration 1	1.600000	0.266667	-1.466667
Iteration 2	0.960000	1.182222	-0.902222
Iteration 3	1.002667	0.965630	-1.017630
Iteration 4	0.999822	1.005995	-0.996973
Iteration 5	1.000012	0.998983	-1.000510

b) Gauss–Seidel iterations

Iteration 1   0.428571   −1.535714   0.875000

Iteration 2   1.211735   −0.925383   0.984375

Iteration 3   0.965789   −1.021012   0.998047

Iteration 4   1.008726   −0.996125   0.999756

Iteration 5   0.998305   −1.000909   0.999969

Note that exact solutions are: a) $[1 \quad 1 \quad -1]^T$ and b) $[1 \quad -1 \quad 1]^T$

12. State the convergence condition for the Gauss–Seidel method, and use pivoting such that the following systems satisfy convergence condition. Then, solve the following systems correct up to three decimal places, starting with the initial approximation $(0, 0, 0)$.

$$40x + 5y + 26z = 19 \qquad\qquad 2x + 6y - z = 9$$
$$\text{a)} \quad 10x + 7y + 20z = -3 \qquad \text{b)} \quad 2x + y + 14z = -11$$
$$21x + 30y + 4z = 47 \qquad\qquad 6x + 4y + z = 9$$

Ans. a) To make the system diagonally dominant (for convergence of Gauss–Seidel iterations), interchange the second and third equations. We have

$$40x + 5y + 26z = 19$$
$$21x + 30y + 4z = 47$$
$$10x + 7y + 20z = -3$$

Applying Gauss–Seidel method to this system, we have

Iteration 1   0.475000   1.234167   −0.819458

Iteration 2   0.853377   1.078564   −0.954186

Iteration 3   0.960400   1.021611   −0.987764

Iteration 4   0.989345   1.005827   −0.996712

Iteration 5   0.997135   1.001567   −0.999116

Iteration 6   0.999229   1.000422   −0.999762

Iteration 7   0.999793   1.000113   −0.999936

b) The system can be rewritten as follows

$$6x + 4y + z = 9$$
$$2x + 6y - z = 9$$
$$2x + y + 14z = -11$$

Applying Gauss–Seidel method to this system, we have

Iteration 1   1.500000   1.000000   −1.071429

Iteration 2   1.011905   0.984127   −1.000567

Iteration 3   1.010677   0.996347   −1.001264

Iteration 4   1.002646   0.998907   −1.000300

Iteration 5   1.000778   0.999691   −1.000089

Iteration 6   1.000221   0.999911   −1.000025

13. Find the solution of the following system of linear equations correct to four decimal places with the help of Gauss–Seidel method. Use initial approximation (0.5, –0.5, 0.5).

$$\begin{bmatrix} 2 & -1 & 4 \\ 1 & 3 & -1 \\ -5 & 1 & -3 \end{bmatrix} X = \begin{bmatrix} 7 \\ -3 \\ -9 \end{bmatrix}$$

Ans. Rewrite the system as follows to convert it into the diagonally dominant system

$$\begin{bmatrix} -5 & 1 & -3 \\ 1 & 3 & -1 \\ 2 & -1 & 4 \end{bmatrix} X = \begin{bmatrix} -9 \\ -3 \\ 7 \end{bmatrix}$$

Iteration 1   1.400000   –1.300000   0.725000

Iteration 2   1.105000   –1.126667   0.915833

Iteration 3   1.025167   –1.036444   0.978306

Iteration 4   1.005728   –1.009141   0.994851

Iteration 5   1.001261   –1.002137   0.998835

Iteration 6   1.000272   –1.000479   0.999745

Iteration 7   1.000057   –1.000104   0.999945

Iteration 8   1.000012   –1.000022   0.999988

Iteration 9   1.000003   –1.000005   0.999998

14. Compute 5–7 iterations of Gauss–Seidel method with initial approximations (0, 0, 0) for the following system of equations

$$2x_1 + 3x_2 + x_3 = -1; \ 3x_1 + 2x_2 + 2x_3 = 1; \ ; \ x_1 + 2x_2 + 2x_3 = 6$$

Conclude that the method diverges for this system.

Ans.

Iteration 1   –0.500000   1.250000   2.000000

Iteration 2   –3.375000   3.562500   1.125000

Iteration 3   –6.406250   8.984375   –2.781250

Iteration 4   –12.585938   22.160156   –12.867188

Iteration 5   –27.306641   54.327148   –37.673828

Iteration 6   –63.153809   132.904541   –98.327637

Iteration 7   –150.692993   324.867126   –246.520630

15. Compute 7 iterations of SOR method with Jacobi iterations for the following system with initial approximation $(0 \ 0 \ 0)^T$.

$$5x_1 + x_2 - x_3 = 7$$
$$x_1 + 2x_2 + x_3 = 2$$
$$3x_1 - x_2 + 5x_3 = -3$$

Use relaxation parameter $\omega = 1.15$

Ans.

Iteration 1	1.610000	1.150000	-0.690000
Iteration 2	0.945300	0.448500	-1.432900
Iteration 3	1.035483	1.363095	-1.024167
Iteration 4	0.905607	0.939029	-0.937346
Iteration 5	1.042593	1.027395	-0.988290
Iteration 6	0.996903	0.947417	-1.029344
Iteration 7	1.005809	1.026541	-1.005556

16. Solve the Exercise 15 using SOR method with Gauss–Seidel Iteration

Ans.

Iteration 1	1.610000	0.224250	-1.749323
Iteration 2	0.914578	1.596340	-0.691502
Iteration 3	0.946609	0.763862	-1.063747
Iteration 4	1.047658	1.044672	-1.013048
Iteration 5	0.979576	1.012546	-0.981065
Iteration 6	1.004533	0.984624	-1.009505
Iteration 7	1.000670	1.007386	-0.997338

17. Consider the following systems of linear equations

a) $\begin{bmatrix} 5 & 3 & 0 \\ 3 & 5 & -2 \\ 0 & -2 & 7 \end{bmatrix} \begin{bmatrix} x_1 \\ x_2 \\ x_3 \end{bmatrix} = \begin{bmatrix} 7 \\ 10 \\ 6 \end{bmatrix}$

b) $\begin{bmatrix} 6 & -2 & 0 \\ -2 & 6 & -3 \\ 0 & -3 & 6 \end{bmatrix} \begin{bmatrix} x_1 \\ x_2 \\ x_3 \end{bmatrix} = \begin{bmatrix} 5 \\ 3 \\ 3 \end{bmatrix}$

Find the optimal relaxation parameter ω_{opt} for SOR method, and use it to compute the solution correct upto 3 decimal places. Start with initial approximation $[0.5\ 1\ 1]^T$.

Ans.

a)
$$T_j = -D^{-1}(L+U) = \begin{bmatrix} 0 & -0.6 & 0 \\ -0.6 & 0 & 0.4 \\ 0 & 0.285714 & 0 \end{bmatrix}$$

Eigenvalues of T_j's are $(\pm 0.688684, 0)$

$$\omega_{opt} = \frac{2}{1+\sqrt{1-\mu^2}} = 1.159379$$

Iteration 1	1.623131	1.189664	1.387831
Iteration 2	0.536875	2.399295	1.567331
Iteration 3	-0.131451	2.754654	1.656436
Iteration 4	-0.272132	2.837201	1.669578

Iteration 5	−0.307132	2.854487	1.673210
Iteration 6	−0.313578	2.857900	1.673762
Iteration 7	−0.314925	2.858549	1.673889
Iteration 8	−0.315162	2.858669	1.673908

The solution upto three decimal places is given by

$x_1 = -0.315162, x_2 = 2.858669, x_3 = 1.673908$

b)

$$T_j = -D^{-1}(L+U) = \begin{bmatrix} 0 & 0.333333 & 0 \\ 0.333333 & 0 & 0.5 \\ 0 & 0.5 & 0 \end{bmatrix}$$

Eigenvalues of T_j's are $(\pm 0.600925, 0)$

$$\omega_{opt} = \frac{2}{1+\sqrt{1-\mu^2}} = 1.111540$$

Iteration 1	0.926283	0.898970	1.055391
Iteration 2	1.156046	1.470384	1.255247
Iteration 3	1.342135	1.586671	1.303328
Iteration 4	1.364464	1.605503	1.303328
Iteration 5	1.368951	1.608258	1.304218
Iteration 6	1.369471	1.608638	1.304330
Iteration 7	1.369554	1.608688	1.304346

The solution correct upto three decimal places is as follows

$x_1 = 1.369554, x_2 = 1.608688, x_3 = 1.304346$

Eigenvalues and Eigenvectors

> *Mathematics is the language with which God wrote the universe.*
>
> **Galileo Galilei**
> (February 15, 1564–January 8, 1642)
> He is known as the 'Father of Science'. He was a physicist, mathematician,
> astronomer, engineer, and philosopher.

6.1 Introduction

The eigenvalue problems play a crucial role in many engineering and scientific applications. For example, consider the following system of linear homogeneous first order differential equations with constant coefficients.

$$\frac{dx_1}{dt} = a_{11}x_1 + a_{12}x_2$$

$$\frac{dx_2}{dt} = a_{21}x_1 + a_{22}x_2$$

This system represents many important physical phenomena like concentrations x_1 and x_2 of two constituents in a chemical process as a function of time t. This system in matrix form is as follows

$$\frac{dx}{dt} = Ax \tag{6.1}$$

where $x = \begin{bmatrix} x_1 \\ x_2 \end{bmatrix}$ and $A = \begin{bmatrix} a_{11} & a_{12} \\ a_{21} & a_{22} \end{bmatrix}$

Let the solution of this system be of the following form

$$x = Xe^{\lambda t}$$

where X is a column vector of two elements and λ is a parameter to be determined. On substituting this solution in the system (6.1), we get

$$\lambda X e^{\lambda t} = A X e^{\lambda t}$$

The quantity $e^{\lambda t}$ is a nonzero scalar; after cancellation, we have $AX = \lambda X$.

The problem to solve the system of differential equations is now the problem to determine a scalar λ and a column vector X such that

$$AX = \lambda X$$

This problem is known as the eigenvalue problem.

Let A be a square matrix of order n, X is a column vector of n elements and λ is scalar quantity, then the eigenvalue problem is defined as follows

$$AX = \lambda X \tag{6.2}$$

The column vector X and matrix A are given by

$$X = \begin{bmatrix} x_1 \\ x_2 \\ \vdots \\ x_n \end{bmatrix} \quad \text{and} \quad A = \begin{bmatrix} a_{11} & a_{12} & \cdots & a_{1n} \\ a_{21} & a_{22} & \cdots & a_{2n} \\ \vdots & & & \\ a_{n1} & a_{n2} & \cdots & a_{nn} \end{bmatrix}$$

Equation (6.2) always has zero solution ($X = 0$) for any scalar λ. But this trivial solution is not an interesting and useful solution from application points. So, the aim is to find the nontrivial solutions X, known as eigenvectors or characteristic vectors. These solutions exist for some particular values of λ called as eigenvalues or characteristic values (or sometimes latent roots or proper values). The set of all eigenvalues is called as a spectrum of A. The radius ρ of the smallest circle, with center at the origin, and which contains all the eigenvalues is called the spectral radius of the A.

In terms of linear transformation, the eigenvalue problem can be defined as follows

we know that any matrix A transform a column vector X into a new column vector Y, i.e.

$$AX = Y$$

where A is a square matrix of order n, and X and Y are column vectors of n elements. If we want our transformed vector Y is a scalar multiple of X, then the problem now transformed to eigenvalue problem

$$AX = \lambda X$$

Only certain special vectors X and scalars λ satisfy this equation, and these are known as eigenvectors and eigenvalues, respectively. The eigenvector X has the property that the transformation A does not change its direction, but only scaled by a factor of λ. In fact, matrix A acts to stretch the vector X without changing the direction.

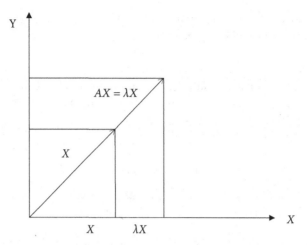

Fig.6.1 Eigenvalue and eigenvector

6.2 Eigenvalues and Eigenvectors

In this section, we will discuss the method to compute the eigenvalues $(\lambda_1, \lambda_2, \ldots \lambda_n)$ and corresponding eigenvectors $(X_1, X_2, \ldots X_n)$ of a square matrix A of order n. Equation (6.2) is given by

$$AX = \lambda X$$

$$AX = \lambda(IX)$$

where the matrix I is the identity matrix. We can rewrite the above equation in the following form

$$(A - \lambda I)X = 0 \tag{6.3}$$

The expression (6.3) is a homogeneous system of n linear equations. The system (6.3) has nontrivial solutions, if

$$\det(A - \lambda I) = \begin{vmatrix} a_{11} - \lambda & a_{12} & \cdots & a_{1n} \\ a_{21} & a_{22} - \lambda & \cdots & a_{2n} \\ \vdots & & & \\ a_{n1} & a_{n2} & \cdots & a_{nn} - \lambda \end{vmatrix} = 0$$

Let

$$p(\lambda) = \det(A - \lambda I) = 0 \tag{6.4}$$

The $p(\lambda)$ is the polynomial of degree n. $p(\lambda)$ is known as the characteristic polynomial, and the equation $p(\lambda) = \det(A - \lambda I) = 0$ is called as a characteristic equation. Roots of Eq.(6.4) are the eigenvalues of matrix A (eigenvalues may be complex). Once we obtain the eigenvalue λ_k, the corresponding eigenvector X_k is derived by solving following homogeneous system for nontrivial solution

$$(A - \lambda_k I) X_k = 0$$

It easy to see that if $X = \begin{bmatrix} x_1 & x_2 & \dots & x_n \end{bmatrix}^T$ is an eigenvector, then for any scalar, α, αX is also an eigenvector.

$$AX = \lambda X$$
$$\alpha AX = \alpha \lambda X$$
$$A(\alpha X) = \lambda(\alpha X)$$

So, we often use the normalized eigenvector $\widehat{X}$ defined by

$$\widehat{X} = \frac{1}{\|X\|} \begin{bmatrix} x_1 \\ x_2 \\ \vdots \\ x_n \end{bmatrix} \tag{6.5}$$

where $\|X\| = \left(X^T X\right)^{1/2} = \left(x_1^2 + x_2^2 + \dots + x_n^2\right)$ is the norm or Euclidean length of the vector X in the space $\mathfrak{R}^n$.

Let any eigenvalue μ be repeated r times with k linearly independent eigenvectors; then r is the algebraic multiplicity, and k is the geometric multiplicity of the eigenvalue, μ. The linearly independent eigenvectors of any matrix A form a basis for the corresponding vector space. This space comprises the all possible linear combinations of these eigenvectors. It is known as solution space or eigenspace. Now, we will discuss different types of eigenvalues problems.

6.2.1 Real Eigenvalues

Example 6.1

Find the eigenvalues and corresponding eigenvectors of the matrix $A = \begin{bmatrix} 1 & 2 \\ 3 & 2 \end{bmatrix}$.

Ans. The characteristic polynomial of matrix A is as follows

$$p(\lambda) = |A - \lambda I| = \begin{vmatrix} 1 - \lambda & 2 \\ 3 & 2 - \lambda \end{vmatrix} = (\lambda - 4)(\lambda + 1)$$

The solution of characteristic equation $p(\lambda) = 0$ provides the two eigenvalues $\lambda_1 = 4$ and $\lambda_2 = -1$. The spectrum of matrix A is the set $\{4, -1\}$ and the spectral radius is 4.

For the eigenvalue, $\lambda_1 = 4$, the corresponding eigenvector $X_1 = \begin{bmatrix} x_1 & x_2 \end{bmatrix}^T$ is the solution of the following homogeneous system

$$(A - 4I)X_1 = 0$$

$$\left(\begin{bmatrix} 1 & 2 \\ 3 & 2 \end{bmatrix} - 4 \begin{bmatrix} 1 & 0 \\ 0 & 1 \end{bmatrix} \right) \begin{bmatrix} x_1 \\ x_2 \end{bmatrix} = \begin{bmatrix} 0 \\ 0 \end{bmatrix}$$

$$-3x_1 + 2x_2 = 0$$

$$3x_1 - 2x_2 = 0$$

These two equations are the same, so effectively we have only one equation in two variables. We can set one variable as arbitrary. Let $x_2 = \alpha$, then $x_1 = \dfrac{2}{3}\alpha$. Therefore, the eigenvector X_1 corresponding to the eigenvalue $\lambda_1 = 4$ is given by

$$X_1 = \alpha \begin{bmatrix} \dfrac{2}{3} \\ 1 \end{bmatrix}$$

The normalized eigenvector is $\widehat{X_1} = \begin{bmatrix} \dfrac{2}{\sqrt{13}} \\ \dfrac{3}{\sqrt{13}} \end{bmatrix}$

Similarly, for the eigenvalue, $\lambda_2 = -1$, the pair of consistent equations $(A+1)X_2 = 0$ is as follows

$$2x_1 + 2x_2 = 0$$

$$3x_1 + 3x_2 = 0$$

The eigenvector X_2 corresponding to the eigenvalue, $\lambda_2 = -1$, is $X_2 = \alpha \begin{bmatrix} 1 \\ -1 \end{bmatrix}$ for any arbitrary scalar α, and corresponding normalized eigenvector is $\widehat{X_2} = \begin{bmatrix} \dfrac{1}{\sqrt{2}} \\ \dfrac{-1}{\sqrt{2}} \end{bmatrix}$. Note that the vectors X_1 and X_2 are linearly independent eigenvectors, so they form a basis for the solution space or eigenspace, which is R^2 space.

6.2.2 Complex Eigenvalues

Example 6.2

Find the eigenvalues and corresponding eigenvectors of the matrix $A = \begin{bmatrix} 1 & -2 \\ 3 & 2 \end{bmatrix}$.

Ans.
The characteristic equation for matrix A is given by

$$p(\lambda) = |A - \lambda I| = \begin{vmatrix} 1-\lambda & -2 \\ 3 & 2-\lambda \end{vmatrix} = \lambda^2 - 3\lambda + 8 = 0$$

Roots of this characteristics equation give eigenvalues $\frac{1}{2}\left(3 \pm i\sqrt{23}\right)$. For the eigenvalue

$\lambda_1 = \frac{1}{2}\left(3 + i\sqrt{23}\right)$, the corresponding eigenvector is the solution of the following homogeneous system

$$-\frac{1}{2}\left(1 + i\sqrt{23}\right)x_1 - 2x_2 = 0$$

$$3x_1 + \frac{1}{2}\left(1 - i\sqrt{23}\right)x_2 = 0$$

The solution of the system is an eigenvector $X_1 = \alpha \begin{bmatrix} 2 \\ \dfrac{-1 - i\sqrt{23}}{2} \end{bmatrix}$, where α is an arbitrary

complex constant. The corresponding normalized eigenvector is $\widehat{X}_1 = \begin{bmatrix} \dfrac{1}{\sqrt{7}} \\ \dfrac{-1 - i\sqrt{23}}{4\sqrt{7}} \end{bmatrix}$.

Similarly, the linear system of equations for the eigenvalue $\lambda_2 = \frac{1}{2}\left(3 - i\sqrt{23}\right)$ is as follow

$$\frac{1}{2}\left(-1 + i\sqrt{23}\right)x_1 - 2x_2 = 0$$

$$3x_1 + \frac{1}{2}\left(1 + i\sqrt{23}\right)x_2 = 0$$

The corresponding eigenvector is $X_2 = \alpha \begin{bmatrix} 2 \\ \dfrac{-1 + i\sqrt{23}}{2} \end{bmatrix}$. The normalized eigenvector is

$$\widehat{X}_1 = \begin{bmatrix} \dfrac{1}{\sqrt{7}} \\ \dfrac{-1 + i\sqrt{23}}{4\sqrt{7}} \end{bmatrix}.$$

6.2.3 Matrix with Real and Distinct Eigenvalues

Example ─── 6.3

Determine the eigenvalues and normalized eigenvectors of the matrix

$$A = \begin{bmatrix} 0 & 2 & 4 \\ 1 & 1 & -2 \\ -2 & 0 & 5 \end{bmatrix}.$$

Ans.

Characteristic equation

$$p(\lambda) = \begin{vmatrix} 0-\lambda & 2 & 4 \\ 1 & 1-\lambda & -2 \\ -2 & 0 & 5-\lambda \end{vmatrix} = -\lambda^3 + 6\lambda^2 - 11\lambda + 6 = 0$$

gives eigenvalues, $\lambda_1 = 1$, $\lambda_2 = 2$, $\lambda_3 = 3$. Since the eigenvalues are distinct, we will get three linearly independent eigenvectors corresponding to each eigenvalue. For the first eigenvalue $\lambda_1 = 1$, the corresponding homogeneous system is as follows

$$(A - I)X = 0$$

In algebraic form, this system is given by

$$-x_1 + 2x_2 + 4x_3 = 0$$
$$x_1 - 2x_3 = 0$$
$$-2x_1 + 4x_3 = 0$$

This system of equations has only two linearly independent equations (as the second equation is a constant multiple of the third equation). We can assume one variable, let x_3, as arbitrary and other variables in terms of this variable. Let $x_3 = \alpha$, then $x_1 = 2\alpha$ and $x_2 = -\alpha$; so the eigenvector is $X_1 = \alpha \begin{bmatrix} 2 \\ -1 \\ 1 \end{bmatrix}$. We can easily deduce the normalized

eigenvector $\widehat{X}_1 = \dfrac{1}{\sqrt{6}} \begin{bmatrix} 2 \\ -1 \\ 1 \end{bmatrix}.$

Similarly, corresponding to eigenvalues $\lambda_2 = 2$ and $\lambda_3 = 3$, the normalized eigenvectors

are $\widehat{X}_2 = \dfrac{1}{\sqrt{14}} \begin{bmatrix} 3 \\ -1 \\ 2 \end{bmatrix}$ and $\widehat{X}_3 = \dfrac{1}{3} \begin{bmatrix} 2 \\ -1 \\ 2 \end{bmatrix}$, respectively. Note that the three linearly independent

eigenvectors form a basis for the solution space of the eigenvalue problem. The solution space is a real three-dimensional vector space.

6.2.4 Matrix with Real and Repeated Eigenvalues

6.2.4.1 Linearly independent eigenvectors

Example **6.4**

Find the eigenvalues and eigenvectors of the matrix $A = \begin{bmatrix} -1 & 3 & -3 \\ -3 & 5 & -3 \\ -6 & 6 & -4 \end{bmatrix}$.

Ans. The characteristics equation is given by

$$|A - \lambda I| = \begin{bmatrix} -1-\lambda & 3 & -3 \\ -3 & 5-\lambda & -3 \\ -6 & 6 & -4-\lambda \end{bmatrix} = 0$$

It gives eigenvalues $\lambda_1 = 2$, $\lambda_2 = 2$, $\lambda_3 = -4$. The eigenvalue $\lambda_1 = \lambda_2 = 2$ is twice repeated (algebraic multiplicity 2). For $\lambda_1 = \lambda_2 = 2$, the associated homogeneous linear system takes the form

$$-3x_1 + 3x_2 - 3x_3 = 0$$
$$-3x_1 + 3x_2 - 3x_3 = 0$$
$$-6x_1 + 6x_2 - 6x_3 = 0$$

This system has only one linearly independent equation. Two variables can assume any arbitrary values. We can easily find two linearly independent solutions, for example

$X_1 = \alpha \begin{bmatrix} 1 \\ 0 \\ -1 \end{bmatrix}$ and $X_2 = \beta \begin{bmatrix} 1 \\ 1 \\ 0 \end{bmatrix}$. Since we have two linearly independent eigenvectors,

hence geometric multiplicity is also two.

For eigenvalue, $\lambda_3 = -4$, the system of equations is as follows

$$3x_1 + 3x_2 - 3x_3 = 0$$
$$-3x_1 + 9x_2 - 3x_3 = 0$$
$$-6x_1 + 6x_2 = 0$$

It is easy to see that, out of three only two equations are independent, as $R_1 + R_3 = R_2$. So, we can neglect any one equation, and can continue with other two equations. A solution

of the system is an eigenvector $X_3 = \gamma \begin{bmatrix} 1 \\ 1 \\ 2 \end{bmatrix}$.

The eigenvalues of the matrix are, $\lambda_1 = 2$, $\lambda_2 = 2$, $\lambda_3 = -4$, and corresponding normalized eigenvectors are $X_1 = \dfrac{1}{\sqrt{2}}\begin{bmatrix} 1 \\ 0 \\ -1 \end{bmatrix}$, $X_2 = \dfrac{1}{\sqrt{2}}\begin{bmatrix} 1 \\ 1 \\ 0 \end{bmatrix}$ and $X_3 = \dfrac{1}{\sqrt{6}}\begin{bmatrix} 1 \\ 1 \\ 2 \end{bmatrix}$, respectively.

Example 6.5

Similarly, the matrix $A = \begin{bmatrix} 1 & 2 & 5 \\ 2 & 4 & 10 \\ 3 & 6 & 15 \end{bmatrix}$ has eigenvalues, $\lambda_1 = 0$, $\lambda_2 = 0$, $\lambda_3 = 20$.

The eigenvalue $\lambda_1 = 0$ is of algebraic as well as geometric multiplicity two with two independent normalized eigenvectors $X_1 = \dfrac{1}{\sqrt{5}}\begin{bmatrix} 2 \\ -1 \\ 0 \end{bmatrix}$ and $X_2 = \dfrac{1}{\sqrt{26}}\begin{bmatrix} 1 \\ 0 \\ -5 \end{bmatrix}$. The normalized eigenvector corresponding to $\lambda_3 = 20$ is $X_3 = \dfrac{1}{\sqrt{14}}\begin{bmatrix} 1 \\ 2 \\ 3 \end{bmatrix}$.

6.2.4.2 Linearly dependent eigenvectors

Example 6.6

Find the eigenvalues and eigenvectors of the matrix $A = \begin{bmatrix} 5 & 6 & 7 \\ 0 & -2 & -2 \\ 0 & 2 & 2 \end{bmatrix}$.

Ans. The characteristics equation is given by

$$|A - \lambda I| = \begin{bmatrix} 5-\lambda & 6 & 7 \\ 0 & -2-\lambda & -2 \\ 0 & 2 & 2-\lambda \end{bmatrix} = 0$$

The eigenvalues are $\lambda_1 = 0$, $\lambda_2 = 0$, $\lambda_3 = 5$. The eigenvalue, $\lambda_1 = \lambda_2 = 0$, has algebraic multiplicity 2, and the associated homogeneous linear system is as follows

$$5x_1 + 6x_2 + 7x_3 = 0$$
$$-2x_2 - 2x_3 = 0$$
$$2x_2 + 2x_3 = 0$$

Its solution gives only one linearly independent normalized eigenvector, $X_1 = \dfrac{1}{\sqrt{51}}\begin{bmatrix} 1 \\ 5 \\ -5 \end{bmatrix}$.

So, eigenvalue $\lambda_1 = \lambda_2 = 0$ has one geometric multiplicity. Similarly, for eigenvalue, $\lambda_3 = 5$,

the normalized eigenvector is $X_3 = \begin{bmatrix} 1 \\ 0 \\ 0 \end{bmatrix}$.

6.3 Bounds on Eigenvalues

In this section, we will discuss the Gerschgorin and Brauer theorems, which will provide us the bounds on eigenvalues of a given square matrix.

6.3.1 Gerschgorin Theorem

Theorem 6.1

Modulus of the eigenvalue of a matrix cannot exceed the largest sum of the modulus of its elements along any row or column. Let λ be any eigenvalue of matrix $A = [a_{ij}]$ of order n, then

$$|\lambda| \leq \max_i \left[\sum_{j=1}^{n} |a_{ij}| \right] \qquad i = 1, 2, ..., n. \qquad \text{(Sum along any row)} \qquad (6.6)$$

and $\qquad |\lambda| \leq \max_j \left[\sum_{i=1}^{n} |a_{ij}| \right] \qquad j = 1, 2, ..., n. \qquad \text{(Sum along any column)} \quad (6.7)$

Proof: Let the matrix A have eigenvectors $X_1, X_2, ..., X_n$ corresponding to eigenvalues $\lambda_1, \lambda_2, ..., \lambda_n$, respectively.

$$AX_i = \lambda_i X_i; \qquad 1 \leq i \leq n$$

This system of linear equations is as follows

$$a_{11}x_{i1} + a_{12}x_{i2} + \cdots + a_{1n}x_{in} = \lambda_i x_{i1}$$
$$a_{21}x_{i1} + a_{22}x_{i2} + \cdots + a_{2n}x_{in} = \lambda_i x_{i2}$$
$$\vdots$$
$$a_{n1}x_{i1} + a_{n2}x_{i2} + \cdots + a_{nn}x_{in} = \lambda_i x_{in}$$

Let the largest of all $|x_{ik}|$ be $|x_{ir}| = \max_k |x_{ik}|$. Then divide the rth equation by the quantity x_{ir}. We obtain the following equation

$$\lambda_i = a_{r1}\left(\frac{x_{i1}}{x_{ir}}\right) + a_{r2}\left(\frac{x_{i2}}{x_{ir}}\right) + \cdots + a_{rr} + \cdots + a_{rn}\left(\frac{x_{in}}{x_{ir}}\right) \tag{6.8}$$

As $|x_{ir}| = \max_k |x_{ik}|$, so we have

$$\left|\frac{x_{ij}}{x_{ir}}\right| \le 1; \ j = 1, 2, ..., n$$

$$\Rightarrow |\lambda_i| \le |a_{r1}| + |a_{r2}| + \cdots + |a_{rn}| = \sum_{j=1}^{n} |a_{rj}|$$

Since r is unknown, so we can write it as

$$|\lambda| \le \max_i \left[\sum_{j=1}^{n} |a_{ij}|\right]; \qquad i = 1, 2, ..., n$$

The matrices A and A^T (transpose of matrix A) have same eigenvalues. So, proceeding in a similar manner, we can obtain the following result

$$|\lambda| \le \max_j \left[\sum_{i=1}^{n} |a_{ij}|\right] \qquad j = 1, 2, ..., n$$

Example ── **6.7**

Find the bounds on eigenvalues of a matrix $A = \begin{bmatrix} 0 & 2 & 4 \\ 1 & -1 & -2 \\ -2 & 0 & 5 \end{bmatrix}$ using Gerschgorin theorem.

Ans. Equation (6.6) of Gerschgorin theorem provides the result, $|\lambda| \le 7$ (the largest sum of modulus of elements along rows).

From Eq.(6.7), we have $|\lambda| \le 11$ (the largest sum of modulus of elements along columns)

Therefore, the intersection of these two inequalities provides the bound on eigenvalues $|\lambda| \le 7$.

6.3.2 Brauer Theorem

Theorem 6.2

Every eigenvalue λ of a matrix A satisfies at least one of the following relations

$$\left|\lambda-a_{rr}\right|\leq\sum_{\substack{j=1\\j\neq r}}^{n}\left|a_{rj}\right| \qquad r=1,2,...,n. \text{ (along rows)} \tag{6.9}$$

Similarly, along columns, we have that every eigenvalue λ of a matrix A satisfies at least one of the following relation

$$\left|\lambda-a_{rr}\right|\leq\sum_{\substack{j=1\\j\neq r}}^{n}\left|a_{jr}\right| \qquad r=1,2,...,n \tag{6.10}$$

(Consider the circles with centers at the diagonal elements of the matrix, and their radiuses are the sum of the modulus of elements along rows excluding diagonal element. Then, every eigenvalue of the matrix will lie inside or on the boundary of at least one of these circles. These circles are known as Gerschgorin circles, and bounds on eigenvalues are known as Gerschgorin bounds)

Proof: In the previous theorem, Eq.(6.8) is given by

$$\lambda_i = a_{r1}\left(\frac{x_{i1}}{x_{ir}}\right)+a_{r2}\left(\frac{x_{i2}}{x_{ir}}\right)+\cdots+a_{rr}+\cdots+a_{rn}\left(\frac{x_{in}}{x_{ir}}\right)$$

$$\lambda_i - a_{rr} = a_{r1}\left(\frac{x_{i1}}{x_{ir}}\right)+a_{r2}\left(\frac{x_{i2}}{x_{ir}}\right)+\cdots+a_{rr-1}\left(\frac{x_{ir-1}}{x_{rr-1}}\right)+a_{rr+1}\left(\frac{x_{ir+1}}{x_{ir+1}}\right)\cdots+a_{rn}\left(\frac{x_{in}}{x_{ir}}\right) \tag{6.11}$$

As $\left|x_{ir}\right| = \max_k\left|x_{ik}\right|$, so we have

$$\left|\frac{x_{ij}}{x_{ir}}\right|\leq 1; \; j=1,2,...,n$$

On using these inequalities in Eq.(6.11), we get

$$\left|\lambda_i - a_{rr}\right|\leq\left|a_{r1}\right|+\left|a_{r2}\right|+\cdots+\left|a_{rr-1}\right|+\left|a_{rr+1}\right|\cdots+\left|a_{rn}\right|$$

$$\Rightarrow\left|\lambda_i - a_{rr}\right|\leq\sum_{\substack{j=1\\j\neq r}}^{n}\left|a_{rj}\right|$$

Since λ_i is not specific, so we can generalize it for every eigenvalue (λ).

$$\left|\lambda-a_{rr}\right|\leq\sum_{\substack{j=1\\j\neq r}}^{n}\left|a_{rj}\right| \qquad r=1,2,...,n$$

Note that the eigenvalues of A and A^T are equal, so we can modify the above result as follows

$$\left|\lambda - a_{rr}\right| \leq \sum_{\substack{j=1 \\ j \neq r}}^{n} \left|a_{jr}\right| \qquad r = 1, 2, \ldots, n$$

Hence proved.

Example   **6.8**

Find the bounds on eigenvalues of a matrix $A = \begin{bmatrix} 0 & 2 & 4 \\ 1 & -1 & -2 \\ -2 & 0 & 5 \end{bmatrix}$ using Brauer theorem.

Ans.
According to Brauer theorem, each and every eigenvalue satisfies at least one of following inequalities (6.9)

$$\left|\lambda - a_{rr}\right| \leq \sum_{\substack{j=1 \\ j \neq r}}^{n} \left|a_{rj}\right| \quad r = 1, 2, \ldots, n. \text{ (along rows)}$$

For matrix A, we have

$$\left|\lambda - a_{11}\right| = \left|\lambda - 0\right| \leq 6$$
$$\left|\lambda - a_{22}\right| = \left|\lambda + 1\right| \leq 3$$
$$\left|\lambda - a_{33}\right| = \left|\lambda - 5\right| \leq 2 \qquad\qquad (6.12)$$

Since each and every eigenvalue satisfies at least one of these inequalities, so in general, we can use the union of these circles for the bounds on eigenvalues. Now, the inequality $\left|\lambda + 1\right| \leq 3$ is included in the inequality, $\left|\lambda - 0\right| \leq 6$. Therefore we can avoid the second inequality of system (6.12).

Also, eigenvalues satisfy at least one of inequalities of system (6.10)

$$\left|\lambda - a_{rr}\right| \leq \sum_{\substack{j=1 \\ j \neq r}}^{n} \left|a_{jr}\right| \quad r = 1, 2, \ldots, n \qquad \text{(along columns)}$$

For matrix A, we can easily have following three inequalities for bounds on eigenvalues along columns.

$$\left|\lambda - a_{11}\right| = \left|\lambda - 0\right| \leq 3$$
$$\left|\lambda - a_{22}\right| = \left|\lambda + 1\right| \leq 2$$
$$\left|\lambda - a_{33}\right| = \left|\lambda - 5\right| \leq 6 \qquad\qquad (6.13)$$

The union of these circles provides the bounds on eigenvalues. The inequality $|\lambda+1|\leq 2$ is included in the inequality, $|\lambda-0|\leq 3$; therefore we can avoid the former inequality. From inequalities (6.12 and 6.13), the eigenvalues lie in the following region

$$\left(|\lambda-5|\leq 2 \cup |\lambda-0|\leq 6\right)\cap\left(|\lambda-0|\leq 3 \cup |\lambda-5|\leq 6\right)$$

Example 6.9

Find the bounds on eigenvalues of the matrix $A = \begin{bmatrix} 0 & 2 & 4 \\ 1 & -1 & -2 \\ -2 & 0 & 5 \end{bmatrix}$.

Ans.

From Example (6.7), the Gerschgorin theorem provides the following bound on eigenvalues

$$|\lambda|\leq 7 \tag{6.14}$$

We have the following region for eigenvalues as derived in Example (6.8) by using Brauer theorem.

$$\left(|\lambda-5|\leq 2 \cup |\lambda-0|\leq 6\right)\cap\left(|\lambda-0|\leq 3 \cup |\lambda-5|\leq 6\right) \tag{6.15}$$

The intersection of (6.14) and (6.15) is the final region for the eigenvalues.

In Section 6.2, we have discussed a direct method for computing eigenvalues of a square matrix. The eigenvalues can be obtained by using this method, but for higher order matrix, expanding the characteristic determinant and obtaining roots from the high-degree characteristic equation are very difficult. Also in the direct methods, the errors committed will remain in final results. In the case of higher order matrices, the numbers of operational counts are large, and the error propagation will cause great damage to the results obtained.

Consequently, we require iterative procedures for the solution of eigenvalue problems. In next sections, we will discuss following iterative procedures to compute eigenvalues and eigenvectors for a square matrix.

1. Rayleigh Power Method
2. Rutishauser (or) LU Decomposition Method

6.4 Rayleigh Power Method

Power method is used to determine the largest eigenvalue (in magnitude) of matrix A of order n. Let $\lambda_1, \lambda_2,..., \lambda_n$ be the eigenvalues of the matrix A, such that

$$\left|\lambda_1\right| > \left|\lambda_2\right| > ... > \left|\lambda_n\right|$$

The aim is to determine the absolutely largest eigenvalue (λ_1).

Let $X_1, X_2,..., X_n$ be the eigenvectors corresponding to the eigenvalues, $\lambda_1, \lambda_2,..., \lambda_n$, respectively. It implies

$$AX_i = \lambda_i X_i; \qquad 1 \leq i \leq n$$

If the matrix A has n-linearly independent eigenvectors, then we can write any vector X (from same vector space) as a linear combination of the vectors, $X_1, X_2,..., X_n$. Therefore, for some scalars, $c_i; 1 \leq i \leq n$, we have

$$X = c_1 X_1 + c_2 X_2 + ... + c_n X_n \qquad (6.16)$$

Pre-multiplying Eq. (6.16) with the matrix A, we get

$$
\begin{aligned}
AX &= A\left(c_1 X_1 + c_2 X_2 + ... + c_n X_n\right) \\
&= c_1 AX_1 + c_2 AX_2 + ... + c_n A X_n && \left(c_i; 1 \leq i \leq n \text{ are scalars}\right) \\
&= c_1 \lambda_1 X_1 + c_2 \lambda_2 X_2 + ... + c_n \lambda_n X_n && \left(AX_i = \lambda X_i; 1 \leq i \leq n\right) \\
&= \lambda_1 \left(c_1 X_1 + c_2 \frac{\lambda_2}{\lambda_1} X_2 + ... + c_n \frac{\lambda_n}{\lambda_1} X_n\right)
\end{aligned}
$$

Again, pre-multiplying with matrix A, we get

$$A^2 X = \lambda_1 \left(c_1 AX_1 + c_2 \frac{\lambda_2}{\lambda_1} AX_2 + ... + c_n \frac{\lambda_n}{\lambda_1} AX_n\right) \quad \left(c_i, \lambda_i; 1 \leq i \leq n \text{ are scalars}\right)$$

$$= \lambda_1^2 \left(c_1 X_1 + c_2 \left(\frac{\lambda_2}{\lambda_1}\right)^2 X_2 + ... + c_n \left(\frac{\lambda_n}{\lambda_1}\right)^2 X_n\right)$$

Repeating this process k-times successively, we obtain

$$A^k X = \lambda_1^k \left(c_1 X_1 + c_2 \left(\frac{\lambda_2}{\lambda_1}\right)^k X_2 + ... + c_n \left(\frac{\lambda_n}{\lambda_1}\right)^k X_n\right)$$

$$A^{k+1} X = \lambda_1^{k+1} \left(c_1 X_1 + c_2 \left(\frac{\lambda_2}{\lambda_1}\right)^{k+1} X_2 + ... + c_n \left(\frac{\lambda_n}{\lambda_1}\right)^{k+1} X_n\right)$$

Since $\left|\lambda_1\right| > \left|\lambda_2\right| > ... > \left|\lambda_n\right|$, it implies

$$\lim_{k \to \infty} \left(\frac{\lambda_i}{\lambda_1}\right)^k \to 0; \, 2 \leq i \leq n$$

$$\Rightarrow \lim_{k \to \infty} \frac{A^{k+1} X}{A^k X} = \lambda_1$$

It provides the largest eigenvalue λ_1.

Theoretically, the method is as follows: first, we take any initial vector X, then we multiply it by matrix A infinitely many times $(k \to \infty)$. At last, we divide the last two vectors. Practically, it is not possible to repeat the process infinite times. So, we can multiply the vector X as many times as feasible, for example 50 times. Then the common ratio $\dfrac{A^{51}X}{A^{50}X}$ is the largest eigenvalue. But, it can create the problem of rounding error, as the elements of the vector $A^{51}X$ become very large. Therefore, the method is applied by taking the largest element (magnitude) common at each iteration (to minimize the round-off error), and then continue with the remaining vector.

A stepwise procedure is as follows

i)   Let $X^{(0)}$ be any non-zero initial vector.

ii)   Multiply $X^{(0)}$ with the matrix A to obtain the vector $Y^{(0)}$ i.e. $Y^{(0)} = AX^{(0)}$.

iii)   Take the absolutely largest element $(\lambda^{(1)})$ common from the vector $Y^{(0)}$. Let remaining vector be $X^{(1)}$.

$$Y^{(0)} = \lambda^{(1)} X^{(1)}$$

iv)   Repeat steps ii) and iii) till the last iteration has the desired accuracy.

$$Y^{(k)} = \lambda^{(k+1)} X^{(k+1)} \qquad k = 0, 1, 2, \ldots$$

v)   At last, $\lambda^{(k+1)}$ and $X^{(k+1)}$ are the approximations to the largest eigenvalue and eigenvector, respectively.

Note that we cannot start with trivial initial vector, i.e., zero vector $X_0 = [0 \ \ 0 \ \ 0]^T$.

Example   **6.10**

Determine the largest eigenvalue and corresponding eigenvector of the matrix

$$A = \begin{bmatrix} 0 & 2 & 4 \\ 1 & 1 & -2 \\ -2 & 0 & 5 \end{bmatrix}$$

Start with the initial vector, $X^{(0)} = [1 \ \ 1 \ \ 1]^T$. Perform the iterations till the eigenvalue and eigenvector are same up to two decimal places, in last two iterations.

Ans. The first iteration of the Power method is given by

$$Y^{(0)} = AX^{(0)} = \begin{bmatrix} 0 & 2 & 4 \\ 1 & 1 & -2 \\ -2 & 0 & 5 \end{bmatrix} \begin{bmatrix} 1 \\ 1 \\ 1 \end{bmatrix} = \begin{bmatrix} 6 \\ 0 \\ 3 \end{bmatrix}$$

On scaling the vector $Y^{(0)}$ with the absolutely largest element, we have

$$Y^{(0)} = 6\begin{bmatrix} 1 & 0 & 0.5 \end{bmatrix} = \lambda^{(1)} X^{(1)}$$

Similarly, the second iteration is computed as follows

$$Y^{(1)} = AX^{(1)} = \begin{bmatrix} 0 & 2 & 4 \\ 1 & 1 & -2 \\ -2 & 0 & 5 \end{bmatrix}\begin{bmatrix} 1 \\ 0 \\ 0.5 \end{bmatrix} = \begin{bmatrix} 2 \\ 0 \\ 0.5 \end{bmatrix} = 2\begin{bmatrix} 1 \\ 0 \\ 0.25 \end{bmatrix}$$

$\lambda^{(2)} = 2$ and $X^{(2)} = \begin{bmatrix} 1 & 0 & 0.25 \end{bmatrix}^T$.

Proceeding in a similar manner, the subsequent iterations of the Power method are as follows

$$\lambda^{(3)} = 1.000000, \quad X^{(3)} = \begin{bmatrix} 1.000000 & 0.500000 & 0.7500003 \end{bmatrix}^T$$

$$\lambda^{(4)} = 5.750000, \quad X^{(4)} = \begin{bmatrix} -0.347826 & 0.521739 & -1.000000 \end{bmatrix}^T$$

$$\lambda^{(5)} = 4.304348 \quad X^{(5)} = \begin{bmatrix} -0.686869 & 0.505050 & -1.000000 \end{bmatrix}^T$$

$$\lambda^{(6)} = 3.626263 \quad X^{(6)} = \begin{bmatrix} -0.824513 & 0.501393 & -1.000000 \end{bmatrix}^T$$

$$\lambda^{(7)} = 3.350975 \quad X^{(7)} = \begin{bmatrix} -0.894431 & 0.500416 & -1.000000 \end{bmatrix}^T$$

$$\lambda^{(8)} = 3.211139 \quad X^{(8)} = \begin{bmatrix} -0.933989 & 0.500129 & -1.000000 \end{bmatrix}^T$$

$$\lambda^{(9)} = 3.132022 \quad X^{(9)} = \begin{bmatrix} -0.957765 & 0.500041 & -1.000000 \end{bmatrix}^T$$

$$\lambda^{(10)} = 3.084470 \quad X^{(10)} = \begin{bmatrix} -0.972588 & 0.500013 & -1.000000 \end{bmatrix}^T$$

$$\lambda^{(11)} = 3.054824 \quad X^{(11)} = \begin{bmatrix} -0.982044 & 0.500004 & -1.000000 \end{bmatrix}^T$$

$$\lambda^{(12)} = 3.035911 \quad X^{(12)} = \begin{bmatrix} -0.988168 & 0.500001 & -1.000000 \end{bmatrix}^T$$

$$\lambda^{(13)} = 3.023663 \quad X^{(13)} = \begin{bmatrix} -0.992173 & 0.500000 & -1.000000 \end{bmatrix}^T$$

The difference in the values at last two iterations (twelfth and thirteenth) are less than 0.005. Therefore, the approximate eigenvalue and eigenvector are $\lambda^{(13)} = 3.023663$ and $X^{(13)} = \begin{bmatrix} -0.992173 & 0.500000 & -1.000000 \end{bmatrix}$, respectively.

From Example 6.3, the exact eigenvalue is 3 and eigenvector is $\begin{bmatrix} -1 & 0.5 & -1 \end{bmatrix}^T$.

Note: In Example 3, the differences between the largest eigenvalue $\lambda_3 = 3$ and other eigenvalues $\lambda_1 = 1$, $\lambda_2 = 2$ are relatively less. Therefore, a large number of iterations are required for higher accuracy. Note that the power method has following restrictions.

1. The largest (in magnitude) eigenvalue of the matrix must be distinct.
2. The matrix A has n-linearly independent eigenvectors.

3. The rate of convergence is proportional to the ratio, $\dfrac{|\lambda_2|}{|\lambda_1|}$, where λ_2 is the second largest (in magnitude) eigenvalue and λ_1 is the largest (in magnitude) eigenvalue of the matrix A.

Example ── **6.11**

Determine the largest eigenvalue and the corresponding eigenvector of the matrix

$$A = \begin{bmatrix} 10 & 6 & 7 \\ 1 & 7 & -2 \\ 2 & 2 & 2 \end{bmatrix}$$

Ans. Let us start the iterations with the initial vector, $X^{(0)} = [1 \quad 1 \quad 1]^T$. The first iteration is as follows

$$Y^{(0)} = AX^{(0)} = \begin{bmatrix} 10 & 6 & 7 \\ 1 & 7 & -2 \\ 2 & 2 & 2 \end{bmatrix} \begin{bmatrix} 1 \\ 1 \\ 1 \end{bmatrix} = \begin{bmatrix} 23 \\ 6 \\ 6 \end{bmatrix} = 23 \begin{bmatrix} 1 \\ 0.260870 \\ 0.260870 \end{bmatrix}$$

The second iteration is given by

$$Y^{(1)} = AX^{(1)} = \begin{bmatrix} 10 & 6 & 7 \\ 1 & 7 & -2 \\ 2 & 2 & 2 \end{bmatrix} \begin{bmatrix} 1 \\ 0.260870 \\ 0.260870 \end{bmatrix} = 13.391304 \begin{bmatrix} 1.000000 \\ 0.172078 \\ 0.227273 \end{bmatrix}$$

$$\lambda^{(2)} = 13.391304 \text{ and } X^{(2)} = \begin{bmatrix} 1.000000 & 0.172078 & 0.227273 \end{bmatrix}^T$$

Similarly, the subsequent iterations are as follows

$\lambda^{(3)} = 12.623377$	$X^{(3)} = \begin{bmatrix} 1.000000 & 0.138632 & 0.2217083 \end{bmatrix}^T$
$\lambda^{(4)} = 12.383745$	$X^{(4)} = \begin{bmatrix} 1.000000 & 0.123307 & 0.2196984 \end{bmatrix}^T$
$\lambda^{(5)} = 12.277726$	$X^{(5)} = \begin{bmatrix} 1.000000 & 0.115963 & 0.2187715 \end{bmatrix}^T$
$\lambda^{(6)} = 12.227172$	$X^{(6)} = \begin{bmatrix} 1.000000 & 0.112389 & 0.2183236 \end{bmatrix}^T$
$\lambda^{(7)} = 12.202589$	$X^{(7)} = \begin{bmatrix} 1.000000 & 0.110638 & 0.2181037 \end{bmatrix}^T$
$\lambda^{(8)} = 12.190553$	$X^{(8)} = \begin{bmatrix} 1.000000 & 0.109779 & 0.2179958 \end{bmatrix}^T$
$\lambda^{(9)} = 12.184639$	$X^{(9)} = \begin{bmatrix} 1.000000 & 0.109356 & 0.2179429 \end{bmatrix}^T$
$\lambda^{(10)} = 12.181730$	$X^{(10)} = \begin{bmatrix} 1.000000 & 0.109148 & 0.2179161 \end{bmatrix}^T$

Note that the eigenvalues of the matrix A are 12.1789083458, 6, 0.8210916542. The difference between the largest and the second largest eigenvalues is large, so comparatively fewer iterations are required.

6.4.1 Inverse Power Method

The inverse power method is used to compute the smallest (in magnitude) eigenvalue of a given square matrix A. Inverse power method is a variation of power method. It involves computing of the largest (in magnitude) eigenvalue of the inverse matrix, A^{-1}.

Theorem 6.3

Let λ_i be an eigenvalue of matrix A, then $\dfrac{1}{\lambda_i}$ is the eigenvalue of the matrix A^{-1}. The eigenvector X_i of matrix A^{-1} remains same as that of matrix A.

Proof: Let λ_i be an eigenvalue and X_i is the corresponding eigenvector of matrix A, then we have

$$AX_i = \lambda_i X_i$$

On pre-multiplying with the matrix A^{-1}, we have

$$A^{-1}(AX_i) = A^{-1}(\lambda_i X_i) \tag{6.17}$$

The matrix multiplication is associative, so we have

$$A^{-1}(AX_i) = \left(A^{-1}A\right)X_i = I\,X_i = X_i \tag{6.18}$$

where the matrix I is the identity matrix. Also, the eigenvalue λ_i is scaler quantity, so

$$A^{-1}(\lambda_i X_i) = \lambda_i \left(A^{-1}X_i\right) \tag{6.19}$$

Equations (6.17–6.19) provide the following result

$$X_i = \lambda_i A^{-1} X_i$$

$$\frac{1}{\lambda_i} X_i = A^{-1} X_i$$

It implies that $\dfrac{1}{\lambda_i}$ is the eigenvalue of A^{-1}, the eigenvector X_i of matrix A^{-1} remains same as that of matrix A.

To find the smallest (in magnitude) eigenvalue of the matrix A, we find the largest eigenvalue (in magnitude) of the matrix A^{-1}, and then the inverse of that eigenvalue is the smallest eigenvalue of the matrix A.

Example ── **6.12**

Determine the smallest eigenvalue and the corresponding eigenvector of the matrix

$$A = \begin{bmatrix} 10 & 6 & 7 \\ 1 & 7 & -2 \\ 2 & 2 & 2 \end{bmatrix}.$$

Ans.

The inverse of matrix A is given by

$$A^{-1} = \frac{1}{60} \begin{bmatrix} 18 & 2 & -61 \\ -6 & 6 & 27 \\ -12 & -8 & 64 \end{bmatrix}$$

To compute the smallest (in magnitude) eigenvalue of matrix A, first we find the largest eigenvalue (in magnitude) of A^{-1}, and then inverse of that eigenvalue is the smallest eigenvalue of A.

To compute the largest eigenvalue of matrix, A^{-1}, let us start the iterations with initial vector $X^{(0)} = [1 \ \ 1 \ \ 1]$. The first iteration of Power method is given by

$$Y^{(0)} = A^{-1} X^{(0)} = \frac{1}{60} \begin{bmatrix} 18 & 2 & -61 \\ -6 & 6 & 27 \\ -12 & -8 & 64 \end{bmatrix} \begin{bmatrix} 1 \\ 1 \\ 1 \end{bmatrix} = \frac{1}{60} \begin{bmatrix} -41 \\ 27 \\ 44 \end{bmatrix} = \frac{44}{60} \begin{bmatrix} -0.931818 \\ 0.613636 \\ 1 \end{bmatrix}$$

$$\lambda^{(1)} = \frac{44}{60} \text{ and } X^{(1)} = \begin{bmatrix} -0.931818 & 0.613636 & 1 \end{bmatrix}^T$$

The second iteration is as follows

$$Y^{(1)} = A^{-1} X^{(1)} = \frac{1}{60} \begin{bmatrix} 18 & 2 & -61 \\ -6 & 6 & 27 \\ -12 & -8 & 64 \end{bmatrix} \begin{bmatrix} -0.931818 \\ 0.613636 \\ 1 \end{bmatrix} = \frac{76.545456}{60} \begin{bmatrix} -1 \\ 0.473872 \\ 0.918052 \end{bmatrix}$$

$$\lambda^{(2)} = \frac{76.545456}{60} \text{ and } X^{(2)} = \begin{bmatrix} -1 & 0.473872 & 0.918052 \end{bmatrix}^T$$

Other iterations are given by

$$\lambda^{(3)} = 73.053444 / 60 \qquad X^{(3)} = \begin{bmatrix} -1.000000 & 0.460357 & 0.9166493 \end{bmatrix}^T$$

$$\lambda^{(4)} = 72.994881 / 60 \qquad X^{(4)} = \begin{bmatrix} -1.000000 & 0.459096 & 0.9176354 \end{bmatrix}^T$$

$$\lambda^{(5)} = 73.057564 / 60 \qquad X^{(5)} = \begin{bmatrix} -1.000000 & 0.458963 & 0.9178505 \end{bmatrix}^T$$

$$\lambda^{(6)} = 73.070930 / 60 \qquad X^{(6)} = \begin{bmatrix} -1.000000 & 0.458948 & 0.9178856 \end{bmatrix}^T$$

$$\lambda^{(7)} = 73.073082 / 60 \qquad X^{(7)} = \begin{bmatrix} -1.000000 & 0.458946 & 0.9178907 \end{bmatrix}^T$$

$$\lambda^{(8)} = 73.073402 / 60 \qquad X^{(8)} = \begin{bmatrix} -1.000000 & 0.458945 & 0.9178918 \end{bmatrix}^T$$

$$\lambda^{(9)} = 73.073441 / 60 \qquad X^{(9)} = \begin{bmatrix} -1.000000 & 0.458945 & 0.9178919 \end{bmatrix}^T$$

$$\lambda^{(10)} = 73.073448 / 60 \qquad X^{(10)} = \begin{bmatrix} -1.000000 & 0.458945 & 0.9178911 \end{bmatrix}^T$$

The approximate value of the largest eigenvalue of A^{-1} is $\lambda^{(10)} = 73.073448 / 60 = 1.2178908$. Hence, the smallest eigenvalue of A is $1/1.2178908 = 0.8210916775$.

6.4.2 Shifted Power Method

Shifted power method is another variation of power method. It is used to compute the eigenvalues which are farthest/nearest from a given scalar k.

Theorem 6.4

Let λ_i be an eigenvalue of matrix A, then $(\lambda_i - k)$ is an eigenvalue of the matrix $(A - k\,I)$ with the same eigenvector as that of matrix A.

Proof: Let λ_i be an eigenvalue and X_i is the corresponding eigenvector of matrix A, then we have

$$AX_i = \lambda_i X_i$$

To compute eigenvalues of $(A - k\,I)$, we have

$$
\begin{aligned}
(A - k\,I)X_i &= AX_i - k\,I\,X_i \\
&= \lambda_i X_i - k\,X_i \\
&= (\lambda_i - k)\,X_i
\end{aligned}
$$

It implies that if λ_i is an eigenvalue of matrix A, then $(\lambda_i - k)$ is an eigenvalue of a matrix, $(A - k\,I)$. The vector X_i is the corresponding eigenvector of matrix A as well as $(A - k\,I)$.

Eigenvalue farthest to a given scalar: To compute eigenvalue of matrix A farthest to a given number k, first we find the largest eigenvalue (in magnitude) of the matrix, $(A - k\,I)$, and then that eigenvalue in addition with k is the desired eigenvalue of matrix A.

For example, let us assume the eigenvalues of a matrix A are -5, 2 and 8 and we want to compute the eigenvalue that is farthest from the scalar 5. The eigenvalues of the matrix $(A - 5\,I)$ are -10, -3 and 3. The computational procedure is to compute the largest (in magnitude) eigenvalue of a matrix $(A - 5\,I)$ (i.e. -10), and then add scalar 5 to that eigenvalue to get the desired eigenvalue (i.e. -5).

Eigenvalue nearest to a given scalar: To compute eigenvalue of matrix A nearest to number k, first, we find the largest eigenvalue (in magnitude) of matrix $(A - k\,I)^{-1}$, and then inverse of that eigenvalue in addition with k is the desired eigenvalue of matrix A.

For example, let us assume the eigenvalues of a matrix A are -1, 4.5 and 7 and we want to compute the eigenvalue that is nearest to 4. We have

Eigenvalues of matrix A are -1, 4.5 and 7

Eigenvalues of matrix $(A - 4\,I)$ are -5, 0.5 and 3

Eigenvalues of matrix $(A - 4\,I)^{-1}$ are $\dfrac{-1}{5}$, 2 and $\dfrac{1}{3}$

The computational procedure is to compute the largest (in magnitude) eigenvalue of a matrix $(A - 4\,I)^{-1}$ (i.e. 2). Then reciprocal (0.5) of that eigenvalue in addition with $k\,(= 4)$ is the desired eigenvalue (4.5) of matrix A.

Example 6.13

Determine the eigenvalue farthest to 4 for the matrix

$$A = \begin{bmatrix} 2 & 6 & -3 \\ 5 & 3 & -3 \\ 5 & -4 & 4 \end{bmatrix}.$$

Start the iterations with the initial vector $X^{(0)} = [1 \quad 0 \quad 1]^T$.

Ans.

To compute the eigenvalue of matrix A which is farthest to 4, we will find the largest (in magnitude) eigenvalue of the matrix $(A - 4I)$, and then add scalar 4 to that eigenvalue.

$$A - 4I = \begin{bmatrix} -2 & 6 & -3 \\ 5 & -1 & -3 \\ 5 & -4 & 0 \end{bmatrix}$$

Proceeding in a similar manner as in previous examples with an initial vector $X^{(0)} = [1 \quad 0 \quad 1]^T$ for matrix $A - 4I$, the largest eigenvalue of this matrix is computed as follows

$$Y^{(0)} = (A - 4I) X^{(0)} = \begin{bmatrix} -2 & 6 & -3 \\ 5 & -1 & -3 \\ 5 & -4 & 0 \end{bmatrix} \begin{bmatrix} 1 \\ 0 \\ 1 \end{bmatrix} = \begin{bmatrix} -5 \\ 1 \\ 1 \end{bmatrix} = 5 \begin{bmatrix} -1 \\ 0.4 \\ 1 \end{bmatrix}$$

$$\lambda^{(1)} = 5 \text{ and } X^{(1)} = \begin{bmatrix} -1 & 0.4 & 1 \end{bmatrix}^T$$

Other iterations are given by

$$\lambda^{(2)} = 8.4 \qquad X^{(2)} = \begin{bmatrix} 0.166667 & -1.000000 & -0.7857142 \end{bmatrix}^T$$

$$\lambda^{(3)} = 4.833333 \quad X^{(3)} = \begin{bmatrix} -0.822660 & 0.866995 & 1.000000 \end{bmatrix}^T$$

$$\lambda^{(4)} = 7.980295 \quad X^{(4)} = \begin{bmatrix} 0.482099 & -1.000000 & -0.9500004 \end{bmatrix}^T$$

$$\lambda^{(5)} = 6.410494 \quad X^{(5)} = \begin{bmatrix} -0.641791 & 0.976601 & 1.000000 \end{bmatrix}^T$$

$$\lambda^{(6)} = 7.185556 \quad X^{(6)} = \begin{bmatrix} 0.576599 & -1.000000 & -0.9902316 \end{bmatrix}^T$$

$$\lambda^{(7)} = 6.882997 \quad X^{(7)} = \begin{bmatrix} -0.607658 & 0.995742 & 1.000000 \end{bmatrix}^T$$

$$\lambda^{(8)} = 7.034031 \quad X^{(8)} = \begin{bmatrix} 0.595643 & -1.000000 & -0.9981848 \end{bmatrix}^T$$

$$\lambda^{(9)} = 6.978212 \quad X^{(9)} = \begin{bmatrix} -0.601405 & 0.999219 & 1.000000 \end{bmatrix}^T$$

$$\lambda^{(10)} = 7.006245 \quad X^{(10)} = \begin{bmatrix} 0.599198 & -1.000000 & -0.99966610 \end{bmatrix}^T$$

These iterations are converging to the eigenvalue 7. Since the elements of eigenvectors are changing the sign alternatively, so the eigenvalue is –7. The largest eigenvalue of matrix $(A - 4\,I)$ is –7, so the eigenvalue of matrix A is –7 + 4 = –3. The eigenvalue $\lambda = -3$ of matrix A is farthest from scalar 4.

Note that the eigenvalues of matrix A are –3, 5 and 7.

Example ———————————————————————————————— **6.14**

Determine the eigenvalue nearest to 5 and the corresponding eigenvector of the matrix

$$A = \begin{bmatrix} 10 & 6 & 7 \\ 1 & 7 & -2 \\ 2 & 2 & 2 \end{bmatrix}.$$

Ans.

First, we find the largest eigenvalue (in magnitude) of the matrix $(A - 5\,I)^{-1}$. Then, by adding number 5 to the inverse of that eigenvalue will produce an eigenvalue of the matrix A (nearest to number 5).

$$A - 5I = \begin{bmatrix} 5 & 6 & 7 \\ 1 & 2 & -2 \\ 2 & 2 & -3 \end{bmatrix}$$

$$(A - 5I)^{-1} = \frac{1}{30}\begin{bmatrix} 2 & -32 & 26 \\ 1 & 29 & -17 \\ 2 & -2 & -4 \end{bmatrix}$$

Now, we have to compute the largest eigenvalue of the matrix, $(A - 5\,I)^{-1}$. Let us start the iterations with the initial vector, $X^{(0)} = [1 \quad 1 \quad 1]^T$. The first iteration is given by

$$Y^{(0)} = (A - 5I)^{-1} X^{(0)} = \frac{1}{30}\begin{bmatrix} 2 & -32 & 26 \\ 1 & 29 & -17 \\ 2 & -2 & -4 \end{bmatrix}\begin{bmatrix} 1 \\ 1 \\ 1 \end{bmatrix} = \frac{1}{30}\begin{bmatrix} -4 \\ 13 \\ -4 \end{bmatrix} = \frac{13}{30}\begin{bmatrix} -0.307692 \\ 1 \\ -0.307692 \end{bmatrix}$$

Similarly, other iterations are as follows

$$\lambda^{(2)} = 40.615383 / 30 \qquad X^{(2)} = \begin{bmatrix} -1.000000 & 0.835227 & -0.034091 \end{bmatrix}^T$$

$$\lambda^{(3)} = 29.613638 / 30 \qquad X^{(3)} = \begin{bmatrix} -1.000000 & 0.835227 & -0.034091 \end{bmatrix}^T$$

$$\lambda^{(4)} = 30.821949 / 30 \qquad X^{(4)} = \begin{bmatrix} -1.000000 & 0.789591 & -0.101554 \end{bmatrix}^T$$

$$\lambda^{(5)} = 29.907299 / 30 \qquad X^{(5)} = \begin{bmatrix} -1.000000 & 0.789926 & -0.106093 \end{bmatrix}^T$$

The largest eigenvalue of the matrix $(A - 5\,I)^{-1}$ is $\dfrac{29.907299}{30} \approx 1$

The smallest eigenvalue of the matrix $(A - 5\,I)$ is $\dfrac{1}{1} = 1$

The eigenvalue of the matrix A nearest to 5 is $1 + 5 = 6$.

Theorem 6.5

Let λ_i be an eigenvalue of matrix A, then λ_i^k is the eigenvalue of A^k. The eigenvector X_i is same for both the matrices A and A^k.

Proof: Let λ_i be an eigenvalue and X_i is the corresponding eigenvector of matrix A, then

$$AX_i = \lambda_i X_i$$

On pre-multiplying with the matrix A, we have

$$A^2 X_i = A\left(\lambda_i X_i\right) = \lambda_i \left(AX_i\right) = \lambda_i \left(\lambda_i X_i\right) = \lambda_i^2 X_i$$

It implies that if λ_i is an eigenvalue of matrix A, then λ_i^2 is the eigenvalue of A^2; and X_i is the corresponding eigenvector of both matrices A and A^2.

On repeating the process of pre-multiplying with matrix A, we can establish that λ_i^k is an eigenvalue of the matrix A^k. We can use this to compute eigenvalues of A^k and for other purposes.

6.5 Rutishauser (or) LU Decomposition Method

In this method, we will generate a convergent sequence of upper triangular matrices, $A_1, A_2, A_3 \ldots$. LU decomposition method discussed in the last chapter is used for this purpose. Initially, we decompose the matrix A in lower and upper triangular matrices $A = L_1 U_1$ with diagonal elements of the lower triangular matrix (L_1) are unity, i.e., $l_{ii} = 1$. Then we create, $A_1 = U_1 L_1$.

We repeat the process to get

$$A_1 = L_2 U_2 \text{ and } A_2 = U_2 L_2$$

Proceeding in a similar manner, decompose the matrix $A_k = L_{k+1} U_{k+1}$ and then create the matrix $A_{k+1} = U_{k+1} L_{k+1}$. At last, we have an upper triangular matrix, A_{k+1}. The diagonal elements of upper triangular matrix are the eigenvalues of the matrix.

Example ———————————————————— 6.15

Find the eigenvalues of a matrix $A = \begin{bmatrix} 10 & 6 & 7 \\ 1 & 7 & -2 \\ 2 & 2 & 2 \end{bmatrix}$ using LU decomposition method.

Ans.

First Iteration

First, we decompose the matrix A into lower triangular matrix (with diagonal element as 1) and upper triangular matrix using LU-decomposition method

$$A = \begin{bmatrix} 10 & 6 & 7 \\ 1 & 7 & -2 \\ 2 & 2 & 2 \end{bmatrix}$$

$$= \underbrace{\begin{bmatrix} 1.000000 & 0.000000 & 0.000000 \\ 0.100000 & 1.000000 & 0.000000 \\ 0.200000 & 0.125000 & 1.000000 \end{bmatrix}}_{L_1} \underbrace{\begin{bmatrix} 10.000000 & 6.000000 & 7.000000 \\ 0.000000 & 6.400000 & -2.700000 \\ 0.000000 & 0.000000 & 0.937500 \end{bmatrix}}_{U_1}$$

Then, we compute the following matrix

$$A_1 = U_1 L_1 = \begin{bmatrix} 12.000000 & 6.875000 & 7.000000 \\ 0.100000 & 6.062500 & -2.700000 \\ 0.187500 & 0.117187 & 0.937500 \end{bmatrix}$$

Second Iteration

We compute $A_1 = L_2 U_2$ as follows

$$A_1 = \underbrace{\begin{bmatrix} 1.000000 & 0.000000 & 0.000000 \\ 0.008333 & 1.000000 & 0.000000 \\ 0.015625 & 0.001626 & 1.000000 \end{bmatrix}}_{L_2} \underbrace{\begin{bmatrix} 12.000000 & 6.875000 & 7.000000 \\ 0.000000 & 6.005208 & -2.758333 \\ 0.000000 & 0.000000 & 0.832610 \end{bmatrix}}_{U_2}$$

On computing $A_2 = U_2 L_2$, we get

$$A_2 = \begin{bmatrix} 12.166667 & 6.886384 & 7.000000 \\ 0.006944 & 6.000723 & -2.758333 \\ 0.013010 & 0.001354 & 0.832610 \end{bmatrix}$$

Third Iteration

$$A_2 = \begin{bmatrix} 1.000000 & 0.000000 & 0.000000 \\ 0.000571 & 1.000000 & 0.000000 \\ 0.001069 & -0.001002 & 1.000000 \end{bmatrix} \begin{bmatrix} 12.166667 & 6.886384 & 7.000000 \\ 0.000000 & 5.996792 & -2.762329 \\ 0.000000 & 0.000000 & 0.822357 \end{bmatrix}$$

$$\qquad\qquad\qquad L_3 \qquad\qquad\qquad\qquad\qquad U_3$$

$$A_3 = U_3 L_3 = \begin{bmatrix} 12.178083 & 6.879369 & 7.000000 \\ 0.000469 & 5.999560 & -2.762329 \\ 0.000879 & -0.000824 & 0.822357 \end{bmatrix}$$

Fourth Iteration

$$A_3 = \begin{bmatrix} 1.000000 & 0.000000 & 0.000000 \\ 0.000039 & 1.000000 & 0.000000 \\ 0.000072 & -0.000220 & 1.000000 \end{bmatrix} \begin{bmatrix} 12.178083 & 6.879369 & 7.000000 \\ 0.000000 & 5.999295 & -2.762599 \\ 0.000000 & 0.000000 & 0.821244 \end{bmatrix}$$

$$\qquad\qquad\qquad L_4 \qquad\qquad\qquad\qquad\qquad U_4$$

$$A_4 = U_4 L_4 = \begin{bmatrix} 12.178854 & 6.877828 & 7.000000 \\ 0.000032 & 5.999904 & -2.762599 \\ 0.000059 & -0.000181 & 0.821244 \end{bmatrix}$$

The matrix A_4 is approximately an upper triangular matrix. So, we can approximate the eigenvalues of matrix A as 12.178854, 5.999904 and 0.821244.

Example ── **6.16**

Find the eigenvalues of a matrix $A = \begin{bmatrix} 4 & 5 & 2 \\ 2 & -3 & 4 \\ 2 & 1 & 3 \end{bmatrix}$ using LU decomposition method.

Ans.

First Iteration

$$A = \begin{bmatrix} 4 & 5 & 2 \\ 2 & -3 & 4 \\ 2 & 1 & 3 \end{bmatrix} = \begin{bmatrix} 1.0000 & 0.0000 & 0.0000 \\ 0.5000 & 1.0000 & 0.0000 \\ 0.5000 & 0.2727 & 1.0000 \end{bmatrix} \begin{bmatrix} 4.0000 & 5.0000 & 2.0000 \\ 0.0000 & -5.5000 & 3.0000 \\ 0.0000 & 0.0000 & 1.1818 \end{bmatrix}$$

$$\qquad\qquad\qquad\qquad\qquad L_1 \qquad\qquad\qquad\qquad\qquad U_1$$

$$A_1 = U_1 L_1 = \begin{bmatrix} 7.5000 & 5.5455 & 2.0000 \\ -1.2500 & -4.6818 & 3.0000 \\ 0.5909 & 0.3223 & 1.1818 \end{bmatrix}$$

Second Iteration

$$A_1 = \begin{bmatrix} 1.0000 & 0.0000 & 0.0000 \\ -0.1667 & 1.0000 & 0.0000 \\ 0.0788 & 0.0305 & 1.0000 \end{bmatrix} \begin{bmatrix} 7.5000 & 5.5455 & 2.0000 \\ 0.0000 & -3.7576 & 3.3333 \\ 0.0000 & 0.0000 & 0.9226 \end{bmatrix}$$

$$\qquad\qquad L_2 \qquad\qquad\qquad\qquad U_2$$

$$A_2 = \begin{bmatrix} 6.7333 & 5.6065 & 2.0000 \\ 0.8889 & -3.6559 & 3.3333 \\ 0.0727 & 0.0281 & 0.9226 \end{bmatrix}$$

Third Iteration

$$A_2 = \begin{bmatrix} 1.0000 & 0.0000 & 0.0000 \\ 0.1320 & 1.0000 & 0.0000 \\ 0.0108 & 0.0074 & 1.0000 \end{bmatrix} \begin{bmatrix} 6.7333 & 5.6065 & 2.0000 \\ 0.0000 & -4.3960 & 3.0693 \\ 0.0000 & 0.0000 & 0.8784 \end{bmatrix}$$

$$\qquad\qquad L_3 \qquad\qquad\qquad\qquad U_3$$

$$A_3 = U_3 L_3 = \begin{bmatrix} 7.4950 & 5.6212 & 2.0000 \\ -0.5472 & -4.3734 & 3.0693 \\ 0.0095 & 0.0065 & 0.8784 \end{bmatrix}$$

Fourth Iteration

$$A_3 = \begin{bmatrix} 1.0000 & 0.0000 & 0.0000 \\ -0.0730 & 1.0000 & 0.0000 \\ 0.0013 & 0.0002 & 1.0000 \end{bmatrix} \begin{bmatrix} 7.4950 & 5.6212 & 2.0000 \\ 0.0000 & -3.9630 & 3.2153 \\ 0.0000 & 0.0000 & 0.8753 \end{bmatrix}$$

$$\qquad\qquad L_4 \qquad\qquad\qquad\qquad U_4$$

$$A_4 = U_4 L_4 = \begin{bmatrix} 7.0872 & 5.6215 & 2.0000 \\ 0.2934 & -3.9625 & 3.2153 \\ 0.0011 & 0.0001 & 0.8753 \end{bmatrix}$$

Fifth Iteration

$$A_4 = \begin{bmatrix} 1.0000 & 0.0000 & 0.0000 \\ 0.0414 & 1.0000 & 0.0000 \\ 0.0002 & 0.0002 & 1.0000 \end{bmatrix} \begin{bmatrix} 7.0872 & 5.6215 & 2.0000 \\ 0.0000 & -4.1952 & 3.1325 \\ 0.0000 & 0.0000 & 0.8745 \end{bmatrix}$$

$$\qquad\qquad L_5 \qquad\qquad\qquad\qquad U_5$$

$$A_5 = U_5 L_5 = \begin{bmatrix} 7.3202 & 5.6219 & 2.0000 \\ -0.1732 & -4.1947 & 3.1325 \\ 0.0001 & 0.0002 & 0.8745 \end{bmatrix}$$

Sixth Iteration

$$A_5 = \begin{bmatrix} 1.0000 & 0.0000 & 0.0000 \\ -0.0237 & 1.0000 & 0.0000 \\ 0.0000 & 0.0000 & 1.0000 \end{bmatrix} \begin{bmatrix} 7.3202 & 5.6219 & 2.0000 \\ 0.0000 & -4.0617 & 3.1798 \\ 0.0000 & 0.0000 & 0.8745 \end{bmatrix}$$

$$\qquad\qquad\qquad L_6 \qquad\qquad\qquad\qquad\qquad U_6$$

$$A_6 = U_6 L_6 = \begin{bmatrix} 7.1873 & 5.6218 & 2.0000 \\ 0.0962 & -4.0617 & 3.1798 \\ 0.0000 & 0.0000 & 0.8745 \end{bmatrix}$$

Seventh Iteration

$$A_6 = \begin{bmatrix} 1.0000 & 0.0000 & 0.0000 \\ 0.0134 & 1.0000 & 0.0000 \\ 0.0000 & 0.0000 & 1.0000 \end{bmatrix} \begin{bmatrix} 7.1873 & 5.6218 & 2.0000 \\ 0.0000 & -4.1369 & 3.1531 \\ 0.0000 & 0.0000 & 0.8744 \end{bmatrix}$$

$$\qquad\qquad\qquad L_7 \qquad\qquad\qquad\qquad\qquad U_7$$

$$A_7 = U_7 L_7 = \begin{bmatrix} 7.2625 & 5.6218 & 2.0000 \\ -0.0553 & -4.1369 & 3.1531 \\ 0.0000 & 0.0000 & 0.8744 \end{bmatrix}$$

The approximate eigenvalues of matrix A are 7.2625, −4.1369 and 0.8744.

Exercise 6

1. Compute the eigenvalues and the corresponding eigenvectors for the following matrices

a) $\begin{bmatrix} -1 & 1 & 2 \\ 1 & -1 & -2 \\ 1 & 1 & 2 \end{bmatrix}$ **Ans.** $\begin{bmatrix} 2 \\ 0 \\ -2 \end{bmatrix}$ $\begin{bmatrix} 1 & 0 & -1 \\ -1 & -2 & 1 \\ 2 & 1 & 0 \end{bmatrix}$

b) $\begin{bmatrix} 0 & 1 & -2 \\ 2 & 1 & 0 \\ 4 & -2 & 5 \end{bmatrix}$ **Ans.** $\begin{bmatrix} 1 \\ 2 \\ 3 \end{bmatrix}$ $\begin{bmatrix} 0 & 1 & -1 \\ 2 & 2 & -1 \\ 1 & 0 & 1 \end{bmatrix}$

c) $\begin{bmatrix} -1 & 3 & 3 \\ -3 & 5 & 3 \\ -3 & 3 & 2 \end{bmatrix}$ **Ans.** $\begin{bmatrix} 2 \\ 2 \\ 2 \end{bmatrix}$ $\begin{bmatrix} 1 & 0 & 0 \\ 1 & 0 & 0 \\ 0 & 0 & 0 \end{bmatrix}$

d) $\begin{bmatrix} 2 & 3 & 3 \\ 3 & 5 & 3 \\ 3 & 3 & 2 \end{bmatrix}$ **Ans.** $\begin{bmatrix} 5+3\sqrt{2} \\ 5-3\sqrt{2} \\ -1 \end{bmatrix}$ $\begin{bmatrix} 1 & 1 & -1 \\ \sqrt{2} & -\sqrt{2} & 0 \\ 1 & 1 & 1 \end{bmatrix}$

e) $\begin{bmatrix} 1 & 1 & -1 \\ 3 & 2 & 4 \\ -1 & 4 & 2 \end{bmatrix}$ **Ans.** $\begin{bmatrix} 2 \\ 6 \\ -3 \end{bmatrix}$ $\begin{bmatrix} -4 & 0 & 9 \\ -1 & 1 & -19 \\ 3 & 1 & 17 \end{bmatrix}$

f) $\begin{bmatrix} 4 & 3 & -3 \\ 1 & 6 & -3 \\ -1 & -3 & 6 \end{bmatrix}$ **Ans.** $\begin{bmatrix} 10 \\ 3 \\ 3 \end{bmatrix}$ $\begin{bmatrix} -1 & -3 & 3 \\ -1 & 1 & 0 \\ 1 & 0 & 1 \end{bmatrix}$

Note that eigenvalues are elements of column vector and eigenvectors are columns of matrix in Answer.

2. Find the bounds on eigenvalues of matrix $A = \begin{bmatrix} 1 & 1 & -1 \\ 3 & 2 & 4 \\ -1 & 4 & 2 \end{bmatrix}$ using Gerschgorin theorem only.

   Ans. $|\lambda| \le 7$

3. Find the bounds on eigenvalues of matrix $A = \begin{bmatrix} 1 & 1 & -1 \\ 3 & 2 & 4 \\ -1 & 4 & 2 \end{bmatrix}$ with the help of Gerschgorin and Brauer theorems.

   Ans. $|\lambda - 2| \le 5$

4. Using Rayleigh power method, find the largest eigenvalue and the corresponding eigenvector of the following matrices with initial approximation $[1, 1, 1]^t$. Perform only five iterations.

   a) $\begin{bmatrix} 1 & 1 & -1 \\ 3 & 2 & 4 \\ -1 & 4 & 2 \end{bmatrix}$ b) $\begin{bmatrix} 4 & 3 & -3 \\ 1 & 6 & -3 \\ -1 & -3 & 6 \end{bmatrix}$

   c) $\begin{bmatrix} 0 & 2 & 4 \\ 1 & 1 & -2 \\ -2 & 0 & 5 \end{bmatrix}$ d) $\begin{bmatrix} -9 & 2 & 6 \\ 5 & 0 & -3 \\ -16 & 4 & 11 \end{bmatrix}$

Ans. a) 9.000000, [0.111111 1.000000 0.555556]^T
   5.000000, [0.111111 0.911111 1.000000]^T
   6.155556, [0.003610 1.000000 0.898917]^T
   5.794224, [0.018069 0.967601 1.000000]^T
   5.989408, [−0.002393 1.000000 0.977114]^T

b) 4.000000, [1.000000   1.000000   0.500000]^T

5.500000, [1.000000   1.000000   -0.1818182]^T

7.545455, [1.000000   1.000000   -0.6746993]^T

9.024096, [1.000000   1.000000   -0.891856]^T

9.675568, [1.000000   1.000000   -0.966470]^T

c) 6.000000, [1.000000   0.000000   0.500000]^T

2.000000, [1.000000   0.000000   0.250000]^T

1.000000, [1.000000   0.500000   -0.750000]^T

5.750000, [-0.347826   0.521739   -1.000000]^T

4.304348, [-0.686869   0.505050   -1.000000]^T

d) 2.000000, [-0.500000   1.000000   -0.500000]^T

6.500000, [0.538462   -0.153846   1.000000]^T

1.769230, [0.478261   -0.173913   1.000000]^T

2.652176, [0.508197   -0.229508   1.000000]^T

1.950821, [0.495798   -0.235294   1.000000]^T

5. Find all the eigenvalues and the corresponding eigenvectors of the matrix $A = \begin{bmatrix} 2 & -1 & 0 \\ -1 & 2 & -1 \\ 0 & -1 & 2 \end{bmatrix}$

with the help of Rayleigh power method.

Ans. All three eigenvalues and corresponding eigenvectors are as follows

3.4142, [0.7071   -1.0000   0.7071]^T

0.5858, [0.7071   1.0000   0.7071]^T

1.9999, [-1.0000   0.0001   1.0000]^T

6. Use LU decomposition method to compute the eigenvalues of the following matrix. Perform only three iterations.

$$\begin{bmatrix} 5 & 4 & 0 \\ 4 & 3 & -1 \\ 0 & -1 & 6 \end{bmatrix}$$

Ans.

First Iteration

Lower triangular matrix L_1:	Upper triangular matrix U_1:	Similarity matrix A_1:
1.0000 0.0000 0.0000	5.0000 4.0000 0.0000	8.2000 4.0000 0.0000
0.8000 1.0000 0.0000	0.0000 -0.2000 -1.0000	-0.1600 -5.2000 -1.0000
0.0000 5.0000 1.0000	0.0000 0.0000 11.0000	0.0000 55.0000 11.0000

Second Iteration

Lower triangular matrix L_2:	Upper triangular matrix U_2:	Similarity matrix A_2:
1.0000　0.0000　0.0000	8.2000　4.0000　0.0000	8.1220　4.0000　0.0000
−0.0195　1.0000　0.0000	0.0000　−5.1220　−1.0000	0.0999　5.6161　−1.0000
0.0000　−10.7381　1.0000	0.0000　0.0000　0.2619	0.0000　−2.8124　0.2619

Third Iteration

The lower triangular matrix L_3:	The upper triangular matrix U_3:	The similarity matrix A_3:
1.0000　0.0000　0.0000	8.1220　4.0000　0.0000	8.1712　4.0000　0.0000
0.0123　1.0000　0.0000	0.0000　5.5669　−1.0000	0.0685　6.0721　−1.0000
0.0000　−0.5052　1.0000	0.0000　0.0000　−0.2433	0.0000　0.1229　−0.2433

7.　Consider the following matrices

$$A = \begin{bmatrix} 1 & 1 & -1 \\ 3 & 2 & 4 \\ -1 & 4 & 2 \end{bmatrix} \text{ and } S = \begin{bmatrix} -4 & 0 & 9 \\ -1 & 1 & -19 \\ 3 & 1 & 17 \end{bmatrix}$$

The matrix $S^{-1} AS$ is a diagonal matrix. Use it to calculate the eigenvalues and the corresponding eigenvectors of matrix A.

Ans. $S^{-1}AS = \begin{bmatrix} 2 & 0 & 0 \\ 0 & 6 & 0 \\ 0 & 0 & -3 \end{bmatrix}$. It implies eigenvalues are 2, 6 and −3, while the corresponding

eigenvectors are the columns of matrix S, respectively.

Eigenvalues and Eigenvectors of Real Symmetric Matrices

Do not worry about your problems with mathematics, I assure you mine are far greater.

Albert Einstein
(March 14, 1879–April 18, 1955)
He was a great theoretical physicist who gave the famous
'mass energy formula $E = mc^{2}$' and the 'theory of relativity'.

7.1 Introduction

In Chapter 6, we have discussed a few methods to solve the eigenvalue problem

$$AX = \lambda X \tag{7.1}$$

where λ is a scalar quantity. The column vector X and matrix A are given by

$$X = \begin{bmatrix} x_1 \\ x_2 \\ \vdots \\ x_n \end{bmatrix} \text{ and } A = \begin{bmatrix} a_{11} & a_{12} & \cdots & a_{1n} \\ a_{21} & a_{22} & \cdots & a_{2n} \\ \vdots & & & \\ a_{n1} & a_{n2} & \cdots & a_{nn} \end{bmatrix}$$

The following properties of eigenvalues and eigenvectors of real matrices have already been discussed in the last chapter. Let λ_i be an eigenvalue of matrix A, then

i. $\dfrac{1}{\lambda_i}$ is the eigenvalue of the matrix A^{-1}.

ii. $\lambda_i - k$ is an eigenvalue of the matrix $(A - k\,I)$.

iii. λ_i^{k} is the eigenvalue of the matrix A^k.

Note that the eigenvector X_i is same for both the matrices in all these three cases.

The matrix A is real symmetric matrix if all its elements are real and transpose matrix A^T equals A. In this chapter, we will discuss some methods to compute eigenvalues and eigenvectors of real symmetric matrices. These methods are easy to implement, and provide all the eigenvalues and eigenvectors. We will first discuss a few definitions and properties of matrices, which are pre-requisites for these methods. Let A be any square matrix, transpose A^T, conjugate $\overline{A}$, and transpose conjugate $A^* = \left(\overline{A}\right)^T$, then

1. Matrix A is symmetric if $A^T = A$
2. Matrix A is skew-symmetric if $A^T = -A$
3. Matrix A is Hermitian if $A^* = A$
4. Matrix A is skew-Hermitian if $A^* = -A$
5. Matrix A is orthogonal if $A^T = A^{-1}$ or $A^T A = AA^T = I$
6. Matrix A is unitary matrix if $A^* = A^{-1}$ or $A^* A = AA^* = I$.

Theorem 7.1

Eigenvalues of a matrix A and its transpose A^T are the same.

Proof: Let λ be an eigenvalue matrix A, then

$$\left|(A - \lambda I)\right| = 0$$

$$\left|(A - \lambda I)^T\right| = 0 \qquad \text{(Determinants of matrix and its transpose are equal)}$$

$$\left|A^T - \lambda I\right| = 0 \qquad \text{(Identity matrix is symmetrical)}$$

It implies that characteristics equations of matrices A and A^T are equal. So, the eigenvalues of these two matrices are equal.

Theorem 7.2

Eigenvectors of a matrix A and its transpose A^T are bi-orthogonal.

Proof: Let us assume that the eigenvalues $\lambda_1, \lambda_2, \cdots, \lambda_n$ of matrix A are distinct. Since the eigenvalues of A and A^T are same, so $\lambda_1, \lambda_2, \cdots, \lambda_n$ are also the eigenvalues of matrix A^T. Corresponding to these eigenvalues, assume the eigenvectors of the matrix A and A^T are $X_1, X_2, \cdots, X_n$ and $Y_1, Y_2, \cdots, Y_n$, respectively. So, we have

$$AX_i = \lambda_i X_i \tag{7.2}$$

$$A^T Y_j = \lambda_j Y_j \tag{7.3}$$

On taking transpose on both sides of Eq. (7.2), we obtain

$$\left(AX_i\right)^T = \left(\lambda_i X_i\right)^T$$

$$X_i^T A^T = \lambda_i X_i^T \quad \left(\lambda_i^T = \lambda_i, \text{ for scaler } \lambda_i\right) \tag{7.4}$$

The expression X_i^T (7.3) – (7.4) Y_j gives the following equation

$$\left(\lambda_i - \lambda_j\right) X_i^T Y_j = 0$$

The eigenvalues are distinct $\left(\lambda_i \neq \lambda_j \text{ if } i \neq j\right)$; thus we have

$$X_i^T Y_j = 0$$

So, the eigenvectors are bi-orthogonal.

Corollary: Eigenvectors of a symmetric matrix are orthogonal.

Proof: Let matrix A be symmetric ($A^T = A$), then eigenvectors of A and A^T are identical, i.e. $X_j = Y_j$. So, for distinct eigenvalues $\left(\lambda_i \neq \lambda_j \text{ if } i \neq j\right)$, we have

$$X_i^T X_j = 0$$

Theorem 7.3

Eigenvalues of a Hermitian matrix are real.

Proof: Let λ be an eigenvalue and X is corresponding eigenvector of a Hermitian matrix A, then

$$AX = \lambda X \tag{7.5}$$

The transpose conjugate (*) provides the following results

$$\left(AX\right)^* = \left(\lambda X\right)^*$$

$$X^* A^* = \bar{\lambda} X^*$$

$$X^* A = \bar{\lambda} X^* \quad \left(A^* = A\right)$$

On post-multiplying with vector X, we get

$$\left(X^* A\right) X = \left(\bar{\lambda} X^*\right) X = \bar{\lambda}\left(X^* X\right) \tag{7.6}$$

On pre-multiplying Eq. (7.5) with X^*, we have

$$X^*\left(AX\right) = X^*\left(\lambda X\right) = \lambda\left(X^* X\right) \tag{7.7}$$

The subtraction of Eq. (7.7) from Eq. (7.6) provides the following equation

$$\left(\overline{\lambda} - \lambda\right) X^* X = 0$$

The product X^*X cannot be zero, as it is the sum of the square of non-negative numbers with at least one nonzero. So, it gives

$\overline{\lambda} = \lambda$, which implies λ is real.

Corollary 1: Eigenvalue of the real symmetric matrix is real, as $A^* = A^T$ for real symmetric matrix.

Corollary 2: Eigenvalue of a skew-Hermitian matrix is either zero or pure imaginary.

Proof: Let λ be an eigenvalue and X is the corresponding eigenvector of a skew-Hermitian matrix A, then

$$AX = \lambda X$$
$$i\left(AX\right) = i\left(\lambda X\right)$$
$$\left(iA\right)X = \left(i\lambda\right)X$$

It shows that $i\lambda$ is an eigenvalue of the matrix iA. If matrix A is skew-Hermitian, then the matrix iA is Hermitian. From theorem 7.3, the eigenvalue $i\lambda$ is real. Therefore eigenvalue λ is either zero or pure imaginary.

Theorem 7.4

Eigenvectors of a Hermitian matrix are orthogonal.

Proof: Let λ_i and λ_j be two eigenvalues and X_i and X_j are corresponding eigenvectors of a Hermitian matrix A, then

$$AX_i = \lambda_i X_i \tag{7.8}$$

$$AX_j = \lambda_j X_j \tag{7.9}$$

On pre-multiplying Eq. (7.8) with the vector X_j^*, we get

$$X_j^* A X_i = X_j^* \lambda_i X_i = \lambda_i \left(X_j^* X_i\right)$$

On taking transpose conjugate(*) of this equation, we have

$$\left(X_j^* A X_i\right)^* = \left(X_j^* \lambda_i X_i\right)^*$$
$$X_i^* A^* \left(X_j^*\right)^* = X_i^* \lambda_i^* \left(X_j^*\right)^*$$
$$X_i^* A^* X_j = X_i^* \lambda_i^* X_j$$

The matrix A is Hermitian $(A^* = A)$ and eigenvalues of Hermitian matrix are real $(\lambda^* = \bar{\lambda} = \lambda)$. So, we have

$$\left(X_i^* A^* X_j\right)^* = \left(X_i^* \lambda_i^* X_j\right)^*$$
$$X_i^* A X_j = X_i^* \lambda_i X_j$$
$$X_i^* A X_j = \lambda_i\left(X_i^* X_j\right) \qquad\qquad (7.10)$$

On pre-multiplying Eq. (7.9) with the vector X_i^* and subtracting from Eq. (7.10), we have

$$0 = \left(\lambda_i - \lambda_j\right)\left(X_i^* X_j\right)$$

For $\lambda_i \neq \lambda_j$, we have

$$X_i^* X_j = 0$$

$\Rightarrow$ Eigenvectors are orthogonal.

<hr>

Theorem 7.5

Eigenvalues of a unitary matrix are of unit modulus.

Proof: Let λ be an eigenvalue and X is the corresponding eigenvector of a unitary matrix A, then

$$AX = \lambda X$$

Transpose conjugate of this equation $(AX)^* = (\lambda X)^*$ provides the following equation

$$X^* A^* = \bar{\lambda} X^*$$

From these two equations, we have

$$X^* A^* A X = \bar{\lambda} X^* \lambda X$$
$$X^*\left(A^* A\right)X = \bar{\lambda}\lambda\left(X^* X\right)$$
$$X^*\left(I\right)X = \bar{\lambda}\lambda\left(X^* X\right) \qquad (A \text{ is unitary matrix}, A^* A = I)$$
$$X^* X = \bar{\lambda}\lambda\left(X^* X\right)$$
$$\left(1 - \bar{\lambda}\lambda\right)\left(X^* X\right) = 0$$

As $X^* X \neq 0$, therefore $\lambda\bar{\lambda} = 1$.

So, the eigenvalues of unitary matrix are of unit modulus.

Theorem 7.6

Eigenvectors of a unitary matrix are orthogonal

Left as an exercise.

7.1.1 Similarity Transformations

The two matrices A and B (of same order) are said to be similar matrices, if a nonsingular matrix S exists, such that

$$B = S^{-1}AS \tag{7.11}$$

The matrix S is called similarity matrix, and transformation (7.11) is called similarity transformation.

Theorem 7.7

Let λ_i be an eigenvalue and X_i is corresponding eigenvector of matrix A. Also, let us assume the matrices A and B are similar matrices through the transformation $B = S^{-1}AS$. Then, the eigenvalues of similar matrices A and B are equal, and eigenvector Y_i of B can be obtained easily from the relation $Y_i = S^{-1}X_i$.

Proof: For matrix A, we have

$$AX_i = \lambda_i X_i$$

(or) $\quad S^{-1}AX_i = \lambda_i S^{-1}X_i \tag{7.12}$

Substituting $X_i = SY_i$ in Eq. (7.12), we have

$$S^{-1}A(SY_i) = \lambda_i S^{-1}(SY_i)$$

$$(S^{-1}AS)Y_i = \lambda_i(S^{-1}S)Y_i$$

Using $B = S^{-1}AS$, we obtain

$$BY_i = \lambda_i Y_i \tag{7.13}$$

It implies that λ_i is eigenvalue and Y_i is eigenvector of matrix B.

Note: Let all the eigenvectors of matrix A be linearly independent, and S is the matrix of eigenvectors, then S^{-1} exists, and it is such that

$$S^{-1}AS = D$$

where the matrix D is a diagonal matrix. The eigenvalues of A are the diagonal elements of diagonal matrix D, and matrix A is said to be diagonalizable.

Example ——————————————————————————— **7.1**

Compute the eigenvalues and corresponding eigenvectors for the following matrix, and prove that the matrix is diagonalizable.

$$A = \begin{bmatrix} 4 & 3 & -3 \\ 1 & 6 & -3 \\ -1 & -3 & 6 \end{bmatrix}$$

Ans. The eigenvalues and corresponding eigenvectors of the matrix A have already obtained in exercise 6.1 (f). The eigenvalues are 10, 3, 3 and eigenvectors are $\begin{bmatrix} -1 \\ -1 \\ 1 \end{bmatrix}$, $\begin{bmatrix} -3 \\ 1 \\ 0 \end{bmatrix}$,

$\begin{bmatrix} 3 \\ 0 \\ 1 \end{bmatrix}$, respectively. On using the eigenvectors as column vectors for similarity matrix, we have

$$S = \begin{bmatrix} -1 & -3 & 3 \\ -1 & 1 & 0 \\ 1 & 0 & 1 \end{bmatrix}$$

Matrix inverse of S is given by

$$S^{-1} = \frac{1}{7} \begin{bmatrix} -1 & -3 & 3 \\ -1 & 4 & 3 \\ 1 & 3 & 4 \end{bmatrix}$$

The transformation $S^{-1}AS$ is given by

$$S^{-1}AS = \frac{1}{7} \begin{bmatrix} -1 & -3 & 3 \\ -1 & 4 & 3 \\ 1 & 3 & 4 \end{bmatrix} \begin{bmatrix} 4 & 3 & -3 \\ 1 & 6 & -3 \\ -1 & -3 & 6 \end{bmatrix} \begin{bmatrix} -1 & -3 & 3 \\ -1 & 1 & 0 \\ 1 & 0 & 1 \end{bmatrix}$$

$$= \begin{bmatrix} 10 & 0 & 0 \\ 0 & 3 & 0 \\ 0 & 0 & 3 \end{bmatrix}$$

The diagonal elements are eigenvalues of matrix A. It shows that the matrix A is diagonalizable.

7.1.2 Orthogonal Transformations

A nonsingular matrix S is said to be an orthogonal matrix if its inverse and transpose are equal, i.e., $S^T = S^{-1}$ or simply $S\,S^T = I = S^T\,S$. Let S be an orthogonal matrix, then the matrices A and $B = S^T\,AS$ are similar matrices.

Product matrix of two orthogonal matrices is also orthogonal.

Proof: Let S_1 and S_2 be two orthogonal matrices, then we have

$$S_1 S_2 \left(S_1 S_2\right)^T = S_1 S_2 \left(S_2^{T} S_1^{T}\right)$$
$$= S_1 \left(S_2 S_2^{T}\right) S_1^{T} \quad \left(\text{Matrix } S_2 \text{ is orthogonal} \Rightarrow S_2 S_2^{T} = I\right)$$
$$= S_1 \left(I\right) S_1^{T}$$
$$= S_1\, S_1^{T} \quad \left(\text{Matrix } S_1 \text{ is orthogonal}\right)$$
$$= I$$

It implies that product matrix $S_1 S_2$ is orthogonal.

We can extend this result to prove that the product of any n orthogonal matrices is also an orthogonal matrix.

Consider the real symmetric matrix A and orthogonal matrices S_r, $r = 1, 2, 3, \ldots$ $\left(S_r^T = S_r^{-1}\right)$ to obtain following successive similarity transformations

$$A_1 = S_1^T A S_1$$
$$A_{r+1} = S_{r+1}^T A_r S_{r+1}, \quad r = 1, 2, 3, \ldots$$

Since the eigenvalues of similar matrices are equal, therefore the eigenvalues of A and A_{r+1} are equal. We will use these successive transformations to discuss following three methods for the eigenvalues of real symmetric matrix.

    i)   Jacobi Method
    ii)   Givens Method
    iii)   Householder Method

In Jacobi method, similarity transformations convert the matrix A into the diagonal matrix (A_{r+1}), so the diagonal elements of the matrix A_{r+1} give the eigenvalues of the matrix A. Jacobi method needs a large number of similarity transformations for a desired accuracy. In Givens and Householder methods, the matrix A is transformed into the tridiagonal matrix (A_{r+1}). The eigenvalues of matrix A_{r+1} are obtained using Strum sequence These methods require fixed number of iterations to obtain tridiagonal matrix.

7.2 Jacobi Method

All the eigenvalues of a real symmetric matrix are real. There exists a real orthogonal matrix S such that $S^{-1}AS$ is a diagonal matrix D, and diagonal elements of D are eigenvalues (as D and A are similar), and column vectors of S are corresponding eigenvectors. We will apply repeated orthogonal transformations to obtain the matrix S. First, we will explain how the orthogonal transformation is used to make non-diagonal elements of a real symmetric matrix zero with an example of a second order real symmetric matrix.

Let $A_1 = \begin{bmatrix} a_{ii} & a_{ij} \\ a_{ij} & a_{jj} \end{bmatrix}$ be any real symmetric matrix. The aim is to make this matrix a

diagonal matrix under orthogonal transformation. Let us consider an orthogonal matrix $\left(S_1^{-1} = S_1^T\right)$ defined by $S_1 = \begin{bmatrix} \cos\theta & -\sin\theta \\ \sin\theta & \cos\theta \end{bmatrix}$. The angle θ is to be determined in such a way

that matrix A_1 becomes diagonal matrix after similarity transformation. On applying the similarity transformation, we have

$$S_1^T A_1 S_1 = \begin{bmatrix} \cos\theta & \sin\theta \\ -\sin\theta & \cos\theta \end{bmatrix}\begin{bmatrix} a_{ii} & a_{ij} \\ a_{ij} & a_{jj} \end{bmatrix}\begin{bmatrix} \cos\theta & -\sin\theta \\ \sin\theta & \cos\theta \end{bmatrix}$$

$$= \begin{bmatrix} a_{ii}\cos^2\theta + a_{ij}\sin 2\theta + a_{jj}\sin^2\theta & (a_{jj}-a_{ii})\sin\theta\cos\theta + a_{ij}\cos 2\theta \\ (a_{jj}-a_{ii})\sin\theta\cos\theta + a_{ij}\cos 2\theta & a_{ii}\sin^2\theta - a_{ij}\sin 2\theta + a_{jj}\cos^2\theta \end{bmatrix}$$

For this matrix to be a diagonal matrix, we have

$$(a_{jj}-a_{ii})\sin\theta\cos\theta + a_{ij}\cos 2\theta = 0$$

$$\frac{1}{2}(a_{jj}-a_{ii})\sin 2\theta + a_{ij}\cos 2\theta = 0$$

$$\tan 2\theta = \frac{2a_{ij}}{(a_{ii}-a_{jj})}$$

$$\theta = \frac{1}{2}\tan^{-1}\left(\frac{2a_{ij}}{(a_{ii}-a_{jj})}\right)$$

The values of the inverse of tangent function lie in the interval $\left[-\dfrac{\pi}{2}, \dfrac{\pi}{2}\right]$, so the value

of $\theta \in \left[-\dfrac{\pi}{4}, \dfrac{\pi}{4}\right]$. So, non-diagonal elements vanish under this orthogonal transformation.

These orthogonal transformations are also known as plane rotations. We can use repeated applications of this transformation in case of a matrix of order ≥ 3. The following steps are involved during application of Jacobi method for a real symmetric matrix A of order n.

i. Find the absolutely largest off-diagonal element a_{ij} of matrix A. Compute
$\theta = \frac{1}{2}\tan^{-1}\left(\dfrac{a_{ij}}{a_{ii} - a_{jj}}\right)$. Note that at $a_{ii} = a_{jj}$, then $\theta = -45°$ or $45°$ according to sign of a_{ij}

ii. Consider an identity matrix of order n, then replace the elements at (i, i), (i, j), (j, i), (j, j) positions with $\cos\theta$, $-\sin\theta$, $\sin\theta$, $\cos\theta$, respectively to create an

orthogonal matrix $S_1 = \begin{bmatrix} 1 & 0 & 0 & \cdots & 0 & \cdots & 0 \\ 0 & 1 & 0 & \cdots & 0 & \cdots & 0 \\ \vdots & & \ddots & \vdots & \cdots & \vdots & \vdots \\ 0 & 0 & \cos\theta & \cdots & -\sin\theta & \cdots & 0 \\ \vdots & & \vdots & \vdots & & \vdots & \vdots \\ 0 & 0 & \sin\theta & \cdots & \cos\theta & \cdots & 0 \\ \vdots & & \vdots & \vdots & & \ddots & \vdots \\ 0 & 0 & 0 & \cdots & 0 & & 1 \end{bmatrix}$.

iii. Perform the similarity transformation $A_1 = S_1^T A S_1$.

iv. Repeat the steps (i–iii) on the matrix A_1 to create the matrix $A_2 = S_2^T A_1 S_2$ by using similarity matrix S_2.

v. Obtain following successive similarity transformations till the matrix A_{r+1} is approximately diagonal matrix as per desired accuracy.

$$A_1 = S_1^T A S_1$$
$$A_{r+1} = S_{r+1}^T A_r S_{r+1}, \quad r = 1, 2, 3, \ldots$$

vi. Since the eigenvalues of similar matrices (A and A_{r+1}) are equal, so the diagonal elements of A_{r+1} are the eigenvalues of matrix A.

vii. The eigenvectors can be calculated from similarity transformations. So, either column of matrix $S = S_1 S_2 \ldots S_r$ or rows of the matrix $S^T = S_r^T S_{r-1}^T \ldots S_1^T$ are the eigenvectors of matrix A.

Example ── **7.2**

Compute all the eigenvalues of the following real symmetric matrix with the aid of Jacobi method

$$A = \begin{bmatrix} 3 & -2 & 1 \\ -2 & -5 & -1 \\ 1 & -1 & 3 \end{bmatrix}$$

Ans.

A maximum off-diagonal element in the magnitude of the matrix A is the element $a_{12} = -2$. The value of θ is given by

$$\theta = \frac{1}{2}\tan^{-1}\left(\frac{2a_{ij}}{(a_{ii} - a_{jj})}\right) = \frac{1}{2}\tan^{-1}\left(\frac{2a_{12}}{(a_{11} - a_{22})}\right) = \frac{1}{2}\tan^{-1}\left(\frac{-4}{(3+5)}\right) = -0.231824$$

On using this value of θ, we can easily obtain following orthogonal matrix

$$S_1 = \begin{bmatrix} \cos\theta & -\sin\theta & 0 \\ \sin\theta & \cos\theta & 0 \\ 0 & 0 & 1 \end{bmatrix} = \begin{bmatrix} 0.973249 & 0.229753 & 0 \\ -0.229753 & 0.973249 & 0 \\ 0 & 0 & 1 \end{bmatrix}$$

The first plane rotation is given by

$$A_1 = S_1^T A S_1 = \begin{bmatrix} 0.973249 & -0.229753 & 0 \\ 0.229753 & 0.973249 & 0 \\ 0 & 0 & 1 \end{bmatrix} \begin{bmatrix} 3 & -2 & 1 \\ -2 & -5 & -1 \\ 1 & -1 & 3 \end{bmatrix}$$

$$\begin{bmatrix} 0.973249 & 0.229753 & 0 \\ -0.229753 & 0.973249 & 0 \\ 0 & 0 & 1 \end{bmatrix}$$

$$A_1 = \begin{bmatrix} 3.472136 & 0 & 1.203002 \\ 0 & -5.472136 & -0.743496 \\ 1.203002 & -0.743496 & 3.000000 \end{bmatrix}$$

The maximum off-diagonal element in the matrix A_1 is $a_{13} = 1.203002$. It gives

$$\theta = \frac{1}{2}\tan^{-1}\left(\frac{2a_{ij}}{(a_{ii} - a_{jj})}\right) = \frac{1}{2}\tan^{-1}\left(\frac{2a_{13}}{(a_{11} - a_{33})}\right) = \frac{1}{2}\tan^{-1}\left(\frac{2.406004}{0.472136}\right) = 0.688513$$

The orthogonal matrix for rotation is as follows

$$S_2 = \begin{bmatrix} \cos\theta & 0 & -\sin\theta \\ 0 & 1 & 0 \\ \sin\theta & 0 & \cos\theta \end{bmatrix} = \begin{bmatrix} 0.772192 & 0 & -0.635390 \\ 0 & 1 & 0 \\ 0.635390 & 0 & 0.772192 \end{bmatrix}$$

Second plane rotation is given by

$$A_2 = S_2^T A_1 S_2$$

$$= \begin{bmatrix} 0.772192 & 0 & 0.635390 \\ 0 & 1 & 0 \\ -0.635390 & 0 & 0.772192 \end{bmatrix} \begin{bmatrix} 3.472136 & 0 & 1.203002 \\ 0 & -5.472136 & -0.743496 \\ 1.203002 & -0.743496 & 3.000000 \end{bmatrix}$$

$$\begin{bmatrix} 0.772192 & 0 & -0.635390 \\ 0 & 1 & 0 \\ 0.635390 & 0 & 0.772192 \end{bmatrix}$$

$$A_2 = \begin{bmatrix} 4.462013 & -0.472410 & 0 \\ -0.472410 & -5.472136 & -0.574121 \\ 0 & -0.574121 & 2.010123 \end{bmatrix}$$

Similarly, other plane rotations are as follows

Third Plane Rotation:

Maximum element in matrix = −0.574121

Value of θ = −0.076137

$$A_3 = \begin{bmatrix} 4.462013 & -0.471041 & 0.035933 \\ -0.471041 & -5.515933 & 0 \\ 0.035933 & 0 & 2.053919 \end{bmatrix}$$

Fourth Plane Rotation:

Maximum element in matrix = −0.471041

Value of θ = −0.047069

$$A_4 = \begin{bmatrix} 4.484200 & 0 & 0.035893 \\ 0 & -5.538120 & 0.001691 \\ 0.035893 & 0.001691 & 2.053919 \end{bmatrix}$$

Fifth Plane Rotation:

Maximum element in matrix = 0.035893

Value of θ = 0.014765

$$A_5 = \begin{bmatrix} 4.484730 & 0.000025 & 0 \\ 0.000025 & -5.538120 & 0.001691 \\ 0 & 0.001690 & 2.053389 \end{bmatrix}$$

Maximum element in matrix = 2.000000

Sixth Plane Rotation:

Maximum element in matrix = 0.001691

Value of θ = −0.000223

$$A_6 = \begin{bmatrix} 4.484730 & 0.000025 & 0 \\ 0.000025 & -5.538121 & 0 \\ 0 & 0 & 2.053390 \end{bmatrix}$$

Seventh Plane Rotation:

Maximum element in matrix = 0.000025

Value of θ = −0.000002

$$A_7 = \begin{bmatrix} 4.484730 & 0 & 0 \\ 0 & -5.538121 & 0 \\ 0 & 0 & 2.053390 \end{bmatrix}$$

The diagonal elements of the matrix A_7 give the eigenvalues of matrix A, which are 4.484730, −5.538121, 2.053390.

Note:

1. The eigenvalues of similar matrices are equal, and the trace of a matrix is equal to sum of the eigenvalues, so

$$\text{Tr}(A) = \sum_{i=1}^{n} \lambda_i = \text{Tr}(A_1) = \text{Tr}(A_2) = \dots$$

2. The eigenvectors of matrix A are columns of the matrix $S = S_1 S_2 \dots S_r$. Here, due to the complexity of calculations, this step is avoided.
3. The computational work is very cumbersome for the method, so programming is helpful this regard.

7.3 Strum Sequence for Real Symmetric Tridiagonal Matrix

In Givens and Householder methods, the matrix A is transformed into the following tridiagonal matrix

$$B = \begin{bmatrix} b_1 & c_1 & 0 & 0 & \dots & 0 & 0 \\ c_1 & b_2 & c_2 & 0 & \dots & 0 & 0 \\ & \ddots & \ddots & \ddots & & & \\ 0 & 0 & 0 & 0 & \dots & b_{n-1} & c_{n-1} \\ 0 & 0 & 0 & 0 & \dots & c_{n-1} & b_n \end{bmatrix}$$

First, we will introduce Strum sequence to obtain the characteristic equation of the tridiagonal matrix B. Consider the following determinant

$$f_n = |\lambda I - B|$$

$$= \begin{vmatrix} \lambda - b_1 & -c_1 & 0 & 0 & \dots & 0 & 0 \\ -c_1 & \lambda - b_2 & -c_2 & 0 & \dots & 0 & 0 \\ & \ddots & \ddots & \ddots & & & \\ 0 & 0 & 0 & 0 & \dots & \lambda - b_{n-1} & -c_{n-1} \\ 0 & 0 & 0 & 0 & \dots & -c_{n-1} & \lambda - b_n \end{vmatrix}$$

On expanding the determinant through minors, we can easily construct the Strum sequence, which is given by

$$f_0 = 1, \quad f_1 = \lambda - b_1$$
$$f_r = (\lambda - b_r) f_{r-1} - c_{r-1}^2 f_{r-2}; \qquad 2 \leq r \leq n \qquad (7.14)$$

At last, we obtain the polynomial f_n, and the equation $f_n = 0$ is required characteristic equation. The roots of characteristics equation are the eigenvalues of the matrix. The Strum sequence $\{f_n\}$ satisfies the properties discussed in subsection 4.5.2. So, we can easily obtain the intervals containing the zeroes of polynomial f_n, and then roots can be computed using any method discussed in Chapters 3, 4.

Example   **7.3**

Find all the eigenvalues of the tridiagonal matrix $A = \begin{bmatrix} 2 & -1 & 0 \\ -1 & 2 & -1 \\ 0 & -1 & 2 \end{bmatrix}$ with the help of Strum sequence.

Ans.

On using Strum sequence (7.14), we have

$$f_0 = 1,$$
$$f_1 = \lambda - b_1 = \lambda - 2$$
$$f_2 = (\lambda - b_2) f_1 - c_1^2 f_0 = (\lambda - 2)(\lambda - 2) - (-1)^2 .1 = \lambda^2 - 4\lambda + 3$$
$$f_3 = (\lambda - b_3) f_2 - c_2^2 f_1 = (\lambda - 2)(\lambda^2 - 4\lambda + 3) - (-1)^2 (\lambda - 2)$$
$$= \lambda^3 - 6\lambda^2 + 10\lambda - 4$$

The characteristic equation is given by

$$\lambda^3 - 6\lambda^2 + 10\lambda - 4 = 0$$

We can compute the roots (0.5858, 2 and 3.4142) of this equation by any method discussed in Chapters 3, 4. The eigenvalues are 0.5858, 2 and 3.4142.

7.4 Givens Method

The major disadvantage of Jacobi method is that, the off-diagonal elements which are made zeroes in an iteration may not remain zeroes in next iteration. In Givens method, the zero off-diagonal elements remain zeroes in subsequent iterations. The similarity transformations used in Givens method can reduce the matrix A into a similar tridiagonal matrix B. The Strum sequence $\{f_n\}$ can be obtained from this tridiagonal matrix and zeroes of polynomial f_n are the eigenvalues of the matrix A. Further, the eigenvectors of matrix B can be used to compute the eigenvectors of matrix A. On a similar pattern as in the Jacobi method, we will use the orthogonal transformations to reduce the matrix A into its tridiagonal similar matrix B.

In this method, we obtain zeroes at (1, 3) positions in first plane rotation, then we annihilate the elements at (1, 4), (1, 5), ..., (1, n) positions in subsequent plane rotations. After, vanishing the off-tridiagonal elements in first row, we proceed to second row and use plane rotations to vanish the elements at (2, 4), (2, 5), ..., (2, n) positions. We will continue this process till all the off-tridiagonal elements are zeroes.

Consider the real symmetric matrix $A = \begin{bmatrix} a_{11} & a_{12} & \cdots & a_{1n} \\ a_{21} & a_{22} & \cdots & a_{2n} \\ \vdots & & & \\ a_{n1} & a_{n2} & \cdots & a_{nn} \end{bmatrix}$. The aim is to

obtain a similar tridiagonal matrix B by using similarity transformations. Givens method has the following steps.

i. Consider an identity matrix of order n, then replace the elements at $(2, 2)$, $(2, 3)$, $(3, 2)$, $(3, 3)$ positions with $\cos\theta$, $-\sin\theta$, $\sin\theta$, $\cos\theta$, respectively to create

    an orthogonal matrix $S_1 = \begin{bmatrix} 1 & 0 & 0 & 0 & \cdots & 0 \\ 0 & \cos\theta & -\sin\theta & 0 & \cdots & 0 \\ 0 & \sin\theta & \cos\theta & 0 & \cdots & 0 \\ & & & \ddots & & \\ 0 & 0 & 0 & & \cdots & 1 \end{bmatrix}$. Now the

    angle θ is to be determined such that the elements $a'_{13} = a'_{31} = 0$ in similar matrix $A_1 = S_1^T A S_1$. We find that

    $$a'_{13} = a'_{31} = -a_{12}\sin\theta + a_{13}\cos\theta = 0$$

    $$\Rightarrow \tan\theta = \frac{a_{13}}{a_{12}}$$

    We obtain zeroes at $(3, 1)$ and $(1, 3)$ positions by performing the plane rotation $A_1 = S_1^T A S_1$ with this value of θ.

ii. On a similar pattern, we will obtain the zeroes at $(4, 1)$ and $(1, 4)$ positions by using similarity transformation on the matrix A_1. For this, we create an orthogonal matrix S_2 by replacing the elements of an identity matrix at $(2, 2)$, $(2, 4)$, $(4, 2)$, $(4, 4)$ positions with $\cos\theta$, $-\sin\theta$, $\sin\theta$, $\cos\theta$ respectively. The value of angle θ is

    given by $\tan\theta = \frac{a_{14}}{a_{12}}$. Then, the plane rotation $A_2 = S_2^T A_1 S_2$ gives zeroes at $(4, 1)$

    and $(1, 4)$ positions. Proceeding in a similar manner, we will obtain zeroes at $(5, 1)$ and $(1, 5)$ positions and so on till zeroes at $(n, 1)$ and $(1, n)$ positions. At last, we will obtain zeroes at $(i, 1)$ and $(1, i)$ positions for all $i = 3, 4, ..., n$.

iii. Create an orthogonal matrix by replacing the elements of an identity matrix at $(3, 3)$, $(3, 4)$, $(4, 3)$, $(4, 4)$ positions with $\cos\theta$, $-\sin\theta$, $\sin\theta$, $\cos\theta$, respectively, and perform the plane rotation to obtain zeroes at $(4, 2)$ and $(2, 4)$ positions. Similarly, we will obtain zeroes at $(i, 2)$ and $(2, i)$ positions for all $i = 4, 5, ..., n$.

iv. Repeat the process till the following tridiagonal matrix is obtained

$$B = \begin{bmatrix} b_1 & c_1 & 0 & 0 & \cdots & 0 & 0 \\ c_1 & b_2 & c_2 & 0 & \cdots & 0 & 0 \\ & \ddots & \ddots & \ddots & & & \\ 0 & 0 & 0 & 0 & \cdots & b_{n-1} & c_{n-1} \\ 0 & 0 & 0 & 0 & \cdots & c_{n-1} & b_n \end{bmatrix}$$

Since all the transformations are orthogonal transformations, so the matrix B obtained from these transformations is similar to matrix A. The eigenvalues of both the matrices are equal. We can compute the Strum sequence $\{f_n\}$ to obtain characteristic polynomial of matrix B. The zeroes of this polynomial provide the eigenvalues of matrix A.

The matrices A and B are similar matrices, so the eigenvectors of these two matrices are related by the following relation

$$X_i = S Y_i$$

where $S = S_1 S_2 \ldots S_j$ is the product of all orthogonal matrices used for plane rotations. The eigenvectors of the matrix A can be obtained with the help of eigenvectors of the matrix B.

Example 7.4

Obtain the similar tridiagonal form for the matrix $A = \begin{bmatrix} 5 & 1 & -2 \\ 1 & 0 & 1 \\ -2 & 1 & -3 \end{bmatrix}$ with the help of Givens method. Then, compute the eigenvalues of matrix A by using Strum sequence.

Ans.

Step 1. To obtain similar tridiagonal matrix

Let us consider an orthogonal transformation defined by $S_1 = \begin{bmatrix} 1 & 0 & 0 \\ 0 & \cos\theta & -\sin\theta \\ 0 & \sin\theta & \cos\theta \end{bmatrix}$

to perform the plane rotation $A_1 = S_1^T A S_1$. The angle θ is to be determined by the following formula

$$\tan\theta = \frac{a_{13}}{a_{12}} = \frac{-2}{1} = -2$$
$$\Rightarrow \theta = -1.107149$$

On using this value of θ, we have

$$S_1 = \begin{bmatrix} 1 & 0 & 0 \\ 0 & 0.447214 & 0.894427 \\ 0 & -0.894427 & 0.447214 \end{bmatrix}$$

The plane rotation $A_1 = S_1^T A S_1$ is given by

$$A_1 = S_1^T A S_1$$

$$= \begin{bmatrix} 1 & 0 & 0 \\ 0 & 0.447214 & -0.894427 \\ 0 & 0.894427 & 0.447214 \end{bmatrix} \begin{bmatrix} 5 & 1 & -2 \\ 1 & 0 & 1 \\ -2 & 1 & -3 \end{bmatrix} \begin{bmatrix} 1 & 0 & 0 \\ 0 & 0.447214 & 0.894427 \\ 0 & -89.4427 & 0.447214 \end{bmatrix}$$

$$= \begin{bmatrix} 5 & 2.236068 & 0 \\ 2.236068 & -3.2 & 0.6 \\ 0 & 0.6 & 0.2 \end{bmatrix}$$

The zeroes are obtained at (3, 1) and (1, 3) positions, so we have following required tridiagonal form

$$A_1 = \begin{bmatrix} 5 & 2.236068 & 0 \\ 2.236068 & -3.2 & 0.6 \\ 0 & 0.6 & 0.2 \end{bmatrix}$$

Step 2: To generate the Strum sequence from tridiagonal matrix

Let the tridiagonal matrix is B.

$$B = A_1 = \begin{bmatrix} 5 & 2.236068 & 0 \\ 2.236068 & -3.2 & 0.6 \\ 0 & 0.6 & 0.2 \end{bmatrix}$$

The characteristic equation of matrix B is given by

$$|\lambda I - B| = \begin{vmatrix} \lambda - 5 & -2.236068 & 0 \\ -2.236068 & \lambda + 3.2 & -0.6 \\ 0 & -0.6 & \lambda - 0.2 \end{vmatrix} = 0$$

Strum sequence can be computed in the following manner

$$f_0 = 1$$
$$f_1 = \lambda - b_1 = \lambda - 5$$
$$f_2 = (\lambda - b_2) f_1 - c_1^2 f_0 = (\lambda + 3.2)(\lambda - 5) - (2.236068)^2 . 1 = \lambda^2 - 1.8\lambda - 21$$
$$f_3 = (\lambda - b_3) f_2 - c_2^2 f_1 = (\lambda - 0.2)(\lambda^2 - 1.8\lambda - 21) - (0.6)^2 (\lambda - 5)$$
$$= \lambda^3 - 2\lambda^2 - 21\lambda + 6$$

Characteristic equation is given by

$$\lambda^3 - 2\lambda^2 - 21\lambda + 6 = 0$$

The roots of this characteristics equation are lies in the intervals $(-4, 3)$, $(0, 1)$, $(5, 6)$.

These roots can be easily determined by any iterative method discussed in Chapters 3 and 4. The roots of the characteristic equation are -3.853556, 0.279322 and 5.574234 correct up to six decimal places. These are the required eigenvalues.

Example ── **7.5**

Reduce the symmetric matrix $A = \begin{bmatrix} 4 & -1 & 1 & 1 \\ -1 & 2 & 0 & 2 \\ 1 & 0 & 3 & 1 \\ 1 & 2 & 1 & -4 \end{bmatrix}$ to the similar tridiagonal matrix

by using Givens method. Then compute the eigenvalues using strum sequence.

Ans. (Step 1) Givens method to obtain tridiagonal matrix

Let us consider an orthogonal transformation defined by $S_1 = \begin{bmatrix} 1 & 0 & 0 & 0 \\ 0 & \cos\theta & -\sin\theta & 0 \\ 0 & \sin\theta & \cos\theta & 0 \\ 0 & 0 & 0 & 1 \end{bmatrix}$,

where the value of θ is given by

$$\tan\theta = \frac{a_{13}}{a_{12}} = \frac{1}{-1} = -1$$

$$\Rightarrow \theta = \frac{-\pi}{4}$$

The orthogonal matrix is as follows

$$S_1 = \begin{bmatrix} 1 & 0 & 0 & 0 \\ 0 & \cos\theta & -\sin\theta & 0 \\ 0 & \sin\theta & \cos\theta & 0 \\ 0 & 0 & 0 & 1 \end{bmatrix} = \begin{bmatrix} 1 & 0 & 0 & 0 \\ 0 & 0.707107 & 0.707107 & 0 \\ 0 & -0.707107 & 0.707107 & 0 \\ 0 & 0 & 0 & 1 \end{bmatrix}$$

The first rotation $A_1 = S_1^T A S_1$ annihilates the following resulting matrix with zeroes at $(3, 1)$ and $(1, 3)$ positions. We have

$A_1 = S_1^T A S_1$

$$= \begin{bmatrix} 1 & 0 & 0 & 0 \\ 0 & 0.707107 & -0.707107 & 0 \\ 0 & 0.707107 & 0.707107 & 0 \\ 0 & 0 & 0 & 1 \end{bmatrix} \begin{bmatrix} 4 & -1 & 1 & 1 \\ -1 & 2 & 0 & 2 \\ 1 & 0 & 3 & 1 \\ 1 & 2 & 1 & -4 \end{bmatrix} \begin{bmatrix} 1 & 0 & 0 & 0 \\ 0 & 0.707107 & 0.707107 & 0 \\ 0 & -0.707107 & 0.707107 & 0 \\ 0 & 0 & 0 & 1 \end{bmatrix}$$

$$= \begin{bmatrix} 4 & -1.414214 & 0 & 1 \\ -1.414214 & 2.5 & -0.5 & 0.707107 \\ 0 & -0.5 & 2.5 & 2.121320 \\ 1 & 0.707107 & 2.121320 & -4 \end{bmatrix}$$

On a similar pattern, we will obtain the zeroes at $(4, 1)$ and $(1, 4)$ positions. The value of angle θ is given by

$$\tan\theta = \frac{a_{14}}{a_{12}} = \frac{1}{-1.414214} = -0.707106$$

$$\Rightarrow \theta = -0.615480$$

The orthogonal matrix S_2 is given by

$$S_2 = \begin{bmatrix} 1 & 0 & 0 & 0 \\ 0 & \cos\theta & 0 & -\sin\theta \\ 0 & 0 & 1 & 0 \\ 0 & \sin\theta & 0 & -\cos\theta \end{bmatrix} = \begin{bmatrix} 1 & 0 & 0 & 0 \\ 0 & 0.816497 & 0 & 0.577350 \\ 0 & 0 & 1 & 0 \\ 0 & -0.577350 & 0 & 0.816497 \end{bmatrix}$$

The second plane rotation $A_2 = S_2^T A_1 S_2$ produces the following similar matrix

$A_2 = S_2^T A_1 S_2$

$$= \begin{bmatrix} 1 & 0 & 0 & 0 \\ 0 & 0.816497 & 0 & -0.577350 \\ 0 & 0 & 1 & 0 \\ 0 & 0.577350 & 0 & 0.816497 \end{bmatrix} \begin{bmatrix} 4 & -1.414214 & 0 & 1 \\ -1.414214 & 2.5 & -0.5 & 0.707107 \\ 0 & -0.5 & 2.5 & 2.121320 \\ 1 & 0.707107 & 2.121320 & -4 \end{bmatrix} \begin{bmatrix} 1 & 0 & 0 & 0 \\ 0 & 0.816497 & 0 & 0.577350 \\ 0 & 0 & 1 & 0 \\ 0 & -0.577350 & 0 & 0.816497 \end{bmatrix}$$

$$= \begin{bmatrix} 4.000000 & -1.732051 & 0.000000 & 0.000000 \\ -1.732051 & 0.333333 & -1.632993 & 3.299832 \\ 0.000000 & -1.632993 & 2.500000 & 1.443376 \\ 0.000000 & 3.299832 & 1.443376 & -1.166667 \end{bmatrix}$$

To obtain the zeroes at $(4, 2)$ and $(2, 4)$ positions, the value of angle θ and orthogonal matrix are given by

$$\tan \theta = \frac{a_{24}}{a_{23}} = \frac{3.299832}{-1.632993} = -2.020726$$

$$\Rightarrow \theta = -1.111260$$

$$S_3 = \begin{bmatrix} 1 & 0 & 0 & 0 \\ 0 & 1 & 0 & 0 \\ 0 & 0 & 0.443533 & -0.896258 \\ 0 & 0 & 0.896258 & 0.443533 \end{bmatrix}$$

The plane rotation is given by

$$S_3^T A_2 S_3 = \begin{bmatrix} 1 & 0 & 0 & 0 \\ 0 & 1 & 0 & 0 \\ 0 & 0 & 0.443533 & -0.896258 \\ 0 & 0 & 0.896258 & 0.443533 \end{bmatrix} \begin{bmatrix} 4.000000 & -1.732051 & 0.000000 & 0.000000 \\ -1.732051 & 0.333333 & -1.632993 & 3.299832 \\ 0.000000 & -1.632993 & 2.500000 & 1.443376 \\ 0.000000 & 3.299832 & 1.443376 & -1.166667 \end{bmatrix} \begin{bmatrix} 1 & 0 & 0 & 0 \\ 0 & 1 & 0 & 0 \\ 0 & 0 & 0.443533 & -0.896258 \\ 0 & 0 & 0.896258 & 0.443533 \end{bmatrix}$$

$$A_3 = \begin{bmatrix} 4.000000 & -1.732051 & 0.000000 & 0.000000 \\ -1.732051 & -0.333333 & -3.681787 & 0.000000 \\ 0.000000 & -3.681787 & -1.592897 & 0.582083 \\ 0.000000 & 0.000000 & 0.582083 & 2.926229 \end{bmatrix}$$

This matrix A_3 is the required tridiagonal matrix.

Step 2: Strum sequence

The tridiagonal matrix is given by

$$B = A_3 = \begin{bmatrix} 4.000000 & -1.732051 & 0.000000 & 0.000000 \\ -1.732051 & -0.333333 & -3.681787 & 0.000000 \\ 0.000000 & -3.681787 & -1.592897 & 0.582083 \\ 0.000000 & 0.000000 & 0.582083 & 2.926229 \end{bmatrix}$$

We have to compute the Strum sequence for the matrix B. The associated characteristic equation is as follows

$$|\lambda I - B| = \begin{vmatrix} \lambda - 4.000000 & 1.732051 & 0.000000 & 0.000000 \\ 1.732051 & \lambda + 0.333333 & 3.681787 & 0.000000 \\ 0.000000 & 3.681787 & \lambda + 1.592897 & -0.582083 \\ 0.000000 & 0.000000 & -0.582083 & \lambda - 2.926229 \end{vmatrix} = 0$$

Strum sequence is given by

$f_0 = 1$

$f_1 = \lambda - b_1 = \lambda - 4$

$f_2 = (\lambda - b_2)f_1 - c_1^2 f_0 = (\lambda + 0.333333)(\lambda - 4) - (1.732051)^2 .1 = \lambda^2 - 3.666667\lambda - 4.333353$

$f_3 = (\lambda - b_3)f_2 - c_2^2 f_1 = (\lambda + 1.592897)(\lambda^2 - 3.666667\lambda - 4.333353) - (3.681787)^2(\lambda - 4)$

$\quad = \lambda^3 - 2.073770\lambda^2 - 23.729511\lambda + 47.319667$

$f_4 = (\lambda - b_4)f_3 - c_3^2 f_2 = (\lambda - 2.926229)(\lambda^3 - 1.073770\lambda^2 - 23.729531\lambda + 47.319637)$

$\quad - (-0.582083)^2(\lambda^2 - 3.666667\lambda - 4.333353)$

$\quad \approx \lambda^4 - 5\lambda^3 - 18\lambda^2 + 118\lambda - 137$

The characteristic equation is given by

$$\lambda^4 - 5\lambda^3 - 18\lambda^2 + 118\lambda - 137 = 0$$

Roots of this equation are -4.867263, 1.859930, 3.057171 and 4.950163, which are eigenvalues of the matrix A.

7.5 Householder Method

Givens method requires $(n-1)(n-2)/2$ numbers of plane rotations to obtain similar tridiagonal matrix. But, Householder method produces tridiagonal matrix only after $(n-2)$ numbers of plane rotations. The following theorem is useful for similarity transformations in Householder method.

Theorem 7.9

Consider the matrix $S = I - 2XX^T$, where $X^T = (x_1, x_2, ..., x_n) \in R^n$ is a vector, and such that

$$X^T X = x_1^2 + x_2^2 + ... + x_n^2 = 1$$

Then, prove that the matrix S is symmetric and orthogonal, i.e., $S^T = S$ and $S^T = S^{-1}$.

Proof: Transpose of the matrix $S = I - 2XX^T$ is given by

$$S^T = (I - 2XX^T)^T$$

$$= I^T - 2(XX^T)^T$$

$$= I - 2(X^T)^T X^T$$

$$= I - 2XX^T = S$$

It proves that the matrix S is symmetric.

$$S^T S = \left(I - 2 X X^T \right) \left(I - 2 X X^T \right)$$
$$= I - 4 X X^T + 4 X X^T X X^T$$
$$= I - 4 X X^T + 4 X (1) X^T$$
$$= I$$
$$\Rightarrow S^T = S^{-1}$$

It proves that the matrix S is orthogonal.

Consider the real symmetric matrix A of order n. The aim is to obtain a similar tridiagonal matrix B. In Householder method, we use the following symmetric and orthogonal matrices for similarity transformations

$$S_r = I - 2 X_r X_r^T, \, r = 1, 2, 3, \dots, n-2;$$

where vector $X_r^T = [0, 0, \dots, 0, \, x_{r+1}, \, x_{r+2}, \dots, x_n]^T$, and it is such that

$$X_r^T X_r = x_{r+1}^2 + x_{r+2}^2 + \dots + x_n^2 = 1$$

From theorem 7.9, it is easy to see that the matrix S_r is symmetric and orthogonal. The following plane rotations are performed successively

$$A_1 = A$$
$$A_{r+1} = S_r^{-1} A_r S_r \quad \text{for } r = 1, 2, 3, \dots, n-2$$
$$= S_r A_r S_r$$

Note that S_r is symmetric and orthogonal, therefore $S_r = S_r^{-1}$.

The aim is to reduce the matrix A into a tridiagonal matrix. The first plane rotation is given by

$$A_2 = S_1 A_1 S_1$$

The vector $X_1^T = \left[0, x_2, x_3, \dots, x_n \right]^T$ is constructed with the aim to obtain zeroes at $(1, 3), (1, 4), \dots, (1, n)$ positions. Similarly, during second plane rotation $A_3 = S_2 A_2 S_2$, zeroes are obtained at $(2, 4), (2, 5), \dots, (2, n)$ positions. Proceeding in a similar manner, the matrix A is reduced to tridiagonal form after $(n-2)$ such plane rotations. From tridiagonal matrix, the eigenvalues are obtained using Strum sequence.

Stepwise Procedure for r^{th} transformation:

i)   Select vector $X_r^T = (0, 0, \dots, 0, \, x_{r+1}, \, x_{r+2}, \dots, x_n)$ for the rth transformation. Like vector $X_1^T = (0, x_2, x_3, \dots, x_n)$ for first rotation, vector $X_2^T = (0, 0, x_3, x_4, \dots, x_n)$ for second rotation and so on.

ii)   Construct orthogonal matrices $S_r = I - 2 X_r X_r^T$; $\, r = 1, 2, 3, \dots, n-2$. For example, let us consider matrix A of order 4, then for first rotation, we have

$$S_1 = I - 2X_1 X_1^T = \begin{bmatrix} 1 & 0 & 0 & 0 \\ 0 & 1 & 0 & 0 \\ 0 & 0 & 1 & 0 \\ 0 & 0 & 0 & 1 \end{bmatrix} - 2 \begin{bmatrix} 0 \\ x_2 \\ x_3 \\ x_4 \end{bmatrix} \begin{bmatrix} 0 & x_2 & x_3 & x_4 \end{bmatrix}$$

$$S_1 = \begin{bmatrix} 1 & 0 & 0 & 0 \\ 0 & 1-2x_2^2 & -2x_2 x_3 & -2x_2 x_4 \\ 0 & -2x_2 x_3 & 1-2x_3^2 & -2x_3 x_4 \\ 0 & -2x_3 x_4 & -2x_3 x_4 & 1-2x_4^2 \end{bmatrix}$$

iii) The rth plane rotation is given by

$$A_{r+1} = S_r A_r S_r; \quad r = 1, 2, ..., n-2$$

In first plane rotation $A_2 = S_1 A_1 S_1$ (with $A_1 = A$), we obtain zeroes at (1, 3), (1, 4), ..., (1, n) positions. Similarly, during the second rotation $A_3 = S_2 A_2 S_2$, zeroes at (2, 4), (2, 5), ..., (2, n) positions. Applying this procedure successively, the matrix A is reduced to tridiagonal form.

For example for first plane rotation $A_2 = S_1 A_1 S_1$, we have

$$x_2^2 = \frac{1}{2}\left(1 + \frac{a_{12}\text{sign}(a_{12})}{s_1}\right), \quad x_3 = \frac{1}{2}\left(\frac{a_{13}\text{sign}(a_{12})}{s_1 x_2}\right), \quad x_4 = \frac{1}{2}\left(\frac{a_{14}\text{sign}(a_{12})}{s_1 x_2}\right),$$

$$x_5 = \frac{1}{2}\left(\frac{a_{15}\text{sign}(a_{12})}{s_1 x_2}\right)....,$$

where $s_1 = \sqrt{a_{12}^2 + a_{13}^2 + ... + a_{1n}^2}$

Similarly, for second plane rotation $A_3 = S_2 A_2 S_2$, we have

$$x_3^2 = \frac{1}{2}\left(1 + \frac{a_{23}\text{sign}(a_{23})}{s_2}\right), \quad x_4 = \frac{1}{2}\left(\frac{a_{24}\text{sign}(a_{23})}{s_2 x_3}\right), \quad x_5 = \frac{1}{2}\left(\frac{a_{25}\text{sign}(a_{23})}{s_2 x_3}\right), ...,$$

Where $s_2 = \sqrt{a_{23}^2 + a_{24}^2 + \cdots + a_{2n}^2}$

$\vdots$

iv) After $(n - 2)$ such rotations, the matrix A is reduced into the tridiagonal matrix. Eigenvalues and corresponding eigenvectors can be obtained from this tridiagonal form as explained earlier.

Example ──────────────────────────────── 7.6 ──

Compute the equivalent similar tridiagonal form for the matrix $A = \begin{bmatrix} 3 & 1 & -1 \\ 1 & 3 & -1 \\ -1 & -1 & 5 \end{bmatrix}$

with the help of Householder method. Then, compute the eigenvalues of matrix A by using Strum sequence.

Ans. The matrix A is given by

$$A = \begin{bmatrix} 3 & 1 & -1 \\ 1 & 3 & -1 \\ -1 & -1 & 5 \end{bmatrix}$$

First, we will compute the elements for the orthogonal matrix as follows

$$s_1 = \sqrt{a_{12}^2 + a_{13}^2} = \sqrt{1+1} = \sqrt{2}$$

$$x_2^2 = \frac{1}{2}\left(1 + \frac{a_{12}\text{sign}(a_{12})}{s_1}\right) = \frac{1}{2}\left(1 + \frac{1}{\sqrt{2}}\right) = 0.853553$$

$$\Rightarrow x_2 = 0.923880$$

$$x_3 = \frac{1}{2}\left(\frac{a_{13}\text{sign}(a_{12})}{s_1 x_2}\right) = \frac{1}{2}\left(\frac{-1}{\sqrt{2}(0.923880)}\right) = -0.382683$$

Using these values in following orthogonal matrix, we get

$$S_1 = \begin{bmatrix} 1 & 0 & 0 \\ 0 & 1 - 2x_2^2 & -2x_2 x_3 \\ 0 & -2x_2 x_3 & 1 - 2x_3^2 \end{bmatrix}$$

$$S_1 = \begin{bmatrix} 1.000000 & 0.000000 & 0.000000 \\ 0.000000 & -0.707107 & 0.707107 \\ 0.000000 & 0.707107 & 0.707107 \end{bmatrix}$$

The similarity matrix is given by

$A_2 = S_1 A_1 S_1$

$$= \begin{bmatrix} 1.000000 & 0.000000 & 0.000000 \\ 0.000000 & -0.707107 & 0.707107 \\ 0.000000 & 0.707107 & 0.707107 \end{bmatrix} \begin{bmatrix} 3 & 1 & -1 \\ 1 & 3 & -1 \\ -1 & -1 & 5 \end{bmatrix} \begin{bmatrix} 1.000000 & 0.000000 & 0.000000 \\ 0.000000 & -0.707107 & 0.707107 \\ 0.000000 & 0.707107 & 0.707107 \end{bmatrix}$$

$$= \begin{bmatrix} 3.000000 & -1.414214 & 0.000000 \\ -1.414214 & 5.00000 & 1.000000 \\ 0.000000 & 1.000000 & 3.000000 \end{bmatrix}$$

This matrix is required tridiagonal matrix. Now, we will compute Strum sequence to compute the eigenvalues of the matrix A. Let the tridiagonal matrix be B.

$$B = A_2 = \begin{bmatrix} 3.000000 & -1.414214 & 0.000000 \\ -1.414214 & 5.00000 & 1.000000 \\ 0.000000 & 1.000000 & 3.000000 \end{bmatrix}$$

The characteristic equation of matrix B is given by

$$|\lambda I - B| = \begin{bmatrix} \lambda-3 & 1.414214 & 0 \\ 1.414214 & \lambda-5 & -1 \\ 0 & -1 & \lambda-3 \end{bmatrix} = 0$$

Strum sequence can be computed in the following manner

$$f_0 = 1$$
$$f_1 = \lambda - b_1 = \lambda - 3$$
$$f_2 = (\lambda - b_2) f_1 - c_1^2 f_0 = (\lambda-5)(\lambda-3) - (1.414214)^2 .1 = \lambda^2 - 8\lambda + 13$$
$$f_3 = (\lambda - b_3) f_2 - c_2^2 f_1 = (\lambda-3)(\lambda^2 - 8\lambda + 13) - (-1)^2 (\lambda-3)$$
$$= \lambda^3 - 11\lambda^2 + 36\lambda - 36$$

Characteristic equation is given by

$$\lambda^3 - 11\lambda^2 + 36\lambda - 36 = 0$$

The roots of this characteristics equation are $\lambda = 2, 3, 6$. These are the required eigenvalues.

Example 7.7

Transform the following symmetric matrix into equivalent tridiagonal similar matrix with the help of Householder method.

$$A = \begin{bmatrix} 4 & 2 & 1 & -1 \\ 2 & 1 & 0 & 3 \\ 1 & 0 & -1 & 4 \\ -1 & 3 & 4 & 0 \end{bmatrix}$$

Ans.

The matrix A is given by

$$A_1 = A = \begin{bmatrix} 4 & 2 & 1 & -1 \\ 2 & 1 & 0 & 3 \\ 1 & 0 & -1 & 4 \\ -1 & 3 & 4 & 0 \end{bmatrix}$$

First plane rotation

First, we will compute the elements for the orthogonal matrix as follows

$$s_1 = \sqrt{a_{12}^2 + a_{13}^2 + a_{14}^2} = \sqrt{4+1+1} = \sqrt{6}$$

$$x_2^2 = \frac{1}{2}\left(1 + \frac{a_{12}\text{sign}(a_{12})}{s_1}\right) = \frac{1}{2}\left(1 + \frac{2}{\sqrt{6}}\right) = 0.9082483$$

$$\Rightarrow x_2 = 0.953021$$

$$x_3 = \frac{1}{2}\left(\frac{a_{13}\text{sign}(a_{12})}{s_1 x_2}\right) = \frac{1}{2}\left(\frac{1}{\sqrt{6}(0.953021)}\right) = 0.214186$$

$$x_4 = \frac{1}{2}\left(\frac{a_{14}\text{sign}(a_{12})}{s_1 x_2}\right) = \frac{1}{2}\left(\frac{-1}{\sqrt{6}(0.953021)}\right) = -0.214186$$

On using these values, we have following orthogonal matrix

$$S_1 = \begin{bmatrix} 1 & 0 & 0 & 0 \\ 0 & 1-2x_2^2 & -2x_2x_3 & -2x_2x_4 \\ 0 & -2x_2x_3 & 1-2x_3^2 & -2x_3x_4 \\ 0 & -2x_3x_4 & -2x_3x_4 & 1-2x_4^2 \end{bmatrix}$$

$$S_1 = \begin{bmatrix} 1 & 0 & 0 & 0 \\ 0 & -0.816497 & -0.408248 & 0.408248 \\ 0 & -0.408248 & 0.908248 & 0.091752 \\ 0 & 0.408248 & 0.091752 & 0.908248 \end{bmatrix}$$

The similarity matrix is given by

$$A_2 = S_1 A_1 S_1 = \begin{bmatrix} 4 & -2.44949 & 0 & 0 \\ -2.44949 & -2.83333 & 1.312713 & -3.353954 \\ 0 & 1.312713 & -0.216326 & 2.083333 \\ 0 & -3.353954 & 2.083333 & 3.049660 \end{bmatrix}$$

Second plane rotation

The elements for the orthogonal transformation are given by

$$s_2 = \sqrt{a_{23}^2 + a_{24}^2} = \sqrt{(1.312713)^2 + (-3.353954)^2} = 3.601697$$

$$x_3^2 = \frac{1}{2}\left(1 + \frac{a_{23}\text{sign}(a_{23})}{s_2}\right) = \frac{1}{2}\left(1 + \frac{1.312713}{3.601697}\right) = 0.682235$$

$$\Rightarrow x_3 = 0.825975$$

$$x_4 = \frac{1}{2}\left(\frac{a_{24}\text{sign}(a_{23})}{s_2 x_3}\right) = \frac{1}{2}\left(\frac{-3.353954}{3.601697(0.825975)}\right) = -0.563706$$

The orthogonal matrix with these elements is as follows

$$S_2 = \begin{bmatrix} 1 & 0 & 0 & 0 \\ 0 & 1 & 0 & 0 \\ 0 & 0 & 1-2x_3^2 & -2x_3x_4 \\ 0 & 0 & -2x_3x_4 & 1-2x_4^2 \end{bmatrix}$$

$$S_2 = \begin{bmatrix} 1 & 0 & 0 & 0 \\ 0 & 1 & 0 & 0 \\ 0 & 0 & -0.364471 & 0.931215 \\ 0 & 0 & 0.931215 & 0.364471 \end{bmatrix}$$

The plane rotation provides the following matrix

$$A_3 = S_2 A_2 S_2 = \begin{bmatrix} 4 & -2.44949 & 0 & 0 \\ -2.44949 & -2.833333 & -3.601697 & 0 \\ 0 & -3.601697 & 1.201642 & 2.638316 \\ 0 & 0 & 2.638316 & 1.631692 \end{bmatrix}$$

This matrix is the required tridiagonal similar matrix.

Step 2: Strum Sequence:

The tridiagonal matrix is as follows

$$B = A_3 = \begin{bmatrix} 4 & -2.44949 & 0 & 0 \\ -2.44949 & -2.833333 & -3.601697 & 0 \\ 0 & -3.601697 & 1.201642 & 2.638316 \\ 0 & 0 & 2.638316 & 1.631692 \end{bmatrix}$$

The characteristic equation for matrix B is given by

$$|\lambda I - B| = \begin{vmatrix} \lambda-4 & 2.44949 & 0 & 0 \\ 2.44949 & \lambda+2.833333 & 3.601697 & 0 \\ 0 & 3.601697 & \lambda-1.201642 & -2.638316 \\ 0 & 0 & -2.638316 & \lambda-1.631692 \end{vmatrix} = 0$$

Strum sequence is given by

$$f_0 = 1$$
$$f_1 = \lambda - b_1 = \lambda - 4$$
$$f_2 = (\lambda - b_2)f_1 - c_1^2 f_0 = (\lambda+2.833333)(\lambda-4) - (2.44949)^2 .1 = \lambda^2 - 1.166667\lambda - 17.33333$$

$$f_3 = (\lambda - b_3)f_2 - c_2^2 f_1 = (\lambda - 1.201642)(\lambda^2 - 1.166667\lambda - 17.33333) - (3.601697)^2(\lambda - 4)$$

$$= \lambda^3 - 2.368309\lambda^2 - 28.903638\lambda + 72.717346$$

$$f_4 = (\lambda - b_4)f_3 - c_3^2 f_2 = (\lambda - 2.926229)(\lambda^3 - 1.073770\lambda^2 - 23.729531\lambda + 47.319637)$$

$$- (-0.582083)^2(\lambda^2 - 3.666667\lambda - 4.333353)$$

$$\approx \lambda^4 - 4\lambda^3 - 32\lambda^2 + 128\lambda + 2$$

On solving the characteristic equation, the eigenvalues are -5.653614, -0.0155646, 4.031504 and 5.637675.

Exercise 7

1. Prove that eigenvectors of a Hermitian matrix are orthogonal.

2. Prove that eigenvectors of a unitary matrix are orthogonal.

3. Compute the eigenvalues and corresponding eigenvectors for the following matrices and prove that these matrices are diagonalizable.

a) $\begin{bmatrix} -1 & 1 & 2 \\ 1 & -1 & -2 \\ 1 & 1 & 2 \end{bmatrix}$
    b) $\begin{bmatrix} 0 & 1 & -2 \\ 2 & 1 & 0 \\ 4 & -2 & 5 \end{bmatrix}$

Ans. The elements of the following column vectors are the eigenvalues, and eigenvectors are given by the columns of the following matrices.

a) $\begin{bmatrix} 2 \\ 0 \\ -2 \end{bmatrix}$ $\begin{bmatrix} 1 & 0 & -1 \\ -1 & -2 & 1 \\ 2 & 1 & 0 \end{bmatrix}$
    b) $\begin{bmatrix} 1 \\ 2 \\ 3 \end{bmatrix}$ $\begin{bmatrix} 0 & 1 & -1 \\ 2 & 2 & -1 \\ 1 & 0 & 1 \end{bmatrix}$

Use these matrices to prove that the given matrices are diagonalizable, by simply using $S^{-1}AS = D$.

4. Apply Jacobi method to compute the eigenvalues of the following matrices

i) $\begin{bmatrix} 4 & -2 & 1 \\ -2 & 3 & 0 \\ 1 & 0 & 2 \end{bmatrix}$
    ii) $\begin{bmatrix} 5 & -1 & 0.5 \\ -1 & 2 & 0.2 \\ 0.5 & 0.2 & -2 \end{bmatrix}$

Use six decimal digits arithmetic till all the off-diagonal elements are less than 0.002.

Ans. 4. i)			Ans. 4. ii)		
First Iteration: $\theta = -0.662909$			First Iteration: $\theta = -0.294001$		
5.561553	-0.000000	0.788205	5.302776	-0.000000	0.420589
0.000000	1.438447	0.615412	0.000000	1.697224	0.336311
0.788205	0.615412	2.000000	0.420589	0.336311	-2.000000

Second Iteration: $\theta = 0.208349$		
5.728193	0.127295	−0.000000
0.127295	1.438447	0.602103
0.000000	0.602103	1.833360
Third Iteration: $\theta = -0.626952$		
5.728193	0.103086	0.074681
0.103086	1.002250	−0.000000
0.074681	−0.000000	2.269557
Fourth Iteration: $\theta = 0.021799$		
5.730441	0.000000	0.074664
−0.000000	1.000002	−0.001628
0.074664	−0.001628	2.269557
Fifth Iteration: $\theta = -0.021560$		
5.732051	−0.000035	−0.000000
−0.000035	1.000002	−0.001627
−0.000000	−0.001628	2.267947

Eigenvalues are 5.732051, 1.000002 and 2.267947

Second Iteration: $\theta = 0.057340$		
5.326919	0.019274	−0.000000
0.019274	1.697224	0.335758
−0.000000	0.335758	−2.024143
Third Iteration: $\theta = 0.089264$		
5.326919	0.019197	−0.001718
0.019197	1.727275	−0.000000
−0.001718	0.000000	−2.054194
Fourth Iteration: $\theta = -0.005333$		
5.327022	−0.000000	−0.001718
−0.000000	1.727173	0.000009
−0.001718	0.000009	−2.054194

Eigenvalues are 5.327022, 1.727173 and −2.054194

5. Construct the Strum sequence of the functions of eigenvalues, and hence find the eigenvalues of the following tridiagonal matrix

$$\begin{bmatrix} 1 & -2 & 0 \\ -2 & 3 & 1 \\ 0 & 1 & -3 \end{bmatrix}$$

Ans. Strum sequence is given by

$f_0 = 1$

$f_1 = \lambda - b_1 = \lambda - 1$

$f_2 = (\lambda - b_2)f_1 - c_1^2 f_0 = (\lambda - 3)(\lambda - 1) - (2)^2 \cdot 1 = \lambda^2 - 4\lambda - 1$

$f_3 = (\lambda - b_3)f_2 - c_2^2 f_1 = (\lambda + 3)(\lambda^2 - 4\lambda - 1) - (1)^2(\lambda - 1)$

$\quad = \lambda^3 - \lambda^2 - 14\lambda - 2$

Characteristic equation is given by

$$\lambda^3 - \lambda^2 - 14\lambda - 2 = 0$$

Roots of this equation are lying in the intervals (−4, −3), (−1, 0) and (4, 5). Computing the roots with the help of Newton Raphson method, we obtain following eigenvalues correct to five decimal places

−3.19097, −0.14456 and 4.33553.

6. Consider the following tridiagonal matrix

$$\begin{bmatrix} 5 & 3 & 0 \\ 3 & 4 & -1 \\ 0 & -1 & 3 \end{bmatrix}$$

a) Construct the Strum sequence of the functions of eigenvalues and hence find the characteristics equations.

b) Obtain the bounds on the eigenvalues of matrices using Gerschgorin and Brauer theorems.

c) Obtain the intervals of unit length in which eigenvalues of the matrices lies by using Strum sequence and further find the eigenvalues correct to four decimal places using NR method.

Ans. a) Strum sequence is as follows

$$f_0 = 1$$

$$f_1 = \lambda - b_1 = \lambda - 5$$

$$f_2 = (\lambda - b_2)f_1 - c_1^2 f_0 = (\lambda - 4)(\lambda - 5) - (3)^2 .1 = \lambda^2 - 9\lambda + 11$$

$$f_3 = (\lambda - b_3)f_2 - c_2^2 f_1 = (\lambda - 3)(\lambda^2 - 9\lambda + 11) - (1)^2(\lambda - 5)$$

$$= \lambda^3 - 12\lambda^2 + 37\lambda - 28$$

Characteristic equation is given by

$$\lambda^3 - 12\lambda^2 + 37\lambda - 28 = 0$$

b) By using Gerschgorin and Brauer theorems, the bounds on eigenvalues are given by

$$|\lambda - 4| \le 4$$

$$\Rightarrow 0 \le \lambda \le 8$$

c) In this range, the roots of the equation lie in the intervals (1, 2), (3, 4) and (7, 8). We can compute the roots of the equation by any iterative method discussed in Chapter 3. The eigenvalues are as follows

1.135115, 3.231460, 7.633425

7. Consider the real symmetric matrix $\begin{bmatrix} 3 & 1 & 1 \\ 1 & 4 & 2 \\ 1 & 2 & 5 \end{bmatrix}$.

a) Obtain the tridiagonal form using Givens and Householder methods.

b) Find the Sturm sequences, and hence characteristics equation of the matrix.

c) Find the intervals of unit length in which eigenvalues of the matrix lies by using Strum sequence and further find the eigenvalues correct to four decimal places using NR method.

Ans. a) The tridiagonal forms using Givens and Householder methods are as follows

Givens method:

$$\begin{bmatrix} 3.000000 & 1.414214 & 0.000000 \\ 1.414214 & 6.500001 & 0.500000 \\ 0.000000 & 0.500000 & 2.500000 \end{bmatrix}$$

Householder method:

$$\begin{bmatrix} 3.000000 & -1.414214 & 0.000000 \\ -1.414214 & 6.500001 & -0.500000 \\ 0.000000 & -0.500000 & 2.500000 \end{bmatrix}$$

b) Using Strum sequence, the final characteristic equation is as follows

$$\lambda^3 - 12\lambda^2 + 41\lambda - 43 = 0$$

c) Eigenvalues are 2.307978, 2.643104 and 7.048917

8. Reduce the following matrices to tridiagonal form using Givens and Householder methods and further obtain the eigenvalues of the matrices

i) $\begin{bmatrix} 2 & -3 & 1 \\ -3 & 2 & -2 \\ 1 & -2 & 5 \end{bmatrix}$
ii) $\begin{bmatrix} 2 & 1 & -1 \\ 1 & 3 & 4 \\ -1 & 4 & 6 \end{bmatrix}$
iii) $\begin{bmatrix} 4 & 2 & -1 & 1 \\ 2 & -2 & 2 & -1 \\ -1 & 2 & 5 & 0 \\ 1 & -1 & 0 & -5 \end{bmatrix}$

Ans.

Givens method				Householder method			
i) $\theta = -0.321751$				i)			
2.000000	-3.162278	0.000000		2.000000	3.162278	0.000000	
-3.162278	3.500000	-2.500000		3.162278	3.500000	2.500000	
0.000000	-2.500000	3.500000		0.000000	2.500000	3.500000	
ii) $\theta = -0.785398$				ii)			
2.000000	1.414214	0.000000		2.000000	-1.414214	0.000000	
1.414214	0.500000	-1.500000		-1.414214	0.500000	1.500000	
0.000000	-1.500000	8.500000		0.000000	1.500000	8.500000	
iii) First Rotation $\theta = -0.463648$				iii) First rotation			
4.000000	2.236068	0.000000	1.000000	4.000000	-2.449490	0.000000	0.000000
2.236068	-2.200000	-1.600000	-0.894427	-2.449490	-3.333333	1.799660	1.466327
0.000000	-1.600000	5.200000	-0.447214	0.000000	1.799660	5.157398	-0.666667
1.000000	-0.894427	-0.447214	-5.000000	0.000000	1.466326	-0.666667	-3.824064
Second Rotation $\theta = 0.420534$				Second Rotation			
4.000000	2.449490	0.000000	0.000000	4.000000	-2.449490	0.000000	0.000000
2.449490	-3.333333	-1.643168	-1.639783	-2.449490	-3.333333	-2.321398	0.000000
0.000000	-1.643168	5.200000	0.244949	0.000000	-2.321398	0.920962	4.532819
0.000000	-1.639783	0.244949	-3.866667	0.000000	0.000000	4.532818	0.412371
Third Rotation $\theta = 0.784367$							
4.000000	2.449490	0.000000	0.000000				
2.449490	-3.333333	-2.321398	0.000000				
0.000000	-2.321398	0.920962	-4.532819				
0.000000	0.000000	-4.532819	0.412371				

Eigenvalues are as follows:

i)　　−1.093373, 2.941164, 7.152208

ii)　　−0.537231, 2.755516, 8.781715

iii)　　−5.649347, −2.603370, 4.514705, 5.738012

Interpolation

> *'Obvious' is the most dangerous word in mathematics.*
>
> **Eric Temple Bell**
> (February 7, 1883–December 21, 1960)
> He was a mathematician, science fiction, and non-fiction writer.
>
> *Mathematics is not only real, but it is the only reality.*
>
> **Martin Gardner**
> (October 21, 1914–May 22, 2010)
> He was a famous mathematician and science writer.

8.1 Introduction

One of the important aspects in numerical methodology is the interpolation. The primary aim of this chapter is to obtain polynomial approximation for a given set of discrete data points. Interpolation deals with the problem of constructing a polynomial $P(x)$ of minimum degree, which passes through a given set of discrete data points, (x_i, y_i), $i = 0, 1, ..., n$. This polynomial $P(x)$ is known as an interpolating polynomial. The interpolating polynomial can be used to estimate the value of the dependent variable y for any intermediate value of the independent variable x.

For example, let us consider data points $(0, -1)$, $(1, 1)$, $(2, 9)$, $(3, 29)$, $(5, 129)$. The aim is to construct a polynomial of minimum degree which passes through all these points. The polynomial $P(x) = x^3 + x - 1$ is the required interpolating polynomial. It is easy to see that the polynomial $P(x)$ passes through every point of given data set $\{(0, -1), (1, 1), (2, 9), (3, 29), (5, 129)\}$ as $P(0) = -1$, $P(1) = 1$,.... Now, we can use this interpolating polynomial to find the approximate value of the dependent variable y for any intermediate value of the independent variable x, say for example $y(4) \simeq P(4) = 67$.

In this chapter, we will discuss how to build such polynomials in an algorithmic manner; Lagrange and Newton divided difference methods are most suitable for such constructions.

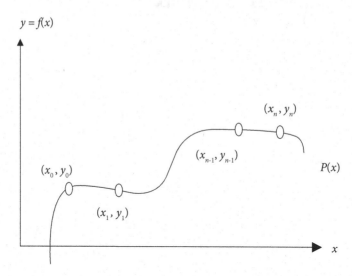

$y = f(x)$

(x_n, y_n)

(x_{n-1}, y_{n-1})

(x_0, y_0)

$P(x)$

(x_1, y_1)

x

Fig. 8.1 Interpolation

Experimental observations contain a set of data points, i.e., inputs (x) and outputs (y). In many practical problems, we have data points instead of the mathematical models for such problems. For example, Indian government carries out national census after a gap of 10 years to speculate about the development of the population of the country. Recently, the Indian government has carried out the census in 2011, before that it was carried out in 2001, 1991, ..., 1951. The population data in these years is as follows

Table 8.1

Years	Population (in crores)
1961	43.9235
1971	54.8160
1981	68.3329
1991	84.6421
2001	102.8737
2011	121.0193

(The data set has population in four decimal points only)
Population data is available only for the years given in the above Table 8.1. To estimate the populations in intermediate years such as 1977, 2010, etc., numerical techniques can be used.

Sometimes, we also require approximating different complicated functions with polynomials. In fact, polynomials have simple structures, and polynomials are easy for fundamental operations of calculus like differentiation and integration etc. Also, the Weierstrass approximation theorem (Section 8.9) states that any continuous function on a finite interval is well approximated by a polynomial function within error limit. So, we have many reasons for polynomial approximations of data sets as well as functions.

Many other applications of numerical analysis involve the polynomial approximation of various functions. For examples, methods for finding roots of nonlinear equations, solutions of integral and differential equations and most importantly numerical approximation for integration and differentiation. The next section is devoted to various forms of the polynomials, and in successive sections, we will discuss the methods for computing the interpolating polynomials.

Readers who are interested only in finding interpolating polynomials, they can skip the next section and jump to interpolating polynomial Section 8.3.

8.2 Polynomial Forms

The following forms of the polynomials are widely used.

    i)   Power Form
    ii)   Shifted Power Form
    iii)   Newton Form
    iv)   Nested Newton Form

8.2.1 Power Form

The most common form of the polynomial of degree n is the following power form

$$P_n(x) = a_0 + a_1 x + a_2 x^2 + \cdots + a_n x^n, \ (a_n \neq 0)$$

The calculus (differentiations, integrations) is easy with this polynomial form.

8.2.2 Shifted Power Form

The shifted power form about the center $x = c$ has the following expression

$$P_n(x) = a_0 + a_1(x-c) + a_2(x-c)^2 + \cdots + a_n(x-c)^n$$

To discuss the advantage of the shifted power form, let us take a very simple example

Let $f(1110) = \dfrac{-5}{6}$ and $f(1111) = \dfrac{1}{6}$

We will use five significant digits floating point rounding arithmetic to compute the polynomial $P(x) = a_0 + a_1 x$, which passes through these two points. For this, we have

$$P(1110) = \frac{-5}{6} \qquad\qquad \Rightarrow a_0 + 1110 a_1 = -.83333$$

$$P(1111) = \frac{1}{6} \qquad\qquad \Rightarrow a_0 + 1111 a_1 = .16667$$

On solving these two equations with five significant digits floating point arithmetic, the polynomial obtained is as follows

$$P(x) = -1110.8 + x \tag{8.1}$$

The polynomial (8.1) gives $P(1110) = -.8$ and $P(1111) = .2$. These estimates are correct up to one significant digit only. So, there is a loss of four significant digits, which can be avoided using the shifted power form $P(x) = a_0 + a_1(x-1110)$ about center $x = 1110$.

The equations are as follows

$$a_0 = -.83333$$
$$a_0 + a_1 = .16667 \tag{8.2}$$

Solution of these two equations gives the following polynomial

$$P(x) = -.83333 + (x-1110)$$

The shifted power form (8.2) provides the following values

$$P(1110) = -.8.3333$$
$$P(1111) = .16667$$

These values are correct up to five significant digits.

The data values are near to the point $x = 1110$. We are analysing the data set from origin $(x = 0)$ in power form (8.1), and the origin is at $x = 1110$ in shifted power form (8.2). This is the reason behind the huge difference in both the approximations. The power form is like to see (analyze) the things (function) from the origin, while the region of study is far from the origin. The major advantage in shifted power form is that we shift the origin to the point in the data set where more information could be drawn for a better approximation. Therefore, it is always better to work with shifted power form with the center in the interval, in which we are discussing the function.

It is also easy to see that all the coefficients a_i's of the polynomial provide the derivative values at point $x = c$ in the shifted power form.

$$\frac{P^i(c)}{i!} = a_i$$

8.2.3 Newton Form

A more generalized form of the shifted power form is Newton form

$$P_n(x) = a_0 + a_1(x-c_1) + a_2(x-c_1)(x-c_2) + \cdots + a_n(x-c_1)(x-c_2)\cdots(x-c_n)$$

This form plays a vital role in the interpolation theory. Newton form becomes shifted power form for the values $c_1 = c_2 = \ldots c_n = c$, and further power form for the value $c = 0$.

8.2.4 Nested Newton Form

In Newton form of the polynomial of degree n, the total number of multiplications and additions are $\frac{n(n+1)}{2}$ and $n + \frac{n(n+1)}{2}$, respectively. It will be better to use the following nested Newton form to minimize the number of arithmetic operations.

$$P_n(x) = a_0 + (x-c_1)\left\{a_1 + (x-c_2)\left\{a_2 + \cdots + (x-c_{n-1})\left\{a_{n-1} + a_n(x-c_n)\right\}\cdots\right\}\right\}$$

The nested Newton form involves only n multiplications and $2n$ additions.

8.2.5 Recursive Algorithm for the Nested Newton Form

Consider the nested Newton polynomial

$$P_n(x) = a_0 + (x-c_1)\left\{a_1 + (x-c_2)\left\{a_2 + \cdots + (x-c_{n-1})\left\{a_{n-1} + a_n(x-c_n)\right\}\cdots\right\}\right\}$$

Let us start with the innermost part of this polynomial, and continue by adding a new term to the resulting expression.

We can easily obtain the following recursive algorithm to get the polynomial $P_n(x)$.

$$a'_n = a_n$$
$$a'_{n-1} = a_{n-1} + (x-c_n)a'_n$$
$$a'_{n-2} = a_{n-2} + (x-c_{n-1})a'_{n-1}$$
$$\vdots$$
$$a'_i = a_i + (x-c_{i+1})a'_{i+1}$$
$$\vdots$$
$$a'_0 = a_0 + (x-c_1)a'_1 = P_n(x) \tag{8.3}$$

The value of polynomial at any point $x = z$ can be computed with this algorithm.

Example — 8.1

Evaluate the following polynomial at $x = 4$ using recursive algorithm (8.3)

$$P_4(x) = x + 2x(x-1) + 3x(x-1)(x-2) + 4x(x-1)(x-2)(x-3)$$

Ans.

The coefficients of the polynomial and centers are given by

$$a_0 = 0, \ a_1 = 1, \ a_2 = 2, \ a_3 = 3, \ a_4 = 4$$
$$c_1 = 0, \ c_2 = 1, \ c_3 = 2, \ c_4 = 3$$

On using recursive algorithm (8.3), we get

$$a_4' = a_4 = 4$$
$$a_3' = a_3 + (z - c_4)a_4' = 3 + (4-3).4 = 7$$
$$a_2' = a_2 + (z - c_3)a_3' = 2 + (4-2).7 = 16$$
$$a_1' = a_1 + (z - c_2)a_2' = 1 + (4-1).16 = 49$$
$$a_0' = a_0 + (z - c_1)a_1' = 0 + (4-0).49 = 196$$

It provides the value $P_4(4) = 196$.

8.2.6 Change of Center in Newton Form

Suppose we want to replace the center c_n with a new center z in the polynomial

$$P_n(x) = a_0 + a_1(x - c_1) + a_2(x - c_1)(x - c_2) + \cdots + a_n(x - c_1)(x - c_2)\cdots(x - c_n) \qquad (8.4)$$

On replacing the old coefficients $a_0, a_1, a_2, \ldots, a_n$ of the polynomial by the new coefficients $a_0', a_1', a_2', \ldots, a_n'$ from the recursive algorithm (8.3), we get the following equivalent Newton form.

$$P_n(x) = \left[a_0' + (c_1 - z)a_1'\right] + \left[a_1' + (c_2 - z)a_2'\right](x - c_1) + \left[a_2' + (c_3 - z)a_3'\right](x - c_1)(x - c_2) +$$
$$\cdots + \left[a_{n-1}' + (c_n - z)a_n'\right](x - c_1)(x - c_2)\cdots(x - c_{n-1}) + \left[a_n'\right](x - c_1)(x - c_2)\cdots(x - c_n)$$

(or)

$$P_n(x) = a_0' + a_1'(x - z) + a_2'(x - z)(x - c_1) + \cdots + a_n'(x - z)(x - c_1)\cdots(x - c_{n-1}) \qquad (8.5)$$

The equivalent form has the centers $z, c_1, \ldots, c_{n-1}$, where $P_n(z) = a_0'$. Note that both the polynomials (8.4) and (8.5) are equal, but in different forms.

Example ─── **8.2**

Replace the center $x = 3$ by new center $x = 4$ in the following polynomial, and write down the equivalent polynomial form using the recursive algorithm

$$P_4(x) = x + 2x(x-1) + 3x(x-1)(x-2) + 4x(x-1)(x-2)(x-3) \qquad (8.6)$$

Ans.

In Example 8.1, we have already computed the coefficients as follows

$$a_4' = 4,\ a_3' = 7,\ a_2' = 16,\ a_1' = 49,\ a_0' = 196$$

On using these values in polynomial (8.5), we have the following equivalent form of the given polynomial with new center $x = 4$.

$$P_4(x) = 196 + 49(x-4) + 16(x-4)(x) + 7(x-4)(x)(x-1) + 4(x-4)(x)(x-1)(x-2) \quad (8.7)$$

It is worth mentioning here that the polynomial (8.7) is equal to polynomial (8.6). The algorithm (8.3) is very helpful in obtaining the equivalent polynomial forms (8.4) and (8.5) with different centers. These equivalent polynomials are otherwise very difficult to obtain. Also, the algorithm (8.3) can be implemented on the computer easily.

Example 8.3

Change the following polynomial

$$P_3(x) = 1 + 2(x-1) + 3(x-1)(x-2) + 4(x-1)(x-2)(x-3)$$

into its equivalent power form using the recursive algorithm.

Ans.

In power form, all the centers are zeroes, i.e., $c_1 = c_2 = c_3 = z = 0$. We will change all the centers one by one. On using recursive algorithm 8.3 for the polynomial with coefficients $a_0 = 1,\ a_1 = 2,\ a_2 = 3,\ a_3 = 4$; and centers $c_1 = 1,\ c_2 = 2,\ c_3 = 3$, we get

$$a_3' = a_3 = 4$$
$$a_2' = a_2 + (z - c_3)a_3' = 3 + (0-3)4 = -9$$
$$a_1' = a_1 + (z - c_2)a_2' = 2 + (0-2)(-9) = 20$$
$$a_0' = a_0 + (z - c_1)a_1' = 1 + (0-1)20 = -19$$

Equivalent polynomial (8.5) with centers $c_1 = 0,\ c_2 = 1,\ c_3 = 2$ is given by

$$P_3(x) = -19 + 20(x-0) - 9(x-0)(x-1) + 4(x-0)(x-1)(x-2)$$

Again applying the recursive algorithm to this new polynomial, we get

$$a_3'' = a_3' = 4$$
$$a_2'' = a_2' + (z - c_3)a_3'' = -9 + (0-2)4 = -17$$
$$a_1'' = a_1' + (z - c_2)a_2'' = 20 + (0-1)(-17) = 37$$
$$a_0'' = a_0' + (z - c_1)a_1'' = 1 + (0-1)20 = -19$$

On using these coefficients in polynomial (8.5), we have

$$P_3(x) = -19 + 37(x-0) - 17(x-0)(x-0) + 4(x-0)(x-0)(x-1)$$

A similar treatment to this polynomial leads us to the polynomial with all new centers as zeroes, and the polynomial is as follows

$$P_3(x) = -19 + 37x - 21x^2 + 4x^3$$

Note: One can easily verify that the various polynomials computed in this example are equivalent forms with different centers, i.e.,

$$\begin{aligned}
P_3(x) &= 1 + 2(x-1) + 3(x-1)(x-2) + 4(x-1)(x-2)(x-3) \\
&= -19 + 20(x-0) - 9(x-0)(x-1) + 4(x-0)(x-1)(x-2) \\
&= -19 + 37(x-0) - 17(x-0)(x-0) + 4(x-0)(x-0)(x-1) \\
&= -19 + 37x - 21x^2 + 4x^3
\end{aligned}$$

Note that the polynomial in Example 8.3 can be expanded directly to get power form. But, we cannot multiply characters (x) with the real values (a_i's and c_i's) on the computer to get the power form directly. While, the recursive algorithm (8.3) requires arithmetic of real numbers (a_i's, c_i's and z) only. So, the algorithm (8.3) is needed to obtain the equivalent polynomial forms by using computer.

Next, we will discuss following two lemmas which show that there exists a unique polynomial of degree $\leq n$, which passes through $(n + 1)$ points.

Lemma 8.1

Let $P(x)$ be any polynomial having k distinct zeroes, $z_1, z_2, ..., z_k$. Then, for some polynomial, $R(x)$, we can write $P(x) = (x-z_1)(x-z_2)\cdots(x-z_k)R(x)$.

Proof: Let $P(x)$ be a polynomial of degree n. On applying the algorithm (8.1) on the polynomial $P(x)$ to introduce the center z, we get

$$P(x) = a_0' + a_1'(x-z) + a_2'(x-z)(x-c_1) + \cdots + a_n'(x-z)(x-c_1)\cdots(x-c_{n-1})$$
$$P(x) = P(z) + (x-z)\{a_1' + a_2'(x-c_1) + \cdots + a_n'(x-c_1)\cdots(x-c_{n-1})\}$$
$$\Rightarrow P(x) = P(z) + (x-z)Q(x), \text{ where } Q(x) \text{ is a polynomial of degree less than } n.$$

Since z_1 is the zero of the polynomial $P(x)$, so $P(z_1) = 0$. The above expression at the point $z = z_1$ is as follows

$$\Rightarrow P(x) = (x-z_1)Q(x)$$

The z_2 is zero of the polynomial $P(x)$ and so that of the polynomial $Q(x)$.

$$\Rightarrow Q(x) = (x-z_2)Q_1(x)$$

On continuing in this way, we arrive at following result

$$P(x) = (x-z_1)(x-z_2)\cdots(x-z_k)R(x), \text{ for some } R(x) \text{ of the degree } \leq (n-k).$$

Lemma 8.2

Let $P(x)$ and $Q(x)$ be two polynomials of degree $\leq k-1$. Let these polynomials match the function $f(x)$ at k distinct points, $z_1, z_2, ..., z_k$. Then $P(x) = Q(x)$.

Proof: Consider the following polynomial

$$M(x) = P(x) - Q(x)$$

Since $P(x)$ and $Q(x)$ are two polynomials which agree with the function $f(x)$ at k distinct points $z_1, z_2, ..., z_k$. This implies

$$P(z_i) = Q(z_i) = f(z_i) \qquad \text{for } i = 1, 2, ..., k$$

Hence $M(x) = P(x) - Q(x)$ has k distinct zeroes $z_1, z_2, ..., z_k$.

According to Lemma 8.1, we have

$$M(x) = (x - z_1)(x - z_2) \cdots (x - z_k) R(x) \tag{8.8}$$

for some polynomial $R(x)$.

Now the polynomials $P(x)$ and $Q(x)$ are of degree $\leq k-1$, so the degree of the polynomial $M(x) = P(x) - Q(x)$ must be $\leq k-1$. But from Eq. (8.8), if $R(x)$ is nonzero, then the degree of $M(x)$ is $\geq k$. Hence, only possibility is $R(x) = 0$. Then, from Eq. (8.8), we have

$$M(x) = 0$$

This result eventually implies that $P(x) = Q(x)$.

Remark: This Lemma has very important consequence in polynomial approximation, that there exist a unique polynomial of degree $\leq n$, which passes through $(n + 1)$ points $\left(x_i, f(x_i)\right)$; $i = 0, 1, ..., n$. This unique polynomial is known as an interpolating polynomial. It is worth mentioning here that there are infinitely many polynomials of degree $> n$, which pass through $(n + 1)$ points $\left(x_i, f(x_i)\right)$; $i = 0, 1, ..., n$.

Definition 8.1

(Interpolating Polynomial): Let $y = f(x)$ be a real-valued function defined on some interval $I = [a, b]$ and $x_0, x_1, ..., x_n$ are $(n+1)$ points in the interval I. Then, interpolating polynomial $P_n(x)$ is a polynomial of degree $\leq n$, which agrees with the function $f(x)$ at these points, i.e.,

$$P_n(x_i) = f(x_i); \; i = 0, 1, ..., n$$

Now, we will discuss Lagrange and Newton divided difference methods to construct interpolating polynomial.

8.3 Lagrange Method

Let $y = f(x)$ be a real-valued function defined on some interval $I = [a, b]$ and $x_0, x_1, ..., x_n$ are $(n+1)$ points in the interval I. We have already proved that there exists a unique interpolating polynomial of degree $\leq n$ which agrees $f(x)$ at these $(n+1)$ points. We will use the following form, also known as the Lagrange form of polynomials, to obtain this interpolating polynomial.

$$P_n(x) = \sum_{i=1}^{n} a_i l_i(x) = a_0 l_0(x) + a_1 l_1(x) + ... + a_n l_n(x)$$

where $l_i(x) = \displaystyle\prod_{\substack{k=0 \\ k \neq i}}^{n} \frac{x - x_k}{x_i - x_k} = \dfrac{(x - x_0)(x - x_1)\cdots(x - x_{i-1})(x - x_{i+1})\cdots(x - x_n)}{(x_i - x_0)(x_i - x_1)\cdots(x_i - x_{i-1})(x_i - x_{i+1})\cdots(x_i - x_n)}$

The expression for polynomial $P_n(x)$ is given by

$$\begin{aligned}
P_n(x) &= a_0 l_0(x) + a_1 l_1(x) + ... + a_n l_n(x) \\
&= a_0 \frac{(x - x_1)(x - x_2)\cdots(x - x_n)}{(x_0 - x_1)(x_0 - x_2)\cdots(x_0 - x_n)} \\
&+ a_1 \frac{(x - x_0)(x - x_2)\cdots(x - x_n)}{(x_1 - x_0)(x_1 - x_2)\cdots(x_1 - x_n)} \\
&\vdots \\
&+ a_n \frac{(x - x_0)(x - x_1)\cdots(x - x_{n-1})}{(x_n - x_0)(x_n - x_1)\cdots(x_n - x_{n-1})}
\end{aligned} \tag{8.9}$$

Since the polynomial $P_n(x)$ is an interpolating polynomial, it must agree $f(x)$ at $(n+1)$ points $(x_i, f(x_i))$, $i = 0, 1, 2, \cdots, n$

$$P_n(x_i) = f(x_i), \, i = 0, 1, 2, \cdots, n \tag{8.10}$$

The values of coefficients a_i, $i = 0, 1, 2, \cdots, n$ can be computed with the help of Eqs. (8.9) and (8.10) as follows

At any point $x = x_i$, Eq. (8.9) implies that

$$P_n(x_i) = a_i$$

So, from Eq. (8.10), we have

$$a_i = f(x_i)$$

On using the values of the constants a_i, $i = 0, 1, 2, \cdots, n$ in Eq. (8.9), the resulting interpolating polynomial is given by

$$P_n(x) = \sum_{i=0}^{n} f(x_i) l_i(x)$$

$$= f(x_0) l_0(x) + f(x_1) l_1(x) + \ldots + f(x_n) l_n(x)$$

$$= f(x_0) \frac{(x-x_1)(x-x_2)\cdots(x-x_n)}{(x_0-x_1)(x_0-x_2)(x_0-x_n)}$$

$$+ f(x_1) \frac{(x-x_0)(x-x_2)\cdots(x-x_n)}{(x_1-x_0)(x_1-x_2)\cdots(x_1-x_n)}$$

$$+ \ldots$$

$$+ f(x_n) \frac{(x-x_0)(x-x_1)\cdots(x-x_{n-1})}{(x_n-x_0)(x_n-x_1)\cdots(x_n-x_{n-1})}$$

$$= \sum_{i=0}^{n} f(x_i) \prod_{\substack{j=0 \\ j \neq i}}^{n} \frac{(x-x_j)}{(x_i-x_j)} \qquad (8.11)$$

Therefore, there exists an interpolating polynomial $P_n(x)$ of degree $\leq n$ which agrees the function $f(x)$ at $(n+1)$ points.

From Lemma 8.2, there is at the most one interpolating polynomial $P_n(x)$ of degree $\leq n$ which agrees $f(x)$ at $(n+1)$ points.

Hence, we can conclude that there exists a unique polynomial of degree $\leq n$ which interpolates $f(x)$ at $(n+1)$ points. (*Existence and Uniqueness of Interpolating Polynomial*)

Example 8.4

We have following data for the population of India from the year 1981 to 2011. Use Lagrange interpolation to find the approximate population in 2006.

Year (x)	1981	1991	2001	2011
Population (y) (in crores)	68.3329	84.6421	102.8737	121.0193

Ans.

The four data points in the table are as follows

$x_0 = 1981$	$x_1 = 1991$	$x_2 = 2001$	$x_3 = 2011$
$f(x_0) = 68.3329$	$f(x_1) = 84.6421$	$f(x_2) = 102.8737$	$f(x_3) = 121.0193$

Lagrange interpolating polynomial (8.11) for $n = 3$, is given by

$$P(x) = f(x_0)\frac{(x - x_1)(x - x_2)(x - x_3)}{(x_0 - x_1)(x_0 - x_2)(x_0 - x_3)} + f(x_1)\frac{(x - x_0)(x - x_2)(x - x_3)}{(x_1 - x_0)(x_1 - x_2)(x_1 - x_3)}$$

$$+ f(x_2)\frac{(x - x_0)(x - x_1)(x - x_3)}{(x_2 - x_0)(x_2 - x_1)(x_2 - x_3)} + f(x_3)\frac{(x - x_0)(x - x_1)(x - x_2)}{(x_3 - x_0)(x_3 - x_1)(x_3 - x_2)}$$

On computing this polynomial at $x = 2006$, we have

$$P(2006) = (68.3329)\frac{(2006 - 1991)(2006 - 2001)(2006 - 2011)}{(1981 - 1991)(1981 - 2001)(1981 - 2011)}$$

$$+ (84.6421)\frac{(2006 - 1981)(2006 - 2001)(2006 - 2011)}{(1991 - 1981)(1991 - 2001)(1991 - 2011)}$$

$$+ (102.8737)\frac{(2006 - 1981)(2006 - 1991)(2006 - 2011)}{(2001 - 1981)(2001 - 1991)(2001 - 2011)}$$

$$+ (121.0193)\frac{(2006 - 1981)(2006 - 1991)(2006 - 2001)}{(2011 - 1981)(2011 - 1991)(2011 - 2001)}$$

$$P(2006) = (68.3329)\left(\frac{1}{16}\right) + (84.6421)\left(\frac{-5}{16}\right) + (102.8737)\left(\frac{15}{16}\right) + (121.0193)\left(\frac{5}{16}\right)$$

$$P(2006) = 112.082775$$

The approximate population in the year 2006 is 112.082775 crores.

Example ——————————————————————— **8.5**

Derive interpolating polynomial for the data points $(0, -1)$, $(1, 1)$, $(2, 9)$, $(3, 29)$, $(5, 129)$ using Lagrange interpolation. Use this polynomial to compute the value of the function $y(4)$.

Ans.

Interpolating polynomial (8.11) is of degree ≤ 4 for following five points. $(0 = x_0, -1 = f(x_0))$, $(1 = x_1, 1 = f(x_1))$, $(2 = x_2, 9 = f(x_2))$, $(3 = x_3, 29 = f(x_3))$, $(5 = x_4, 129 = f(x_4))$ and is given by

$$P(x) = f(x_0)\frac{(x - x_1)(x - x_2)(x - x_3)(x - x_4)}{(x_0 - x_1)(x_0 - x_2)(x_0 - x_3)(x_0 - x_4)} + f(x_1)\frac{(x - x_0)(x - x_2)(x - x_3)(x - x_4)}{(x_1 - x_0)(x_1 - x_2)(x_1 - x_3)(x_1 - x_4)}$$

$$+ f(x_2)\frac{(x - x_0)(x - x_1)(x - x_3)(x - x_4)}{(x_2 - x_0)(x_2 - x_1)(x_2 - x_3)(x_2 - x_4)} + f(x_3)\frac{(x - x_0)(x - x_1)(x - x_2)(x - x_4)}{(x_3 - x_0)(x_3 - x_1)(x_3 - x_2)(x_3 - x_4)}$$

$$+ f(x_4)\frac{(x - x_0)(x - x_1)(x - x_2)(x - x_3)}{(x_4 - x_0)(x_4 - x_1)(x_4 - x_2)(x_4 - x_3)}$$

$$P(x) = -1\frac{(x-1)(x-2)(x-3)(x-5)}{(0-1)(0-2)(0-3)(0-5)} + 1\frac{(x-0)(x-2)(x-3)(x-5)}{(1-0)(1-2)(1-3)(1-5)}$$

$$+9\frac{(x-0)(x-1)(x-3)(x-5)}{(2-0)(2-1)(2-3)(2-5)} + 29\frac{(x-0)(x-1)(x-2)(x-5)}{(3-0)(3-1)(3-2)(3-5)}$$

$$+129\frac{(x-0)(x-1)(x-2)(x-3)}{(5-0)(5-1)(5-2)(5-3)}$$

$$P(x) = x^3 + x - 1$$

So, the polynomial $P(x) = x^3 + x - 1$ is the required interpolating polynomial.

We have $P(4) = 67$.

Note: It is easy to see that this polynomial passes through all these points $(0, -1)$, $(1, 1)$, $(2, 9)$, $(3, 29)$, $(5, 129)$ as $P(0) = -1$, $P(1) = 1$,....

8.4 Newton Divided Difference (NDD) Method

In Section 8.3, we have obtained Lagrange form of the interpolating polynomial. In this section, we will obtain another form of interpolating polynomial known as Newton divided difference form. We have already proved the importance of Newton form over other polynomial forms in Section 8.2. Let us consider the interpolating polynomial in the following Newton form with centers $x_0, x_1, ..., x_{n-1}$.

$$P_n(x) = a_0 + a_1(x-x_0) + a_2(x-x_0)(x-x_1) + \cdots + a_n(x-x_0)(x-x_1)\cdots(x-x_{n-1}) \quad (8.12)$$

This polynomial is interpolating polynomial, if it passes through the $(n+1)$ points $(x_0, f(x_0))$, $(x_1, f(x_1))$,..., $(x_n, f(x_n))$, i.e.,

$$P_n(x_i) = f(x_i); \quad i = 0, 1, ..., n$$

At point $x = x_0$, Eq. (8.12) provides the following result

$$P_n(x_0) = f(x_0) = a_0 \quad (8.13)$$

Equation (8.12) at point $x = x_1$ is given by

$$P_n(x_1) = f(x_1) = f(x_0) + a_1(x_1 - x_0)$$

$$\Rightarrow a_1 = \frac{f(x_1) - f(x_0)}{x_1 - x_0} \quad (8.14)$$

The coefficient a_1 depends on the points x_0 and x_1 only. For this reason, we take $f\left[x_0, x_1\right]$ notation for this first divided difference. Similarly, we have following formula for other first divided differences

$$f\left[x_i, x_j\right] = \frac{f(x_i) - f(x_j)}{x_i - x_j} = f\left[x_j, x_i\right]$$

Equation (8.12) at point $x = x_2$ gives the following equation

$$P_n(x_2) = f(x_2) = a_0 + a_1(x - x_0) + a_2(x - x_0)(x - x_1)$$

On using results (8.13) and (8.14) in this expression, and making some rearrangements, we have

$$a_2 = \frac{f\left[x_1, x_2\right] - f\left[x_0, x_1\right]}{x_2 - x_0} \tag{8.15}$$

Now, a_2 depends on the points x_0, x_1 and x_2 only, this motivates the notation for second divided difference as $f[x_0, x_1, x_2]$, and so on. In the next section, we will prove that the second and higher order divided differences are defined in terms of lower order divided differences by the following formulae

$$f\left[x_i, x_{i+1}, x_{i+2}\right] = \frac{f\left[x_{i+1}, x_{i+2}\right] - f\left[x_i, x_{i+1}\right]}{x_{i+2} - x_i}$$

$$f\left[x_i, x_{i+1}, x_{i+2}, x_{i+3}, ..., x_{i+k}\right] = \frac{f\left[x_{i+1}, x_{i+2}, x_{i+3}, ..., x_{i+k}\right] - f\left[x_i, x_{i+1}, x_{i+2}, x_{i+3}, ..., x_{i+k-1}\right]}{x_{i+k} - x_i} \tag{8.16}$$

On using Eqs. 8.13–8.16 in Eq. 8.12, the interpolating polynomial is given by the following formula

$$P_n(x) = f(x_0) + f[x_0, x_1](x - x_0) + f[x_0, x_1, x_2](x - x_0)(x - x_1) + \cdots$$
$$+ f[x_0, x_1, \cdots, x_n](x - x_0)(x - x_1)\cdots(x - x_{n-1}) \tag{8.17}$$

This formula is known as Newton divided difference (NDD) formula for interpolating polynomial.

Divided Difference Table

x	$f(x)$	[,]	[, ,]	[, , ,]
x_0	$f(x_0)$			
		$f\left[x_0, x_1\right] = \frac{f(x_1) - f(x_0)}{x_1 - x_0}$		
x_1	$f(x_1)$		$f\left[x_0, x_1, x_2\right] = \frac{f[x_1, x_2] - f[x_0, x_1]}{x_2 - x_0}$	
		$f\left[x_1, x_2\right] = \frac{f(x_2) - f(x_1)}{x_2 - x_1}$		$f\left[x_0, x_1, x_2, x_3\right] = \frac{f[x_1, x_2, x_3] - f[x_0, x_1, x_2]}{x_3 - x_0}$
x_2	$f(x_2)$		$f\left[x_1, x_2, x_3\right] = \frac{f[x_2, x_3] - f[x_1, x_2]}{x_3 - x_1}$	
		$f\left[x_2, x_3\right] = \frac{f(x_3) - f(x_2)}{x_3 - x_2}$		$\vdots$
x_3	$f(x_3)$		$\vdots$	
		$\vdots$		
$\vdots$	$\vdots$			

Example ————————————————————— **8.6**

Use Newton divided difference (NDD) formula to derive interpolating polynomial for the data points $(0, -1)$, $(1, 1)$, $(2, 9)$, $(3, 29)$, $(5, 129)$, and hence compute the value of the point $y(4)$.

Ans.

We have following five points

$$\left(0 = x_0,\ -1 = f(x_0)\right), \left(1 = x_1,\ 1 = f(x_1)\right), \left(2 = x_2,\ 9 = f(x_2)\right), \left(3 = x_3,\ 29 = f(x_3)\right)$$
$$\left(5 = x_4,\ 129 = f(x_4)\right)$$

To compute interpolating polynomial, first we will construct divided difference table for these data points as follows

x	$f(x)$	$[\ ,\ ]$	$[\ ,\ ,\ ]$	$[\ ,\ ,\ ,\ ]$	$[\ ,\ ,\ ,\ ,\ ]$
$0 = x_0$	$-1 = f(x_0)$				
		$2 = f\left[x_0, x_1\right]$			
$1 = x_1$	$1 = f(x_1)$		$3 = f\left[x_0, x_1, x_2\right]$		
		$8 = f\left[x_1, x_2\right]$		$1 = f\left[x_0, x_1, x_2, x_3\right]$	
$2 = x_2$	$9 = f(x_2)$		$6 = f\left[x_1, x_2, x_3\right]$		$0 = f\left[x_0, x_1, x_2, x_3, x_4\right]$
		$20 = f\left[x_2, x_3\right]$		$1 = f\left[x_1, x_2, x_3, x_4\right]$	
$3 = x_3$	$29 = f(x_3)$		$10 = f\left[x_2, x_3, x_4\right]$		
		$50 = f\left[x_3, x_4\right]$			
$5 = x_4$	$129 = f(x_4)$				

Newton divided difference formula (8.17) for interpolating polynomial is given by

$$P_n(x) = f(x_0) + f[x_0,\ x_1](x - x_0) + f[x_0,\ x_1,\ x_2](x - x_0)(x - x_1) + \cdots$$
$$+ f[x_0,\ x_1,\ \cdots, x_n](x - x_0)(x - x_1) \cdots (x - x_{n-1})$$
$$P(x) = -1 + 2(x - 0) + 3(x - 0)(x - 1) + 1(x - 0)(x - 1)(x - 2) + 0.(x - 0)(x - 1)(x - 2)(x - 3)$$
$$P(x) = x^3 + x - 1$$

Therefore, $x^3 + x - 1$ is the required interpolating polynomial, and $P(4) = 67$.

Note: The polynomials obtained from Lagrange and Newton divided difference formulas are same. In fact, we have already established in Lemma 2, that a unique polynomial of degree $\leq n$ exists, which passes through $(n + 1)$ points $\left(x_i, f(x_i)\right)$; $i = 0, 1, ..., n$. In fact, Lagrange and NDD polynomials are different polynomial forms of an identical polynomial.

Example — 8.7

Following table gives the population data of India from the year 1981 to 2011. Use Newton divided difference formula to find the approximate population in 2006.

Table 8.2

Year (x)	1981	1991	2001	2011
Population (y) (in crores)	68.3329	84.6421	102.8737	121.0193

Ans.

The data points in the Table 8.2 are as follows

$x_0 = 1981$ $x_1 = 1991$ $x_2 = 2001$ $x_3 = 2011$

$f(x_0) = 68.3329$ $f(x_1) = 84.6421$ $f(x_3) = 102.8737$ $f(x_3) = 121.0193$

The divided difference table for these data points is given by

Years (x)	Population(y)	$[\,,\,]$	$[\,,\,,\,]$	$[\,,\,,\,,\,]$
1981	68.3329			
		1.63092		
1991	84.6421		0.009612	
		1.82316		−0.000334733
2001	102.8737		−0.00043	
		1.81456		
2011	121.0193			

Newton divided difference formula (8.17) with $n = 3$ is of the following form

$$P(x) = f(x_0) + f[x_0, x_1](x - x_0) + f[x_0, x_1, x_2](x - x_0)(x - x_1)$$
$$+ f[x_0, x_1, x_2, x_3](x - x_0)(x - x_1)(x - x_2)$$
$$P(2006) = 68.3329 + (1.63092)(2006 - 1981) + (0.009612)(2006 - 1981)(2006 - 1991)$$
$$+ (-0.000334733)(2006 - 1981)(2006 - 1991)(2006 - 2001)$$
$$P(2006) = 112.082775$$

8.4.1 Proof for Higher Order Divided Differences

In many available textbooks on numerical methods, the proofs for general k^{th} order divided differences are not available. The kth divided difference is given by following formula

$$f\left[x_0, x_1, x_2, x_3, \ldots x_k\right] = \frac{f\left[x_1, x_2, x_3, \ldots x_k\right] - f\left[x_0, x_1, x_2, x_3, \ldots x_{k-1}\right]}{x_k - x_0}$$

Here, we are providing the proof for general k^{th} order divided differences.

Proof:

Let us take three polynomials $P_{k-1}(x)$, $Q_{k-1}(x)$ and $P_k(x)$ of degree $\leq k-1$, $k-1$ and k, respectively, such that $P_{k-1}(x)$ interpolates x_0, x_1, ..., x_{k-1}, $Q_{k-1}(x)$ interpolates x_1, x_2, ..., x_k, and $P_k(x)$ interpolates x_0, x_1, ..., x_k.

Newton divided difference formulae for these three polynomials are as follows

$$P_{k-1}(x) = f(x_0) + f[x_0, x_1](x-x_0) + f[x_0, x_1, x_2](x-x_0)(x-x_1) + \cdots$$
$$+ f[x_0, x_1, \cdots, x_{k-1}](x-x_0)(x-x_1)\cdots(x-x_{k-2})$$
$$Q_{k-1}(x) = f(x_1) + f[x_1, x_2](x-x_1) + f[x_1, x_2, x_3](x-x_1)(x-x_2) + \cdots$$
$$+ f[x_1, x_2, \cdots, x_k](x-x_1)(x-x_2)\cdots(x-x_{k-1})$$
$$P_k(x) = f(x_0) + f[x_0, x_1](x-x_0) + f[x_0, x_1, x_2](x-x_0)(x-x_1) + \cdots$$
$$+ f[x_0, x_1, \cdots, x_k](x-x_0)(x-x_1)\cdots(x-x_{k-1})$$

Consider the polynomial $H_k(x)$ of the following form

$$H_k(x) = \frac{(x-x_0)}{(x_k-x_0)}Q_{k-1}(x) + \frac{(x_k-x)}{(x_k-x_0)}P_{k-1}(x)$$

It is easy to see that $H_k(x)$ interpolates all the points, x_0, x_1, ..., x_k. We have used the interpolating polynomials $P_{k-1}(x)$ and $Q_{k-1}(x)$ for this result. Therefore, we have

$$H_k(x_i) = f(x_i), \quad i = 0, 1, 2, \cdots, k$$

Note that a unique polynomial of degree $\leq k$ exists such that it agrees with the function $f(x)$ at $(k+1)$ points [Refer Lemma 8.2], so $H_k(x) = P_k(x)$. It implies

$$P_k(x) = \frac{(x-x_0)}{(x_k-x_0)}Q_{k-1}(x) + \frac{(x_k-x)}{(x_k-x_0)}P_{k-1}(x)$$

On comparing the coefficients of x^k from both sides of the above equation, we get

$$f\left[x_0, x_1, x_2, x_3, \ldots x_k\right] = \frac{f\left[x_1, x_2, x_3, \ldots x_k\right] - f\left[x_0, x_1, x_2, x_3, \ldots x_{k-1}\right]}{x_k - x_0}$$

Hence, proved.

8.4.2 Advantages of NDD Interpolation over Lagrange Interpolation

There are following two major disadvantages of using Lagrange method for interpolation.

1. Lagrange interpolation requires more arithmetic operations than NDD. In Lagrange method, more repetitions of calculations occurred as compared to NDD. Further, nested Newton form can be used in NDD formula to avoid repetitions of computational work.

2. If we want to add a new point to the data set, then Lagrange method requires starting afresh and repeating all computation. But in NDD, we can simply add a new point without repeating all the arithmetic. Let $P_n(x)$ be the NDD interpolating polynomial of degree $\leq n$, for the points $x_0, x_1, \cdots, x_n$, i.e.,

$$P_n(x) = f(x_0) + f[x_0, x_1](x - x_0) + f[x_0, x_1, x_2](x - x_0)(x - x_1) + \cdots$$
$$+ f[x_0, x_1, \cdots, x_n](x - x_0)(x - x_1) \cdots (x - x_{n-1})$$

Let any point $\tilde{x}$ other than the points, $x_0, x_1, \cdots, x_n$. Let $P_{n+1}(x)$ be polynomial of degree $\leq (n+1)$, which interpolates the points $x_0, x_1, \cdots, x_n$ and $\tilde{x}$, then we have

$$P_{n+1}(x) = f(x_0) + f[x_0, x_1](x - x_0) + f[x_0, x_1, x_2](x - x_0)(x - x_1) + \cdots$$
$$+ f[x_0, x_1, \cdots, x_n](x - x_0)(x - x_1) \cdots (x - x_{n-1})$$
$$+ f[x_0, x_1, \cdots, x_n, \tilde{x}](x - x_0)(x - x_1) \cdots (x - x_n)$$
$$P_{n+1}(x) = P_n(x) + f[x_0, x_1, \cdots, x_n, \tilde{x}](x - x_0)(x - x_1) \cdots (x - x_n)$$

In another way, if we want to add a data point to existing data, then we have to add only a last term to the existing polynomial from the previous data set. It is easy to see that, if we want to interpolate only one point $(x_0, f(x_0))$, then first term of the interpolating polynomial $P_n(x)$ is sufficient. Similarly for two points $(x_0, f(x_0))$ and $(x_1, f(x_1))$, only first two terms of NDD polynomial are required. In general, first $(k+1)$ terms are sufficient for $(k+1)$ points, $(x_0, f(x_0)), (x_1, f(x_1)), \dots, (x_k, f(x_k))$; as the higher terms are zero on these $(k+1)$ points $x_0, x_1, \dots, x_k$.

8.4.3 Properties of Divided Differences

i) The Newton divided difference is symmetrical in operation, since

$$f\left[x_i, x_{i+1}\right] = \frac{f(x_{i+1}) - f(x_i)}{x_{i+1} - x_i} = f\left[x_{i+1}, x_i\right]$$

ii) The kth generalization of Newton divided difference formula is as follows

$$f\left[x_0, x_1, x_2, x_3, \dots x_k\right] = \frac{f\left[x_1, x_2, x_3, \dots x_k\right] - f\left[x_0, x_1, x_2, x_3, \dots x_{k-1}\right]}{x_k - x_0}$$

iii) Divided differences are independent of the order of data points, i.e., if we interchange the position of the points in the divided difference table, then the last divided difference remains same. This result can be proved using the fact that there exists a unique interpolating polynomial $P_n(x)$ for given $(n+1)$ points. So, we get the desired result by equating the coefficients of x^n in both the polynomials.

For example, the coefficient of x^n in the NDD interpolating polynomial (8.17) is $f\left[x_0, x_1, x_2, \dots x_n\right]$. Let us interchange the position of x_0 and x_1, then the coefficient of x^n in new interpolating polynomial is $f\left[x_1, x_0, x_2, x_3, \ldots x_n\right]$. Since, the polynomial is unique, so we have

$$f\left[x_0, x_1, x_2, x_3, \ldots x_n\right] = f\left[x_1, x_0, x_2, x_3, \ldots x_n\right]$$

iv) The mth divided difference in compact form is as follows

$$f\left[x_0, x_1, x_2, x_3, \ldots x_m\right] = \frac{f^m(\xi)}{m!}$$

for some ξ in the given interval. The proof of the result is given in following theorem.

Theorem 8.1

Let $f(x)$ be a well-defined real-valued function on the interval $[a, b]$ containing $(m+1)$ – points, $x_0, x_1, \cdots, x_m$. If the function $f(x)$ is m-times differentiable in (a, b), then there exists a point, $\xi \in (a, b)$, such that

$$f\left[x_0, x_1, x_2, x_3, \ldots x_m\right] = \frac{f^m(\xi)}{m!}$$

Proof: For $m = 1$, we can prove the result using Lagrange mean value theorem as follows

$$f\left[x_0, x_1\right] = \frac{f(x_1) - f(x_0)}{x_1 - x_0} = f'(\xi), \text{ at some point } \xi \in (a, b)$$

For $m > 1$, the interpolating polynomial for points $x_0, x_1, \cdots, x_m$ is given by

$$P_m(x) = f(x_0) + f[x_0, x_1](x - x_0) + f[x_0, x_1, x_2](x - x_0)(x - x_1) + \cdots$$
$$+ f[x_0, x_1, \cdots, x_n](x - x_0)(x - x_1) \cdots (x - x_{m-1})$$

We have

$$P_m(x_i) = f(x_i), \quad i = 0, 1, 2, \cdots, m$$

Consider the error function as follows

$$\varepsilon(x) = f(x) - P_m(x)$$

It has at least $(m+1)$ zeroes, $x_0, x_1, \cdots, x_m$, as $\varepsilon(x_0) = \varepsilon(x_1) = \cdots = \varepsilon(x_m) = 0$.

According to Rolle theorem, the function $\varepsilon'(x)$ has at least m zeroes in the intervals, $(x_0, x_1), (x_1, x_2), \cdots, (x_{m-1}, x_m) \in (a, b)$.

Similarly, the function $\varepsilon''(x)$ has at least $m-1$ zeroes in the interval (a, b).

$\vdots$

The function $\varepsilon'''(x)$ has at least one zero in the interval (a, b), let that zero be ξ. So, we have

$$\varepsilon'''(\xi) = 0.$$
$$\Rightarrow \varepsilon'''(\xi) = f'''(\xi) - P_m'''(\xi) = 0 \tag{8.18}$$

On using interpolating polynomial $P_m(x)$, we get

$$P_m'''(x) = f\left[x_0, x_1, x_2, x_3, \ldots x_m\right] m! \tag{8.19}$$

The last two equations (8.18) and (8.19) provide the desired result

$$f\left[x_0, x_1, x_2, x_3, \ldots x_m\right] = \frac{f'''(\xi)}{m!}$$

8.5 Error in Interpolating Polynomial

Let $P_n(x)$ be the interpolating polynomial of degree $\leq n$, for the points $x_0, x_1, \cdots, x_n$ in the interval (a, b)

$$P_n(x) = f(x_0) + f[x_0, x_1](x - x_0) + f[x_0, x_1, x_2](x - x_0)(x - x_1) + \cdots$$
$$+ f[x_0, x_1, \cdots, x_n](x - x_0)(x - x_1) \cdots (x - x_{n-1})$$

Hence $P_n(x_i) = f(x_i), \quad i = 0, 1, 2, \cdots, n.$

At any point $\tilde{x}$ other than $x_0, x_1, \cdots, x_n$, the error in interpolating polynomial is given by
$$\varepsilon(\tilde{x}) = f(\tilde{x}) - P_n(\tilde{x})$$

Let $P_{n+1}(x)$ be polynomial of degree $\leq (n+1)$, which interpolates the points $x_0, x_1, \cdots, x_n$ and $\tilde{x}$ also. We have

$$P_{n+1}(x) = f(x_0) + f[x_0, x_1](x - x_0) + f[x_0, x_1, x_2](x - x_0)(x - x_1) + \cdots$$
$$+ f[x_0, x_1, \cdots, x_n](x - x_0)(x - x_1) \cdots (x - x_{n-1})$$
$$+ f[x_0, x_1, \cdots, x_n, \tilde{x}](x - x_0)(x - x_1) \cdots (x - x_n)$$
$$P_{n+1}(x) = P_n(x) + f[x_0, x_1, \cdots, x_n, \tilde{x}](x - x_0)(x - x_1) \cdots (x - x_n)$$

The polynomial $P_{n+1}(x)$ interpolates $\tilde{x}$, so $P_{n+1}(\tilde{x}) = f(\tilde{x})$. This implies

$$f(\tilde{x}) = P_{n+1}(\tilde{x}) = P_n(\tilde{x}) + f[x_0, x_1, \cdots, x_n, \tilde{x}](\tilde{x} - x_0)(\tilde{x} - x_1) \cdots (\tilde{x} - x_n)$$

The error term is given by

$$\varepsilon_n(\tilde{x}) = f(\tilde{x}) - P_n(\tilde{x}) = f[x_0, x_1, \cdots, x_n, \tilde{x}](\tilde{x} - x_0)(\tilde{x} - x_1) \cdots (\tilde{x} - x_n) \tag{8.20}$$

The error is likely to be the next term in the interpolating polynomial.

For $(n + 2)$ points $x_0, x_1, \cdots, x_n$ and $\tilde{x}$, the theorem 8.1 implies that

$$f[x_0, x_1, \cdots, x_n, \tilde{x}] = \frac{f^{n+1}(\xi)}{(n+1)!}$$

On using this expression in Eq. (8.20), we get error term as follows

$$\varepsilon_n(\tilde{x}) = \frac{f^{n+1}(\xi)}{(n+1)!}(\tilde{x} - x_0)(\tilde{x} - x_1)\cdots(\tilde{x} - x_n) = \frac{f^{n+1}(\xi)}{(n+1)!}\prod_{i=0}^{n}(\tilde{x} - x_i) \qquad (8.21)$$

where ξ is any point in the interval (a, b).

Let $\left|f^{n+1}(\xi)\right| \le M$ for all $\xi \in (a, b)$.

Then, error bound (8.21) for interpolating polynomial is given by

$$\left|\varepsilon_n(\tilde{x})\right| \le \frac{M}{(n+1)!}\prod_{i=0}^{n}\left|\tilde{x} - x_i\right| \qquad (8.22)$$

Note: It is worth emphasizing here that we cannot compute the error bound in interpolation for the data points only. For example, we cannot compute an error bound in the examples 8.6, 8.7, etc. But if we are approximating a sufficiently differentiable function with an interpolating polynomial, then the error bound can be computed like in next example.

Example   **8.8**

From an equally spaced table of the function $\sin(x)$ with spacing $h = 0.1$, we have following data set

x	0.1	0.2	0.3
$\sin x$	0.0998334	0.198669	0.295520

Compute the value of $\sin(x)$ at $x = 0.123$ using Lagrange method. Also, compute the error bound.

Ans.
Lagrange interpolation formula (8.11) for 3 data points is as follows

$$P(x) = f(x_0)\frac{(x - x_1)(x - x_2)}{(x_0 - x_1)(x_0 - x_2)} + f(x_1)\frac{(x - x_0)(x - x_2)}{(x_1 - x_0)(x_1 - x_2)} + f(x_2)\frac{(x - x_0)(x - x_1)}{(x_2 - x_0)(x_2 - x_1)}$$

$$P(0.123) = (0.0998334)\frac{(0.123 - 0.2)(0.123 - 0.3)}{(0.1 - 0.2)(0.1 - 0.3)} + (0.198669)\frac{(0.123 - 0.1)(0.123 - 0.3)}{(0.2 - 0.1)(0.2 - 0.3)}$$

$$\qquad + (0.295520)\frac{(0.123 - 0.1)(0.123 - 0.2)}{(0.3 - 0.1)(0.3 - 0.2)}$$

$$P(0.123) = 0.122741$$

Error term (8.22) for a quadratic interpolation is given by

$$\left|\varepsilon_2(\tilde{x})\right| \le \frac{M}{(3)!} \prod_{i=0}^{2} \left|\tilde{x} - x_i\right|, \text{ where } M = \max_{0.1 \le x \le 0.3} \left|f'''(\xi)\right| \tag{8.23}$$

As ξ is not known to us, we simply compute its upper bound for $f(x) = \sin(x)$,

$$M = \max_{0.1 \le x \le 0.3} \left|f'''(x)\right| = \max_{0.1 \le x \le 0.3} \left|\cos(x)\right| = 0.9950$$

The error bound (8.23) is given by

$$\left|\varepsilon_2(\tilde{x})\right| \le \frac{0.995}{6} \left|(0.123 - 0.1)(0.123 - 0.2)(0.123 - 0.3)\right|$$

$$\left|\varepsilon_2(\tilde{x})\right| \le 0.000010517$$

Example ──────────────────────────────────── 8.9

Suppose we want to create an equally spaced table with spacing h for the function $f(x) = \cos(x)$ in the interval $\left(0, \dfrac{\pi}{2}\right)$. Compute the total number of entries required in the table so that interpolation with a quadratic polynomial will yield minimum accuracy of six decimal places.

Ans.
Let the table contain equally spaced points with spacing h, i.e.,

$$x_i = 0 + ih, \ i = 1, 2, \cdots N$$

where N is the number of entries in the table and given by

$$N = \frac{\dfrac{\pi}{2} - 0}{h} = \frac{\pi}{2h}$$

Let the polynomial approximation be computed at any point $\tilde{x} \in (x_{i-1}, x_{i+1})$. Consider $P_2(x)$ is the quadratic polynomial, which interpolates the function $f(x)$ at points x_{i-1}, x_i, x_{i+1}. Then the error bound (8.21) is given by

$$\left|f(\tilde{x}) - P_2(\tilde{x})\right| = \frac{\left|f'''(\xi)\right|}{3!} \left|(\tilde{x} - x_{i-1})(\tilde{x} - x_i)(\tilde{x} - x_{i+1})\right|, \text{ for some } \xi \in \left(x_{i-1}, x_{i+1}\right)$$

As ξ is not known to us, we simply compute its upper bound,

$$f'''(\xi) \le \max_{0 \le x \le \pi/2} \left|\sin(x)\right| = 1$$

Let $\tilde{x} - x_i = y$, then we have

$$(\tilde{x} - x_{i-1})(\tilde{x} - x_i)(\tilde{x} - x_{i+1}) = (y-h)y(y+h)$$

The maximum of value of the function $|(y-h)y(y+h)|$ is at the point $y = \pm\dfrac{h}{\sqrt{3}}$, and that

value is $\dfrac{2h^3}{3\sqrt{3}}$. So, the maximum error in interpolating polynomial is given by

$$\left|f(\tilde{x}) - P_2(\tilde{x})\right| = \frac{\left|f'''(\xi)\right|}{3!}\left|(\tilde{x} - x_{i-1})(\tilde{x} - x_i)(\tilde{x} - x_{i+1})\right| \le \frac{1}{6}\frac{2h^3}{3\sqrt{3}} \le \frac{h^3}{9\sqrt{3}}$$

For an accuracy of six decimal places, the error term must be less than .0000005. Therefore, we have

$$\frac{h^3}{9\sqrt{3}} \le 5 \times 10^{-7}$$

$$\Rightarrow h \le .019827 \text{ (or) } N \ge \frac{\dfrac{\pi}{2} - 0}{.019827} \approx 80$$

8.6 Discussion

It is worth emphasizing here that the interpolating polynomial is unique. However, we use different forms of polynomials depending on their use in different applications (so far we examined only Lagrange and Newton forms). The computations in Lagrange method are repetitive, so the number of arithmetic operations in Lagrange method is greater than the NDD-method. Consequently, the result we obtain from Lagrange method is more error-prone to the round-off error than the NDD-method. But, Lagrange method has its applications in some theoretical concepts.

Polynomials are the best approximation for the smooth experiments. But sometimes if the data size is large, then it is very difficult to obtain a polynomial of very high degree. For examples, if we have a set of 50 data points, then the interpolating polynomial is of degree 49. Also, it is very difficult to handle the round-off error during the calculations of values at any intermediate points. Let there be an intermediate value up to 6-significant digits, then, to compute 49th power of this value without round-off requires $6(2^{49})$ significant digits, i.e., something unrealistic to compute with normal computers so we have to switch over to some other methodology.

Also, the global nature of the interpolating polynomial is a major concern because if we change the position of one point then the whole polynomial gets changed. So, we cannot use the interpolation for designing/graphics purpose.

The Lagrange and NDD interpolation methods are applicable for equally as well as unequally spaced data sets. But if our input variable (x) is equally spaced $\left(x_i = x_0 + ih, \quad i = 1, 2, 3, \cdots, n\right)$, then our interpolation can be simpler. In Chapter 10, we will obtain certain simplified interpolation formulas, which are applicable only for equally spaced data sets.

8.7 Hermite Interpolation

We have discussed the interpolating polynomial of degree $\leq n$; which passes through $(n+1)$ points $\left(x_i, f(x_i)\right)$; $i = 0, 1, 2,...,n$. Now, let us derive interpolating polynomial for a function $f(x)$ such that the values of the function $f(x)$ and its derivative $f'(x)$ match with this polynomial at $(n + 1)$ points x_i; $i = 0, 1, 2,...,n$. The polynomial of degree $\leq 2n+1$ is required to satisfies $2n+2$ conditions. Let us consider an interpolating polynomial of degree $\leq 2n+1$ which satisfies the following $2(n+1)$ restrictions at $(n+1)$ points x_i; $i = 0, 1, 2,...,n$.

$$\left. \begin{array}{l} P_{2n+1}(x_i) = f(x_i) \\ P'_{2n+1}(x_i) = f'(x_i) \end{array} \right\}, \qquad i = 0, 1, 2, \cdots, n \tag{8.24}$$

We have to express the polynomial $P_{2n+1}(x)$ in terms of $(n+1)$ points, x_i; $i = 0, 1, 2,...,n$. Therefore, let the polynomial $P_{2n+1}(x)$ be of the following form

$$\begin{aligned} P_{2n+1}(x) &= \sum_{i=0}^{n} u_i(x) P_{2n+1}(x_i) + \sum_{i=0}^{n} v_i(x) P'_{2n+1}(x_i) \\ &= \sum_{i=0}^{n} u_i(x) f(x_i) + \sum_{i=0}^{n} v_i(x) f'(x_i) \end{aligned} \tag{8.25}$$

where $u_i(x)$ and $v_i(x)$ are polynomials of degree $\leq 2n+1$. Let us rewrite these polynomials in terms of Lagrange polynomial coefficients $l_i(x)$ as follows

$$\left. \begin{array}{l} u_i(x) = (a_i x + b_i) l_i^2(x) \\ v_i(x) = (c_i x + d_i) l_i^2(x) \end{array} \right., \qquad i = 0, 1, 2, \cdots, n \tag{8.26}$$

where a_i, b_i, c_i, d_i are constants to be determined. The coefficients $l_i(x)$ are given by

$$l_i(x) = \prod_{\substack{k=0 \\ k \neq i}}^{n} \frac{x - x_k}{x_i - x_k} = \frac{(x-x_0)(x-x_1)\cdots(x-x_{i-1})(x-x_{i+1})\cdots(x-x_n)}{(x_i-x_0)(x_i-x_1)\cdots(x_i-x_{i-1})(x_i-x_{i+1})\cdots(x_i-x_n)}$$

with property $l_i(x_j) = \begin{cases} 1 & i = j \\ 0 & i \neq j \end{cases}$ \tag{8.27}

The polynomial (8.25) is interpolating polynomial if it satisfies the conditions (8.24). For this, we have

$$u_i(x_j) = \begin{cases} 1 & i=j \\ 0 & i \neq j \end{cases} \qquad v_i'(x_j) = \begin{cases} 1 & i=j \\ 0 & i \neq j \end{cases}$$

$$v_i(x_j) = 0 \; \forall \, j = 0, 1, \cdots, n \qquad \text{and} \qquad u_i'(x_j) = 0 \; \forall \, j = 0, 1, \cdots, n \qquad (8.28)$$

On using Eqs. (8.26–8.28), we have

$$u_i(x_j) = \begin{cases} 1 & i=j \\ 0 & i \neq j \end{cases} \qquad \Rightarrow \qquad (a_i x_j + b_i) l_i^2(x_j) = \begin{cases} 1 & i=j \\ 0 & i \neq j \end{cases}$$

$$a_i x_i + b_i = 1$$

$$v_i(x_j) = 0 \; \forall \, j = 0, 1, \cdots, n \qquad \Rightarrow \qquad (c_i x_j + d_i) l_i^2(x_j) = 0$$

$$c_i x_i + d_i = 0$$

$$v_i'(x_j) = \begin{cases} 1 & i=j \\ 0 & i \neq j \end{cases} \qquad \Rightarrow \qquad (c_i x_j + d_i) 2 l_i(x_j) l_i'(x_j) + c_i l_i^2(x_j) = \begin{cases} 1 & i=j \\ 0 & i \neq j \end{cases}$$

$$(c_i x_i + d_i) 2 l_i'(x_i) + c_i = 1$$

$$u_i'(x_j) = 0 \; \forall \, j = 0, 1, \cdots, n \qquad \Rightarrow \qquad (a_i x_j + b_i) 2 l_i(x_j) l_i'(x_j) + a_i l_i^2(x_j) = 0$$

$$(a_i x_i + b_i) 2 l_i'(x_i) + a_i = 0$$

Therefore, we have following four sets of equations in the variables, a_i, b_i, c_i, d_i; $i = 0, 1, \cdots, n$.

$$\left. \begin{aligned} a_i x_i + b_i &= 1 \\ c_i x_i + d_i &= 0 \\ (c_i x_i + d_i) 2 l_i'(x_i) + c_i &= 1 \\ (a_i x_i + b_i) 2 l_i'(x_i) + a_i &= 0 \end{aligned} \right\} \Rightarrow \begin{cases} a_i = -2 l_i'(x_i) \\ b_i = 1 + 2 x_i l_i'(x_i) \\ c_i = 1 \\ d_i = -x_i \end{cases}$$

On using these values of constants, a_i, b_i, c_i, d_i; $i = 0, 1, \cdots, n$ in Eqs. (8.26), we get

$$u_i(x) = (a_i x + b_i) l_i^2(x) = \left(-2 l_i'(x_i) x + 1 + 2 x_i l_i'(x_i) \right) l_i^2(x)$$

$$v_i(x) = (c_i x + d_i) l_i^2(x) = (x - x_i) l_i^2(x)$$

Use these values in Eq. (8.25) to get the following Hermite interpolating polynomial

$$P_{2n+1}(x) = \sum_{i=0}^{n} u_i(x) f(x_i) + \sum_{i=0}^{n} v_i(x) f'(x_i)$$

$$= \sum_{i=0}^{n} \left(-2 l_i'(x_i) x + 1 + 2 x_i l_i'(x_i) \right) l_i^2(x) f(x_i) + \sum_{i=0}^{n} (x - x_i) l_i^2(x) f'(x_i) \qquad (8.29)$$

Example ─────────────────────────────────────── **8.10** ───────

Compute the Hermite interpolating polynomial and then the value of the function $f(0.5)$ from the following data set.

x	-1	0	1	2
$f(x)$	2	2	2	26
$f'(x)$	2	0	2	68

Ans: We have $4(=n+1)$ points, $x_0 = -1$, $x_1 = 0$, $x_2 = 1$, $x_3 = 2$; therefore, the Hermite polynomial (8.29) is of degree $\leq 7 (= 2n+1)$. It is given by

$$P(x) = \sum_{i=0}^{3}\left(-2l_i'(x_i)x + 1 + 2x_i l_i'(x_i)\right)l_i^2(x)f(x_i) + \sum_{i=0}^{3}(x-x_i)l_i^2(x)f'(x_i)$$

We have to calculate Lagrange coefficients polynomials $l_i(x)$ and their derivatives $l_i'(x_i)$ to compute the interpolating polynomial. On using $n = 3$ in the following formula, we have

$$l_i(x) = \prod_{\substack{k=0 \\ k\neq i}}^{n}\frac{x-x_k}{x_i-x_k} = \frac{(x-x_0)(x-x_1)\cdots(x-x_{i-1})(x-x_{i+1})\cdots(x-x_n)}{(x_i-x_0)(x_i-x_1)\cdots(x_i-x_{i-1})(x_i-x_{i+1})\cdots(x_i-x_n)}$$

For $i = 0, 1, 2, 3$, we have

$$l_0(x) = \prod_{\substack{k=0 \\ k\neq 0}}^{3}\frac{x-x_k}{x_0-x_k} = \frac{(x-x_1)(x-x_2)(x-x_3)}{(x_0-x_1)(x_0-x_2)(x_0-x_3)} = \frac{-1}{6}(x-0)(x-1)(x-2)$$

$$\Rightarrow l_0'(x_0) = \frac{-11}{6}$$

$$l_1(x) = \prod_{\substack{k=0 \\ k\neq 1}}^{3}\frac{x-x_k}{x_1-x_k} = \frac{(x-x_0)(x-x_2)(x-x_3)}{(x_1-x_0)(x_1-x_2)(x_1-x_3)} = \frac{1}{2}(x+1)(x-1)(x-2)$$

$$\Rightarrow l_1'(x_1) = \frac{-1}{2}$$

$$l_2(x) = \prod_{\substack{k=0 \\ k\neq 2}}^{3}\frac{x-x_k}{x_2-x_k} = \frac{(x-x_0)(x-x_1)(x-x_3)}{(x_2-x_0)(x_2-x_1)(x_2-x_3)} = \frac{-1}{2}(x+1)(x-0)(x-2)$$

$$l_2'(x_2) = \frac{1}{2}$$

$$l_3(x) = \prod_{\substack{k=0 \\ k\neq 3}}^{3}\frac{x-x_k}{x_3-x_k} = \frac{(x-x_0)(x-x_1)(x-x_2)}{(x_3-x_0)(x_3-x_1)(x_3-x_2)} = \frac{1}{6}(x+1)(x-0)(x-1)$$

$$l_3'(x_3) = \frac{11}{6}$$

The interpolating polynomial is given by

$$P(x) = \sum_{i=0}^{3} \left(-2l_i'(x_i)x + 1 + 2x_i l_i'(x_i) \right) l_i^2(x) f(x_i) + \sum_{i=0}^{3} (x - x_i) l_i^2(x) f'(x_i)$$

$$P(x) = \left(-2l_0'(x_0)x + 1 + 2x_0 l_0'(x_0) \right) l_0^2(x) f(x_0) + \left(-2l_1'(x_1)x + 1 + 2x_1 l_1'(x_1) \right) l_1^2(x) f(x_1)$$

$$+ \left(-2l_2'(x_2)x + 1 + 2x_2 l_2'(x_2) \right) l_2^2(x) f(x_2) + \left(-2l_3'(x_3)x + 1 + 2x_3 l_3'(x_3) \right) l_3^2(x) f(x_3)$$

$$+ (x - x_0) l_0^2(x) f'(x_0) + (x - x_1) l_1^2(x) f'(x_1) + (x - x_2) l_2^2(x) f'(x_2) + (x - x_3) l_3^2(x) f'(x_3)$$

$$P(x) = \left(\frac{11}{3}x + \frac{14}{3} \right) \left(\frac{-1}{6}(x-0)(x-1)(x-2) \right)^2 (2) + (x+1) \left(\frac{1}{2}(x+1)(x-1)(x-2) \right)^2 (2)$$

$$+ (-x+2) \left(\frac{-1}{2}(x+1)(x-0)(x-2) \right)^2 (2) + \left(\frac{-11}{3}x + \frac{25}{3} \right) \left(\frac{1}{6}(x+1)(x-0)(x-1) \right)^2 (26)$$

$$+ (x+1) \left(\frac{-1}{6}(x-0)(x-1)(x-2) \right)^2 (2) + (x-0) \left(\frac{1}{2}(x+1)(x-1)(x-2) \right)^2 (0)$$

$$+ (x-1) \left(\frac{-1}{2}(x+1)(x-0)(x-2) \right)^2 (2) + (x-2) \left(\frac{1}{6}(x+1)(x-0)(x-1) \right)^2 (68)$$

$$P(x) = \left(\frac{28}{3}x + \frac{34}{3} \right) \left(\frac{-1}{6}(x-0)(x-1)(x-2) \right)^2 + (2x+2) \left(\frac{1}{2}(x+1)(x-1)(x-2) \right)^2$$

$$+ 2 \left(\frac{-1}{2}(x+1)(x-0)(x-2) \right)^2 + \left(\frac{-82}{3}x + \frac{244}{3} \right) \left(\frac{1}{6}(x+1)(x-0)(x-1) \right)^2$$

$$P(x) = x^5 - x^3 + 2$$

$$P(0.5) = (0.5)^5 - (0.5)^3 + 2 = 1.90625$$

Note: We can easily verify that the polynomial satisfies all the conditions

x	−1	0	1	2
$P(x) = x^5 - x^3 + 2$	2	2	2	26
$P'(x) = 5x^4 - 3x^2$	2	0	2	68

The polynomial $x^5 - x^3 + 2$ is a unique polynomial of degree $\leq 7 (= 2n+1)$, and it satisfies the conditions above. Again, it is worth to mentioning here that there are an infinite number of polynomials of degree > 7 which satisfying above conditions.

8.8 Piecewise Interpolation

If the data points are large in number, then interpolating polynomial is of very high degree which is difficult to handle. Oscillations may occur for a polynomial of large degree, and also the global nature of interpolating polynomial is a major issue. To overcome these problems, we can divide the large set of data points into small sets. For example, consider data points,

(x_i, y_i), $i = 0, 1, ..., 10$, then instead of fitting a single polynomial of degree 10, we can fit lower degree polynomials by subdividing the interval. For example, quadratic polynomials can be used for following five different sets of points.

$$\{(x_0, y_0), (x_1, y_1), (x_2, y_2)\}, \ \{(x_2, y_2), (x_3, y_3), (x_4, y_4)\}, ... \{(x_8, y_8), (x_9, y_9), (x_{10}, y_{10})\}$$

We can fit five different quadratic polynomials to these five sets.

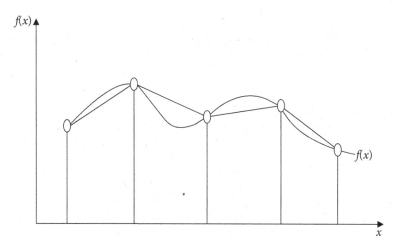

Fig. 8.2 Linear piecewise interpolation

In piecewise interpolation, we have different polynomials in subintervals. So, smoothness is a major issue at nodal points which is also visible from the above figure.

Example ── **8.11**

Fit piecewise quadratic polynomials in the following data set

x:	−2	−1	0	1	3	4	5
y:	10	5	2	5	23	43	69

Hence, compute $y(-0.5)$, $y(2)$ and $y(4.2)$.

Ans.

To fit quadratic polynomial, divide the interval $(-2, 5)$ into following subintervals and compute the interpolating polynomials for each subinterval

x:	−2	−1	0
y:	10	5	2

The interpolating polynomial is $x^2 - 2x + 2$

$$x: \quad 0 \quad 1 \quad 3$$
$$y: \quad 2 \quad 5 \quad 23$$

The interpolating polynomial is $2x^2 + x + 2$

$$x: \quad 3 \quad 4 \quad 5$$
$$y: \quad 23 \quad 43 \quad 69$$

The interpolating polynomial is $3x^2 - x - 1$

Hence, piecewise quadratic polynomial is given by

$$P_2(x) = \begin{cases} x^2 - 2x + 2 & -2 \le x \le 0 \\ 2x^2 + x + 2 & 0 \le x \le 3 \\ 3x^2 - x - 1 & 3 \le x \le 5 \end{cases}$$

One can easily compute the desired values from this piecewise polynomial as follows

$$y(-0.5) = P_2(-0.5) = 3.25, \quad y(2) = P_2(2) = 12 \text{ and } y(4.2) = P_2(4.2) = 47.72$$

8.9 Weierstrass Approximation Theorem

In interpolation, we use polynomials as approximating functions. Polynomials are most widely used approximations from theoretical as well as applications point of view. Other commonly used approximating functions are trigonometric functions, exponential/ hyperbolic functions and rational functions. Any continuous function on a finite interval is well approximated by a polynomial function within error limit, and the existence of such a polynomial is given by Weierstrass approximation theorem, which states.

Weierstrass Approximation Theorem: Consider a continuous real-valued function defined on a real finite interval [a, b], then for every ε>0, there exists a polynomial P(x), such that

$$\left| f(x) - P(x) \right| < \varepsilon \text{ for all } x \in [a, b]$$

In fact, Weierstrass approximation theorem has far reaching consequences as it implies that we can approximate any continuous function with a polynomial function as close as required. The polynomials have simple structures and can easily be computed with a computer. Further, polynomials are convenient for analysis like differentiation, integration, etc. Therefore, polynomial approximations are most commonly and widely used structures.

Exercise 8

1. Prove that there exists at most one polynomial of degree less than equal to k which takes on specified values at (k+1) distinct points.

2. Use nested multiplication algorithm to convert the polynomial $1 + 2x + 3x^2$ into its equivalent Newton form whose centers are 1 and 2, respectively.

    Ans. $6 + 11(x-1) + 3(x-1)(x-2)$

3. Derive nested multiplication algorithm for converting a Newton form into another Newton form with different centers. Obtain the first derivative at $x = 1$ by converting the polynomial $P(x) = 1 + 2(x-1) + 3(x-1)(x-2) + 4(x-1)(x-2)(x-3)$ to shifted power form with center 1.

Ans. $1 + 7(x-1) - 9(x-1)^2 + 4(x-1)^3$, $P'(1) = 7$

4. Use Lagrange interpolating polynomial to compute the value of $f(1.3)$ for data points $(0, -1)$, $(1, 1)$, $(2, 9)$, $(3, 29)$.

Ans. 2.497

5. Obtain interpolating polynomial for the following table, and hence compute the value of $f(2.5)$ with the help of Lagrange formula.

x	1	2	3	4
$f(x)$	1	8	27	64

Ans. x^3, 15.625

6. Using Lagrange interpolation formula, find the value of $\sin\left(\dfrac{\pi}{6}\right)$ from the following set of data points

x:	0	$\pi/4$	$\pi/2$
$\sin x$:	0	0.707107	1.0

Also, estimate the error limit in the solution.

Ans. $\sin\left(\dfrac{\pi}{6}\right) = 0.517428$, $\quad$ Error $\leq \dfrac{\pi^3}{1296}$

7. In an experiment, the current (I) in a wire for various imposed voltages (V) is as follows

V(Volts):	5	10	20	40
I(Amperes):	7.4	12.9	21.7	45.8

Use Lagrange interpolation to compute the approximate current in wire for a voltage of 15 V.

Ans. 17.4845 Ampere

8. Find a polynomial of degree 3 or less, such that it interpolates the data set $(0,1)$, $(1,3)$, $(2,13)$ and $(4,81)$ by using Newton divided difference interpolation technique.

Ans. $x^3 + x^2 + 1$

9. Construct the Lagrange interpolating polynomial for the following data set

x	0	0.1	0.2	0.4
$f(x)$	0.74	1.2	1.87	5.15

Ans. $54.583x^3 - 5.875x^2 + 4.642x + 0.74$

10. The tensile strengths of stainless steel cables of different diameters are investigated to give following results.

Cable diameter (x):	1.2	1.6	2.4	3.6
Tensile Strength(y):	1.32	2.53	5.28	8.34

Use Lagrange's interpolation to compute the tensile strength of a cable of diameter 2.

Ans. 3.892

11. Given

x	1.0	2.5	3.0
$f(x)$	0.34	0.42	0.53

Ans. 2.7255

Find the value of x for $f(x) = 0.44$ by using inverse Lagrange method.

12. In an experimental setup, we have following input and output values. Compute the approximation for the output value if the input value is 1.35. Find the input value for maximum output. Also find that maximum output value. Use Newton divided difference method for the calculations.

Input	0.5	1.0	1.5	2.0
Output	2.4	5.7	12.8	6.3

Ans. 11.3063, Maximum output is 13.1453 at the input 1.604

13. The following data reveals the tensile strength of a given plastic, and the time it is heat treated. Use Lagrange interpolation to determine the tensile strength at a time of 40 min. and 65 min.

Time (In Min.):   15   30   45   60   70

Tensile Strength: 18   28   26   37   48

Ans. 25.734, 42.817

14. The growth of cell culture (optical density) at various pH levels are tabulated in the following table.

pH: 4   4.5   5   5.5   6

Optical density: 0.28   0.35   0.41   0.46   0.52

Compute the optical density at pH level 5.8.

Ans. 0.492688

15. The following data set represents the resistivity of a given metal with temperature. Predict the resistivity at 350 K temperature

Temperature (K):   100   200   300   400   500

Resistivity (Ω-cm, $\times 10^6$):   4.2   8.1   12.8   16.6   20.0

Ans. 14.8328×10^6 Ω-cm

16. The population of a certain town (as obtained from census data) is shown in the following table

Year   1981   1991   2001   2011

Population   12.92   16.46   21.14   25.35 (in millions)

Find the estimate for the population in the year 2008 using the Lagrange method.

Ans. 24.232145 millions

17. The tensile strengths of stainless steel cables of different diameters are investigated to give following results.

Cable diameter (x):   1.2   1.6   2.4   3.6   5.0

Tensile Strength(y):   1.32   2.53   5.28   8.34   12.83

Use Newton divided difference formula to compute the tensile strength of a cable of diameter 2.

Ans. 3.920725

18. The function values $f(x)$ for given values of x are as follows

$$f(0) = 0, f(1) = -1, f(2) = 0, f(4) = 32$$

Use Lagrange formula to compute the value of x for which $f(x)$ is minimum.

Ans. The interpolating polynomial is $x^3 - 2x^2$, and point of minima is $x = 1.333$

19. By the mean of Newton divided difference formula, find the value of $f(5)$ from the following table

x	2	3	4	6	7
$f(x)$	25	80	189	641	1020

Ans. $f(5) = 370$

20. Derive interpolating polynomial for $\log_{10} 1 = 0$, $\log_{10} 2 = 0.3010$, $\log_{10} 3 = 0.4771$, and hence calculate $\log_{10}(15)$ using Newton divided difference formula.

Ans. $-0.06245x^2 + 0.48835x - 0.42590$; $\log_{10}(1.5) = 0.16611250$

$\log_{10}(15) = \log_{10}(10 \times 1.5) = \log_{10}(10) \times \log_{10}(1.5) = 1 + 0.16611250 = 1.16611250$

21. Approximate the exponential value $e^{1.3}$ by Newton divided difference polynomial from the data set $e^0 = 1$, $e^1 = 2.7183$, $e^2 = 7.3891$. What is the estimate for the error bound?

Ans. $e^{1.3} = 3.8095$, Error bound $= 0.3362$

22. The following values are obtained from the function $f(x) = 2^x$

$x:$	1	2	3	5	7
$f(x):$	2	4	8	32	128

Ans. 15.65

Use NDD formula to obtain $f(4)$, and explain why it differs from the function 2^4.

23. Let $f(x)$ be a real-valued function defined on the interval $[a, b]$ and n times differentiable in the interval (a, b). If $x_0, x_1, ..., x_n$ are $(n+1)$ distinct points in interval $[a, b]$, then show that there exists

a point $c \in [a, b]$, such that $f[x_0, x_1, ..., x_n] = \dfrac{f^{(n)}(c)}{n!}$.

24. Derive an error bound in the linear approximation of a function $f(x)$.

25. Determine the spacing h in a table of equally spaced values of the function $f(x) = \sin(x)$ in the interval, $\left(0, \dfrac{\pi}{2}\right)$, so that the interpolation with a second degree polynomial in the table will yield the desired accuracy of six decimal points.

Ans. $h \leq .019827$ or $N \geq \dfrac{\dfrac{\pi}{2} - 0}{.019827} \approx 80$

26. From the following table, find $f(0.5)$, $f'(1.5)$ using Hermite interpolation

x	0	2	3
$f(x)$	2	16	80
$f'(x)$	−1	31	107

Ans. $f(0.5) = 1.5625$, $f'(1.5) = 12.5$ $\left(x^4 - x + 2 \text{ is the Hermite polynomial}\right)$

27. Apply Hermite interpolation formula to obtain a cubic polynomial which satisfies the following specifications

x	:	0.1	0.2
$f(x)$	:	0.201	0.408
$f'(x)$	:	2.03	2.12

Ans. $x^3 + 2x$

28. Calculate $f(1.2)$ by approximating the following values with cubic polynomial

$f(1) = 0, f'(1) = 1, f(2) = 0.693147, f'(2) = 0.5$

Ans. 0.0792

Finite Operators

> *Emptiness is everywhere, and it can be calculated, which gives us a great opportunity.*
> *I know how to control the universe. So, tell me, why should I run for a million.*
>
> **Grigori Yakovlevich Perelman**
> (Born: June 13, 1966)
> A mathematician who declined the field medal; he has made great contributions to
> Riemannian geometry and geometric topology.

9.1 Introduction

The finite difference operators play a central role in obtaining the interpolating polynomials for equally spaced points. Difference equations are also solved using finite differences. Most importantly, the finite difference methods for the numerical solutions of ordinary and partial differential equations with boundary conditions are based on the finite difference approximations of derivatives. This chapter presents various finite difference operators and their properties.

Definition 9.1

Consider a function $y = f(x)$ defined on an interval containing points $x + a$ and $x + b$. The finite difference is a mathematical expression of the form $f(x+b) - f(x+a)$.

The most commonly used finite differences are forward difference (Δ), backward difference (∇), and central difference (δ). We will discuss these finite differences along with average (μ), shift (E), and differential (D) operators.

Let $x_0, x_1, x_2, \ldots, x_n$ be $(n + 1)$ equidistant points in the given interval. The spacing of these points is equal and let it be given by

$$h = x_{i+1} - x_i$$
$$\Rightarrow x_i = x_0 + ih; \quad i = 1, 2, \cdots, n$$

Let $y_0 = f(x_0)$, $y_1 = f(x_1)$, $y_2 = f(x_2), \ldots, y_n = f(x_n)$ be corresponding values for the function, $y = f(x)$. The differences $y_1 - y_0$, $y_2 - y_1, \ldots, y_n - y_{n-1}$ are called finite differences.

9.2 Finite Difference Operators

In this section, we will discuss finite difference operators. Let us consider a function $y = f(x)$ defined on the interval $[a, b]$. Let x_0, x_1, $x_2, \ldots, x_n$ be a set of $(n + 1)$ equidistant points with spacing h in the given interval, and let $y_0 = f(x_0)$, $y_1 = f(x_1)$, $y_2 = f(x_2), \ldots, y_n = f(x_n)$ be corresponding values for the curve $y = f(x)$. Then, the finite difference operators, viz. forward difference (Δ), backward difference (∇), and central difference (δ), are defined as follows.

9.2.1 Forward Difference Operator (Δ)

The first forward difference operator is denoted by Δ, and defined as follows
$$\Delta y_i = y_{i+1} - y_i; \qquad i = 0, 1, \ldots, n-1$$
(or) $\quad \Delta y_0 = y_1 - y_0, \ \Delta y_1 = y_2 - y_1, \ \ldots \Delta y_{n-1} = y_n - y_{n-1}$

The forward differences Δy_0, $\Delta y_1, \ldots, \Delta y_{n-1}$ are first forward differences of y.

Further, the forward differences of first forward differences are called second forward differences and are denoted by Δ^2. The second forward differences are defined as follows

$$\Delta^2 y_i = \Delta(\Delta y_i) = \Delta y_{i+1} - \Delta y_i$$
$$= (y_{i+2} - y_{i+1}) - (y_{i+1} - y_i)$$
$$= y_{i+2} - 2y_{i+1} + y_i \qquad i = 0, 1, \ldots, n-2$$

For example, let $i = 0$, then
$$\Delta^2 y_0 = \Delta(\Delta y_0) = \Delta y_1 - \Delta y_0$$
$$= (y_2 - y_1) - (y_1 - y_0)$$
$$= y_2 - 2y_1 + y_0$$

Similarly, the forward differences of second forward differences are third forward differences Δ^3 and so on, in general

$$\Delta^m y_i = \Delta^{m-1} y_{i+1} - \Delta^{m-1} y_i; \qquad i = 0, 1, \ldots, n-m$$

9.2.2 Backward Difference Operator (∇)

The first backward difference operator (∇) is defined as follows

$$\nabla y_i = y_i - y_{i-1}; \qquad i = 1, 2, \ldots, n$$

(or) $\nabla y_1 = y_1 - y_0$, $\nabla y_2 = y_2 - y_1$, ..., $\nabla y_n = y_n - y_{n-1}$

The backward differences $\nabla y_1, \nabla y_2, \ldots, \nabla y_n$ are first backward differences of y.

The second backward differences (∇^2) are defined as follows

$$\begin{aligned}
\nabla^2 y_i &= \nabla(\nabla y_i) = \nabla y_i - \nabla y_{i-1} \\
&= (y_i - y_{i-1}) - (y_{i-1} - y_{i-2}) \\
&= y_i - 2y_{i-1} + y_{i-2} \qquad i = 2, 3, \ldots, n
\end{aligned}$$

For example

$$\begin{aligned}
\nabla^2 y_2 &= \nabla(\nabla y_2) = \nabla y_2 - \nabla y_1 \\
&= (y_2 - y_1) - (y_1 - y_0) \\
&= y_2 - 2y_1 + y_0
\end{aligned}$$

In general, we have m^{th} backward differences as follows

$$\nabla^m y_i = \nabla^{m-1} y_i - \nabla^{m-1} y_{i-1}; \qquad i = m, m+1, \ldots, n$$

9.2.3 Central Difference Operator (δ)

The first central differences (δ) are defined as follows

$$\delta y_{i+1/2} = y_{i+1} - y_i; \qquad i = 0, 1, \ldots, n-1$$

(or) $\delta y_{1/2} = y_1 - y_0$, $\delta y_{3/2} = y_2 - y_1$, ..., $\delta y_{n-1/2} = y_n - y_{n-1}$

Similarly, the second central differences (δ^2) are defined in terms of first central differences, and given by

$$\begin{aligned}
\delta^2 y_i &= \delta(\delta y_i) = \delta(y_{i+1/2} - y_{i-1/2}) \\
&= \delta y_{i+1/2} - \delta y_{i-1/2} \\
&= (y_{i+1} - y_i) - (y_i - y_{i-1}) \\
&= y_{i+1} - 2y_i + y_{i-1}; \qquad i = 1, 2, \ldots, n-1
\end{aligned}$$

For example

$$\begin{aligned}
\delta^2 y_1 &= \delta(\delta y_1) = \delta(y_{3/2} - y_{1/2}) \\
&= \delta y_{3/2} - \delta y_{1/2} \\
&= (y_2 - y_1) - (y_1 - y_0) \\
&= y_2 - 2y_1 + y_0
\end{aligned}$$

The higher order central differences are as follows

$$\delta^m y_{i-1/2} = \delta^{m-1} y_i - \delta^{m-1} y_{i-1}; \qquad \text{if } m \text{ is odd}$$

$$\delta^m y_i = \delta^{m-1} y_{i+1/2} - \delta^{m-1} y_{i-1/2}; \qquad \text{if } m \text{ is even}$$

Remark: For a given function, $y = f(x)$, the first differences are given by

$$\Delta f(x) = f(x+h) - f(x)$$
$$\nabla f(x) = f(x) - f(x-h)$$
$$\delta f(x) = f(x+h/2) - f(x-h/2)$$

Higher order differences are as follows

$$\Delta^m f(x) = \Delta^{m-1} f(x+h) - \Delta^{m-1} f(x)$$
$$\nabla^m f(x) = \nabla^{m-1} f(x) - \nabla^{m-1} f(x-h)$$
$$\delta^m f(x) = \delta^{m-1} f(x+h/2) - \delta^{m-1} f(x-h/2)$$

9.3 Average, Shift and Differential Operators

In this section, we will discuss average (μ), shift (E) and differential (D) operators. These operators are not finite difference operators, but these are useful for many applications in science and engineering.

9.3.1 Mean or Average Operator (μ)

The average operator is denoted by μ, and is given by

$$\mu y_i = \frac{1}{2}\left(y_{i+1/2} + y_{i-1/2}\right)$$

If h is the spacing in the two consecutive values of equidistant points, $x_0, x_1, x_2, ..., x_n$, then for a given function $y = f(x)$, the average operator is as follows

$$\mu f(x) = \frac{1}{2}\left(f(x+h/2) + f(x-h/2)\right)$$

9.3.2 Shift Operator (E)

The shift operator is denoted by E and is defined as follows

$$E y_i = y_{i+1}$$

If h is the spacing in the two consecutive values of equidistant points, $x_0, x_1, x_2, ..., x_n$, then for a given function $y = f(x)$, the shift operator is as follows

$$E f(x) = f(x+h)$$

Similarly, second order shift operator is defined as follows

$$E^2 f(x) = E\big(Ef(x)\big) = E\big(f(x+h)\big) = f(x+2h)$$

In general, the shift operator is given by

$$E^m f(x) = E^{m-1}\big(Ef(x)\big) = f(x+mh)$$

(or) $E^m(y_i) = y_{i+m}$

Similarly, we can define inverse shift operator as follows

$$E^{-1} y_i = y_{i-1}$$

For a given function $y = f(x)$, the inverse shift operator is as follows

$$E^{-1} f(x) = f(x-h)$$

In general, we have

$$E^{-m} f(x) = f(x-mh) \ (or) E^{-m}(y_i) = y_{i-m}$$

9.3.3 Differential Operator (D)

The differential operator is denoted by D, and defined as follows

$$Df(x) = \frac{d}{dx} f(x) = f'(x)$$

$$D^2 f(x) = \frac{d}{dx^2} f(x) = f''(x) \quad \text{etc}$$

The differential is not a finite operator, but we can relate it with the finite operator through Taylor series; we will discuss it later in this chapter.

Table 9.1 Finite Differences and Other Operators

Forward Difference Operator (Δ) $\Delta f(x) = f(x+h) - f(x)$	$\Delta f(x_i) = f(x_i + h) - f(x_i)$ $\Delta f(x_i) = f(x_{i+1}) - f(x_i)$ $\Delta y_i = y_{i+1} - y_i$
Backward Difference Operator (∇) $\nabla f(x) = f(x) - f(x-h)$	$\nabla f(x_i) = f(x_i) - f(x_i - h)$ $\nabla f(x_i) = f(x_i) - f(x_{i-1})$ $\nabla y_i = y_i - y_{i-1}$
Central Difference Operator (δ) $\delta f(x) = f(x+h/2) - f(x-h/2)$	$\delta f(x_i) = f(x_i + h/2) - f(x_i - h/2)$ $\delta f(x_i) = f(x_{i+1/2}) - f(x_{i-1/2})$ $\delta y_i = y_{i+1/2} - y_{i-1/2}$

Mean or Average Operator (μ)	
$\mu f(x) = \dfrac{1}{2}[f(x+h/2)+f(x-h/2)]$	$\mu f(x_i) = \dfrac{1}{2}\left[f(x_i+h/2)+f(x_i-h/2)\right]$ $\mu f(x_i) = \dfrac{1}{2}\left[f(x_{i+1/2})+f(x_{i-1/2})\right]$ $\mu y_i = \dfrac{1}{2}\left(y_{i+1/2}+y_{i-1/2}\right)$
Shift Operator (E) $Ef(x) = f(x+h)$	$Ef(x_i) = f(x_i+h)$ $Ef(x_i) = f(x_{i+1})$ $Ey_i = y_{i+1}$
Differential Operator (D) $Df(x) = \dfrac{d}{dx}f(x)$	$Df(x) = \dfrac{d}{dx}f(x)$

It is worth mentioning here that these are only first order differences, higher order differences are further obtained on a similar pattern. Like second order forward difference is $\Delta^2 y_i = \Delta y_{i+1} - \Delta y_i$ and so on.

It is worth mentioning here that only three major differences are used to derive the interpolating polynomials, these are forward difference (Δ), backward difference (∇) and central difference (δ). Hence, we will primarily concentrate here on these three operators (Δ, ∇ and δ). Other operators (shift, average, and differential) are not the finite differences, but these operators are used for various other applications.

9.4 Properties and Interrelations of Finite Operators

This section contains the linearity and commutative properties satisfied by various operators. Also, we will tabulate the relations among the operators and some identities.

9.4.1 Linearity and Commutative Properties

The operators defined in this chapter, (viz. forward difference (Δ), backward difference $\left(\nabla\right)$, central difference (δ), average (μ), shift (E) and differential (D) operators) satisfy the linearity and commutative properties.

Let us assume that $a_1, a_2, ..., a_n$ are constants; $L, L_1, L_2, ..., L_n$ are operators; $f(x), f_1(x), f_2(x), ..., f_n(x)$ are given functions, then following axioms hold

i) Linearity:

$$L\left[a_1 f_1(x)+a_2 f_2(x)+\cdots+a_n f_n(x)\right] = a_1 L\left[f_1(x)\right]+a_2 L\left[f_2(x)\right]+\cdots+a_n L\left[f_n(x)\right]$$
$$\left(a_1 L_1+a_2 L_2+\cdots+a_n L_n\right)f(x) = a_1 L_1\left[f(x)\right]+a_2 L_2\left[f(x)\right]+\cdots+a_n L_n\left[f(x)\right]$$

ii) Commutative:

$$L_1 L_2 f(x) = L_2 L_1 f(x)$$

$$\left[L_1 + L_2'\right]f(x) = \left[L_2 + L_1\right]f(x)$$

iii) Exponentiation:

$$L^m L^n f(x) = L^{m+n} f(x) = L^n L^m f(x)$$

9.4.2 Interrelations of Finite Operators

This subsection is mainly concerned with the interrelations of various finite operators discussed in this chapter.

Example   **9.1**

Prove the following relations

i) $E \equiv 1 + \Delta$,   ii) $\nabla \equiv 1 - E^{-1}$,   iii) $\Delta \equiv E\nabla \equiv \nabla E \equiv \delta E^{\frac{1}{2}}$,

iv) $\Delta \nabla \equiv \nabla \Delta \equiv \delta^2$,   v) $(1+\Delta)(1-\nabla) \equiv 1$,   vi) $\Delta \equiv \nabla(1-\nabla)^{-1}$,

vii) $(1+\Delta) \equiv (E-1)\nabla^{-1}$,   viii) $\delta \equiv E^{\frac{1}{2}} - E^{\frac{-1}{2}}$,   ix) $\mu \equiv \dfrac{1}{2}\left(E^{\frac{1}{2}} + E^{\frac{-1}{2}}\right)$

x) $\mu^2 \equiv 1 + \dfrac{1}{4}\delta^2$,   xi) $\mu\delta \equiv \dfrac{\Delta + \nabla}{2}$ xii) $E \equiv e^{hD}$,

xiii) $\mu \equiv \cosh\left(\dfrac{hD}{2}\right)$,   xiv) $\delta \equiv 2\sinh\left(\dfrac{hD}{2}\right)$,   xv) $\mu\delta \equiv \sinh(hD)$

xvi) $\Delta^3 y_2 = \nabla^3 y_5$.

Ans.

i) $(1+\Delta)f(x) = f(x) + \Delta f(x) = f(x) + \left(f(x+h) - f(x)\right) = f(x+h) = Ef(x)$

   $\Rightarrow E = 1 + \Delta$

ii) L.H.S $= \nabla f(x) = f(x) - f(x-h)$

   R.H.S $= \left(1 - E^{-1}\right)f(x) = f(x) - E^{-1}f(x) = f(x) - f(x-h)$

   $\Rightarrow \nabla \equiv 1 - E^{-1}$

iii) $\Delta f(x) = f(x+h) - f(x)$

   $E\left(\nabla f(x)\right) = E\left(f(x) - f(x-h)\right) = f(x+h) - f(x)$

   $\nabla\left(Ef(x)\right) = \nabla\left(f(x+h)\right) = f(x+h) - f(x)$

$$\delta\left(E^{\frac{1}{2}}f(x)\right)=\delta\big(f(x+h/2)\big)=f(x+h)-f(x)$$

$$\Rightarrow \Delta \equiv E\nabla \equiv \nabla E \equiv \delta E^{\frac{1}{2}}$$

iv) $\Delta\big(\nabla f(x)\big)=\Delta\big(f(x)-f(x-h)\big)=\Delta f(x)-\Delta f(x-h)=f(x+h)-2f(x)+f(x-h)$

$\nabla\big(\Delta f(x)\big)=\nabla\big(f(x+h)-f(x)\big)=\nabla f(x+h)-\nabla f(x)=f(x+2h)-2f(x)+f(x-h)$

$\delta^2\big(f(x)\big)=\delta\big(\delta f(x)\big)=\delta\big(f(x+h/2)-f(x-h/2)\big)=\delta\big(f(x+h/2)\big)-\delta\big(f(x-h/2)\big)$

$\qquad = f(x+2h)-2f(x)+f(x-h)$

$\qquad \Rightarrow \Delta\nabla \equiv \nabla\Delta \equiv \delta^2$

v) $(1+\Delta)(1-\nabla)\big(f(x)\big)=(1+\Delta)\big(f(x)-\nabla f(x)\big)=(1+\Delta)\big(f(x-h)\big)$

$= f(x-h)+\Delta f(x-h)=f(x)=1.f(x)$

$\Rightarrow (1+\Delta)(1-\nabla)\equiv 1$

vi) $\qquad \Delta \equiv \nabla(1-\nabla)^{-1}$ (or) $\Delta(1-\nabla)\equiv \nabla$

$\text{L.H.S.} = \Delta(1-\nabla)f(x)=\Delta\big(f(x)-\nabla f(x)\big)=\Delta\big(f(x-h)\big)$

$\qquad = f(x)-f(x-h)=\nabla f(x)=\text{R.H.S.}$

$\qquad \Rightarrow \Delta \equiv \nabla(1-\nabla)^{-1}$

vii) $\qquad (1+\Delta)\equiv (E-1)\nabla^{-1}$ (or) $(1+\Delta)\nabla \equiv (E-1)$

$(1+\Delta)\nabla f(x)=(1+\Delta)\big(f(x)-f(x-h)\big)=\big(f(x)-f(x-h)\big)+\Delta\big(f(x)-f(x-h)\big)$

$\qquad = \big(f(x)-f(x-h)\big)+\big(f(x+h)-f(x)\big)-\big(f(x)-f(x-h)\big)$

$\qquad = f(x+h)-f(x)$

$(E-1)f(x)=Ef(x)-f(x)=f(x+h)-f(x)$

$\qquad \Rightarrow (1+\Delta)\equiv (E-1)\nabla^{-1}$

viii) $\delta f(x)=f(x+h/2)-f(x-h/2)=\left(E^{\frac{1}{2}}-E^{\frac{-1}{2}}\right)f(x)$

$$\Rightarrow \delta \equiv E^{\frac{1}{2}}-E^{\frac{-1}{2}}$$

ix) $\mu f(x)=\dfrac{1}{2}\big(f(x+h/2)+f(x-h/2)\big)=\dfrac{1}{2}\left(E^{\frac{1}{2}}+E^{\frac{-1}{2}}\right)f(x)$

$$\Rightarrow \mu \equiv \dfrac{1}{2}\left(E^{\frac{1}{2}}+E^{\frac{-1}{2}}\right)$$

x) $\mu^2 f(x)=\dfrac{1}{4}\left(E^{\frac{1}{2}}+E^{\frac{-1}{2}}\right)^2 f(x)=\dfrac{1}{4}\left(\left(E^{\frac{1}{2}}+E^{\frac{-1}{2}}\right)^2+4\right)f(x)=\left(1+\dfrac{1}{4}\delta^2\right)f(x)$

$$\Rightarrow \mu^2 \equiv 1+\dfrac{1}{4}\delta^2$$

xi) L.H.S. $= \mu\delta f(x) = \mu\big(f(x+h/2) - f(x-h/2)\big) = \mu\big(f(x+h/2)\big) - \mu\big(f(x-h/2)\big)$

$$= \frac{1}{2}\big(f(x+h) + f(x)\big) - \frac{1}{2}\big(f(x) + f(x-h)\big) = \frac{1}{2}\big(f(x+h) - f(x-h)\big)$$

R.H.S. $= \dfrac{\Delta + \nabla}{2} f(x) = \dfrac{1}{2}\big(\Delta f(x) + \nabla f(x)\big) = \dfrac{1}{2}\big(f(x+h) - f(x-h)\big)$

$$\Rightarrow \mu\delta \equiv \frac{\Delta + \nabla}{2}$$

xii) L.H.S. $= Ef(x) = f(x+h)$

On using Taylor series, we have

$$Ef(x) = f(x+h) = f(x) + hf'(x) + \frac{h^2}{2!}f''(x) + \frac{h^3}{3!}f'''(x) + \cdots$$

$$= \left(1 + hD + \frac{h^2}{2!}D^2 + \frac{h^3}{3!}D^3 + \cdots\right)f(x)$$

$$= e^{hD} f(x) = \text{R.H.S.}$$

$$\Rightarrow E \equiv e^{hD}$$

xiii) $\mu f(x) = \dfrac{1}{2}\left(E^{\frac{1}{2}} + E^{\frac{-1}{2}}\right)f(x) = \dfrac{1}{2}\left(e^{\frac{hD}{2}} + e^{\frac{-hD}{2}}\right)f(x) = \cosh\left(\dfrac{hD}{2}\right)f(x)$

$$\Rightarrow \mu \equiv \cosh\left(\frac{hD}{2}\right)$$

xiv) $\delta f(x) = \left(E^{\frac{1}{2}} - E^{\frac{-1}{2}}\right)f(x) = \left(e^{\frac{hD}{2}} - e^{\frac{-hD}{2}}\right)f(x) = 2\sinh\left(\dfrac{hD}{2}\right)f(x)$

$$\Rightarrow \delta \equiv 2\sinh\left(\frac{hD}{2}\right)$$

xv) Using results in xiii) and xiv), we have

$$\mu\delta \equiv 2\cosh\left(\frac{hD}{2}\right)\sinh\left(\frac{hD}{2}\right) \equiv \sinh(hD)$$

xvi) $\Delta^3 y_2 = \nabla^3 y_5$

$$\Delta^3 y_2 = \Delta^2\big(\Delta y_2\big) = \Delta^2\big(y_3 - y_2\big) = \Delta\big(\Delta(y_3 - y_2)\big) = \Delta\big(\Delta y_3 - \Delta y_2\big)$$

$$= \Delta\big(y_4 - 2y_3 + y_2\big) = \Delta y_4 - 2\Delta y_3 + \Delta y_2 = y_5 - 3y_4 + 3y_3 - y_2$$

$$\nabla^3 y_5 = \nabla^2\big(\nabla y_5\big) = \nabla^2\big(y_5 - y_4\big) = \nabla\big(\nabla(y_5 - y_4)\big) = \nabla\big(\nabla y_5 - \nabla y_4\big)$$

$$= \nabla\big(y_5 - 2y_4 + y_3\big) = \nabla y_5 - 2\nabla y_4 + \nabla y_3 = y_5 - 3y_4 + 3y_3 - y_2$$

$$\Rightarrow \Delta^3 y_2 = \nabla^3 y_5$$

Table 9.2 Relations between the Operators

	E	Δ	∇	δ	μ	hD
E	E	$1+\Delta$	$(1-\nabla)^{-1}$	$1+\dfrac{\delta^2}{2}+\delta\sqrt{1+\dfrac{\delta^2}{4}}$	$\left(\mu+\sqrt{\mu^2-1}\right)^2$	e^{hD}
Δ	$E-1$	Δ	$(1-\nabla)^{-1}-1$	$\dfrac{\delta^2}{2}+\delta\sqrt{1+\dfrac{\delta^2}{4}}$	$\left(\mu+\sqrt{\mu^2-1}\right)^2-1$	$e^{hD}-1$
∇	$1-E^{-1}$	$1-(1+\Delta)^{-1}$	∇	$-\dfrac{\delta^2}{2}+\delta\sqrt{1+\dfrac{\delta^2}{4}}$	$1-\left(\mu+\sqrt{\mu^2-1}\right)^{-2}$	$1-e^{-hD}$
δ	$E^{\frac{1}{2}}-E^{-\frac{1}{2}}$	$\Delta(1+\Delta)^{\frac{1}{2}}$	$\nabla(1-\nabla)^{\frac{1}{2}}$	δ	$2\sqrt{\mu^2-1}$	$2\sinh(hD/2)$
μ	$\dfrac{1}{2}\left(E^{\frac{1}{2}}+E^{-\frac{1}{2}}\right)$	$\left(1+\dfrac{\Delta}{2}\right)(1+\Delta)^{\frac{1}{2}}$	$\left(1-\dfrac{\nabla}{2}\right)(1-\nabla)^{\frac{1}{2}}$	$\sqrt{1+\dfrac{\delta^2}{4}}$	μ	$\cosh(hD/2)$
hD	$\log E$	$\log(1+\Delta)$	$-\log(1-\nabla)$	$2\sinh^{-1}(\delta/2)$	$\sinh^{-1}\left(2\mu\sqrt{\mu^2-1}\right)$	hD

9.5 Operators on Some Functions

In this section, we will apply finite operators on some functions especially on polynomials. Forward difference operator has been considered for most of the cases, but similar kind of results can also be proved for backward and central differences.

Example   **9.2**

Prove that

$$\Delta^n \ln(ax) = \ln(a(x+nh)) - {}^nC_1 \ln(a(x+(n-1)h)) + {}^nC_2 \ln(a(x+(n-2)h)) - \cdots$$
$$+ (-1)^r \, {}^nC_r \ln(a(x+(n-r)h)) + \cdots + (-1)^{n-1} \, {}^nC_{n-1} \ln(a(x+h)) + (-1)^n \ln(ax)$$

Ans.

$$\Delta^n \ln(ax) = \Delta^{n-1}\left(\Delta \ln(ax)\right) = \Delta^{n-1}\left(\ln(a(x+h)) - \ln(ax)\right) = \Delta^{n-2}\left(\Delta \ln(a(x+h)) - \Delta \ln(ax)\right)$$

$$= \Delta^{n-2}\left(\ln(a(x+2h)) - 2\ln(a(x+h)) + \ln(ax)\right)$$

$$= \Delta^{n-3}\left(\ln(a(x+3h)) - 3\ln(a(x+2h)) + 3\ln(a(x+h)) - \ln(ax)\right)$$

$$= \Delta^{n-4}\left(\ln(a(x+4h)) - 4\ln(a(x+3h)) + 6\ln(a(x+2h)) - 4\ln(a(x+h)) + \ln(ax)\right)$$

$$\vdots$$

$$= \ln(a(x+nh)) - {}^nC_1 \ln(a(x+(n-1)h)) + {}^nC_2 \ln(a(x+(n-2)h)) - \cdots$$
$$+ (-1)^r \, {}^nC_r \ln(a(x+(n-r)h)) + \cdots + (-1)^{n-1} \, {}^nC_{n-1} \ln(a(x+h)) + (-1)^n \ln(ax)$$

Example   **9.3**

Prove that

$$\nabla\left[f(x)g(x)\right] = f(x)\nabla g(x) + g(x)\nabla f(x) - \nabla f(x)\nabla g(x)$$

Ans.

$$\nabla\left[f(x)g(x)\right] = f(x)g(x) - f(x-h)g(x-h)$$

$$= f(x)g(x) - f(x-h)g(x) + f(x-h)g(x) - f(x-h)g(x-h)$$

$$= g(x)\left(f(x) - f(x-h)\right) + f(x-h)\left(g(x) - g(x-h)\right)$$

$$= g(x)\nabla f(x) + f(x-h)\nabla g(x)$$

$$= g(x)\nabla f(x) + \left(f(x) - \nabla f(x)\right)\nabla g(x)$$

$$= f(x)\nabla g(x) + g(x)\nabla f(x) - \nabla f(x)\nabla g(x)$$

Similarly, we can easily prove the result

$$\Delta\left[f(x)g(x)\right] = f(x)\Delta g(x) + g(x)\Delta f(x) + \Delta f(x)\Delta g(x)$$

Example ———————————————————————— 9.4

Prove the following results

a) $\Delta(c) = 0$, where c is a constant

b) $\Delta(cx) = c\Delta(x) = ch$ and $\Delta^2(cx) = 0$

c) $\Delta(cx^2) = c\Delta(x^2) = c(2hx + h^2)$, $\Delta^2(cx^2) = 2ch^2$ and $\Delta^3(cx^2) = 0$

d) $\Delta(cx^n) = c\Delta(x^n) = c\left({}^nC_1 x^{n-1}h + {}^nC_2 x^{n-2}h^2 + {}^nC_3 x^{n-3}h^3 + \ldots + {}^nC_{n-1}x\, h^{n-1} + h^n\right)$

e) $\Delta^n(cx^n) = ch^n n!$

f) Use all these results to find the nth finite differences of a polynomial of degree n.

Ans.

a) $\Delta(c) = c - c = 0$

b) $\Delta(cx) = c\Delta(x)$ (linearity property)

$c\Delta(x) = c(x + h - x) = ch$

$\Delta^2(cx) = \Delta\Delta(cx) = \Delta(ch) = ch - ch = 0$

Similarly, third and higher order differences are zero.

c) $\Delta(cx^2) = c\Delta(x^2) = c\left((x+h)^2 - x^2\right) = c\left(2hx + h^2\right)$

$\Delta^2(cx^2) = c\Delta\Delta(x^2) = c\Delta\left(2hx + h^2\right) = c\left(2h(x+h) + h^2 - 2hx - h^2\right) = 2ch^2$

$\Delta^3(cx^2) = \Delta\Delta^2(cx^2) = \Delta\left(2ch^2\right) = 0$

Fourth and higher order differences are also zero.

d) $\Delta(cx^n) = c\Delta(x^n)$

$= c\left((x+h)^n - x^n\right)$

$= c\left((x^n + {}^nC_1 x^{n-1}h + {}^nC_2 x^{n-2}h^2 + {}^nC_3 x^{n-3}h^3 + \ldots + {}^nC_{n-1}x\, h^{n-1} + h^n) - x^n\right)$

$= c\left({}^nC_1 x^{n-1}h + {}^nC_2 x^{n-2}h^2 + {}^nC_3 x^{n-3}h^3 + \ldots + {}^nC_{n-1}x\, h^{n-1} + h^n\right)$

e) We will use the principle of mathematical induction to prove the result $\Delta^n(cx^n) = ch^n n!$.

From part b) and c), the result is true for $n = 1, 2$.

Let this be true for $n - 1$, i.e.

$$\Delta^{n-1}(cx^{n-1}) = ch^{n-1}(n-1)!$$

Also, we have

$$\Delta^k(x^j) = 0 \text{ for all } k > j$$

It means all differences of order $k + 1$ and higher, are zeroes for the polynomial x^k.

For n, we have

$$\Delta^n(cx^n) = c\Delta^n(x^n)$$

$$= \Delta^{n-1}\Delta(cx^n)$$

$$= c\Delta^{n-1}({}^nC_1 x^{n-1}h + {}^nC_2 x^{n-2}h^2 + {}^nC_3 x^{n-3}h^3 + \ldots + {}^nC_{n-1}x\ h^{n-1} + h^n)$$

$$= c({}^nC_1 h\ \Delta^{n-1}(x^{n-1}) + {}^nC_2 h^2\Delta^{n-1}(x^{n-2}) + {}^nC_3 h^3\Delta^{n-1}(x^{n-3}) + \ldots$$

$$+ {}^nC_{n-1}h^{n-1}\Delta^{n-1}(x) + \Delta^{n-1}(h^n))$$

Since all the differences of order $k + 1$ and higher are zeroes for x^k, therefore, we have

$$\Delta^n(cx^n) = c\left({}^nC_1 h\ \Delta^{n-1}(x^{n-1})\right)$$

$$= c\left({}^nC_1 h\ \left(h^{n-1}(n-1)!\right)\right)$$

$$= ch^n n!$$

Hence, it is also true for all n by the principle of mathematical induction.

f)   Let $P_n(x) = a_0 + a_1 x + a_2 x^2 + \cdots + a_n x^n$

$$\Delta^n\left(P_n(x)\right) = \Delta^n\left(a_0 + a_1 x + a_2 x^2 + \cdots + a_n x^n\right)$$

$$= \Delta^n a_0 + a_1\Delta^n x + a_2\Delta^n x^2 + \cdots + a_n\Delta^n x^n$$

Since all the differences of order $k + 1$ and higher are zeroes for x^k, thus we have

$$\Delta^n\left(P_n(x)\right) = a_n\Delta^n x^n,$$

On using the result $\Delta^n(cx^n) = ch^n n!$ of part d), we have

$$\Delta^n\left(P_n(x)\right) = a_n h^n n!$$

Note: On a similar pattern, it is easy to prove that

i) $\nabla^n\left(P_n(x)\right) = a_n h^n n!$

ii) $\delta^n\left(P_n(x)\right) = a_n h^n n!$

iii)   All the backward and central differences of order $k + 1$ and higher are zeroes for x^k,

---

Example   **9.5**

Evaluate $\delta^6\left[(1-x)(1-x^2)(1-x^3)\right]$.

Ans.

Since all the central differences of order $k + 1$ and higher are zeroes for x^k, hence we have

$$\delta^6\left[(1-x)(1-x^2)(1-x^3)\right] = \delta^6(-x^6) = -6!\,h^6$$

9.6 Newton Divided Differences and Other Finite Differences

In this section, we will derive the relations between the divided differences and the other finite differences. In the next chapter, we will use these relations to obtain various interpolation formulas for equally spaced points.

> **Lemma 9.1**
>
> For equispaced points $x_i = x_0 + ih$; $i = 1, 2, \cdots, n$, the following relation holds between forward differences and divided differences

$$f\left[x_i, x_{i+1}, x_{i+2}, \ldots x_{i+m}\right] = \frac{1}{m! \, h^m} \Delta^m f(x_i) \qquad (9.1)$$

Proof: We will prove the above result with the aid of mathematical induction:
For $m = 1$:

$$f\left[x_i, x_{i+1}\right] = \frac{f(x_{i+1}) - f(x_i)}{x_{i+1} - x_i} = \frac{f(x_{i+1}) - f(x_i)}{h} = \frac{1}{h} \Delta f(x_i)$$

$$\Rightarrow f\left[x_0, x_1\right] = \frac{1}{h} \Delta f(x_0)$$

So, it is true for $m = 1$.
For $m = 2$:

$$f\left[x_i, x_{i+1}, x_{i+2}\right] = \frac{f\left[x_{i+1}, x_{i+2}\right] - f\left[x_i, x_{i+1}\right]}{x_{i+2} - x_i}$$

$$= \frac{f\left[x_{i+1}, x_{i+2}\right] - f\left[x_i, x_{i+1}\right]}{2h}$$

$$= \frac{1}{2h}\left(\frac{f(x_{i+2}) - f(x_{i+1})}{x_{i+2} - x_{i+1}} - \frac{f(x_{i+1}) - f(x_i)}{x_{i+1} - x_i}\right)$$

$$= \frac{1}{2h}\left(\frac{1}{h} \Delta f(x_{i+1}) - \frac{1}{h} \Delta f(x_i)\right) = \frac{1}{2! \, h^2} \Delta^2 f(x_i)$$

$$\Rightarrow f\left[x_0, x_1, x_2\right] = \frac{1}{2! \, h^2} \Delta^2 f(x_0)$$

So, it is also true for $m = 2$.

[Many authors simply generalize the above result up to any finite number of indices, but here we shall prove the generalization in a logical manner by using mathematical induction]

Let the relation be true for $m = k$, i.e.,

$$f\left[x_i, x_{i+1}, x_{i+2}, \ldots x_{i+k}\right] = \frac{1}{k! \, h^k} \Delta^k f(x_i) \qquad (9.2)$$

Since we do not have any restriction on the subscript i (it may take any value from 0 to m), the following relation is also true by simply replacing i with $i + 1$:

$$f\left[x_{i+1}, x_{i+2}, x_{i+3}, \ldots x_{i+k+1}\right] = \frac{1}{k!\, h^k} \Delta^k f(x_{i+1}) \tag{9.3}$$

Now, we have to prove that the relation is true for $m = k + 1$. The $k + 1$ divided differences are given by

$$f\left[x_i, x_{i+1}, x_{i+2}, \ldots x_{i+k+1}\right] = \frac{f\left[x_{i+1}, x_{i+2}, \ldots x_{i+k+1}\right] - f\left[x_i, x_{i+1}, x_{i+2}, \ldots x_{i+k}\right]}{x_{i+k+1} - x_i}$$

$$= \frac{\dfrac{1}{k!\, h^k} \Delta^k f(x_{i+1}) - \dfrac{1}{k!\, h^k} \Delta^k f(x_i)}{(k+1)h}$$

$$= \frac{1}{(k+1)!\, h^{k+1}} \left(\Delta^k f(x_{i+1}) - \Delta^k f(x_i)\right)$$

$$= \frac{1}{(k+1)!\, h^{k+1}} \Delta^{k+1} f(x_i)$$

The relation is true for $m = k + 1$, whenever it is true for $m = k$. Therefore, by the principle of mathematical induction, it is true for each m

$$\Rightarrow f\left[x_0, x_1, x_2, \ldots x_m\right] = \frac{1}{m!\, h^m} \Delta^m f(x_0)$$

Note: On a similar pattern, it is easy to prove the following relations between backward differences and divided differences

$$f\left[x_i, x_{i+1}, x_{i+2}, \ldots x_{i+m}\right] = \frac{1}{m!\, h^m} \nabla^m f(x_{i+m})$$

$$\Rightarrow f\left[x_0, x_1, x_2, \ldots x_m\right] = \frac{1}{m!\, h^m} \nabla^m f(x_m)$$

Similarly, relations between central differences and divided differences are as follows

$$f\left[x_i, x_{i+1}, x_{i+2}, \ldots x_{i+2m}\right] = \frac{1}{2m!\, h^{2m}} \delta^{2m} f\left(x_{i+m}\right)$$

$$f\left[x_i, x_{i+1}, x_{i+2}, \ldots x_{i+2m+1}\right] = \frac{1}{(2m+1)!\, h^{2m+1}} \delta^{2m+1} f\left(x_{i+m+\frac{1}{2}}\right)$$

Let $f(x)$ be a well-defined real-valued function on the interval $[a, b]$ containing $(m+1)$ – points, $x_0, x_1, \cdots, x_m$. If the function $f(x)$ is m-times differentiable in (a, b), then there exists a point, $\xi \in (a, b)$, such that

$$\frac{1}{h^m} \Delta^m f(x_0) = f^m(\xi)$$

Proof: From theorem 8.1, we have

$$f\left[x_0, x_1, x_2, x_3, \ldots x_m\right] = \frac{f^m(\xi)}{m!}$$

On using expression (9.1), we can easily obtain the required result, i.e.,

$$\frac{1}{h^m} \Delta^m f(x_0) = f^m(\xi)$$

Note: Similarly, following results can also be proved for backward and central differences:

$$\frac{1}{h^m} \nabla^m f\left(x_m\right) = f^m(\xi)$$

$$\frac{1}{h^{2m}} \delta^{2m} f\left(x_{i+m}\right) = f^{2m}(\xi)$$

$$\frac{1}{h^{2m+1}} \delta^{2m+1} f\left(x_{i+m+\frac{1}{2}}\right) = f^{2m+1}(\xi)$$

9.7 Finite Difference Tables and Error Propagation

In this section, we will focus on the construction of the finite difference tables for three finite differences (Δ, ∇ and δ) and then examine the error propagation in these tables.

It is easy to construct various difference tables with the aid of the finite difference formulas given the Table 9.1, like forward difference, $\Delta y_i = y_{i+1} - y_i$, backward difference, $\nabla y_i = y_i - y_{i-1}$, and central difference, $\delta y_i = y_{i+1/2} - y_{i-1/2}$. We have similar formulas for higher order differences like $\Delta^2 y_i = \Delta y_{i+1} - \Delta y_i$, etc.

Table 9.3 Forward Differences

x	$y = f(x)$	First forward differences (Δ)	Second forward differences (Δ^2)	Third forward differences (Δ^3)	Fourth forward differences (Δ^4)
x_0	y_0				
		$y_1 - y_0 = \Delta y_0$			
x_1	y_1		$\Delta y_1 - \Delta y_0 = \Delta^2 y_0$		
		$y_2 - y_1 = \Delta y_1$		$\Delta^2 y_1 - \Delta^2 y_0 = \Delta^3 y_0$	
x_2	y_2		$\Delta y_2 - \Delta y_1 = \Delta^2 y_1$		$\Delta^3 y_1 - \Delta^3 y_0 = \Delta^4 y_0$
		$y_3 - y_2 = \Delta y_2$		$\Delta^2 y_2 - \Delta^2 y_1 = \Delta^3 y_1$	$\vdots$
x_3	y_3		$\Delta y_3 - \Delta y_2 = \Delta^2 y_2$	$\vdots$	
		$y_4 - y_3 = \Delta y_3$	$\vdots$		
x_4	y_4	$\vdots$			
$\vdots$	$\vdots$				
x_{n-1}	y_{n-1}				
		$y_n - y_{n-1} = \Delta y_{n-1}$			
x_n	y_n				

Table 9.4 Backward Differences

x	$y = f(x)$	First backward differences (∇)	Second backward differences (∇^2)	Third backward differences (∇^3)	Fourth backward differences (∇^4)
x_0	y_0				
		$y_1 - y_0 = \nabla y_1$			
x_1	y_1		$\nabla y_2 - \nabla y_1 = \nabla^2 y_2$		
		$y_2 - y_1 = \nabla y_2$		$\nabla^2 y_3 - \nabla^2 y_2 = \nabla^3 y_3$	
x_2	y_2		$\nabla y_3 - \nabla y_2 = \nabla^2 y_3$		$\nabla^3 y_4 - \nabla^3 y_3 = \nabla^4 y_4$
		$y_3 - y_2 = \nabla y_3$		$\nabla^2 y_4 - \nabla^2 y_3 = \nabla^3 y_4$	$\vdots$
x_3	y_3		$\nabla y_4 - \nabla y_3 = \nabla^2 y_4$	$\vdots$	
		$y_4 - y_3 = \nabla y_4$	$\vdots$		
x_4	y_4	$\vdots$			
$\vdots$	$\vdots$				
x_{n-1}	y_{n-1}				
		$y_n - y_{n-1} = \nabla y_n$			
x_n	y_n				

Table 9.5 Central Differences

x	$y = f(x)$	δ	δ^2	δ^3	δ^4
x_0	y_0				
		$y_1 - y_0 = \delta y_{1/2}$			
x_1	y_1		$\delta y_{3/2} - \delta y_{1/2} = \delta^2 y_1$		
		$y_2 - y_1 = \delta y_{3/2}$		$\delta^2 y_2 - \delta^2 y_1 = \delta^3 y_{3/2}$	
x_2	y_2		$\delta y_{5/2} - \delta y_{3/2} = \delta^2 y_2$		$\delta^3 y_{5/2} - \delta^3 y_{3/2} = \delta^4 y_2$
		$y_3 - y_2 = \delta y_{5/2}$		$\delta^2 y_3 - \delta^2 y_2 = \delta^3 y_{5/2}$	
x_3	y_3		$\delta y_{7/2} - \delta y_{5/2} = \delta^2 y_4$	$\vdots$	
		$y_4 - y_3 = \delta y_{7/2}$	$\vdots$		
x_4	y_4	$\vdots$			
$\vdots$	$\vdots$				

From these three Tables (9.3–9.5), it is easy to notice that the values in the difference table are same and expressed in terms of different differences, like

$$y_1 - y_0 = \Delta y_0 = \nabla y_1 = \delta y_{1/2}$$

Hence, we have only one difference table expressed in terms of forward, backward and central differences.

Example ────────────────────────────────────── 9.6

Construct the finite difference table for the following data set

x	0	15	30	45	60	75	90
y	0	0.258819	0.5	0.707107	0.866025	0.965926	1

Ans.

We have following values of variables x and y

$$x_0 = 0, \; x_1 = 15, \; x_2 = 30, \ldots, x_6 = 90$$
$$y_0 = 0, \; y_1 = 0.258819, \; y_2 = 0.5, \ldots, y_6 = 1$$

Different finite differences can be obtained as follows

First difference, $d_{i,1} = y_{i+1} - y_i$, $i = 0, 1, 2, \ldots, 6$

Second difference, $d_{i,2} = d_{i+1,1} - d_{i,1}$, $i = 0, 1, 2, \ldots, 5$

Third difference, $d_{i,3} = d_{i+1,2} - d_{i,2}$, $i = 0, 1, 2, \ldots, 4$ and so on.

x	y	$\Delta(\nabla, \delta)$	$\Delta^2(\nabla^2, \delta^2)$	$\Delta^3(\nabla^3, \delta^3)$	$\Delta^4(\nabla^4, \delta^4)$	$\Delta^5(\nabla^5, \delta^5)$	$\Delta^6(\nabla^6, \delta^6)$
0	0						
		0.258819					
15	0.258819		−0.017638				
		0.241181		−0.016436			
30	0.500000		−0.034074		0.002321		
		0.207107		−0.014115		0.000966	
45	0.707107		−0.048189		0.003287		−0.000235
		0.158918		−0.010828		0.000731	
60	0.866025		−0.059017		0.004018		
		0.099901		−0.006810			
75	0.965926		−0.065827				
		0.034074					
90	1						

All three finite differences can be represented by one table, and we can treat it as forward, backward or central difference table according to our requirement. More precisely the difference between these finite differences can only be viewed from their respective relative positions in the table, but they all are same in the absolute form.

This table as forward, backward and central difference table can be represented as follows

Forward Difference Table:

x	y	Δ	Δ^2	Δ^3	Δ^4	Δ^5	Δ^6
$0(x_0)$	$0(y_0)$						
		$0.258819(\Delta y_0)$					
$15(x_1)$	$0.258819(y_1)$		$-0.017638(\Delta^2 y_0)$				
		$0.241181(\Delta y_1)$		$-0.016436(\Delta^3 y_0)$			
$30(x_2)$	$0.500000(y_2)$		$-0.034074(\Delta^2 y_1)$		$0.002321(\Delta^4 y_0)$		
		$0.207107(\Delta y_2)$		$-0.014115(\Delta^3 y_1)$		$0.000966(\Delta^5 y_0)$	
$45(x_3)$	$0.707107(y_3)$		$-0.048189(\Delta^2 y_2)$		$0.003287(\Delta^4 y_1)$		$-0.000235(\Delta^6 y_0)$
		$0.158918(\Delta y_3)$		$-0.010828(\Delta^3 y_2)$		$0.000731(\Delta^5 y_1)$	
$60(x_4)$	$0.866025(y_4)$		$-0.059017(\Delta^2 y_3)$		$0.004018(\Delta^4 y_2)$		
		$0.099901(\Delta y_4)$		$-0.006810(\Delta^3 y_3)$			
$75(x_5)$	$0.965926(y_5)$		$-0.065827(\Delta^2 y_4)$				
		$0.034074(\Delta y_5)$					
$90(x_6)$	$1(y_6)$						

Backward Difference Table:

x	y	∇	∇^2	∇^3	∇^4	∇^5	∇^6
$0(x_0)$	$0(y_0)$						
		$0.258819(\nabla y_1)$					
$15(x_1)$	$0.258819(y_1)$		$-0.017638(\nabla^2 y_2)$				
		$0.241181(\nabla y_2)$		$-0.016436(\nabla^3 y_3)$			
$30(x_2)$	$0.500000(y_2)$		$-0.034074(\nabla^2 y_3)$		$0.002321(\nabla^4 y_4)$		
		$0.207107(\nabla y_3)$		$-0.014115(\nabla^3 y_4)$		$0.000966(\nabla^5 y_5)$	
$45(x_3)$	$0.707107(y_3)$		$-0.048189(\nabla^2 y_4)$		$0.003287(\nabla^4 y_5)$		$-0.000235(\nabla^6 y_6)$
		$0.158918(\nabla y_4)$		$-0.010828(\nabla^3 y_5)$		$0.000731(\nabla^5 y_6)$	
$60(x_4)$	$0.866025(y_4)$		$-0.059017(\nabla^2 y_5)$		$0.004018(\nabla^4 y_6)$		
		$0.099901(\nabla y_5)$		$-0.006810(\nabla^3 y_6)$			
$75(x_5)$	$0.965926(y_5)$		$-0.065827(\nabla^2 y_6)$				
		$0.034074(\nabla y_6)$					
$90(x_6)$	$1(y_6)$						

Central Difference Table:

x	y	δ	δ^2	δ^3	δ^4	δ^5	δ^6
$0(x_0)$	$0(y_0)$						
		$0.258819(\delta y_{1/2})$					
$15(x_1)$	$0.258819(y_1)$		$-0.017638(\delta^2 y_1)$				
		$0.241181(\delta y_{3/2})$		$-0.016436(\delta^3 y_{3/2})$			
$30(x_2)$	$0.500000(y_2)$		$-0.034074(\delta^2 y_2)$		$0.002321(\delta^4 y_2)$		
		$0.207107(\delta y_{5/2})$		$-0.014115(\delta^3 y_{5/2})$		$0.000966(\delta^5 y_{5/2})$	
$45(x_3)$	$0.707107(y_3)$		$-0.048189(\delta^2 y_3)$		$0.003287(\delta^4 y_3)$		$-0.000235(\delta^6 y_3)$
		$0.158918(\delta y_{7/2})$		$-0.010828(\delta^3 y_{7/2})$		$0.000731(\delta^5 y_{7/2})$	
$60(x_4)$	$0.866025(y_4)$		$-0.059017(\delta^2 y_4)$		$0.004018(\delta^4 y_4)$		
		$0.099901(\delta y_{9/2})$		$-0.006810(\delta^3 y_{9/2})$			
$75(x_5)$	$0.965926(y_5)$		$-0.065827(\delta^2 y_5)$				
		$0.034074(\delta y_{11/2})$					
$90(x_6)$	$1(y_6)$						

It is easy to see that the underline values can be defined in terms of different finite differences as follows

$$0.207107 = \Delta y_2 = \nabla y_3 = \delta y_{5/2}$$

$$-0.034074 = \Delta^2 y_1 = \nabla^2 y_3 = \delta^2 y_2$$

$$0.003287 = \Delta^4 y_1 = \nabla^4 y_5 = \delta^4 y_3$$

$$-0.000235 = \Delta^6 y_0 = \nabla^6 y_6 = \delta^6 y_3$$

Note: It is easy to verify that the data set is taken for the function, $y = \sin x$, where x is in degree.

Example ————————————————————————————————————— 9.7

Construct the finite difference table for the function $y = \ln x$ on interval $[1, 2]$ by taking step size equal to 0.25.

Ans.

x	$\ln(x)$	$\Delta(\nabla\ \delta)$	$\Delta^2(\nabla^2\delta^2)$	$\Delta^3(\nabla^3\delta^3)$	$\Delta^4(\nabla^4\delta^4)$
1	0				
		0.223143551			
1.25	0.223143551		−0.040821995		
		0.182321557		0.012651118	
1.5	0.405465108		−0.028170877		−0.005099528
		0.15415068		0.00755159	
1.75	0.559615788		−0.020619287		
		0.133531393			
2	0.693147181				

Example ————————————————————————————————————— 9.8

The input and output values of a certain experiment are given below except one output value.

$$\text{Input}: \quad 50 \quad 100 \quad 150 \quad 200 \quad 250$$
$$\text{Output}: \quad 4.52 \quad 6.78 \quad 9.31 \quad ? \quad 16.54$$

Compute that value by constructing finite difference table.

Ans.
We have following values of x and y

$$x_0 = 50, \ x_1 = 100, \ x_2 = 150, \ x_3 = 200, \ x_4 = 250$$
$$y_0 = 4.52, \ y_1 = 6.78, \ y_2 = 9.31, \ y_3 = ?, \ y_4 = 16.54$$

A polynomial of degree three can be fitted to these four data points. For polynomial of degree three, the fourth forward difference $\Delta^4 y_0$ must be zero, this implies

$$\Delta^4 y_0 = (E-1)^4 y_0$$
$$= \left(E^4 - 4E^3 + 6E^2 - 4E + 1\right) y_0$$
$$= y_4 - 4y_3 + 6y_2 - 4y_1 + y_0$$
$$= 16.54 - 4y_3 + 6(9.31) - 4(6.78) + 4.52$$
$$= 49.8 - 4y_3$$
$$\Rightarrow 49.8 - 4y_3 = 0 \text{ or } y_3 = \frac{49.8}{4} = 12.45$$

Error in Difference Table:

It is easy to see that, once an error is committed in tabulated value, it will propagate and amplify in subsequent entries. An illustration (by introducing error ε in the value of y_3) from forward difference table can be seen as follows

x	$y = f(x)$	Δ	Δ^2	Δ^3	Δ^4
x_0	y_0				
		Δy_0			
x_1	y_1		$\Delta^2 y_0$		
		Δy_1		$\Delta^3 y_0 + \varepsilon$	
x_2	y_2		$\Delta^2 y_1 + \varepsilon$		$\Delta^4 y_0 - 4\varepsilon$
		$\Delta y_2 + \varepsilon$		$\Delta^3 y_1 - 3\varepsilon$	
x_3	$y_3 + \varepsilon$		$\Delta^2 y_2 - 2\varepsilon$		$\Delta^4 y_1 + 6\varepsilon$
		$\Delta y_3 - \varepsilon$		$\Delta^3 y_2 + 3\varepsilon$	
x_4	y_4		$\Delta^2 y_3 + \varepsilon$		$\Delta^4 y_2 - 4\varepsilon$
		Δy_4		$\Delta^3 y_3 - \varepsilon$	$\vdots$
x_5	y_5		$\Delta^2 y_4$	$\vdots$	
		Δy_5	$\vdots$		
x_6	y_6	$\vdots$			
$\vdots$	$\vdots$				

It is easy to see that the error (ε) propagates in the difference table in a Binomial manner i.e. $[(1-1)^n]$. For example, the error in first difference is given by $\varepsilon(1, -1)$. So, binomially, it is $\varepsilon(1, -2, 1)$ in second difference; $\varepsilon(1, -3, 3, -1)$ in third difference; $\varepsilon(1, -4, 6, -4, 1)$ in fourth difference, and so on.

Example   9.9

The following data is calculated from a polynomial of degree three, but there is some error in data, locate the error and remove it to obtain correct data set.

x:	-1	0	1	2	3	4
y:	-2	1	4	15	34	73

Ans.

Since the data is obtained from a polynomial of degree 3, therefore all the fourth and higher order differences must be zero. We will construct the forward difference table for the data set to locate and correct the error.

x	y	Δ	Δ^2	Δ^3	Δ^4
−1	−2				
		3			
0	1		0		
		3		8	
1	4		8		−8
		11		0	
2	15		8		12
		19		12	
3	34		20		
		39			
4	73				

The fourth differences must vanish for a polynomial of degree 3. The error terms propagate in a Binomial manner (for fourth differences $\varepsilon(1, -4, 6, -4, 1)$). It is easy to see that in fourth differences, we have $(-8, 12) \sim 2(\ldots, -4, 6, \ldots, \ldots)$. Hence we have an error, $\varepsilon = 2$, which is corresponding to $x = 2$.

Hence, the correct value for y_2 is given by

$$y_2 = 15 - 2 = 13$$

On using this value of y_2, we have a correct table as follows

x:	−1	0	1	2	3	4
y:	−2	1	4	13	34	73

Exercise 9

1. Use the definitions of various finite differences to prove the following results

   a) $\Delta^3 f_1 = \nabla^3 f_4 = \delta^3 f_{5/2}$,

   b) $1 + \delta^2 \mu^2 \equiv \left(1 + \dfrac{\delta^2}{2}\right)^2$,

   c) $E^{1/2} \equiv \mu + \dfrac{\delta}{2}$,

   d) $\Delta + \nabla = \dfrac{\Delta}{\nabla} - \dfrac{\nabla}{\Delta}$.

2. Find the value of the expression, $\left(\dfrac{\Delta^2}{E}\right)x^2$. **Ans.** $2h^2$

3. Prove that, if the polynomial $P_n(x)$ is a polynomial of degree n, then $\Delta^2 P_n(x)$ is of degree $\leq n-2$.

4. Show that for the polynomial, $P_n(x) = a_0 + a_1 x + a_2 x^2 + \cdots\cdots + a_n x^n$; the nth order difference is constant and higher order differences are zero. Is the converse true?

5. Prepare the finite difference table for the polynomial $P_3(x) = x^3 + 3x - 1$ for the points $x = -1, 0, 1, 2, 3, 4$ and 5. Verify that the third differences are constants and fourth and higher differences are zeroes.

Ans.

x	y	$\Delta(\nabla,\delta)$	$\Delta^2(\nabla^2,\delta^2)$	$\Delta^3(\nabla^3,\delta^3)$	$\Delta^4(\nabla^4,\delta^4)$	$\Delta^5(\nabla^5,\delta^5)$	$\Delta^6(\nabla^6,\delta^6)$
-1	-5	4	0	6	0	0	0
0	-1	4	6	6	0	0	
1	3	10	12	6	0		
2	13	22	18	6			
3	35	40	24				
4	75	64					
5	139						

6. Prove that $\Delta\left[\dfrac{f(x)}{g(x)}\right] = \dfrac{g(x)\Delta f(x) - f(x)\Delta g(x)}{g(x)\,g(x+h)}$ and use this result to verify that

$$\Delta\left[\dfrac{1}{g(x)}\right] = \dfrac{-\Delta g(x)}{g(x)g(x+h)} \text{ provided } g(x) \neq 0.$$

7. Find the nth forward differences of the functions $\sin(ax+b)$, $\cos(ax+b)$ and e^{ax+b}.

Ans.

$$\left(2\sin\left(\dfrac{ah}{2}\right)\right)^n \sin\left(ax+b+\dfrac{n}{2}(ah+\pi)\right)$$

$$\left(2\sin\left(\dfrac{ah}{2}\right)\right)^n \cos\left(ax+b+\dfrac{n}{2}(ah+\pi)\right)$$

$$\left(e^{ah}-1\right)^n e^{ax+b}$$

8. Evaluate $\nabla^6\left[(1-x)(1-2x^2)(1-3x^3)\right]$ at $h = 2$.

Ans. $-6(6!2^6) = -276480$

9. Evaluate $\dfrac{\Delta^2}{E^2}(\sin(x+h))$.

Ans. $\sin(x+h) - 2\sin(x) + \sin(x-h)$

10. Prepare the finite difference table for the function $f(x) = \dfrac{1}{1+x^2}$ at $x = -2\,(0.5)\,2$.

Ans.

x	$f(x)$	$\Delta(\delta,\nabla)$	$\Delta^2(\delta^2,\nabla^2)$	$\Delta^3(\delta^3,\nabla^3)$	$\Delta^4(\delta^4,\nabla^4)$	$\Delta^5(\delta^5,\nabla^5)$	$\Delta^6(\delta^6,\nabla^6)$	$\Delta^7(\delta^7,\nabla^7)$	$\Delta^8(\delta^8,\nabla^8)$
−2	0.2								
−1.5	0.307692	0.107692							
−1	0.5	0.192308	0.084615						
−0.5	0.8	0.3	0.107692	0.023077					
0	1	0.2	−0.1	−0.20769	−0.23077				
0.5	0.8	−0.2	−0.4	−0.3	−0.09231	0.138462			
1	0.5	−0.3	−0.1	0.3	0.6	0.692308	0.553846		
1.5	0.307692	−0.19231	0.107692	0.207692	−0.09231	−0.69231	−1.38462	−1.93846	
2	0.2	−0.10769	0.084615	−0.02308	−0.23077	−0.13846	0.553846	1.938462	3.876923

11. The population of Haryana state during census years is given below

Years	Population (in Lakh)
1951	52.8
1961	75.9
1971	100.4
1981	129.2
1991	?
2001	211.4
2011	253.5

Compute the missing population data and compare with the exact population of Haryana in 1991.

Ans. 164.6 Lakh.

12. Change the following data, such that the data follows a certain fourth-degree polynomial.

$x:$	−2	−1	0	1	2	3	4
$y:$	25	8	6	4	17	80	253

Ans. Change the value of variable y at $x = 0$, from 6 to 5.

$x:$	−2	−1	0	1	2	3	4
$y:$	25	8	5	4	17	80	253

Note that we can also compute the interpolating polynomial $x^4 - 2x + 5$ from Lagrange or NDD formula.

Interpolation for Equal Intervals and Bivariate Interpolation

Chapter 10

> *Mathematics is the supreme judge; from its decisions there is no appeal.*
>
> **Tobias Dantzig**
> (February 19, 1884–August 9, 1956)
> He is the author of *Number: The Language of Science (A critical survey written for the cultured non-mathematician)* (1930) and *Aspects of Science* (1937).
>
> *The saddest aspect of life right now is that science gathers knowledge faster than society gathers wisdom.*
>
> **Isaac Asimov**
> (January 2, 1920–April 6, 1992)
> He was a biochemist, professor, and prolific writer, who wrote or edited approximately 500 books on popular science and science fiction.

In Chapter 8, we have discussed the Lagrange and Newton divided difference interpolation methods. These methods are applicable to the data set of equally as well as unequally spaced points. We have already discussed that the interpolating polynomial is unique for a given set of data points. But, for equally spaced points, the calculations become simpler with lesser number of arithmetic operations. So, various errors and most importantly rounding error tend to decrease. A wide range of interpolation problems falls in the class of equally spaced points, for examples, the experiments with equispaced inputs. In this chapter, we will discuss the following interpolation methods for equally spaced points:

Forward difference formula

i) Newton forward difference formula

Backward difference formula

ii) Newton backward difference formula

Central difference formulas

  iii) Gauss formulas (forward central and backward central)
  iv) Stirling formula
  v) Bessel formula
  vi) Everett formula
  vii) Steffensen formula

Note: The motivation for introducing so many interpolation methods will be discussed later in this chapter. The position of the interpolation point in the data set defines the suitability of all these interpolation methods. If we want to compute the value of function $y = f(x)$ at the beginning of the table, Newton forward difference formula is suitable. If we want to perform interpolation for a point lying near to the end of the data set, Newton backward difference formula is suitable for interpolation. If the value is in the mid of the table, central difference formulas are suitable.

10.1 Gregory–Newton Forward Difference Formula

Let us consider a function, $y = f(x)$ defined on the interval, $[a, b]$. Let x_0, x_1, x_2,...,x_n be $(n + 1)$ equidistant points in the given interval. Let h be the spacing between two consecutive values of x, i.e.,

$$h = x_{i+1} - x_i$$

$$\Rightarrow x_i = x_0 + ih; \quad i = 1, 2, \cdots, n \tag{10.1}$$

Newton divided difference interpolating polynomial (8.12) is given by

$$P_n(x) = f(x_0) + f[x_0, x_1](x - x_0) + f[x_0, x_1, x_2](x - x_0)(x - x_1) + \cdots$$
$$+ f[x_0, x_1, \cdots, x_n](x - x_0)(x - x_1)\cdots(x - x_{n-1}) \tag{10.2}$$

The relation between forward differences and divided differences (Lemma 9.1) is as follows

$$f\left[x_0, x_1, x_2, \ldots x_m\right] = \frac{1}{m! \, h^m} \Delta^m f(x_0); \quad m = 1, 2, \ldots, n \tag{10.3}$$

On using the relation (10.3) in the expression (10.2), we get following Newton forward difference formula for interpolating polynomial

$$P_n(x) = f(x_0) + \frac{1}{h}\Delta f(x_0)(x - x_0) + \frac{1}{2! \, h^2}\Delta^2 f(x_0)(x - x_0)(x - x_1) + \cdots$$

$$+ \frac{1}{n! \, h^n}\Delta^n f(x_0)(x - x_0)(x - x_1)\cdots(x - x_{n-1}) \tag{10.4}$$

The formula (10.4) can be modified to compute the value at any intermediate point. For this, consider the following transformation

$$x = x_0 + sh \tag{10.5}$$

On using expressions (10.1) and (10.5) in Eq. (10.4), we have

$$P_n(x) = f(x_0) + \Delta f(x_0)(s) + \frac{\Delta^2 f(x_0)}{2!}(s)(s-1) + \cdots + \frac{\Delta^n f(x_0)}{n!}(s)(s-1)\cdots(s-n+1) \tag{10.6}$$

The expression (10.6) provides another form of Newton forward difference formula.

Note: The formula (10.4) produces the interpolating polynomial to approximate the function $f(x)$. While, the formula (10.6) gives the approximate value of function $f(x)$ at any given point $x = x_0 + sh$.

Alternative:
The formula (10.6) can be obtained alternatively using shift operator. Let $x = x_0 + sh$

$$f(x) = f(x_0 + sh)$$
$$= E^s f(x_0)$$
$$= (1 + \Delta)^s f(x_0) \tag{10.7}$$

The binomial expansion for $(1 + \Delta)^s$ (only symbolic expansion) is as follows

$$(1 + \Delta)^s = 1 + s\Delta + \frac{s(s-1)}{2!}\Delta^2 + \cdots \tag{10.8}$$

If we approximate the function $f(x)$ with interpolating polynomial $P_n(x)$ of degree $\leq n$, then finite differences of order $n + 1$ and higher are zero. We can easily obtain the Newton forward difference formula from (10.7) and (10.8)

$$P_n(x) = f(x_0) + \Delta f(x_0)(s) + \frac{\Delta^2 f(x_0)}{2!}(s)(s-1) + \cdots + \frac{\Delta^n f(x_0)}{n!}(s)(s-1)\cdots(s-n+1)$$

We can use the expression $s = \dfrac{x - x_0}{h}$ in the above formula to get the interpolating polynomial formula (10.4)

It is worth mentioning here that tabular points for Newton forward difference formula are given by following table (from Chapter 9)

x	$f(x)$	Δ	Δ^2	Δ^3
x_0	$f(x_0)$			
		$\Delta f(x_0)$		
x_1	$f(x_1)$		$\Delta^2 f(x_0)$	
		$\Delta f(x_1)$		$\Delta^3 f(x_0)$
x_2	$f(x_2)$		$\Delta^2 f(x_1)$	$\vdots$
		$\Delta f(x_2)$	$\vdots$	
x_3	$f(x_3)$	$\vdots$		
$\vdots$	$\vdots$			

Note: The Newton forward difference formula is used to compute the values at the points which are at the beginning of the table. Therefore, the absolute value of variable $s = \dfrac{x - x_0}{h}$ is small in magnitude. It will reduce the round-off error during computations.

Example ── **10.1**

Use Newton forward difference formula to find the value of $f(1.2)$ for the following data set

x:	1	1.25	1.5	1.75	2.0
$f(x)$	0	0.223144	0.405465	0.559616	0.693147

Ans.
The data set is equispaced with spacing $h = 0.25$.

$x_0 = 1,\ x_1 = 1.25,\ x_2 = 1.5,\ x_3 = 1.75,\ x_4 = 2.0$
$f(x_0) = 0,\ f(x_1) = 0.223144,\ f(x_2) = 0.405465,\ f(x_3) = 0.559616,\ f(x_4) = 0.693147$

We simply use the forward difference formulas, $\Delta f(x_i) = f(x_{i+1}) - f(x_i)$, $\Delta^2 f(x_i) = \Delta f(x_{i+1}) - \Delta f(x_i)$ and so on, to make the following forward difference table for the given data set

x	$f(x)$	Δ	Δ^2	Δ^3	Δ^4
$1(x_0)$	$0(f(x_0))$				
		$0.223144(\Delta f(x_0))$			
$1.25(x_1)$	$0.223144(f(x_1))$		$-0.040823(\Delta^2 f(x_0))$		
		$0.182321(\Delta f(x_1))$		$0.012653(\Delta^3 f(x_0))$	
$1.5(x_2)$	$0.405465(f(x_2))$		$-0.028170(\Delta^2 f(x_1))$		$-0.005103(\Delta^4 f(x_0))$
		$0.154151(\Delta f(x_2))$		$0.007550(\Delta^3 f(x_1))$	
$1.75(x_3)$	$0.559616(f(x_3))$		$-0.020620(\Delta^2 f(x_2))$		
		$0.133531(\Delta f(x_3))$			
$2(x_4)$	$0.693147(f(x_4))$				

To compute value at $x = 1.2$, we have

$$s = \frac{x - x_0}{h} = \frac{1.2 - 1}{0.25} = 0.8$$

On using Newton forward difference formula (10.6), we have

$$P(1.2) = f(x_0) + \Delta f(x_0)(s) + \frac{\Delta^2 f(x_0)}{2!}(s)(s-1) + \cdots + \frac{\Delta^n f(x_0)}{n!}(s)(s-1)\cdots(s-n+1)$$

$$= 0 + 0.223144(0.8) + \frac{-0.040823}{2}(0.8)(-0.2) + \frac{0.012653}{6}(0.8)(-0.2)(-1.2)$$

$$+ \frac{-0.005103}{24}(0.8)(-0.2)(-1.2)(-2.2)$$

$$P(1.2) = 0.178515 + 0.003266 + 0.000405 + 0.000090 = 0.182276$$

Hence, the approximate value of $f(1.2) = 0.182276$.

10.1.1 Error in Newton Forward Difference Formula

Since the interpolating polynomial obtained through any method is unique, hence the error in each method is the same. But the error formula becomes simpler for equispaced points. As discussed in Chapter 8, the error term, (8.14), is as follows

$$\varepsilon_n(x) = \frac{f^{n+1}(\xi)}{(n+1)!}(x - x_0)(x - x_1)\cdots(x - x_n) = \frac{f^{n+1}(\xi)}{(n+1)!}\prod_{i=0}^{n}(x - x_i) \qquad (10.9)$$

where the point ξ is a point in the given interval. The points are equally spaced points, i.e., $x_i = x_0 + ih;\ i = 1, 2, \cdots, n$.

On using the transformation $x = x_0 + sh$ in the error formula (10.9), we have

$$\varepsilon_n(x) = \frac{f^{n+1}(\xi)}{(n+1)!}(s)(h)(s-1)h\cdots(s-n)h$$

$$= \frac{f^{n+1}(\xi)}{(n+1)!}h^{n+1}(s)(s-1)\cdots(s-n) \tag{10.10}$$

Example ──────────────────────────────────── **10.2**

The tabulated values in Example 10.1 are obtained from the function $f(x) = \ln(x)$. Compute the error bound in the approximation for $f(1.2)$ from Newton forward difference formula.

Ans.
In Example 10.1, the points are equispaced points in the interval $[1, 2]$ with spacing $h = 0.25$. Also, we have $s = \dfrac{x - x_0}{h} = \dfrac{1.2 - 1}{0.25} = 0.8$. The polynomial of degree 4 is used for interpolation, so the error term from Eq. (10.10) is given by

$$\varepsilon = \frac{f^{(v)}(\xi)}{5!}h^5(s)(s-1)(s-2)(s-3)(s-4) \tag{10.11}$$

Since the function $f(x) = \ln(x)$, we have

$$f^{(v)}(x) = \frac{24}{x^5}$$

Its absolutely maximum value in the interval $[1, 2]$ is at $x = 1$.

$$\left|f^{(v)}(x)\right| \le 24 \quad \forall x \in [1, 2]$$

Hence, from Eq. (10.11), maximum absolute error is given by

$$\varepsilon = \frac{24}{5!}(0.25)^5(0.8)(-0.2)(-1.2)(-2.2)(-3.2) = 0.000264$$

Note: The approximate value obtained from the Newton forward difference formula is

$$f(1.2) = P(1.2) = 0.182276$$

Note that, the exact value is $f(1.2) = \ln(1.2) = 0.182322$.
The error in obtained value is 0.000046.

10.2 Gregory–Newton Backward Difference Formula

Consider the transformation, $x = x_n + sh$, then we have

$$f(x) = f(x_n + sh) = E^s f(x_n) = (1 - \nabla)^{-s} f(x_n)$$

On using the expansion

$$(1 - \nabla)^{-s} = 1 + s\nabla + \frac{s(s+1)}{2!} \nabla^2 + \cdots,$$

and the fact that for a polynomial of degree n, the finite differences of order $n + 1$ and higher are zero, we have

$$P_n(x) = f(x_n) + \nabla f(x_n)(s) + \frac{\nabla^2 f(x_n)}{2!}(s)(s+1) + \frac{\nabla^3 f(x_n)}{3!}(s)(s+1)(s+2) + \cdots$$

$$+ \frac{\nabla^n f(x_n)}{n!}(s)(s+1)(s+2)\cdots(s+n-1) \tag{10.12}$$

This formula is known as Gregory–Newton backward difference interpolation formula. The interpolating polynomial can be obtained by using $s = \dfrac{x - x_n}{h}$ in the above formula. We have

$$P_n(x) = f(x_n) + \frac{\nabla f(x_n)}{h}(x - x_n) + \frac{\nabla^2 f(x_n)}{2!h^2}(x - x_n)(x - x_{n-1}) + \frac{\nabla^3 f(x_n)}{3!h^3}(x - x_n)(x - x_{n-1})$$

$$(x - x_{n-2}) + \cdots + \frac{\nabla^n f(x_n)}{n!h^n}(x - x_n)(x - x_{n-1})\cdots(x - x_1) \tag{10.13}$$

The following table contains the tabular points for Newton backward difference formula by using table 9.4.

x	$f(x)$	∇	∇^2	∇^3
$\vdots$	$\vdots$			
x_{n-3}	$f(x_{n-3})$	$\vdots$		
		$\nabla f(x_{n-2})$	$\vdots$	
x_{n-2}	$f(x_{n-2})$		$\nabla^2 f(x_{n-1})$	$\vdots$
		$\nabla f(x_{n-1})$		$\nabla^3 f(x_n)$
x_{n-1}	$f(x_{n-1})$		$\nabla^2 f(x_n)$	
		$\nabla f(x_n)$		
x_n	$f(x_n)$			

Note that Newton backward difference formula will be used to compute the values at the points which are at the end of the table.

Example ─────────────────────────────────── **10.3** ─

Use Newton backward difference formula to find the value of $f(1.9)$ for the following data set

x:	1	1.25	1.5	1.75	2.0
$f(x)$	0	0.223144	0.405465	0.559616	0.693147

Ans.
Similar to Example 10.1, the data set is equispaced with spacing $h = 0.25$.

$x_0 = 1$, $x_1 = 1.25$, $x_2 = 1.5$, $x_3 = 1.75$, $x_4 = 2.0$
$f(x_0) = 0$, $f(x_1) = 0.223144$, $f(x_2) = 0.405465$, $f(x_3) = 0.559616$, $f(x_4) = 0.693147$

We will use the backward difference formulas, $\nabla f(x_i) = f(x_i) - f(x_{i-1})$, $\nabla^2 f(x_i) = \nabla f(x_i) - \nabla f(x_{i-1})$ and so on, to make the following backward difference table for the given data set

x	$f(x)$	∇	∇^2	∇^3	∇^4
$1(x_0)$	$0(f(x_0))$				
		$0.223144(\nabla f(x_1))$			
$1.25(x_1)$	$0.223144(f(x_1))$		$-0.040823(\nabla^2 f(x_2))$		
		$0.182321(\nabla f(x_2))$		$0.012653(\nabla^3 f(x_3))$	
$1.5(x_2)$	$0.405465(f(x_2))$		$-0.028170(\nabla^2 f(x_3))$		$-0.005103(\nabla^4 f(x_4))$
		$0.154151(\nabla f(x_3))$		$0.007550(\nabla^3 f(x_4))$	
$1.75(x_3)$	$0.559616(f(x_3))$		$-0.020620(\nabla^2 f(x_4))$		
		$0.133531(\nabla f(x_4))$			
$2(x_4)$	$0.693147(f(x_4))$				

At $x = 1.9$, $n = 4$, we have

$$s = \frac{x - x_n}{h} = \frac{1.9 - 2}{0.25} = -0.4$$

On using Newton backward difference formula, (10.12), we have

$$P(1.9) = f(x_4) + \nabla f(x_4)(s) + \frac{\nabla^2 f(x_4)}{2!}(s)(s+1) + \frac{\nabla^3 f(x_4)}{3!}(s)(s+1)(s+2) + \cdots$$

$$+ \frac{\nabla^4 f(x_4)}{4!}(s)(s+1)(s+2)(s+3)$$

$$P(1.9) = 0.693147 + 0.133531(-0.4) + \frac{-0.020620}{2}(-0.4)(0.6) + \frac{0.007550}{6}(-0.4)(0.6)(1.6) + \cdots$$

$$+ \frac{-0.005103}{24}(-0.4)(0.6)(1.6)(2.6)$$

$$P(1.9) = 0.693147 - 0.053412 + 0.002474 - 0.000483 + 0.000212 = 0.641938$$

10.2.1 Error in Newton Backward Difference Formula

On using $x = x_n + sh$ and $x_i = x_0 + ih$; $i = 1, 2, \cdots, n$ in Eq. (10.9), we have following error term for Newton backward difference formula

$$\varepsilon_n(x) = \frac{f^{n+1}(\xi)}{(n+1)!}(x - x_0)(x - x_1) \cdots (x - x_n)$$

$$\varepsilon_n(x) = \frac{f^{n+1}(\xi)}{(n+1)!}(s+n)(h)(s+n-1)(h) \cdots (s)h$$

$$= \frac{f^{n+1}(\xi)}{(n+1)!}h^{n+1}(s)(s+1) \cdots (s+n) \tag{10.14}$$

Example ————————————————————————————————— **10.4** ——

Obtain the error bound for the approximate value of $f(1.9)$ in Example 10.3.

Ans.

The error term from Eq. (10.14) is given by

$$\varepsilon = \frac{f^{(v)}(\xi)}{5!}h^5(s)(s+1)(s+2)(s+3)(s+4) \tag{10.15}$$

The maximum value of $f^{(v)}(x)$ in the interval $[1, 2]$ is 24 (Example 10.2). The value of s is given by

$$s = \frac{x - x_n}{h} = \frac{1.9 - 2}{0.25} = -0.4$$

On substituting the values of various constants in Eq. (10.15), we have the following upper bound for error

$$\varepsilon = \frac{24}{5!}(0.25)^5(-0.4)(0.6)(1.6)(2.6)(3.6) = 0.000702$$

Note: The approximate value obtained from the Newton backward difference formula is

$$f(1.9) = P(1.9) = 0.641938$$

The exact value is $f(1.9) = \ln(1.9) = 0.641854$.
Therefore, the error in obtained value is 0.000084.

10.3 Central Difference Formulas

The interpolating polynomial is unique for a given data set. But round-off error can be reduced by using the appropriate formula for computation of function value at any intermediate point. In case, if we have to find the value at a point, which is in beginning of the table (near x_0), it is always better to use Newton forward difference formula. Similarly, for the points at the end of the table (near x_n), the Newton backward difference formula is suitable. If the point is in the middle of data set, then central difference formulas are used. The following central difference formulas will be discussed in this chapter

    i)   Gauss forward central difference formula
    ii)   Gauss backward central difference formula
    iii)   Stirling central difference formula
    iv)   Bessel central difference formula
    v)   Everett central difference formula
    vi)   Steffensen central difference formula

As the interpolating polynomial is unique, therefore, the error term is also same for all the interpolation formulas. The only difference is in their representations. To avoid repetitions of cumbersome computations and lengthy formulas, we will not discuss error formulas for all these central difference interpolation formulas.

The tabular points from table 9.5 for all these central difference formulas are as follows

x	$f(x)$	δ	δ^2	δ^3
$\vdots$	$\vdots$			
x_{-2}	$f(x_{-2})$	$\vdots$		
		$\delta f(x_{-3/2})$	$\vdots$	
x_{-1}	$f(x_{-1})$		$\delta^2 f(x_{-1})$	$\vdots$
		$\delta f(x_{-1/2})$		$\delta^3 f(x_{-1/2})$
x_0	$f(x_0)$		$\delta^2 f(x_0)$	
		$\delta f(x_{1/2})$		$\delta^3 f(x_{1/2})$
x_1	$f(x_1)$		$\delta^2 f(x_1)$	$\vdots$
		$\delta f(x_{3/2})$	$\vdots$	
x_2	$f(x_2)$	$\vdots$		
$\vdots$	$\vdots$			

10.4 Gauss Forward Central Difference Formula

The Newton forward difference interpolation formula, (10.6), is as follows

$$P_n(x) = f(x_0) + \Delta f(x_0)(s) + \frac{\Delta^2 f(x_0)}{2!}(s)(s-1) + \cdots + \frac{\Delta^n f(x_0)}{n!}(s)(s-1)\cdots(s-n+1) \quad (10.16)$$

We have the following formulae for forward differences

$$\Delta^2 f(x_0) - \Delta^2 f(x_{-1}) = \Delta^3 f(x_{-1})$$

$$\Rightarrow \Delta^2 f(x_0) = \Delta^2 f(x_{-1}) + \Delta^3 f(x_{-1})$$

Similarly, we have

$$\Delta^3 f(x_0) = \Delta^3 f(x_{-1}) + \Delta^4 f(x_{-1})$$
$$\Delta^4 f(x_0) = \Delta^4 f(x_{-1}) + \Delta^5 f(x_{-1})$$
$$\vdots$$

On substituting these results into Newton forward difference formula, (10.16), we have

$$P_n(x) = f(x_0) + s\Delta f(x_0) + \frac{s(s-1)}{2!}\left(\Delta^2 f(x_{-1}) + \Delta^3 f(x_{-1})\right)$$

$$+ \frac{s(s-1)(s-2)}{3!}\left(\Delta^3 f(x_{-1}) + \Delta^4 f(x_{-1})\right) + \frac{s(s-1)(s-2)(s-3)}{4!}\left(\Delta^4 f(x_{-1}) + \Delta^5 f(x_{-1})\right) + \cdots$$

$$P_n(x) = f(x_0) + s\Delta f(x_0) + \frac{s(s-1)}{2!}\Delta^2 f(x_{-1}) + \frac{(s+1)s(s-1)}{3!}\Delta^3 f(x_{-1})$$

$$+ \frac{(s+1)s(s-1)(s-2)}{4!}\Delta^4 f(x_{-1}) + \frac{(s+1)s(s-1)(s-2)(s-3)}{5!}\Delta^5 f(x_{-1}) + \cdots \quad (10.17)$$

Also, we have

$$\Delta^4 f(x_{-1}) = \Delta^4 f(x_{-2}) + \Delta^5 f(x_{-2})$$

$$\Delta^5 f(x_{-1}) = \Delta^5 f(x_{-2}) + \Delta^6 f(x_{-2})$$

$$\vdots$$

On using these values in Eq. (10.17), we get

$$P_n(x) = f(x_0) + s\Delta f(x_0) + \frac{s(s-1)}{2!}\Delta^2 f(x_{-1}) + \frac{(s+1)s(s-1)}{3!}\Delta^3 f(x_{-1})$$

$$+ \frac{(s+1)s(s-1)(s-2)}{4!}\left(\Delta^4 f(x_{-2}) + \Delta^5 f(x_{-2})\right)$$

$$+ \frac{(s+1)s(s-1)(s-2)(s-3)}{5!}\left(\Delta^5 f(x_{-2}) + \Delta^6 f(x_{-2})\right) + \cdots$$

$$P_n(x) = f(x_0) + s\Delta f(x_0) + \frac{s(s-1)}{2!}\Delta^2 f(x_{-1}) + \frac{(s+1)s(s-1)}{3!}\Delta^3 f(x_{-1})$$

$$+ \frac{(s+1)s(s-1)(s-2)}{4!}\Delta^4 f(x_{-2}) + \frac{(s+2)(s+1)s(s-1)(s-2)}{5!}\Delta^5 f(x_{-2}) + \cdots$$

Similarly, we will continue these steps to get the following formula

$$P_n(x) = f(x_0) + s\Delta f(x_0) + \frac{s(s-1)}{2!}\Delta^2 f(x_{-1}) + \frac{(s+1)s(s-1)}{3!}\Delta^3 f(x_{-1})$$

$$+ \frac{(s+1)s(s-1)(s-2)}{4!}\Delta^4 f(x_{-2}) + \frac{(s+2)(s+1)s(s-1)(s-2)}{5!}\Delta^5 f(x_{-2}) + \cdots \quad (10.18)$$

On using the results $\Delta f(x_0) = \delta f(x_{1/2})$, $\Delta^2 f(x_{-1}) = \delta^2 f(x_0)$, ..., we have

$$P_n(x) = f(x_0) + \delta f(x_{1/2})(s) + \frac{\delta^2 f(x_0)}{2!}(s)(s-1) + \frac{\delta^3 f(x_{1/2})}{3!}(s+1)(s)(s-1)$$

$$+ \frac{\delta^4 f(x_0)}{4!}(s+1)(s)(s-1)(s-2) + \frac{\delta^5 f(x_{1/2})}{5!}(s+2)(s+1)(s)(s-1)(s-2) + \cdots \quad (10.19)$$

The formula (10.19) is known as Gauss forward central difference formula.

Example   **10.5**

Use Gauss forward central difference formula to find the value of $f(1.55)$ for the following data set

x:	1	1.25	1.5	1.75	2.0
$f(x)$	0	0.223144	0.405465	0.559616	0.693147

Ans.

The data set is equispaced with spacing $h = 0.25$. Contrary to Newton forward and backward differences, we will now assume the points as follows

$$x_{-2} = 1, \ x_{-1} = 1.25, \ x_0 = 1.5, \ x_1 = 1.75, \ x_2 = 2.0$$
$$f(x_{-2}) = 0, \ f(x_{-1}) = 0.223144, \ f(x_0) = 0.405465, \ f(x_1) = 0.559616, \ f(x_2) = 0.693147$$

We will use the central difference formulas for the given data set to construct the following central difference table.

x	$f(x)$	δ	δ^2	δ^3	δ^4
$1(x_{-2})$	$0(f(x_{-2}))$				
		$0.223144(\delta f(x_{-3/2}))$			
$1.25(x_{-1})$	$0.223144(f(x_{-1}))$		$-0.040823(\delta^2 f(x_{-1}))$		
		$0.182321(\delta f(x_{-1/2}))$		$0.012653(\delta^3 f(x_{-1/2}))$	
$1.5(x_0)$	$0.405465(f(x_0))$		$-0.028170(\delta^2 f(x_0))$		$-0.005103(\delta^4 f(x_0))$
		$0.154151(\delta f(x_{1/2}))$		$0.007550(\delta^3 f(x_{1/2}))$	
$1.75(x_1)$	$0.559616(f(x_1))$		$-0.020620(\delta^2 f(x_1))$		
		$0.133531(\delta f(x_{3/2}))$			
$2(x_2)$	$0.693147(f(x_2))$				

To compute value at $x = 1.55$, we have

$$s = \frac{x - x_0}{h} = \frac{1.55 - 1.5}{0.25} = 0.2$$

On using Gauss forward central difference formula (10.19), we have

$$P(1.55) = f(x_0) + \delta f(x_{1/2})(s) + \frac{\delta^2 f(x_0)}{2!}(s)(s-1) + \frac{\delta^3 f(x_{1/2})}{3!}(s+1)(s)(s-1)$$

$$+ \frac{\delta^4 f(x_0)}{4!}(s+1)(s)(s-1)(s-2) + \frac{\delta^5 f(x_{1/2})}{5!}(s+2)(s+1)(s)(s-1)(s-2) + \cdots$$

$$P(1.55) = 0.405465 + 0.154151(0.2) + \frac{-0.028170}{2}(0.2)(-0.8) + \frac{0.007550}{6}(1.2)(0.2)(-0.8)$$

$$+ \frac{-0.005103}{24}(1.2)(0.2)(-0.8)(-1.8)$$

$$P(1.55) = 0.405465 + 0.030830 + 0.002254 - 0.000242 - 0.000073 = 0.438234$$

10.5 Gauss Backward Central Difference Formula

The Newton forward interpolation formula is given as follows

$$P_n(x) = f(x_0) + \Delta f(x_0)(s) + \frac{\Delta^2 f(x_0)}{2!}(s)(s-1) + \cdots$$

$$+ \frac{\Delta^n f(x_0)}{n!}(s)(s-1)\cdots(s-n+1) \tag{10.20}$$

On using forward difference, we have

$$\Delta f(x_0) - \Delta f(x_{-1}) = \Delta^2 f(x_{-1})$$

$$\Rightarrow \Delta f(x_0) = \Delta f(x_{-1}) + \Delta^2 f(x_{-1})$$

Similarly, the following results can be obtained easily

$$\Delta^2 f(x_0) = \Delta^2 f(x_{-1}) + \Delta^3 f(x_{-1})$$

$$\Delta^3 f(x_0) = \Delta^3 f(x_{-1}) + \Delta^4 f(x_{-1})$$

$$\vdots$$

On substituting these results into Newton forward difference formula, (10.20), we get

$$P_n(x) = f(x_0) + s\left(\Delta f(x_{-1}) + \Delta^2 f(x_{-1})\right) + \frac{s(s-1)}{2!}\left(\Delta^2 f(x_{-1}) + \Delta^3 f(x_{-1})\right)$$

$$+ \frac{s(s-1)(s-2)}{3!}\left(\Delta^3 f(x_{-1}) + \Delta^4 f(x_{-1})\right) + \frac{s(s-1)(s-2)(s-3)}{4!}\left(\Delta^4 f(x_{-1}) + \Delta^5 f(x_{-1})\right) + \cdots$$

$$P_n(x) = f(x_0) + s\Delta f(x_{-1}) + \frac{(s+1)s}{2!}\Delta^2 f(x_{-1}) + \frac{(s+1)s(s-1)}{3!}\left(\Delta^3 f(x_{-2}) + \Delta^4 f(x_{-2})\right)$$

$$+ \frac{s(s+1)(s-1)(s-2)}{4!}\left(\Delta^4 f(x_{-2}) + \Delta^5 f(x_{-2})\right) + \cdots \tag{10.21}$$

The higher order forward differences provide the following results

$$\Delta^3 f(x_{-1}) - \Delta^3 f(x_{-2}) = \Delta^4 f(x_{-2})$$

$$\Rightarrow \Delta^3 f(x_{-1}) = \Delta^3 f(x_{-2}) + \Delta^4 f(x_{-2})$$

$$\Delta^4 f(x_{-1}) = \Delta^4 f(x_{-2}) + \Delta^5 f(x_{-2})$$

$$\Delta^5 f(x_{-1}) = \Delta^5 f(x_{-2}) + \Delta^6 f(x_{-2})$$

$$\vdots$$

On using these forward differences in Eq. (10.21), we get

$$P_n(x) = f(x_0) + s\Delta f(x_{-1}) + \frac{(s+1)s}{2!}\Delta^2 f(x_{-1}) + \frac{(s+1)s(s-1)}{3!}\Delta^3 f(x_{-2})$$

$$+ \frac{(s+2)(s+1)s(s-1)}{4!}\Delta^4 f(x_{-2}) + \frac{(s+2)(s+1)s(s-1)(s-2)}{5!}\Delta^5 f(x_{-3}) + \cdots \tag{10.22}$$

On using the results, $\Delta f(x_{-1}) = \delta f(x_{-1/2})$, $\Delta^2 f(x_{-1}) = \delta^2 f(x_0)$, ..., we get

$$P_n(x) = f(x_0) + \delta f(x_{-1/2})(s) + \frac{\delta^2 f(x_0)}{2!}(s+1)(s) + \frac{\delta^3 f(x_{-1/2})}{3!}(s+1)(s)(s-1)$$

$$+ \frac{\delta^4 f(x_0)}{4!}(s+2)(s+1)(s)(s-1) + \frac{\delta^5 f(x_{-1/2})}{5!}(s+2)(s+1)(s)(s-1)(s-2) + \cdots \tag{10.23}$$

The formulas (10.22) and (10.23) are known as the Gauss backward central difference interpolation formula.

Example ——————————————————————————— **10.6**

Use Gauss backward difference formula to find the value of $f(1.45)$ for the data set of Example 10.5.

Ans.

The data set is equispaced with spacing $h = 0.25$. We will now assume the points as follows

$$x_{-2} = 1, \ x_{-1} = 1.25, \ x_0 = 1.5, \ x_1 = 1.75, \ x_2 = 2.0$$

$$f(x_{-2}) = 0, \ f(x_{-1}) = 0.223144, \ f(x_0) = 0.405465, \ f(x_1) = 0.559616, \ f(x_2) = 0.693147.$$

To compute value at $x = 1.45$, we have

$$s = \frac{x - x_0}{h} = \frac{1.45 - 1.5}{0.25} = -0.2$$

On using Gauss backward difference formula (10.23) and values from central difference table of Example 10.5, we have

$$P(1.45) = f(x_0) + \delta f(x_{-1/2})(s) + \frac{\delta^2 f(x_0)}{2!}(s+1)(s) + \frac{\delta^3 f(x_{-1/2})}{3!}(s+1)(s)(s-1)$$

$$+ \frac{\delta^4 f(x_0)}{4!}(s+2)(s+1)(s)(s-1) + \frac{\delta^5 f(x_{-1/2})}{5!}(s+2)(s+1)(s)(s-1)(s-2) + \cdots$$

$$P(1.45) = 0.405465 + 0.182321(-0.2) + \frac{-0.028170}{2}(0.8)(-0.2) + \frac{0.012653}{6}(0.8)(-0.2)(-1.2)$$

$$+ \frac{-0.005103}{24}(1.8)(0.8)(-0.2)(-1.2)$$

$$P(1.45) = 0.405465 - 0.036464 + 0.002254 + 0.000405 - 0.000073 = 0.371587$$

10.6 Stirling Formula

Gauss forward central difference interpolation formula, (10.19), is given by

$$P_n(x) = f(x_0) + s\Delta f(x_0) + \frac{s(s-1)}{2!}\Delta^2 f(x_{-1}) + \frac{(s+1)s(s-1)}{3!}\Delta^3 f(x_{-1})$$

$$+ \frac{(s+1)s(s-1)(s-2)}{4!}\Delta^4 f(x_{-2}) + \frac{(s+2)(s+1)s(s-1)(s-2)}{5!}\Delta^5 f(x_{-2}) + \cdots$$

$$P_n(x) = f(x_0) + \delta f(x_{1/2})(s) + \frac{\delta^2 f(x_0)}{2!}(s)(s-1) + \frac{\delta^3 f(x_{1/2})}{3!}(s+1)(s)(s-1)$$

$$+ \frac{\delta^4 f(x_0)}{4!}(s+1)(s)(s-1)(s-2) + \frac{\delta^5 f(x_{1/2})}{5!}(s+2)(s+1)(s)(s-1)(s-2) + \cdots \quad (10.24)$$

Gauss backward central difference interpolation formula, (10.23), is given by

$$P_n(x) = f(x_0) + s\Delta f(x_{-1}) + \frac{(s+1)s}{2!}\Delta^2 f(x_{-1}) + \frac{(s+1)s(s-1)}{3!}\Delta^3 f(x_{-2})$$

$$+ \frac{(s+2)(s+1)s(s-1)}{4!}\Delta^4 f(x_{-2}) + \frac{(s+2)(s+1)s(s-1)(s-2)}{5!}\Delta^5 f(x_{-3}) + \cdots$$

$$P_n(x) = f(x_0) + \delta f(x_{-1/2})(s) + \frac{\delta^2 f(x_0)}{2!}(s+1)(s) + \frac{\delta^3 f(x_{-1/2})}{3!}(s+1)(s)(s-1)$$

$$+ \frac{\delta^4 f(x_0)}{4!}(s+2)(s+1)(s)(s-1) + \frac{\delta^5 f(x_{-1/2})}{5!}(s+2)(s+1)(s)(s-1)(s-2) + \cdots \quad (10.25)$$

On taking the mean of (10.24) and (10.25), we have

$$P_n(x) = f(x_0) + s\left[\frac{\Delta f(x_0) + \Delta f(x_{-1})}{2}\right] + \frac{s^2}{2!}\Delta^2 f(x_{-1}) + \frac{s(s^2-1)}{3!}\left[\frac{\Delta^3 f(x_{-1}) + \Delta^3 f(x_{-2})}{2}\right]$$

$$+ \frac{s^2(s^2-1)}{4!}\Delta^4 f(x_{-2}) + \frac{s(s^2-1)(s^2-2^2)}{5!}\left[\frac{\Delta^5 f(x_{-2}) + \Delta^5 f(x_{-3})}{2}\right] + \cdots$$

$$P_n(x) = f(x_0) + (s)\left[\frac{\delta f(x_{1/2}) + \delta f(x_{-1/2})}{2}\right] + \frac{(s^2)}{2!}\delta^2 f(x_0)$$

$$+ \frac{(s)(s^2-1)}{3!}\left[\frac{\delta^3 f(x_{1/2}) + \delta^3 f(x_{-1/2})}{2}\right] + \frac{(s^2)(s^2-1)}{4!}\delta^4 f(x_0)$$

$$+ \frac{(s)(s^2-1)(s^2-2^2)}{5!}\left[\frac{\delta^5 f(x_{1/2}) + \delta^5 f(x_{-1/2})}{2}\right] + \cdots \quad (10.26)$$

The formula (10.26) is known as Stirling central difference formula.

Example ─── **10.7**

Use Stirling central difference formula to find the value of $f(1.55)$ for the data points of Example 10.5.

Ans.
Given
$h = 0.25$

$$x_{-2} = 1, \ x_{-1} = 1.25, \ x_0 = 1.5, \ x_1 = 1.75, \ x_2 = 2.0$$
$$f(x_{-2}) = 0, \ f(x_{-1}) = 0.223144, \ f(x_0) = 0.405465, \ f(x_1) = 0.559616, \ f(x_2) = 0.693147$$

At $x = 1.55$, we have

$$s = \frac{x - x_0}{h} = \frac{1.55 - 1.5}{0.25} = 0.2$$

On using the values from central difference table of Example 10.5 and Stirling central difference formula (10.26), we have

$$P(1.55) = f(x_0) + (s)\left[\frac{\delta f(x_{1/2}) + \delta f(x_{-1/2})}{2}\right] + \frac{(s^2)}{2!}\delta^2 f(x_0)$$

$$+ \frac{(s)(s^2 - 1)}{3!}\left[\frac{\delta^3 f(x_{1/2}) + \delta^3 f(x_{-1/2})}{2}\right] + \frac{(s^2)(s^2 - 1)}{4!}\delta^4 f(x_0) + \cdots$$

$$P(1.55) = 0.405465 + (0.2)\left[\frac{0.154151 + 0.182321}{2}\right] + \frac{(.04)}{2!}(-0.028170)$$

$$+ \frac{(0.2)(-0.96)}{3!}\left[\frac{0.007550 + 0.012653}{2}\right] + \frac{(.04)(-0.96)}{4!}(-0.005103)$$

$$P(1.55) = 0.405465 + 0.033647 - 0.000563 - 0.000323 + 0.000008 = 0.438234$$

10.7 Bessel Formula

Gauss forward interpolation formula (10.19) is as follows

$$P_n(x) = f(x_0) + s\Delta f(x_0) + \frac{s(s-1)}{2!}\Delta^2 f(x_{-1}) + \frac{(s+1)s(s-1)}{3!}\Delta^3 f(x_{-1})$$

$$+ \frac{(s+1)s(s-1)(s-2)}{4!}\Delta^4 f(x_{-2}) + \frac{(s+2)(s+1)s(s-1)(s-2)}{5!}\Delta^5 f(x_{-2}) + \cdots \quad (10.27)$$

The formula (10.27) can be rewritten as follows

$$P_n(x) = \left[\frac{1}{2}f(x_0) + \frac{1}{2}f(x_0)\right] + s\Delta f(x_0) + \frac{s(s-1)}{2!}\left[\frac{1}{2}\Delta^2 f(x_{-1}) + \frac{1}{2}\Delta^2 f(x_{-1})\right]$$

$$+ \frac{(s+1)s(s-1)}{3!}\Delta^3 f(x_{-1}) + \frac{(s+1)s(s-1)(s-2)}{4!}\left[\frac{1}{2}\Delta^4 f(x_{-2}) + \frac{1}{2}\Delta^2 f(x_{-1})\right]$$

$$+ \frac{(s+2)(s+1)s(s-1)(s-2)}{5!}\Delta^5 f(x_{-2}) + \cdots \quad (10.28)$$

The forward differences provide the following results

$$f(x_1) - f(x_0) = \Delta f(x_0) \qquad \Rightarrow f(x_1) - \Delta f(x_0) = f(x_0)$$

$$\Delta^2 f(x_0) - \Delta^2 f(x_{-1}) = \Delta^3 f(x_{-1}) \qquad \Rightarrow \Delta^2 f(x_{-1}) = \Delta^2 f(x_0) - \Delta^3 f(x_{-1})$$

Similarly, we have

$$\Delta^4 f(x_{-2}) = \Delta^4 f(x_{-1}) - \Delta^5 f(x_{-2}) \text{ etc.,}$$

On using these results in Eq. (10.28), we have

$$P_n(x) = \frac{1}{2}f(x_0) + \frac{1}{2}\left[f(x_1) - \Delta f(x_0)\right] + s\Delta f(x_0) + \frac{1}{2}\frac{s(s-1)}{2!}\Delta^2 f(x_{-1})$$

$$+ \frac{1}{2}\frac{s(s-1)}{2!}\left[\Delta^2 f(x_0) - \Delta^3 f(x_{-1})\right] + \frac{(s+1)s(s-1)}{3!}\Delta^3 f(x_{-1}) + \frac{1}{2}\frac{(s+1)s(s-1)(s-2)}{4!}\Delta^4 f(x_{-2})$$

$$+ \frac{1}{2}\frac{(s+1)s(s-1)(s-2)}{4!}\left[\Delta^4 f(x_{-1}) - \Delta^5 f(x_{-2})\right] + \frac{(s+2)(s+1)s(s-1)(s-2)}{5!}\Delta^5 f(x_{-2}) + \cdots$$

$$P_n(x) = \frac{1}{2}\left[f(x_0) + f(x_1)\right] + \left(s - \frac{1}{2}\right)\Delta f(x_0) + \frac{s(s-1)}{2!}\frac{1}{2}\left(\Delta^2 f(x_{-1}) + \Delta^2 f(x_0)\right)$$

$$+ \frac{1}{3!}\left(s - \frac{1}{2}\right)s(s-1)\Delta^3 f(x_{-1}) + \frac{(s+1)s(s-1)(s-2)}{4!}\frac{1}{2}\left(\Delta^4 f(x_{-2}) + \Delta^4 f(x_{-1})\right)$$

$$+ \frac{1}{5!}\left(s - \frac{1}{2}\right)(s+1)s(s-1)(s-2)\Delta^5 f(x_{-2}) + \cdots$$

(or)

$$P_n(x) = \frac{1}{2}\left(f(x_0) + f(x_1)\right) + \left(s - \frac{1}{2}\right)\delta f(x_{1/2}) + \frac{(s)(s-1)}{2!}\frac{1}{2}\left(\delta^2 f(x_0) + \delta^2 f(x_1)\right)$$

$$+ \left(s - \frac{1}{2}\right)\frac{(s)(s-1)}{3!}\delta^3 f(x_{1/2}) + \frac{(s+1)(s)(s-1)(s-2)}{4!}\frac{1}{2}\left(\delta^4 f(x_0) + \delta^4 f(x_1)\right)$$

$$+ \left(s - \frac{1}{2}\right)\frac{(s+1)(s)(s-1)(s-2)}{5!}\delta^5 f(x_{1/2})$$

$$+ \frac{(s+2)(s+1)(s)(s-1)(s-2)(s-3)}{6!}\frac{1}{2}\left(\delta^6 f(x_0) + \delta^6 f(x_1)\right) + \cdots \qquad (10.29)$$

The formula (10.29) is known as Bessel central difference formula.

Example ――――――――――――――――――――――――――――――― 10.8

Compute the value of $f(1.65)$ by using Bessel central difference formula to the data set of Example 10.5.

Ans.
At $x = 1.65$, we have

$$s = \frac{x - x_0}{h} = \frac{1.65 - 1.5}{0.25} = 0.6$$

Bessel central difference formula (10.29) provides the following approximation to the value of $f(1.65)$

$$P(1.65) = \frac{1}{2}\left(f(x_0) + f(x_1)\right) + \left(s - \frac{1}{2}\right)\delta f(x_{1/2}) + \frac{(s)(s-1)}{2!}\frac{1}{2}\left(\delta^2 f(x_0) + \delta^2 f(x_1)\right)$$

$$+ \left(s - \frac{1}{2}\right)\frac{(s)(s-1)}{3!}\delta^3 f(x_{1/2}) + \cdots$$

$$P(1.65) = \frac{1}{2}\left(0.405465 + 0.559616\right) + \left(0.6 - \frac{1}{2}\right)(0.154151)$$

$$+ \frac{(0.6)(-0.4)}{2}\frac{1}{2}\left(-0.028170 - 0.020620\right) + \left(0.6 - \frac{1}{2}\right)\frac{(0.6)(-0.4)}{6}(0.007550)$$

$$P(1.65) = 0.482540 + 0.015415 + 0.002927 - 0.000030 = 0.500852$$

10.8 Everett Formula

In Everett formula, we will retain only even order central differences, and eliminate all the odd order central differences.

Gauss forward interpolation formula is as follows

$$P_n(x) = f(x_0) + s\Delta f(x_0) + \frac{s(s-1)}{2!}\Delta^2 f(x_{-1}) + \frac{(s+1)s(s-1)}{3!}\Delta^3 f(x_{-1})$$

$$+ \frac{(s+1)s(s-1)(s-2)}{4!}\Delta^4 f(x_{-2}) + \frac{(s+2)(s+1)s(s-1)(s-2)}{5!}\Delta^5 f(x_{-2}) + \cdots \quad (10.30)$$

On eliminating odd differences with the following formulas

$$\Delta f(x_0) = f(x_1) - f(x_0)$$

$$\Delta^3 f(x_{-1}) = \Delta^2 f(x_0) - \Delta^2 f(x_{-1})$$

$$\Delta^5 f(x_{-2}) = \Delta^4 f(x_{-1}) - \Delta^4 f(x_{-2}) \text{ etc.}$$

We have

$$P_n(x) = f(x_0) + s\left[f(x_1) - f(x_0)\right] + \frac{s(s-1)}{2!}\Delta^2 f(x_{-1}) + \frac{(s+1)s(s-1)}{3!}\left[\Delta^2 f(x_0) - \Delta^2 f(x_{-1})\right]$$

$$+ \frac{(s+1)s(s-1)(s-2)}{4!}\Delta^4 f(x_{-2}) + \frac{(s+2)(s+1)s(s-1)(s-2)}{5!}\left[\Delta^4 f(x_{-1}) - \Delta^4 f(x_{-2})\right] + \cdots$$

On rearranging the terms, we get

$$P_n(x) = (1-s)f(x_0) + s\left[f(x_1)\right] + \frac{(s+1)s(s-1)}{3!}\left[\Delta^2 f(x_0)\right] + \frac{s(s-1)(2-s)}{3!}\Delta^2 f(x_{-1})$$

$$\frac{(s+2)(s+1)s(s-1)(s-2)}{5!}\left[\Delta^4 f(x_{-1})\right] + \frac{(3-s)(s+1)s(s-1)(s-2)}{5!}\Delta^4 f(x_{-2}) + \cdots$$

Let $r = 1-s$, then

$$P_n(x) = \left[r f(x_0) + s f(x_1)\right] + \left[\frac{(r+1)r(r-1)}{3!}\Delta^2 f(x_{-1}) + \frac{(s+1)s(s-1)}{3!}\left[\Delta^2 f(x_0)\right]\right]$$

$$\left[\frac{(r+2)(r+1)r(r-1)(r-2)}{5!}\Delta^4 f(x_{-2}) + \frac{(s+2)(s+1)s(s-1)(s-2)}{5!}\left[\Delta^4 f(x_{-1})\right]\right] + \cdots$$

$$(10.31)$$

On using relations between forward and central difference formulas, we have

$$P_n(x) = \left(r f(x_0) + s f(x_1)\right) + \left(\frac{(r+1)r(r-1)}{3!}\delta^2 f(x_0) + \frac{(s+1)s(s-1)}{3!}\delta^2 f(x_1)\right)$$

$$+ \left(\frac{(r+2)(r+1)r(r-1)(r-2)}{5!}\delta^4 f(x_0) + \frac{(s+2)(s+1)s(s-1)(s-2)}{5!}\delta^4 f(x_1)\right) + \cdots$$

$$(10.32)$$

The formula (10.32) is known as Everett central difference formula.

Example ── **10.9** ──

Consider Example 10.5, use Everett central difference formula to obtain the value of $f(1.70)$.

Ans.
At $x = 1.70$, we have

$$s = \frac{x - x_0}{h} = \frac{1.70 - 1.5}{0.25} = 0.8$$

$$r = 1 - s = 0.2$$

On using Everett central difference formula (10.32), we have

$$P(1.70) = \left(r f(x_0) + s f(x_1)\right) + \left(\frac{(r+1)r(r-1)}{3!}\delta^2 f(x_0) + \frac{(s+1)s(s-1)}{3!}\delta^2 f(x_1)\right) + \cdots$$

$$P(1.70) = \big((0.2)(0.405465) + 0.8(0.559616)\big)$$

$$+ \left(\frac{(1.2)(0.2)(-0.8)}{6}(-0.028170) + \frac{(1.8)(0.8)(-0.2)}{6}(-0.020620)\right)$$

$$P(1.70) = (0.081093 + 0.447693) + (0.000901 + 0.000990) = 0.530677$$

10.9 Steffensen Formula

In Steffensen formula, we will retain only odd order central differences and eliminate all the even order differences.

Gauss forward interpolation formula is as follow

$$P_n(x) = f(x_0) + s\Delta f(x_0) + \frac{s(s-1)}{2!}\Delta^2 f(x_{-1}) + \frac{(s+1)s(s-1)}{3!}\Delta^3 f(x_{-1})$$

$$+ \frac{(s+1)s(s-1)(s-2)}{4!}\Delta^4 f(x_{-2}) + \frac{(s+2)(s+1)s(s-1)(s-2)}{5!}\Delta^5 f(x_{-2}) + \cdots \qquad (10.33)$$

On eliminating even differences with the formulas

$$\Delta^2 f(x_{-1}) = \Delta f(x_0) - \Delta f(x_{-1})$$

$$\Delta^4 f(x_{-2}) = \Delta^3 f(x_{-1}) - \Delta^3 f(x_{-2})$$

$$\Delta^6 f(x_{-3}) = \Delta^5 f(x_{-2}) - \Delta^5 f(x_{-3}) \text{ etc.}$$

We have

$$P_n(x) = f(x_0) + s\Delta f(x_0) + \frac{s(s-1)}{2!}\big[\Delta f(x_0) - \Delta f(x_{-1})\big] + \frac{(s+1)s(s-1)}{3!}\Delta^3 f(x_{-1})$$

$$+ \frac{(s+1)s(s-1)(s-2)}{4!}\big[\Delta^3 f(x_{-1}) - \Delta^3 f(x_{-2})\big] + \frac{(s+2)(s+1)s(s-1)(s-2)}{5!}\Delta^5 f(x_{-2}) + \cdots$$

On rearranging the terms, we get

$$P_n(x) = f(x_0) + \left[\frac{(s+1)s}{2!}\Delta f(x_0) - \frac{s(s-1)}{2!}\Delta f(x_{-1})\right]$$

$$+ \left[\frac{(s+2)(s+1)s(s-1)}{4!}\Delta^3 f(x_{-1}) - \frac{(s+1)s(s-1)(s-2)}{4!}\Delta^3 f(x_{-2})\right]$$

$$+\left[\frac{(s+3)(s+2)(s+1)s(s-1)(s-2)}{6!}\Delta^5 f(x_{-2})-\frac{(s+2)(s+1)s(s-1)(s-2)(s-3)}{6!}\Delta^5 f(x_{-3})\right]+\cdots$$

$$(10.34)$$

On rewriting the above formula in central differences, we have

$$P_n(x)=f(x_0)+\left(\frac{(s+1)s}{2!}\delta f(x_{1/2})-\frac{s(s-1)}{2!}\delta f(x_{-1/2})\right)$$

$$+\left(\frac{(s+2)(s+1)s(s-1)}{4!}\delta^3 f(x_{1/2})-\frac{(s+1)s(s-1)(s-2)}{4!}\delta^3 f(x_{-1/2})\right)$$

$$+\left(\frac{(s+3)(s+2)(s+1)s(s-1)(s-2)}{6!}\delta^5 f(x_{1/2})-\frac{(s+2)(s+1)s(s-1)(s-2)(s-3)}{6!}\delta^5 f(x_{-1/2})\right)$$

$$+\cdots$$

$$(10.35)$$

The formula (10.35) is known as Steffensen central difference formula.

Example **10.10**

Find the value of $f(1.40)$ for the data set of Example 10.5. Use Steffensen central difference formula.

Ans.

At point $x = 1.40$, we have

$$s=\frac{x-x_0}{h}=\frac{1.4-1.5}{0.25}=-0.4$$

On using Steffensen central difference formula (10.35) and central differences from Example 10.5, we have

$$P(1.40)=f(x_0)+\left(\frac{(s+1)s}{2!}\delta f(x_{1/2})-\frac{s(s-1)}{2!}\delta f(x_{-1/2})\right)$$

$$+\left(\frac{(s+2)(s+1)s(s-1)}{4!}\delta^3 f(x_{1/2})-\frac{(s+1)s(s-1)(s-2)}{4!}\delta^3 f(x_{-1/2})\right)+\cdots$$

$$P(1.40)=0.405465+\left(\frac{(0.6)(-0.4)}{2}(0.154151)-\frac{(-0.4)(-1.4)}{2}(0.182321)\right)$$

$$+\left(\frac{(1.6)(0.6)(-0.4)(-1.4)}{24}(0.007550)-\frac{(0.6)(-0.4)(-1.4)(-2.4)}{24}(0.012653)\right)$$

$$P(1.40)=0.405465+(-0.018498-0.051050)+(0.000169+0.000425)=0.336511$$

Table 10.1 Finite Differences Formulas

Newton forward difference formula: $(0 < s < 1)$ $(x = x_0 + sh)$

$$P_n(x) = f(x_0) + \Delta f(x_0)(s) + \frac{\Delta^2 f(x_0)}{2!}(s)(s-1) + \frac{\Delta^3 f(x_0)}{3!}(s)(s-1)(s-2) + \cdots$$

x	$f(x)$	Δ	Δ^2	Δ^3
x_0	$f(x_0)$			
		$\Delta f(x_0)$		
x_1	$f(x_1)$		$\Delta^2 f(x_0)$	
		$\Delta f(x_1)$		$\Delta^3 f(x_0)$
x_2	$f(x_2)$		$\Delta^2 f(x_1)$	$\cdots$
		$\Delta f(x_2)$	$\cdots$	
x_3	$f(x_3)$			
$\cdots$	$\cdots$			

Newton backward difference formula: $(-1 < s < 0)$ $(x = x_n + sh)$

$$P_n(x) = f(x_n) + \nabla f(x_n)(s) + \frac{\nabla^2 f(x_n)}{2!}(s)(s+1) + \frac{\nabla^3 f(x_n)}{3!}(s)(s+1)(s+2) + \cdots$$

x	$f(x)$	∇	∇^2	∇^3
$\cdots$	$\cdots$			
x_{n-3}	$f(x_{n-3})$			
		$\nabla f(x_{n-2})$		
x_{n-2}	$f(x_{n-2})$	$\cdots$	$\nabla^2 f(x_{n-1})$	
		$\nabla f(x_{n-1})$		$\cdots$
x_{n-1}	$f(x_{n-1})$		$\nabla^2 f(x_n)$	$\nabla^3 f(x_n)$
		$\nabla f(x_n)$		
x_n	$f(x_n)$			

Gauss forward central difference formula: $(0 < s < 1/2)$ $(x = x_0 + sh)$

$$P_n(x) = f(x_0) + \delta f(x_{1/2})(s) + \frac{\delta^2 f(x_0)}{2!}(s)(s-1) + \frac{\delta^3 f(x_{1/2})}{3!}(s+1)(s)(s-1)$$

$$+ \frac{\delta^4 f(x_0)}{4!}(s+1)(s)(s-1)(s-2)$$

$$+ \frac{\delta^5 f(x_{1/2})}{5!}(s+2)(s+1)(s)(s-1)(s-2) + \cdots$$

x	$f(x)$	δ	δ^2	δ^3
...	...			
x_{-2}	$f(x_{-2})$		...	
		$\delta f(x_{-3/2})$		
x_{-1}	$f(x_{-1})$		$\delta^2 f(x_{-1})$	...
		$\delta f(x_{-1/2})$		$\delta^3 f(x_{-1/2})$
x_0	$f(x_0)$		$\delta^2 f(x_0)$	
		$\delta f(x_{1/2})$		$\delta^3 f(x_{1/2})$
x_1	$f(x_1)$		$\delta^2 f(x_1)$	...
		$\delta f(x_{3/2})$	...	
x_2	$f(x_2)$			
...	...			

Gauss backward central difference formula: $(-1/2 < s < 0)$ $(x = x_0 + sh)$

$$P_n(x) = f(x_0) + \delta f(x_{-1/2})(s) + \frac{\delta^2 f(x_0)}{2!}(s+1)(s) + \frac{\delta^3 f(x_{-1/2})}{3!}(s+1)(s)(s-1)$$

$$+ \frac{\delta^4 f(x_0)}{4!}(s+2)(s+1)(s)(s-1)$$

$$+ \frac{\delta^5 f(x_{-1/2})}{5!}(s+2)(s+1)(s)(s-1)(s-2) + \cdots$$

x	$f(x)$	δ	δ^2	δ^3
...	...			
x_{-2}	$f(x_{-2})$		...	
		$\delta f(x_{-3/2})$		
x_{-1}	$f(x_{-1})$		$\delta^2 f(x_{-1})$	...
		$\delta f(x_{-1/2})$		$\delta^3 f(x_{-1/2})$
x_0	$f(x_0)$		$\delta^2 f(x_0)$	
		$\delta f(x_{1/2})$		$\delta^3 f(x_{1/2})$
x_1	$f(x_1)$		$\delta^2 f(x_1)$	...
		$\delta f(x_{3/2})$	...	
x_2	$f(x_2)$			
...	...			

Stirling formula: $(-1/4 < s < 1/4)$ $(x = x_0 + sh)$

$$P_n(x) = f(x_0) + (s)\mu\delta f(x_0) + (s^2)\frac{1}{2!}\delta^2 f(x_0) + \frac{1}{3!}(s)(s^2 - 1^2)\mu\delta^3 f(x_0)$$

$$+ \frac{1}{4!}(s^2)(s^2 - 1^2)\delta^4 f(x_0) + \frac{1}{5!}(s)(s^2 - 1^2)(s^2 - 2^2)\mu\delta^5 f(x_0) + \cdots.$$

(OR)

$$P_n(x) = f(x_0) + (s)\frac{1}{2}\left(\delta f(x_{-1/2}) + \delta f(x_{1/2})\right) + (s^2)\frac{1}{2!}\delta^2 f(x_0)$$

$$+ \frac{1}{3!}(s)(s^2 - 1^2)\frac{1}{2}\left(\delta^3 f(x_{-1/2}) + \delta^3 f(x_{1/2})\right) + \frac{1}{4!}(s)(s^2 - 1^2)\delta^4 f(x_0)$$

$$+ \frac{1}{5!}(s)(s^2 - 1^2)(s^2 - 2^2)\frac{1}{2}\left(\delta^5 f(x_{-1/2}) + \delta^5 f(x_{1/2})\right) + \cdots.$$

x	$f(x)$	δ	δ^2	δ^3
...	...	...		
x_{-2}	$f(x_{-2})$		...	
		$\delta f(x_{-3/2})$		$\delta^3 f(x_{-1/2})$
x_{-1}	$f(x_{-1})$		$\delta^2 f(x_{-1})$	
		$\delta f(x_{-1/2})$		$\delta^3 f(x_{1/2})$
x_0	$f(x_0)$		$\delta^2 f(x_0)$	
		$\delta f(x_{1/2})$		...
x_1	$f(x_1)$		$\delta^2 f(x_1)$	
		$\delta f(x_{3/2})$	...	
x_2	$f(x_2)$	...		
...				

Bessel formula: $(1/4 < s < 3/4)$ $(x = x_0 + sh)$

$$P_n(x) = \mu f(x_{1/2}) + \left(s - \frac{1}{2}\right)\delta f(x_{1/2}) + \frac{(s)(s-1)}{2!}\mu\delta^2 f(x_{1/2}) + \left(s - \frac{1}{2}\right)\frac{(s)(s-1)}{3!}\delta^3 f(x_{1/2})$$

$$+ \frac{(s+1)(s)(s-1)(s-2)}{4!}\mu\delta^4 f(x_{1/2}) + \left(s - \frac{1}{2}\right)\frac{(s+1)(s)(s-1)(s-2)}{5!}\delta^5 f(x_{1/2})$$

$$+ \frac{(s+2)(s+1)(s)(s-1)(s-2)(s-3)}{6!}\mu\delta^6 f(x_{1/2}) + \cdots.$$

(OR)

$$P_n(x) = \frac{1}{2}(f(x_0) + f(x_1)) + \left(s - \frac{1}{2}\right)\delta f(x_{1/2}) + \frac{(s)(s-1)}{2!}\frac{1}{2}(\delta^2 f(x_0) + \delta^2 f(x_1))$$

$$+ \left(s - \frac{1}{2}\right)\frac{(s)(s-1)}{3!}\delta^3 f(x_{1/2}) + \frac{(s+1)(s)(s-1)(s-2)}{4!}\frac{1}{2}(\delta^4 f(x_0) + \delta^4 f(x_1))$$

$$+ \left(s - \frac{1}{2}\right)\frac{(s+1)(s)(s-1)(s-2)}{5!}\delta^5 f(x_{1/2})$$

$$+ \frac{(s+2)(s+1)(s)(s-1)(s-2)(s-3)}{6!}\frac{1}{2}(\delta^6 f(x_0) + \delta^6 f(x_1)) + \cdots.$$

x	$f(x)$	δ	δ^2	δ^3
...	...	...		
x_{-2}	$f(x_{-2})$		...	
		$\delta f(x_{-3/2})$		$\delta^3 f(x_{-1/2})$
x_{-1}	$f(x_{-1})$		$\delta^2 f(x_{-1})$	
		$\delta f(x_{-1/2})$		$\delta^3 f(x_{1/2})$
x_0	$f(x_0)$		$\delta^2 f(x_0)$	
		$\delta f(x_{1/2})$		...
x_1	$f(x_1)$		$\delta^2 f(x_1)$	
		$\delta f(x_{3/2})$	...	
x_2	$f(x_2)$	...		
...				

Everett formula: $(0 < s < 1)$ $(x = x_0 + sh)$ $(r = 1 - s)$

$$P_n(x) = (rf(x_0) + sf(x_1)) + \left(\frac{(r+1)r(r-1)}{3!}\delta^2 f(x_0) + \frac{(s+1)s(s-1)}{3!}\delta^2 f(x_1)\right)$$

$$+ \left(\frac{(r+2)(r+1)r(r-1)(r-2)}{5!}\delta^4 f(x_0) + \frac{(s+2)(s+1)s(s-1)(s-2)}{5!}\delta^4 f(x_1)\right)$$

$$+ \cdots$$

x	$f(x)$	δ	δ^2	δ^3
$\ldots$	$\ldots$	$\ldots$		
x_{-2}	$f(x_{-2})$			
		$\delta f(x_{-3/2})$		
x_{-1}	$f(x_{-1})$		$\delta^2 f(x_{-1})$	
		$\delta f(x_{-1/2})$		$\delta^3 f(x_{-1/2})$
x_0	$f(x_0)$		$\delta^2 f(x_0)$	
		$\delta f(x_{1/2})$		$\delta^3 f(x_1/2)$
x_1	$f(x_1)$		$\delta^2 f(x_1)$	
		$\delta f(x_{3/2})$	$\ldots$	$\ldots$
x_2	$f(x_2)$	$\ldots$		

Steffensen formula: $(-1/2 < s < 1/2)$ $(x = x_0 + sh)$

$$P_n(x) = f(x_0) + \left(\frac{(s+1)s}{2!}\delta f(x_{1/2}) - \frac{s(s-1)}{2!}\delta f(x_{-1/2})\right)$$

$$+ \left(\frac{(s+2)(s+1)s(s-1)}{4!}\delta^3 f(x_{1/2}) - \frac{(s+1)s(s-1)(s-2)}{4!}\delta^3 f(x_{-1/2})\right)$$

$$+ \left(\frac{(s+3)(s+2)(s+1)s(s-1)(s-2)}{6!}\delta^5 f(x_{1/2})\right.$$

$$\left. - \frac{(s+2)(s+1)s(s-1)(s-2)(s-3)}{6!}\delta^5 f(x_{-1/2})\right) + \cdots$$

x	$f(x)$	δ	δ^2	δ^3
$\ldots$	$\ldots$	$\ldots$		
x_{-2}	$f(x_{-2})$			
		$\delta f(x_{-3/2})$		
x_{-1}	$f(x_{-1})$		$\delta^2 f(x_{-1})$	
		$\delta f(x_{-1/2})$		$\delta^3 f(x_{-1/2})$
x_0	$f(x_0)$		$\delta^2 f(x_0)$	
		$\delta f(x_{1/2})$		$\delta^3 f(x_{1/2})$
x_1	$f(x_1)$		$\delta^2 f(x_1)$	
		$\delta f(x_{3/2})$	$\ldots$	$\ldots$
x_2	$f(x_2)$	$\ldots$		

Motivation and Few Remarks:

The interpolating polynomial is unique for a given set of data points. All the interpolation methods produce same function value at any intermediate point theoretically. But, we have different sources of errors in practical implications of all these interpolation methods to find function values. The main motivation to discuss so many interpolation methods is to minimize these errors. There are following three major sources of errors in approximating the given function $f(x)$ with the interpolating polynomial, $P_n(x)$.

1. **Approximation Error:** We have already discussed that the error term in interpolating polynomial of degree n for a given function $f(x)$ is given by

$$\varepsilon_n(x) = \frac{f^{n+1}(\xi)}{(n+1)!}(x-x_0)(x-x_1)\cdots(x-x_n) = \frac{f^{n+1}(\xi)}{(n+1)!}\prod_{i=0}^{n}(x-x_i)$$

This error can be reduced by taking appropriate numbers of data points, such that interpolating polynomial is of sufficient degree. But, we should keep it in mind that a polynomial of very large degree suffers from the polynomial oscillations. For example, consider the following polynomial of degree 13

$$P_{13}(x) = (x-6)(x-5)(x-4)(x-3)(x-2)(x-1)(x)(x+1)(x+2)(x+3)(x+4)$$
$$(x+5)(x+6)$$

It is easy to see from the Figure 10.1 that this polynomial has very large values (up to 10^7) as we move away from the zeroes of this polynomial and the origin. It is due to use of large degree of polynomial. As degree of polynomial increases, the oscillation increases wildly. So, we have to use polynomial approximation up to a reasonable degree.

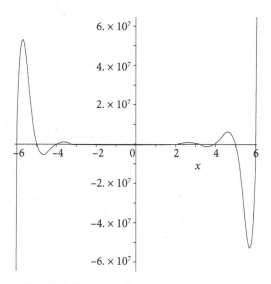

Fig. 10.1 Example of polynomial oscillations

2. **Round-off Error**: We have already discussed that the rounding error will increase with the number of computations. It is worth to remind here that the numbers 10.13 and 0.1013 have same significant digits, but the round off errors in these numbers are 0.005 and 0.00005 respectively. So, it is always recommended to work with the number having absolute value less than 1 to reduce rounding error.

We have to select the appropriate difference formula for the computation of interpolated value with less rounding error. For example, Newton forward difference formula is used to compute the value at the beginning of the table. The question is why to use Newton forward difference formula? If the value of x is near the starting point x_0, then the value of $s = \dfrac{x - x_0}{h}$ is in the interval $(0, 1)$. So, the rounding error is less in the Newton forward difference formula which contains s in each term except the first term. Similar arguments authenticate our assertions that Newton forward difference formula is suitable for the values at the upper end of the table, Newton backward difference formula is suitable for the values at the lower end of the table, and finally if the value is in the mid of the table, central difference formulas are suitable.

Newton forward difference formula: $(0 < s < 1)$
Newton backward difference formula: $(-1 < s < 0)$
Gauss forward central difference formula: $(0 < s < 1/2)$
Gauss backward central difference formula: $(-1/2 < s < 0)$
Stirling formula: $(-1/4 < s < 1/4)$
Bessel formula: $(1/4 < s < 3/4)$
Everett formula: $(0 < s < 1)$
Steffensen formula: $(-1/2 < s < 1/2)$

3. **Truncation Error**: We have already discussed that polynomial approximation of very large degree is not suitable for practical applications. So, we can neglect higher order differences by truncating the interpolation formula. But, it will produce the truncation error. An appropriate selection is required to minimize the truncation error. For example, in second order approximations of function values, the truncation terms in various central difference formulas are given by

i. Gauss forward: $\dfrac{\delta^3 f(x_{1/2})}{3!}(s+1)(s)(s-1)$

ii. Gauss backward: $\dfrac{\delta^3 f(x_{-1/2})}{3!}(s+1)(s)(s-1)$

iii. Stirling: $\dfrac{(s)(s^2-1)}{3!}\left[\dfrac{\delta^3 f(x_{1/2}) + \delta^3 f(x_{-1/2})}{2}\right]$

iv. Bessel:
$$\left(s-\frac{1}{2}\right)\frac{(s)(s-1)}{3!}\delta^3 f(x_{1/2})$$

v. Everett:
$$\frac{(r+2)(r+1)r(r-1)(r-2)}{5!}\delta^4 f(x_0)+\frac{(s+2)(s+1)s(s-1)(s-2)}{5!}\delta^4 f(x_1)$$

vi. Steffensen:
$$\frac{(s+2)(s+1)s(s-1)}{4!}\delta^3 f(x_{1/2})-\frac{(s+1)s(s-1)(s-2)}{4!}\delta^3 f(x_{-1/2})$$

It is clear from the truncation terms that the truncation terms are of order 3 except Everett formula. In Everett formula the approximation is exact up to polynomial of degree 3, and it has truncation term of order 4. Therefore, Everett formula has minimum truncation error for second order approximations of function values. Similarly, Steffensen formula has minimum truncation error for third order approximations.

To conclude, we have to keep in mind the round-off error, truncation error and degree of polynomial approximations to get the desired accuracy and optimize the computational results.

Example   **10.11**

In a given table of $\sin(\theta)$, we have following values

θ(In degree)	0	15	30	45	60	75	90
$\sin(\theta)$	0	0.258819	0.5	0.707107	0.866025	0.965926	1

Compute the values of the function $\sin(\theta)$ at the values $\theta=5°, 40°, 48°, 50°, 55°, 57°, 87°$.

Ans.
First, we construct the finite difference table by simply subtracting the values of $\sin(\theta)$, and then we proceed further. Finite difference table for the given values of $\sin(\theta)$ is as follows

θ	$\sin(\theta)$	$\Delta(\nabla,\delta)$	$\Delta^2(\nabla^2,\delta^2)$	$\Delta^3(\nabla^3,\delta^3)$	$\Delta^4(\nabla^4,\delta^4)$	$\Delta^5(\nabla^5,\delta^5)$	$\Delta^6(\nabla^6,\delta^6)$
0	0						
		0.258819					
15	0.258819		-0.017638				
		0.241181		-0.016436			
30	0.500000		-0.034074		0.002321		
		0.207107		-0.014115		0.000966	
45	0.707107		-0.048189		0.003287		-0.000235
		0.158918		-0.010828		0.000731	
60	0.866025		-0.059017		0.004018		
		0.099901		-0.006810			
75	0.965926		-0.065827				
		0.034074					
90	1						

The values of $\theta = 5^\circ, 40^\circ, 48^\circ, 50^\circ, 55^\circ, 57^\circ, 87^\circ$ are at starting, middle and at the end of the table, so we will use appropriate forward, backward and central difference formulas to compute values of the function $\sin(\theta)$. We have following values of s corresponding to these values of θ.

$$s = \frac{x - x_0}{h} = \frac{5 - 0}{15} = \frac{1}{3}; \text{ Newton forward difference formula}$$

$$s = \frac{x - x_0}{h} = \frac{40 - 45}{15} = \frac{-1}{3}; \text{ Gauss backward difference formula}$$

$$s = \frac{x - x_0}{h} = \frac{48 - 45}{15} = \frac{1}{5}; \text{ Stirling central difference formula}$$

$$s = \frac{x - x_0}{h} = \frac{50 - 45}{15} = \frac{1}{3}; \text{ Gauss forward difference (or) Steffensen formula}$$

$$s = \frac{x - x_0}{h} = \frac{55 - 45}{15} = \frac{2}{3}; \text{ Bessel formula}$$

$$s = \frac{x - x_0}{h} = \frac{57 - 45}{15} = \frac{4}{5}; \text{ Everett formula}$$

$$s = \frac{x - x_6}{h} = \frac{87 - 90}{15} = -\frac{1}{5}; \text{ Newton backward difference formula}$$

Note that in Newton forward difference formula, the point $x_0 = 0$, while in all central difference formulas $x_0 = 45$. It is due to the fact that the tabular points are given different subscripts in forward and central difference tables.

The value of $\sin(\theta)$ **at** $\theta = 5°$: $s = \dfrac{x - x_0}{h} = \dfrac{5 - 0}{15} = \dfrac{1}{3}$

Newton forward difference formula:

θ	$\sin(\theta)$	Δ	Δ^2	Δ^3	Δ^4	Δ^5	Δ^6
$0(x_0)$	$0(y_0)$						
		$0.258819(\Delta y_0)$					
$15(x_1)$	$0.258819(y_1)$		$-0.017638(\Delta^2 y_0)$				
		$0.241181(\Delta y_1)$		$-0.016436(\Delta^3 y_0)$			
$30(x_2)$	$0.500000(y_2)$		$-0.034074(\Delta^2 y_1)$		$0.002321(\Delta^4 y_0)$		
		$0.207107(\Delta y_2)$		$-0.014115(\Delta^3 y_1)$		$0.000966(\Delta^5 y_0)$	
$45(x_3)$	$0.707107(y_3)$		$-0.048189(\Delta^2 y_2)$		$0.003287(\Delta^4 y_1)$		$-0.000235(\Delta^6 y_0)$
		$0.158918(\Delta y_3)$		$-0.010828(\Delta^3 y_2)$		$0.000731(\Delta^5 y_1)$	
$60(x_4)$	$0.866025(y_4)$		$-0.059017(\Delta^2 y_3)$		$0.004018(\Delta^4 y_2)$		
		$0.099901(\Delta y_4)$		$-0.006810(\Delta^3 y_3)$			
$75(x_5)$	$0.965926(y_5)$		$-0.065827(\Delta^2 y_4)$				
		$0.034074(\Delta y_5)$					
$90(x_6)$	$1(y_6)$						

Newton forward difference formula is given by

$$P(x) = f(x_0) + \Delta f(x_0)(s) + \frac{\Delta^2 f(x_0)}{2!}(s)(s-1) + \cdots + \frac{\Delta^n f(x_0)}{n!}(s)(s-1)\cdots(s-n+1)$$

$$P(5°) = 0 + \left(\frac{1}{3}\right)(0.258819) + \left(\frac{1}{2}\right)\left(\frac{1}{3}\right)\left(\frac{-2}{3}\right)(-0.017638) + \left(\frac{1}{6}\right)\left(\frac{1}{3}\right)\left(\frac{-2}{3}\right)\left(\frac{-5}{3}\right)(-0.016436)$$

$$+ \left(\frac{1}{24}\right)\left(\frac{1}{3}\right)\left(\frac{-2}{3}\right)\left(\frac{-5}{3}\right)\left(\frac{-8}{3}\right)(0.002321) + \left(\frac{1}{120}\right)\left(\frac{1}{3}\right)\left(\frac{-2}{3}\right)\left(\frac{-5}{3}\right)\left(\frac{-8}{3}\right)\left(\frac{-11}{3}\right)(0.000966)$$

$$+ \left(\frac{1}{720}\right)\left(\frac{1}{3}\right)\left(\frac{-2}{3}\right)\left(\frac{-5}{3}\right)\left(\frac{-8}{3}\right)\left(\frac{-11}{3}\right)\left(\frac{-14}{3}\right)(-0.000235)$$

$$P(5°) = 0 + 0.086273 + 0.001960 - 0.001014 - 0.000094 + 0.000031 + 0.000008$$

$$= 0.087164 \textbf{ (Approximate value of } \sin(\theta) \textbf{ at } \theta = 5°)$$

Note that the exact value of $\sin(\theta)$ up to six decimal points at $\theta = 5°$ is 0.087156.

Error = Exact value – Approximate value

$$= - 0.000008$$

Value of $\sin(\theta)$ **at** $\theta = 40°$: $s = \dfrac{x - x_0}{h} = \dfrac{40 - 45}{15} = \dfrac{-1}{3}$;

Gauss backward difference formula:

θ	$\sin(\theta)$	δ	δ^2	δ^3	δ^4	δ^5	δ^6
$0(x_{-3})$	$0(y_{-3})$						
		$0.258819(\delta y_{-5/2})$					
$15(x_{-2})$	$0.258819(y_{-2})$		$-0.017638(\delta^2 y_{-2})$				
		$0.241181(\delta y_{-3/2})$		$-0.016436(\delta^3 y_{-3/2})$			
$30(x_{-1})$	$0.500000(y_{-1})$		$-0.034074(\delta^2 y_{-1})$		$0.002321(\delta^4 y_{-1})$		
		$0.207107(\delta y_{-1/2})$		$-0.014115(\delta^3 y_{-1/2})$		$0.000966(\delta^5 y_{-1/2})$	
$45(x_0)$	$0.707107(y_0)$		$-0.048189(\delta^2 y_0)$		$0.003287(\delta^4 y_0)$		$-0.000235(\delta^6 y_0)$
		$0.158918(\delta y_{1/2})$		$-0.010828(\delta^3 y_{1/2})$		$0.000731(\delta^5 y_{1/2})$	
$60(x_1)$	$0.866025(y_1)$		$-0.059017(\delta^2 y_1)$		$0.004018(\delta^4 y_1)$		
		$0.099901(\delta y_{3/2})$		$-0.006810(\delta^3 y_{3/2})$			
$75(x_2)$	$0.965926(y_2)$		$-0.065827(\delta^2 y_2)$				
		$0.034074(\delta y_{5/2})$					
$90(x_3)$	$1(y_3)$						

Gauss backward difference formula

$$P(x) = f(x_0) + \delta f(x_{-1/2})(s) + \frac{\delta^2 f(x_0)}{2!}(s+1)(s) + \frac{\delta^3 f(x_{-1/2})}{3!}(s+1)(s)(s-1)$$

$$+ \frac{\delta^4 f(x_0)}{4!}(s+2)(s+1)(s)(s-1) + \frac{\delta^5 f(x_{-1/2})}{5!}(s+2)(s+1)(s)(s-1)(s-2) + \cdots$$

$$P(40°) = 0.707107 + \left(\frac{-1}{3}\right)(0.207107) + \left(\frac{1}{2}\right)\left(\frac{2}{3}\right)\left(\frac{-1}{3}\right)(-0.048189)$$

$$+ \left(\frac{1}{6}\right)\left(\frac{2}{3}\right)\left(\frac{-1}{3}\right)\left(\frac{-4}{3}\right)(-0.014115)$$

$$+ \left(\frac{1}{24}\right)\left(\frac{5}{3}\right)\left(\frac{2}{3}\right)\left(\frac{-1}{3}\right)\left(\frac{-4}{3}\right)(0.003287) + \left(\frac{1}{120}\right)\left(\frac{5}{3}\right)\left(\frac{2}{3}\right)\left(\frac{-1}{3}\right)\left(\frac{-4}{3}\right)\left(\frac{-7}{3}\right)(0.000966)$$

$$+ \left(\frac{1}{720}\right)\left(\frac{8}{3}\right)\left(\frac{5}{3}\right)\left(\frac{2}{3}\right)\left(\frac{-1}{3}\right)\left(\frac{-4}{3}\right)\left(\frac{-7}{3}\right)(-0.000235)$$

$$P(40°) = 0.707107 - 0.069036 + 0.005354 - 0.000697 + 0.000068 - 0.000009 + 0.000003$$

$$= 0.642790 \text{ (\textbf{Approximate value of} } \sin(\theta) \text{ \textbf{at} } \theta = 40°)$$

Exact value of $\sin(\theta)$ up to six decimal points at $\theta = 40°$ is 0.642788

Value of $\sin(\theta)$ **at** $\theta = 48°$: $s = \dfrac{x - x_0}{h} = \dfrac{48 - 45}{15} = \dfrac{1}{5}$;

Stirling central difference formula:

θ	$\sin(\theta)$	δ	δ^2	δ^3	δ^4	δ^5	δ^6
$0(x_{-3})$	$0(y_{-3})$						
		$0.258819(\delta y_{-5/2})$					
$15(x_{-2})$	$0.258819(y_{-2})$		$-0.017638(\delta^2 y_{-2})$				
		$0.241181(\delta y_{-3/2})$		$-0.016436(\delta^3 y_{-3/2})$			
$30(x_{-1})$	$0.500000(y_{-1})$		$-0.034074(\delta^2 y_{-1})$		$0.002321(\delta^4 y_{-1})$		
		$0.207107(\delta y_{-1/2})$		$-0.014115(\delta^3 y_{-1/2})$		$0.000966(\delta^5 y_{-1/2})$	
$45(x_0)$	$0.707107(y_0)$		$-0.048189(\delta^2 y_0)$		$0.003287(\delta^4 y_0)$		$-0.000235(\delta^6 y_0)$
		$0.158918(\delta y_{1/2})$		$-0.010828(\delta^3 y_{1/2})$		$0.000731(\delta^5 y_{1/2})$	
$60(x_1)$	$0.866025(y_1)$		$-0.059017(\delta^2 y_1)$		$0.004018(\delta^4 y_1)$		
		$0.099901(\delta y_{3/2})$		$-0.006810(\delta^3 y_{3/2})$			
$75(x_2)$	$0.965926(y_2)$		$-0.065827(\delta^2 y_2)$				
		$0.034074(\delta y_{5/2})$					
$90(x_3)$	$1(y_3)$						

Stirling central difference formula

$$P(x) = f(x_0) + (s)\frac{1}{2}\left(\delta f(x_{-1/2}) + \delta f(x_{1/2})\right) + (s^2)\frac{1}{2!}\delta^2 f(x_0) + \frac{1}{3!}(s)(s^2 - 1^2)\frac{1}{2}\left(\delta^3 f(x_{-1/2}) + \delta^3 f(x_{1/2})\right)$$

$$+ \frac{1}{4!}(s^2)(s^2 - 1^2)\delta^4 f(x_0) + \frac{1}{5!}(s)(s^2 - 1^2)(s^2 - 2^2)\frac{1}{2}\left(\delta^5 f(x_{-1/2}) + \delta^5 f(x_{1/2})\right) + \cdots$$

$$P(48°) = 0.707107 + \left(\frac{1}{5}\right)\left(\frac{1}{2}\right)(0.207107 + 0.158918) + \left(\frac{1}{5}\right)\left(\frac{1}{5}\right)\left(\frac{1}{2}\right)(-0.048189)$$

$$+ \left(\frac{1}{6}\right)\left(\frac{1}{5}\right)\left(\frac{-24}{25}\right)\left(\frac{1}{2}\right)(-0.014115 - 0.010828) + \left(\frac{1}{24}\right)\left(\frac{1}{5}\right)\left(\frac{1}{5}\right)\left(\frac{-24}{25}\right)(0.003287)$$

$$+ \left(\frac{1}{120}\right)\left(\frac{1}{5}\right)\left(\frac{-24}{25}\right)\left(\frac{-99}{25}\right)\left(\frac{1}{2}\right)(0.000966 + 0.000731)$$

$$+ \left(\frac{1}{720}\right)\left(\frac{1}{5}\right)\left(\frac{1}{5}\right)\left(\frac{-24}{25}\right)\left(\frac{-99}{25}\right)(-0.000235)$$

$$P(48°) = 0.707107 + 0.036602 - 0.000964 + 0.000399 - 0.000005 + 0.000005 - 0$$

$$P(48°) = 0.743144 \text{ (\textbf{Approximate value of} } \sin(\theta) \text{ \textbf{at} } \theta = 48°)$$

Exact value of $\sin(\theta)$ up to six decimal points at $\theta = 48°$ is 0.743145

Value of $\sin(\theta)$ **at** $\theta = 50°$: $s = \dfrac{x - x_0}{h} = \dfrac{50 - 45}{15} = \dfrac{1}{3}$;

Gauss forward difference formula (or) Steffensen formula:

θ	$\sin(\theta)$	δ	δ^2	δ^3	δ^4	δ^5	δ^6
$0(x_{-3})$	$0(y_{-3})$						
		$0.258819(\delta y_{-5/2})$					
$15(x_{-2})$	$0.258819(y_{-2})$		$-0.017638(\delta^2 y_{-2})$				
		$0.241181(\delta y_{-3/2})$		$-0.016436(\delta^3 y_{-3/2})$			
$30(x_{-1})$	$0.500000(y_{-1})$		$-0.034074(\delta^2 y_{-1})$		$0.002321(\delta^4 y_{-1})$		
		$0.207107(\delta y_{-1/2})$		$-0.014115(\delta^3 y_{-1/2})$		$0.000966(\delta^5 y_{-1/2})$	
$45(x_0)$	$0.707107(y_0)$		$-0.048189(\delta^2 y_0)$		$0.003287(\delta^4 y_0)$		$-0.000235(\delta^6 y_0)$
		$0.158918(\delta y_{1/2})$		$-0.010828(\delta^3 y_{1/2})$		$0.000731(\delta^5 y_{1/2})$	
$60(x_1)$	$0.866025(y_1)$		$-0.059017(\delta^2 y_1)$		$0.004018(\delta^4 y_1)$		
		$0.099901(\delta y_{3/2})$		$-0.006810(\delta^3 y_{3/2})$			
$75(x_2)$	$0.965926(y_2)$		$-0.065827(\delta^2 y_2)$				
		$0.034074(\delta y_{5/2})$					
$90(x_3)$	$1(y_3)$						

Gauss forward difference formula

$$P(x) = f(x_0) + \delta f(x_{1/2})(s) + \frac{\delta^2 f(x_0)}{2!}(s)(s-1) + \frac{\delta^3 f(x_{1/2})}{3!}(s+1)(s)(s-1)$$

$$+ \frac{\delta^4 f(x_0)}{4!}(s+1)(s)(s-1)(s-2) + \frac{\delta^5 f(x_{1/2})}{5!}(s+2)(s+1)(s)(s-1)(s-2) + \cdots$$

$$P(50°) = 0.707107 + \left(\frac{1}{3}\right)(0.158918) + \left(\frac{1}{2}\right)\left(\frac{1}{3}\right)\left(\frac{-2}{3}\right)(-0.048189) + \left(\frac{1}{6}\right)\left(\frac{4}{3}\right)\left(\frac{1}{3}\right)\left(\frac{-2}{3}\right)(-0.010828)$$

$$+ \left(\frac{1}{24}\right)\left(\frac{4}{3}\right)\left(\frac{1}{3}\right)\left(\frac{-2}{3}\right)\left(\frac{-5}{3}\right)(0.003287) + \left(\frac{1}{120}\right)\left(\frac{7}{3}\right)\left(\frac{4}{3}\right)\left(\frac{1}{3}\right)\left(\frac{-2}{3}\right)\left(\frac{-5}{3}\right)(0.000731)$$

$$+ \left(\frac{1}{720}\right)\left(\frac{7}{3}\right)\left(\frac{4}{3}\right)\left(\frac{1}{3}\right)\left(\frac{-2}{3}\right)\left(\frac{-5}{3}\right)\left(\frac{-8}{3}\right)(-0.000235)$$

$$P(50°) = 0.707107 + 0.052973 - 0.005354 + 0.000535 + 0.000068 + 0.000007 + 0.000003$$

$P(50°) = 0.766047$ **(Approximate value of** $\sin(\theta)$ **at** $\theta = 50°$**)**

Exact value of $\sin(\theta)$ up to six decimal points at $\theta = 50°$ is 0.766044

Steffensen formula

$$P(x) = f(x_0) + \left(\frac{(s+1)s}{2!}\delta f(x_{1/2}) - \frac{s(s-1)}{2!}\delta f(x_{-1/2})\right)$$

$$+ \left(\frac{(s+2)(s+1)s(s-1)}{4!}\delta^3 f(x_{1/2}) - \frac{(s+1)s(s-1)(s-2)}{4!}\delta^3 f(x_{-1/2})\right)$$

$$+ \left(\frac{(s+3)(s+2)(s+1)s(s-1)(s-2)}{6!}\delta^5 f(x_{1/2}) - \frac{(s+2)(s+1)s(s-1)(s-2)(s-3)}{6!}\delta^5 f(x_{-1/2})\right)$$

$$+ \cdots$$

$$P(50°) = 0.707107 + \left(\frac{4}{3}\right)\left(\frac{1}{3}\right)\left(\frac{1}{2}\right)(0.158918) - \left(\frac{1}{3}\right)\left(\frac{-2}{3}\right)\left(\frac{1}{2}\right)(0.207107)$$

$$+ \left(\frac{7}{3}\right)\left(\frac{4}{3}\right)\left(\frac{1}{3}\right)\left(\frac{-2}{3}\right)\left(\frac{1}{24}\right)(-0.010828) - \left(\frac{4}{3}\right)\left(\frac{1}{3}\right)\left(\frac{-2}{3}\right)\left(\frac{-5}{3}\right)\left(\frac{1}{24}\right)(-0.014115)$$

$$+ \left(\frac{10}{3}\right)\left(\frac{7}{3}\right)\left(\frac{4}{3}\right)\left(\frac{1}{3}\right)\left(\frac{-2}{3}\right)\left(\frac{-5}{3}\right)\left(\frac{1}{720}\right)(0.000731)$$

$$- \left(\frac{7}{3}\right)\left(\frac{4}{3}\right)\left(\frac{1}{3}\right)\left(\frac{-2}{3}\right)\left(\frac{-5}{3}\right)\left(\frac{-8}{3}\right)\left(\frac{1}{720}\right)(0.000966)$$

$$P(50°) = 0.707107 + 0.035315 + 0.023012 + 0.000312 + 0.000290 + 0.000004 + 0.000004$$

$$P(50°) = 0.766044 \text{ (\textbf{Approximate value of} } \sin(\theta) \text{ \textbf{at} } \theta = 50°)$$

Exact value of $\sin(\theta)$ up to six decimal points at $\theta = 50°$ is 0.766044

Value of $\sin(\theta)$ **at** $\theta = 55°$: $s = \dfrac{x - x_0}{h} = \dfrac{55 - 45}{15} = \dfrac{2}{3}$;

Bessel formula:

$$P(x) = \frac{1}{2}\left(f(x_0) + f(x_1)\right) + \left(s - \frac{1}{2}\right)\delta f(x_{1/2}) + \frac{(s)(s-1)}{2!}\frac{1}{2}\left(\delta^2 f(x_0) + \delta^2 f(x_1)\right)$$

$$+ \left(s - \frac{1}{2}\right)\frac{(s)(s-1)}{3!}\delta^3 f(x_{1/2}) + \frac{(s+1)(s)(s-1)(s-2)}{4!}\frac{1}{2}\left(\delta^4 f(x_0) + \delta^4 f(x_1)\right)$$

$$+ \left(s - \frac{1}{2}\right)\frac{(s+1)(s)(s-1)(s-2)}{5!}\delta^5 f(x_{1/2})$$

$$+ \frac{(s+2)(s+1)(s)(s-1)(s-2)(s-3)}{6!}\frac{1}{2}\left(\delta^6 f(x_0) + \delta^6 f(x_1)\right) + \cdots$$

$$P(55°) = \frac{1}{2}(0.707107 + 0.866025) + \left(\frac{1}{6}\right)(0.158918) + \left(\frac{1}{2}\right)\left(\frac{2}{3}\right)\left(\frac{-1}{3}\right)\left(\frac{1}{2}\right)(-0.048189 - 0.059017)$$

$$+ \left(\frac{1}{6}\right)\left(\frac{2}{3}\right)\left(\frac{-1}{3}\right)\left(\frac{1}{6}\right)(-0.010828) + \left(\frac{1}{24}\right)\left(\frac{5}{3}\right)\left(\frac{2}{3}\right)\left(\frac{-1}{3}\right)\left(\frac{-4}{3}\right)\left(\frac{1}{2}\right)(0.003287 + 0.004018)$$

$$+ \left(\frac{1}{6}\right)\left(\frac{5}{3}\right)\left(\frac{2}{3}\right)\left(\frac{-1}{3}\right)\left(\frac{-4}{3}\right)\left(\frac{1}{120}\right)(0.000731)$$

$$P(55°) = 0.786566 + 0.026486 + 0.005956 + 0.000067 + 0.000075$$

$$P(55°) = 0.819152 \text{ (\textbf{Approximate value of} } \sin(\theta) \text{ \textbf{at} } \theta = 55°)$$

Exact value of $\sin(\theta)$ up to six decimal points at $\theta = 55°$ is 0.819152

Value of $\sin(\theta)$ **at** $\theta = 57°$: $s = \dfrac{x - x_0}{h} = \dfrac{57 - 45}{15} = \dfrac{4}{5}$;

Everett formula:

$$P(x) = \left(r\,f(x_0) + s\,f(x_1)\right) + \left(\frac{(r+1)r(r-1)}{3!}\delta^2 f(x_0) + \frac{(s+1)s(s-1)}{3!}\delta^2 f(x_1)\right)$$

$$+ \left(\frac{(r+2)(r+1)r(r-1)(r-2)}{5!}\delta^4 f(x_0) + \frac{(s+2)(s+1)s(s-1)(s-2)}{5!}\delta^4 f(x_1)\right)$$

$$+ \dots$$

$$P(57°) = \left(\frac{1}{5}\right)(0.707107) + \left(\frac{4}{5}\right)(0.866025) + \left(\frac{1}{6}\right)\left(\frac{6}{5}\right)\left(\frac{1}{5}\right)\left(\frac{-4}{5}\right)(-0.048189)$$

$$+ \left(\frac{1}{6}\right)\left(\frac{9}{5}\right)\left(\frac{4}{5}\right)\left(\frac{-1}{5}\right)(-0.059017) + \left(\frac{1}{120}\right)\left(\frac{11}{5}\right)\left(\frac{6}{5}\right)\left(\frac{1}{5}\right)\left(\frac{-4}{5}\right)\left(\frac{-9}{5}\right)(0.003287)$$

$$+ \left(\frac{1}{120}\right)\left(\frac{14}{5}\right)\left(\frac{9}{5}\right)\left(\frac{4}{5}\right)\left(\frac{-1}{5}\right)\left(\frac{-6}{5}\right)(0.004018)$$

$$P(57°) = 0.141421 + 0.692820 + 0.001542 + 0.002833 + 0.000021 + 0.000032$$

$$P(57°) = 0.838669 \ (\textbf{Approximate value of } \sin(\theta) \textbf{ at } \theta = 57°)$$

Exact value of $\sin(\theta)$ up to six decimal points at $\theta = 57°$ is 0.838670

Value of $\sin(\theta)$ **at** $\theta = 87°$: $s = \dfrac{x - x_6}{h} = \dfrac{87 - 90}{15} = -\dfrac{1}{5}$;

Newton backward difference formula:

θ	$\sin(\theta)$	∇	∇^2	∇^3	∇^4	∇^5	∇^6
$0(x_0)$	$0(y_0)$						
		$0.258819(\nabla y_1)$					
$15(x_1)$	$0.258819(y_1)$		$-0.017638(\nabla^2 y_2)$				
		$0.241181(\nabla y_2)$		$-0.016436(\nabla^3 y_3)$			
$30(x_2)$	$0.500000(y_2)$		$-0.034074(\nabla^2 y_3)$		$0.002321(\nabla^4 y_4)$		
		$0.207107(\nabla y_3)$		$-0.014115(\nabla^3 y_4)$		$0.000966(\nabla^5 y_5)$	
$45(x_3)$	$0.707107(y_3)$		$-0.048189(\nabla^2 y_4)$		$0.003287(\nabla^4 y_5)$		$-0.000235(\nabla^6 y_6)$
		$0.158918(\nabla y_4)$		$-0.010828(\nabla^3 y_5)$		$0.000731(\nabla^5 y_6)$	
$60(x_4)$	$0.866025(y_4)$		$-0.059017(\nabla^2 y_5)$		$0.004018(\nabla^4 y_6)$		
		$0.099901(\nabla y_5)$		$-0.006810(\nabla^3 y_6)$			
$75(x_5)$	$0.965926(y_5)$		$-0.065827(\nabla^2 y_6)$				
		$0.034074(\nabla y_6)$					
$90(x_6)$	$1(y_6)$						

$$P(x) = f(x_n) + \nabla f(x_n)(s) + \frac{\nabla^2 f(x_n)}{2!}(s)(s+1) + \frac{\nabla^3 f(x_n)}{3!}(s)(s+1)(s+2) + \cdots$$

$$P(87°) = 1 + \left(\frac{-1}{5}\right)(0.034074) + \left(\frac{1}{2}\right)\left(\frac{-1}{5}\right)\left(\frac{4}{5}\right)(-0.065827) + \left(\frac{1}{6}\right)\left(\frac{-1}{5}\right)\left(\frac{4}{5}\right)\left(\frac{9}{5}\right)(-0.006810)$$

$$+ \left(\frac{1}{24}\right)\left(\frac{-1}{5}\right)\left(\frac{4}{5}\right)\left(\frac{9}{5}\right)\left(\frac{14}{5}\right)(0.004018) + \left(\frac{1}{120}\right)\left(\frac{-1}{5}\right)\left(\frac{4}{5}\right)\left(\frac{9}{5}\right)\left(\frac{14}{5}\right)\left(\frac{19}{5}\right)(0.000731)$$

$$+ \left(\frac{1}{720}\right)\left(\frac{-1}{5}\right)\left(\frac{4}{5}\right)\left(\frac{9}{5}\right)\left(\frac{14}{5}\right)\left(\frac{19}{5}\right)\left(\frac{24}{5}\right)(-0.000235)$$

$$P(87°) = 1 - 0.006815 + 0.005266 + 0.000327 - 0.000135 - 0.000018 + 0.000007$$

$$P(87°) = 0.998632 \text{ (\textbf{Approximate value of} } \sin(\theta) \textbf{ at } \theta = 87°)$$

Exact value of $\sin(\theta)$ up to six decimal points at $\theta = 87°$ is 0.998630

Example   10.12

Indian government carries out census after a gap of 10 years. According to census details conducted from 1961 to 2011, the populations up to four decimal digits in these years are as follows

Years	Population (in crores)
1961	43.9235
1971	54.8160
1981	68.3329
1991	84.6421
2001	102.8737
2011	121.0193

Use appropriate finite difference technique to find the approximate population in 1966, 1985, 1996 and 2009.

Ans.
The finite difference table for given data set is as follows

Years (x)	Population(y)	$\Delta(\nabla,\delta)$	$\Delta^2(\nabla,\delta)$	$\Delta^3(\nabla,\delta)$	$\Delta^4(\nabla,\delta)$	$\Delta^5(\nabla,\delta)$
1961	43.9235					
		10.8925				
1971	54.8160		2.6244			
		13.5169		0.1679		
1981	68.3329		2.7923		-1.0378	
		16.3092		-0.8699		-0.1007
1991	84.6421		1.9224		-1.1385	
		18.2316		-2.0084		
2001	102.8737		-0.0860			
		18.1456				
2011	121.0193					

The years 1966, 1985 and 2009 are at starting, middle and at the end of the table. The values of s for these years are as follows

$$s = \frac{x-x_0}{h} = \frac{1966-1961}{10} = \frac{1}{2} = 0.5 \text{ ; Newton forward difference formula}$$

$$s = \frac{x-x_0}{h} = \frac{1985-1981}{10} = \frac{2}{5} = 0.4 \text{ ; Bessel formula}$$

$$s = \frac{x-x_5}{h} = \frac{2009-2011}{10} = -\frac{1}{5} = -0.2 \text{ ; Newton backward difference formula}$$

The year 1996 is not at the end of the table, therefore here we are using 2001 as x_n, while discarding the value at 2011.

$$s = \frac{x-x_4}{h} = \frac{1996-2001}{10} = -\frac{1}{2} = -0.5 \text{ ; Newton backward difference formula.}$$

Note that this exclusion of data point will create error in interpolation value. Rather, here we are excluding the data point, but is recommended to use all the data points.

Population at year (x) = 1966:
Newton forward difference formula:

Years (x)	Population(y)	Δ	Δ^2	Δ^3	Δ^4	Δ^5
1961(x_0)	43.9235(y_0)					
		10.8925(Δy_0)				
1971(x_1) .	54.8160(y_1)		2.6244($\Delta^2 y_0$)			
		13.5169(Δy_1)		0.1679($\Delta^3 y_0$)		
1981(x_2)	68.3329(y_2)		2.7923($\Delta^2 y_1$)		−1.0378($\Delta^4 y_0$)	
		16.3092(Δy_2)		−0.8699($\Delta^3 y_1$)		−0.1007($\Delta^5 y_0$)
1991(x_3)	84.6421(y_3)		1.9224($\Delta^2 y_2$)		−1.1385($\Delta^4 y_1$)	
		18.2316(Δy_3)		−2.0084($\Delta^3 y_2$)		
2001(x_4)	102.8737(y_4)		−0.0860($\Delta^2 y_3$)			
		18.1456(Δy_4)				
2011(x_5)	121.0193(y_5)					

$$s = \frac{x - x_0}{h} = \frac{1966 - 1961}{10} = \frac{1}{2} = 0.5$$

Newton forward difference formula is given by

$$P(x) = f(x_0) + \Delta f(x_0)(s) + \frac{\Delta^2 f(x_0)}{2!}(s)(s-1) + \cdots + \frac{\Delta^n f(x_0)}{n!}(s)(s-1)\cdots(s-n+1)$$

$$P(1966) = 43.9235 + \left(\frac{1}{2}\right)(10.8925) + \left(\frac{1}{2}\right)\left(\frac{1}{2}\right)\left(-\frac{1}{2}\right)(2.6244) + \left(\frac{1}{6}\right)\left(\frac{1}{2}\right)\left(-\frac{1}{2}\right)\left(-\frac{3}{2}\right)(0.1679)$$

$$+ \left(\frac{1}{24}\right)\left(\frac{1}{2}\right)\left(-\frac{1}{2}\right)\left(-\frac{3}{2}\right)\left(-\frac{5}{2}\right)(-1.0378) + \left(\frac{1}{120}\right)\left(\frac{1}{2}\right)\left(-\frac{1}{2}\right)\left(-\frac{3}{2}\right)\left(-\frac{5}{2}\right)\left(-\frac{7}{2}\right)(-0.1007)$$

$$P(1966) = 43.9235 + 5.44625 - 0.32805 + 0.010494 + 0.040539 - 0.0027535$$

$$P(1966) = 49.0899795 \text{ (\textbf{Approximate population of India in 1966})}$$

Population at year (x) = 1985:
Bessel formula:

$$s = \frac{x - x_0}{h} = \frac{1985 - 1981}{10} = \frac{2}{5} = 0.4; \text{ Bessel formula}$$

Years (x)	Population(y)	δ	δ^2	δ^3	δ^4	δ^5
$1961(x_{-2})$	$43.9235(y_{-2})$					
		$10.8925(\delta y_{-3/2})$				
$1971(x_{-1})$	$54.8160(y_{-1})$		$2.6244(\delta^2 y_{-1})$			
		$13.5169(\delta y_{-1/2})$		$0.1679(\delta^3 y_{-1/2})$		
$1981(x_0)$	$68.3329(y_0)$		$2.7923(\delta^2 y_0)$		$-1.0378(\delta^4 y_0)$	
		$16.3092(\delta y_{1/2})$		$-0.8699(\delta^3 y_{1/2})$		$-0.1007(\delta^5 y_{1/2})$
$1991(x_1)$	$84.6421(y_1)$		$1.9224(\delta^2 y_1)$		$-1.1385(\delta^4 y_1)$	
		$18.2316(\delta y_{3/2})$		$-2.0084(\delta^3 y_{3/2})$		
$2001(x_2)$	$102.8737(y_2)$		$-0.0860(\delta^2 y_2)$			
		$18.1456(\delta y_{5/2})$				
$2011(x_3)$	$121.0193(y_3)$					

Bessel formula

$$P(x) = \frac{1}{2}\left(f(x_0) + f(x_1)\right) + \left(s - \frac{1}{2}\right)\delta f(x_{1/2}) + \frac{(s)(s-1)}{2!}\frac{1}{2}\left(\delta^2 f(x_0) + \delta^2 f(x_1)\right)$$

$$+ \left(s - \frac{1}{2}\right)\frac{(s)(s-1)}{3!}\delta^3 f(x_{1/2}) + \frac{(s+1)(s)(s-1)(s-2)}{4!}\frac{1}{2}\left(\delta^4 f(x_0) + \delta^4 f(x_1)\right)$$

$$+ \left(s - \frac{1}{2}\right)\frac{(s+1)(s)(s-1)(s-2)}{5!}\delta^5 f(x_{1/2})$$

$$+ \frac{(s+2)(s+1)(s)(s-1)(s-2)(s-3)}{6!}\frac{1}{2}\left(\delta^6 f(x_0) + \delta^6 f(x_1)\right) + \cdots$$

$$P(1985) = \frac{1}{2}(68.3329 + 84.6421) - \frac{1}{10}(16.3092) + \left(\frac{1}{4}\right)\left(\frac{2}{5}\right)\left(\frac{-3}{5}\right)(2.7923 + 1.9224)$$

$$+ \left(\frac{-1}{10}\right)\left(\frac{2}{5}\right)\left(\frac{-3}{5}\right)\left(\frac{1}{6}\right)(-0.8699) + \left(\frac{1}{48}\right)\left(\frac{7}{5}\right)\left(\frac{2}{5}\right)\left(\frac{-3}{5}\right)\left(\frac{-8}{5}\right)(-1.0378 - 1.1385)$$

$$+ \left(\frac{-1}{10}\right)\left(\frac{7}{5}\right)\left(\frac{2}{5}\right)\left(\frac{-3}{5}\right)\left(\frac{-8}{5}\right)\left(\frac{1}{120}\right)(-0.1007)$$

$$P(1985) = 76.4875 - 1.63092 - 0.282876 - 0.0034796 - 0.0243746 + 0.00004511$$

$$= 74.545895 \textbf{ (Approximate population of India in 1985)}$$

Population at year (x) = 2009:
Newton backward difference formula:

$$s = \frac{x - x_5}{h} = \frac{2009 - 2011}{10} = -\frac{1}{5} = -0.2$$

Years (x)	Population(y)	∇	∇^2	∇^3	∇^4	∇^5
1961(x_0)	43.9235(y_0)					
		10.8925(∇y_1)				
1971(x_1)	54.8160(y_1)		2.6244$(\nabla^2 y_2)$			
		13.5169(∇y_2)		0.1679$(\nabla^3 y_3)$		
1981(x_2)	68.3329(y_2)		2.7923$(\nabla^2 y_3)$		$-1.0378(\nabla^4 y_4)$	
		16.3092(∇y_3)		$-0.8699(\nabla^3 y_4)$		$-0.1007(\nabla^5 y_5)$
1991(x_3)	84.6421(y_3)		1.9224$(\nabla^2 y_4)$		$-1.1385(\nabla^4 y_5)$	
		18.2316(∇y_4)		$-2.0084(\nabla^3 y_5)$		
2001(x_4)	102.8737(y_4)		$-0.0860(\nabla^2 y_5)$			
		18.1456(∇y_5)				
2011(x_5)	121.0193(y_5)					

Newton backward difference formula with $n = 5$

$$P(x) = f(x_5) + \nabla f(x_5)(s) + \frac{\nabla^2 f(x_5)}{2!}(s)(s+1) + \frac{\nabla^3 f(x_5)}{3!}(s)(s+1)(s+2) + \cdots$$

$$P(2009) = 121.0193 + \left(\frac{-1}{5}\right)(18.1456) + \left(\frac{1}{2}\right)\left(\frac{-1}{5}\right)\left(\frac{4}{5}\right)(-0.0860) + \left(\frac{1}{6}\right)\left(\frac{-1}{5}\right)\left(\frac{4}{5}\right)\left(\frac{9}{5}\right)(-2.0084)$$

$$+ \left(\frac{1}{24}\right)\left(\frac{-1}{5}\right)\left(\frac{4}{5}\right)\left(\frac{9}{5}\right)\left(\frac{14}{5}\right)(-1.1385) + \left(\frac{1}{120}\right)\left(\frac{-1}{5}\right)\left(\frac{4}{5}\right)\left(\frac{9}{5}\right)\left(\frac{14}{5}\right)\left(\frac{19}{5}\right)(-0.1007)$$

$$P(2009) = 121.0193 - 3.62912 + 0.00688 + 0.0964032 + 0.0382536 + 0.0257148$$

$$P(2009) = 117.5574316 \text{ (Approximate population of India in 2009)}$$

Population at year $(x) = 1996$:
Newton backward difference formula:
The value (1996) is not at the end of the table, so we can either continue with the formula as for the year 2009 or skip the last tabulated value i.e. for 2011 and continue with the $n = 4$. i.e.

$$s = \frac{x - x_4}{h} = \frac{1996 - 2001}{10} = -\frac{1}{2} = -0.5$$

Years (x)	Population(y)	∇	∇^2	∇^3	∇^4	∇^5
$1961(x_0)$	$43.9235(y_0)$					
		$10.8925(\nabla y_1)$				
$1971(x_1)$	$54.8160(y_1)$		$2.6244(\nabla^2 y_2)$			
		$13.5169(\nabla y_2)$		$0.1679(\nabla^3 y_3)$		
$1981(x_2)$	$68.3329(y_2)$		$2.7923(\nabla^2 y_3)$		$-1.0378(\nabla^4 y_4)$	
		$16.3092(\nabla y_3)$		$-0.8699(\nabla^3 y_4)$		
$1991(x_3)$	$84.6421(y_3)$		$1.9224(\nabla^2 y_4)$			
		$18.2316(\nabla y_4)$				
$2001(x_4)$	$102.8737(y_4)$					

Newton backward difference formula with $n = 4$

$$P(x) = f(x_4) + \nabla f(x_4)(s) + \frac{\nabla^2 f(x_4)}{2!}(s)(s+1) + \frac{\nabla^3 f(x_4)}{3!}(s)(s+1)(s+2) + \cdots$$

$$P(1996) = 102.8737 + \left(\frac{-1}{2}\right)(18.2316) + \left(\frac{1}{2}\right)\left(\frac{-1}{2}\right)\left(\frac{1}{2}\right)(1.9224)$$

$$+ \left(\frac{1}{6}\right)\left(\frac{-1}{2}\right)\left(\frac{1}{2}\right)\left(\frac{3}{2}\right)(-0.8699) + \left(\frac{1}{24}\right)\left(\frac{-1}{2}\right)\left(\frac{1}{2}\right)\left(\frac{3}{2}\right)\left(\frac{5}{2}\right)(-1.0378)$$

$$P(1996) = 102.8737 - 9.1158 - 0.2403 + 0.05436875 + 0.04053906$$

$$= 93.61250781 \text{ (Approximate population of India in 1996)}$$

10.10 Bivariate Interpolation

So far, we have discussed interpolation for one independent variable (x); we can extend it to multivariate interpolation for data set where we have more than one independent variable. In our current discussion, we will restrict ourselves to bivariate interpolation only; the extension to more number of independent variables is straightforward but cumbersome.

10.10.1 Lagrange Bivariate Interpolation

Let us consider the function $f(x, y)$ of two independent variables x and y defined at $(m+1)$ $(n+1)$ distinct points, (x_i, y_j), $i = 0,1,...,m, j = 0,1,...,n$. Our aim is to obtain an interpolating polynomial $P_{mn}(x, y)$ of degree at most m in x and n in y, such that

$$P_{mn}(x_i, y_j) = f(x_i, y_j); \quad i = 0,1,...,m, j = 0,1,...,n.$$

First, we define two different Lagrange polynomials of degree m in variable x and degree n in variable y as follows.

$$X_i(x) = \prod_{\substack{k=0 \\ k \neq i}}^{m} \frac{x - x_k}{x_i - x_k} = \frac{(x - x_0)(x - x_1) \cdots (x - x_{i-1})(x - x_{i+1}) \cdots (x - x_m)}{(x_i - x_0)(x_i - x_1) \cdots (x_i - x_{i-1})(x_i - x_{i+1}) \cdots (x_i - x_m)}$$

$$i = 0, 1, \ldots, m$$

$$Y_j(y) = \prod_{\substack{l=0 \\ l \neq j}}^{n} \frac{y - y_l}{y_j - y_l} = \frac{(y - y_0)(y - y_1) \cdots (y - y_{j-1})(y - y_{j+1}) \cdots (y - y_n)}{(y_j - y_0)(y_j - y_1) \cdots (y_j - y_{j-1})(y_j - y_{j+1}) \cdots (y_j - y_n)}$$

$$j = 0, 1, \ldots, n$$

We have

$$X_i(x_k) = \delta_{ik} \text{ and } Y_j(y_k) = \delta_{jk}$$

where δ_{ij} is Kronecker delta function, such that

$$\delta_{ij} = \begin{cases} 0 & i \neq j \\ 1 & i = j \end{cases}$$

Lagrange interpolating polynomial is given by

$$P_{mn}(x, y) = \sum_{i=0}^{m} \sum_{j=0}^{n} X_i(x) Y_j(y) f(x_i, y_j) \tag{10.36}$$

Since $X_i(x)$ is polynomial degree m in variable x and $Y_j(y)$ is polynomial degree n in variable y, therefore the interpolating polynomial $P_{mn}(x, y)$ is of degree at most m in x and n in y. Note that interpolating polynomial $P_{mn}(x, y)$ satisfies the conditions

$$P_{mn}(x_i, y_j) = f(x_i, y_j); \quad i = 0, 1, \ldots, m, j = 0, 1, \ldots, n$$

Example   **10.13**

Compute the bivariate interpolating polynomial for the following set of data points for the function $f(x, y)$. Hence find the value of $f(-0.5, 4)$

x \ y	1	2	5
-1	-1	-7	-49
0	2	1	-2

Ans.

The data points are given by

$$x_0 = -1, \; x_1 = 0$$

$$y_0 = 1, \; y_1 = 2, \; y_2 = 5$$

$$f(x_0, y_0) = -1, \quad f(x_0, y_1) = -7, f(x_0, y_2) = -49$$
$$f(x_1, y_0) = 2, \quad\;\; f(x_1, y_1) = 1, \quad\;\; f(x_1, y_2) = -2$$

The bivariate interpolating polynomial (10.36) is given by

$$P_{mn}(x, y) = \sum_{i=0}^{m} \sum_{j=0}^{n} X_i(x) Y_j(y) f(x_i, y_j)$$

For $m = 1$ and $n = 2$, we have

$$P_{12}(x, y) = \sum_{i=0}^{1} \sum_{j=0}^{2} X_i(x) Y_j(y) f(x_i, y_j)$$

$$= \sum_{i=0}^{1} X_i(x) Y_0(y) f(x_i, y_0) + \sum_{i=0}^{1} X_i(x) Y_1(y) f(x_i, y_1) + \sum_{i=0}^{1} X_i(x) Y_2(y) f(x_i, y_2)$$

$$P_{12}(x, y) = X_0(x) Y_0(y) f(x_0, y_0) + X_1(x) Y_0(y) f(x_1, y_0)$$
$$+ X_0(x) Y_1(y) f(x_0, y_1) + X_1(x) Y_1(y) f(x_1, y_1)$$
$$+ X_0(x) Y_2(y) f(x_0, y_2) + X_1(x) Y_2(y) f(x_1, y_2)$$

$$P_{12}(x, y) = \frac{(x - x_1)}{(x_0 - x_1)} \frac{(y - y_1)(y - y_2)}{(y_0 - y_1)(y_0 - y_2)} f(x_0, y_0) + \frac{(x - x_0)}{(x_1 - x_0)} \frac{(y - y_1)(y - y_2)}{(y_0 - y_1)(y_0 - y_2)} f(x_1, y_0)$$

$$+ \frac{(x - x_1)}{(x_0 - x_1)} \frac{(y - y_0)(y - y_2)}{(y_1 - y_0)(y_1 - y_2)} f(x_0, y_1) + \frac{(x - x_0)}{(x_1 - x_0)} \frac{(y - y_0)(y - y_2)}{(y_1 - y_0)(y_1 - y_2)} f(x_1, y_1)$$

$$+ \frac{(x - x_1)}{(x_0 - x_1)} \frac{(y - y_0)(y - y_1)}{(y_2 - y_0)(y_2 - y_1)} f(x_0, y_2) + \frac{(x - x_0)}{(x_1 - x_0)} \frac{(y - y_0)(y - y_1)}{(y_2 - y_0)(y_2 - y_1)} f(x_1, y_2)$$

On using the data values in this expression, we have

$$P_{12}(x, y) = \frac{(x - 0)}{(-1 - 0)} \frac{(y - 2)(y - 5)}{(1 - 2)(1 - 5)} (-1) + \frac{(x - (-1))}{(0 - (-1))} \frac{(y - 2)(y - 5)}{(1 - 2)(1 - 5)} (2)$$

$$+ \frac{(x - 0)}{(-1 - 0)} \frac{(y - 1)(y - 5)}{(2 - 1)(2 - 5)} (-7) + \frac{(x - (-1))}{(0 - (-1))} \frac{(y - 1)(y - 5)}{(2 - 1)(2 - 5)} (1)$$

$$+ \frac{(x - 0)}{(-1 - 0)} \frac{(y - 1)(y - 2)}{(5 - 1)(5 - 2)} (-49) + \frac{(x - (-1))}{(0 - (-1))} \frac{(y - 1)(y - 2)}{(5 - 1)(5 - 2)} (-2)$$

$$P_{12}(x,y)=\frac{1}{4}x(y-2)(y-5)+\frac{1}{2}(x+1)(y-2)(y-5)-\frac{7}{3}x(y-1)(y-5)$$

$$-\frac{1}{3}(x+1)(y-1)(y-5)+\frac{49}{12}x(y-1)(y-2)-\frac{1}{6}(x+1)(y-1)(y-2)$$

$$P_{12}(x,y)=2xy^2-xy+2x-y+3$$

It is required interpolating polynomial.
Approximate value of $f(-0.5, 4)$ is computed with this polynomial, and given by

$$f(-0.5,\ 4)=P_{12}(-0.5,4)=2(-0.5)(4)^2-(-0.5)4+2(-0.5)-4+3=-16$$

Example   10.14

Obtain the bivariate interpolating polynomial for the following values of function $f(x, y)$, and hence find the value of $f(1.5, 1)$

x \ y	-1	0	2
0	1	2	4
1	5	2	2
2	17	2	-4

Ans.
We have

$$m=n=2 \qquad \begin{aligned} &x_0=0,\ x_1=1,\ x_2=2\\ &y_0=-1,\ y_1=0,\ y_2=2 \end{aligned}$$

$$\begin{aligned} &f(x_0,y_0)=1, & f(x_0,y_1)=2, & \quad f(x_0,y_2)=4,\\ &f(x_1,y_0)=5, & f(x_1,y_1)=2, & \quad f(x_1,y_2)=2,\\ &f(x_2,y_0)=17, & f(x_2,y_1)=2, & \quad f(x_2,y_2)=-4 \end{aligned}$$

For $m = n = 2$, the bivariate interpolating polynomial (10.36) is given by

$$P_{22}(x,y)=\sum_{i=0}^{2}\sum_{j=0}^{2}X_i(x)Y_j(y)f(x_i,y_j)$$

$$=\sum_{i=0}^{2}X_i(x)Y_0(y)f(x_i,y_0)+\sum_{i=0}^{2}X_i(x)Y_1(y)f(x_i,y_1)+\sum_{i=0}^{2}X_i(x)Y_2(y)f(x_i,y_2)$$

$$\begin{aligned} P_{22}(x,y)=&X_0(x)Y_0(y)f(x_0,y_0)+X_1(x)Y_0(y)f(x_1,y_0)+X_2(x)Y_0(y)f(x_2,y_0)\\ &+X_0(x)Y_1(y)f(x_0,y_1)+X_1(x)Y_1(y)f(x_1,y_1)+X_2(x)Y_1(y)f(x_2,y_1)\\ &+X_0(x)Y_2(y)f(x_0,y_2)+X_1(x)Y_2(y)f(x_1,y_2)+X_2(x)Y_2(y)f(x_2,y_2) \end{aligned}$$

$$P_{22}(x,y) = \frac{(x-x_1)(x-x_2)}{(x_0-x_1)(x_0-x_2)} \frac{(y-y_1)(y-y_2)}{(y_0-y_1)(y_0-y_2)} f(x_0,y_0) + \frac{(x-x_0)(x-x_2)}{(x_1-x_0)(x_1-x_2)} \frac{(y-y_1)(y-y_2)}{(y_0-y_1)(y_0-y_2)} f(x_1,y_0)$$

$$+ \frac{(x-x_0)(x-x_1)}{(x_2-x_0)(x_2-x_1)} \frac{(y-y_1)(y-y_2)}{(y_0-y_1)(y_0-y_2)} f(x_2,y_0) + \frac{(x-x_1)(x-x_2)}{(x_0-x_1)(x_0-x_2)} \frac{(y-y_0)(y-y_2)}{(y_1-y_0)(y_1-y_2)} f(x_0,y_1)$$

$$+ \frac{(x-x_0)(x-x_2)}{(x_1-x_0)(x_1-x_2)} \frac{(y-y_0)(y-y_2)}{(y_1-y_0)(y_1-y_2)} f(x_1,y_1) + \frac{(x-x_0)(x-x_1)}{(x_2-x_0)(x_2-x_1)} \frac{(y-y_0)(y-y_2)}{(y_1-y_0)(y_1-y_2)} f(x_2,y_1)$$

$$+ \frac{(x-x_1)(x-x_2)}{(x_0-x_1)(x_0-x_2)} \frac{(y-y_0)(y-y_1)}{(y_2-y_0)(y_2-y_1)} f(x_0,y_2) + \frac{(x-x_0)(x-x_2)}{(x_1-x_0)(x_1-x_2)} \frac{(y-y_0)(y-y_1)}{(y_2-y_0)(y_2-y_1)} f(x_1,y_2)$$

$$+ \frac{(x-x_0)(x-x_1)}{(x_2-x_0)(x_2-x_1)} \frac{(y-y_0)(y-y_1)}{(y_2-y_0)(y_2-y_1)} f(x_2,y_2)$$

On using the data values and after simplifying the expression, we have

$$P_{22}(x,y) = x^2 y^2 - 3x^2 y + y + 2$$

Approximate value of $f(1.5, 1)$ is as follows

$$f(1.5, 1) = P_{22}(1.5,1) = -1.5$$

10.10.2 Newton Bivariate Interpolation for Equi-spaced Points

Consider the forward differences for two variables with equispaced points, with spacing h in x and k in y as follows

$$\Delta_x f(x,y) = f(x+h,y) - f(x,y)$$

$$= (E_x - 1) f(x,y)$$

$$\Delta_y f(x,y) = f(x,y+k) - f(x,y)$$

$$= (E_y - 1) f(x,y)$$

Similarly, second and higher order differences are given by

$$\Delta_{xx} f(x,y) = \Delta_x f(x+h,y) - \Delta_x f(x,y)$$

$$= (E_x - 1)^2 f(x,y)$$

$$\Delta_{yy} f(x,y) = \Delta_y f(x,y+k) - \Delta_y f(x,y)$$

$$= (E_y - 1)^2 f(x,y)$$

$$\Delta_{xy} f(x,y) = \Delta_x \left[f(x,y+k) - f(x,y) \right] = \Delta_x \Delta_y f(x,y)$$

$$= (E_x - 1)(E_y - 1) f(x,y)$$

$$= (E_y - 1)(E_x - 1) f(x,y)$$

$$= \Delta_y \Delta_x f(x,y) = \Delta_{yx} f(x,y)$$

Let $x = x_0 + rh$ and $y = y_0 + sk$, then we have

$$f(x, y) = f(x_0 + rh, y_0 + sk) = E_x^r E_y^s f(x_0, y_0)$$

$$= (1 + \Delta_x)^r (1 + \Delta_y)^s f(x_0, y_0)$$

$$= \left[1 + r\Delta_x + \frac{r(r-1)}{2!} \Delta_{xx} + \frac{r(r-1)(r-2)}{3!} \Delta_{xxx} + \cdots \right]$$

$$\times \left[1 + s\Delta_y + \frac{s(s-1)}{2!} \Delta_{yy} + \frac{s(s-1)(s-2)}{3!} \Delta_{yyy} + \cdots \right] f(x_0, y_0)$$

$$= \left[1 + r\Delta_x + s\Delta_y + rs\Delta_x \Delta_y + \frac{r(r-1)}{2!} \Delta_{xx} + \frac{s(s-1)}{2!} \Delta_{yy} + \cdots \right] f(x_0, y_0)$$

On applying the operator, we have

$$f(x, y) = f(x_0, y_0) + \left[r\Delta_x + s\Delta_y \right] f(x_0, y_0)$$

$$+ \frac{1}{2!} \left[2rs\Delta_x \Delta_y + r(r-1)\Delta_{xx} + s(s-1)\Delta_{yy} \right] f(x_0, y_0) + \cdots$$

$$f(x, y) = f(x_0, y_0) + \left[r\Delta_x f(x_0, y_0) + s\Delta_y f(x_0, y_0) \right]$$

$$+ \frac{1}{2!} \left[2rs\Delta_x \Delta_y \left(f(x_0, y_0) \right) + r(r-1)\Delta_{xx} \left(f(x_0, y_0) \right) + s(s-1)\Delta_{yy} \left(f(x_0, y_0) \right) \right] + \cdots$$

$$(10.37)$$

The formula (10.37) is known as Newton bivariate interpolation.
It is worth mentioning here that if the number of points in variable x is m+1 and in y is n+1;
then we must consider terms up to order (m)(n).

If we replace $r = \dfrac{x - x_0}{h}$ and $s = \dfrac{y - y_0}{k}$, we will get the Newton bivariate interpolating

polynomial for equispaced points in x and y as follows

$$P(x, y) = f(x_0, y_0) + \left[\frac{x - x_0}{h} \Delta_x + \frac{y - y_0}{k} \Delta_y \right] f(x_0, y_0)$$

$$+ \frac{1}{2!} \left[\frac{1}{h^2} (x - x_0)(x - x_1) \Delta_{xx} + \frac{2}{hk} (x - x_0)(y - y_0) \Delta_{xy} + \frac{1}{k^2} (y - y_0)(y - y_1) \Delta_{yy} \right] f(x_0, y_0)$$

$$+ \cdots$$

Example 10.15

Given the following set of data points for the values of $f(x, y)$, compute the value of $f(1.5, 0.25)$

x \ y	0	1	2
1	2	5	10
3	4	11	20

Ans.

We have
$$
\begin{array}{llll}
x_0 = 1, \; x_1 = 3 & f(x_0, y_0) = 2, & f(x_0, y_1) = 5, & f(x_0, y_2) = 10, \\
y_0 = 0, \; y_1 = 1, \; y_2 = 2 & f(x_1, y_0) = 4, & f(x_1, y_1) = 11, & f(x_1, y_2) = 20
\end{array}
\tag{10.38}
$$

Also, the step size for x is $h = 2$, and for y is $k = 1$.

The data points are up to x_1 and y_2, so all second and higher differences are zero for x, and all third and higher differences are zero for y. Therefore, we will consider the formula (10.37) that is up to first order differences in x and second order differences in y. For this, we have

$$f(x, y) = f(x_0 + rh, y_0 + sk) = E_x^r E_y^s f(x_0, y_0)$$

$$= (1 + \Delta_x)^r (1 + \Delta_y)^s f(x_0, y_0)$$

$$f(x, y) = \left[1 + r\Delta_x \right] \times \left[1 + s\Delta_y + \frac{s(s-1)}{2!} \Delta_{yy} \right] f(x_0, y_0)$$

$$f(x, y) = \left[1 + r\Delta_x + s\Delta_y + rs\Delta_x \Delta_y + \frac{s(s-1)}{2!} \Delta_{yy} + r\frac{s(s-1)}{2!} \Delta_{yy}\Delta_x \right] f(x_0, y_0) \tag{10.39}$$

We have to compute the value of $f(1.5, 0.25)$, so

$$r = \frac{x - x_0}{h} = \frac{1.5 - 1}{2} = 0.25$$

$$s = \frac{y - y_0}{k} = \frac{0.25 - 0}{1} = 0.25$$

On using these values of r and s in Eq. (10.39), we have

$$f(x, y) = \begin{bmatrix} 1 + (0.25)\Delta_x + (0.25)\Delta_y + (0.25)(0.25)\Delta_x \Delta_y \\ + \dfrac{(0.25)(-0.75)}{2!}\Delta_{yy} + (0.25)\dfrac{(0.25)(-0.75)}{2!}\Delta_{yy}\Delta_x \end{bmatrix} f(x_0, y_0) \qquad (10.40)$$

Now, we have to find forward differences for the given data set as follows

$$\Delta_x f(x_0, y_0) = f(x_0 + h, y_0) - f(x_0, y_0)$$
$$= f(x_1, y_0) - f(x_0, y_0) = 4 - 2 = 2$$

$$\Delta_y f(x_0, y_0) = f(x_0, y_0 + k) - f(x_0, y_0)$$
$$= f(x_0, y_1) - f(x_0, y_0) = 5 - 2 = 3$$

Similarly, second forward differences are given by

$$\Delta_{yy} f(x_0, y_0) = (E_y - 1)^2 f(x_0, y_0)$$
$$= (E_y^2 + 1 - 2E_y) f(x_0, y_0)$$
$$= f(x_0, y_2) + f(x_0, y_0) - 2f(x_0, y_1)$$
$$= 10 + 2 - 2(5) = 2$$

$$\Delta_{xy} f(x_0, y_0) = (E_y - 1)(E_x - 1) f(x_0, y_0)$$
$$= (E_y E_x - E_y - E_x + 1) f(x_0, y_0)$$
$$= f(x_1, y_1) - f(x_0, y_1) - f(x_1, y_0) + f(x_0, y_0)$$
$$= 11 - 5 - 4 + 2 = 4$$

Similarly, we have

$$\Delta_x \Delta_{yy} f(x_0, y_0) = (E_x - 1)(E_y - 1)^2 f(x_0, y_0)$$
$$= (E_x - 1)(f(x_0, y_2) + f(x_0, y_0) - 2f(x_0, y_1))$$
$$= [f(x_1, y_2) + f(x_1, y_0) - 2f(x_1, y_1)] - [f(x_0, y_2) + f(x_0, y_0) - 2f(x_0, y_1)]$$
$$= 20 + 4 - 2(11) - (10 + 2 - 2(5)) = 0$$

On using all these forward differences in Eq. (10.40), we have

$$f(1.5, 0.25) = \begin{bmatrix} f(x_0, y_0) + (0.25)\Delta_x \left(f(x_0, y_0) \right) + (0.25)\Delta_y \left(f(x_0, y_0) \right) \\ + (0.25)(0.25)\Delta_x \Delta_y \left(f(x_0, y_0) \right) + \dfrac{(0.25)(-0.75)}{2!}\Delta_{yy} \left(f(x_0, y_0) \right) \\ + (0.25)\dfrac{(0.25)(-0.75)}{2!}\Delta_{yy}\Delta_x \left(f(x_0, y_0) \right) \end{bmatrix}$$

$$f(1.5, 0.25) = \begin{bmatrix} 2+(0.25)(2)+(0.25)(3)+(0.25)(0.25)(4) \\ +\dfrac{(0.25)(-0.75)}{2!}(2)+(0.25)\dfrac{(0.25)(-0.75)}{2!}(0) \end{bmatrix}$$

$$f(1.5, 0.25) = \begin{bmatrix} 2+(0.25)(2)+(0.25)(3)+(0.25)(0.25)(4) \\ +\dfrac{(0.25)(-0.75)}{2!}(2)+(0.25)\dfrac{(0.25)(-0.75)}{2!}(0) \end{bmatrix}$$

$$f(1.5, 0.25) = 3.3125$$

(Result is correct as data from polynomial $y^2 + 2xy + x + 1$)

Example 10.16

Consider the following set of data points for the values of $f(x, y)$

x \ y	0	1	2
-1	0	1	0
0	-1	1	3
1	0	3	8

Compute the interpolating polynomial using Newton bivariate formula, and hence find the value of $f(-0.5, 0.5)$.

Ans.

We have $x_0 = -1, x_1 = 0, x_2 = 1$
$y_0 = 0, y_1 = 1, y_2 = 2$

$$f(x_0, y_0) = 0, \quad f(x_0, y_1) = 1, \quad f(x_0, y_2) = 0,$$
$$f(x_1, y_0) = -1, \quad f(x_1, y_1) = 1, \quad f(x_1, y_2) = 3, \quad (10.41)$$
$$f(x_2, y_0) = 0, \quad f(x_2, y_1) = 3, \quad f(x_2, y_2) = 8$$

Step sizes for x and y are $h = k = 1$.

The data points are up to x_2 and y_2, so the Newton bivariate formula (10.37) must be up to second order in x and second order in y. For this, we have

$$f(x, y) = \left[1 + r\Delta_x + \frac{r(r-1)}{2!}\Delta_{xx}\right] \times \left[1 + s\Delta_y + \frac{s(s-1)}{2!}\Delta_{yy}\right] f(x_0, y_0)$$

$$f(x, y) = \begin{bmatrix} 1 + r\Delta_x + s\Delta_y + rs\Delta_x\Delta_y + \dfrac{r(r-1)}{2!}\Delta_{xx} + \dfrac{s(s-1)}{2!}\Delta_{yy} \\[2mm] r\dfrac{s(s-1)}{2!}\Delta_{yy}\Delta_x + s\dfrac{r(r-1)}{2!}\Delta_{xx}\Delta_y + \dfrac{r(r-1)}{2!}\dfrac{s(s-1)}{2!}\Delta_{xx}\Delta_{yy} \end{bmatrix} f(x_0, y_0) \quad (10.42)$$

On using $r = \dfrac{x - x_0}{h}$, $s = \dfrac{y - y_0}{k}$, and $h = k = 1$ in Eq. (10.42), we have

$$P(x, y) = f(x_0, y_0) + \left[(x - x_0)\Delta_x + (y - y_0)\Delta_y\right]f(x_0, y_0)$$

$$+ \frac{1}{2!}\left[(x - x_0)(x - x_1)\Delta_{xx} + 2(x - x_0)(y - y_0)\Delta_{xy} + (y - y_0)(y - y_1)\Delta_{yy}\right]f(x_0, y_0)$$

$$+ \left[\begin{array}{c} (x - x_0)\dfrac{(y - y_0)(y - y_1)}{2!}\Delta_{yy}\Delta_x + (y - y_0)\dfrac{(x - x_0)(x - x_1)}{2!}\Delta_{xx}\Delta_y \\[2mm] + \dfrac{(x - x_0)(x - x_1)}{2!}\dfrac{(y - y_0)(y - y_1)}{2!}\Delta_{xx}\Delta_{yy} \end{array}\right]f(x_0, y_0)$$

$$P(x, y) = f(x_0, y_0) + \left[(x - x_0)\Delta_x + (y - y_0)\Delta_y\right]f(x_0, y_0)$$

$$+ \frac{1}{2}\left[(x - x_0)(x - x_1)\Delta_{xx} + 2(x - x_0)(y - y_0)\Delta_{xy} + (y - y_0)(y - y_1)\Delta_{yy}\right]f(x_0, y_0)$$

$$+ \frac{1}{4}\left[\begin{array}{c} 2(x - x_0)(y - y_0)(y - y_1)\Delta_{yy}\Delta_x + 2(y - y_0)(x - x_0)(x - x_1)\Delta_{xx}\Delta_y \\[2mm] + (x - x_0)(x - x_1)(y - y_0)(y - y_1)\Delta_{xx}\Delta_{yy} \end{array}\right]f(x_0, y_0)$$

On using the values of variables x_i, y_j, and $f(x_i, y_j)$, we have

$$P(x, y) = 0 + \left[(x + 1)\Delta_x + (y)\Delta_y\right]f(x_0, y_0)$$

$$+ \frac{1}{2}\left[(x + 1)(x)\Delta_{xx} + 2(x + 1)(y)\Delta_{xy} + (y)(y - 1)\Delta_{yy}\right]f(x_0, y_0)$$

$$+ \frac{1}{4}\left[\begin{array}{c} 2(x + 1)(y)(y - 1)\Delta_{yy}\Delta_x + 2(y)(x + 1)(x)\Delta_{xx}\Delta_y \\[2mm] + (x + 1)(x)(y)(y - 1)\Delta_{xx}\Delta_{yy} \end{array}\right]f(x_0, y_0) \qquad (10.43)$$

The first forward differences for the given data set are as follows

$$\Delta_x f(x_0, y_0) = f(x_0 + h, y_0) - f(x_0, y_0)$$
$$= f(x_1, y_0) - f(x_0, y_0) = -1 - 0 = -1$$

$$\Delta_y f(x_0, y_0) = f(x_0, y_0 + k) - f(x_0, y_0)$$
$$= f(x_0, y_1) - f(x_0, y_0) = 1 - 0 = 1$$

Similarly, second forward differences are given by

$$\Delta_{xx} f(x_0, y_0) = (E_x - 1)^2 f(x_0, y_0)$$
$$= (E_x^2 + 1 - 2E_x)f(x_0, y_0)$$
$$= f(x_2, y_0) + f(x_0, y_0) - 2f(x_1, y_0)$$
$$= 0 + 0 - 2(-1) = 2$$

$$\Delta_{yy} f(x_0, y_0) = (E_y - 1)^2 f(x_0, y_0)$$
$$= (E_y^2 + 1 - 2E_y) f(x_0, y_0)$$
$$= f(x_0, y_2) + f(x_0, y_0) - 2f(x_0, y_1)$$
$$= 0 + 0 - 2(1) = -2$$

$$\Delta_{xy} f(x_0, y_0) = (E_y - 1)(E_x - 1) f(x_0, y_0)$$
$$= (E_y E_x - E_y - E_x + 1) f(x_0, y_0)$$
$$= f(x_1, y_1) - f(x_0, y_1) - f(x_1, y_0) + f(x_0, y_0)$$
$$= 1 - 1 + 1 + 0 = 1$$

The other differences which are to be used in the interpolation are given by

$$\Delta_x \Delta_{yy} f(x_0, y_0) = (E_x - 1)(E_y - 1)^2 f(x_0, y_0)$$
$$= (E_x - 1)(f(x_0, y_2) + f(x_0, y_0) - 2f(x_0, y_1))$$
$$= [f(x_1, y_2) + f(x_1, y_0) - 2f(x_1, y_1)] - [f(x_0, y_2) + f(x_0, y_0) - 2f(x_0, y_1)]$$
$$= 3 - 1 - 2 - (0 + 0 - 2) = 2$$

$$\Delta_y \Delta_{xx} f(x_0, y_0) = (E_y - 1)(E_x - 1)^2 f(x_0, y_0)$$
$$= (E_y - 1)(f(x_2, y_0) + f(x_0, y_0) - 2f(x_1, y_0))$$
$$= [f(x_2, y_1) + f(x_0, y_1) - 2f(x_1, y_1)] - [f(x_2, y_0) + f(x_0, y_0) - 2f(x_1, y_0)]$$
$$= 3 + 1 - 2 - (0 + 0 + 2) = 0$$

$$\Delta_{xx} \Delta_{yy} f(x_0, y_0) = (E_x - 1)^2 (E_y - 1)^2 f(x_0, y_0)$$
$$= (E_x^2 + 1 - 2E_x)(f(x_0, y_2) + f(x_0, y_0) - 2f(x_0, y_1))$$
$$= [f(x_2, y_2) + f(x_2, y_0) - 2f(x_2, y_1)] + [f(x_0, y_2) + f(x_0, y_0) - 2f(x_0, y_1)]$$
$$\quad - 2[(f(x_1, y_2) + f(x_1, y_0) - 2f(x_1, y_1))]$$
$$= 8 + 0 - 6 + 0 + 0 - 2 - 2(3 - 1 - 2) = 0$$

On using different values of forward differences in Eq. (10.43), we get

$$P(x, y) = 0 + \left[(x+1)\Delta_x + (y)\Delta_y \right] f(x_0, y_0)$$
$$+ \frac{1}{2} \left[(x+1)(x)\Delta_{xx} + 2(x+1)(y)\Delta_{xy} + (y)(y-1)\Delta_{yy} \right] f(x_0, y_0)$$
$$+ \frac{1}{4} \left[\begin{array}{c} 2(x+1)(y)(y-1)\Delta_{yy}\Delta_x + 2(y)(x+1)(x)\Delta_{xx}\Delta_y \\ + (x+1)(x)(y)(y-1)\Delta_{xx}\Delta_{yy} \end{array} \right] f(x_0, y_0)$$

$$P(x,\ y)=\Big[(x+1)(-1)+(y)(1)\Big]+\frac{1}{2!}\Big[(x+1)(x)(2)+2(x+1)(y)(1)+(y)(y-1)(-2)\Big]$$

$$+\frac{1}{4}\Big[2(x+1)(y)(y-1)(2)+2(y)(x+1)(x)(0)+(x+1)(x)(y)(y-1)(0)\Big]$$

$$P(x,\ y)=xy^2+x^2+2y-1.$$

So, the interpolating polynomial is $P(x,\ y)=xy^2+x^2+2y-1.$

The approximate value of $f(-0.5,\ 0.5)$ is as follows

$$f(-0.5,\ 0.5)=(-0.5)(0.5)^2+(-0.5)^2+2(0.5)-1=0.125$$

Exercise 10

1. Use Newton forward, Newton backward, Lagrange and NDD methods to find the cubic polynomial which takes the following values, and verify the result that the polynomial is unique

x	0	1	2	3
f(x)	1	2	1	10

Ans. $2x^3-7x^2+6x+1$

2. The populations (in millions) of Punjab state up to two decimal points in the census years are given below. Estimate the populations for the years 1973, 1995, and 2007.

year	1971	1981	1991	2001	2011
Population (in millions)	13.55	16.79	20.28	24.36	27.70

Ans. 14.25, 21.88, 26.61

3. Compute the value of log(2875) from the following values of log(x) with the help of Newton backward difference formula

x	25	30	35	40	45
log x	1.39794	1.47712	1.54407	1.60206	1.65321

Ans. 3.45863

4. Given:

$\theta°$:	0	5	10	15	20	25	30
$\tan\theta$:	0	0.0875	0.1763	0.2679	0.364	0.4663	0.5774

Find the value of tan 16° using Stirling formula.

Ans. 0.2867

5. The growth of cell culture (optical density) at various pH levels are tabulated in the following table.

pH:	4	4.5	5	5.5	6
Optical density:	0.28	0.35	0.41	0.46	0.52

   Compute the optical density at pH level 5.8.

   Ans. 0.492688

6. The following data set represents the resistivity of a given metal with temperature. Predict the resistivity at 350 K temperature

Temperature (K):	100	200	300	400	500
Resistivity (Ω-cm, $\times 10^6$):	4.2	8.1	12.8	16.6	20.0

   Ans. 14.8328×10^6 Ω-cm

7. Prepare the finite difference table for the function $f(x) = \dfrac{1}{1+x^2}$ at $x = -2\,(0.5)\,2$. Use this table to compute $f(-1.7)$, $f(0.2)$, $f(1.8)$, and compare the results with exact values.

   Ans.

x	-2	-1.5	-1	-0.5	0	0.5	1	1.5	2
$f(x) = \dfrac{1}{1+x^2}$	0.2	0.307692	0.5	0.8	1	0.8	0.5	0.307692	0.2

   Approximate values $f(-1.6) = 0.252424$, $f(0.25) = 0.943369$, $f(1.9) = 0.127260$

   Exact values $f(-1.6) = 0.280899$, $f(0.25) = 0.941176$; $f(1.9) = 0.216920$

   (Error due to polynomial approximation of a rational function)

8. Use Stirling formula to find the value of y_{18} for the given data set

   $y_{10} = 362$ $y_{15} = 487$ $y_{20} = 642$ $y_{25} = 897$ $y_{30} = 1206$

   Ans. 569.32

9. From an equally spaced table of $\sin(x)$ with spacing $h = 0.1$, we have following data

x	0.1	0.2	0.3	0.4	0.5
$\sin x$	.0998334	.198669	.295520	.389418	.479426

   Compute the value of $\sin(x)$ at $x = 0.17$ by using Newton forward difference formula.

   Ans. 0.169182

10. Use Newton forward and backward difference formulas to find the values of $\sin 48°$ and $\sin 62°$ respectively from the given table

$\theta°$	45	50	55	60	65
$\sin \theta$	0.7071	0.7660	0.8192	0.8660	0.9063

    Ans. $\sin 48° = 0.7431$, $\sin 62° = 0.8829$

11. Apply Stirling and Steffensen formulas to find the value of $f(37.5)$ from the following table

x	35	36	37	38	39	40
$f(x)$	4.1	4.146	4.204	4.321	4.448	4.583

Ans. 4.258

12. Use appropriate formulas to compute the values of $f(1.2)$, $f(3.7)$ and $f(5.8)$ from the data set $(0,1)$ $(1,3)$ $(2,11)$ $(3,31)$ $(4,69)$ $(5,131)$ $(6,223)$.

Ans. 3.928, 55.353, 201.912

13. Compute the value of $f(-0.5, 4)$ for the following set of data points for the function $f(x, y)$.

x \ y	1	2	5
-1	2	5	26
0	1	4	25

Ans. 16.5

14. Given the following set of data points for the values of $f(x, y)$

x \ y	-1	1
0	-2	2
2	-4	8
4	-14	22

Compute the interpolating polynomial using Newton bivariate formula, and hence find the value of $f(0.5, -0.25)$

Ans. Interpolating polynomial is $x^2y + 2y + x$, and $f(0.5, -0.25) = -0.0625$

15. Compute the value of $f(1, -1)$ from the following data points

x \ y	-2	0	2
0	4	0	4
2	24	4	24
4	84	16	84

Ans. $f(1, -1) = 3$

(Data was generated from interpolating polynomial $x^2y^2 + y^2 + x^2$)

16. Obtain the Lagrange bivariate interpolating polynomial from the following tabulated values

x \ y	-1	1	2
-1	-4	-4	-7
0	-1	-1	-1
1	2	2	5

Ans. $xy^2 + 2x - 1$

Splines, Curve Fitting, and Other Approximating Curves

Chapter 11

> *The essence of mathematics lies in its freedom.*
>
> **Georg Ferdinand Ludwig Philipp Cantor**
> (March 3, 1845–January 6, 1918)
> The mathematician who developed the set theory.
>
> *Mathematics is the science of what is clear by itself.*
>
> **Carl Gustav Jacob Jacobi**
> (December 10, 1804–February 18, 1851)
> He made landmark contributions to the fields of differential equations,
> elliptic functions, and number theory.

11.1 Introduction

Experimental observations produce data sets of the type $\{(x_0, y_0), (x_1, y_1), (x_2, y_2), \cdots, (x_n, y_n)\}$ with $(n + 1)$ readings. Here, the variable "x" can serve as input, and variable "y" serves as the output of experiment. At any intermediate value of a variable, the corresponding value of another variable can be approximated by some numerical techniques. The problem of constructing an approximating function $y(x)$ that estimates value of dependent variable y at any intermediate value of independent variable x for a given set of discrete data points is very important. The data values (x_i, y_i), $i = 0, 1, ..., n$ are values of some well-defined function or may be some experimental observations. Some important approximating techniques are as follows

i)   Interpolation
ii) Piecewise Interpolation
iii) Spline Interpolation
iv) Bézier Curves
v)   B-Spline Curves

vi)   Least Squares Approximation
vii) Chebyshev Polynomial Approximation
viii) Padé Approximation

We have already discussed the interpolation and piecewise interpolation for a given set of data points in Chapters 8 and 10. Polynomials are the best approximation if the data size is small and observations are without any singularity. However, if the data size is large, then it is very difficult to obtain a polynomial of large degree. For examples, consider a data set of 50 points, the interpolating polynomial is of degree 49. The difficulty is not only to compute the polynomial of degree 49 but also to handle the round-off error during the approximation of values at any intermediate points. For example, if we have an intermediate value, say, up to 6-significant digits. Then, the 49th power of this value contains approximately $6(2^{49})$ significant digits. It is very difficult to compute such value without round-off error. Therefore we have to switch over to some other methodology. Also, higher order polynomials produce large oscillations which may give erroneous results compare to its lower order counterpart. Also, the global nature of the interpolating polynomial is a major concern as if we change the position of one point then the whole polynomial has to be changed. Therefore, for designing or graphic purposes, where large sets of data points are involved, we cannot carry out polynomial approximations. Rather these problems can be handled using piecewise interpolation, but in the case of piecewise interpolation, the smoothness is a major issue at the nodal points, as two different polynomials are joining at nodal points.

In this chapter, we will discuss the cubic spline interpolation, Bězier and B-Spline curves for the smooth and elegant fitting of the polynomial for a given data set. Oscillations and rounding error in large degree interpolation and problems of smooth fitting in piecewise interpolation are two major reasons for popularity of cubic spline for a large data set. Bězier and B-Spline curves serve as mandatory basic tools for computer graphics and computer aided designing. Then, we will discuss the least squares fitting to fit a curve of our choice that needs not to be a polynomial. The Chebyshev polynomial approximation to find the best lower order polynomial approximation is also a part of this chapter. At last, approximation by rational function in polynomials known as Padé approximation is also discussed which will be helpful to introduce singularity in experiments.

11.2 Spline Interpolation

A sequence of continuous curves that are connected to form a single continuous curve is called as a spline curve. Consider a given set of data points (x_i, y_i), $i = 0, 1, ..., n$ such that $x_i < x_{i+1}$ for all $i = 0, 1, ..., n-1$. In general, a mth degree spline $P_s(x)$ for this data set is a piecewise polynomial of degree m, which satisfies following two conditions.

1. It is of degree $\leq m$ for each interval (x_i, x_{i+1}), $i = 0, 1, ..., n-1$ and of degree m in at least one such interval.
2. The spline $P_s(x)$ and its first $m-1$ derivatives are continuous at each node points x_i, $i = 1, ..., n-1$ in the interval (x_0, x_n).

Example ———————————————————————————— **11.1**

The following function is a spline of degree 2 (quadratic spline)

$$f(x) = \begin{cases} x^2 - 5x + 1, & 0 \le x \le 1 \\ 2x^2 - 7x + 2, & 1 \le x \le 2 \\ x^2 - 3x - 2 & 2 \le x \le 3 \end{cases}$$

The function is a polynomial of degree 2 over each interval. Also, the function and its first derivative are continuous throughout the interval [0, 3]. We get following values of the function and its derivatives from both sides

$$f(1) = -3, f(2) = -4$$
$$f'(1) = -3, \ f'(2) = 1$$

The linear spline provides only continuity of the function; the quadratic spline is also continuous for first order derivative. In cubic spline approximation, the approximating cubic polynomials are such that the function values, first derivatives and second derivatives are continuous at nodal points. So, the cubic spline curve is continuous curve such that the gradient and curvature also remain same everywhere in the whole domain. Therefore, cubic spline provides sufficient smoothness on the approximating curve. For example, we have following cubic spline polynomials in different intervals for the data set (0, 1), (0.5, –2), (1, –5), (1.5, 7), (2, 10) and (2.5, 8).

$$P_{cs}(x) = \begin{cases} 1 - 3.55x - 9.80x^3 & x \in (0, 0.5) \\ -6.35 + 40.54x - 88.19x^2 + 48.99x^3 & x \in (0.5, 1) \\ 108.83 - 304.99x + 257.34x^2 - 66.18x^3 & x \in (1, 1.5) \\ -194.63 + 301.93x - 147.27x^2 + 23.73x^3 & x \in (1.5, 2) \\ -30.80 + 56.19x - 24.40x^2 + 3.25x^3 & x \in (2, 2.5) \end{cases}$$

The graphs of these polynomials are as follows. We can easily see the smoothness in the graph at nodal points.

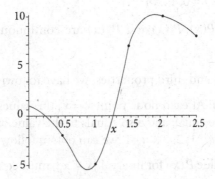

Fig. 11.1 Cubic spline curve

Among all the spline interpolations, cubic spline is one of the best approximations and most widely used as it ensures the smooth approximation with a less computational error. Here, we will discuss the construction of cubic spline polynomials only. The linear and quadratics splines are easy to obtain, and splines of higher degree (>3) require much computation and very difficult to handle.

11.2.1 Cubic Spline Interpolation

Let us approximate the function $f(x)$ by different cubic polynomials $P_i(x) = a_i x^3 + b_i x^2 + c_i x + d_i$, $i = 1, 2, ..., n$ for each subinterval $\left[x_{i-1}, x_i \right]$ in the given interval $[x_0, x_n]$.

$$P(x) = \begin{cases} P_1(x) = a_1 x^3 + b_1 x^2 + c_1 x + d_1 & x_0 \le x \le x_1 \\ P_2(x) = a_2 x^3 + b_2 x^2 + c_2 x + d_2 & x_1 \le x \le x_2 \\ \vdots & \\ P_n(x) = a_n x^3 + b_n x^2 + c_n x + d_n & x_{n-1} \le x \le x_n \end{cases}$$

In cubic spline approximation, the polynomials and their first and second derivatives are continuous at node points. A cubic spline polynomial $P(x)$ satisfies the following three properties

1. On each subinterval $\left[x_{i-1}, x_i \right]$, $1 \le i \le n$, $P(x)$ is a third-degree polynomial, i.e.,

$$P_i(x) = a_i x^3 + b_i x^2 + c_i x + d_i, \quad i = 1, 2, ..., n$$

   We have to find $4n$ unknowns: a_i, b_i, c_i, d_i; $i = 1, 2, ..., n$.
2. The values of the cubic spline at node points equal the values of the function at these points.

$$P(x_i) = f_i, \ i = 0, 1, ..., n$$

3. The polynomials $P(x)$, $P'(x)$ and $P''(x)$ are continuous throughout the interval (x_0, x_n).

On using the above second and third properties, we have following results

a) Continuity of $P(x)$: At each node point $x = x_i$, the values of two polynomials $P_i(x)$ and $P_{i+1}(x)$ must be equal, and also equals to the value of the function $f(x_i)$. At any node point $x = x_i$, $i = 1, 2, ..., n-1$, we can obtain following equations.

   The polynomial value $P_i(x_i)$ for interval $[x_{i-1}, x_i]$, must equals the function value $f(x_i)$

$$P_i(x_i) = a_i x^3_i + b_i x^2_i + c_i x_i + d_i = f(x_i) = f_i$$

Similarly, polynomial for interval $[x_i, x_{i+1}]$ gives following equations

$$P_{i+1}(x_i) = a_{i+1}x_i^3 + b_{i+1}x_i^2 + c_{i+1}x_i + d_{i+1} = f(x_i) = f_i$$

So, we have following set of equations

$$a_i x_i^3 + b_i x_i^2 + c_i x_i + d_i = f_i$$
$$a_{i+1}x_i^3 + b_{i+1}x_i^2 + c_{i+1}x_i + d_{i+1} = f_i \qquad (11.1)$$

b) At the end points x_0 and x_n of the interval, the values of splines must be equal to the values of the function.

$$f_0 = a_1 x_0^3 + b_1 x_0^2 + c_1 x_0 + d_1$$
$$f_n = a_n x_n^3 + b_n x_n^2 + c_n x_n + d_n \qquad (11.2)$$

c) Continuity of $P'(x)$ and $P''(x)$: At each node point $x = x_i$; the values of polynomials $P'_i(x)$ and $P'_{i+1}(x)$ are equal, and the values of polynomials $P''_i(x)$ and $P''_{i+1}(x)$ are also equal. At node points $x = x_i$, $i = 1, 2, \ldots, n-1$; we must have

$$3a_i x_i^2 + 2b_i x_i + c_i = 3a_{i+1}x_i^2 + 2b_{i+1}x_i + c_{i+1} \qquad (11.3)$$

$$6a_i x_i + 2b_i = 6a_{i+1}x_i + 2b_{i+1} \qquad (11.4)$$

We have $2(n-1)$ equations from system (11.1); two equations from system (11.2); and $2(n-1)$ equations from systems (11.3) and (11.4). So, we have total $4n-2$ equations, while the number of arbitrary constants to be determined is $4n$ $(a_i, b_i, c_i, d_i; i = 1, 2, \ldots, n)$. Hence, we need two more equations for the polynomials to be unique.

Let us take the notation $P'(x_i) = m_i$ and $P''(x_i) = M_i$. In general, we assign some values to the polynomial $P''(x)$ at the end points, that is $P''(x_0) = M_0$ and $P''(x_n) = M_n$. If the end conditions are $M_0 = 0$ and $M_n = 0$, then our spline is called as a natural spline (As the drafting of the spline always behaves in this manner).

At last, we have 4n equations in 4n variables; which can be easily solved to obtain the required cubic spline. But to reduce the computational work, we use an alternative method to obtain the cubic spline interpolation described below.

Alternative Method for Cubic Spline:
Since the function $P(x)$ is a cubic polynomial, so the function $P''(x)$ is linear function of x in the interval $x_{i-1} \leq x \leq x_i$ and can be written as

$$P''(x) = \frac{x_i - x}{x_i - x_{i-1}} P''(x_{i-1}) + \frac{(x - x_{i-1})}{x_i - x_{i-1}} P''(x_i)$$

Let us assume that the length of the interval (x_{i-1}, x_i) is h_i i.e. $h_i = x_i - x_{i-1}$. Also, assume $M_i = P''(x_i)$

$$P''(x) = \frac{x_i - x}{h_i} M_{i-1} + \frac{(x - x_{i-1})}{h_i} M_i$$

On integrating this equation twice on x, we have

$$P(x) = \frac{(x_i - x)^3}{6h_i} M_{i-1} + \frac{(x - x_{i-1})^3}{6h_i} M_i + k_1 x + k_2 \tag{11.5}$$

where k_1 and k_2 are arbitrary constants. The values of cubic spline polynomials must equal to function values at nodal points; therefore, we have

$$P(x_{i-1}) = f(x_{i-1}) = f_{i-1} \text{ and } P(x_i) = f(x_i) = f_i$$

On using these conditions in Eq. (11.5), we have

$$P(x_{i-1}) = f_{i-1} = \frac{1}{6} h_i^2 M_{i-1} + k_1 x_{i-1} + k_2$$

$$P(x_i) = f_i = \frac{1}{6} h_i^2 M_i + k_1 x_i + k_2$$

Solution of these two equations for k_1 and k_2 is given by

$$k_1 = \frac{1}{h_i}(f_i - f_{i-1}) - \frac{1}{6}(M_i - M_{i-1})h_i$$

$$k_2 = \frac{1}{h_i}(x_i f_{i-1} - x_{i-1} f_i) - \frac{1}{6}(x_i M_{i-1} - x_{i-1} M_i)h_i$$

On substituting these values of k_1 and k_2 in Eq. (11.5), we have

$$P(x) = \frac{1}{6h_i}(x_i - x)^3 M_{i-1} + \frac{1}{6h_i}(x - x_{i-1})^3 M_i + \frac{x}{h_i}(f_i - f_{i-1})$$

$$-\frac{x}{6}(M_i - M_{i-1})h_i + \frac{1}{h_i}(x_i f_{i-1} - x_{i-1} f_i) - \frac{1}{6}(x_i M_{i-1} - x_{i-1} M_i)h_i \qquad (x_{i-1} \le x \le x_i)$$

$$= \frac{1}{6h_i}\left[(x_i - x)\left\{(x_i - x)^2 - h_i^2\right\}\right]M_{i-1} + \frac{1}{6h_i}\left[(x - x_{i-1})\left\{(x - x_{i-1})^2 - h_i^2\right\}\right]M_i$$

$$+\frac{1}{h_i}(x_i - x)f_{i-1} + \frac{1}{h_i}(x - x_{i-1})f_i \quad i = 1, 2, ..., n \tag{11.6}$$

To compute values of M_{i-1} and M_i, we will use continuity of the polynomial $P'(x)$. On differentiating the Eq. (11.6) w.r.t. x, we get

$$P'(x) = -\frac{(x_i - x)^2}{2h_i} M_{i-1} + \frac{(x - x_{i-1})^2}{2h_i} M_i - \frac{(M_i - M_{i-1})h_i}{6} + \frac{f_i - f_{i-1}}{h_i} \quad x_{i-1} \le x \le x_i \quad (11.7)$$

Similarly, we can obtain $P'(x)$ for the interval $x_i \leq x \leq x_{i+1}$, by simply changing $i = i+1$ in Eq. (11.7).

$$P'(x) = -\frac{(x_{i+1}-x)^2}{2h_{i+1}} M_i + \frac{(x-x_i)^2}{2h_{i+1}} M_{i+1} - \frac{1}{6}(M_{i+1}-M_i)h_{i+1} + \frac{f_{i+1}-f_i}{h_{i+1}} \quad (x_i \leq x \leq x_{i+1})$$

$$(11.8)$$

The continuity of the derivatives implies that the derivatives $P'(x)$ in both the intervals $x_{i-1} \leq x \leq x_i$ and $x_i \leq x \leq x_{i+1}$ must be equal at the node point $x = x_i$. We have

$$\Rightarrow \frac{1}{2}h_i M_i - \frac{(M_i - M_{i-1})h_i}{6} + \frac{f_i - f_{i-1}}{h_i} = -\frac{1}{2}h_{i+1}M_i - \frac{1}{6}(M_{i+1}-M_i)h_{i+1} + \frac{f_{i+1}-f_i}{h_{i+1}}$$

On rewriting this equation, we have

$$\frac{h_i}{6}M_{i-1} + \frac{h_i + h_{i+1}}{3}M_i + \frac{h_{i+1}}{6}M_{i+1} = \frac{1}{h_{i+1}}(f_{i+1}-f_i) - \frac{1}{h_i}(f_i - f_{i-1}) \quad i = 1,2,...,n-1$$

$$x_{i-1} \leq x \leq x_i \quad (11.9)$$

The system (11.9) will produce a linear system of $(n-1)$ equations in $(n+1)$ unknowns $M_0, M_1,..., M_n$. We can use any two additional conditions for unique solution of the system. The spline is a natural spline in case of end conditions $M_0 = 0$ and $M_n = 0$.

We solve the system (11.9), and then use the values of $M_0, M_1,..., M_n$ in system (11.6) to obtain the following cubic spline as desired.

$$P(x) = \frac{1}{6h_i}(x_i - x)^3 M_{i-1} + \frac{1}{6h_i}(x - x_{i-1})^3 M_i + \frac{x}{h_i}(f_i - f_{i-1}) - \frac{x}{6}(M_i - M_{i-1})$$

$$h_i + \frac{1}{h_i}(x_i f_{i-1} - x_{i-1} f_i) - \frac{1}{6}(x_i M_{i-1} - x_{i-1} M_i)h_i \quad i = 1,2,...,n \quad (11.10)$$

11.2.2 Cubic Spline for Equi-spaced Points

The interval length of all the intervals is same in case of equi-spaced points, i.e.,

$$h_1 = h_2 = ... = h_n = h$$

So, our system of Eqs. (11.9) and (11.10) reduces to following Eqs. (11.11) and (11.12), respectively

$$M_{i-1} + 4M_i + M_{i+1} = \frac{6}{h^2}(f_{i+1} - 2f_i + f_{i-1}) \quad i = 1,2,...,n-1 \quad (11.11)$$

$$P(x) = \frac{1}{6h}\left[(x_i - x)^3 M_{i-1} + (x - x_{i-1})^3 M_i \right] + \frac{1}{h}(x_i - x)\left(f_{i-1} - \frac{h^2}{6} M_{i-1} \right)$$

$$+ \frac{1}{h}(x - x_{i-1})\left(f_i - \frac{h^2}{6} M_i \right) \qquad\qquad i = 1, 2, \ldots, n \qquad\qquad (11.12)$$

These systems can be solved to obtain the desired cubic spline.

Example ———————————————————————————————— **11.2** ———

Determine the cubic spline polynomial for the following data set and hence compute the values of $f(0.3)$ and $f(2.6)$.

x	0	1	2	3
$f(x)$	1	-8	-30	-59

Use natural spline conditions. Also, verify the interpolation and continuity conditions of the cubic spline at $x = 2$.

Ans.

Here, we have equispaced points with spacing $h = 1$, and the points are

$$x_0 = 0, \quad x_1 = 1, \quad x_2 = 2, \quad x_3 = 3$$
$$f_0 = 1, \quad f_1 = -8, \quad f_2 = -30, \quad f_3 = -59$$

The natural spline condition implies $M_0 = M_3 = 0$.
First, we will use the Eqs. (11.11) to compute the values of M_1 and M_2.

$$M_{i-1} + 4M_i + M_{i+1} = \frac{6}{h^2}\left(f_{i+1} - 2f_i + f_{i-1} \right)$$

$$i = 1 \quad M_0 + 4M_1 + M_2 = \frac{6}{h^2}\left(f_2 - 2f_1 + f_0 \right)$$

$$4M_1 + M_2 = -78 \qquad\qquad (11.13)$$

$$i = 2 \quad M_1 + 4M_2 + M_3 = \frac{6}{h^2}\left(f_3 - 2f_2 + f_1 \right) = -42$$

$$M_1 + 4M_2 = -42 \qquad\qquad (11.14)$$

On solving Eqs. (11.13) and (11.14) for M_1 and M_2, we get

$$M_1 = -18 \text{ and } M_2 = -6$$

We will use these values of M_1 and M_2 in Eq. (11.12) to obtain the cubic spline curves as follows

$$P(x) = \frac{1}{6h}\left[(x_i - x)^3 M_{i-1} + (x - x_{i-1})^3 M_i\right] + \frac{1}{h}(x_i - x)\left(f_{i-1} - \frac{h^2}{6}M_{i-1}\right)$$

$$+ \frac{1}{h}(x - x_{i-1})\left(f_i - \frac{h^2}{6}M_i\right) \qquad\qquad x_{i-1} \le x \le x_i$$

$i = 1$

$$P(x) = \frac{1}{6h}\left[(x_1 - x)^3 M_0 + (x - x_0)^3 M_1\right] + \frac{1}{h}(x_1 - x)\left(f_0 - \frac{h^2}{6}M_0\right) + \frac{1}{h}(x - x_0)\left(f_1 - \frac{h^2}{6}M_1\right)$$

$$P(x) = -3x^3 - 6x + 1; \text{ for } x_0 \le x \le x_1 \text{ (or) } 0 \le x \le 1 \qquad\qquad (11.15)$$

$i = 2$

$$P(x) = \frac{1}{6h}\left[(x_2 - x)^3 M_1 + (x - x_1)^3 M_2\right] + \frac{1}{h}(x_2 - x)\left(f_1 - \frac{h^2}{6}M_1\right)$$

$$+ \frac{1}{h}(x - x_1)\left(f_2 - \frac{h^2}{6}M_2\right)$$

$$P(x) = 2x^3 - 15x^2 + 9x - 4; \text{ for } x_1 \le x \le x_2 \text{ (or) } 1 \le x \le 2 \qquad\qquad (11.16)$$

$i = 3$

$$P(x) = \frac{1}{6h}\left[(x_3 - x)^3 M_2 + (x - x_2)^3 M_3\right] + \frac{1}{h}(x_3 - x)\left(f_2 - \frac{h^2}{6}M_2\right)$$

$$+ \frac{1}{h}(x - x_2)\left(f_3 - \frac{h^2}{6}M_3\right)$$

$$P(x) = x^3 - 9x^2 - 3x + 4; \text{ for } x_2 \le x \le x_3 \text{ (or) } 2 \le x \le 3 \qquad\qquad (11.17)$$

On combining the results (11.15), (11.16) and (11.17), we finally have cubic spline as follows

$$P(x) = \begin{cases} -3x^3 - 6x + 1 & 0 \le x \le 1 \\ 2x^3 - 15x^2 + 9x - 4 & 1 \le x \le 2 \\ x^3 - 9x^2 - 3x + 4 & 2 \le x \le 3 \end{cases} \qquad\qquad (11.18)$$

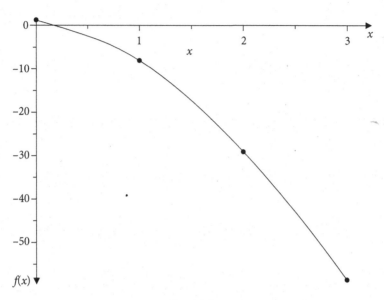

Fig. 11.2

On using the polynomials (11.18), the values of $f(0.3)$ and $f(2.6)$ are given by

$$f(0.3) = -0.881$$
$$f(2.6) = -47.064$$

Verification: We can easily check the authenticity of cubic spline by simply verifying the interpolatory conditions $P(x_i) = f_i$ at $x = 0, 1, 2, 3$ and the continuity conditions of $P(x)$, $P'(x)$ and $P''(x)$ at the nodes $x = 1, 2$, say for example at the point $x = 2$.

$$\text{Interpolatory Condition: } P(2) = \begin{cases} -30 & 1 \le x \le 2 \\ -30 & 2 \le x \le 3 \end{cases} = f_2$$

$$\text{Continuity of } \quad P(x): P(2) = \begin{cases} -30 & 1 \le x \le 2 \\ -30 & 2 \le x \le 3 \end{cases}$$

$$\text{Continuity of } P'(x): P'(x) = \begin{cases} 6x^2 - 30x + 9 & 1 \le x \le 2 \\ 3x^2 - 18x - 3 & 2 \le x \le 3 \end{cases} \Rightarrow P'(2) = \begin{cases} -27 & 1 \le x \le 2 \\ -27 & 2 \le x \le 3 \end{cases}$$

$$\text{Continuity of } P''(x): P''(x) = \begin{cases} 12x - 30 & 1 \le x \le 2 \\ 6x - 18 & 2 \le x \le 3 \end{cases} \Rightarrow P''(2) = \begin{cases} -6 & 1 \le x \le 2 \\ -6 & 2 \le x \le 3 \end{cases}$$

Example ——————————————————————————— 11.3

Obtain cubic spline approximation for $e^{0.2}$ from the following values of e^x correct up to six significant digits. Use natural spline conditions.

x	0	0.1	0.3	0.4
e^x	1	1.10517	1.34986	1.49182

Ans.

Given

$$h_1 = 0.1,\ h_2 = 0.2,\ h_3 = 0.1$$

$$x_0 = 0,\ x_1 = 0.1,\ x_2 = 0.3,\ x_3 = 0.4$$

$$f_0 = 1,\ f_1 = 1.10517,\ f_2 = 1.34986,\ f_3 = 1.49182$$

Natural spline conditions are $M_0 = M_3 = 0$.

Since the points are not equispaced; we have to use Eq. (11.9) for the values of M_1 and M_2

$$\frac{h_i}{6}M_{i-1} + \frac{h_i + h_{i+1}}{3}M_i + \frac{h_{i+1}}{6}M_{i+1} = \frac{1}{h_{i+1}}(f_{i+1} - f_i) - \frac{1}{h_i}(f_i - f_{i-1})$$

$i = 1$

$$\frac{h_1}{6}M_0 + \frac{h_1 + h_2}{3}M_1 + \frac{h_2}{6}M_2 = \frac{1}{h_2}(f_2 - f_1) - \frac{1}{h_1}(f_1 - f_0)$$

$$6M_1 + 2M_2 = 10.305 \qquad (11.19)$$

$i = 2$

$$\frac{h_2}{6}M_1 + \frac{h_2 + h_3}{3}M_2 + \frac{h_3}{6}M_3 = \frac{1}{h_3}(f_3 - f_2) - \frac{1}{h_2}(f_2 - f_1)$$

$$2M_1 + 6M_2 = 11.769 \qquad (11.20)$$

On solving Eqs. (11.19) and (11.20) for M_1 and M_2, we get

$$M_1 = 1.196625 \text{ and } M_2 = 1.562625$$

We have to compute the value of $e^{0.2}$, that is in the interval (x_1, x_2). Therefore, we will use these values of M_1 and M_2 in the Eq. (11.10) for $i = 1$ to obtain the cubic spline approximation of the value $e^{0.2}$.

$$P(x) = \frac{1}{6h_i}(x_i - x)^3 M_{i-1} + \frac{1}{6h_i}(x - x_{i-1})^3 M_i + \frac{x}{h_i}(f_i - f_{i-1})$$

$$- \frac{x}{6}(M_i - M_{i-1})h_i + \frac{1}{h_i}(x_i f_{i-1} - x_{i-1} f_i) - \frac{1}{6}(x_i M_{i-1} - x_{i-1} M_i)h_i$$

$i = 1$

$$P(x) = \frac{1}{6h_1}(x_1 - x)^3 M_0 + \frac{1}{6h_1}(x - x_0)^3 M_1 + \frac{x}{h_1}(f_1 - f_0)$$

$$- \frac{x}{6}(M_1 - M_0)h_1 + \frac{1}{h_1}(x_1 f_0 - x_0 f_1) - \frac{1}{6}(x_1 M_0 - x_0 M_1)h_1$$

On using different values and $x = 0.2$, we obtain following cubic spline approximation for $e^{0.2}$

$$P(0.2) = 1.22088$$

While the exact value of $e^{0.2}$ is 1.22140.

We can also compute the following cubic spline polynomials for the data set

$$\begin{cases} 1 + 1.032x + 1.994x^3 & 0 \le x \le 0.1 \\ 1.002 + 0.981x + 0.507x^2 + 0.305x^3 & 0.1 \le x \le 0.3 \\ 1.080 + 0.196x + 3.125x^2 - 2.604x^3 & 0.3 \le x \le 0.4 \end{cases}$$

11.3 Bězier Curve

In addition to splines, we have other important curves like Bězier and B-Spline curves. These curves are not like the interpolating splines; as these curves, in general, do not pass through all the data points (in this section and next section, we will use the words control points instead of data points). However, these curves stay within the convex polygon formed by the data points. These curves are commonly used in computer-aided designs and computer graphics. These curves are also used in interior designing, animation, graphics, etc. (e.g. shape of the interior of a car) where we need aesthetical approach over accuracy.

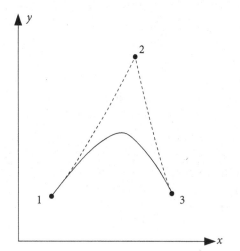

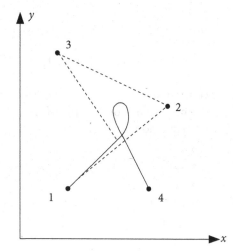

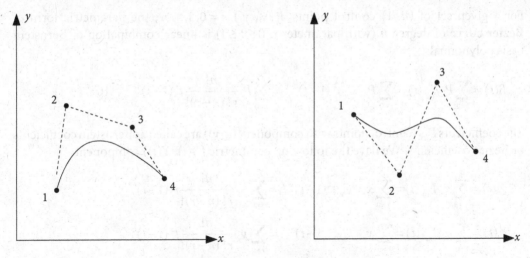

Fig. 11.3 Graphs of some Bězier curves

Before we discuss Bězier and B-Spline curves, it is better to introduce convex hull and Bernstein polynomials.

a) **Convex Polygon and Convex Hull**

Let us consider n – points of a set $P_1(x_1, y_1), P_2(x_2, y_2),..., P_n(x_n, y_n)$. Then draw these points on the drawing board and stick pins to these points. Then take a rubber band and stretch it over a minimum number of pins such that all pins are either touching the rubber band or inside the rubber band. Such polygon formed by the rubber band is a convex polygon of these points and interior of the polygon is convex hull. Bězier and B-Spline curves always remain in the convex hull formed by the control points.

b) **Bernstein Basis Polynomials**

The $(n + 1)$ Bernstein basis polynomials of degree n are defined as follows

$$J_{n,r}(t) = {}^nC_r\, t^r (1-t)^{n-r} = \frac{n!}{r!\,(n-r)!} t^r (1-t)^{n-r} \qquad r = 0, 1, ..., n$$

For first few n, Bernstein polynomials are as follows

$n = 0$ $\quad J_{0,0} = 1$

$n = 1$ $\quad J_{1,0} = (1-t), \qquad J_{1,1} = t$

$n = 2$ $\quad J_{2,0} = (1-t)^2, \quad J_{2,1} = 2t(1-t), J_{2,2} = t^2$

$n = 3$ $\quad J_{3,0} = (1-t)^3, \quad J_{3,1} = 3t(1-t)^2, J_{3,2} = 3t^2(1-t), J_{3,3} = t^3$

$n = 4$ $\quad J_{4,0} = (1-t)^4, \quad J_{4,1} = 4t(1-t)^3, J_{4,2} = 6t^2(1-t)^2, J_{4,3} = 4t^3(1-t), J_{4,4} = t^4$

These $(n + 1)$ Bernstein basis polynomials of degree n form a basis of the vector space $(\wp_n)$ of polynomials of degree at most n.

For a given set of $(n+1)$ control points $P_i(x_i, y_i)$, $i = 0, 1, ..., n$, the parametric form of **Bězier curve** of degree n (with parameter t, $0 \le t \le 1$) is linear combination of Bernstein basis polynomials

$$B(t) = \sum_{r=0}^{n} P_r J_{n,r}(t) = \sum_{r=0}^{n} P_r \, {}^nC_r \, t^r (1-t)^{n-r} = \sum_{r=0}^{n} P_r \frac{n!}{r!\,(n-r)!} t^r (1-t)^{n-r} \quad (0 \le t \le 1)$$

The coefficients P_r's (control points with component (x, y)) are called as Bernstein coefficients or Bězier coefficients. We have the following parametric $(X(t), Y(t))$ components

$$X(t) = \sum_{r=0}^{n} x_r J_{n,r}(t) = \sum_{r=0}^{n} x_r \, {}^nC_r \, t^r (1-t)^{n-r} = \sum_{r=0}^{n} x_r \frac{n!}{r!\,(n-r)!} t^r (1-t)^{n-r}$$

$$Y(t) = \sum_{r=0}^{n} y_r J_{n,r}(t) = \sum_{r=0}^{n} y_r \, {}^nC_r \, t^r (1-t)^{n-r} = \sum_{r=0}^{n} y_r \frac{n!}{r!\,(n-r)!} t^r (1-t)^{n-r}$$

$$(0 \le t \le 1) \qquad (11.21)$$

Example   **11.4**

Obtain the quartic Bězier curve for the following set of control points $P_i(x_i, y_i)$, $i = 0, 1, ..., 4$.

x	-1	2	3	5	6
y	1	3	-4	8	11

Ans.

The control points are given by

$$x_0 = -1, \; x_1 = 2, \; x_2 = 3, \quad x_3 = 5, \; x_4 = 6$$
$$y_0 = 1, \quad y_1 = 3, \; y_2 = -4, \; y_3 = 8, \; y_4 = 11$$

On using these values in the Eq. (11.21), we have following Bězier curve

$$X(t) = \sum_{r=0}^{4} x_r J_{4,r}(t) = x_0 J_{4,0}(t) + x_1 J_{4,1}(t) + x_2 J_{4,2}(t) + x_3 J_{4,3}(t) + x_4 J_{4,4}(t)$$

$$= (-1)(1-t)^4 + (2)4t(1-t)^3 + (3)6t^2(1-t)^2 + (5)4t^3(1-t) + (6)t^4$$

$$= -5t^4 + 12t^3 - 12t^2 + 12t - 1$$

$$Y(t) = \sum_{r=0}^{4} y_r J_{4,r}(t) = y_0 J_{4,0}(t) + y_1 J_{4,1}(t) + y_2 J_{4,2}(t) + y_3 J_{4,3}(t) + y_4 J_{4,4}(t)$$

$$= (1)(1-t)^4 + (3)4t(1-t)^3 + (-4)6t^2(1-t)^2 + (8)4t^3(1-t) + (11)t^4$$

$$= -56t^4 + 112t^3 - 54t^2 + 8t + 1$$

$$(0 \le t \le 1)$$

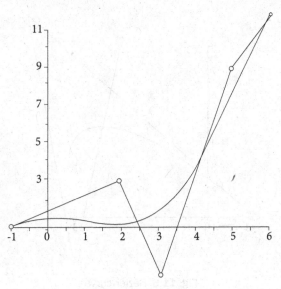

Fig. 11.4 Bězier curve

Example ———————————————————————————————————— **11.5**

Obtain the cubic Bězier curve for the following set of control points

x	−1	5	3	2
y	1	11	−4	8

Ans.

We have the following set of control points

$$x_0 = -1, \quad x_1 = 5, \quad x_2 = 3, \quad x_3 = 2$$
$$y_0 = 1, \quad y_1 = 11, \quad y_2 = -4, \quad y_3 = 8$$

The Bězier curve (11.21) for these control points is given by

$$X(t) = \sum_{r=0}^{3} x_r \, J_{3,r}(t) = x_0 J_{3,0}(t) + x_1 J_{3,1}(t) + x_2 \, J_{3,2}(t) + x_3 J_{3,3}(t)$$
$$= (-1)(1-t)^3 + (5)3t(1-t)^2 + (3)3t^2(1-t) + (2)t^3$$
$$= 9t^3 - 24t^2 + 18t - 1$$

$$Y(t) = \sum_{r=0}^{3} y_r \, J_{3,r}(t) = y_0 J_{3,0}(t) + y_1 J_{3,1}(t) + y_2 \, J_{3,2}(t) + y_3 J_{3,3}(t)$$
$$= (1)(1-t)^3 + (11)3t(1-t)^2 + (-4)3t^2(1-t) + (8)t^3$$
$$= 52t^3 - 75t^2 + 30t + 1$$

$$0 \le t \le 1$$

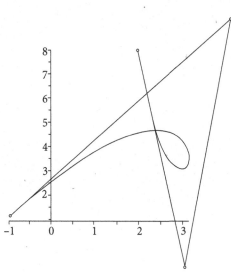

Fig. 11.5 Bězier curve

Matrix Form of Bězier Curve

It is convenient to represent the Bězier curve in the matrix form. Here, we are presenting the matrix forms of Bězier curves up to order 4.

$n = 0$	$J_{0,0} = 1$ $P(t) = P_0$
$n = 1$	$J_{1,0} = (1-t),\ J_{1,1} = t$ $P(t) = \begin{bmatrix} t & 1 \end{bmatrix} \begin{bmatrix} -1 & 1 \\ 1 & 0 \end{bmatrix} \begin{bmatrix} P_0 \\ P_1 \end{bmatrix}$
$n = 2$	$J_{2,0} = (1-t)^2,\ J_{2,1} = 2t(1-t),\ J_{2,2} = t^2$ $P(t) = \begin{bmatrix} t^2 & t & 1 \end{bmatrix} \begin{bmatrix} 1 & -2 & 1 \\ -2 & 2 & 0 \\ 1 & 0 & 0 \end{bmatrix} \begin{bmatrix} P_0 \\ P_1 \\ P_2 \end{bmatrix}$
$n = 3$	$J_{3,0} = (1-t)^3,\ J_{3,1} = 3t(1-t)^2,\ J_{3,2} = 3t^2(1-t),\ J_{3,3} = t^3$ $P(t) = \begin{bmatrix} t^3 & t^2 & t & 1 \end{bmatrix} \begin{bmatrix} -1 & 3 & -3 & 1 \\ 3 & -6 & 3 & 0 \\ -3 & 3 & 0 & 0 \\ 1 & 0 & 0 & 0 \end{bmatrix} \begin{bmatrix} P_0 \\ P_1 \\ P_2 \\ P_3 \end{bmatrix}$

$$n = 4 \quad J_{4,0} = (1-t)^4, \; J_{4,1} = 4t(1-t)^3, \; J_{4,2} = 6t^2(1-t)^2, \; J_{4,3} = 4t^3(1-t), \; J_{4,4} = t^4$$

$$P(t) = \begin{bmatrix} t^4 & t^3 & t^2 & t & 1 \end{bmatrix} \begin{bmatrix} 1 & -4 & 6 & -4 & 1 \\ -4 & 12 & -12 & 4 & 0 \\ 6 & -12 & 6 & 0 & 0 \\ -4 & 4 & 0 & 0 & 0 \\ 1 & 0 & 0 & 0 & 0 \end{bmatrix} \begin{bmatrix} P_0 \\ P_1 \\ P_2 \\ P_3 \\ P_4 \end{bmatrix} \quad (11.22)$$

Properties of Bernstein Polynomials and Bézier Curves

1. We have different Bézier curves for the same set of control points but not in same order. For example, the Bézier curves for following three sets are different

    a)
x	-1	5	3	2
y	1	11	-4	8

    b)
x	-1	2	3	5
y	1	8	-4	11

    c)
x	5	3	2	-1
y	11	-4	8	1

2. In general, the Bézier curve does not pass through the control points except the first and last points.
3. The summation of all the Bernstein Polynomials is equal to 1 for any given t, i.e.,

$$\sum_{r=0}^{n} J_{n,r}(t) = \sum_{r=0}^{n} {}^nC_r \, t^r (1-t)^{n-r} = \sum_{r=0}^{n} \big((1-t)+t\big)^r = 1$$

4. The Bézier curve always lies within the convex hull of the control points. So, it remains bounded by the control points.
5. For a given set of $(n+1)$ control points $P_i(x_i, y_i)$, $i = 0, 1,...,n$, Bézier curve is always of degree n. Therefore, two control points result in a straight line; similarly, Bézier curve for three control points is a parabola, etc.
6. A closed curve can be generated with equal first and last control points.
7. If the data size is large, then it is very difficult to obtain a large degree of Bézier curve. For examples, if we have a dataset of 20 control points, then the Bézier curve is of degree 19. The difficulty is not only to compute the polynomial of degree 19 but also to handle the round-off error during the calculations of values at any intermediate points. Therefore, we will derive different sets of Bézier curves in this case, and club together these Bézier curves to form a single curve. Since the Bézier curve passes through first and last control points, we have continuity of the curve. Also, first order continuity can be achieved by simply ensuring that the tangents between first two points and last two points are equal. In general, it is not possible to achieve second order continuity.

11.4 B-Spline Curve

B-spline curves are commonly used in computer-aided designs and computer graphics. We have a single Bězier curve of degree n for $(n+1)$ control points. But B-Spline curves are different for each interval, and may be of any degree. B-Spline is similar to cubic spline in the sense that, we have different B-Spline curves for each subinterval. But contrary to cubic splines, the B-Spline curves need not pass through the control points. Also, it is worth mentioning here that B-Spline curves have first and second order continuity (like cubic spline). *Without going into further details, here we will discuss only the **cubic B-Spline** curve.*

For a given set of control points $P_i(x_i, y_i)$, $i = 0, 1,...,m$, we have following cubic B-Spline curves $B_j(t)$ for the intervals (P_j, P_{j+1}), $j = 1, 2,...,m-1$ in parametric form (with parameter t)

$$B_j(t) = \sum_{r=-1}^{2} P_{j+r} \, \beta_r(t); \quad 0 \le t \le 1$$

where the set $\beta_r(t)$ is the basis, and it remains same for every set of intervals (P_j, P_{j+1}), $j = 1, 2,...,m-1$. The basis polynomials $\beta_r(t)$ $r = -1, 0, 1, 2$ are given by

$$\beta_{-1}(t) = \frac{1}{6}(1-t)^3$$

$$\beta_0(t) = \frac{t^3}{2} - t^2 + \frac{2}{3}$$

$$\beta_1(t) = -\frac{t^3}{2} + \frac{t^2}{2} + \frac{t}{2} + \frac{1}{6}$$

$$\beta_2(t) = \frac{t^3}{6}$$

$$\Rightarrow B_j(t) = \sum_{r=-1}^{2} P_{j+r} \, \beta_r(t) = P_{j-1}\beta_{-1}(t) + P_j\beta_0(t) + P_{j+1}\beta_1(t) + P_{j+2}\beta_2(t)$$

$$= P_{j-1}\left(\frac{1}{6}(1-t)^3\right) + P_j\left(\frac{t^3}{2} - t^2 + \frac{2}{3}\right) + P_{j+1}\left(-\frac{t^3}{2} + \frac{t^2}{2} + \frac{t}{2} + \frac{1}{6}\right) + P_{j+2}\left(\frac{t^3}{6}\right)$$

The parametric components $\left(X_j(t),\, Y_j(t)\right)$ are as follows

$$X_j(t) = \sum_{r=-1}^{2} x_{j+r} \, \beta_r(t) = x_{j-1}\beta_{-1}(t) + x_j\beta_0(t) + x_{j+1}\beta_1(t) + x_{j+2}\beta_2(t)$$

$$= x_{j-1}\left(\frac{1}{6}(1-t)^3\right) + x_j\left(\frac{t^3}{2} - t^2 + \frac{2}{3}\right) + x_{j+1}\left(-\frac{t^3}{2} + \frac{t^2}{2} + \frac{t}{2} + \frac{1}{6}\right) + x_{j+2}\left(\frac{t^3}{6}\right)$$

$$0 \le t \le 1$$

$$Y_j(t) = \sum_{r=-1}^{2} y_{j+r} \, \beta_r(t) = y_{j-1}\beta_{-1}(t) + y_j\beta_0(t) + y_{j+1}\beta_1(t) + y_{j+2}\beta_2(t)$$

$$= y_{j-1}\left(\frac{1}{6}(1-t)^3\right) + y_j\left(\frac{t^3}{2} - t^2 + \frac{2}{3}\right) + y_{j+1}\left(-\frac{t^3}{2} + \frac{t^2}{2} + \frac{t}{2} + \frac{1}{6}\right) + y_{j+2}\left(\frac{t^3}{6}\right)$$

$$0 \leq t \leq 1$$

(Set $P_{-1} = P_0$ and $P_{m+1} = P_m$ for calculations)

Example 11.6

Obtain the cubic B-Spline curve for the following set of control points $P_i(x_i, y_i)$, $i = 0, 1, 2$.

$$x: \quad -3 \quad 1 \quad 5$$
$$y: \quad 1 \quad -2 \quad 6$$

Ans.

We have the following set of control points (keeping all the control points in order) $P_i(x_i, y_i)$, $i = 0,1,2$.

$$x_0 = -3, \; x_1 = 1, \quad x_2 = 5$$
$$y_0 = 1, \quad y_1 = -2, \; y_2 = 6$$

To start with the computation, we set the following conditions

$$P_{-1}(x_{-1}, y_{-1}) = P_0(x_0, y_0) = (-3, 1)$$
$$P_3(x_3, y_3) = P_2(x_2, y_2) = (5, 6)$$

The cubic B-Spline curve $B_j(t)$ for the interval (P_j, P_{j+1}), $j = 1, 2, \ldots, m-1$ is given by

$$X(t) = \sum_{r=-1}^{2} x_{j+r} \, \beta_r(t) = x_{j-1}\beta_{-1}(t) + x_j\beta_0(t) + x_{j+1}\beta_1(t) + x_{j+2}\beta_2(t)$$

$$= x_{j-1}\left(\frac{1}{6}(1-t)^3\right) + x_j\left(\frac{t^3}{2} - t^2 + \frac{2}{3}\right) + x_{j+1}\left(-\frac{t^3}{2} + \frac{t^2}{2} + \frac{t}{2} + \frac{1}{6}\right) + x_{j+2}\left(\frac{t^3}{6}\right)$$

$$Y(t) = \sum_{r=-1}^{2} y_{j+r} \, \beta_r(t) = y_{j-1}\beta_{-1}(t) + y_j\beta_0(t) + y_{j+1}\beta_1(t) + y_{j+2}\beta_2(t)$$

$$= y_{j-1}\left(\frac{1}{6}(1-t)^3\right) + y_j\left(\frac{t^3}{2} - t^2 + \frac{2}{3}\right) + y_{j+1}\left(-\frac{t^3}{2} + \frac{t^2}{2} + \frac{t}{2} + \frac{1}{6}\right) + y_{j+2}\left(\frac{t^3}{6}\right)$$

B-Spline curve for the interval (P_0, P_1) $[j = 0]$ is given by

$$X(t) = (-3)\left(\frac{1}{6}(1-t)^3\right) + (-3)\left(\frac{t^3}{2} - t^2 + \frac{2}{3}\right) + (1)\left(-\frac{t^3}{2} + \frac{t^2}{2} + \frac{t}{2} + \frac{1}{6}\right) + (5)\left(\frac{t^3}{6}\right)$$

$$= -\frac{7}{3} + 2t + 2t^2 - \frac{2}{3}t^3$$

$$Y(t) = (1)\left(\frac{1}{6}(1-t)^3\right) + (1)\left(\frac{t^3}{2} - t^2 + \frac{2}{3}\right) + (-2)\left(-\frac{t^3}{2} + \frac{t^2}{2} + \frac{t}{2} + \frac{1}{6}\right) + (6)\left(\frac{t^3}{6}\right)$$

$$= \frac{1}{2} - \frac{3}{2}t - \frac{3}{2}t^2 + \frac{7}{3}t^3$$

B-Spline curve for the interval (P_1, P_2) $[j = 0]$ is as follows

$$X(t) = (-3)\left(\frac{1}{6}(1-t)^3\right) + (1)\left(\frac{t^3}{2} - t^2 + \frac{2}{3}\right) + (5)\left(-\frac{t^3}{2} + \frac{t^2}{2} + \frac{t}{2} + \frac{1}{6}\right) + (5)\left(\frac{t^3}{6}\right)$$

$$= 1 + 4t - \frac{2}{3}t^3$$

$$Y(t) = (1)\left(\frac{1}{6}(1-t)^3\right) + (-2)\left(\frac{t^3}{2} - t^2 + \frac{2}{3}\right) + (6)\left(-\frac{t^3}{2} + \frac{t^2}{2} + \frac{t}{2} + \frac{1}{6}\right) + (6)\left(\frac{t^3}{6}\right)$$

$$= -\frac{1}{6} + \frac{5}{2}t + \frac{11}{2}t^2 - \frac{19}{6}t^3$$

We can combine these curves for each subintervals to get the final B-Spline curve.

Example ——————————————————————— 11.7

Obtain the cubic B-Spline curve for the following set of control points $P_i(x_i, y_i)$, $i = 0, 1, ..., 4$

x	−1	2	3	5	6
y	1	3	−4	8	11

Ans.
The set of control points $P_i(x_i, y_i)$, $i = 0, 1, 2, 3, 4$ is as follows

$$x_0 = -1, \quad x_1 = 2, \quad x_2 = 3, \quad x_3 = 5, \quad x_4 = 6$$
$$y_0 = 1, \quad y_1 = 3, \quad y_2 = -4, y_3 = 8, \quad y_4 = 11$$

Let $P_{-1}(x_{-1}, y_{-1}) = P_0(x_0, y_0) = (-1, 1)$

$$P_5(x_5, y_5) = P_4(x_4, y_4) = (6, 11)$$

Proceeding in a similar manner as in the previous example, we can easily obtain the following B–Spline curves for the intervals (P_j, P_{j+1}), $j = 1, 2, ..., m - 1$ in parametric form (with parameter t).

The B-Spline curve for the interval (P_0, P_1)

$$X(t) = (-1)\left(\frac{1}{6}(1-t)^3\right) + (-1)\left(\frac{t^3}{2} - t^2 + \frac{2}{3}\right) + (2)\left(-\frac{t^3}{2} + \frac{t^2}{2} + \frac{t}{2} + \frac{1}{6}\right) + (3)\left(\frac{t^3}{6}\right)$$

$$= -\frac{1}{2} + \frac{3}{2}t + \frac{3}{2}t^2 - \frac{5}{6}t^3$$

$$Y(t) = (1)\left(\frac{1}{6}(1-t)^3\right) + (1)\left(\frac{t^3}{2} - t^2 + \frac{2}{3}\right) + (3)\left(-\frac{t^3}{2} + \frac{t^2}{2} + \frac{t}{2} + \frac{1}{6}\right) + (-4)\left(\frac{t^3}{6}\right)$$

$$= \frac{4}{3} + t + t^2 - \frac{11}{6}t^3$$

The B-Spline curve for the interval (P_1, P_2)

$$X(t) = (-1)\left(\frac{1}{6}(1-t)^3\right) + (2)\left(\frac{t^3}{2} - t^2 + \frac{2}{3}\right) + (3)\left(-\frac{t^3}{2} + \frac{t^2}{2} + \frac{t}{2} + \frac{1}{6}\right) + (5)\left(\frac{t^3}{6}\right)$$

$$= \frac{5}{3} + 2t - t^2 + \frac{1}{2}t^3$$

$$Y(t) = (1)\left(\frac{1}{6}(1-t)^3\right) + (3)\left(\frac{t^3}{2} - t^2 + \frac{2}{3}\right) + (-4)\left(-\frac{t^3}{2} + \frac{t^2}{2} + \frac{t}{2} + \frac{1}{6}\right) + (8)\left(\frac{t^3}{6}\right)$$

$$= \frac{3}{2} - \frac{5}{2}t - \frac{9}{2}t^2 + \frac{14}{3}t^3$$

The B-Spline curve for the interval (P_2, P_3)

$$X(t) = (2)\left(\frac{1}{6}(1-t)^3\right) + (3)\left(\frac{t^3}{2} - t^2 + \frac{2}{3}\right) + (5)\left(-\frac{t^3}{2} + \frac{t^2}{2} + \frac{t}{2} + \frac{1}{6}\right) + (6)\left(\frac{t^3}{6}\right)$$

$$= \frac{19}{6} + \frac{3}{2}t + \frac{1}{2}t^2 - \frac{1}{3}t^3$$

$$Y(t) = (3)\left(\frac{1}{6}(1-t)^3\right) + (-4)\left(\frac{t^3}{2} - t^2 + \frac{2}{3}\right) + (8)\left(-\frac{t^3}{2} + \frac{t^2}{2} + \frac{t}{2} + \frac{1}{6}\right) + (11)\left(\frac{t^3}{6}\right)$$

$$= -\frac{5}{6} + \frac{5}{2}t + \frac{19}{2}t^2 - \frac{14}{3}t^3$$

The B-Spline curve for the interval (P_3, P_4)

$$X(t) = (3)\left(\frac{1}{6}(1-t)^3\right) + (5)\left(\frac{t^3}{2} - t^2 + \frac{2}{3}\right) + (6)\left(-\frac{t^3}{2} + \frac{t^2}{2} + \frac{t}{2} + \frac{1}{6}\right) + (6)\left(\frac{t^3}{6}\right)$$

$$= \frac{29}{6} + \frac{3}{2}t - \frac{1}{2}t^2$$

$$Y(t) = (-4)\left(\frac{1}{6}(1-t)^3\right) + (8)\left(\frac{t^3}{2} - t^2 + \frac{2}{3}\right) + (11)\left(-\frac{t^3}{2} + \frac{t^2}{2} + \frac{t}{2} + \frac{1}{6}\right) + (11)\left(\frac{t^3}{6}\right)$$

$$= \frac{13}{2} + \frac{15}{2}t - \frac{9}{2}t^2 + t^3$$

On combining all these curves, we get following final B-Spline curve

$$B(t) = \begin{cases} \begin{aligned} X(t) &= -\frac{1}{2} + \frac{3}{2}t + \frac{3}{2}t^2 - \frac{5}{6}t^3 \\ Y(t) &= \frac{4}{3} + t + t^2 - \frac{11}{6}t^3 \end{aligned} & \big((-1,1),\,(2,3)\big) \\[3mm] \begin{aligned} X(t) &= \frac{5}{3} + 2t - t^2 + \frac{1}{2}t^3 \\ Y(t) &= \frac{3}{2} - \frac{5}{2}t - \frac{9}{2}t^2 + \frac{14}{3}t^3 \end{aligned} & \big((2,3),\,(3,-4)\big) \\[3mm] \begin{aligned} X(t) &= \frac{19}{6} + \frac{3}{2}t + \frac{1}{2}t^2 - \frac{1}{3}t^3 \\ Y(t) &= -\frac{5}{6} + \frac{5}{2}t + \frac{19}{2}t^2 - \frac{14}{3}t^3 \end{aligned} & \big((3,-4),\,(5,8)\big) \\[3mm] \begin{aligned} X(t) &= \frac{29}{6} + \frac{3}{2}t - \frac{1}{2}t^2 \\ Y(t) &= \frac{13}{2} + \frac{15}{2}t - \frac{9}{2}t^2 + t^3 \end{aligned} & \big((5,8),\,(6,11)\big) \end{cases}$$

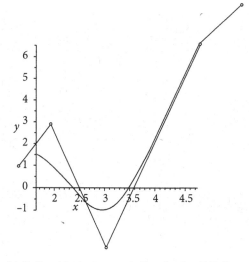

Fig. 11.6 Graphical representation for the B-Spline curve

Properties of B-Spline Curves:

1. If order of control points is altered, then B-Spline curve get changed. For example, we have different B-Spline curves for the following two sets

    a)
x	−1	5	3	2
y	1	11	−4	8

    b)
x	−1	2	3	5
y	1	8	−4	11

2. In general, the B-Spline curves do not pass through any of the control points.
3. Like Bězier curve, B-Spline curve also lies within the convex hull of the control points.
4. For a set of $(n+1)$ control points $P_i(x_i, y_i)$, $i = 0, 1,...,n$, the B-Spline curve can be of any degree.
5. B-Spline curves have first and second order continuity at nodal points.
6. B-Spline curve has a local effect, i.e. if we change one control point P_j, then the B-Spline is changing only for the points $P_i(x_i, y_i)$, $i = j-1, j, j+1, j+2$.

11.5 Least Squares Curve

In the least squares approximation, we approximate given data set by some standard functions e.g. straight line, polynomial, exponential function, etc. Suppose we have to fit a function $g(x)$ to n points (x_i, y_i); $i = 1, 2,...,n$. At any given point x_i, the function value $g(x_i)$ is approximate value and y_i is an exact value. Therefore, the error at this point is given by

$$e_i = y_i - g(x_i)$$

The aim is to find $g(x)$, such that the error at points (x_i, y_i), $i = 1, 2,...,n$ is minimum.

To find a total effective error (E), we cannot simply add these errors as some errors may be of opposite signs and hence during addition total error could be less than effective error. To overcome this difficulty, we can use absolute values of errors i.e. $E = \sum_{i=1}^{n} |e_i|$. But in that case, it is difficult to perform analysis like differentiation, etc. on modulus function.

Keeping in mind all these, we can use the sum of squares of errors i.e. $E = \sum_{i=1}^{n} e_i^2$. The aim is to find the function $g(x)$, such that this error E is least. It is the reason that the method is known as least squares method. Now, we will obtain equations for some possible fittings like a straight line, parabola, etc.

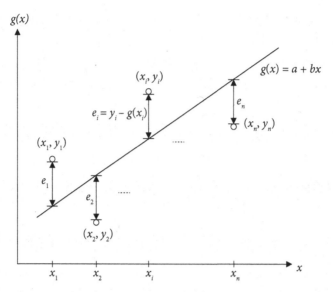

Fig. 11.7 Approximation by a straight line g(x) = a+bx. The dots are exact data points
(xᵢ, yᵢ); i = 1,2,...,n, therefore the vertical differences between straight line and dots
are errors.

11.5.1 Linear Curve (or) Straight Line Fitting

Let us consider a straight line $g(x) = a + bx$. The aim is to find a and b such that the error
(E) is minimum

$$E = \sum_{i=1}^{n} e_i^2 = \sum_{i=1}^{n}\left(y_i - g(x_i)\right)^2 = \sum_{i=1}^{n}\left(y_i - (a+bx_i)\right)^2$$

To find the extreme points of the variable E, the partial derivatives with respect to variables
a and b must vanish

$$\frac{\partial E}{\partial a} = \sum_{i=1}^{n}\frac{\partial}{\partial a}\left(y_i - (a+bx_i)\right)^2 = 0$$

$$\sum_{i=1}^{n} -2\left(y_i - (a+bx_i)\right) = 0$$

$$\sum_{i=1}^{n}\left(y_i - (a+bx_i)\right) = 0$$

$$\sum_{i=1}^{n} y_i - \sum_{i=1}^{n} a - \sum_{i=1}^{n} bx_i = 0$$

$$\Rightarrow na + b\sum_{i=1}^{n} x_i = \sum_{i=1}^{n} y_i \qquad\qquad (11.21)$$

$$\frac{\partial E}{\partial b} = \sum_{i=1}^{n} \frac{\partial}{\partial b}\left(y_i - (a + bx_i)\right)^2 = 0$$

$$\sum_{i=1}^{n} 2\left(y_i - (a + bx_i)\right)(-x_i) = 0$$

$$\sum_{i=1}^{n} x_i y_i - a \sum_{i=1}^{n} x_i - b \sum_{i=1}^{n} x_i^2 = 0$$

$$\Rightarrow a \sum_{i=1}^{n} x_i + b \sum_{i=1}^{n} x_i^2 = \sum_{i=1}^{n} x_i y_i \qquad (11.22)$$

Equations (11.21) and (11.22) are called as normal equations for straight line fitting.

$$na + b \sum_{i=1}^{n} x_i = \sum_{i=1}^{n} y_i$$

$$a \sum_{i=1}^{n} x_i + b \sum_{i=1}^{n} x_i^2 = \sum_{i=1}^{n} x_i y_i \qquad (11.23)$$

We can solve these equations for variables a and b. The straight line $g(x) = a + bx$ with these values of a and b is the straight line with minimum least square error.

Example ————————————————————————— **11.8**

Fit a straight line to the following data set

x: 1   1.2   1.4   1.6   1.8   2.0

y: 0   0.182322 0.336472 0.470004 0.587787 0.693147

Ans.

Normal Eqs. (11.23) for the straight line fitting are as follows

$$na + b \sum_{i=1}^{n} x_i = \sum_{i=1}^{n} y_i$$

$$a \sum_{i=1}^{n} x_i + b \sum_{i=1}^{n} x_i^2 = \sum_{i=1}^{n} x_i y_i$$

We have to compute the values $\sum_{i=1}^{n} x_i, \sum_{i=1}^{n} y_i, \sum_{i=1}^{n} x_i^2$ and $\sum_{i=1}^{n} x_i y_i$ for solution of these equations

x	y	x^2	xy
1	0	1	0
1.2	0.182322	1.44	0.218786
1.4	0.336472	1.96	0.471061
1.6	0.470004	2.56	0.752006
1.8	0.587787	3.24	1.058017
2.0	0.693147	4.0	1.386294
$\sum x = 9.0$	$\sum y = 2.269732$	$\sum x^2 = 14.2$	$\sum xy = 3.886164$

On using these values and $n = 6$ in normal equations (11.23), we get

$$6a + 9b = 2.269732$$
$$9a + 14.2b = 3.886164$$

Solution of these equations is given by

$$a = -0.653638, \ b = 0.687951$$

Therefore, the linear curve with minimum least square error is as follows

$$y = -0.653638 + 0.687951 \, x$$

Note: It is worth mentioning here that the data set is generated by the function $y = \ln(x)$.

11.5.2 Nonlinear Curve Fitting by Linearization of Data

There are some nonlinear curves, which are equivalent to linear fitting after some transformations in the dependent and independent variables. For example, if we want to fit a curve of the type $y = ae^{bx}$ to a data set. Then taking natural log on both sides, we have

$$\ln(y) = \ln(a) + b \, x$$

This expression is equivalent to following linear expression

$$Y = A + BX$$

where $Y = \ln(y)$, $A = \ln(a)$, $B = b$, $X = x$

In the next table, we are summarizing some nonlinear curves, which with simple operations and transformations can be converted into linear curve fitting.

Table 11.1 Linearization of Nonlinear Curves

Sr. No.	Function $y = f(x)$	Operations	Linearization $Y = A + BX$	New Variables and Constants $Y = A + BX$
1	$y = ae^{bx}$	Take Log	$\ln(y) = \ln(a) + bx$	$Y = \ln(y),\ A = \ln(a), B = b,\ X = x$
	$y = a\,xe^{bx}$		$\ln\left(\dfrac{y}{x}\right) = \ln(a) + bx$	$Y = \ln\left(\dfrac{y}{x}\right),\ A = \ln(a), B = b,\ X = x$
	$y = ax^{b}$		$\ln(y) = \ln(a) + b\ln(x)$	$Y = \ln(y),\ A = \ln(a), B = b,\ X = \ln(x)$
2	$y = \dfrac{x}{ax + b}$	Inverse	$\dfrac{1}{y} = \dfrac{ax + b}{x} = a + b\dfrac{1}{x}$	$Y = \dfrac{1}{y}, A = a,\ B = b,\ X = \dfrac{1}{x}$
	$y = \dfrac{1}{a + bx}$		$\dfrac{1}{y} = a + bx$	$Y = \dfrac{1}{y}, A = a,\ B = b,\ X = x$
	$y = \dfrac{a}{x + b}$		$\dfrac{1}{y} = \dfrac{1}{a}x + \dfrac{b}{a}$	$Y = \dfrac{1}{y}, A = \dfrac{b}{a},\ B = \dfrac{1}{a},\ X = x$
3	$y = \dfrac{c}{1 + ae^{bx}}$	Inverse and Log	$\ln\left(\dfrac{c}{y} - 1\right) = \ln(a) + bx$	$Y = \ln\left(\dfrac{c}{y} - 1\right),\ A = \ln(a), B = b,\ X = x$
4	$y = (a + bx)^{m}$	m^{th} root (m fixed)	$(y)^{-m} = a + bx$	$Y = (y)^{-m},\ A = a, B = b,\ X = x$
5	$y = a + b\,g(x)$	Let $X = g(x)$	$y = a + b\,g(x)$	$Y = y,\ A = a, B = b,\ X = g(x)$
	$y = a + b\dfrac{1}{x}$	$X = \dfrac{1}{x}$	$y = a + b\dfrac{1}{x}$	$Y = y,\ A = a, B = b,\ X = \dfrac{1}{x}$
	$y = a + b\ln(x)$	$X = \ln(x)$	$y = a + b\ln(x)$	$Y = y,\ A = a, B = b,\ X = \ln(x)$
	Etc.			

Example   11.9

Fit a curve $y = ax^{b}$ to the following data

$x:$	1	2	3	5	6
$y:$	1	9	29	129	221

Ans.

On taking log on both sides of the curve $y = ax^{b}$, we get

$$\ln(y) = \ln(a) + b\ln(x)$$

So, the curve fitting of type $y = ax^{b}$ is equivalent to fit a straight line $Y = A + bX$, where

$Y = \ln(y),\ A = \ln(a),\ X = \ln(x)$.

Normal equations for straight line $Y = A + bX$ fitting are as follows

$$nA + b\sum_{i=1}^{n} X_i = \sum_{i=1}^{n} Y_i$$

$$A\sum_{i=1}^{n} X_i + b\sum_{i=1}^{n} X_i^2 = \sum_{i=1}^{n} X_i Y_i \tag{11.24}$$

On computing various terms in normal equations

x	$X = \ln(x)$	y	$Y = \ln(y)$	X^2	XY
1	0	1	0	0	0
2	0.693147181	9	2.197224577	0.480453014	1.523000021
3	1.098612289	29	3.36729583	1.206948961	3.699352578
5	1.609437912	129	4.859812404	2.590290394	7.821566331
6	1.791759469	221	5.398162702	3.210401996	9.672209137
	5.192956851		15.82249551	7.488094364	22.71612807

The normal Eqs. (11.24) are as follows

$$5A + 5.192956851b = 15.82249551$$

$$5.192956851A + 7.488094364b = 22.71612807$$

On solving these equations for A and b, we obtain

$$A = 0.04931094359, b = 2.99943582$$

$$a = e^A = 1.050546961, b = 2.99943582$$

Hence, our curve is $y = ax^b = 1.050546961\ x^{2.99943582}$.

Example   **11.10**

Following are census details for the population of India from the year 1961 to 2011. Fit an exponential curve $y = ae^{bx}$ to these data, and hence find the approximate population in the years 1966, 1985, 1996 and 2009.

Year (x)	1961	1971	1981	1991	2001	2011
Population (y) (in crores)	43.9235	54.8160	68.3329	84.6421	102.8737	121.0193

Ans.

We have to fit an exponential curve $y = ae^{bx}$ for years (x) from 1961 to 2011. To avoid lengthy calculations (like $y = ae^{bx} = a(2.718)^{(1961)b}$), we can shift the origin and rescale the data as follows

$$X = \frac{x - 1981}{10}$$

Now, we have to fit the exponential curve $y = ae^{bX}$ to the following data

X:	-2	-1	0	1	2	3
y:	43.9235	54.8160	68.3329	84.6421	102.8737	121.0193

On taking logarithmic on both side to the equation $y = ae^{bX}$, we get

$$\ln y = \ln a + bX$$

By replacing $Y = \ln(y)$, $A = \ln(a)$, we have following straight line

$$Y = A + bX$$

Normal equations for this straight line are as follows

$$nA + b\sum_{i=1}^{n} X_i = \sum_{i=1}^{n} Y_i$$

$$A\sum_{i=1}^{n} X_i + b\sum_{i=1}^{n} X_i^2 = \sum_{i=1}^{n} X_i Y_i$$

X	y	$Y = \ln(y)$	X^2	XY
-2	43.9235	3.78245	4	-7.5649
-1	54.8160	4.00398	1	-4.00398
0	68.3329	4.22439	0	0
1	84.6421	4.43843	1	4.43843
2	102.8737	4.63350	4	9.26700
3	121.0193	4.79595	9	14.38785
3		25.87870	19	16.52440

The normal equations are given by

$$6A + 3b = 25.87870$$

$$3A + 19b = 16.52440$$

On solving these equations, we have

$$A = 4.21069, \ b = 0.20486$$

Since $A = \ln(a) \Rightarrow a = 67.40281$

Hence, the fitted curve is given by

$$y = 67.40281e^{0.20486\,X}$$

Now, we have to compute populations in the years $(x) = 1966, 1985, 1996$ and 2009. Corresponding to these years, the variable X is given by

$$X = \frac{x - 1981}{10} = -1.5, 0.4, 1.5., 2.8$$

So, the populations are given by

$$y(1966) = 67.40281e^{0.20486(-1.5)} = 49.570533$$

$$y(1985) = 67.40281e^{0.20486(0.4)} = 73.158664$$

$$y(1996) = 67.40281e^{0.20486(1.5)} = 91.649962$$

$$y(2009) = 67.40281e^{0.20486(2.8)} = 119.616951$$

Note: Try the example without shifting of origin and scaling. We will see that the calculations are almost impossible to even with a good computer.

11.5.3 Quadratic Curve Fitting

So far, we have discussed the straight line fitting and fitting of those nonlinear curves which are reducible to linear curve fitting. Now, we will discuss fitting of a quadratic curve. Proceeding in a similar manner as in the case of straight line fitting; we will find the normal equations for quadratic fit. Let $g(x) = a + bx + cx^2$ be a required quadratic curve. The error term is given by

$$E = \sum_{i=1}^{n} e_i^2 = \sum_{i=1}^{n} \left(y_i - g(x_i)\right)^2 = \sum_{i=1}^{n} \left(y_i - (a + bx_i + cx_i^2)\right)^2$$

On equating the partial derivatives of function E with respect to variables a, b and c with zero, we can produce the following normal equations for the quadratic fitting

$$\frac{\partial E}{\partial a} = \sum_{i=1}^{n} \frac{\partial}{\partial a} \left(y_i - \left(a + bx_i + cx_i^2 \right) \right)^2 = 0$$

$$\sum_{i=1}^{n} -2 \left(y_i - \left(a + bx_i + cx_i^2 \right) \right) = 0$$

$$\sum_{i=1}^{n} \left(y_i - \left(a + bx_i + cx_i^2 \right) \right) = 0$$

$$\sum_{i=1}^{n} y_i - \sum_{i=1}^{n} a - \sum_{i=1}^{n} bx_i - \sum_{i=1}^{n} cx_i^2 = 0$$

$$\Rightarrow na + b \sum_{i=1}^{n} x_i + c \sum_{i=1}^{n} x_i^2 = \sum_{i=1}^{n} y_i \tag{11.25}$$

$$\frac{\partial E}{\partial b} = \sum_{i=1}^{n} \frac{\partial}{\partial b} \left(y_i - \left(a + bx_i + cx_i^2 \right) \right)^2 = 0$$

$$\sum_{i=1}^{n} 2 \left(y_i - \left(a + bx_i + cx_i^2 \right) \right) \left(-x_i \right) = 0$$

$$\sum_{i=1}^{n} x_i y_i = a \sum_{i=1}^{n} x_i + b \sum_{i=1}^{n} x_i^2 + c \sum_{i=1}^{n} x_i^3$$

$$\Rightarrow a \sum_{i=1}^{n} x_i + b \sum_{i=1}^{n} x_i^2 + c \sum_{i=1}^{n} x_i^3 = \sum_{i=1}^{n} x_i y_i \tag{11.26}$$

$$\frac{\partial E}{\partial c} = \sum_{i=1}^{n} \frac{\partial}{\partial c} \left(y_i - \left(a + bx_i + cx_i^2 \right) \right)^2 = 0$$

$$\sum_{i=1}^{n} 2 \left(y_i - \left(a + bx_i + cx_i^2 \right) \right) \left(-x_i^2 \right) = 0$$

$$\sum_{i=1}^{n} x_i^2 y_i = a \sum_{i=1}^{n} x_i^2 + b \sum_{i=1}^{n} x_i^3 + c \sum_{i=1}^{n} x_i^4$$

$$\Rightarrow a \sum_{i=1}^{n} x_i^2 + b \sum_{i=1}^{n} x_i^3 + c \sum_{i=1}^{n} x_i^4 = \sum_{i=1}^{n} x_i^2 y_i \tag{11.27}$$

Equations (11.25), (11.26) and (11.27) are three normal equations for quadratic fitting

$$na + b \sum_{i=1}^{n} x_i + c \sum_{i=1}^{n} x_i^2 = \sum_{i=1}^{n} y_i$$

$$a \sum_{i=1}^{n} x_i + b \sum_{i=1}^{n} x_i^2 + c \sum_{i=1}^{n} x_i^3 = \sum_{i=1}^{n} x_i y_i$$

$$a \sum_{i=1}^{n} x_i^2 + b \sum_{i=1}^{n} x_i^3 + c \sum_{i=1}^{n} x_i^4 = \sum_{i=1}^{n} x_i^2 y_i \tag{11.28}$$

Example   11.11

Fit a quadratic curve to the following data, and compute the value of variable y at point $x = 3$

x:	0	1	2	4	5
y:	-2	0	10	78	148

Ans.

The normal equations (11.28) for quadratic fitting are given by

$$na + b\sum_{i=1}^{n} x_i + c\sum_{i=1}^{n} x_i^2 = \sum_{i=1}^{n} y_i$$

$$a\sum_{i=1}^{n} x_i + b\sum_{i=1}^{n} x_i^2 + c\sum_{i=1}^{n} x_i^3 = \sum_{i=1}^{n} x_i y_i$$

$$a\sum_{i=1}^{n} x_i^2 + b\sum_{i=1}^{n} x_i^3 + c\sum_{i=1}^{n} x_i^4 = \sum_{i=1}^{n} x_i^2 y_i$$

For these equations, we have to compute the following values

x	y	x^2	x^3	x^4	xy	$x^2 y$
0	-2	0	0	0	0	0
1	0	1	1	1	0	0
2	10	4	8	16	20	40
4	78	16	64	256	312	1248
5	148	25	125	625	740	3700
12	234	46	198	898	1072	4988

The following normal equations, are obtained

$$5a + 12b + 46c = 234$$
$$12a + 46b + 198c = 1072$$
$$46a + 198b + 898c = 4988$$

On solving these equations, we have

$$a = 5/11, \ b = -269/22, \ c = 181/22$$

The desired quadratic curve is as follows

$$y = \frac{5}{11} - \frac{269}{22}x + \frac{181}{22}x^2$$

$$\Rightarrow y(3) = \frac{416}{11} = 37.818182$$

Example 11.12

Following are the values of $\ln(x)$ for $x = 1\,(0.2)\,2$

x:	1	1.2	1.4	1.6	1.8	2.0
$\ln(x)$	0	0.182322	0.336472	0.470004	0.587787	0.693147

Fit a quadratic curve to these values, and use the result to compute the value of $\ln(1.7)$.

Ans.

Proceeding in a similar manner as in the previous example, compute the following values

x	$y = \ln(x)$	x^2	x^3	x^4	xy	$x^2 y$
1	0	1	1	1	0	0
1.2	0.182322	1.44	1.728	2.0736	0.218786	0.262544
1.4	0.336472	1.96	2.744	3.8416	0.471061	0.659485
1.6	0.470004	2.56	4.096	6.5536	0.752006	1.203210
1.8	0.587787	3.24	5.832	10.4976	1.058017	1.904430
2.0	0.693147	4.0	8.00	16.00	1.386294	2.772588
9.0	2.269732	14.2	23.4	39.9664	3.886164	6.802257

The normal equations, Eqs. (11.28) are given by

$$6a + 9b + 14.2c = 2.269732$$
$$9a + 14.2b + 23.4c = 3.886164$$
$$14.2a + 23.4b + 39.9664c = 6.802257$$

We get $a = -1.158640$, $b = 1.398110$, $c = -0.236720$, hence our quadratic curve is as follows

$$y = -1.158640 + 1.398110x - 0.236720x^2$$

$$\Rightarrow y(1.7) = -1.158640 + 1.398110(1.7) - 0.236720(1.7)^2 = 0.534026$$

The exact value of $\ln(1.7) = 0.530628$.

Note: In general, the least squares curves do not pass through any data point. The least squares curves have global effects (if we change the position of one point, it will change the whole curve).

Interpolating polynomial is best suitable for a data set having less number of points, as it has zero least squares error. But, for large data set, we have already discussed the disadvantages of higher order polynomials. Least squares fitting are suitable for large data set having global patterns like a straight line, exponential, parabolic, etc. But selection of appropriate curve is very difficult task, as it is not possible to predict the suitable curve by just looking at the data set. Scatter diagram will be helpful in this regard.

11.6 Chebyshev Polynomials Approximation

In this section, we are mainly concerned with the approximation by Chebyshev polynomials. To begin, here we will briefly discuss some basic aspects of Chebyshev polynomials.

Chebyshev differential equation of degree n is given by

$$(1-x^2)\frac{d^2 y}{dx^2} - x\frac{dy}{dx} + n^2 y = 0 \quad -1 \leq x \leq 1$$

Two linearly independent solutions of this differential equation are Chebyshev polynomial of the first kind $T_n(x) = \cos(n\cos^{-1} x)$ and Chebyshev polynomial of the second kind $U_n(x) = \sin(n\cos^{-1} x)$. Here, we will concentrate only on Chebyshev polynomials of the first kind $T_n(x)$ and use its minimax property to obtain best lower approximation for a given polynomial.

Now, we will discuss important forms, recurrence relation and orthogonal property of Chebyshev polynomials of the first kind $T_n(x)$ of degree n.

1. *Forms of Chebyshev polynomial*

$$T_n(x) = \cos(n\cos^{-1} x), \qquad -1 \leq x \leq 1$$

On replacing $x = \cos(\theta)$ or $\theta = \cos^{-1} x$, we have

$$T_n(\cos\theta) = \cos(n\theta)$$

On using the de Moivre's formula $(\cos\theta \pm i\sin\theta)^n = \cos(n\theta) \pm i\sin(n\theta)$, we have

$$T_n(\cos\theta) = \cos(n\theta)$$

$$= \frac{1}{2}\left[(\cos\theta + i\sin\theta)^n + (\cos\theta - i\sin\theta)^n\right]$$

$$= \frac{1}{2}\left[\left(\cos\theta + i\sqrt{1-\cos^2\theta}\right)^n + \left(\cos\theta - i\sqrt{1-\cos^2\theta}\right)^n\right]$$

$$= \frac{1}{2}\left[\left(\cos\theta + \sqrt{\cos^2\theta - 1}\right)^n + \left(\cos\theta - \sqrt{\cos^2\theta - 1}\right)^n\right]$$

Substitute $x = \cos(\theta)$ to get the following form

$$T_n(x) = \frac{1}{2}\left[\left(x + \sqrt{x^2 - 1}\right)^n + \left(x - \sqrt{x^2 - 1}\right)^n\right]$$

2. *Chebyshev polynomial in terms of Gauss hypergeometric function*

If we put $x = 1 - 2t$ in Chebyshev differential equation, then it will transform into the following differential equation

$$\left(t-t^2\right)\frac{d^2y}{dt^2}+\left(\frac{1}{2}-t\right)\frac{dy}{dt}+n^2y=0$$

This differential equation is equivalent to following Gauss hypergeometric equation

$$\left(t-t^2\right)\frac{d^2y}{dt^2}+\left(\gamma-(\alpha+\beta+1)t\right)\frac{dy}{dt}-\alpha\beta y=0,\text{ with }\alpha=n,\ \beta=-n\text{ and }\gamma=\frac{1}{2}$$

Therefore, the Chebyshev polynomial can also be written in terms of Gauss hypergeometric function $F(\alpha,\ \beta;\ \gamma;\ t)$ as follows

$$T_n\left(x\right)=F\left(n,\ -n;\ \frac{1}{2};\ \frac{1-x}{2}\right)$$

3. *Polynomial expansion for Chebyshev polynomials*

De Moivre's formula is given by

$$\cos(n\theta)+i\sin(n\theta)=\left(\cos\theta+i\sin\theta\right)^n=\sum_{m=0}^{n}{}^nC_m\left(\cos\theta\right)^{n-m}\left(i\sin\theta\right)^m$$

In this expansion, real terms exist only for $m=2k$ as

$$\left(i\sin\theta\right)^m=\left(i\sin\theta\right)^{2k}=(-1)^k\left(\sin^2\theta\right)^k=\left(\cos^2\theta-1\right)^k$$

On equating real terms, we have

$$\cos(n\theta)=\sum_{k=0}^{[n/2]}{}^nC_{2k}\left(\cos\theta\right)^{n-2k}\left(\cos^2\theta-1\right)^k$$

Use $T_n\left(\cos\theta\right)=\cos\left(n\theta\right)$ and $x=\cos(\theta)$ to get following form

$$T_n\left(x\right)=\sum_{k=0}^{[n/2]}\frac{n!}{2k!(n-2k)!}x^{n-2k}\left(x^2-1\right)^k$$

4. *Recurrence relation for Chebyshev polynomials*

Consider the following trigonometric identies

$$\cos\left(n\theta\right)=\cos\left(\theta+(n-1)\theta\right)$$
$$=\cos\left(\theta\right)\cos\left((n-1)\theta\right)-\sin\left(\theta\right)\sin\left((n-1)\theta\right)$$

$$\cos\left((n-2)\theta\right)=\cos\left(-\theta+(n-1)\theta\right)$$
$$=\cos\left(\theta\right)\cos\left((n-1)\theta\right)+\sin\left(\theta\right)\sin\left((n-1)\theta\right)$$

On adding these equations, we get

$$\cos(n\theta) + \cos((n-2)\theta) = 2\cos(\theta)\cos((n-1)\theta)$$

On using $T_n(\cos\theta) = \cos(n\theta)$ and $x = \cos(\theta)$, we have

$$T_n(x) + T_{n-2}(x) = 2xT_{n-1}(x)$$

$$\text{(or)} \quad T_n(x) = 2xT_{n-1}(x) - T_{n-2}(x) \qquad (11.29)$$

Since, we have $T_n(x) = \cos(n\cos^{-1}x)$, $\Rightarrow T_0(x) = 1$, $T_1(x) = x$

Higher degree Chebyshev polynomials can be obtained using recurrence relation (11.29), and Chebyshev polynomials up to degree six are listed in following table

Chebyshev Polynomial in Power of *x*	Power of *x* in Chebyshev Polynomial
$T_0(x) = 1$	$1 = T_0(x)$
$T_1(x) = x$	$x = T_1(x)$
$T_2(x) = 2xT_1(x) - T_0(x) = 2x^2 - 1$	$x^2 = \dfrac{1}{2}(T_0(x) + T_2(x))$
$T_3(x) = 2xT_2(x) - T_1(x) = 4x^3 - 3x$	$x^3 = \dfrac{1}{4}(3T_1(x) + T_3(x))$
$T_4(x) = 8x^4 - 8x^2 + 1$	$x^4 = \dfrac{1}{8}(3T_0(x) + 4T_2(x) + T_4(x))$
$T_5(x) = 16x^5 - 20x^3 + 5x$	$x^5 = \dfrac{1}{16}(10T_1(x) + 5T_3(x) + T_5(x))$
$T_6(x) = 32x^6 - 48x^4 + 18x^2 - 1$	$x^6 = \dfrac{1}{32}(10T_0(x) + 15T_2(x) + 6T_4(x) + T_6(x))$
$\vdots$	$\vdots$

5. *Orthogonal property of Chebyshev polynomials*

Chebyshev polynomials $T_n(x)$ are orthogonal w.r.t. weight function $\dfrac{1}{\sqrt{1-x^2}}$ over the interval $[-1, 1]$, i.e.

$$\int_{-1}^{1} \frac{T_m(x)T_n(x)}{\sqrt{1-x^2}}\,dx = 0; \quad n \neq m \qquad (11.30)$$

When $n = m$, we have $\displaystyle\int_{-1}^{1} \frac{T_n^{\,2}(x)}{\sqrt{1-x^2}}\,dx = \begin{cases} \pi & n = 0 \\ \dfrac{\pi}{2} & n \neq 0 \end{cases}$

6. Minimax property of Chebyshev polynomials

One of the most important properties of Chebyshev polynomials is minimax property. We will use this property to obtain lower order polynomial approximation for a given polynomial.

The coefficient of x^n in the polynomial $T_n(x)$ is 2^{n-1}, therefore $2^{1-n}T_n(x)$ is a polynomial with coefficient of x^n is 1. It means leading coefficient in polynomial $2^{1-n}T_n(x)$ is 1. Since $T_n(x) = \cos(n\cos^{-1}x)$ so its maximum absolute value is 1.

$$\max_{-1 \le x \le 1} |T_n(x)| = 1$$

On using these facts, we can state the minimax property of Chebyshev polynomial as follows

Consider polynomials with leading coefficients 1 (known as a monic polynomial) and of degree $n > 0$, i.e.,

$$P_n(x) = x^n + a_{n-1}x^{n-1} + a_{n-2}x^{n-2} + \cdots + a_1 x + a_0$$

Then, following relations hold in the domain $-1 \le x \le 1$,

$$\max_{-1 \le x \le 1} |P_n(x)| \ge \max_{-1 \le x \le 1} |2^{1-n}T_n(x)| = 2^{1-n} \qquad (11.31)$$

The minimax property implies that among all the monic polynomials of degree n, the $2^{1-n}T_n(x)$ has smallest least upper bound for its absolute value in the domain $-1 \le x \le 1$. Thus if we approximate a given polynomial by lower order polynomial, then by Chebyshev polynomial we can minimize the maximum absolute error.

Example   **11.13**

Use Chebyshev polynomials to compute the best lower order approximation for the polynomial $3x^4 + 5x^3 - x + 1$ in the domian $-1 \le x \le 1$. Also, compute the error bound in this approximation.

Ans.

First, we replace the highest order term in the polynomial with the help of Chebyshev polynomial as follows

$$3x^4 + 5x^3 - x + 1 = \frac{3}{8}\left(3T_0(x) + 4T_2(x) + T_4(x)\right) + 5x^3 - x + 1$$

$$= \frac{3}{8}T_4(x) + \frac{3}{2}T_2(x) + \frac{9}{8}T_0(x) + 5x^3 - x + 1$$

$$= \frac{3}{8}T_4(x) + \frac{3}{2}(2x^2 - 1) + \frac{9}{8} + 5x^3 - x + 1$$

$$= \frac{3}{8}T_4(x) + 5x^3 + 3x^2 - x + \frac{5}{8}$$

On neglecting the term $\frac{3}{8}T_4(x)$, the lower order approximation is as follows

$$3x^4 + 5x^3 - x + 1 \approx 5x^3 + 3x^2 - x + \frac{5}{8}$$

The maximum absolute error (11.31) in this approximation is given by

$$\frac{3}{2^3}T_4(x) = \frac{3}{2^3} = \frac{3}{8} = 0.375$$

Note that if we directly neglect the term $3x^4$ from the given polynomial, then the maximum possible error in the interval $-1 \le x \le 1$ is 3.

Example 11.14

Economize the Taylor series expansion $\cos x = 1 - \frac{x^2}{2!} + \frac{x^4}{4!} - \frac{x^6}{6!} + O(x^8)$ to lower order approximation over the interval $-1 \le x \le 1$. Also, compute the error bound.

Ans.
On using the value of x^6 in terms of Chebyshev polynomials, we have

$$1 - \frac{x^2}{2!} + \frac{x^4}{4!} - \frac{x^6}{6!} = 1 - \frac{x^2}{2!} + \frac{x^4}{4!} - \frac{1}{6!}\left(\frac{1}{32}\left(10T_0(x) + 15T_2(x) + 6T_4(x) + T_6(x)\right)\right)$$

$$= 1 - \frac{x^2}{2!} + \frac{x^4}{4!} - \frac{1}{6!}\left(\frac{1}{32}\left(10 + 15(2x^2 - 1) + 6(8x^4 - 8x^2 + 1) + T_6(x)\right)\right)$$

On neglecting the term containing $T_6(x)$, we have

$$1 - \frac{x^2}{2!} + \frac{x^4}{4!} - \frac{x^6}{6!} \approx 1 - \frac{x^2}{2!} + \frac{x^4}{4!} - \frac{1}{6!}\left(\frac{1}{32}\left(10 + 15(2x^2 - 1) + 6(8x^4 - 8x^2 + 1)\right)\right)$$

$$= \frac{23039}{23040} - \frac{639}{1280}x^2 + \frac{19}{480}x^4$$

This polynomial is the lower order economized approximation for the function cos x. The error in this approximation is given by

$$\frac{1}{6!}2^{1-n}T_n(x) = \frac{1}{6!}\frac{1}{2^5}T_6(x) = \frac{1}{6!}\frac{1}{2^5} = 0.000043402777$$

Example ———————————————————————————— **11.15**

Approximate the polynomial $x^3 + 5x^2 + 2x - 1$ to a quadratic polynomial with minimum error in the interval (3, 4).

Ans.
To apply the Chebyshev approximation, first of all, we have to change the variable x to variable t, such that the interval converts from (3, 4) to (−1, 1). Let our new variable be $t = ax + b$. At $x = 3$ and $x = 4$, we want $t = -1$ and $t = 1$ respectively, i.e.,

$$-1 = 3a + b$$

$$1 = 4a + b$$

On solving these two equations for a and b and using these values, we have

$$t = 2x - 7 \quad \text{or} \quad x = \frac{1}{2}(t + 7)$$

On using this expression for x in given polynomial, we have

$$x^3 + 5x^2 + 2x - 1 = \left(\frac{1}{2}(t+7)\right)^3 + 5\left(\frac{1}{2}(t+7)\right)^2 + 2\left(\frac{1}{2}(t+7)\right) - 1$$

$$= \frac{1}{8}t^3 + \frac{31}{8}t^2 + \frac{295}{8}t + \frac{881}{8}$$

Now, we have to convert $\frac{1}{8}t^3 + \frac{31}{8}t^2 + \frac{295}{8}t + \frac{881}{8}$ to a quadratic polynomial over the domain (−1, 1).

$$\frac{1}{8}t^3 + \frac{31}{8}t^2 + \frac{295}{8}t + \frac{881}{8} = \frac{1}{8}\left(\frac{1}{4}\left(3T_1(t) + T_3(t)\right)\right) + \frac{31}{8}t^2 + \frac{295}{8}t + \frac{881}{8}$$

$$= \frac{1}{32}T_3(t) + \frac{3}{32}T_1(t) + \frac{31}{8}t^2 + \frac{295}{8}t + \frac{881}{8}$$

$$= \frac{1}{32}T_3(t) + \frac{3}{32}t + \frac{31}{8}t^2 + \frac{295}{8}t + \frac{881}{8}$$

$$= \frac{1}{32}T_3(t) + \frac{31}{8}t^2 + \frac{1183}{32}t + \frac{881}{8}$$

Lower order approximation is given by

$$\frac{31}{8}t^2 + \frac{1183}{32}t + \frac{881}{8} \quad \text{over the interval } (-1, 1)$$

(or) $\frac{31}{8}(2x-7)^2 + \frac{1183}{32}(2x-7) + \frac{881}{8} = 15.5x^2 - 34.5625x + 41.21875$ over the interval

(3, 4).

11.7 Approximation by Rational Function of Polynomials (Padé Approximation)

So far, we have discussed only polynomial approximations, except in Section 11.5 of curve fitting where other functions are also discussed. Polynomials are best approximations for the functions which are smooth in behavior and for the data sets of experimental observations without singularity. But if our function/experiment behaves in chaos or singular manner (i.e., tends to infinity at some points), then we have to approximate with some other functions, one of that function is rational function of polynomials

$$R_{m,n}(x) = \frac{a_0 + a_1 x + a_2 x^2 + \cdots + a_m x^m}{1 + b_1 x + b_2 x^2 + \cdots + b_n x^n}$$

This approximation by the rational function of polynomials is called Padé approximation.

Example ─────────────────────────────────── **11.16**

Fit a rational function $R_{2,2}(x) = \dfrac{a_0 + a_1 x + a_2 x^2}{1 + b_1 x + b_2 x^2}$ to the function e^x.

Ans.
We require total five equations to compute values of five arbitrary constants a_0, a_1, a_2, b_1, b_2 in the function $R_{2,2}(x)$. Consider the first five terms of the Taylor series expansion of the function e^x as follows

$$e^x = 1 + x + \frac{x^2}{2!} + \frac{x^3}{3!} + \frac{x^4}{4!}$$

The approximation of the function e^x with $R_{2,2}(x)$ implies that

$$R_{2,2}(x) = e^x$$

$$\Rightarrow \frac{a_0 + a_1 x + a_2 x^2}{1 + b_1 x + b_2 x^2} = 1 + x + \frac{x^2}{2} + \frac{x^3}{6} + \frac{x^4}{24}$$

$$a_0 + a_1 x + a_2 x^2 = \left(1 + x + \frac{x^2}{2} + \frac{x^3}{6} + \frac{x^4}{24}\right)\left(1 + b_1 x + b_2 x^2\right)$$

On comparing various coefficients of different powers of x up to x^4, we have the following set of equations

$$a_0 = 1$$

$$a_1 = b_1 + 1$$

$$a_2 = \frac{1}{2} + b_1 + b_2$$

$$0 = \frac{1}{6} + \frac{b_1}{2} + b_2$$

$$0 = \frac{1}{24} + \frac{b_1}{6} + \frac{b_2}{2}$$

On solving this system of equations, we get

$$a_0 = 1, \ a_1 = \frac{1}{2}, \ a_2 = \frac{1}{12}, \ b_1 = \frac{-1}{2}, \ b_2 = \frac{1}{12}$$

Therefore, Padé approximation to e^x is as follows

$$R_{2,2} = \frac{1 + \dfrac{1}{2}x + \dfrac{1}{12}x^2}{1 + \dfrac{-1}{2}x + \dfrac{1}{12}x^2}$$

Example ── **11.17**

Fit a rational function $R_{2,2}(x) = \dfrac{a_0 + a_1 x + a_2 x^2}{1 + b_1 x + b_2 x^2}$ to the following data set, and use the result to compute the output at an input 2.5.

Input :	−1	0	1	2	3
Output :	−1	−1	2	13/5	13/5

Ans.

The approximation $R_{2,2}(x) = \dfrac{a_0 + a_1 x + a_2 x^2}{1 + b_1 x + b_2 x^2}$ should match with the data set at given

points. Therefore, we have

$R_{22}(-1) = y(-1) = -1$

$R_{22}(0) = y(0) = -1$

$R_{22}(1) = y(1) = 2$

$R_{22}(2) = y(2) = \dfrac{13}{5}$

$R_{22}(3) = y(3) = \dfrac{13}{5}$

On using the value of $R_{2,2}(x) = \dfrac{a_0 + a_1 x + a_2 x^2}{1 + b_1 x + b_2 x^2}$ in these equations and simplifying the

expressions, we have following five equations in five unknowns a_0, a_1, a_2, b_1, b_2.

$a_0 - a_1 + a_2 - b_1 + b_2 = -1$

$a_0 = -1$

$a_0 + a_1 + a_2 - 2b_1 - 2b_2 = 2$

$5a_0 + 10a_1 + 20a_2 - 26b_1 - 52b_2 = 13$

$5a_0 + 15a_1 + 45a_2 - 39b_1 - 117b_2 = 13$

On solving these five equations, we have

$a_0 = -1, \, a_1 = 3, a_2 = 2, b_1 = 0, b_2 = 1$

Therefore, the Padé approximation is given by

$R_{2,2} = \dfrac{-1 + 3x + 2x^2}{1 + x^2}$

Example 11.18

Obtain a rational approximation $R_{2,4}(x) = \dfrac{a_0 + a_1 x + a_2 x^2}{1 + b_1 x + b_2 x^2 + b_3 x^3 + b_4 x^4}$ for the Taylor series expansion of the function

$$\cos x = 1 - \frac{x^2}{2!} + \frac{x^4}{4!} - \frac{x^6}{6!} + \cdots$$

Ans.

We have seven arbitrary constants $a_0, a_1, a_2, b_1, b_2, b_3, b_4$ in rational approximating function $R_{2,4}(x)$. Therefore, we will consider first seven terms of the Taylor series expansion of function

$$\cos x = 1 - \frac{x^2}{2!} + \frac{x^4}{4!} - \frac{x^6}{6!}$$

$$R_{2,4}(x) = 1 - \frac{x^2}{2!} + \frac{x^4}{4!} - \frac{x^6}{6!}$$

$$\frac{a_0 + a_1 x + a_2 x^2}{1 + b_1 x + b_2 x^2 + b_3 x^3 + b_4 x^4} = 1 - \frac{x^2}{2!} + \frac{x^4}{4!} - \frac{x^6}{6!}$$

$$a_0 + a_1 x + a_2 x^2 = \left(1 - \frac{x^2}{2!} + \frac{x^4}{4!} - \frac{x^6}{6!}\right)\left(1 + b_1 x + b_2 x^2 + b_3 x^3 + b_4 x^4\right)$$

On comparing various coefficients of different powers of x up to x^6, we have the following set of linear equations

$$a_0 = 1, \qquad a_1 = b_1, \quad a_2 = \frac{-1}{2} + b_2, \qquad\qquad 0 = \frac{-1}{2} b_1 + b_3,$$

$$0 = \frac{1}{24} - \frac{1}{2} b_2 + b_4, \qquad 0 = \frac{1}{24} b_1 - \frac{1}{2} b_3, \quad 0 = -\frac{1}{720} + \frac{1}{24} b_2 - \frac{1}{2} b_4.$$

Solution of this system provides the following result

$$a_0 = 1, \, a_1 = 0, \, a_2 = \frac{-61}{150}, \, b_1 = 0, \, b_2 = \frac{7}{75}, \, b_3 = 0, \, b_4 = \frac{1}{200}$$

So, Padé approximation to $\cos x$ is as follows

$$R_{2,4} = \frac{1 - \dfrac{61}{150} x^2}{1 + \dfrac{7}{75} x^2 + \dfrac{1}{200} x^4}$$

Table 11.2 Summary and Comparison

Approximation techniques for a given data set of $(n+1)$ points (x_i, y_i); $i = 0,1,2,...,n$

	Interpolation	Piecewise Interpolation	Cubic Spline	Bézier	B-Spline	Least Squares	Padé Approximation
Approximating function	A single polynomial of degree $\leq n$ for complete interval (x_0, x_n)	Divide (x_0, x_n) into m subintervals, and then fit different polynomials of degree n/m to each subinterval	Cubic polynomials to each subinterval (x_0, x_1), (x_1, x_2), ..., (x_{n-1}, x_n)	A single Polynomial of degree exactly n in parametric form	polynomials of any degree to each subinterval (x_0, x_1), (x_1, x_2), ..., (x_{n-1}, x_n)	Fit any single suitable function not necessarily polynomial	Rational function of polynomials
Continuity	Interpolating polynomial and all derivatives are continuous	Piecewise polynomials are continuous, but derivatives are not continuous at nodal points	Interpolating polynomial, its first and second derivatives are continuous	Bézier curve and its first derivative are continuous	B-Spline curve and its first and second derivatives are continuous	Depend on the approximating curve	Discontinuous at the points where denominator is zero
Pass through	All the data points	All the data points	All the data points	First and last points (in general)	no points (in general)	no points (in general)	All the data points
Effect; If we change the value of one data point (x_j, y_j)	Global, complete polynomial will change	Local (only on the intervals containing the changed point, i.e. (x_{j-1}, x_j), (x_j, x_{j+1}))	Global, all cubic spline will change	Global, complete polynomial will change	Local (only on the intervals (x_{j-1}, x_j) (x_j, x_{j+1}) and (x_{j+1}, x_{j+2})	Global, complete approximating curve will change	Global
No. of curves	one	n/m	n	One	n	one	one

Chebyshev polynomial computes best lower order approximation for a given polynomial.

Exercise 11

1. Check the following functions, that they are splines or not

(i) $f(x) = \begin{cases} 5x, & 0 \le x \le 1 \\ 11x - 6, & 1 \le x \le 2 \\ -4 + 10x, & 2 \le x \le 3 \end{cases}$

(ii) $f(x) = \begin{cases} 12x - 7x^2, & 0 \le x \le 1 \\ 1 + 10x - 6x^2, & 1 \le x \le 2 \end{cases}$

Ans. i) Linear spline, satisfying interpolating conditions at nodes

   ii) Quadratic spline, satisfying interpolating condition as well as continuity of first derivatives at nodes.

2. Find the cubic spline fit for the following data points

x:	-1	0	1
$f(x)$:	2	5	9

   Use natural spline conditions $f''(-1) = 0$ and $f''(1) = 0$.

   Ans. $f(x) = \begin{cases} 5 + \dfrac{7}{2}x + \dfrac{3}{4}x^2 + \dfrac{1}{4}x^3, & -1 \le x \le 0 \\ 5 + \dfrac{7}{2}x + \dfrac{3}{4}x^2 - \dfrac{1}{4}x^3, & 0 \le x \le 1 \end{cases}$

3. Fit a cubic spline curve to the following data set

x:	-1	1	3	5
y:	2	4	11	23

   Use end conditions $f''(-1) = 0$ and $f''(5) = 0$.

   Ans. $f(x) = \begin{cases} \dfrac{21}{8} + \dfrac{7}{8}x + \dfrac{3}{8}x^2 + \dfrac{1}{8}x^3, & -1 \le x \le 1 \\ \dfrac{11}{4} + \dfrac{1}{2}x + \dfrac{3}{4}x^2, & 1 \le x \le 3 \\ \dfrac{49}{8} - \dfrac{23}{8}x + \dfrac{15}{8}x^2 - \dfrac{1}{8}x^3, & 3 \le x \le 5 \end{cases}$

4. Construct the cubic spline with the natural end conditions that passes through the points $(-1, 0)$, $(0, 1)$, $(2, 5)$ and $(3, 2)$.

   Ans. $f(x) = \begin{cases} 1 + 2x + \dfrac{3}{2}x^2 + \dfrac{1}{2}x^3, & -1 \le x \le 0 \\ 1 + 2x + \dfrac{3}{2}x^2 - \dfrac{3}{4}x^3, & 0 \le x \le 2 \\ -13 + 23x - 9x^2 + x^3, & 2 \le x \le 3 \end{cases}$

5. Find the values of $y(1.5)$ and $y'(3)$ by using the cubic spline fit for the following data with the end conditions $y'(0) = -1, y'(4) = 2$

$x:$	0	1	3	4
$y:$	-10	1	8	-5

Ans. $y(x) = \begin{cases} -10 - x + \dfrac{39}{2}x^2 - \dfrac{15}{2}x^3, & 0 \le x \le 1 \\[2mm] -16 + 17x + \dfrac{3}{2}x^2 - \dfrac{3}{2}x^3, & 1 \le x \le 3 \\[2mm] -421 + 422x - \dfrac{267}{2}x^2 + \dfrac{27}{2}x^3, & 3 \le x \le 4 \end{cases}$

$y(1.5) = 7.8125, \quad y'(3) = -14.5$

6. Obtain the cubic Bézier curve for the following set of control points

x	2	0	-1	1.5
y	1	-3	4	8

Ans. $\left(2 - 6t + 3t^2 + 2.5t^3, \ 1 - 12t + 33t^2 - 14t^3\right)$

7. Obtain the cubic Bézier curve for the following set of control points $P_i(x_i, y_i)$, $i = 0, 1, \ldots, 3$

x	-1	2	3	5
y	1	8	-4	11

Compare the obtained curve with the cubic Bézier curve of Example 5. It establishes the fact that we have different Bézier curves for the same data sets but data points are in different order.

Ans. $\left(-1 + 9t - 6t^2 + 3t^3, \ 1 + 21t - 57t^2 + 46t^3\right)$

8. Obtain the quartic Bézier curve for the following set of control points

x	-2	0	1	3	2
y	1	-3	4	7	9

Ans. $-2 + 8t - 6t^2 + 8t^3 - 6t^4, \ 1 - 16t + 66t^2 - 60t^3 + 18t^4$

9. Obtain the cubic B-Spline curve for the following set of control points

x	-2	-1	0	3
y	-5	-1	-2	4

Ans. B-Spline curve for the interval (P_0, P_1)

$$X(t) = (-2)\left(\frac{1}{6}(1-t)^3\right) + (-2)\left(\frac{t^3}{2} - t^2 + \frac{2}{3}\right) + (-1)\left(-\frac{t^3}{2} + \frac{t^2}{2} + \frac{t}{2} + \frac{1}{6}\right) + (0)\left(\frac{t^3}{6}\right)$$

$$-\frac{11}{6} + \frac{1}{2}t + \frac{1}{2}t^2 - \frac{1}{6}t^3$$

$$Y(t) = (-5)\left(\frac{1}{6}(1-t)^3\right) + (-5)\left(\frac{t^3}{2} - t^2 + \frac{2}{3}\right) + (-1)\left(-\frac{t^3}{2} + \frac{t^2}{2} + \frac{t}{2} + \frac{1}{6}\right) + (-2)\left(\frac{t^3}{6}\right)$$

$$-\frac{13}{2} + 2t + 2t^2 - \frac{3}{2}t^3$$

The B-Spline curve for the interval (P_1, P_2)

$$X(t) = (-2)\left(\frac{1}{6}(1-t)^3\right) + (-1)\left(\frac{t^3}{2} - t^2 + \frac{2}{3}\right) + (0)\left(-\frac{t^3}{2} + \frac{t^2}{2} + \frac{t}{2} + \frac{1}{6}\right) + (3)\left(\frac{t^3}{6}\right)$$

$$-1 + t + \frac{1}{3}t^3$$

$$Y(t) = (-5)\left(\frac{1}{6}(1-t)^3\right) + (-1)\left(\frac{t^3}{2} - t^2 + \frac{2}{3}\right) + (-2)\left(-\frac{t^3}{2} + \frac{t^2}{2} + \frac{t}{2} + \frac{1}{6}\right) + (4)\left(\frac{t^3}{6}\right)$$

$$-\frac{11}{6} + \frac{3}{2}t - \frac{5}{2}t^2 + 2t^3$$

The B-Spline curve for the interval (P_2, P_3)

$$X(t) = (-1)\left(\frac{1}{6}(1-t)^3\right) + (0)\left(\frac{t^3}{2} - t^2 + \frac{2}{3}\right) + (3)\left(-\frac{t^3}{2} + \frac{t^2}{2} + \frac{t}{2} + \frac{1}{6}\right) + (3)\left(\frac{t^3}{6}\right)$$

$$\frac{1}{3} + 2t + t^2 - \frac{5}{6}t^3$$

$$Y(t) = (-1)\left(\frac{1}{6}(1-t)^3\right) + (-2)\left(\frac{t^3}{2} - t^2 + \frac{2}{3}\right) + (4)\left(-\frac{t^3}{2} + \frac{t^2}{2} + \frac{t}{2} + \frac{1}{6}\right) + (4)\left(\frac{t^3}{6}\right)$$

$$-\frac{5}{6} + \frac{5}{2}t + \frac{7}{2}t^2 - \frac{13}{6}t^3$$

Hence, the final B-Spline curve is as follows

$$B(t) = \begin{cases} \begin{cases} X(t) = -\frac{11}{6} + \frac{1}{2}t + \frac{1}{2}t^2 - \frac{1}{6}t^3 \\ Y(t) = -\frac{13}{3} + 2t + 2t^2 - \frac{3}{2}t^3 \end{cases} & ((-2,-5),(-1,-1)) \\[2em] \begin{cases} X(t) = -1 + t + \frac{1}{3}t^3 \\ Y(t) = -\frac{11}{6} + \frac{3}{2}t - \frac{5}{2}t^2 + 2t^3 \end{cases} & ((-1,-1),(0,-2)) \\[2em] \begin{cases} X(t) = \frac{1}{3} + 2t + t^2 - \frac{5}{6}t^3 \\ Y(t) = -\frac{5}{6} + \frac{5}{2}t + \frac{7}{2}t^2 - \frac{13}{6}t^3 \end{cases} & ((0,-2),(3,4)) \end{cases}$$

10. Find the least squares curve $y = a + bx$ for the data set

x:	50	70	100	120
y:	12	15	21	25

Ans. $y = 2.276 + 0.118x$

11. Fit a straight line $y = a + bx$ to the following data

x:	1	1.1	2.2	3.3	4.4
y:	2.8	4.7	6.4	8.5	10.3

Ans. $y = 1.751 + 1.993x$

12. The growth of cell culture (optical density) at various pH levels are tabulated in the following table.

pH (x):	3.5	4	4.5	5	5.5	6	6.5	7
Optical density(y):	0.20	0.28	0.35	0.41	0.46	0.52	0.55	0.62

Fit a linear curve $y = a + bx$ to data set, and compute the optical density at pH level 5.8.
Ans. $y = -0.183 + 0.115\,x$, 0.484

13. The following data represents the heat capacity (σ) at different temperatures (T) for a given gas. Use the method of least square to determine heat capacity as a linear function of temperature.

T:	−20	0	30	70	120
σ:	1180	1235	1310	1470	1675

Ans. $\sigma = 1231.778 + 3.556T$

14. The tensile strengths of stainless steel cables of different diameters are investigated to give following results.

Cable diameter (x):	1.2	1.6	2.4	3.6	5.0	8.0
Tensile Strength(y):	1.32	2.53	5.28	8.34	12.83	23.92

Fit linear and exponential curves to the data set. Find the least square error estimates for both the curves.

Ans. Linear curve $y = -2.902 + 3.286x$, Least square error = 1.146
Exponential curve $y = 1.449\,e^{0.390x}$, Least square error = 95.5996

15. An empirical formula for the effect of temperature on viscosity of a liquid is given by Andrade's equation

$$\mu = ae^{bT}$$

where μ is the dynamic viscosity of the liquid, T is absolute temperature, and a, b are constants. Compute the best fitted Andrade's model to the following data for a given liquid.

T:	10	20	30	40	50
μ:	4.67	3.84	3.17	2.71	2.53

Ans. $\mu = 5.288\,e^{-0.016T}$

16. Find the least squares curve $y = a + bx + cx^2$ for the following data points

x:	−3	−1	1	3
y:	15	5	1	5

Ans. $y = 2.125 - 1.7x + 0.875x^2$

17. Obtain the least squares fit of the form $y = ax^2 + bx + c$ for the following data set.

x:	10	12	15	23	20
y:	14	17	23	25	21

Solve the system of normal equations with the aid of Gauss elimination method.
Ans. $y = 0.069 + 3.010x - 8.728x^2$

18. Obtain the quadratic approximation for the following discrete data points

x:	−1	0	1	2	3
$f(x)$:	1.4	1	1.5	3.5	12

Ans. $y = 0.131 - 0.387x + 1.379x^2$

19. It is expected from theoretical consideration, that the rate of flow is proportional to some power of the pressure at the nozzle of a fire hose. Get the least squares values for exponent and proportionality factor.

Flow rate (F):	90	110	130	150	170
Pressure (P):	10	18	28	41	53.

Ans. $F = 37.306\,P^{0.377}$

20. Fit a curve of the type $y = ae^{bx}$ to the following data

x:	0	1	3	5
y:	1.8	6.3	10.1	15.2

Ans. $y = 2.751e^{0.379x}$

21. Fit a curve $y = \dfrac{c_0}{x} + c_1\sqrt{x}$ to the following data values

x:	1	1.2	1.4	1.5	1.7	2
y:	1	2.3	5.5	6.3	6.6	5.3

Hint: The given expression is as follows

$$yx = c_0 + c_1(x)^{\frac{3}{2}}$$

Let it is $Y = c_0 + c_1 X$ with $Y = yx$ and $X = (x)^{\frac{3}{2}}$

Ans. $y = \dfrac{-3.334}{x} + 5.780\sqrt{x}$

22. Fit a curve of the form $y = ae^{bx}$ to the following data points

x	0	0.1	0.2	0.3
y	1.8	2.7	3.6	5.0

How far does the fit agree with the data?

Ans. $y = 1.85\,e^{3.353x}$, least square error estimate $= 0.019015$

23. Economize Taylor series expansion $e^x = 1 + x + \dfrac{x^2}{2!} + \dfrac{x^3}{3!} + \dfrac{x^4}{4!} + O(x^5)$ to lower order Chebyshev approximation over the interval $-1 \le x \le 1$.

Ans. $0.9946 + 0.9973x + 0.5430x^2 + 0.1772x^3$

24. Use the Chebyshev polynomials to obtain the approximations of second degree for the following polynomials

(i) $2x^4 + 3x^3 - x + 2$ on $[-1, 1]$

(ii) $x^3 + 2x^2 - 5x + 3$ on $[2, 3]$

Ans. i) $3x^3 + 2x^2 - x + 1.75,$

ii) $x = \dfrac{t+5}{2}, \ x^3 + 2x^2 - 5x + 3 = \dfrac{t^3}{8} + \dfrac{19t^2}{8} + \dfrac{95t}{8} + \dfrac{149}{8}, \ t \in [-1, 1]$

Lower order approximation $= \dfrac{19t^2}{8} + \dfrac{383t}{32} + \dfrac{149}{8}$

$$= \dfrac{19x^2}{2} - \dfrac{377x}{16} + \dfrac{581}{32} = 9.5x^2 - 23.5625x + 18.15625)$$

25. From the Maclaurin expansion $e^{-x} = 1 - x + \dfrac{x^2}{2!} - \dfrac{x^3}{3!} + \dfrac{x^4}{4!},$ obtain an approximation of the form

$$\dfrac{a_0 + a_1 x + a_2 x^2}{1 + b_1 x + b_2 x^2}$$

Ans. $\dfrac{1 - \dfrac{1}{2}x + \dfrac{1}{12}x^2}{1 + \dfrac{1}{2}x + \dfrac{1}{12}x^2}$

26. Fit a rational function $R_{2,2}(x) = \dfrac{a_0 + a_1 x + a_2 x^2 + a_3 x^3}{1 + b_1 x + b_2 x^2}$ to the function $\sinh(x)$.

Ans. $\dfrac{\dfrac{7}{60}x^3 + x}{1 - \dfrac{1}{20}x^2}$. Hint: Use the following series expansion for the hyperbolic sine function

$$\sinh x = \sum_{n=0}^{\infty} \dfrac{x^{2n+1}}{(2n+1)!} = x + \dfrac{x^3}{3!} + \dfrac{x^5}{5!} + \dfrac{x^7}{7!} + \cdots$$

27. Following are the experimental data points

Input :	−1	0	1	2
Output :	−15	−1000	2000	150

Fit a rational function $R_{2,1}(x) = \dfrac{a_0 + a_1 x + a_2 x^2}{1 + b_1 x}$ to these data points, and use the result to compute

the output at an input 0.5.

Ans. $R_{2,1}(x) = \dfrac{-1000 - \dfrac{200725}{381}x + \dfrac{165725}{381}x^2}{1 - \dfrac{589}{381}x}$

$R_{2,1}(0.5) = -5085.910405$

28. Use Maclaurin series expansion $\tan x = x + \dfrac{x^3}{3} + \dfrac{2x^5}{15} + \dfrac{17x^7}{315}$ to obtain the following expression

for the function tan (x)

$$\tan x = \dfrac{x - x^3/15}{1 - 2x^2/5}$$

Numerical Differentiation

> *No mathematician should ever allow him to forget that mathematics, more than any other art or science, is a young man's game. ... Galois died at twenty one, Abel at twenty seven, Ramanujan at thirty three, Riemann at forty. There have been men who have done great work later... [but] I do not know a single instance of a major mathematical advance initiated by a man past fifty... A mathematician may still be competent enough at sixty, but it is useless to expect him to have original ideas.*
>
> **Godfrey Harold Hardy**
> (February 7, 1877–December 1, 1947)
> He had made influential contributions in mathematical analysis and number theory.

12.1 Introduction

Differentiation and integration are two important concepts of mathematics. Many physical phenomena involve differentiation of the original function. Continuous change produces differentiation and it has many applications in science and engineering. Very few but important examples are Newton second law, $F = ma = m\dfrac{d^2 x}{dt^2}$; heat flow, $q = -k\dfrac{dT}{dx}$; current flow, $J = -\sigma\dfrac{dV}{dx}$, etc.

In fact, differentiation is an integral part of almost each branch of science and engineering. We have already mentioned that in the case of experimental results, data sets are obtained, and no analytical differentiation is possible in these cases. So, numerical differentiation provides alternative ways to deal such problems.

In this chapter, numerical differentiation techniques based on interpolation are discussed. We have already discussed in previous chapters that the interpolating polynomial is unique. So for each method, the same interpolating polynomial is obtained. But, each numerical method has its advantages and disadvantages.

In this chapter, we will discuss various examples for equally as well as unequally spaced points. Lagrange and Newton divided difference methods are used for unequally spaced data sets.

Example ── **12.1** ──

Derive interpolating polynomial for the data points $(0, -1)$, $(1, 1)$, $(2, 9)$, $(3, 29)$, $(5, 129)$ using Newton divided difference formula, and hence compute the value of $y'(4)$ and $y''(4)$.

x	$f(x)$	$[\,,]$	$[\,,,]$	$[\,,,,]$	$[\,,,,,]$
$0 = x_0$	$-1 = f(x_0)$				
		$2 = f[x_0, x_1]$			
$1 = x_1$	$1 = f(x_1)$		$3 = f[x_0, x_1, x_2]$		
		$8 = f[x_1, x_2]$		$1 = f[x_0, x_1, x_2, x_3]$	
$2 = x_2$	$9 = f(x_2)$		$6 = f[x_1, x_2, x_3]$		$0 = f[x_0, x_1, x_2, x_3, x_4]$
		$20 = f[x_2, x_3]$		$1 = f[x_1, x_2, x_3, x_4]$	
$3 = x_3$	$29 = f(x_3)$		$10 = f[x_2, x_3, x_4]$		
		$50 = f[x_3, x_4]$			
$5 = x_4$	$129 = f(x_4)$				

$$P_n(x) = f(x_0) + f[x_0, x_1](x - x_0) + f[x_0, x_1, x_2](x - x_0)(x - x_1) + \cdots\cdots$$
$$+ f[x_0, x_1, \cdots, x_n](x - x_0)(x - x_1)\cdots(x - x_{n-1})$$
$$P(x) = -1 + 2(x - 0) + 3(x - 0)(x - 1) + 1(x - 0)(x - 1)(x - 2) + 0.(x - 0)(x - 1)(x - 2)(x - 3)$$
$$P(x) = x^3 + x - 1$$

It is required interpolating polynomial. To compute $y'(4)$ and $y''(4)$, we have

$$P'(x) = x^3 + x - 1 = 3x^2 + 1$$
$$\Rightarrow P'(4) = 49 \quad \text{(Approximate value of } y'(4))$$

Similarly, the approximate value of $y''(4)$ is given by

$$P''(x) = x^3 + x - 1 = 6x$$
$$\Rightarrow P''(4) = 24.$$

In the case of equally spaced data, the computational work becomes easier. Therefore we will also discuss following numerical differentiation techniques for the equally spaced data set.

Forward and Backward Difference Formulas

  i)   Newton Forward Difference Formula
  ii)   Newton Backward Difference Formula

Central Difference Formulas

i)   Gauss Formulas (Forward and Backward Differences)
ii) Stirling Formula
iii) Bessel Formula
iv) Everett Formula
v)   Steffensen Formula

All these formulas are used according to our choices for interpolating points. Newton forward difference formula is suitable for the points which are at the beginning of the table. If we want to perform numerical differentiation for a point lying at the end of the data set, Newton backward difference formula is suitable. For the numerical differentiation at the point in the mid of the table, central difference formulas are suitable.

12.2 Numerical Differentiation Formulas

In this section, we will compute numerical differentiation formula using Newton forward difference interpolation, and then list all the formulas in Table 12.1.

Let $P_n(x)$ be the interpolating polynomial of degree $\leq n$, for the points $x_0, x_1, \cdots, x_n$. Newton forward difference formula [10.4] for interpolating polynomial is as follows

$$P_n(x) = f(x_0) + \frac{1}{h}\Delta f(x_0)(x-x_0) + \frac{1}{2!\,h^2}\Delta^2 f(x_0)(x-x_0)(x-x_1) + \cdots$$

$$+ \frac{1}{n!\,h^n}\Delta^n f(x_0)(x-x_0)(x-x_1)\cdots(x-x_{n-1})$$

This formula is used to compute the interpolating polynomial for numerical differentiation. If we want to compute the numerical differentiation at any intermediate point, then following form of Newton forward difference formula [10.6] is suitable.

$$P_n(x) = f(x_0) + \Delta f(x_0)(s) + \frac{\Delta^2 f(x_0)}{2!}(s)(s-1) + \cdots + \frac{\Delta^n f(x_0)}{n!}(s)(s-1)\cdots(s-n+1)$$

where $s = \dfrac{x - x_0}{h}$ (12.1)

Differentiating with respect to x, we get

$$P_n'(x) = \frac{d}{dx}\left[f(x_0) + \Delta f(x_0)(s) + \frac{\Delta^2 f(x_0)}{2!}(s)(s-1) + \cdots + \frac{\Delta^n f(x_0)}{n!}(s)(s-1)\cdots(s-n+1) \right]$$

$$P_n'(x) = \frac{d}{ds}\left[f(x_0) + \Delta f(x_0)(s) + \frac{\Delta^2 f(x_0)}{2!}(s)(s-1) + \cdots + \frac{\Delta^n f(x_0)}{n!}(s)(s-1)\cdots(s-n+1) \right]\frac{ds}{dx}$$

We have, $x = x_0 + sh$. It implies $dx = h\,ds$ or $\dfrac{ds}{dx} = \dfrac{1}{h}$.

$$P_n'(x) = \frac{1}{h}\left(\Delta f(x_0) + \frac{2s-1}{2}\Delta^2 f(x_0) + \frac{3s^2 - 6s + 2}{6}\Delta^3 f(x_0) \right.$$

$$\left. + \frac{2s^3 - 9s^2 + 11s - 3}{12}\Delta^4 f(x_0) + \cdots \right) \tag{12.2}$$

Similarly, again differentiating this formula with respect to x, we have

$$P_n''(x) = \frac{1}{h^2}\left(\Delta^2 f(x_0) + (s-1)\Delta^3 f(x_0) + \frac{6s^2 - 18s + 11}{12}\Delta^4 f(x_0) \right.$$

$$\left. + \frac{2s^3 - 12s^2 + 21s - 10}{12}\Delta^5 f(x_0) + \cdots \right) \tag{12.3}$$

The formulae (12.2) and (12.3) are forward difference formulae for first and second derivatives, respectively.

On a similar pattern, we can easily obtain the various numerical differentiation formula for equally spaced points listed in the following table

Table 12.1 Summary Table for Numerical Differentiation Formulas

Newton Forward Difference: $(0 < s < 1)$ $(x = x_0 + sh)$

$$P_n(x) = f(x_0) + \Delta f(x_0)(s) + \frac{\Delta^2 f(x_0)}{2!}(s)(s-1) + \frac{\Delta^3 f(x_0)}{3!}(s)(s-1)(s-2) + \frac{\Delta^4 f(x_0)}{4!}(s)(s-1)(s-2)(s-3)$$

$$+ \frac{\Delta^5 f(x_0)}{5!}(s)(s-1)(s-2)(s-3)(s-4) + \cdots$$

$$P_n'(x) = \frac{1}{h}\left(\Delta f(x_0) + \frac{2s-1}{2}\Delta^2 f(x_0) + \frac{3s^2 - 6s + 2}{6}\Delta^3 f(x_0) + \frac{2s^3 - 9s^2 + 11s - 3}{12}\Delta^4 f(x_0) \right.$$

$$\left. + \frac{5s^4 - 40s^3 + 105s^2 - 100s + 24}{120}\Delta^5 f(x_0) + \frac{6s^5 - 75s^4 + 340s^3 - 675s^2 + 548s - 120}{720}\Delta^6 f(x_0) + \cdots \right)$$

$$P_n''(x) = \frac{1}{h^2}\left(\Delta^2 f(x_0) + (s-1)\Delta^3 f(x_0) + \frac{6s^2 - 18s + 11}{12}\Delta^4 f(x_0) + \frac{2s^3 - 12s^2 + 21s - 10}{12}\Delta^5 f(x_0) \right.$$

$$\left. + \frac{15s^4 - 150s^3 + 510s^2 - 675s + 274}{360}\Delta^6 f(x_0) + \cdots \right)$$

Newton Backward Difference: $(-1 < s < 0)$ $(x = x_n + sh)$

$$P_n(x) = f(x_n) + \nabla f(x_n)(s) + \frac{\nabla^2 f(x_n)}{2!}(s)(s+1) + \frac{\nabla^3 f(x_n)}{3!}(s)(s+1)(s+2) + \frac{\nabla^4 f(x_n)}{4!}(s)(s+1)(s+2)(s+3)$$

$$+ \frac{\nabla^5 f(x_n)}{5!}(s)(s+1)(s+2)(s+3)(s+4) + \cdots$$

$$P_n'(x) = \frac{1}{h}\left(\nabla f(x_n) + \frac{2s+1}{2}\nabla^2 f(x_n) + \frac{3s^2+6s+2}{6}\nabla^3 f(x_n) + \frac{2s^3+9s^2+11s+3}{12}\nabla^4 f(x_n) \right.$$

$$\left. + \frac{5s^4+40s^3+105s^2+100s+24}{120}\nabla^5 f(x_n) + \frac{6s^5+75s^4+340s^3+675s^2+548s+120}{720}\nabla^6 f(x_0) + \cdots \right)$$

$$P_n''(x) = \frac{1}{h^2}\left(\nabla^2 f(x_n) + (s+1)\nabla^3 f(x_n) + \frac{6s^2+18s+11}{12}\nabla^4 f(x_n) + \frac{2s^3+12s^2+21s+10}{12}\nabla^5 f(x_n) \right.$$

$$\left. + \frac{15s^4+150s^3+510s^2+675s+274}{360}\nabla^6 f(x_0) + \cdots \right)$$

Gauss Forward Difference: $(0 < s < 1/2)$ $(x = x_0 + sh)$

$$P_n(x) = f(x_0) + \delta f(x_{1/2})(s) + \frac{\delta^2 f(x_0)}{2!}(s)(s-1) + \frac{\delta^3 f(x_{1/2})}{3!}(s+1)(s)(s-1) + \frac{\delta^4 f(x_0)}{4!}(s+1)(s)(s-1)(s-2)$$

$$+ \frac{\delta^5 f(x_{1/2})}{5!}(s+2)(s+1)(s)(s-1)(s-2) + \cdots$$

$$P_n'(x) = \frac{1}{h}\left(\delta f(x_{1/2}) + \frac{2s-1}{2}\delta^2 f(x_0) + \frac{3s^2-1}{6}\delta^3 f(x_{1/2}) + \frac{2s^3-3s^2-s+1}{12}\delta^4 f(x_0) \right.$$

$$\left. + \frac{5s^4-15s^2+4}{120}\delta^5 f(x_{1/2}) + \frac{6s^5-15s^4-20s^3+45s^2+8s-12}{720}\delta^6 f(x_0) + \cdots \right)$$

$$P_n''(x) = \frac{1}{h^2}\left(\delta^2 f(x_0) + s\delta^3 f(x_{1/2}) + \frac{6s^2-6s-1}{12}\delta^4 f(x_0) + \frac{2s^3-3s}{12}\delta^5 f(x_{1/2}) \right.$$

$$\left. + \frac{30s^4-60s^3-60s^2+90s+8}{720}\delta^6 f(x_0) + \cdots \right)$$

Gauss Backward Difference: $(-1/2 < s < 0)$ $(x = x_0 + sh)$

$$P_n(x) = f(x_0) + \delta f(x_{-1/2})(s) + \frac{\delta^2 f(x_0)}{2!}(s+1)(s) + \frac{\delta^3 f(x_{-1/2})}{3!}(s+1)(s)(s-1)$$

$$+ \frac{\delta^4 f(x_0)}{4!}(s+2)(s+1)(s)(s-1) + \frac{\delta^5 f(x_{-1/2})}{5!}(s+2)(s+1)(s)(s-1)(s-2) + \cdots$$

$$P_n'(x) = \frac{1}{h}\left(\delta f(x_{-1/2}) + \frac{2s+1}{2}\delta^2 f(x_0) + \frac{3s^2-1}{6}\delta^3 f(x_{-1/2}) + \frac{2s^3+3s^2-s-1}{12}\delta^4 f(x_0) \right.$$

$$\left. + \frac{5s^4-15s^2+4}{120}\delta^5 f(x_{-1/2}) + \frac{6s^5+15s^4-20s^3-45s^2+8s+12}{720}\delta^6 f(x_0) + \cdots \right)$$

$$P_n''(x) = \frac{1}{h^2}\left(\delta^2 f(x_0) + s\delta^3 f(x_{-1/2}) + \frac{6s^2+6s-1}{12}\delta^4 f(x_0) + \frac{2s^3-3s}{12}\delta^5 f(x_{-1/2}) \right.$$

$$\left. + \frac{30s^4+60s^3-60s^2-90s+8}{720}\delta^6 f(x_0) + \cdots \right)$$

Stirling Formula: $(-1/4 < s < 1/4)$ $(x = x_0 + sh)$

$$P_n(x) = f(x_0) + (s)\mu\delta f(x_0) + (s^2)\frac{1}{2!}\delta^2 f(x_0) + \frac{1}{3!}(s)(s^2 - 1^2)\mu\delta^3 f(x_0) + \frac{1}{4!}(s^2)(s^2 - 1^2)\delta^4 f(x_0)$$

$$+ \frac{1}{5!}(s)(s^2 - 1^2)(s^2 - 2^2)\mu\delta^5 f(x_0) + \cdots$$

$$P_n'(x) = \frac{1}{h}\left(\mu\delta f(x_0) + s\,\delta^2 f(x_0) + \frac{3s^2 - 1}{6}\mu\delta^3 f(x_0) + \frac{2s^3 - s}{12}\delta^4 f(x_0) + \frac{5s^4 - 15s^2 + 4}{120}\mu\delta^5 f(x_0)\right.$$

$$\left. + \frac{6s^5 - 20s^3 + 8s}{720}\delta^6 f(x_0) + \cdots\right)$$

$$P_n''(x) = \frac{1}{h^2}\left(\delta^2 f(x_0) + (s)\mu\delta^3 f(x_0) + \frac{6s^2 - 1}{12}\delta^4 f(x_0) + \frac{2s^3 - 3s}{12}\mu\delta^5 f(x_0)\right.$$

$$\left. + \frac{15s^4 - 30s^2 + 4}{360}\delta^6 f(x_0) + \cdots\right)$$

Bessel Formula: $(1/4 < s < 3/4)$ $(x = x_0 + sh)$

$$P_n(x) = \mu f(x_{1/2}) + \left(s - \frac{1}{2}\right)\delta f(x_{1/2}) + \frac{(s)(s-1)}{2!}\mu\delta^2 f(x_{1/2}) + \left(s - \frac{1}{2}\right)\frac{(s)(s-1)}{3!}\delta^3 f(x_{1/2})$$

$$+ \frac{(s+1)(s)(s-1)(s-2)}{4!}\mu\delta^4 f(x_{1/2}) + \left(s - \frac{1}{2}\right)\frac{(s+1)(s)(s-1)(s-2)}{5!}\delta^5 f(x_{1/2})$$

$$+ \frac{(s+2)(s+1)(s)(s-1)(s-2)(s-3)}{6!}\mu\delta^6 f(x_{1/2}) + \cdots$$

$$P_n'(x) = \frac{1}{h}\left(\delta f(x_{1/2}) + \frac{(2s-1)}{2}\mu\delta^2 f(x_{1/2}) + \frac{6s^2 - 6s + 1}{12}\delta^3 f(x_{1/2}) + \frac{2s^3 - 3s^2 - s + 1}{12}\mu\delta^4 f(x_{1/2})\right.$$

$$\left. + \frac{5s^4 - 10s^3 + 5s - 1}{120}\delta^5 f(x_{1/2}) + \frac{6s^5 - 15s^4 - 20s^3 + 45s^2 + 8s - 12}{720}\mu\delta^6 f(x_{1/2}) + \cdots\right)$$

$$P_n''(x) = \frac{1}{h^2}\left(\mu\delta^2 f(x_{1/2}) + \left(s - \frac{1}{2}\right)\delta^3 f(x_{1/2}) + \frac{6s^2 - 6s - 1}{12}\mu\delta^4 f(x_{1/2}) + \frac{4s^3 - 6s^2 + 1}{24}\delta^5 f(x_{1/2})\right.$$

$$\left. + \frac{15s^4 - 30s^3 - 30s^2 + 45s + 4}{360}\mu\delta^6 f(x_{1/2}) + \cdots\right)$$

Everett Formula: $(0 < s < 1)$ $(x = x_0 + sh)$ $(r = 1 - s)$

$$P_n(x) = \left(rf(x_0) + sf(x_1)\right) + \left(\frac{(r+1)r(r-1)}{3!}\delta^2 f(x_0) + \frac{(s+1)s(s-1)}{3!}\delta^2 f(x_1)\right)$$

$$+ \left(\frac{(r+2)(r+1)r(r-1)(r-2)}{5!}\delta^4 f(x_0) + \frac{(s+2)(s+1)s(s-1)(s-2)}{5!}\delta^4 f(x_1)\right) + \cdots$$

$$P_n'(x) = \frac{1}{h}\left(\left(f(x_1) - f(x_0)\right) + \frac{1}{6}\left((3s^2 - 1)\delta^2 f(x_1) - (3s^2 - 6s + 2)\delta^2 f(x_0)\right) + \right.$$

$$\left. \frac{1}{120}\left((5s^4 - 15s^2 + 4)\delta^4 f(x_1) - (5s^4 - 20s^3 + 15s^2 + 10s - 6)\delta^4 f(x_0)\right) + \cdots\right)$$

$$P_n''(x) = \frac{1}{h^2}\left(\left((s)\delta^2 f(x_1) - (s-1)\delta^2 f(x_0)\right) + \frac{1}{12}\left((2s^3 - 3s)\delta^4 f(x_1) - (2s^3 - 6s^2 + 3s + 10)\delta^4 f(x_0)\right) + \cdots\right)$$

Steffensen Formula: $(-1/2 < s < 1/2)$ $(x = x_0 + sh)$

$$P_n(x) = f(x_0) + \left(\frac{(s+1)s}{2!} \delta f(x_{1/2}) - \frac{s(s-1)}{2!} \delta f(x_{-1/2}) \right)$$

$$+ \left(\frac{(s+2)(s+1)s(s-1)}{4!} \delta^3 f(x_{1/2}) - \frac{(s+1)s(s-1)(s-2)}{4!} \delta^3 f(x_{-1/2}) \right)$$

$$+ \left(\frac{(s+3)(s+2)(s+1)s(s-1)(s-2)}{6!} \delta^5 f(x_{1/2}) - \frac{(s+2)(s+1)s(s-1)(s-2)(s-3)}{6!} \delta^5 f(x_{-1/2}) \right) + \cdots$$

$$P_n'(x) = \frac{1}{h} \left(\frac{1}{2} \left((2s+1)\delta f(x_{1/2}) - (2s-1)\delta f(x_{-1/2}) \right) \right.$$

$$+ \frac{1}{12} \left((2s^3 + 3s^2 - s - 1)\delta^3 f(x_{1/2}) - (2s^3 - 3s^2 - s + 1)\delta^3 f(x_{-1/2}) \right)$$

$$+ \frac{1}{720} \left((6s^5 + 15s^4 - 20s^3 - 45s^2 + 8s + 12) \delta^5 f(x_{1/2}) - (6s^5 - 15s^4 - 20s^3 + 45s^2 + 8s - 12)\delta^5 f\left(x_{-1/2}\right) + \cdots \right)$$

$$P_n''(x) = \frac{1}{h^2} \left(\left(\delta f(x_{1/2}) - \delta f(x_{-1/2}) \right) + \frac{1}{12} \left((6s^2 + 6s - 1)\delta^3 f(x_{1/2}) - (6s^2 - 6s - 1)\delta^3 f(x_{-1/2}) \right) \right.$$

$$+ \frac{1}{360} \left((15s^4 + 30s^3 - 30s^2 - 45s + 4)\delta^5 f(x_{1/2}) - (15s^4 - 30s^3 - 30s^2 + 45s + 4)\delta^5 f\left(x_{-1/2}\right) + \cdots \right)$$

The formulas in Table 12.1 can be simplified to compute derivatives at a tabulated point. For example, consider forward difference formula for first order derivative

$$P_n'(x) = \frac{1}{h} \left(\Delta f(x_0) + \frac{2s-1}{2} \Delta^2 f(x_0) + \frac{3s^2 - 6s + 2}{6} \Delta^3 f(x_0) + \frac{2s^3 - 9s^2 + 11s - 3}{12} \Delta^4 f(x_0) + \cdots \right)$$

To compute derivative at $x = x_0$, we have $s = 0$. This implies

$$P_n'(x_0) = \frac{1}{h} \left(\Delta f(x_0) - \frac{1}{2} \Delta^2 f(x_0) + \frac{1}{3} \Delta^3 f(x_0) - \frac{1}{4} \Delta^4 f(x_0) + \frac{1}{5} \Delta^5 f(x_0) - \cdots \right)$$

Similarly, the second derivative is given by

$$P_n''(x) = \frac{1}{h^2} \left(\Delta^2 f(x_0) - \Delta^3 f(x_0) + \frac{11}{12} \Delta^4 f(x_0) - \frac{5}{6} \Delta^5 f(x_0) + \cdots \right)$$

In a given table of $\sin(\theta)$, we have following values

Example _____ **12.2**

θ(In degree)	0	15	30	45	60	75	90
$\sin(\theta)$	0	0.258819	0.5	0.707107	0.866025	0.965926	1

compute the values of y' and y'' at the angles, $\theta = 20°, 40°, 48°, 50°, 55°, 57°, 87°$.

Ans.

Finite difference table for the given values of $\sin(\theta)$ is as follows

θ	$\sin(\theta)$	$\Delta(\nabla,\delta)$	$\Delta^2(\nabla,\delta)$	$\Delta^3(\nabla,\delta)$	$\Delta^4(\nabla,\delta)$	$\Delta^5(\nabla,\delta)$	$\Delta^6(\nabla,\delta)$
0	0						
		0.258819					
15	0.258819		−0.017638				
		0.241181		−0.016436			
30	0.500000		−0.034074		0.002321		
		0.207107		−0.014115		0.000966	
45	0.707107		−0.048189		0.003287		−0.000235
		0.158918		−0.010828		0.000731	
60	0.866025		−0.059017		0.004018		
		0.099901		−0.006810			
75	0.965926		−0.065827				
		0.034074					
90	1						

First, we should change the step size $h = 15°$ from degree to radian, $h = 15 \times \dfrac{\pi}{180} = 0.261799$, such that we can use it to various finite difference formulas.

1. **Newton Forward Difference for y' at angle $\theta = 20°$.**

Let $x_0 = 15°$. We will use following finite differences table to compute the value at $\theta = 20°$.

θ	$\sin(\theta)$	Δ	Δ^2	Δ^3	Δ^4	Δ^5	Δ^6
15	0.258819						
		0.241181					
30	0.500000		−0.034074				
		0.207107		−0.014115			
45	0.707107		−0.048189		0.003287		
		0.158918		−0.010828		0.000731	
60	0.866025		−0.059017		0.004018		
		0.099901		−0.006810			
75	0.965926		−0.065827				
		0.034074					
90	1						

We have, $s = \dfrac{20-15}{15} = \dfrac{1}{3}$. Forward difference formula for the first derivative is given by

$$P_n'(x) = \frac{1}{h}\left(\begin{array}{l} \Delta f(x_0) + \dfrac{2s-1}{2}\Delta^2 f(x_0) + \dfrac{3s^2-6s+2}{6}\Delta^3 f(x_0) + \dfrac{2s^3-9s^2+11s-3}{12}\Delta^4 f(x_0) \\ + \dfrac{5s^4-40s^3+105s^2-100s+24}{120}\Delta^5 f(x_0) \end{array} \right)$$

$$P_n'(x) = \frac{1}{h}\left(\begin{array}{l} \Delta f(x_0) + \dfrac{2s-1}{2}\Delta^2 f(x_0) + \dfrac{3s^2-6s+2}{6}\Delta^3 f(x_0) + \dfrac{2s^3-9s^2+11s-3}{12}\Delta^4 f(x_0) \\ + \dfrac{5s^4-40s^3+105s^2-100s+24}{120}\Delta^5 f(x_0) \end{array} \right)$$

$$P_n'(x) = \frac{1}{0.261799}\left(\begin{array}{l} 0.241181 + \dfrac{1}{2}\left(\dfrac{2}{3}-1\right)(-0.34074) + \dfrac{1}{6}\left(\dfrac{1}{3}-2+2\right)(-0.014115) \\ + \dfrac{1}{12}\left(\dfrac{2}{27}-1+\dfrac{11}{3}-3\right)(0.003287) + \dfrac{1}{120}\left(\dfrac{5}{81}-\dfrac{40}{27}+\dfrac{105}{9}-\dfrac{100}{3}+24\right)(0.000731) \end{array} \right)$$

$$P_n'(x) = \frac{1}{0.261799}\left(0.241181 + \frac{1}{6}(0.34074) + \frac{1}{18}(-0.014115) + \frac{-7}{324}(0.003287) + \frac{74}{9720}(0.000731) \right)$$

$$P_n'(x) = \frac{1}{0.261799}(0.241181 + .005679 - 0.000784 - 0.000071 + 0.000006)$$

$$P_n'(x) = 0.939694 \quad \text{(Approximate value of } y' = \frac{d}{dx}(\sin x) = \cos x \text{ at } \theta = 20°)$$

While exact value of $\cos 20° = 0.939693$.

2. Newton Backward Difference for y'' at angle $\theta = 70°$.

Since we have to compute the value at $\theta = 70°$, hence we can skip the last value $\theta = 90°$. Let $x_n = 75°$. Finite differences table is given by

θ	$\sin(\theta)$	∇	∇^2	∇^3	∇^4	∇^5	∇^6
0	0						
		0.258819					
15	0.258819		−0.017638				
		0.241181		−0.016436			
30	0.500000		−0.034074		0.002321		
		0.207107		−0.014115		0.000966	
45	0.707107		−0.048189		0.003287		
		0.158918		−0.010828			

60	0.866025		−0.059017
		0.099901	
75	0.965926		

We have, $s = \dfrac{70-75}{15} = \dfrac{-1}{3}$. Backward difference formula for the second derivative is given by

$$P_n''(x) = \frac{1}{h^2}\left[\nabla^2 f(x_n) + (s+1)\nabla^3 f(x_n) + \frac{6s^2+18s+11}{12}\nabla^4 f(x_n) + \frac{2s^3+12s^2+21s+10}{12}\nabla^5 f(x_n)\right]$$

$$P_n''(x) = \frac{1}{(0.261799)^2}\left(\begin{array}{l} -0.059017 + \left(\dfrac{-1}{3}+1\right)(-0.10828) + \dfrac{1}{12}\left(\dfrac{2}{3}-6+11\right)(0.003287) \\[2mm] +\dfrac{1}{12}\left(\dfrac{-2}{27}+\dfrac{4}{3}-7+10\right)(0.000966) \end{array}\right)$$

$P_n''(x) = -0.938754$ (Approximate value of $y'' = \dfrac{d^2}{dx^2}(\sin x) = -\sin x$ at $\theta = 70°$)

The exact value is given by

$\quad -\sin 70° = -0.939693$.

3. **Stirling Central Difference for y' at angle $\theta = 50°$.**

θ	$\sin(\theta)$	δ	δ^2	δ^3	δ^4	δ^5	δ^6
0	0						
		0.258819					
15	0.258819		−0.017638				
		0.241181		−0.016436			
30	0.500000		−0.034074		0.002321		
		0.207107		−0.014115		0.000966	
45	0.707107		−0.048189		0.003287		−0.000235
		0.158918		−0.010828		0.000731	
60	0.866025		−0.059017		0.004018		
		0.099901		−0.006810			
75	0.965926		−0.065827				
		0.034074					
90	1						

We have, $s = \dfrac{50-45}{15} = \dfrac{1}{3}$, and Stirling central difference formula for the first derivative is as follows

$$P_n'(x) = \frac{1}{h}\left(\begin{array}{l} \mu\delta f(x_0) + s\delta^2 f(x_0) + \dfrac{3s^2-1}{6}\mu\delta^3 f(x_0) + \dfrac{2s^3-s}{12}\delta^4 f(x_0) \\[2mm] + \dfrac{5s^4-15s^2+4}{120}\mu\delta^5 f(x_0) + \dfrac{6s^5-20s^3+8s}{720}\delta^6 f(x_0) + \cdots \end{array}\right)$$

$$P_n'(x) = \frac{1}{0.261799}\left(\begin{array}{l} \dfrac{1}{2}(0.207107+0.158918) + \dfrac{1}{3}(-0.048189) + \dfrac{1}{6}\left(\dfrac{-2}{3}\right)\left(\dfrac{1}{2}\right) \\[2mm] (-0.014115-0.010828) + \dfrac{1}{12}\left(\dfrac{2}{27}-\dfrac{1}{3}\right)(0.003287) + \dfrac{1}{120}\left(\dfrac{5}{81}-\dfrac{5}{3}+4\right) \\[2mm] \dfrac{1}{2}(0.000966+0.000731) + \dfrac{1}{720}\left(\dfrac{6}{243}-\dfrac{20}{27}+\dfrac{8}{3}\right)(-0.000235) \end{array}\right)$$

$$P_n'(x) = \frac{1}{0.261799}(0.183012-0.016063+0.001386-0.000071+0.000017-0.000001)$$

$P_n'(x) = 0.642783$ (Approximate value of $y' = \dfrac{d}{dx}(\sin x) = \cos x$ at $\theta = 50°$)

While exact value of $\cos 50° = 0.642788$

4. Gauss Backward Central Difference Formula for y'' at angle $\theta = 40°$.

We have, $s = \dfrac{40-45}{15} = \dfrac{-1}{3}$, and Gauss backward central difference formula for second derivative is as follows

$$P_n''(x) = \frac{1}{h^2}\left(\begin{array}{l} \delta^2 f(x_0) + s\delta^3 f(x_{-1/2}) + \dfrac{6s^2+6s-1}{12}\delta^4 f(x_0) + \dfrac{2s^3-3s}{12}\delta^5 f(x_{-1/2}) \\[2mm] + \dfrac{30s^4+60s^3-60s^2-90s+8}{720}\delta^6 f(x_0) + \cdots \end{array}\right)$$

$$P_n''(x) = \frac{1}{(0.261799)^2}\left(\begin{array}{l} -0.048189 + \left(\dfrac{-1}{3}\right)(-0.014115) + \dfrac{1}{12}\left(\dfrac{2}{3}-2-1\right)(0.003287) \\[2mm] + \dfrac{1}{12}\left(\dfrac{-2}{27}+1\right)(0.000966) + \dfrac{1}{720}\left(\dfrac{10}{27}-\dfrac{20}{9}-\dfrac{20}{3}+30+8\right)(-0.000235) \end{array}\right)$$

$$P_n''(x) = \frac{1}{(0.261799)^2}(-0.048189+0.004705-0.000639+0.000080-0.000010)$$

$P_n''(x) = -0.642746$ (Approximate value of $y'' = \dfrac{d^2}{dx^2}(\sin x) = -\sin x$ at $\theta = 40°$)

The exact value is given by

$$-\sin 40° = -0.642787$$

5. **Steffensen Central Difference Formula for** y' **at angle** $\theta = 50°$.

We have, $s = \dfrac{50-45}{15} = \dfrac{1}{3}$, and Steffensen central difference formula for the first derivative is as follows

$$P_n'(x) = \frac{1}{h}\left(\begin{array}{l} \frac{1}{2}\big((2s+1)\delta f(x_{1/2}) - (2s-1)\delta f(x_{-1/2})\big) + \frac{1}{12}\big((2s^3+3s^2-s-1)\delta^3 f(x_{1/2}) - (2s^3-3s^2-s+1)\delta^3 f(x_{-1/2})\big) \\ + \frac{1}{720}\big((6s^5+15s^4-20s^3-45s^2+8s+12)\delta^5 f(x_{1/2}) - (6s^5-15s^4-20s^3+45s^2+8s-12)\delta^5 f(x_{-1/2})\big)+\cdots \end{array}\right)$$

$$P_n'(x) = \frac{1}{(0.261799)}\left[\begin{array}{l} \frac{1}{2}\left(\left(\frac{2}{3}+1\right)(0.158918) - \left(\frac{2}{3}-1\right)(0.207107)\right) \\ + \frac{1}{12}\left(\left(\frac{2}{27}+\frac{1}{3}-\frac{1}{3}-1\right)(-0.010828) - \left(\frac{2}{27}-\frac{1}{3}-\frac{1}{3}+1\right)(-0.014115)\right) \\ + \frac{1}{720}\left(\left(\frac{2}{81}+\frac{5}{27}-\frac{20}{27}-5+\frac{8}{3}+12\right)(0.000731) - \left(\frac{2}{81}-\frac{5}{27}-\frac{20}{27}+5+\frac{8}{3}-12\right)(0.000966)\right) \end{array}\right]$$

$$P_n'(x) = \frac{1}{0.261799}\big(0.132432+0.034518+0.000835+0.000479+0.000009+0.000007\big)$$

$$P_n'(x) = 0.642783 \ (\text{Approximate value of } y' = \frac{d}{dx}(\sin x) = \cos x \text{ at } \theta = 50°)$$

The exact value is given by

$$\cos 50° = 0.642788$$

Steffensen formula for second derivative is as follows

$$P_n''(x) = \frac{1}{h^2}\left(\begin{array}{l} \big(\delta f(x_{1/2}) - \delta f(x_{-1/2})\big) + \frac{1}{12}\big((6s^2+6s-1)\delta^3 f(x_{1/2}) - (6s^2-6s-1)\delta^3 f(x_{-1/2})\big) \\ + \frac{1}{360}\big((15s^4+30s^3-30s^2-45s+4)\delta^5 f(x_{1/2}) - (15s^4-30s^3-30s^2+45s+4)\delta^5 f(x_{-1/2})\big) \end{array}\right)$$

$$P_n''(x) = \frac{1}{(0.261799)^2}\left[\begin{array}{l} \big(0.158918-0.207107\big) + \frac{1}{12}\left(\left(\frac{2}{3}+2-1\right)(-0.010828) - \left(\frac{2}{3}-2-1\right)(-0.014115)\right) \\ + \frac{1}{360}\left(\left(\frac{5}{27}+\frac{10}{9}-\frac{10}{3}-15+4\right)(0.000731) - \left(\frac{5}{27}-\frac{10}{9}-\frac{10}{3}+15+4\right)(0.000966)\right) \end{array}\right]$$

$$P_n''(x) = \frac{1}{(0.261799)^2}\big(-0.048189-0.001504-0.002744-0.000026-0.000040\big)$$

$$P_n''(x) = -0.766034 \ (\text{Approximate value of } y'' = \frac{d^2}{dx^2}(\sin x) = -\sin x \text{ at } \theta = 50°)$$

The exact value is given by

$$-\sin 50° = -0.766044$$

Exercise 12

1. Derive interpolating polynomial for $(-1, -1)$, $(0, 1)$, $(1, 3)$, $(3, 31)$, $(5, 131)$ using Newton divided difference formula, and hence compute the value of $y'(2.5)$ and $y''(1.2)$.

   Ans. $P(x) = x^3 + x + 1$, 19.75, 7.2

2. Approximate the values $f'(1.2)$ and $f''(1.2)$ for the following data set.

x:	1	1.25	1.5	1.75	2.0
$f(x)$	0	0.223144	0.405465	0.559616	0.693147

   Ans. 0.834355, -0.696651

3. Obtain the value of $f'(2.2)$ using Stirling formula to the following data

x:	2	2.1	2.2	2.3	2.4
$f(x)$:	0.135335	0.122456	0.110803	0.100259	0.090718

   Ans. -0.110799

4. The following table provides compressive strength of an alloy as a function of concentration of some additive

Concentration:	5	7.5	10	12.5	15
Compressive Strength:	26.4	29.8	31.5	30.8	27.6

   Estimate the concentration that corresponds to maximum compressive strength.

   Ans. 10.568867

5. The following data represents thermal resistance for a certain metal tube as a function of insulation thickness.

Thickness:	1	3	7	10	15
Thermal Resistance:	7.45	6.38	5.23	6.19	6.48

   Use interpolation to compute the thickness for which thermal resistance is minimum.

   Ans. 6.616718

6. Derive interpolating polynomial for following data and hence compute the maximum and minimum values for the function $f(x)$.

x:	0	0.5	1	1.5	2.5
$f(x)$:	2.53	5.38	6.34	4.79	3.21

   Ans. $P_4(x) = 1.537333x^4 - 5.438667x^3 + 1.687667x^2 + 6.023667x + 2.53$

   Extreme points $(x) = 0.929753, 2.202004$

   Maximum value of $f(x) = 6.367049$ at $x = 0.929753$

   Minimum value of $f(x) = 2.052364$ at $x = 2.2020004$

7. Find the first and second derivatives of $y = f(x)$ at $x = 0$ and 0.3, from the following table

x	0	0.1	0.2	0.3
y	1	1.105170	1.221403	1.349859

   Consider that the values are obtained from the function $f(x) = e^x$ correct to 6 decimal places.

   Hence estimate the errors in the values of $\dfrac{dy}{dx}$ and $\dfrac{d^2y}{dx^2}$ at $x = 0.3$.

Ans. $f'(0) = 1.000252, f'(0.3) = 1.349542$

     $f''(0) = 0.990300, f''(0.3) = 1.338300$

8. Compute the growth rate of population in years, 1971, 1991 and 2011 from the following population data and interpret the results.

Years	Population (in crores)
1971	54.8160
1981	68.3329
1991	84.6421
2001	102.8737
2011	121.0193

                                Ans. 1.2113, 1.7509, 1.7147 (aligned with 1991 row)

9. A car traveled on a straight road, and it is found on the certain milestones (distance x) after a certain time (t) as follows

t (InHrs):	0	0.48	0.78	1.10	1.42	1.82
x (InKms):	0	50	100	150	200	250

Compute the approximate time of maximum velocity.

Ans. 0.552554

10. A machine has inputs and their corresponding outputs as follows

Input:	0	5	10	15	20
Output:	3.05	11.43	8.57	5.32	6.58

Find out the maximum output of the machine.

Ans. 11.439852

11. In a given table of $y = \cos(x)$, we have following values

x:	1	1.1	1.2	1.3	1.4	1.5
$\cos(x)$:	0.540302	0.453596	0.362358	0.2674988	0.169967	0.070737

Compute the values of y' and y'' at the point $x = 1$. Also, estimate the errors.

Ans. – 0.841476, –0.540075

12. Find the first and second derivatives of $f(x)$ at $x = 1.2$ and $x = 2.0$ from the following table

x	1.0	1.2	1.4	1.6	1.8	2.0
$f(x)$	2.7183	3.3201	4.0552	4.9530	6.0496	7.3891

These values are obtained from the function $y = e^x$, Use this function to estimate the errors in the values of $\dfrac{dy}{dx}$ and $\dfrac{d^2y}{dx^2}$ at $x = 1.2$ and 2.0.

Ans. $f'(1.2) = 3.3202, f'(2) = 7.3896$

     $f''(1.2) = 3.3212, f''(2) = 7.3854$

13. The following table provides values of the function $f(x) = 1 + \cos^2 x$. Use interpolation to estimate the values of $f'(1.32)$ and $f''(1.65)$.

x:	1.3	1.4	1.5	/ 1.6	1.7
$f(x)$:	1.071556	1.028889	1.005004	1.000853	1.016601

Ans. $f'(1.32) = -0.480808, f''(1.65) = 1.975196.$

Many scientific and engineering applications require integration of a function. For example, arc length of the curve $y = f(x)$ in the interval $[a, b]$ is given by the following formula

$$\text{Arc length} = \int_a^b \sqrt{1 + \left(\frac{dy}{dx}\right)^2}\, dx$$

Consider the arc length of a curve $y = \sin x$ in the interval $0 \le x \le \pi$. It requires computing the integral $\int_0^\pi \sqrt{1 + \cos^2 x}\, dx$. Similarly, surface area and volume of solid of revolution obtained by rotating the curve $y = f(x)$, $a \le x \le b$ about the x-axis are given by

$$S = \int_a^b 2\pi y \sqrt{1 + \left(\frac{dy}{dx}\right)^2}\, dx \text{ and } V = \int_a^b \pi y^2\, dx, \text{ respectively. Integrals are used to compute the}$$

center of gravity of irregular objects, the total amount of given physical quantity, work done, kinetic energy and many more. The normal or Gaussian distribution involves the integral $\int_a^b e^{-z^2}\, dz$, and it is most commonly used continuous probability distribution in statistics.

The integrals are evaluated analytically for simple functions. In general, it is very difficult and some times not possible to integrate analytically. There are two main reasons to study numerical integration

i)   Analytical integration may be impossible or infeasible. For example, we do not have analytical solutions of the integrals like $\int_0^{\pi} \sqrt{1+\cos^2 x} \, dx$, $\int_1^2 \frac{\sin x}{x} dx$ and $\int_0^2 e^{-x^2} dx$, etc. In these cases; numerical methods can be used for the approximate solutions.

ii)   We wish to integrate tabulated data. For example, to compute the value of integral $\int_0^5 y(x)dx$ from data set $(0, -1)$, $(1, 1)$, $(2, 9)$, $(3, 29)$, $(5, 129)$.

This chapter deals with the techniques for numerical integration. In numerical integration, we replace function $f(x)$ or data set with an approximating curve to compute the given integral. Since, the approximating curve has an error, that error needs to be evaluated in case of integration also. This chapter contains following numerical integration techniques

1.   Newton–Cotes quadrature formulas
     1.1   Trapezoidal rule
     1.2   Simpson 1/3 rule
     1.3   Simpson 3/8 rule
     1.4   Boole rule
     1.5   Weddle rule
2.   Gauss Quadrature formulas
     2.1   Gauss–Legendre method
     2.2   Gauss–Chebyshev method
     2.3   Gauss–Laguerre method
     2.4   Gauss–Hermite method
3.   Euler–Maclaurin formula
4.   Romberg integration

13.1 Newton–Cotes Quadrature Formulas (Using Lagrange Method)

Let $f(x)$ be a real-valued function defined on some interval $[a, b]$ containing $(n+1)$ points $x_0, x_1, \cdots, x_n$. Let the points be equispaced points with spacing h, i.e.,

$$x_i = x_0 + ih; \; i = 1, 2, \cdots, n$$

We want to evaluate the integral $\int_{x_0}^{x_n} f(x)dx$.

Let $P_n(x)$ be the interpolating polynomial, which interpolates function $f(x)$ at $(n+1)$ points $x_0, x_1, \cdots, x_n$. Also, let $\varepsilon_n(x)$ be the error in the interpolation. When we approximate the function $f(x)$ with interpolating polynomial $P_n(x)$, then let I_n be the approximate value of integration and EI_n be the error term. The integral can be computed as follows

$$I = \int_{x_0}^{x_n} f(x)dx = \int_{x_0}^{x_n} \left[P_n(x) + \varepsilon_n(x) \right] dx = \int_{x_0}^{x_n} P_n(x)dx + \int_{x_0}^{x_n} \varepsilon_n(x)dx = I_n + EI_n \qquad (13.1)$$

First, we will compute the integration (I_n) and then error (EI_n) in numerical integration. Consider Lagrange form of interpolating polynomial $P_n(x)$ to get the following Newton–Cotes quadrature formula

$$
\begin{aligned}
I_n &= \int_{x_0}^{x_n} P_n(x)dx \\
&= \int_{x_0}^{x_n} \left[f(x_0)l_0(x) + f(x_1)l_1(x) + \cdots + f(x_n)l_n(x) \right] dx \\
&= \int_{x_0}^{x_n} \left(\sum_{i=0}^{n} f(x_i)l_i(x) \right) dx \\
&= \sum_{i=0}^{n} f(x_i) \int_{x_0}^{x_n} l_i(x)dx = \sum_{i=0}^{n} f(x_i)\lambda_i \qquad (13.2)
\end{aligned}
$$

where $\lambda_i = \int_{x_0}^{x_n} l_i(x)dx$; $i = 0, 1, 2, \cdots, n$ are the constants to be determined. The points are equispaced, i.e.,

$$x_i = x_0 + ih; \quad i = 1, 2, \cdots, n,$$

Let $x = x_0 + sh$, we have

$$
\begin{aligned}
l_i(x) &= \frac{(x - x_0)(x - x_1) \cdots (x - x_{i-1})(x - x_{i+1}) \cdots (x - x_n)}{(x_i - x_0)(x_i - x_1) \cdots (x_i - x_{i-1})(x_i - x_{i+1}) \cdots (x_i - x_n)} \\
&= \frac{(s)h(s-1)h \cdots (s-i+1)h(s-i-1)h \cdots (s-n)h}{(i)h(i-1)h \cdots (1)h(-1)h \cdots (i-n)h} \\
&= \frac{(s)(s-1) \cdots (s-i+1)(s-i-1) \cdots (s-n)}{(-1)^{n-i} i! \, (n-i)!}
\end{aligned}
$$

On using the expression for $l_i(x)$ in $\lambda_i = \int_{x_0}^{x_n} l_i(x)dx$, we have

$$
\begin{aligned}
\lambda_i &= \int_{x_0}^{x_n} l_i(x)dx \\
&= \int_{0}^{n} \frac{(s)(s-1) \cdots (s-i+1)(s-i-1) \cdots (s-n)}{(-1)^{n-i} i! \, (n-i)!} h \, ds \qquad i = 0, 1, 2, \cdots, n \qquad (13.3)
\end{aligned}
$$

We will use these expressions in Eq. (13.2) to get the approximate value of the integral for different values of n.

13.1.1 Trapezoidal Rule ($n = 1$)

For $n = 1$, we have only two points x_0 and x_1. Therefore the integral $\int_{x_0}^{x_1} f(x)dx$ can be computed as follows from Eq. (13.2)

$$\int_{x_0}^{x_1} f(x)dx \approx I_T = \sum_{i=0}^{1} f(x_i)\lambda_i = \lambda_0 f(x_0) + \lambda_1 f(x_1)$$

where λ_0 and λ_1 are computed with the aid of formula (13.3), and given by

$$\lambda_0 = \int_0^1 -(s-1)h\,ds = \frac{h}{2}$$

$$\lambda_1 = \int_0^1 (s)h\,ds = \frac{h}{2}$$

Hence, the value of integral I_T is given by

$$I_T = \lambda_0 f(x_0) + \lambda_1 f(x_1) = \frac{h}{2}\left[f(x_0) + f(x_1)\right] \tag{13.4}$$

Geometrical Interpretation:

In Trapezoidal rule, we approximate the function $f(x)$ with a straight line joining $(x_0, f(x_0))$ and $(x_1, f(x_1))$. Therefore, the integral $\int_{x_0}^{x_1} f(x)dx$, which is the area under the curve $f(x)$ in $[x_0, x_1]$ is approximated by the area under the straight line joining $(x_0, f(x_0))$ and $(x_1, f(x_1))$ in $[x_0, x_1]$.

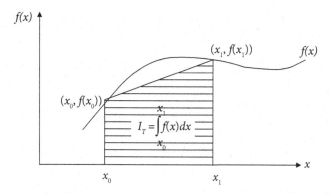

Fig. 13.1 Trapezoidal rule

Since this formula is equivalent to the area of a trapezium, that's why this method is known as Trapezoidal rule.

13.1.2 Simpson 1/3 Rule ($n = 2$)

For $n = 2$, Eq. (13.2) gives following approximation for the integral

$$\int_{x_0}^{x_2} f(x)dx \simeq I_{S1/3} = \sum_{i=0}^{2} f(x_i)\lambda_i = \lambda_0 f(x_0) + \lambda_1 f(x_1) + \lambda_2 f(x_2)$$

where λ_0, λ_1 and λ_2 can be computed easily from Eq. (13.3). We have

$$\lambda_0 = \int_{0}^{2} \frac{(s-1)(s-2)}{2} h\, ds = \frac{h}{3}$$

$$\lambda_1 = \int_{0}^{2} -(s)(s-2) h\, ds = \frac{4h}{3}$$

$$\lambda_2 = \int_{0}^{2} \frac{(s)(s-1)}{2} h\, ds = \frac{h}{3}$$

So, the value of integral is given by

$$I_{S1/3} = \lambda_0 f(x_0) + \lambda_1 f(x_1) + \lambda_2 f(x_2) = \frac{h}{3}\left[f(x_0) + 4f(x_1) + f(x_2)\right] \qquad (13.5)$$

Geometrical Interpretation:
The function $f(x)$ is approximated with a parabola passing through the points $(x_0, f(x_0))$, $(x_1, f(x_1))$ and $(x_2, f(x_2))$. Hence, the integral $\int_{x_0}^{x_2} f(x)dx$, which is the area under the curve $f(x)$ in $[x_0, x_2]$, is approximated by the area under this parabola.

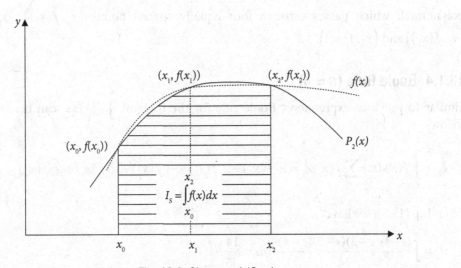

Fig. 13.2 Simpson 1/3 rule

13.1.3 Simpson 3/8 Rule ($n = 3$)

For $n = 3$, Eq. (13.2) provides the following rule

$$\int_{x_0}^{x_3} f(x)dx \approx I_{S3/8} = \sum_{i=0}^{3} f(x_i)\lambda_i = \lambda_0 f(x_0) + \lambda_1 f(x_1) + \lambda_2 f(x_2) + \lambda_3 f(x_3)$$

From Eq. (13.3), we have

$$\lambda_0 = \int_0^3 \frac{-(s-1)(s-2)(s-3)}{6} h\, ds = \frac{3h}{8}$$

$$\lambda_1 = \int_0^3 \frac{(s)(s-2)(s-3)}{2} h\, ds = \frac{9h}{8}$$

$$\lambda_2 = \int_0^3 \frac{-(s)(s-1)(s-3)}{2} h\, ds = \frac{9h}{8}$$

$$\lambda_3 = \int_0^3 \frac{(s)(s-1)(s-2)}{6} h\, ds = \frac{3h}{8}$$

The approximate integral ($I_{S3/8}$) is given by

$$I_{S3/8} = \lambda_0 f(x_0) + \lambda_1 f(x_1) + \lambda_2 f(x_2) = \frac{3h}{8}\left[f(x_0) + 3f(x_1) + 3f(x_2) + f(x_3)\right] \qquad (13.6)$$

Geometrical Interpretation:

In Simpson 3/8 rule, the integral $\int_{x_0}^{x_3} f(x)dx$ is approximated by the area under a cubic polynomial, which passes through four equally spaced points $\left(x_0, f(x_0)\right)$, $\left(x_1, f(x_1)\right)$, $\left(x_2, f(x_2)\right)$ and $\left(x_3, f(x_3)\right)$.

13.1.4 Boole Rule ($n = 4$)

Similar to previous expressions, Boole rule for the integral $\int_{x_0}^{x_4} f(x)dx$ can be derived as follows

$$I_B = \int_{x_0}^{x_4} f(x)dx \approx \sum_{i=0}^{4} f(x_i)\lambda_i = \lambda_0 f(x_0) + \lambda_1 f(x_1) + \lambda_2 f(x_2) + \lambda_3 f(x_3) + \lambda_4 f(x_4)$$

From Eq. (13.3), we have

$$\lambda_0 = \int_0^4 \frac{(s-1)(s-2)(s-3)(s-4)}{24} h\, ds = \frac{14}{45} h$$

$$\lambda_1 = \int_0^4 \frac{-(s)(s-2)(s-3)(s-4)}{6} h\,ds = \frac{64}{45}h$$

$$\lambda_2 = \int_0^4 \frac{(s)(s-1)(s-3)(s-4)}{4} h\,ds = \frac{24}{45}h$$

$$\lambda_3 = \int_0^4 \frac{-(s)(s-1)(s-2)(s-4)}{6} h\,ds = \frac{64}{45}h$$

$$\lambda_4 = \int_0^4 \frac{(s)(s-1)(s-2)(s-3)}{24} h\,ds = \frac{14}{45}h$$

The value of integral is given by

$$I_B = \lambda_0 f(x_0) + \lambda_1 f(x_1) + \cdots + \lambda_4 f(x_4)$$

$$= \frac{2}{45}h\Big[7f(x_0) + 32f(x_1) + 12f(x_2) + 32f(x_3) + 7f(x_4)\Big] \qquad (13.7)$$

Geometrically, the integral $\int_{x_0}^{x_4} f(x)dx$ is approximated by the area under a quartic polynomial (degree 4), which passes through five equally spaced points $(x_0, f(x_0))$, $(x_1, f(x_1))$, $(x_2, f(x_2))$, $(x_3, f(x_3))$ and $(x_4, f(x_4))$.

13.1.5 Weddle Rule ($n = 6$)

Similar to previous expressions, Weddle rule for the integral $\int_{x_0}^{x_6} f(x)dx$ can be derived as follows

$$I_W = \int_{x_0}^{x_6} f(x)dx \approx \sum_{i=0}^{6} f(x_i)\lambda_i = \lambda_0 f(x_0) + \lambda_1 f(x_1) + \cdots + \lambda_6 f(x_6)$$

$$= \frac{h}{140}\Big[41f(x_0) + 216f(x_1) + 27f(x_2) + 272f(x_3) + 27f(x_4) + 216f(x_5) + 41f(x_6)\Big]$$

$$(13.8)$$

In Weddle Rule, the integral $\int_{x_0}^{x_6} f(x)dx$ is approximated by the area under a polynomial of degree 6, which passes through $(x_i, f(x_i))$; $i = 0,1,2,\ldots,6$.

Example ── 13.1

Compute the value of integral $\int\limits_{1}^{2} e^{-x^2} dx$ with the aid of Trapezoidal, Simpson 1/3 and Simpson 3/8 formulas.

Ans.

The value of integral $\int\limits_{1}^{2} e^{-x^2} dx$ from different formulas are given by

   i)   Trapezoidal rule (13.4)

$$x_0 = 1, x_1 = 2 \text{ and } h = 1$$

$$I_T = \frac{h}{2}\left[f(x_0) + f(x_1)\right] = \frac{1}{2}\left[e^{-1} + e^{-4}\right] = \frac{1}{2}[0.367879 + 0.018316] = 0.193098$$

   ii)   Simpson 1/3 rule (13.5)

$$x_0 = 1, \ x_1 = \frac{3}{2}, \ x_2 = 2 \text{ and } h = \frac{1}{2}$$

$$I_{S1/3} = \frac{h}{3}\left[f(x_0) + 4f(x_1) + f(x_2)\right] = \frac{1}{6}\left[e^{-1} + 4e^{-9/4} + e^{-4}\right] = 0.134632$$

   iii)   Simpson 3/8 Rule (13.6)

$$x_0 = 1, \ x_1 = \frac{4}{3}, \ x_2 = \frac{5}{3}, \ x_3 = 2 \text{ and } h = \frac{1}{3}$$

$$I_{S3/8} = \frac{3h}{8}\left[f(x_0) + 3f(x_1) + 3f(x_2) + f(x_3)\right] = \frac{1}{8}\left[e^{-1} + 3e^{-16/9} + 3e^{-25/9} + e^{-4}\right] = 0.134971$$

Example ── 13.2

We have following input and output from an experiment

Input x:	0	0.5	1	2	2.5	3	3.5
Output $f(x)$:	2.3	3.8	4.7	5.4	6.0	6.4	6.7

Use Weddle rule to compute the integral $\int\limits_{0}^{3.5} f(x)dx$.

Ans.

The points and step size are as follows

$$x_0 = 0, \ x_1 = 0.5, ..., \ x_6 = 3.5 \text{ and } h = 0.5.$$

On applying the Weddle rule (13.8), we have

$$I_W = \int_{x_0}^{x_6} f(x)dx$$

$$= \frac{h}{140}\left[41f(x_0)+216f(x_1)+27f(x_2)+272f(x_3)+27f(x_4)+216f(x_5)+41f(x_6)\right]$$

$$= \frac{0.5}{140}\left[41(2.3)+216(3.8)+27(4.7)+272(5.4)+27(6.0)+216(6.4)+41(6.7)\right]$$

$$= 15.463928$$

13.2 Composite Newton–Cotes Quadrature Rules

In case of large intervals, the higher order approximating polynomials are very difficult to apply practically. Also, the lower order polynomial approximations produce large error in such cases. So, it is better to compute the integral by using composite rules. We subdivide the interval in composite rules, and then apply a particular lower order method to each of these subintervals.

13.2.1 Composite Trapezoidal Rule

The interval $[a, b]$ is subdivided into m equal sub-intervals with $a = x_0$ and $b = x_m$. The spacing for variable x is given by

$$h = \frac{x_m - x_0}{m}$$

The integral $\int_a^b f(x)dx$ can be computed as follows

$$\int_a^b f(x)dx = \int_{x_0}^{x_m} f(x)dx = \int_{x_0}^{x_1} f(x)dx + \int_{x_1}^{x_2} f(x)dx + \cdots + \int_{x_{m-1}}^{x_m} f(x)dx$$

On using Trapezoidal rule (13.4) for each interval, we have following approximation for the integral $\int_{x_0}^{x_m} f(x)dx$.

$$\int_{x_0}^{x_m} f(x)dx \approx I_{CT} = \frac{h}{2}\left[f(x_0)+f(x_1)\right]+\frac{h}{2}\left[f(x_1)+f(x_2)\right]+\cdots+\frac{h}{2}\left[f(x_{m-1})+f(x_m)\right]$$

$$= \frac{h}{2}\left\{f(x_0)+f(x_m)+2\left[f(x_1)+f(x_2)+\cdots+f(x_{m-1})\right]\right\} \tag{13.9}$$

Geometrical Interpretation:

In each interval, we approximate the function $f(x)$ with a straight line joining end points of the interval.

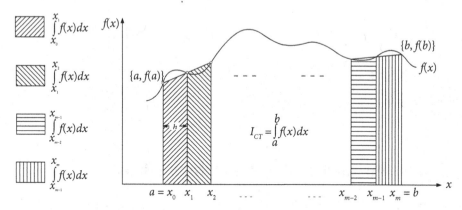

Fig. 13.3 Composite Trapezoidal rule

13.2.2 Composite Simpson 1/3 Rule

Similar to Trapezoidal rule, we divide the interval $[a, b]$ into m equal parts with $a = x_0$ and $b = x_m$. Let spacing for x be $h = \dfrac{x_m - x_0}{m}$. The integral $\displaystyle\int_a^b f(x)dx$ is given by

$$\int_a^b f(x)dx = \int_{x_0}^{x_m} f(x)dx = \int_{x_0}^{x_2} f(x)dx + \int_{x_2}^{x_4} f(x)dx + \cdots + \int_{x_{m-2}}^{x_m} f(x)dx$$

Note that m must be an even number, as the integrands are over the intervals $(x_0 \cdots x_2), (x_2 \cdots x_4), \cdots, (x_{m-2} \cdots x_m)$. We apply Simpson 1/3 rule (13.5) on each interval to get the following composite rule

$$\int_{x_0}^{x_m} f(x)dx = \int_{x_0}^{x_2} f(x)dx + \int_{x_2}^{x_4} f(x)dx + \cdots + \int_{x_{m-2}}^{x_m} f(x)dx$$

$$I_{CS1/3} = \frac{h}{3}\left[f(x_0) + 4f(x_1) + f(x_2)\right] + \frac{h}{3}\left[f(x_2) + 4f(x_3) + f(x_4)\right] +$$

$$\cdots + \frac{h}{3}\left[f(x_{m-2}) + 4f(x_{m-1}) + f(x_m)\right]$$

$$= \frac{h}{3}\left\{ \begin{array}{l} f(x_0) + f(x_m) + 2\left[f(x_2) + f(x_4) + \cdots + f(x_{m-2})\right] \\ \qquad + 4\left[f(x_1) + f(x_3) + \cdots + f(x_{m-1})\right] \end{array} \right\}$$

$$= \frac{h}{3}\left\{ \begin{array}{l} \text{first term} + \text{last term} + 2\left[\text{sum of even terms}\right] \\ \qquad\qquad + 4\left[\text{sum of odd terms}\right] \end{array} \right\} \qquad (13.10)$$

This method is known as composite Simpson 1/3 rule or simply Simpson 1/3 rule with m equal sub-intervals.

Geometrical Interpretation:

Geometrically, we approximate the function $f(x)$ with quadratic polynomials in subintervals, $(x_0 \cdots x_2), (x_2 \cdots x_4), \cdots, (x_{m-2} \cdots x_m)$.

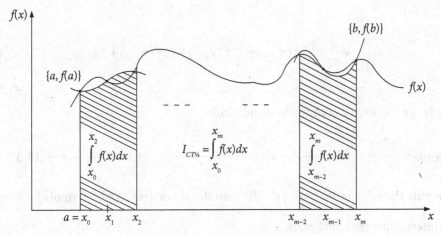

Fig. 13.4 Composite Simpson 1/3 rule

13.2.3 Composite Simpson 3/8 rule

Simpson 3/8 rule (13.6) (m to be a multiple of 3) can be used to produce following composite rule

$$\int_{x_0}^{x_m} f(x)dx = \int_{x_0}^{x_3} f(x)dx + \int_{x_3}^{x_6} f(x)dx + \cdots + \int_{x_{m-3}}^{x_m} f(x)dx$$

$$\approx \frac{3h}{8}\left[f(x_0)+3f(x_1)+3f(x_2)+f(x_3)\right]+\frac{3h}{8}\left[f(x_3)+3f(x_4)+3f(x_5)+f(x_6)\right]+$$

$$\cdots+\frac{3h}{8}\left[f(x_{m-3})+3f(x_{m-2})+3f(x_{m-1})+f(x_m)\right]$$

$$I_{CS3/8} = \frac{3h}{8}\left\{\begin{array}{c} f(x_0)+f(x_m)+3\left[f(x_1)+f(x_2)+f(x_4)+f(x_5)\cdots+f(x_{m-2})+f(x_{m-1})\right] \\ +2\left[f(x_3)+f(x_6)+\cdots+f(x_{m-3})\right] \end{array}\right\}$$

$$(13.11)$$

13.2.4 Composite Boole Rule

Like Trapezoidal and Simpson rules, we can derive following composite Boole rule (for m to be a multiple of 4).

$$\int_{x_0}^{x_m} f(x)dx = \int_{x_0}^{x_4} f(x)dx + \int_{x_4}^{x_8} f(x)dx + \cdots + \int_{x_{m-4}}^{x_m} f(x)dx \approx I_{CB}$$

$$I_{CB} = \frac{2}{45}h\left[7f(x_0)+32f(x_1)+12f(x_2)+32f(x_3)+7f(x_4)\right]$$

$$+\frac{2}{45}h\left[7f(x_4)+32f(x_5)+12f(x_6)+32f(x_7)+7f(x_8)\right]$$

$$+$$

$$\vdots$$

$$+\frac{2}{45}h\left[7f(x_{m-4})+32f(x_{m-3})+12f(x_{m-2})+32f(x_{m-1})+7f(x_m)\right]$$

(13.12)

Similarly, we can derive composite Weddle rule.

Example ─── **13.3**

Compute the value of integral $\int_1^2 e^{-x^2} dx$ with the aid of Trapezoidal formula by dividing the interval into ten equal parts.

Ans.
We have to divide the interval [1, 2] into ten equal subintervals, so the spacing $h = \frac{2-1}{10} = 0.1$. The function $f(x) = e^{-x^2}$ is tabulated below on the points $x_0 = 1$, $x_1 = 1.1$, $x_2 = 1.2$, $x_3 = 1.3$, ..., $x_{10} = 2$ correct to six decimal places

x	$f(x) = e^{-x^2}$
1	0.367879
1.1	0.298197
1.2	0.236928
1.3	0.18452
1.4	0.140858
1.5	0.105399
1.6	0.077305
1.7	0.055576
1.8	0.039164
1.9	0.027052
2	0.018316

Composite Trapezoidal Rule (13.9) gives the following result

$$\int_{x_0}^{x_{10}} f(x)dx \approx I_{CT} = \frac{h}{2}\left\{f(x_0)+f(x_{10})+2\left[f(x_1)+f(x_2)+\cdots+f(x_9)\right]\right\}$$

$$I_{CT} = \frac{0.1}{2}\left\{0.367879+0.018316+2\begin{bmatrix}0.298197+0.236928+0.184520+0.140858+0.105399\\ +0.077305+0.055576+0.039164+0.027052\end{bmatrix}\right\}$$

$$= 0.135810$$

The approximate value of integral $\int_{1}^{2} f(x)dx = 0.135810$.

Example ———————————————————————————————— **13.4**

Evaluate the integral $\int_{1}^{2.2}\dfrac{1}{1+2x+x^2}dx$ by Trapezoidal and Simpson rules by dividing the interval into 12 equal subintervals.

Use six decimal places round-off arithmetic.

Ans.
The step size is given by

$$h = \frac{2.2-1}{12} = 0.1$$

We have

$$x_0 = 1, \quad x_1 = 1.1, \quad x_2 = 1.2, \quad x_3 = 1.3, \dots, \quad x_{12} = 2.2$$

Values of the function $f(x) = \dfrac{1}{1+2x+x^2}$ at these points are as follows

x	1	1.1	1.2	1.3	
$f(x)$	0.250000	0.226757	0.206612	0.189036	
x	1.4	1.5	1.6	1.7	
$f(x)$	0.173611	0.160000	0.147929	0.137174	
x	1.8	1.9	2.0	2.1	2.2
$f(x)$	0.127551	0.118906	0.111111	0.104058	0.097656

We compute the integral with the Trapezoidal and Simpson rules.

i) Composite Trapezoidal Rule (13.9)

$$\int_{x_0}^{x_{12}} f(x)dx \approx I_T = \frac{h}{2}\{f(x_0)+f(x_{12})+2[f(x_1)+f(x_2)+\cdots+f(x_{11})]\}$$

$$I_{CT} = \frac{0.1}{2}\{0.25+0.097656+2[0.226757+0.206612+0.189036+\cdots+0.111111+0.104058]\}$$

$$I_{CT} = 0.187657$$

ii) Composite Simpson 1/3 Rule (13.10)

$$\int_{x_0}^{x_{12}} f(x)dx \approx I_{CS1/3} = \frac{h}{3}\left\{\begin{array}{l} f(x_0)+f(x_{12})+2[f(x_2)+f(x_4)+\cdots+f(x_{10})] \\ \qquad\qquad +4[f(x_1)+f(x_3)+\cdots+f(x_{11})]\end{array}\right\}$$

$$I_{CS1/3} = \frac{0.1}{3}\left\{\begin{array}{l} 0.25+0.097656+2[0.206612+0.173611+0.147929+0.127551+0.111111] \\ +4[0.226757+0.189036+0.160000+0.137174+0.118906+0.104058]\end{array}\right\}$$

$$I_{CS1/3} = 0.187500$$

iii) Composite Simpson 3/8 Rule (13.11)

$$\int_{x_0}^{x_{12}} f(x)dx = \frac{3h}{8}\left\{\begin{array}{l} f(x_0)+f(x_{12})+3\left[\begin{array}{l} f(x_1)+f(x_2)+f(x_4)+f(x_5) \\ +f(x_7)+f(x_8)+f(x_{10})+f(x_{11})\end{array}\right] \\ \qquad\qquad +2[f(x_3)+f(x_6)+f(x_9)]\end{array}\right\}$$

$$I_{CS3/8} = \frac{0.3}{8}\left\{\begin{array}{l} 0.25+0.097656+3\left[\begin{array}{l} 0.206612+0.226757+0.173611+0.160000 \\ +0.137174+0.127551+0.111111+0.104058\end{array}\right] \\ \qquad\qquad +2[0.189036+0.147929+0.118906]\end{array}\right\}$$

$$= 0.187500$$

Note: *The exact value of integral is given by*

$$\int_{1}^{2.2} \frac{1}{1+2x+x^2}dx = \int_{1}^{2.2} \frac{1}{(1+x)^2}dx = \left.\frac{-1}{1+x}\right|_{1}^{2.2} = \frac{-1}{3.2}+\frac{1}{2} = 0.1875$$

The composite Simpson rules produce an exact answer.

Example ———————————————————— **13.5**

Use Trapezoidal rule to compute the integral $\int\limits_0^1 2xe^{x^2}dx$ by dividing the interval into 50, 200, 1000, 5000, 9000, 40000, 50000 and 100000 equal subintervals with the aid of C-programming by declaring variables as float variables. Solve the integral analytically to fine exact answer, and hence errors in the values by Trapezoidal rule. Interpret the results.

Ans.

On using C-Program of composite Trapezoidal rule (13.9) for the integral $\int\limits_0^1 2xe^{-x^2}dx$ by declaring variables as float variables, we have following results

Sr. No.	No. of intervals	Value of Integral from Trapezoidal Rule
1	50	1.718759
2	200	1.718311
3	1000	1.718283
4	5000	1.718283
5	9000	1.718284
6	40000	– 0.000106
7	50000	– 0.000175
8	100000	– 0.000087

Note that the Turbo C/C++ compiler is used for the computation.

Exact value of the integral $\int\limits_0^1 2xe^{x^2}dx$ is given by

$$\int\limits_0^1 2xe^{x^2}dx = e^{x^2}\Big|_0^1 = e - 1 = 2.718282 - 1 = 1.718282$$

It is easy to see from the table that if we increase numbers of points beyond a certain limit, our rounding error starts dominating the results. At 5000 points, our result is 1.718283, while at 9000 points, the result is 1.718183. After 40000 points the result is the totally wrong result.

Example ———————————————————— **13.6**

Evaluate the integral $\int\limits_1^2 \sqrt{1+\cos^2 x}\, dx$ with the aid of Trapezoidal and Simpson 1/3 rule with spacing $h = 0.1$.

Ans.
The values of function $f(x) = \sqrt{1+\cos^2 x}$ in the interval [1, 2] with $h = 0.1$ are given below.

x	$1+\cos^2(x)$	$f(x)=\sqrt{1+\cos^2 x}$
1	1.291927	1.136629
1.1	1.205749	1.098066
1.2	1.131303	1.063627
1.3	1.071556	1.03516
1.4	1.028889	1.014342
1.5	1.005004	1.002499
1.6	1.000853	1.000426
1.7	1.016601	1.008266
1.8	1.051621	1.025486
1.9	1.104516	1.05096
2	1.173178	1.083134

It is again worth to mention here that for calculation of trigonometric functions, our calculator/computer must be in radian mode.

i) Trapezoidal Rule (13.9)

$$\int_{x_0}^{x_{10}} f(x)dx \approx I_{CT} = \frac{h}{2}\{f(x_0)+f(x_{10})+2[f(x_1)+f(x_2)+\cdots+f(x_9)]\}$$

$$I_{CT}=\frac{0.1}{2}\left\{1.136629+1.083134+2\begin{bmatrix}1.098066+1.063627+1.03516+1.014342+1.002499\\+1.000426+1.008266+1.025486+1.05096\end{bmatrix}\right\}$$

$$I_{CT}=1.040871$$

ii) Simpson 1/3 Rule (13.10)

$$\int_{x_0}^{x_{10}} f(x)dx \approx I_{CS1/3} = \frac{h}{3}\left\{\begin{matrix}f(x_0)+f(x_{10})+2[f(x_2)+f(x_4)+\cdots+f(x_8)]\\+4[f(x_1)+f(x_3)+\cdots+f(x_9)]\end{matrix}\right\}$$

$$I_{CS1/3}=\frac{0.1}{3}\left\{\begin{matrix}1.136629+1.083134+2[1.063627+1.014342+1.000426+1.025486]\\+4[1.098066+1.03516+1.002499+1.008266+1.05096]\end{matrix}\right\}$$

$$I_{CS1/3}=1.040244$$

Note: *The integral* $\int_1^2 \sqrt{1+\cos^2 x}\ dx$ *provides the arc length* * *of the curve* sin(x) *from x = 1 to x = 2 and its correct value up to seven decimal points is 1.0402463.*

Formula to compute the arc length of a curve $y=f(x)$ from $x=a$ to $x=b$ is $\int_a^b \sqrt{1+\left(\dfrac{dy}{dx}\right)^2}\ dx$.

Example ——————————————————————————— **13.7**

A car is running on a straight road and the velocity of car at a regular interval of 5 minutes is given below

Time (in minutes):	0	5	10	15	20	25	30
Velocity (in km/hr.):	80	88	90	95	93	85	83

Apply Simpson 1/3 rule and Weddle rule to find the distance covered by the car in these 30 minutes.

Ans.

The step size (h) is $\dfrac{1}{12}$ hrs. The distance covered is integral of velocity, so the distance covered by the car is given by

i) Composite Simpson 1/3 rule (13.10)

$$\int_{x_0}^{x_6} f(x)dx = I_{CS1/3} = \frac{h}{3}\left\{f(x_0)+f(x_6)+2\left[f(x_2)+f(x_4)\right]+4\left[f(x_1)+f(x_3)+f(x_5)\right]\right\}$$

$$\int_{x_0}^{x_6} f(x)dx = I_{CS1/3} = \frac{1}{36}\left\{80+83+2\left[90+93\right]+4\left[88+95+85\right]\right\}$$

$$= 44.472222$$

ii) Weddle rule (13.8)

$$\int_{x_0}^{x_6} f(x)dx = I_W$$

$$= \frac{h}{140}\left[41f(x_0)+216f(x_1)+27f(x_2)+272f(x_3)+27f(x_4)+216f(x_5)+41f(x_6)\right]$$

$$= \frac{1}{1680}\left[3280+19008+2430+25840+2511+18360+3403\right]$$

$$= 44.542857$$

Example ——————————————————————————— **13.8**

The volume of the solid of revolution obtained by rotating the curve $y = f(x)$, $a \le x \le b$ about the x-axis is given by

$$V = \int_{a}^{b} \pi y^2 \, dx$$

Determine the volume of the solid obtained by rotating the region bounded by $y = e^x$, $x = 0$, $x = 2$ and the x-axis about the x-axis. Use Simpson 1/3 rule with nine nodes.

Ans.

The volume of solid of revolution is given by

$$V = \int_a^b \pi y^2 \, dx = \int_0^2 \pi \left(e^x\right)^2 dx = \int_0^2 \pi e^{2x} \, dx$$

The values of function $y = e^{2x}$ in the interval [0, 2] with $h = 0.25$ are given below.

x	e^x	e^{2x}
0	1	1
0.25	1.284025	1.648721
0.5	1.648721	2.718282
0.75	2.117	4.481689
1	2.718282	7.389056
1.25	3.490343	12.18249
1.5	4.481689	20.08554
1.75	5.754603	33.11545
2	7.389056	54.59815

On using Simpson 1/3 Rule (13.10), we have

$$\int_0^2 \pi e^{2x} \, dx \approx \frac{0.25}{3} \left\{ \begin{array}{l} f(x_0) + f(x_8) + 2\left[f(x_2) + f(x_4) + f(x_6)\right] \\ \quad + 4\left[f(x_1) + f(x_3) + f(x_5) + f(x_7)\right] \end{array} \right\}$$

$$= \frac{0.25\pi}{3} \left\{ \begin{array}{l} 1 + 54.59815 + 2\left[2.718282 + 7.389056 + 20.08554\right] \\ \quad + 4\left[1.648721 + 4.481689 + 12.18249 + 33.11545\right] \end{array} \right\}$$

$$= 26.808111\pi$$

Example ————————————————————————————————— **13.8**

We have following width measurements of 80 meters wide lake

distance :	0	10	20	30	40	50	60	70	80
width :	0	18	35	54	65	48	32	26	0

Compute the area of lake by using Simpson 1/3 and Boole rules. Assume that the lake is symmetrical about mid-line.

Ans.

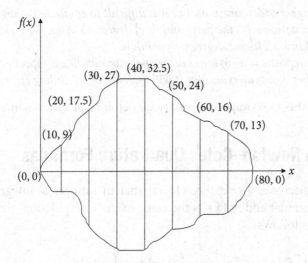

The integral $\int\limits_a^b f(x)dx$ is the area under the curve $f(x)$ from a to b. Now, we approximate the area of lake by computing the half area. For this, we have

x:	0	10	20	30	40	50	60	70	80
$f(x)$:	0	9	17.5	27	32.5	24	16	13	0

i) Simpson 1/3 Rule (13.10)

$$\int\limits_{x_0}^{x_8} f(x)dx \approx I_{s1/3} = \frac{h}{3}\left\{ \begin{array}{l} f(x_0)+f(x_8)+2\left[f(x_2)+f(x_4)+f(x_6)\right] \\ \qquad +4\left[f(x_1)+f(x_3)+f(x_5)+f(x_7)\right] \end{array} \right\}$$

$$\int\limits_0^{80} f(x)dx \approx I_{s1/3} = \frac{10}{3}\left\{0+0+2\left[17.5+32.5+16\right]+4\left[9+27+24+13\right]\right\}$$

$$= 1413.3333$$

The area of the lake is approximately $1413.3333 \times 2 = 2826.667$.

ii) Composite Boole rule (13.12)

$$\int\limits_0^{80} f(x)dx = \frac{2}{45}h\left[\begin{array}{l} 7f(x_0)+32f(x_1)+12f(x_2)+32f(x_3)+14f(x_4) \\ \qquad +32f(x_5)+12f(x_6)+32f(x_7)+7f(x_8) \end{array} \right]$$

$$\int\limits_0^{80} f(x)dx = \frac{4}{9}\left[\begin{array}{l} 0+288+210+864+455 \\ \qquad +768+192+416+0 \end{array} \right]$$

$$= 1419.111$$

The area of the lake is approximately $1419.111 \times 2 = 2838.222$.

Note: *There are following two ways to increase the accuracy during the numerical integration*

1. *To apply higher order formulas, but it is difficult to compute the higher order formulas say approximation with the polynomials of degree 15. Also, high amplitude oscillations are present in such higher degree polynomials.*
2. *To apply composite rules of lower degree polynomials like composite Simpson etc. with a large number of subintervals. But it will increase the rounding error.*

Hence, a balance of these two ways is necessary for better results from numerical integration.

13.3 Errors in Newton–Cotes Quadrature Formulas

We have already discussed [Refer Eq. (13.1)] that in numerical integration, if $P_n(x)$ is interpolating polynomial and $\varepsilon_n(x)$ is the error in the interpolation, then the integration can be computed as follows

$$I = \int_{x_0}^{x_n} f(x)dx = \int_{x_0}^{x_n} \left[P_n(x) + \varepsilon_n(x) \right] dx = \int_{x_0}^{x_n} P_n(x)dx + \int_{x_0}^{x_n} \varepsilon_n(x)dx = I_n + EI_n$$

where I_n is the approximate value of integration, when we approximate the function $f(x)$ with interpolating polynomial $P_n(x)$. The term EI_n gives error in the value of integration. Now, we will compute the errors in numerical integration formulas.

If we approximate a function $f(x)$ with a polynomial of degree n, then the error in interpolating polynomial at any point x is given by Eq. (8.14)

$$\varepsilon_n(x) = \frac{f^{n+1}(\xi)}{(n+1)!}(x-x_0)(x-x_1)\cdots(x-x_n) = \frac{f^{n+1}(\xi)}{(n+1)!}\prod_{i=0}^{n}(x-x_i) \qquad (13.13)$$

where ξ is any point in the given interval. Hence, error in numerical integration formula can be computed as follows

$$EI_n = \int_{x_0}^{x_n} \varepsilon_n(x)dx$$

$$= \int_{x_0}^{x_n} \frac{f^{n+1}(\xi)}{(n+1)!}(x-x_0)(x-x_1)\cdots(x-x_n)dx$$

$$= \frac{f^{n+1}(\xi)}{(n+1)!} \int_{x_0}^{x_n}(x-x_0)(x-x_1)\cdots(x-x_n)dx$$

On using $x = x_0 + sh$ and $x_i = x_0 + ih$, we have

$$dx = h\,ds$$

at $x = x_0$, $s = 0$; and at $x = x_n$, $s = n$.

Error term is given by

$$EI_n = \frac{f^{n+1}(\xi)}{(n+1)!} h^{n+2} \int_0^n s(s-1)\cdots(s-n)ds \qquad (13.14)$$

Now, we will compute the error formulas for different values of n.

13.3.1 Error in Trapezoidal Rule ($n = 1$)

For $n = 1$, Eq. (13.14) is given by

$$EI_T = \frac{f''(\xi)}{2!} h^3 \int_0^1 (s)(s-1)ds = \frac{-h^3 f''(\xi)}{12} \qquad (13.15)$$

This equation gives error in Trapezoidal rule for the integral, $\int_{x_0}^{x_1} f(x)dx$. To compute error in composite Trapezoidal rule, divide the interval into m equal parts, i.e.,

$$\int_a^b f(x)dx = \int_{x_0}^{x_m} f(x)dx = \int_{x_0}^{x_1} f(x)dx + \int_{x_1}^{x_2} f(x)dx + \cdots + \int_{x_{m-1}}^{x_m} f(x)dx$$

Error in the value of integral $\int_{x_0}^{x_m} f(x)dx$ is sum of errors for all these subintervals. The error in the integral $\int_{-x_i}^{x_{i+1}} f(x)dx$ is given by

$$EI_{T_i} = \frac{-h^3 f''(\xi)}{12}$$

where ξ is any point in the given interval $\left[x_i, x_{i+1}\right]$. The exact value of ξ for different intervals is not known, so we can consider only one value of ξ for which the $f''(\xi)$ is maximum. Summing up all the errors and let assuming η is the point of maxima for the function $f''(\xi)$, we have

$$EI_{CT} = \frac{-mh^3 f''(\eta)}{12}$$

On using the number of intervals $m = \dfrac{b-a}{h}$, we have

$$EI_{CT} = \frac{-mh^3 f''(\xi)}{12} = \frac{-(b-a)h^2 f''(\xi)}{12} \qquad (13.16)$$

13.3.2 Error in Simpson 1/3 Rule ($n = 2$)

Equation (13.14) for $n = 2$ is given by

$$EI_{S1/3} = \frac{f'''(\xi)}{3!} h^4 \int_0^2 (s)(s-1)(s-2)ds = 0$$

This result implies that Simpson 1/3 rule is not only exact for quadratic polynomial, but also it is exact for cubic polynomials. So the error term is due to next order polynomial. To compute error term, we have to consider four points such that error term is zero for these points. Without loss of generality, let one more point be, again, x_1 (for simplification only, we can also take a point x_0 or x_2).

$$\varepsilon_3(x) = \frac{f^{iv}(\xi)}{4!}(x-x_0)(x-x_1)^2(x-x_2)$$

$$EI_{S1/3} = \frac{f^{iv}(\xi)}{4!}\int_{x_0}^{x_2}(x-x_0)(x-x_1)^2(x-x_2)dx$$

$$= \frac{f^{iv}(\xi)}{4!}h^5\int_0^2(s)(s-1)^2(s-2)ds$$

$$= \frac{-f^{iv}(\xi)}{90}h^5 \qquad\qquad (13.17)$$

This equation gives error in Trapezoidal rule for the integral, $\int_{x_0}^{x_2} f(x)dx$. To compute error in composite Simpson 1/3 rule, let us divide the interval (x_0, x_m) in m (even) equal parts

$$\int_{x_0}^{x_m} f(x)dx = \int_{x_0}^{x_2} f(x)dx + \int_{x_2}^{x_4} f(x)dx + \cdots + \int_{x_{m-2}}^{x_m} f(x)dx$$

On adding errors for all these intervals, we have

$$EI_{CS1/3} = \frac{-mh^5 f^{iv}(\xi)}{180} = \frac{-(b-a)h^4 f^{iv}(\xi)}{180} \qquad\qquad (13.18)$$

13.3.3 Error in Simpson 3/8 Rule ($n = 3$)

In Simpson 3/8 rule, we approximate our function with cubic interpolation ($n = 3$). From Eq. (13.14), the error is given by

$$EI_{S3/8} = \frac{f^{iv}(\xi)}{4!}h^5\int_0^3(s)(s-1)(s-2)(s-3)ds$$

$$= \frac{-3f^{iv}(\xi)}{80}h^5 \qquad\qquad (13.19)$$

Since the error term in Simpson 3/8 is of equal order as in Simpson 1/3 rule, hence Simpson 3/8 rule has no extra advantage over Simpson 1/3 rule.

Proceeding in a similar manner as in composite Simpson 1/3 rule, we can easily obtain following error term for composite Simpson 3/8 rule

$$EI_{CS3/8} = \frac{-m f^{iv}(\xi)}{80}h^5 \qquad\qquad (13.20)$$

Note that the error terms in Simpson 1/3 rule and Simpson 3/8 rule are of same order (h^4). So, both the formulas are correct up to polynomials of degree 3. Therefore, the Simpson 3/8 rule has no extra advantage over Simpson 1/3 rule.

13.3.4 Error in Boole Rule ($n = 4$)

In Boole rule, we approximate our function with quartic interpolation ($n = 4$). Error Eq. (13.14) is as follows

$$\frac{f^{(v)}(\xi)}{5!}h^6\int_0^4 (s)(s-1)(s-2)(s-3)(s-4)ds = 0$$

This result implies that Boole rule is also exact for a polynomial of degree 5. Consequently, the error term is due to next order polynomial. We can consider six points such that error term is zero for these points. So, without loss of generality, let one more point be x_2. Accordingly, error in Boole rule is given by

$$EI_B = \frac{f^{(vi)}(\xi)}{6!}h^7\int_0^4 (s)(s-1)(s-2)^2(s-3)(s-4)ds$$

$$= \frac{f^{(vi)}(\xi)}{6!}h^7\left(\frac{-128}{21}\right)$$

$$= \left(\frac{-8f^{(vi)}(\xi)}{945}\right)h^7 \tag{13.21}$$

Boole rule is correct up to polynomial of degree 5, so there is no need to derive the formula for $n = 5$.

13.3.5 Error in Weddle Rule ($n = 6$)

In Weddle rule, the error term is as follows

$$\frac{f^{(vii)}(\xi)}{7!}h^8\int_0^6 (s)(s-1)(s-2)(s-3)(s-4)(s-5)(s-6)ds = 0$$

The error term is due to next order polynomial. Let one more point in this case be x_3. The error in Weddle rule is given by

$$EI_W = \frac{f^{(viii)}(\xi)}{8!}h^9\int_0^6 (s)(s-1)(s-2)(s-3)^2(s-4)(s-5)(s-6)ds$$

$$= \frac{f^{(viii)}(\xi)}{8!}h^9\left(\frac{-1296}{5}\right)$$

$$= \left(\frac{-9f^{(viii)}(\xi)}{1400}\right)h^9 \tag{13.22}$$

Example ── **13.9**

Compute the value of integral $\int\limits_{1}^{2} e^{-x^2} dx$ with the aid of Trapezoidal formula by dividing the interval into ten equal parts and find the error bound in the value of the integral.

Ans.

We have already computed the value of integral by the composite Trapezoidal rule (Example 13.3)

$$\int\limits_{1}^{2} e^{-x^2} dx = 0.135810$$

Error in the integral value by composite Trapezoidal rule is given by Eq. (13.16)

$$EI_{CT} = \frac{-mh^3 f''(\xi)}{12}$$

Here, the number of intervals are $m = 10$ and step size $h = 0.1$.

Also $\qquad f(x) = e^{-x^2}$

$$f'(x) = -2xe^{-x^2}$$

$$f''(x) = -2e^{-x^2}(1 - 2x^2)$$

Now, we have to compute maximum value of $f''(x)$ in the interval $[1, 2]$, which is maximum at $x = 1$.

$$\left| f''(x) \right| \leq 2e^{-1}$$

Error bound is given by

$$\left| EI_{CT} \right| = \left| \frac{-mh^3 f''(\xi)}{12} \right| \leq \frac{10(0.1)^3}{12} 2e^{-1} = \frac{1}{600e}$$

Example ── **13.10**

Find the minimum number of interval required to evaluate the integral $\int\limits_{0}^{1} \frac{1}{1 + 2x + x^2} dx$ correct to 4 decimal places by i) Trapezoidal rule and ii) Simpson 1/3 rule.

Ans.

i) Error in the composite Trapezoidal rule (13.16) is as follows

$$EI_{CT} = \frac{-mh^3 f''(\xi)}{12} = \frac{-(b-a)h^2 f''(\xi)}{12}$$

Now $f(x) = \dfrac{1}{(1+x)^2}$

$f'(x) = \dfrac{-2}{(1+x)^3}$

$f''(x) = \dfrac{6}{(1+x)^4}$

The maximum value of $f''(x)$ in the interval $[0, 1]$ is at the point $x = 0$.

$|f''(\xi)| \le 6$

Error bound is given by

$$|EI_{CT}| = \left|\dfrac{-(b-a)h^2 f''(\xi)}{12}\right| \le \left|\dfrac{-(1-0)h^2 6}{12}\right| = \dfrac{h^2}{2}$$

The error bound must be less than 0.00005 for an accuracy of 4 decimal places. It implies

$\dfrac{h^2}{2} \le 0.00005$ or $h \le 0.01$

Hence number of intervals required in Trapezoidal method for accuracy of four decimal places is given by

$$m = \dfrac{b-a}{h} = \dfrac{1-0}{0.01} = 100$$

ii) Error in composite Simpson 1/3 Rule (13.18) is given by

$$EI_{CS1/3} = \dfrac{-mh^5 f^{iv}(\xi)}{180} = \dfrac{-(b-a)h^4 f^{iv}(\xi)}{180}$$

We have, $f^{iv}(x) = \dfrac{120}{(1+x)^6}$, which is maximum at $x = 0$ in the interval $[0, 1]$; this implies

$|f^{iv}(\xi)| \le 120$

The error is given by

$$|EI_{CS1/3}| = \left|\dfrac{-(b-a)h^4 f^{iv}(\xi)}{180}\right| \le \dfrac{2}{3}h^4 \le 0.00005$$

$\Rightarrow h \le 0.09306$

The number of intervals required for an accuracy of four decimal places is given by

$$m = \dfrac{b-a}{h} \ge 11$$

Table 13.1 Newton–Cotes Quadrature Formulas

Newton–Cotes Formulas

Method	Formulation $\left(\int_a^b f(x)dx\right)$	Error Term
Trapezoidal	$\displaystyle\int_{x_0}^{x_1} f(x)dx = \frac{h}{2}\big[f(x_0)+f(x_1)\big]$	$-\dfrac{h^3 f''(\xi)}{12}$
Composite Trapezoidal	$\displaystyle\int_{x_0}^{x_m} f(x)dx = \frac{h}{2}\Big\{f(x_0)+f(x_m)+2\big[f(x_1)+f(x_2)+\cdots+f(x_{m-1})\big]\Big\}$	$-\dfrac{mh^3 f''(\eta)}{12}$
Simpson 1/3 Rule	$\displaystyle\int_{x_0}^{x_2} f(x)dx = \frac{h}{3}\big[f(x_0)+4f(x_1)+f(x_2)\big]$	$-\dfrac{f^{iv}(\xi)}{90}h^5$
Composite Simpson 1/3 Rule	$\displaystyle\int_{x_0}^{x_m} f(x)dx = \frac{h}{3}\left\{ \begin{matrix} f(x_0)+f(x_m)+2\big[f(x_2)+f(x_4)+\cdots+f(x_{m-2})\big] \\ +4\big[f(x_1)+f(x_3)+\cdots+f(x_{m-1})\big] \end{matrix} \right\} \;\; m \text{ must be multiple of 2.}$	$-\dfrac{mh^5 f^{iv}(\xi)}{180}$
Simpson 3/8 Rule	$\displaystyle\int_{x_0}^{x_3} f(x)dx = \frac{3h}{8}\big[f(x_0)+3f(x_1)+3f(x_2)+f(x_3)\big]$	$-\dfrac{3f^{iv}(\xi)}{80}h^5$
Composite Simpson 3/8 Rule	$\displaystyle\int_{x_0}^{x_m} f(x)dx = \frac{3h}{8}\left\{ \begin{matrix} f(x_0)+f(x_m)+3\big[f(x_1)+f(x_2)+f(x_4)+f(x_5)\cdots+f(x_{m-2})+f(x_{m-1})\big] \\ +2\big[f(x_3)+f(x_6)+\cdots+f(x_{m-3})\big] \end{matrix} \right\} \;\; m \text{ must be multiple of 3.}$	$-\dfrac{mf^{iv}(\xi)}{80}h^5$
Boole Rule	$\displaystyle\int_{x_0}^{x_4} f(x)dx = \frac{2}{45}h\big[7f(x_0)+32f(x_1)+12f(x_2)+32f(x_3)+7f(x_4)\big]$	$\left(\dfrac{-8f^{(vi)}(\xi)}{945}\right)h^7$
Weddle Rule	$\displaystyle\int_{x_0}^{x_6} f(x)dx = \frac{h}{140}\big[41f(x_0)+216f(x_1)+27f(x_2)+272f(x_3)+27f(x_4)+216f(x_5)+41f(x_6)\big]$	$\left(\dfrac{-9f^{(viii)}(\xi)}{1400}\right)h^9$

13.4 Gauss Quadrature Formulas

In numerical integration, the value of integral, $\int_a^b f(x)dx$, depends on the values of function $f(x)$ at suitable number of points. It can be written as follows

$$I = \int_a^b f(x)dx \approx \sum_{i=0}^n \lambda_i f(x_i)$$

where $x_0, x_1, \cdots, x_n$ are $(n+1)$ node points in the interval $[a, b]$, and λ_i's are weights given to the values of function $f(x)$ at these node points.

In the case of Newton–Cotes quadrature formula, the node points are already defined (i.e., $x_i = x_0 + ih$; $i = 1, 2, \cdots, n$) and λ_i's are to be computed. A polynomial of degree n is used for approximation to compute these $(n + 1)$ weights λ_i's. Let, there be no restriction on the points x_i's also, then there are total $2n+2$ arbitrary constants [$(n + 1)$ weights λ_i's and $(n+1)$ node points x_i's]. For these $2n+2$ constants, a polynomial of degree $2n+1$ can be utilized to approximate the function. So, higher accuracy can be achieved by these formulas. These methods are known as Gauss quadrature methods. Here, we will discuss following four Gauss quadrature methods

 i)   Gauss–Legendre formula
 ii) Gauss–Chebyshev formula
 iii) Gauss–Laguerre formula
 iv) Gauss–Hermite formula

13.4.1 Gauss–Legendre Formula

In this method, we assume the integral is of the form, $\int_{-1}^1 f(x)dx$. Note that any definite integral, $\int_a^b g(x)dx$ can be converted to the form, $\int_{-1}^1 f(x)dx$ by substituting following formula

$$x = \frac{b-a}{2}t + \frac{b+a}{2}$$

Let the function $f(x)$ in the integral $\int_{-1}^1 f(x)dx$ be approximated by the following polynomial of degree $2n + 1$

$$f(x) \approx a_0 + a_1 x + a_2 x^2 + \cdots + a_{2n+1} x^{2n+1}$$

The integral is approximated by following expression

$$\int_{-1}^1 f(x)dx = \sum_{i=0}^n \lambda_i f(x_i) \tag{13.23}$$

L.H.S. and R.H.S. of Eq. (13.23) are as follows

$$\text{L.H.S.} = \int_{-1}^{1} f(x)dx = \int_{-1}^{1} \left(a_0 + a_1 x + a_2 x^2 + \cdots + a_{2n+1} x^{2n+1} \right) dx$$

$$= 2a_0 + \frac{2}{3} a_2 + \frac{2}{5} a_4 + \cdots$$

$$\text{R.H.S.} = \sum_{i=1}^{n} \lambda_i f(x_i) = \lambda_0 \left(a_0 + a_1 x_0 + a_2 x_0^2 + \cdots + a_{2n+1} x_0^{2n+1} \right)$$

$$+ \lambda_1 \left(a_0 + a_1 x_1 + a_2 x_1^2 + \cdots + a_{2n+1} x_1^{2n+1} \right)$$

$$+ \lambda_2 \left(a_0 + a_1 x_2 + a_2 x_2^2 + \cdots + a_{2n+1} x_2^{2n+1} \right)$$

$$\vdots$$

$$+ \lambda_n \left(a_0 + a_1 x_n + a_2 x_n^2 + \cdots + a_{2n+1} x_n^{2n+1} \right)$$

On comparing both sides, we get

$$\lambda_0 + \lambda_1 + \lambda_2 + \cdots + \lambda_n = 2$$

$$\lambda_0 x_0 + \lambda_1 x_1 + \lambda_2 x_2 + \cdots + \lambda_n x_n = 0$$

$$\lambda_0 x_0^2 + \lambda_1 x_1^2 + \lambda_2 x_2^2 + \cdots + \lambda_n x_n^2 = \frac{2}{3}$$

$$\vdots$$

$$\lambda_0 x_0^{2n+1} + \lambda_1 x_1^{2n+1} + \lambda_2 x_2^{2n+1} + \cdots + \lambda_n x_n^{2n+1} = 0 \qquad (13.24)$$

In general, it is very difficult to solve these $2n + 2$ nonlinear equations. But fortunately, the values of x_i's are zeroes of Legendre orthogonal polynomials (discussed later in the chapter). Once the values of x_i's are known, we can use these values in the system (13.24). We will get linear system for λ_i's, which can be easily solved. Here, we are considering only some particular cases.

1-Point Formula ($n = 0$)
For $n = 0$, we have following two equations from system (13.24)

$$\lambda_0 = 2$$

$$\lambda_0 x_0 = 0$$

Solution is $\lambda_0 = 2$, $x_0 = 0$.

On using these values in Eq. (13.23), we have following Gauss–Legendre 1-point formula

$$\int_{-1}^{1} f(x)dx = \lambda_0 f(x_0) = 2f(0) \qquad (13.25)$$

2-Points Formula ($n = 1$)

For $n = 1$, we have following four Eqs. (13.24)

$$\lambda_0 + \lambda_1 = 2$$
$$\lambda_0 x_0 + \lambda_1 x_1 = 0$$
$$\lambda_0 x_0^2 + \lambda_1 x_1^2 = \frac{2}{3}$$
$$\lambda_0 x_0^3 + \lambda_1 x_1^3 = 0 \qquad\qquad \text{[13.26 (a), (b), (c), (d)]}$$

(d) $- x_1^2$ (b) implies

$$\lambda_0 x_0 \left(x_0^2 - x_1^2 \right) = 0 \qquad \text{or} \qquad \lambda_0 x_0 \left(x_0 - x_1 \right)\left(x_0 + x_1 \right) = 0$$

Now if we select $\lambda_0 = 0$, or $x_0 = 0$, or $x_0 = x_1$, then remaining equations do not hold. Therefore, we have $x_0 = -x_1$. On using this in Eq. (13.26 b) and solving Eqs. (13.26 a) and (13.26 b) simultaneously, we have

$$\lambda_0 = \lambda_1 = 1$$

On substituting the values $x_0 = -x_1$ and $\lambda_0 = \lambda_1 = 1$ in Eq. (13.26 c), we get

$$x_0 = \frac{1}{\sqrt{3}}, \ x_1 = \frac{-1}{\sqrt{3}}$$

Equation (13.23) provides the following Gauss–Legendre 2-points formula

$$\int_{-1}^{1} f(x)dx = \lambda_0 f(x_0) + \lambda_1 f(x_1) = f\left(\frac{-1}{\sqrt{3}}\right) + f\left(\frac{1}{\sqrt{3}}\right) \qquad (13.27)$$

3-Points Formula

For $n = 2$, we have following six equations from system (13.24)

$$\lambda_0 + \lambda_1 + \lambda_2 = 2$$
$$\lambda_0 x_0 + \lambda_1 x_1 + \lambda_2 x_2 = 0$$
$$\lambda_0 x_0^2 + \lambda_1 x_1^2 + \lambda_2 x_2^2 = \frac{2}{3}$$
$$\lambda_0 x_0^3 + \lambda_1 x_1^3 + \lambda_2 x_2^3 = 0$$
$$\lambda_0 x_0^4 + \lambda_1 x_1^4 + \lambda_2 x_2^4 = \frac{2}{5}$$
$$\lambda_0 x_0^5 + \lambda_1 x_1^5 + \lambda_2 x_2^5 = 0$$

Here, we are presenting solution directly without giving any computational details. Readers interested in finding the solutions can continue with the next section.

$$x_0 = -\sqrt{\frac{3}{5}},\ x_1 = 0,\ x_2 = \sqrt{\frac{3}{5}},\qquad \lambda_0 = \frac{5}{9},\ \lambda_1 = \frac{8}{9},\ \lambda_2 = \frac{5}{9}$$

Gauss–Legendre 3-points formula is given by

$$\int_{-1}^{1} f(x)dx = \lambda_0 f(x_0) + \lambda_1 f(x_1) + \lambda_2 f(x_2) = \frac{5}{9} f\left(-\sqrt{\frac{3}{5}}\right) + \frac{8}{9} f(0) + \frac{5}{9} f\left(\sqrt{\frac{3}{5}}\right) \qquad (13.28)$$

On a similar pattern, we can obtain Gauss–Legendre formulas for higher values of n.

Number of points ($n + 1$)	Points (x_i)	Weights (λ_i)	Gauss–Legendre formula $\int_{-1}^{1} f(x)dx = \sum_{i=0}^{n} \lambda_i f(x_i)$
1	0	2	$= 2f(0)$
2	$\pm\sqrt{\frac{1}{3}}$	1	$= f\left(-\sqrt{\frac{1}{3}}\right) + f\left(\sqrt{\frac{1}{3}}\right)$
3	0	$\frac{8}{9}$	$= \frac{5}{9} f\left(-\sqrt{\frac{3}{5}}\right) + \frac{8}{9} f(0) + \frac{5}{9} f\left(\sqrt{\frac{3}{5}}\right)$
	$\pm\sqrt{\frac{3}{5}}$	$\frac{5}{9}$	
4	$\pm\sqrt{\frac{3 - 2\sqrt{6/5}}{7}}$	$\frac{18 + \sqrt{30}}{36}$	$= \frac{18 + \sqrt{30}}{36} f\left(\sqrt{\frac{3 - 2\sqrt{6/5}}{7}}\right) + \frac{18 + \sqrt{30}}{36}$
	$\pm\sqrt{\frac{3 + 2\sqrt{6/5}}{7}}$	$\frac{18 - \sqrt{30}}{36}$	$f\left(-\sqrt{\frac{3 - 2\sqrt{6/5}}{7}}\right) + \frac{18 - \sqrt{30}}{36} f\left(\sqrt{\frac{3 + 2\sqrt{6/5}}{7}}\right)$ $+ \frac{18 - \sqrt{30}}{36} f\left(-\sqrt{\frac{3 + 2\sqrt{6/5}}{7}}\right)$
5	0	$\frac{128}{225}$	$= \frac{322 + 13\sqrt{70}}{900} f\left(-\frac{1}{3}\sqrt{5 - 2\sqrt{\frac{10}{7}}}\right) + \frac{322 + 13\sqrt{70}}{900}$
	$-\frac{1}{3}\sqrt{5 - 2\sqrt{\frac{10}{7}}}$	$\frac{322 + 13\sqrt{70}}{900}$	$f\left(\frac{1}{3}\sqrt{5 - 2\sqrt{\frac{10}{7}}}\right) + \frac{128}{225} f(0) + \frac{322 - 13\sqrt{70}}{900}$
	$\pm\frac{1}{3}\sqrt{5 + 2\sqrt{\frac{10}{7}}}$	$\frac{322 - 13\sqrt{70}}{900}$	$f\left(-\frac{1}{3}\sqrt{5 + 2\sqrt{\frac{10}{7}}}\right) + \frac{322 - 13\sqrt{70}}{900} f\left(\frac{1}{3}\sqrt{5 + 2\sqrt{\frac{10}{7}}}\right)$

Example 13.11

Compute the integral $\int_{-1}^{1}\dfrac{1}{1+x^2}dx$ with the help of Gauss–Legendre 1, 2 and 3-points formulas. Compare the results with exact value.

Ans.

i) Gauss–Legendre 1-point formula (13.25)

$$\int_{-1}^{1} f(x)dx = 2f(0)$$

$$\int_{-1}^{1}\frac{1}{1+x^2}dx = 2(1) = 2$$

ii) Gauss–Legendre 2-points formula (13.27)

$$\int_{-1}^{1} f(x)dx = f\left(\frac{-1}{\sqrt{3}}\right) + f\left(\frac{1}{\sqrt{3}}\right)$$

$$\int_{-1}^{1}\frac{1}{1+x^2}dx = \frac{1}{1+\left(\dfrac{-1}{\sqrt{3}}\right)^2} + \frac{1}{1+\left(\dfrac{1}{\sqrt{3}}\right)^2} = \frac{3}{2} = 1.5$$

iii) Gauss–Legendre 3-points formula (13.28)

$$\int_{-1}^{1} f(x)dx = \frac{5}{9}f\left(-\sqrt{\frac{3}{5}}\right) + \frac{8}{9}f(0) + \frac{5}{9}f\left(\sqrt{\frac{3}{5}}\right)$$

$$\int_{-1}^{1}\frac{1}{1+x^2}dx = \frac{5}{9}\frac{1}{1+\left(-\sqrt{\dfrac{3}{5}}\right)^2} + \frac{8}{9}\frac{1}{1+0} + \frac{5}{9}\frac{1}{1+\left(\sqrt{\dfrac{3}{5}}\right)^2} = \frac{114}{72} = 1.58333$$

Exact solution is given by

$$\int_{-1}^{1}\frac{1}{1+x^2}dx = \tan^{-1}(x)\Big|_{-1}^{1} = \frac{\pi}{2} = 1.571$$

Hence, 3-points formula gives better approximation.

Example —— **13.12**

Use Gauss–Legendre 2-points formula to compute the approximate value of the integral

$$\int_1^2 \sqrt{1+\cos^2 x}\ dx\ .$$

Ans.

To apply Gauss–Legendre formula, we first convert the interval $[1, 2]$ to $[-1, 1]$. For this, we use the following formula

$$x = \frac{b-a}{2}t + \frac{b+a}{2} = \frac{1}{2}t + \frac{3}{2}$$

On substituting this transformation in the integral, we have

$$\int_1^2 \sqrt{1+\cos^2 x}\ dx = \int_{-1}^1 \sqrt{1+\cos^2\left(\frac{1}{2}t+\frac{3}{2}\right)}\ \left(\frac{1}{2}dt\right)$$

$$= \frac{1}{2}\left(\int_{-1}^1 \sqrt{1+\cos^2\left(\frac{1}{2}t+\frac{3}{2}\right)}\ dt\right)$$

This resulting integral is calculated with the help of Gauss–Legendre 2-points formula as follows

$$\frac{1}{2}\left(\int_{-1}^1 \sqrt{1+\cos^2\left(\frac{1}{2}t+\frac{3}{2}\right)}\ dt\right) = \frac{1}{2}\left(f\left(\frac{-1}{\sqrt{3}}\right)+f\left(\frac{1}{\sqrt{3}}\right)\right)$$

$$= \frac{1}{2}\left(\sqrt{1+\cos^2\left(\frac{1}{2}\left(\frac{-1}{\sqrt{3}}\right)+\frac{3}{2}\right)} + \sqrt{1+\cos^2\left(\frac{1}{2}\left(\frac{1}{\sqrt{3}}\right)+\frac{3}{2}\right)}\right)$$

$$= \frac{1}{2}(1.023095667 + 1.060070203)$$

$$= 1.041582935$$

Example —— **13.13**

Solve the integral $\int_0^{0.5} \exp(-x^2)\ dx$ numerically with the help of Gauss–Legendre 3-points formula.

Ans.

To convert the interval $[0, 0.5]$ in to interval $[-1, 1]$, the transformation is given by

$$x = \frac{b-a}{2}t + \frac{b+a}{2} = \frac{1}{4}t + \frac{1}{4}$$

On using this expression in the given integral, we get

$$\int_0^{0.5} \exp(-x^2)\, dx = \int_{-1}^{1} \exp\left(-\left(\frac{t}{4}+\frac{1}{4}\right)^2\right)\frac{1}{4}\, dt$$

$$= \frac{1}{4}\left(\int_{-1}^{1} \exp\left(-\left(\frac{t}{4}+\frac{1}{4}\right)^2\right) dt\right)$$

On applying Gauss–Legendre 3-points formula, we have

$$\frac{1}{4}\left(\int_{-1}^{1} \exp\left(-\left(\frac{t}{4}+\frac{1}{4}\right)^2\right) dt\right) = \frac{1}{4}\left(\frac{5}{9}f\left(-\sqrt{\frac{3}{5}}\right)+\frac{8}{9}f(0)+\frac{5}{9}f\left(\sqrt{\frac{3}{5}}\right)\right)$$

$$= \frac{1}{4}\left(\frac{5}{9}\left(\exp\left(-\left(\frac{1}{4}\left(-\sqrt{\frac{3}{5}}\right)+\frac{1}{4}\right)^2\right)\right)+\frac{8}{9}\left(\exp\left(-\left(\frac{1}{4}\right)^2\right)\right)+\frac{5}{9}\left(\exp\left(-\left(\frac{1}{4}\left(\sqrt{\frac{3}{5}}\right)+\frac{1}{4}\right)^2\right)\right)\right)$$

$$= \frac{1}{4}\left(\frac{5}{9}(0.9968296200)+\frac{8}{9}(0.9394130628)+\frac{5}{9}(0.8213346963)\right)$$

$$= 0.4612812800$$

We have derived composite rules for Trapezoidal and Simpson methods. Similarly, composite rule for Gauss quadrature method can also be obtained. For this, we divide the given interval into subintervals and then apply Gauss quadrature formula to each interval. Here, we are not discussing the composite Gauss quadrature to avoid repetitions.

Now, we will discuss computation of nodes and weights for Gauss-Legendre quadrature formula with the help of Legendre polynomials.

Computation of Weights and Nodes using Legendre Polynomials

Since our major concern is to compute the weights λ_i and nodes x_i for Gauss–Legendre integration $\int_{-1}^{1} f(x)dx = \sum_{i=0}^{n} \lambda_i f(x_i)$ with the help of Legendre polynomials, hence here we are discussing the various properties of Legendre polynomials in brief only.

1. Legendre polynomial $L_n(x)$ of degree n is solution of following second order differential equation

$$(1-x^2)\frac{d^2 y}{dx^2}-2x\frac{dy}{dx}+n(n+1)y = 0$$

2. *Rodrigues formula*
    The Legendre polynomials can be obtained using Rodrigues formula

$$L_n(x)=\frac{1}{2^n\, n!}\frac{d^n}{dx^n}\left((x^2-1)^n\right)$$

3. *Recurrence relation for Legendre polynomials*

We have following recurrence relation for Legendre polynomials

$$(n+1)L_{n+1}(x) = (2n+1)xL_n(x) - nL_{n-1}(x)$$

From Rodrigues formula, we can easily compute following Legendre polynomials

$$L_0(x) = 1 \text{ and } L_1(x) = x$$

On using the recurrence relation for $n = 1$, we have

$$(2)L_2(x) = (3)xL_1(x) - L_0(x) = 3x^2 - 1$$

$$L_2(x) = \frac{1}{2}(3x^2 - 1)$$

Similarly, the recurrence relation provides higher order Legendre polynomials for $n = 2, 3, 4\ldots$ The Legendre polynomials up to order 6 are as follows

$$L_0(x) = 1 \qquad L_1(x) = x \qquad\qquad L_2(x) = \frac{1}{2}(3x^2 - 1)$$

$$L_3(x) = \frac{1}{2}(5x^3 - 3x) \qquad\qquad L_4(x) = \frac{1}{8}(35x^4 - 30x^2 + 3)$$

$$L_5(x) = \frac{1}{8}(63x^5 - 70x^3 + 15x) \qquad\qquad L_6(x) = \frac{1}{16}(231x^6 - 315x^4 + 105x^2 - 5)$$

$$\vdots$$

4. *Orthogonal property of Legendre polynomials*

Here, without going in details, we will only state the orthogonal property of Legendre polynomials.

Legendre polynomials $L_n(x)$ are orthogonal over the interval $[-1, 1]$

$$\int_{-1}^{1} L_m(x)L_n(x)dx = 0; \quad n \neq m$$

When $n = m$, we have

$$\int_{-1}^{1} L_n^2(x)dx = \frac{2}{2n+1}$$

Let us consider that orthogonal polynomials with weight functions $w(x)$ over the interval $[a, b]$. If x_i's $i = 0, 1, 2, \cdots, n$ are zeroes of orthogonal polynomials, then the integral

$\int_a^b w(x)f(x)dx = \sum_{i=0}^n \lambda_i f(x_i)$ is exact for polynomials of degree $\leq (2n+1)$.

Proof:

Let the function $f(x)$ be a polynomial of degree $\leq (2n+1)$.

Let $P_n(x)$ be the interpolating polynomial of degree $\leq n$ which agrees $f(x)$ at $(n+1)$ points

$P_n(x_i) = f(x_i), i = 0, 1, 2, \cdots, n$

Therefore, the function $f(x) - P_n(x)$ has $(n + 1)$ zeroes $x_i, i = 0, 1, 2, \cdots, n$. Let $Q_{n+1}(x)$ be polynomial of degree $(n+1)$ having zeroes x_i's.

We can write $f(x) - P_n(x)$ as product of two polynomials $Q_{n+1}(x)$ and $R_n(x)$, where $R_n(x)$ is a polynomial of degree at most n, i.e.,

$$f(x) - P_n(x) = Q_{n+1}(x)R_n(x) \qquad (13.29)$$

On multiplying (13.29) with $w(x)$ and then integrating from a to b, we have

$$\int_a^b w(x)f(x)dx - \int_a^b w(x)P_n(x)dx = \int_a^b w(x)Q_{n+1}(x)R_n(x)dx$$

The integral on right hand side is zero, if the function $Q_{n+1}(x)$ is orthogonal over the interval $[a, b]$ with respect to weight function, $w(x)$, to all polynomials of degree $\leq n$. Then, we have

$$\int_a^b w(x)f(x)dx = \int_a^b w(x)P_n(x)dx \qquad (13.30)$$

Consider the interpolating polynomial $P_n(x)$ of Lagrange form

$$P_n(x) = \sum_{i=1}^n f(x_i)l_i(x)$$

From Eq. (13.30), we have

$$\int_a^b w(x)f(x)dx = \int_a^b w(x)P_n(x)dx = \int_a^b w(x)\left(\sum_{i=1}^n f(x_i)l_i(x)\right)dx = \sum_{i=0}^n \lambda_i f(x_i) \qquad (13.31)$$

where $\lambda_i = \int_a^b w(x)l_i(x)dx$ are the weights.

As we start with the assumption, that $f(x)$ is a polynomial of degree $\leq (2n+1)$. It proves that the formula has an accuracy of $(2n + 1)$ degree polynomial.

Note: *We prove that if x_i's are zeroes of orthogonal polynomials, then the integral*

$$\int_a^b w(x)f(x)dx = \sum_{i=0}^n \lambda_i f(x_i) \text{ is exact for polynomials of degree } \leq (2n+1).$$

Now, Legendre polynomials are orthogonal with respect to weight function $w(x)=1$ over the interval $[a, b]=[-1, 1]$. So, we will use Legendre polynomials for the calculation of the weights λ_i and nodes x_i for Gauss–Legendre integration $\int_{-1}^1 f(x)dx = \sum_{i=0}^n \lambda_i f(x_i)$

1-Point Formula ($n = 0$)

The nodes x_i's are zeroes of orthogonal polynomials. For $n = 0$, we have Legendre polynomial, $L_1(x) = x$. Therefore, the node x_0 is zero of this polynomial, and it follows $x_0 = 0$. Weight $\lambda_i = \int_a^b w(x)l_i(x)dx$ is given by

$$\lambda_0 = \int_{-1}^1 1\,dx = 2$$

Hence $\int_{-1}^1 f(x)dx = \sum_{i=0}^n \lambda_i f(x_i) = \lambda_0 f(x_0) = 2f(0)$

2-Points Formula ($n = 1$)

For $n = 1$, nodes are zeroes of the Legendre polynomial

$$L_2(x) = \frac{1}{2}\left(3x^2 - 1\right)$$

Nodes are $x_0 = \dfrac{-1}{\sqrt{3}}$, $x_1 = \dfrac{1}{\sqrt{3}}$. The weights are given by the formula $\lambda_i = \int_a^b w(x)l_i(x)dx$, we have

$$\lambda_0 = \int_{-1}^1 l_0(x)dx = \int_{-1}^1 \frac{x-x_1}{x_0-x_1}dx = \frac{-\sqrt{3}}{2}\int_{-1}^1\left(x-\frac{1}{\sqrt{3}}\right)dx = 1$$

$$\lambda_1 = \int_{-1}^1 l_1(x)dx = \int_{-1}^1 \frac{x-x_0}{x_1-x_0}dx = \frac{\sqrt{3}}{2}\int_{-1}^1\left(x+\frac{1}{\sqrt{3}}\right)dx = 1$$

So, Gauss–Legendre 2-points formula is given by

$$\int_{-1}^1 f(x)dx = \lambda_0 f(x_0) + \lambda_1 f(x_1) = f\left(\frac{-1}{\sqrt{3}}\right) + f\left(\frac{1}{\sqrt{3}}\right)$$

3-Points Formula ($n = 2$)

For $n = 2$, we have following equation for Legendre polynomial

$$L_3(x) = \frac{1}{2}\left(5x^3 - 3x\right) = 0$$

Nodes are $x_0 = -\sqrt{\frac{3}{5}},\ x_1 = 0,\ x_2 = \sqrt{\frac{3}{5}}$.

Weights are given by

$$\lambda_0 = \int_{-1}^{1} l_0(x)\,dx = \int_{-1}^{1} \frac{(x - x_1)(x - x_2)}{(x_0 - x_1)(x_0 - x_2)}\,dx = \frac{5}{6}\int_{-1}^{1} x\left(x - \sqrt{\frac{3}{5}}\right)\,dx = \frac{5}{9}$$

$$\lambda_1 = \int_{-1}^{1} l_1(x)\,dx = \int_{-1}^{1} \frac{(x - x_0)(x - x_2)}{(x_1 - x_0)(x_1 - x_2)}\,dx = \frac{-5}{3}\int_{-1}^{1}\left(x^2 - \frac{3}{5}\right)\,dx = \frac{8}{9}$$

$$\lambda_2 = \int_{-1}^{1} l_2(x)\,dx = \int_{-1}^{1} \frac{(x - x_0)(x - x_1)}{(x_2 - x_0)(x_2 - x_1)}\,dx = \frac{5}{6}\int_{-1}^{1} x\left(x + \sqrt{\frac{3}{5}}\right)\,dx = \frac{5}{9}$$

Gauss–Legendre 3-points formula is given by

$$\int_{-1}^{1} f(x)\,dx = \lambda_0 f(x_0) + \lambda_1 f(x_1) + \lambda_2 f(x_2) = \frac{5}{9}f\left(-\sqrt{\frac{3}{5}}\right) + \frac{8}{9}f(0) + \frac{5}{9}f\left(\sqrt{\frac{3}{5}}\right)$$

Similarly, for 4-points Gauss–Legendre formula, we have

$$L_4(x) = \frac{1}{8}\left(35x^4 - 30x^2 + 3\right) = 0$$

Roots of this equation are $\pm\sqrt{\dfrac{3 - 2\sqrt{6/5}}{7}}$, $\pm\sqrt{\dfrac{3 + 2\sqrt{6/5}}{7}}$. Corresponding to these nodes, the weights are $\dfrac{18 + \sqrt{30}}{36}$, $\dfrac{18 + \sqrt{30}}{36}$, $\dfrac{18 - \sqrt{30}}{36}$ and $\dfrac{18 - \sqrt{30}}{36}$ respectively. Gauss–Legendre 4-points formula is as follows

$$\int_{-1}^{1} f(x)\,dx = \frac{18 + \sqrt{30}}{36} f\left(\sqrt{\frac{3 - 2\sqrt{6/5}}{7}}\right) + \frac{18 + \sqrt{30}}{36} f\left(-\sqrt{\frac{3 - 2\sqrt{6/5}}{7}}\right)$$

$$+ \frac{18 - \sqrt{30}}{36} f\left(\sqrt{\frac{3 + 2\sqrt{6/5}}{7}}\right) + \frac{18 - \sqrt{30}}{36} f\left(-\sqrt{\frac{3 + 2\sqrt{6/5}}{7}}\right)$$

We can easily compute similar formulas from other orthogonal functions. Without going into details, here we are discussing some other important integration formulas in the following section.

13.4.2 Gauss–Chebyshev Formula

We have already discussed Chebyshev polynomials in details in Section 11.6.

i)   Chebyshev polynomials $T_n(x) = \cos(n\cos^{-1} x)$, are orthogonal with weight function $w(x) = \dfrac{1}{\sqrt{1-x^2}}$ over the interval $[-1, 1]$, i.e.,

$$\int_{-1}^{1} \frac{T_m(x)T_n(x)}{\sqrt{1-x^2}}\,dx = 0; \quad n \neq m$$

When $n = m$, we have

$$\int_{-1}^{1} \frac{T_n^{\,2}(x)}{\sqrt{1-x^2}}\,dx = \begin{cases} \pi & n = 0 \\[2mm] \dfrac{\pi}{2} & n \neq 0 \end{cases}$$

ii)   Chebyshev polynomials satisfy recurrence relation

$$T_n(x) = 2xT_{n-1}(x) - T_{n-2}(x) \text{ with } T_0(x) = 1 \text{ and } T_1(x) = x$$

Higher order Chebyshev polynomials can be computed as follows

$T_0(x) = 1$

$T_1(x) = x$

$T_2(x) = 2xT_1(x) - T_0(x) = 2x^2 - 1$

$T_3(x) = 2xT_2(x) - T_1(x) = 4x^3 - 3x$

$T_4(x) = 8x^4 - 8x^2 + 1$

$T_5(x) = 16x^5 - 20x^3 + 5x$

$T_6(x) = 32x^6 - 48x^4 + 18x^2 - 1$

$\vdots$

On using the orthogonal property of Chebyshev polynomials, we have Gauss–Chebyshev integration formula as follows

$$\int_{-1}^{1} \frac{1}{\sqrt{1-x^2}} f(x)\,dx = \sum_{i=0}^{n} \lambda_i f(x_i)$$

where nodes x_i are zeroes of orthogonal polynomials $T_i(x)$, which are as follows

1-Point formula: $\qquad T_1(x) = x = 0 \Rightarrow x_0 = 0$

2-Points formula: $\qquad T_2(x) = 2x^2 - 1 = 0 \Rightarrow x_0 = \dfrac{-1}{\sqrt{2}}, \; x_1 = \dfrac{1}{\sqrt{2}}$

3-Points formula: $\qquad T_3(x) = 4x^3 - 3x = 0 \Rightarrow x_0 = \dfrac{-\sqrt{3}}{2}, \; x_1 = 0, \; x_2 = \dfrac{\sqrt{3}}{2}$

$\vdots$

Weights are given by the formula $\lambda_i = \displaystyle\int_{-1}^{1} \dfrac{1}{\sqrt{1-x^2}} l_i(x)dx.$

It can be obtained that all the weights for an n-points Gauss–Chebyshev formula are equal and given by $\dfrac{\pi}{n}$

Using the values of nodes and weights, we can easily have following Gauss–Chebyshev integration formulas.

Number of points $(n+1)$	Points (x_i)	Weights (λ_i)	Gauss–Chebyshev formula $\displaystyle\int_{-1}^{1} \dfrac{1}{\sqrt{1-x^2}} f(x)dx = \sum_{i=0}^{n} \lambda_i f(x_i)$
1	0	π	$= \pi f(0)$
2	$\pm\dfrac{1}{2}\sqrt{2}$	$\dfrac{\pi}{2}$	$= \dfrac{\pi}{2}\left(f\left(-\dfrac{1}{\sqrt{2}}\right) + f\left(\dfrac{1}{\sqrt{2}}\right) \right)$
3	0	$\dfrac{\pi}{3}$	
	$\pm\dfrac{1}{2}\sqrt{3}$		$= \dfrac{\pi}{3}\left(f\left(-\dfrac{\sqrt{3}}{2}\right) + f(0) + f\left(\dfrac{\sqrt{3}}{2}\right) \right)$
4	$\pm\dfrac{1}{2}\sqrt{2-\sqrt{2}}$	$\dfrac{\pi}{4}$	$= \dfrac{\pi}{4}\left(\begin{array}{c} f\left(\dfrac{-1}{2}\sqrt{2-\sqrt{2}}\right) + f\left(\dfrac{1}{2}\sqrt{2-\sqrt{2}}\right) + f\left(\dfrac{-1}{2}\sqrt{2+\sqrt{2}}\right) \\ + f\left(\dfrac{1}{2}\sqrt{2+\sqrt{2}}\right) \end{array} \right)$
	$\pm\dfrac{1}{2}\sqrt{2+\sqrt{2}}$		
5	0	$\dfrac{\pi}{5}$	$= \dfrac{\pi}{5}\left(\begin{array}{c} f\left(\dfrac{-1}{2}\sqrt{\dfrac{1}{2}(5-\sqrt{5})}\right) + f\left(\dfrac{1}{2}\sqrt{\dfrac{1}{2}(5-\sqrt{5})}\right) + f(0) \\ + f\left(\dfrac{-1}{2}\sqrt{\dfrac{1}{2}(5+\sqrt{5})}\right) + f\left(\dfrac{1}{2}\sqrt{\dfrac{1}{2}(5+\sqrt{5})}\right) \end{array} \right)$
	$\pm\dfrac{1}{2}\sqrt{\dfrac{1}{2}(5-\sqrt{5})}$		
	$\pm\dfrac{1}{2}\sqrt{\dfrac{1}{2}(5+\sqrt{5})}$		

Example   **13.14**

Solve the integral $\int_{-1}^{1} \dfrac{x^2}{\sqrt{1-x^2}} dx$ numerically with Gauss–Chebyshev 2-points formula.

Ans.

Gauss–Chebyshev 2-points formula is given by

$$\int_{-1}^{1} \frac{1}{\sqrt{1-x^2}} f(x)dx = \frac{\pi}{2}\left(f\left(-\frac{1}{\sqrt{2}}\right) + f\left(\frac{1}{\sqrt{2}}\right) \right)$$

On using $f(x) = x^2$, we have

$$\int_{-1}^{1} \frac{x^2}{\sqrt{1-x^2}} dx = \frac{\pi}{2}\left(\left(-\frac{1}{\sqrt{2}}\right)^2 + \left(\frac{1}{\sqrt{2}}\right)^2 \right) = \frac{\pi}{2}$$

Example   **13.14**

Approximate the integral $\int_{2}^{4} \dfrac{e^x}{\sin(x)+x} dx$ using Gauss–Chebyshev 3-points formula.

Ans.

To change the limit of the integral, we have

$$x = \frac{b-a}{2}t + \frac{b+a}{2} = t+3$$

On substituting this expression in the given integral, we have

$$\int_{2}^{4} \frac{e^x}{\sin(x)+x} dx = \int_{-1}^{1} \frac{e^{t+3}}{\sin(t+3)+t+3} dt$$

$$= \int_{-1}^{1} \frac{e^{t+3}}{\sin(t+3)+t+3} \frac{\sqrt{1-t^2}}{\sqrt{1-t^2}} dt$$

$$= \int_{-1}^{1} \frac{f(t)}{\sqrt{1-t^2}} dt$$

where $f(t) = \dfrac{e^{t+3}\sqrt{1-t^2}}{\sin(t+3)+t+3}$.

On applying Gauss–Chebyshev 3-points formula, we get

$$\int_{-1}^{1} \frac{f(t)}{\sqrt{1-t^2}} dt = \frac{\pi}{3}\left(f\left(-\frac{\sqrt{3}}{2}\right) + f(0) + f\left(\frac{\sqrt{3}}{2}\right) \right) = 15.986186$$

13.4.3 Gauss–Laguerre Formula

Without going into detail, here we will discuss only a few properties of Laguerre polynomials

$$G_{n+1}(x) = (-1)^{n+1} e^x \frac{d^{n+1}}{dx^{n+1}}\left(e^{-x}x^{n+1}\right)$$

i) Laguerre polynomials are orthogonal with weight function $w(x) = e^{-x}$ over the interval $[0, \infty)$

$$\int\limits_0^\infty e^{-x}G_m(x)G_n(x)dx = 0; \quad n \neq m$$

ii) Laguerre polynomials satisfy recurrence relation

$$(n+1)G_{n+1}(x) = (2n+1-x)G_n(x) - nG_{n-1}(x) \text{ with } G_0(x) = 1 \text{ and } G_1(x) = -x+1.$$

On using this recurrence relation, following Laguerre polynomials can be computed easily

$$G_0(x) = 1$$
$$G_1(x) = -x+1$$
$$G_2(x) = \frac{1}{2}\left(x^2 - 4x + 2\right)$$
$$G_3(x) = \frac{1}{6}\left(-x^3 + 9x^2 - 18x + 6\right)$$
$$\vdots$$

Gauss–Laguerre integration formula is given by

$$\int\limits_0^\infty e^{-x}f(x)dx = \sum_{i=0}^n \lambda_i f(x_i)$$

where nodes x_i are zeroes of orthogonal polynomials $G_i(x)$, which are as follows

1-Point formula: $\quad G_1(x) = -x+1 = 0 \implies x_0 = 1$

2-Points formula: $\quad G_2(x) = \frac{1}{2}\left(x^2 - 4x + 2\right) = 0 \implies x_0 = 0.585786,\ x_1 = 3.41421$

3-Points formula: $\quad G_3(x) = \frac{1}{6}\left(-x^3 + 9x^2 - 18x + 6\right) = 0$

$$\implies x_0 = 0.415775,\ x_1 = 2.29428,\ x_2 = 6.28995$$

$\vdots$

The following formula gives the weights

$$\lambda_i = \int\limits_0^\infty e^{-x}l_i(x)dx$$

In the following table, we are listing nodes and weights for Gauss–Laguerre formulas as well as associated Gauss–Laguerre integration formulas

Number of points $(n + 1)$	Points (x_i)	Weights (λ_i)	Gauss–Laguerre formula $\int_0^\infty e^{-x} f(x)dx = \sum_{i=0}^n \lambda_i f(x_i)$
1	1	1	$= f(1)$
2	0.585786	0.853553	$= 0.853553 f(0.585786) + 0.146447 f(3.41421)$
	3.41421	0.146447	
3	0.415774	0.711093	$= 0.711093 f(0.415774) + 0.278518 f(2.29428)$
	2.29428	0.278518	$+0.0103892 f(6.28994)$
	6.28994	0.0103892	
4	0.322548	0.603154	$= 0.603154 f(0.322548) + 0.357419 f(1.74576)$
	1.74576	0.357419	$+0.0388879 f(4.53662) + 0.00053929 f(9.39507)$
	4.53662	0.0388879	
	9.39507	0.00053929	
5	0.263560	0.521756	$= 0.521756 f(0.263560) + 0.398667 f(1.41340)$
	1.41340	0.398667	$+0.075942 f(3.59642) + 0.0036118 f(7.08581)$
	3.59642	0.075942	$+0.0000234 f(12.6408)$
	7.08581	0.0036118	
	12.6408	0.0000234	

Example ———————————————————————— **13.15** ——

Compute the integral $\int_0^\infty x e^{-x} \sin(x)dx$ with Gauss–Laguerre 2-points formula.

Ans.
Gauss–Laguerre 2-points formula

$$\int_0^\infty e^{-x} f(x)dx = 0.853553\, f\left(0.585786\right) + 0.146447\, f\left(3.41421\right)$$

Here, the function is $f(x) = x\sin(x)$

$$\int_0^\infty e^{-x}\left(x\sin(x)\right)dx = 0.853553\left(0.585786\sin(0.585786)\right) + 0.146447\left(3.41421\sin(3.41421)\right)$$

$$= 0.141800$$

Example ─── **13.16**

Compute the integral $\int_{0}^{\infty} \dfrac{\sin(x)}{x^2} dx$ with Gauss–Laguerre 3-points formula.

Ans.

Gauss–Laguerre 3-points formula

$$\int_{0}^{\infty} e^{-x} f(x)dx = 0.711093 f(0.415774) + 0.278518 f(2.29428) + 0.0103892 f(6.28994)$$

Here, the function $f(x) = \dfrac{e^x \sin(x)}{x^2}$, and approximate value of the integral is given by

$$\int_{0}^{\infty} \dfrac{\sin(x)}{x^2} dx = 2.912237505$$

13.4.4 Gauss–Hermite Formula

Following properties of Hermite polynomials will be useful for the derivation of Gauss–Hermite integration formula.

i)   Hermite polynomials $H_{n+1}(x) = (-1)^{n+1} e^{x^2} \dfrac{d^{n+1}}{dx^{n+1}}\left(e^{-x^2}\right)$, are orthogonal with

weight function $w(x) = e^{-x^2}$ over the interval $(-\infty, \infty)$

$$\int_{-\infty}^{\infty} e^{-x^2} H_m(x)H_n(x)dx = 0; \quad n \neq m$$

ii)   Hermite polynomials satisfy recurrence relation

$$H_{n+1}(x) = 2x H_n(x) - 2nH_{n-1}(x) \text{ with } H_0(x) = 1 \text{ and } H_1(x) = 2x$$

From the recurrence relation, higher-order Hermite polynomials are given by

$$H_2(x) = 4x^2 - 2$$
$$H_3(x) = 8x^3 - 12x$$
$$H_4(x) = 16x^4 - 48x^2 + 12$$
$$H_5(x) = 32x^5 - 160x^3 + 120x$$
$$\vdots$$

On using the orthogonal property of Hermite polynomials, we have Gauss–Hermite integration formula as follows

$$\int_{-\infty}^{\infty} e^{-x^2} f(x)dx = \sum_{i=0}^{n} \lambda_i f(x_i)$$

where nodes x_i are zeroes of orthogonal polynomials $H_i(x)$, and weights are given by the

formula $\lambda_i = \int\limits_{-\infty}^{\infty} e^{-x^2} l_i(x)dx.$

The following table contains the Gauss–Hermite integration formulas up to 5-points

Number of points $(n+1)$	Points (x_i)	Weights (λ_i)	Gauss–Hermite formula $\int\limits_{-\infty}^{\infty} e^{-x^2} f(x)dx = \sum\limits_{i=0}^{n} \lambda_i f(x_i)$
1	0	$\sqrt{\pi}$	$= \sqrt{\pi}f(0)$
2	±0.707107	0.886227	$= 0.886227(f(-0.707107)+f(0.707107))$
3	0	1.18164	$= 0.295409f(-1.22474)+1.18164f(0)$
	±1.22474	0.295409	$+0.295409f(1.22474)$
4	±0.524648	0.804914	$= 0.804914f(-0.524648)+0.804914f(0.524648)$
	±1.65068	0.081313	$+0.081313f(-1.65068)+0.081313f(1.65068)$
5	0	0.945309	$= 0.945309f(0)+0.393619f(-0.958572)$
	±0.958572	0.393619	$+0.393619f(0.958572)+0.019953f(-2.02018)$
	±2.02018	0.019953	$+0.019953f(2.02018)$

Example ────────────────────────────────────── **13.17**

Compute the value of the integral $\int\limits_{-\infty}^{\infty} x^2 e^{-x^2} dx$ with the help of Gauss–Hermite 3-points formula.

Ans.
Gauss–Hermite 3-points formula is as follows

$$\int\limits_{-\infty}^{\infty} e^{-x^2} f(x)dx = 0.295409 f(-1.22474)+1.18164 f(0)+0.295409 f(1.22474)$$

For $f(x) = x^2$, we have

$$\int\limits_{-\infty}^{\infty} e^{-x^2} x^2 dx = 0.295409(-1.22474)^2 +1.18164(0)+0.295409(1.22474)^2$$

$$= 0.88621995$$

13.5 Euler–Maclaurin Formula

Euler–Maclaurin formula is used to compute numerical quadrature and to approximate the sum of finite and infinite series. Some Newton–Cotes formulas like Trapezoidal and Simpson 1/3 can also be deduced using this formula.

Let us derive Euler–Maclaurin formula with the help of Binomial expansion and shift operator $(Ef(x) = f(x+h))$.

$$\frac{1}{E-1}f(x) = \frac{1}{e^{hD}-1}f(x)$$

$$= \frac{1}{hD + \dfrac{(hD)^2}{2!} + \dfrac{(hD)^3}{3!} + \dfrac{(hD)^4}{4!} + \cdots}f(x) \qquad \left(\text{using } E = e^{hD}\right)$$

$$= \frac{1}{hD\left[1 + \dfrac{hD}{2} + \dfrac{(hD)^2}{6} + \dfrac{(hD)^3}{24} + \cdots\right]}f(x)$$

$$= \frac{1}{hD(1+z)}f(x)$$

where $z = \dfrac{hD}{2} + \dfrac{(hD)^2}{6} + \dfrac{(hD)^3}{24} + \cdots$.

On using the expression, $\dfrac{1}{1+z} = 1 - z + z^2 - z^3 + \cdots$, we have

$$\frac{1}{E-1}f(x) = \frac{1}{hD}\left(1 - z + z^2 - z^3 + \cdots\right)f(x)$$

$$= \frac{1}{hD}\left[1 - \left(\frac{hD}{2} + \frac{(hD)^2}{6} + \frac{(hD)^3}{24}\right) + \left(\frac{hD}{2} + \frac{(hD)^2}{6} + \frac{(hD)^3}{24}\right)^2\right.$$

$$\left. - \left(\frac{hD}{2} + \frac{(hD)^2}{6} + \frac{(hD)^3}{24}\right)^3 + \cdots\right]f(x)$$

$$= \frac{1}{hD}\left[1 - \frac{hD}{2} + \frac{(hD)^2}{12} - \frac{(hD)^4}{720} + \frac{(hD)^6}{30240} - \cdots\right]f(x)$$

$$\frac{1}{E-1}f(x) = \left(\frac{1}{hD} - \frac{1}{2} + \frac{hD}{12} - \frac{(hD)^3}{720} + \frac{(hD)^5}{30240} - \cdots\right)f(x) \qquad (13.32)$$

Consider the following expression

$$\frac{E^n-1}{E-1}f(x_0)=\frac{1}{E-1}\Big((E^n-1)f(x_0)\Big)$$

$$=\frac{1}{E-1}\big(f(x_n)-f(x_0)\big)$$

On using Eq. (13.32) in this expression, we get

$$\frac{E^n-1}{E-1}f(x_0)=\left(\frac{1}{hD}-\frac{1}{2}+\frac{hD}{12}-\frac{(hD)^3}{720}+\frac{(hD)^5}{30240}-\cdots\right)\big(f(x_n)-f(x_0)\big)$$

$$=\frac{1}{hD}\big(f(x_n)-f(x_0)\big)-\frac{1}{2}\big(f(x_n)-f(x_0)\big)+\frac{hD}{12}\big(f(x_n)-f(x_0)\big)$$

$$-\frac{(hD)^3}{720}\big(f(x_n)-f(x_0)\big)+\cdots$$

$$=\frac{1}{h}\int_{x_0}^{x_n}f(x)dx-\frac{1}{2}\big(f(x_n)-f(x_0)\big)+\frac{h}{12}\big(f'(x_n)-f'(x_0)\big)$$

$$-\frac{h^3}{720}\big(f'''(x_n)-f'''(x_0)\big)+\cdots \tag{13.33}$$

Also, we have the following expression

$$\frac{E^n-1}{E-1}f(x_0)=\big(1+E+E^2+\cdots+E^{n-1}\big)f(x_0)$$

$$=f(x_0)+f(x_1)+f(x_2)+\cdots+f(x_{n-1})$$

$$=\sum_{i=0}^{n-1}f(x_i) \tag{13.34}$$

On equating Eq(13.33) and Eq(13.34), we have

$$\sum_{i=0}^{n-1}f(x_i)=\frac{1}{h}\int_{x_0}^{x_n}f(x)dx-\frac{1}{2}\big(f(x_n)-f(x_0)\big)+\frac{h}{12}\big(f'(x_n)-f'(x_0)\big)$$

$$-\frac{h^3}{720}\big(f'''(x_n)-f'''(x_0)\big)+\frac{h^5}{30240}\big(f^{(v)}(x_n)-f^{(v)}(x_0)\big)-\cdots$$

$$\sum_{i=0}^{n}f(x_i)=\frac{1}{h}\int_{x_0}^{x_n}f(x)dx+\frac{1}{2}\big(f(x_n)+f(x_0)\big)+\frac{h}{12}\big(f'(x_n)-f'(x_0)\big)$$

$$-\frac{h^3}{720}\big(f'''(x_n)-f'''(x_0)\big)+\frac{h^5}{30240}\big(f^{(v)}(x_n)-f^{(v)}(x_0)\big)-\cdots \tag{13.35}$$

Equation (13.35) can be used to compute the series expansion. But to compute the integral value, rewrite the Eq. (13.35) as follows

$$\int_{x_0}^{x_n} f(x)dx = h\left(\sum_{i=1}^{n-1} f(x_i) + \frac{1}{2}\left(f(x_n) + f(x_0)\right)\right) - \frac{h^2}{12}\left(f'(x_n) - f'(x_0)\right)$$

$$+ \frac{h^4}{720}\left(f'''(x_n) - f'''(x_0)\right) - \frac{h^6}{30240}\left(f^{(v)}(x_n) - f^{(v)}(x_0)\right) + \cdots \qquad (13.36)$$

Note: It is worth mentioning here that from the integral formula (13.36), we can easily derive composite Trapezoidal and Simpson 1/3 rules. In formula (13.36), on neglecting all the derivative terms, we have

$$\int_{x_0}^{x_n} f(x)dx = h\left(\sum_{i=1}^{n-1} f(x_i) + \frac{1}{2}\left(f(x_n) + f(x_0)\right)\right)$$

It is nothing but the composite Trapezoidal rule. Similarly, we can derive Simpson 1/3 rule.

Example 13.18

Find the sum of cubes of first n natural numbers using Euler–Maclaurin formula.

Ans.
Euler–Maclaurin formula for the sum of finite series (Eq. 13.35) is as follows

$$\sum_{i=0}^{n} f(x_i) = \frac{1}{h}\int_{x_0}^{x_n} f(x)dx + \frac{1}{2}\left(f(x_n) + f(x_0)\right) + \frac{h}{12}\left(f'(x_n) - f'(x_0)\right)$$

$$- \frac{h^3}{720}\left(f'''(x_n) - f'''(x_0)\right) + \frac{h^5}{30240}\left(f^{(v)}(x_n) - f^{(v)}(x_0)\right) - \cdots$$

To find the sum of cubes of first n natural numbers, let $f(x) = x^3$ with $x_0 = 0$, $x_n = n$ and $x_i = i$, $i = 0, 1, 2, \ldots, n$. We have $f'(x) = 3x^2$, $f'''(x) = 6$ and higher derivatives terms are zeroes. Also, the step size is $h = 1$. Now, using all these values in Eq. (13.35), we have

$$\sum_{i=0}^{n} x_i^3 = \frac{x^4}{4}\Big|_{x_0}^{x_n} + \frac{1}{2}\left(n^3 + 0\right) + \frac{1}{12}\left(3x_n^2 - 3x_0^2\right) - \frac{1}{720}\left(6 - 6\right)$$

$$\sum_{i=0}^{n} x_i^3 = \frac{n^4}{4} + \frac{1}{2}\left(n^3\right) + \frac{1}{12}\left(3n^2\right) = \left(\frac{n(n+1)}{2}\right)^2$$

Example ———————————————————————————————— **13.19**

Use Euler–Maclaurin formula to prove that

$$\cos(0) + \cos\left(\frac{\pi}{100}\right) + \cos\left(\frac{2\pi}{100}\right) + \cdots + \cos(2\pi) = 1$$

Ans.

We have to prove that

$$\sum_{i=0}^{200} \cos\left(\frac{i\pi}{100}\right) = 1$$

The function is $f(x) = \cos(x)$ with step size $h = \dfrac{\pi}{100}$.

On using the Euler–Maclaurin formula (13.35), we have

$$\sum_{i=0}^{n} f(x_i) = \frac{1}{h}\int_{x_0}^{x_n} f(x)dx + \frac{1}{2}\left(f(x_n) + f(x_0)\right) + \frac{h}{12}\left(f'(x_n) - f'(x_0)\right)$$

$$- \frac{h^3}{720}\left(f'''(x_n) - f'''(x_0)\right) + \frac{h^5}{30240}\left(f^{(v)}(x_n) - f^{(v)}(x_0)\right) - \cdots$$

$$\sum_{i=0}^{200} \cos\left(\frac{i\pi}{100}\right) = \frac{100}{\pi}\int_{0}^{2\pi} \cos(x)dx + \frac{1}{2}\left(\cos(2\pi) + \cos(0)\right) + \frac{\pi}{1200}\left(\sin(2\pi) - \sin(0)\right)$$

$$- \frac{h^3}{720}\left(-\sin(2\pi) + \sin(0)\right) + \frac{h^5}{30240}\left(\sin(2\pi) - \sin(0)\right) - \cdots$$

$$\sum_{i=0}^{200} \cos\left(\frac{i\pi}{100}\right) = \frac{1}{2}(1+1) = 1$$

Example ———————————————————————————————— **13.20**

Use Euler–Maclaurin formula to compute the value of the integral, $\int_{1}^{2} e^{-x^2} dx$. Divide the interval into ten equal parts and use up to third derivative terms only.

Ans.

The spacing is $h = 0.1$, and we have to compute the function $f(x) = e^{-x^2}$ at 11 node points $x_0 = 1$, $x_1 = 1.1$, $x_2 = 1.2$, $x_3 = 1.3$, ... , $x_{10} = 2$

x	$f(x) = e^{-x^2}$
1	0.367879
1.1	0.298197
1.2	0.236928
1.3	0.18452
1.4	0.140858
1.5	0.105399
1.6	0.077305
1.7	0.055576
1.8	0.039164
1.9	0.027052
2	0.018316

For $n = 10$, Euler–Maclaurin formula (13.36) is given by

$$\int_{x_0}^{x_{10}} f(x)dx = h\left(\sum_{i=1}^{9} f(x_i) + \frac{1}{2}\left(f(x_{10}) + f(x_0)\right)\right) - \frac{h^2}{12}\left(f'(x_{10}) - f'(x_0)\right)$$

$$+ \frac{h^4}{720}\left(f'''(x_{10}) - f'''(x_0)\right) - \frac{h^6}{30240}\left(f^{(v)}(x_n) - f^{(v)}(x_0)\right) + \cdots \tag{13.37}$$

The derivative terms up to third order are as follows

$$f(x) = e^{-x^2}$$
$$f'(x) = -2xe^{-x^2} \qquad\qquad f'(1) = -0.735758 \qquad f'(2) = -0.073264$$
$$f''(x) = -2e^{-x^2} + 4x^2 e^{-x^2}$$
$$f'''(x) = 12xe^{-x^2} - 8x^3 e^{-x^2} \quad f'''(1) = 1.471516 \qquad f'''(2) = -0.732640$$

On using these values of derivative terms and $h = 0.1$ in Eq. (13.37), we have

$$\int_{x_0}^{x_{10}} f(x)dx = (0.1)\left(\sum_{i=1}^{9} f(x_i) + \frac{1}{2}\left(f(2) + f(1)\right)\right) - \frac{(0.1)^2}{12}\left(f'(2) - f'(1)\right)$$

$$+ \frac{(0.1)^4}{720}\left(f'''(2) - f'''(1)\right)$$

$$\int_{1}^{2} f(x)dx = (0.1)\left(1.164999 + \frac{1}{2}(0.018316 + 0.367879)\right) - \frac{(0.1)^2}{12}\left(-0.073264 - (-0.735758)\right)$$

$$+ \frac{(0.1)^4}{720}\left(-0.732640 - 1.471516\right)$$

$$\int_{1}^{2} f(x)dx = 0.135810 - 0.000552 - 0.0000003 = 0.1352577$$

Note: *It is worth mentioning here that, if we neglect all the derivative terms then the approximate value of the integral is 0.135810, and the integration formula is the Trapezoidal rule. In this example, we use derivative terms up to third order and result obtained is 0.1352577. The exact value of the integral correct up to nine decimal places is 0.1352572579.*

13.6 Richardson Extrapolation

Richardson extrapolation techniques are used to improve the order of numerical techniques. We consider suitable numerical method with different spacing to improve the accuracy of the method. Here, we will discuss Richardson extrapolation for numerical integration.

Consider a numerical method for the value of the integral $I = \int_a^b f(x)\, dx$ with spacing h has an accuracy of order k. A method is said to be of order k if the order of error term is $k + 1$. Let the approximate value computed by this method be I_1. Hence, we can write the method as follows

$$I = I_1 + a_1 h^{k+1} + a_2 h^{k+2} + \cdots \tag{13.38}$$

where a_i's are the asymptotic error constants.

Suppose we use the same method with spacing $\dfrac{h}{2}$ and computed value is I_2. Then, in that case, we have

$$I = I_2 + a_1 \left(\frac{h}{2}\right)^{k+1} + a_2 \left(\frac{h}{2}\right)^{k+2} + \cdots \tag{13.39}$$

To increase the order of method, multiply Eq. (13.39) with 2^{k+1} and then subtract it from Eq. (13.38), we get

$$I = \frac{\left(2^{k+1} I_2 - I_1\right)}{\left(2^{k+1} - 1\right)} + b_2 h^{k+2} + b_3 h^{k+3} + \cdots$$

The value of the integral is given by

$$I = \int_a^b f(x)\, dx = \frac{\left(2^{k+1} I_2 - I_1\right)}{\left(2^{k+1} - 1\right)} \tag{13.40}$$

This scheme is of order at least $k + 1$.

This process of finding higher-order formula from two different spacing is called Richardson extrapolation.

Here we elaborate the technique for numerical integration, but we can apply this technique to other numerical methods also.

Example ———————————————— 13.21

Use Richardson extrapolation to obtain higher order technique from Trapezoidal rule. Compute the value of the integral $\int_1^2 e^{-x^2} dx$ with the help of this improved technique. Further prove that this improved scheme is equivalent to Simpson 1/3 rule.

Ans.
We have already obtained that the error term in the Trapezoidal rule with spacing h is of order h^2. Let the approximate value computed by the Trapezoidal method be T_1. We can write the method as follows

$$I = T_1 + a_1 h^2 + a_2 h^4 + \cdots \qquad (13.41)$$

Consider the method with spacing, $\dfrac{h}{2}$, and let computed value be T_2

$$I = T_2 + a_1 \left(\frac{h}{2}\right)^2 + a_2 \left(\frac{h}{2}\right)^4 + \cdots \qquad (13.42)$$

Eliminating the term h^2 from Eqs. (13.41) and (13.42), we have

$$I = \frac{2^2 T_2 - T_1}{2^2 - 1} + b_2 h^4 + \cdots \qquad (13.43)$$

The formula, $I = \dfrac{2^2 T_2 - T_1}{2^2 - 1}$ is a higher-order formula, and the order of error term is at least four.

Now, we will use the formula (13.43) to compute the value of the integral, $\int_1^2 e^{-x^2} dx$.

$$T_1 = \frac{h}{2}\big(f(1) + f(2)\big) = \frac{1}{2}(0.367879 + 0.018316) = 0.193098 \qquad (h = 1)$$

$$T_2 = \frac{h}{2}\big(f(1) + 2f(1.5) + f(2)\big)$$

$$= \frac{1}{4}(0.367879 + 2(0.105399) + 0.018316) \qquad (h = 1/2)$$

$$= 0.149248$$

Value of integral with improved scheme is as follows

$$I = \frac{2^2 T_2 - T_1}{2^2 - 1} = \frac{4(0.149248) - 0.193098}{3} = 0.134631$$

Now, we will deduce that the formula (13.43) is equivalent to Simpson 1/3 rule. Trapezoidal rules with spacing h and $h/2$ are given by

$$T_1 = \frac{h}{2}(f(a)+f(b)) \qquad\qquad \text{(interval spacing is } h)$$

$$T_2 = \frac{h}{4}\left(f(a)+2f\left(\frac{a+b}{2}\right)+f(b)\right) \quad \text{(interval spacing is } h/2)$$

The improved formula (13.43) is given by

$$I = \frac{2^2 T_2 - T_1}{2^2 - 1} = \frac{1}{3}\left(4\left(\frac{h}{4}\left(f(a)+2f\left(\frac{a+b}{2}\right)+f(b)\right)\right)-\left(\frac{h}{2}(f(a)+f(b))\right)\right)$$

$$= \frac{h}{6}\left(f(a)+4f\left(\frac{a+b}{2}\right)+f(b)\right)$$

This expression is Simpson 1/3 formula with spacing $h/2$.

Note: We have already discussed higher order integration formula by interpolating several data points. For example, we interpolate seven data points with 6th order polynomial in Weddle rule. The order of global truncation error for Weddle rule is much higher, but we rarely use it for numerical integration. Such types of rules do not produce good numerical approximations, as the rounding error is large and often they are a victim of polynomial wiggle (oscillate widely between the samples).

Therefore, instead of using higher order numerical integration formulas, we use iterative approaches such as Romberg integration for lower order integration techniques like Trapezoidal and Simpson, etc. to reduce the truncation errors of these techniques.

13.7 Romberg Integration

Romberg integration technique is an iterative technique. It uses repeated applications of Richardson extrapolation for numerical integration. In Romberg integration, we use a numerical method with different spacing to improve the accuracy of the method.

The Richardson formula (13.40) is given by

$$I = \int_a^b f(x)dx = \frac{\left(2^{k+1} I_2 - I_1\right)}{\left(2^{k+1} - 1\right)}$$

where I_1 is numerical integration with spacing h and I_2 is numerical integration with spacing $h/2$. In Romberg integration, we will use Richardson scheme successively to obtain further higher order scheme.

For example, we will compute the integral with spacing h, $h/2$, $h/4$, $h/8$, ... from any method like Trapezoidal or Simpson method. Let these values be I_1^0, I_2^0, I_3^0, I_4^0, ... Here the subscript denotes the integration with different spacing (subscript is 1 for spacing h, subscript 2 for spacing $h/2$, so on), and superscript denotes the iteration number (the superscript 0 denotes initial approximation for Romberg integration).

We have error formulas for composite Newton Cotes and Gauss quadrature formulas, which can be used to determine the number of subintervals needed to guarntee a given accuracy (e.g. 13.10). But, in many cases, the computations required to determine number of subintervals is cumbersome and difficult enough to avoid. Also the upper bound of error sometimes provides much greater number of subintervals than actually necessary. We can use Romberg integration for higher accuracy, and also we can stop iterations as the desired accuracy is obtained.

We apply Richardson scheme for each set $(h, h/2)$, $(h/2, h/4)$, $(h/4, h/8)$, $(h/8, h/16)$ ... to obtained the values of I_1^1, I_2^1, $(I, ...$ Then, Richardson scheme is applied further by using these obtained values. This process is repeated till only one value is remained.

For easy understanding and to keep all these computations at one place, we can build a table of the form

I_1^0

$I_2^0 \quad I_2^1$

$I_3^0 \quad I_3^1 \quad I_3^2$

$I_4^0 \quad I_4^1 \quad I_4^2 \quad I_4^3$

$I_5^0 \quad I_5^1 \quad I_5^2 \quad I_5^3 \quad I_5^4$

Let us discuss this scheme for composite Trapezoidal and Simpson schemes for numerical integrations.

Trapezoidal Rule:
The composite Trapezoidal scheme is given by

$$I = I_T + a_1 h^2 + a_2 h^4 + \cdots$$

Let the computed values for the integral with spacing h, $h/2$, $h/4$, $h/8$, ... using composite Trapezoidal scheme be I_1^0, I_2^0, I_3^0, I_4^0, ... As discussed in Example 13.21, the Richardson scheme for composite trapezoidal rule (13.43) provides first iteration as follows

$$I_{k+1}^1 = \frac{\left(4 I_{k+1}^0 - I_k^0\right)}{(4-1)} + b_2 h^4 + \cdots, \quad k = 1, 2, 3, \cdots$$

This expression provides the values of the first approximations $\left(I_1^1, I_2^1, I_3^1, ...\right)$ of Romberg integration. We can further use these values to obtain the higher approximations. In general, the jth iteration is given by

$$I_{k+1}^j = \frac{\left(4^j I_{k+1}^{j-1} - I_k^{j-1}\right)}{\left(4^j - 1\right)}, \quad k = 1, 2, 3, ... \tag{13.44}$$

Simpson Rule:
The composite Simpson scheme is given by

$$I = I_S + a_1 h^4 + a_2 h^6 + \cdots$$

Proceeding in a similar manner as in Trapezoidal method, the jth iteration of Romberg integration for composite Simpson rule is given by

$$I_{k+1}^{j} = \frac{\left(4^{j+1} I_{k+1}^{j-1} - I_{k}^{j-1}\right)}{\left(4^{j+1} - 1\right)}, \quad k = 1, 2, 3, \dots \tag{13.45}$$

Example ——————————————————————————— 13.22 —

Compute the value of integral $I = \int\limits_0^1 \dfrac{1}{1+x} dx$ with the help of Romberg integration. Use only four initial values of integral with the Trapezoidal rule.

Ans.

First, we will compute the four initial approximation to the integral $I = \int\limits_0^1 \dfrac{1}{1+x} dx$ by using Trapezoidal rule with spacing $h = 1$, $h/2 = 0.5$, $h/4 = 0.25$, $h/8 = 0.125$. These values are listed in following Table.

$n = 1$ $h = 1$	$I_1^0 = \dfrac{h}{2}\left(f(0) + f(1)\right) = \dfrac{1}{2}\left(1 + \dfrac{1}{2}\right) = 0.75$
$n = 2$ $h/2 = 0.5$	$I_2^0 = \dfrac{h}{2}\left(f(0) + 2f(0.5) + f(1)\right) = \dfrac{0.5}{2}\left(1 + 2\left(\dfrac{1}{1.5}\right) + \dfrac{1}{2}\right) = 0.708333$
$n = 4$ $h/4 = 0.25$	$I_3^0 = \dfrac{h}{2}\left(f(0) + 2\left(f(0.25) + f(0.5) + f(0.75)\right) + f(1)\right) = 0.697024$
$n = 8$ $h/8 = 0.125$	$I_4^0 = 0.694122$

The Romberg integration formula (13.44) is given by

$$I_{k+1}^{j} = \frac{\left(4^{j} I_{k+1}^{j-1} - I_{k}^{j-1}\right)}{\left(4^{j} - 1\right)}, \quad k = 1, 2, 3, \dots$$

On using this formula and initial values from the table, we can easily compute the iterations of Romberg integration. For the first iteration, we have

$$I_2^1 = \frac{\left(4 I_2^0 - I_1^0\right)}{3} = 0.694444$$

$$I_3^1 = \frac{\left(4 I_3^0 - I_2^0\right)}{3} = 0.693254$$

$$I_4^1 = \frac{\left(4 I_4^0 - I_3^0\right)}{3} = 0.693155$$

Similarly, the second iteration is given by

$$I_3^2 = \frac{\left(16I_3^1 - I_2^1\right)}{15} = 0.693175$$

$$I_4^2 = \frac{\left(16I_4^1 - I_3^1\right)}{15} = 0.693148$$

The last iteration is as follows

$$I_4^3 = \frac{\left(64I_4^2 - I_3^2\right)}{63} = 0.693147$$

This value of the integral is correct up to 6 decimal places. In table form, we can list the iterations as follows

Spacing	Value of integral using Trapezoidal rule	1st iteration of Romberg Integration	2nd iteration of Romberg Integration	3rd iteration of Romberg Integration
$h = 1$	0.750000			
$h/2 = 0.5$	0.708333	0.694444		
$h/4 = 0.25$	0.697024	0.693254	0.693175	
$h/8 = 0.125$	0.694122	0.693155	0.693148	0.693147

Note that last two iterations (0.693148 and 0.693147) matches upto five decimal points. So the result 0.693147 is at least correct upto five decimal places.

Example 13.23

Use Trapezoidal formula to compute the value of integral $\int_1^2 e^{-x^2} dx$ with $n = 1, 2$ and 4.

Then use the Romberg integration to improve these values.

Ans.
First, we will compute initial values with the help of Trapezoidal formula as follows

$n = 1$	$I_1^0 = \dfrac{1}{2}\big(f(1) + f(2)\big) = 0.193098$
$n = 2$	$I_2^0 = \dfrac{0.5}{2}\big(f(1) + 2f(1.5) + f(2)\big) = 0.149248$
$n = 4$	$I_3^0 = \dfrac{0.25}{2}\big(f(1) + 2\big(f(1.25) + f(1.5) + f(1.75)\big) + f(2)\big) = 0.138720$

The following table contains the iterations of the Romberg integration (13.44)

Spacing	Value of integral using Trapezoidal rule	1st iteration of Romberg Integration	2nd iteration of Romberg Integration
$h = 1$	0.193098		
$h/2 = 0.5$	0.149248	0.134632	
$h/4 = 0.25$	0.138720	0.135210	0.135249

Hence, the value of integral $\int_{1}^{2} e^{-x^2} dx$ from Romberg integration is 0.135249.

Example ── **13.24**

Use Simpson formula to compute the value of integral $\int_{1}^{2} e^{-x^2} dx$ with $n = 2, 4$ and 8. Then use the Romberg integration to improve these values.

Ans.

First, we will compute initial values with the help of Simpson formula as follows

$n = 2$	$I_1^0 = \dfrac{0.5}{2}\big(f(1) + 4f(1.5) + f(2)\big) = 0.134632$
$n = 4$	$I_2^0 = \dfrac{0.25}{2}\big(f(1) + 4f(1.25) + 4f(1.75) + 2f(1.5) + f(2)\big) = 0.135210$
$n = 8$	$I_3^0 = \dfrac{0.125}{2}\begin{pmatrix} f(1) + 4\big(f(1.125) + f(1.375) + f(1.625) + f(1.8755)\big) \\ +2\big(f(1.25) + f(1.5) + f(1.75)\big) + f(2) \end{pmatrix} = 0.135254$

The following table contains the iterations of the Romberg integration (13.45)

Spacing	Values of integral using Simpson rule	1st iteration of Romberg Integration	2nd iteration of Romberg Integration
$h = 0.5$	0.134632		
$h/2 = 0.25$	0.135210	0.135249	
$h/4 = 0.125$	0.135254	0.135257	0.135257

Hence, the value of integral $\int_{1}^{2} e^{-x^2} dx$ from Romberg integration is 0.135257. It is correct up to six decimal points. Note that the correct value of the integral up to ten decimal points is 0.1352572579.

Table 13.2 Numerical Techniques for Integration

Method	Newton–Cotes Formulas — Formulation $\left(\int_a^b f(x)\,dx\right)$	Order of Approximating Polynomial	Order of Error term	Programming
Trapezoidal	$\displaystyle\int_{x_0}^{x_1} f(x)dx \approx I_T = \frac{h}{2}\left[f(x_0)+f(x_1)\right]$	1	2	Easy
Composite Trapezoidal	$\displaystyle\int_{x_0}^{x_m} f(x)dx \approx I_{CT} = \frac{h}{2}\left\{f(x_0)+f(x_m)+2\left[f(x_1)+f(x_2)+\cdots+f(x_{m-1})\right]\right\}$	1	2	Easy
Simpson 1/3 Rule	$\displaystyle\int_{x_0}^{x_2} f(x)dx \approx I_{S1/3} = \frac{h}{3}\left[f(x_0)+4f(x_1)+f(x_2)\right]$	3	4	Easy
Composite Simpson 1/3 Rule	$\displaystyle\int_{x_0}^{x_m} f(x)dx \approx I_{CS1/3} = \frac{h}{3}\left\{f(x_0)+f(x_m)+2\left[f(x_2)+f(x_4)+\cdots+f(x_{m-2})\right]\right.$ $\left.+4\left[f(x_1)+f(x_3)+\cdots+f(x_{m-1})\right]\right\}$ m must be multiple of 2.	3	4	Easy
Simpson 3/8 Rule	$\displaystyle\int_{x_0}^{x_3} f(x)dx \approx I_{S3/8} = \frac{3h}{8}\left[f(x_0)+3f(x_1)+3f(x_2)+f(x_3)\right]$	3	4	Easy
Composite Simpson 3/8 Rule	$\displaystyle\int_{x_0}^{x_m} f(x)dx \approx I_{CS3/8} = \frac{3h}{8}\left\{f(x_0)+f(x_m)+3\left[f(x_1)+f(x_2)+f(x_4)+f(x_5)\cdots\right.\right.$ $\left.\left.+f(x_{m-2})+f(x_{m-1})\right]+2\left[f(x_3)+f(x_6)+\cdots+f(x_{m-3})\right]\right\}$ m must be multiple of 3.	3	4	Easy
Boole Rule	$\displaystyle\int_{x_0}^{x_4} f(x)dx \approx I_B = \frac{2}{45}h\left[7f(x_0)+32f(x_1)+12f(x_2)+32f(x_3)+7f(x_4)\right]$	5	6	Easy
Weddle Rule	$\displaystyle\int_{x_0}^{x_6} f(x)dx \approx I_W = \frac{h}{140}\left[41f(x_0)+216f(x_1)+27f(x_2)+272f(x_3)+27f(x_4)\right.$ $\left.+216f(x_5)+41f(x_6)\right]$	7	8	Easy

Gauss–Legendre Formulas

Method	Formulation $\left(\int\limits_{-1}^{1} f(x)dx \approx I_{GL} \right)$	Order of method	Order of Error term	Programming
1 - point	$\int\limits_{-1}^{1} f(x)dx \approx I_{GL1} = 2f(0)$	1	2	Easy
2 - point	$\int\limits_{-1}^{1} f(x)\,dx \approx I_{GL2} = f\left(-\sqrt{\dfrac{1}{3}}\right) + f\left(\sqrt{\dfrac{1}{3}}\right)$	3	4	Easy
3 - point	$\int\limits_{-1}^{1} f(x)\,dx \approx I_{GL3} = \dfrac{5}{9}f\left(-\sqrt{\dfrac{3}{5}}\right) + \dfrac{8}{9}f(0) + \dfrac{5}{9}f\left(\sqrt{\dfrac{3}{5}}\right)$	5	6	Easy
4 - point	$\int\limits_{-1}^{1} f(x)dx \approx I_{GL4} = \dfrac{18+\sqrt{30}}{36}f\left(\sqrt{\dfrac{3-2\sqrt{6/5}}{7}}\right) + \dfrac{18+\sqrt{30}}{36}f\left(-\sqrt{\dfrac{3-2\sqrt{6/5}}{7}}\right)$ $+\dfrac{18-\sqrt{30}}{36}f\left(\sqrt{\dfrac{3+2\sqrt{6/5}}{7}}\right) + \dfrac{18-\sqrt{30}}{36}f\left(-\sqrt{\dfrac{3+2\sqrt{6/5}}{7}}\right)$	7	8	Easy

Euler–Maclaurin Formula

Method	Formulation	Order of method	Order of Error term	Programming
For Series Summation	$\sum\limits_{i=0}^{n} f(x_i) = \dfrac{1}{h}\int\limits_{x_0}^{x_n} f(x)dx + \dfrac{1}{2}\left(f(x_n) + f(x_0)\right) + \dfrac{h}{12}\left(f'(x_n) - f'(x_0)\right)$ $-\dfrac{h^3}{720}\left(f'''(x_n) - f'''(x_0)\right) + \dfrac{h^5}{30240}\left(f^{(v)}(x_n) - f^{(v)}(x_0)\right) - \cdots$	General	General	Programming not possible as it requires derivative terms
For Integration	$\int\limits_{x_0}^{x_n} f(x)dx = h\left(\sum\limits_{i=1}^{n-1} f(x_i) + \dfrac{1}{2}\left(f(x_n) + f(x_0)\right)\right) - \dfrac{h^2}{12}\left(f'(x_n) - f'(x_0)\right)$ $+\dfrac{h^4}{720}\left(f'''(x_n) - f'''(x_0)\right) - \dfrac{h^6}{30240}\left(f^{(v)}(x_n) - f^{(v)}(x_0)\right) + \cdots$	General	General	Programming not possible as it requires derivative terms

13.8 Double Integrals

The numerical

In this section, we will extend Trapezoidal and Simpson rules to evaluate numerical approximation for double integral, $I = \int_c^d \int_a^b f(x,y)dxdy$.

Let $f(x, y)$ be a real-valued function defined on the region, $a \le x \le b$, $c \le y \le d$. Let the interval $[a, b]$ contain $(m+1)$ equispaced points $a = x_0, x_1, \cdots, x_m = b$ with spacing h and the interval $[c, d]$ contains $(n+1)$ equispaced points $c = y_0, y_1, \cdots, y_n = d$ with spacing k.

13.8.1 Trapezoidal Rule

To estimate the integral, $I = \int_c^d \int_a^b f(x,y)dxdy$, first we will apply Trapezoidal rule for variable x keeping variable y constant. Then, Trapezoidal rule is used for variable y.

$$I = \int_c^d \int_a^b f(x,y)\,dxdy = \int_{y_0}^{y_n}\left(\int_{x_0}^{x_m} f(x,y)dx \right)dy \approx I_T$$

$$I_T = \int_{y_0}^{y_n}\left(\frac{h}{2}\left[f(x_0,y)+f(x_m,y)+2\{f(x_1,y)+f(x_2,y)+\cdots+f(x_{m-1},y)\} \right] \right)dy$$

$$= \frac{h}{2}\left[\int_{y_0}^{y_n} f(x_0,y)dy + \int_{y_0}^{y_n} f(x_m,y)dy +2\left\{ \int_{y_0}^{y_n} f(x_1,y)dy + \int_{y_0}^{y_n} f(x_2,y)dy +\cdots \int_{y_0}^{y_n} f(x_{m-1},y)dy \right\} \right]$$

$$= \frac{hk}{4}\left[\begin{array}{l} f(x_0,y_0)+f(x_0,y_n)+2\{f(x_0,y_1)+f(x_0,y_2)+\cdots+f(x_0,y_{n-1})\} \\ +f(x_m,y_0)+f(x_m,y_n)+2\{f(x_m,y_1)+f(x_m,y_2)+\cdots+f(x_m,y_{n-1})\} \\ \left[\begin{array}{l} f(x_1,y_0)+f(x_1,y_n)+2\{f(x_1,y_1)+f(x_1,y_2)+\cdots+f(x_1,y_{n-1})\}+ \\ f(x_2,y_0)+f(x_2,y_n)+2\{f(x_2,y_1)+f(x_2,y_2)+\cdots+f(x_2,y_{n-1})\}+ \\ \vdots \\ +f(x_{m-1},y_0)+f(x_{m-1},y_n)+2\{f(x_{m-1},y_1)+f(x_{m-1},y_2)+\cdots+f(x_{m-1},y_{n-1})\} \end{array} \right] \end{array} \right]$$

$$\qquad\qquad\qquad\qquad (13.46)$$

Example ── **13.25**

Apply Trapezoidal rule to compute the integral $\int_0^{0.6}\int_0^{0.5} e^{x+y}dx\,dy$ by taking two equal subintervals for the variable x and three equal subintervals for the variable y. Also, find the exact solution.

Ans.

We have

$$f(x,y) = e^{x+y}$$
$$m = 2, n = 3$$
$$x_0 = 0, \; x_1 = 0.25, \; x_2 = 0.5$$
$$y_0 = 0, \; y_1 = 0.2, \; y_2 = 0.4, \; y_3 = 0.6$$

On using Trapezoidal rule (13.46) with $h = 0.25$ and $k = 0.2$, we have the following approximation for the integral

$$\int_0^{0.6} \left(\int_0^{0.5} e^{x+y} \, dx \right) dy = \frac{1}{80} \begin{bmatrix} f(x_0,y_0) + f(x_0,y_3) + 2\{f(x_0,y_1) + f(x_0,y_2)\} \\ + f(x_2,y_0) + f(x_2,y_3) + 2\{f(x_2,y_1) + f(x_2,y_2)\} \\ +2\left[f(x_1,y_0) + f(x_1,y_3) + 2\{f(x_1,y_1) + f(x_1,y_2)\} \right] \end{bmatrix}$$

$$= 0.537887$$

The exact solution is given by

$$\int_0^{0.6} \left(\int_0^{0.5} e^{x+y} \, dx \right) dy = \left(\int_0^{0.5} e^x \, dx \right) \left(\int_0^{0.6} e^y \, dy \right) = \left(e^{0.5} - 1 \right) \left(e^{0.6} - 1 \right) = 0.533325953$$

Example ———————————————————————————————— **13.26**

Calculate the approximation to the double integral $\int_0^1 \int_1^2 \dfrac{2y}{\left(1+x^2\right)} \, dx \, dy$ with the aid of Trapezoidal method by dividing both the intervals in 4 equal sub-intervals ($h = k = 0.25$)

Ans.

Dividing both the intervals in 4 equal parts (i.e., $m = n = 4$), we have

$$x_0 = 1, \; x_1 = 1.25, \; x_2 = 1.5, \; x_3 = 1.75, \; x_4 = 2$$
$$y_0 = 0, \; y_1 = 0.25, \; y_2 = 0.5, \; y_3 = 0.75, \; y_4 = 1$$

By using Trapezoidal rule (13.46) with $h = k = 0.25$, we have

$$\int_0^1 \left(\int_1^2 \frac{2y}{\left(1+x^2\right)} \, dx \right) dy = \frac{1}{64} \begin{bmatrix} f(x_0,y_0) + f(x_0,y_4) + 2\{f(x_0,y_1) + f(x_0,y_2) + f(x_0,y_3)\} \\ + f(x_4,y_0) + f(x_4,y_4) + 2\{f(x_4,y_1) + f(x_4,y_2) + f(x_4,y_3)\} \\ +2 \begin{bmatrix} f(x_1,y_0) + f(x_1,y_4) + 2\{f(x_1,y_1) + f(x_1,y_2) + f(x_1,y_3)\} \\ + f(x_2,y_0) + f(x_2,y_4) + 2\{f(x_2,y_1) + f(x_2,y_2) + f(x_2,y_3)\} \\ + f(x_3,y_0) + f(x_3,y_4) + 2\{f(x_3,y_1) + f(x_3,y_2) + f(x_3,y_3)\} \end{bmatrix} \end{bmatrix}$$

$$= 0.323523$$

Note: We can easily separate the integral to get the exact integral as follows

$$\int_0^1\left(\int_1^2 \frac{2y}{(1+x^2)}dx\right)dy = \left(\int_1^2 \frac{1}{(1+x^2)}dx\right)\left(\int_0^1 2y\,dy\right) = \left(\tan^{-1}2 - \tan^{-1}1\right)(1) = 0.3217505546.$$

In some cases, it is possible to separate the terms and integrate. But, in general, it is not possible to find the exact solutions, on that account, we require numerical approximation.

13.8.2 Simpson 1/3 Rule

Simpson 1/3 rule to approximate double integral $I = \int_{y_0}^{y_n}\int_{x_0}^{x_m} f(x,y)dxdy$ is given by

$$I = \int_{y_0}^{y_n}\int_{x_0}^{x_m} f(x,y)dxdy = \int_{y_0}^{y_n}\left(\int_{x_0}^{x_m} f(x,y)dx\right)dy$$

$$= \int_{y_0}^{y_n}\left(\frac{h}{3}\left[\begin{array}{l} f(x_0,y)+f(x_m,y)+4\{f(x_1,y)+f(x_3,y)+\cdots+f(x_{m-1},y)\} \\ +2\{f(x_2,y)+f(x_4,y)+\cdots+f(x_{m-2},y)\}\end{array}\right]\right)dy$$

$$= \frac{h}{3}\left[\begin{array}{l}\left\{\int_{y_0}^{y_n} f(x_0,y)dy+\int_{y_0}^{y_n} f(x_m,y)dy+4\left\{\int_{y_0}^{y_n} f(x_1,y)dy+\int_{y_0}^{y_n} f(x_3,y)dy+\cdots\int_{y_0}^{y_n} f(x_{m-1},y)dy\right\}\right. \\ \left.+2\left\{\int_{y_0}^{y_n} f(x_2,y)dy+\int_{y_0}^{y_n} f(x_4,y)dy+\cdots\int_{y_0}^{y_n} f(x_{m-2},y)dy\right\}\right\}\end{array}\right]$$

$$= \frac{hk}{9}\left[\begin{array}{l}\left[\begin{array}{l} f(x_0,y_0)+f(x_0,y_n)+4\{f(x_0,y_1)+f(x_0,y_3)+\cdots+f(x_0,y_{n-1})\} \\ +2\{f(x_0,y_2)+f(x_0,y_4)+\cdots+f(x_0,y_{n-2})\} \\ +f(x_m,y_0)+f(x_m,y_n)+4\{f(x_m,y_1)+f(x_m,y_3)+\cdots+f(x_m,y_{n-1})\} \\ +2\{f(x_m,y_2)+f(x_m,y_4)+\cdots+f(x_m,y_{n-2})\}\end{array}\right] \\ +4\left[\begin{array}{l}\left[\begin{array}{l} f(x_1,y_0)+f(x_1,y_n)+4\{f(x_1,y_1)+f(x_1,y_3)+\cdots+f(x_1,y_{n-1})\} \\ +2\{f(x_1,y_2)+f(x_1,y_4)+\cdots+f(x_1,y_{n-2})\}\end{array}\right] \\ + \\ \vdots \\ +f(x_{m-1},y_0)+f(x_{m-1},y_n)+4\{f(x_{m-1},y_1)+f(x_{m-1},y_3)+\cdots+f(x_{m-1},y_{n-1})\} \\ +2\{f(x_{m-1},y_2)+f(x_{m-1},y_4)+\cdots+f(x_{m-1},y_{n-2})\}\end{array}\right] \\ +2\left[\begin{array}{l}\left[\begin{array}{l} f(x_2,y_0)+f(x_2,y_n)+4\{f(x_2,y_1)+f(x_2,y_3)+\cdots+f(x_2,y_{n-1})\} \\ +2\{f(x_2,y_2)+f(x_2,y_4)+\cdots+f(x_2,y_{n-2})\}\end{array}\right] \\ + \\ \vdots \\ +f(x_{m-2},y_0)+f(x_{m-2},y_n)+4\{f(x_{m-2},y_1)+f(x_{m-2},y_3)+\cdots+f(x_{m-2},y_{n-1})\} \\ +2\{f(x_{m-2},y_2)+f(x_{m-2},y_4)+\cdots+f(x_{m-2},y_{n-2})\}\end{array}\right]\end{array}\right] \tag{13.47}$$

Note: Similar kind of expressions can be computed for other integral formulas like Simpson 3/8 rule and Gauss quadrature formulas. Also for triple integral, we can derive expressions with the help of Trapezoidal and Simpson rules. Here, we are avoiding all these formulas due to the complexity.

Example ———————————————————————————————————— **13.27** ———

Calculate the approximation to the double integral $\int_1^2 \int_1^2 \dfrac{1}{(1+x+y)} \, dx \, dy$ with the aid of Simpson method by taking $h = 0.5$ and $k = 0.25$.

Ans.

We have

$$f(x, y) = \frac{1}{1+x+y}$$

$h = 0.5, k = 0.25 \ (m = 2, n = 4)$

$x_0 = 1, \ x_1 = 1.5, \ x_2 = 2$

$y_0 = 1, \ y_1 = 1.25, \ y_2 = 1.5, \ y_3 = 1.75, \ y_4 = 2$

Using Simpson rule (13.47), we get

$$\int_1^2 \left(\int_1^2 \frac{1}{(1+x+y)} \, dx \right) dy = \frac{1}{72} \begin{bmatrix} f(x_0, y_0) + f(x_0, y_4) + 4\{f(x_0, y_1) + f(x_0, y_3)\} + 2f(x_0, y_2) \\ + f(x_2, y_0) + f(x_2, y_4) + 4\{f(x_2, y_1) + f(x_2, y_3)\} + 2f(x_2, y_2) \\ + 4\left[f(x_1, y_0) + f(x_1, y_4) + 4\{f(x_1, y_1) + f(x_1, y_3)\} + 2f(x_1, y_2) \right] \end{bmatrix}$$

$$= 0.252681$$

Example ———————————————————————————————————— **13.28** ———

Use Simpson method to estimate the double integral in the Example 13.25.

Ans.

Simpson rule (13.47) provides following approximation to the given integral

$$\int_0^1 \left(\int_1^2 \frac{2y}{(1+x^2)} \, dx \right) dy = \frac{1}{144} \begin{bmatrix} f(x_0, y_0) + f(x_0, y_4) + 4\{f(x_0, y_1) + f(x_0, y_3)\} + 2f(x_0, y_2) \\ + f(x_4, y_0) + f(x_4, y_4) + 4\{f(x_4, y_1) + f(x_4, y_3)\} + 2f(x_4, y_2) \\ + 4\left[f(x_1, y_0) + f(x_1, y_4) + 4\{f(x_1, y_1) + f(x_1, y_3)\} + 2f(x_1, y_2) \\ + f(x_3, y_0) + f(x_3, y_4) + 4\{f(x_3, y_1) + f(x_3, y_3)\} + 2f(x_3, y_2) \right] \\ + 2\left[f(x_2, y_0) + f(x_2, y_4) + 4\{f(x_2, y_1) + f(x_2, y_3)\} + 2f(x_2, y_2) \right] \end{bmatrix}$$

$$= 0.321748$$

It is easy to see that the result is very close to exact value 0.321750 up to six decimal places.

Exercise 13

1. Use Simpson 1/3 rule to estimate the integral $\int_0^1 e^{-x^2} dx$ by taking five ordinates.

   Ans. 0.746855

2. Find the approximate value of the integral $\int_0^{\pi/2} \sqrt{1+\cos\theta}\, d\theta$ by dividing the interval into six equal parts. Use the composite Trapezoidal and Simpson 1/3 rules.

   Ans. 1.997143, 2.000003

3. Integrate the function $f(x)$ between $x = 2.0$ to $x = 2.8$, using the Trapezoidal rule.

x	2	2.1	2.2	2.3	2.4	2.5	2.6	2.7	2.8
$f(x)$	1.53	1.97	2.64	3.34	4.10	4.97	5.86	5.43	4.21

   Ans. 3.118

4. We have many integrals in different branches of sciences and engineering, which cannot be solved analytically. Consider following examples

   a) $\int_0^{\pi} \sqrt{1+\cos^2 x}\, dx$, b) $\int_1^2 \frac{\sin x}{x} dx$, c) $\int_0^2 e^{-x^2} dx$

   Approximate these integrals by Trapezoidal and Simpson 1/3 rules. Divide the intervals into ten equal parts.

   Ans. a) 3.820197, 3.820188
   b) 0.659218, 0.659330
   c) 0.881839, 0.882075

5. The arc length of the curve $y = f(x)$ in the interval $[a, b]$ is given by following formula

   $$\text{Arc length} = \int_a^b \sqrt{1+\left(\frac{dy}{dx}\right)^2}\, dx$$

   Approximate the arc lengths for the following functions. Use composite Trapezoidal and composite Simpson 1/3 rules with nine nodes.

   a) $y = \cos(x)$ $0 \le x \le \pi/2$ (Ans. 1.910099, 1.910099)
   b) $y = e^x$ $0 \le x \le 1$ (Ans. 2.005898, 2.003499)
   c) $y = x^2$ $0 \le x \le 2$ (Ans. 4.656889, 4.646834)

6. The surface area of the solid of revolution obtained by rotating the curve $y = f(x)$, $a \le x \le b$ about the x-axis is given by

   $$S = \int_a^b 2\pi y \sqrt{1+\left(\frac{dy}{dx}\right)^2}\, dx$$

   Approximate the surface area of solid of revolution using composite Simpson 1/3 rule with 11 nodes for the following curves

   a) $y = \cos(x)$ $0 \le x \le \pi/2$ (Ans. 7.211876)
   b) $y = e^x$ $0 \le x \le 1$ (Ans. 22.943201)
   c) $y = x^2$ $0 \le x \le 2$ (Ans. 53.226620)

7. The volume of the solid of revolution obtained by rotating the curve $y = f(x)$, $a \leq x \leq b$ about the x-axis is given by

$$V = \int_a^b \pi y^2 \, dx$$

Approximate the volume of solid of revolution obtained by rotating the following curves about the x-axis. Use Simpson 1/3 rule with 11 nodes.

a) $y = \cos(x)$ $0 \leq x \leq \pi/2$ (Ans. 2.467401)

b) $y = e^x$ $0 \leq x \leq 1$ (Ans. 10.035995)

c) $y = x^2$ $0 \leq x \leq 2$ (Ans. 20.107536)

8. We recorded the car velocity at regular intervals of 10 seconds during its travel of 2 minutes. The velocities (converted in metre/second) are given in following table

Time	0	10	20	30	40	50	60	70	80	90	100	110	120
Velocities	0	2	4	5	7	5	6	4	3	4	3	2	0

Compute the total distance traveled by the car. Use Simpson 1/3 rule.

Ans. 446.67 metre

9. A solid of revolution is formed by rotating about the x-axis, the lines $x = 0$ and $x = 1$ and curve through the points with the following coordinates

$x:$ 0   0.25   0.5   0.75   1

$y:$ 0   0.247404   0.479426   0.681639   0.841471

Estimate the volume of the solid formed using composite Simpson rule.

Ans. 0.856380

10. Estimate the average value of the function $f(x) = \sin(x)$ on the interval $\left[0, \dfrac{\pi}{2}\right]$ by using

Trapezoidal and Simpson 1/3 rules with step size $h = \dfrac{\pi}{12}$.

Ans. 0.632980, 0.636636

11. Find an upper bound of the error in estimation of the integral $\int_0^{\frac{\pi}{2}} \sin(x) dx$ with Trapezoidal and

Simpson 1/3 rules for number of subintervals $n = 6$.

Ans. 0.00897172, 0.0000409941

12. Compute the errors in approximate values of the integral $\int_1^2 e^{-x} dx$ by Trapezoidal, Simpson 1/3

and Simpson 3/8 formulas with 11 nodes.

Ans. $\dfrac{1}{1200\,e}$, $\dfrac{1}{1800000\,e}$, $\dfrac{1}{800000\,e}$

13. Estimate the integral, $I = \int_0^1 \dfrac{dx}{1+x}$, using Trapezoidal and Simpson 1/3 formulas with eight equal

subintervals and compare the result with exact value. Compute the upper bounds of the errors in both the formulas.

Ans. $I_{CT} = 0.694122$, $I_{CS1/3} = 0.693155$, $El_{CT} = 0.00260417$, $El_{CS1/3} = 0.0000325521$

14. Compute the value of integral, $\int_0^{\frac{\pi}{2}} \sin^2(x)dx$ with the aid of Trapezoidal formula by dividing the interval into ten equal parts and find the error bound in the value of the integral.

Ans. 0.785577, 0.00645964

15. Find the minimum number of intervals required to evaluate the integral, $\int_0^1 \sin(x)dx$ correct to 4 decimal places by i) Trapezoidal rule and ii) Simpson 1/3 rule.

Ans. 41, 4

16. Find the minimum number of intervals required to evaluate the integral, $\int_0^1 e^{-x^4} dx$ correct to 4 decimal places by i) Trapezoidal rule ii) Simpson 1/3 rule.

Ans. 77, 11; use $\max\limits_{0\le x\le 1}|f''(\xi)| < 3.5$ and $\max\limits_{0\le x\le 1}|f^{iv}(\xi)| < 95$

17. Evaluate the following integrals by using Gauss–Legendre 2-points and 3-points formulas.

a) $\int_1^2 (x^2 - \ln x)\, dx$ c) $\int_0^{\frac{\pi}{2}} \sqrt{1 + \sin^2 x}\, dx$

b) $\int_0^1 x^2 e^{-x^2}\, dx$ d) $\int_0^1 \dfrac{e^x}{1+\sin(x)}dx$

Ans. a) 1.946738, 1.947033;   c) 1.906879, 1.910344
       b) 0.188321, 0.189539;   d) 1.154197, 1.154765

18. Solve the integral $\int_{-1}^1 \dfrac{x - \sin^2(x)}{\sqrt{1-x^2}}dx$ numerically with Gauss–Chebyshev 2 and 3-points formula.

Ans. −1.506160, −1.427621

19. Approximate the integral $\int_2^4 \dfrac{e^x - \cos(x)}{x^2}dx$ with the help of Gauss–Chebyshev 2 and 3-points formula.

Ans. 5.592587, 5.184131; Hint: change the limit by using transformation $x = t + 3$ and then use function $f(t) = \dfrac{e^{t+3} - \cos(t+3)}{(t+3)^2}\sqrt{1-t^2}$

20. Compute the integral $\int_0^{\infty} \dfrac{(\ln(x)+x^2)\sin(x)}{e^x}dx$ with Gauss–Laguerre 2-points formula. (−0.598504)

21. Use Gauss–Laguerre 3-points formula to compute the numerical approximation for the integral $\int_0^{\infty} \dfrac{x-3}{x^3 + \cos(x)}dx.$

Ans. −2.919626

22. Apply Gauss–Hermite 2 and 3-points formula to obtain the approximate values of the following integral

    a) $\int_{-\infty}^{\infty} \left(x^2 + \sin(x)\right) e^{-x^2} \, dx$ **Ans.** 0.886228, 0.886220

    b) $\int_{-\infty}^{\infty} \left(3x^2 + e^x\right) e^{-x^2} \, dx$ **Ans.** 4.893024, 4.932461

    c) $\int_{-\infty}^{\infty} \frac{\cos(x)}{4x^2 - 3} \, dx$ **Ans.** −1.347500, −0.327079

23. Compute the value of the integral $\int_{-\infty}^{\infty} \frac{1}{1+x^2} dx$ with the help of Gauss–Hermite 3-points formula.

    Ans. 2.240777

24. Calculate the value of $\log_e 2$ from the integral $\int_0^1 \frac{dx}{1+x}$ by using Euler–Maclaurin formula.

    Ans. 0.693147

25. Find the value of the integral $\int_2^3 \frac{dx}{1+2x^2}$ by using Euler–Maclaurin formula.

    Ans. 0.0766218

26. Use Euler–Maclaurin formula to prove that

    $$\sin(0) + \sin\left(\frac{\pi}{100}\right) + \sin\left(\frac{2\pi}{100}\right) + \cdots + \sin(2\pi) = 0.$$

27. Prove the following results with the help of Euler–Maclaurin formula

    a) $\sum_{x=1}^{n} x = \frac{n(n+1)}{2}$ b) $\sum_{x=1}^{n} x^2 = \frac{n(n+1)(2n+1)}{6}$

28. Use Euler–Maclaurin formula to compute the value of the series $\sum_{x=1}^{100} \frac{1}{x}$. Use derivative terms up to order 5.

    Ans. 5.189130

29. Compute the value of integration $I = \int_1^{1.75} e^{-x^2} dx$ with the help of Romberg integration. Use only three initial values of integral with the Trapezoidal rule.

    Ans.

0.155494		
0.134363	0.127320	
0.129271	0.127574	0.127591

30. Compute the value of integration $I = \int_0^1 \frac{1}{1+x^2} dx$ with help of Romberg integration. Use only 4 initial values of integral with Trapezoidal rule.

    Ans.

0.750000			
0.775000	0.783333		
0.782794	0.785392	0.785529	
0.784747	0.785398	0.785399	0.785396

31. Find the approximate value of the double integral $\int_0^1 \int_0^1 \sin(x+y)\,dx\,dy$ with the help of Trapezoidal

and Simpson 1/3 methods by taking $h = 0.5$ and $k = 0.25$.

Ans. 0.753510, 0.773938

32. Calculate the approximation to the double integral $\int_1^2 \int_1^2 \dfrac{1}{\left(1+x^2+y^2\right)}\,dx\,dy$ with the help of

Trapezoidal and Simpson 1/3 methods by taking $h = 0.25$ and $k = 0.25$.

Ans. 0.185930, 0.185470

33. Apply Trapezoidal rule to compute the integral $\int_0^1 \int_0^{0.5} \dfrac{x+y}{\cos(x+y)}\,dx\,dy$ by taking two equal

subintervals for x and four equal subintervals for y.

Ans. 0.988098

First Order Ordinary Differential Equations: Initial Value Problems

Mathematics is the tool specially suited for dealing with abstract concepts of any kind and there is no limit to its power in this field.

Paul Adrien Maurice Dirac
(August 8, 1902–October 20, 1984)
A physicist who had made fundamental contributions in the fields of quantum mechanics and quantum electrodynamics.

The theories of modern physics, generally involve a mathematical model, as far as possible, it is a set of differential equations. In last few decades, the differential equations become the most important tool to study various applied sciences as well as engineering phenomena. We first obtain solutions of the mathematical models, and then come to the mathematical and physical interpretations of these solutions. Various aspects of physical phenomena are well described by the solutions of their respective differential equations.

An equation containing one or more independent variables, dependent variables, and their derivatives is known as a differential equation. For examples

i) $\dfrac{dy}{dx} + xy^{1/2} = \cos x$

ii) $x\dfrac{\partial z}{\partial x} + y\dfrac{\partial z}{\partial y} = z$

iii) $x\dfrac{d^2 y}{dx^2} + \left(\dfrac{dy}{dx}\right)^2 + xy = e^x$

iv) $y^2\left(\dfrac{\partial^2 z}{\partial x^2}\right)^2 + x\dfrac{\partial z}{\partial y} + z\left(\dfrac{\partial z}{\partial x}\right)^3 + \sin y = 0$

v) $\dfrac{\partial^2 z}{\partial t^2} = x^{1/3}\dfrac{\partial^2 z}{\partial y^2} + \dfrac{\partial^2 z}{\partial x^2}$

vi) $\dfrac{d^2 y}{dx^2} + \sin(y) = 0$

vii) $y\dfrac{d^2 y}{dx^2} + x\dfrac{dy}{dx} = 3$ (or) $yy'' + xy' = 3$

viii) $y^2\dfrac{\partial^3 z}{\partial x^3} + \sin x\left(\dfrac{\partial z}{\partial y}\right)^2 + y^{1/2} = 0$.   (or) $y^2 z_{xxx} + \sin x\, z_y^2 + y^{1/2} = 0$

In Section 14.1, we will discuss some basics of differential equations. Subsequent sections present some numerical methods for first order ordinary differential equations with the initial conditions.

14.1 Some Important Classifications and Terms

In this section, we will discuss some basic definitions, classifications, and terms; those are important in the theories of differential equations.

14.1.1 Ordinary and Partial Differential Equations

An ordinary differential equation (ODE) is a relation between an independent variable x, dependent variables, and their derivatives. In case of only one dependent variable and its derivatives, we have a single ODE of the following form

$$\phi\left(x, y, \frac{dy}{dx}, \frac{d^2 y}{dx^2}, \cdots, \frac{d^n y}{dx^n}\right) = 0$$

Equations i), iii), vi) and vii) are ODEs with one independent variable x and only one dependent variable y. We have a system of ODEs for more than one dependent variable (but only one independent variable). For example, the system of ODEs

$$x\frac{d^2 y}{dx^2} + \frac{dz}{dx} + \sin x = 0$$

$$\frac{dy}{dx} + \cos x \frac{dz}{dx} + z = 0$$

consists of two dependent variables $y(x)$ and $z(x)$ and one independent variable x.

A partial differential equation (PDE) is a relation between two or more independent variables (let $x_1, x_2, \cdots, x_m$), dependent variable and its derivatives. In case of only one dependent variable and its derivatives, we have single PDE of the following form

$$\psi\left(x_1, x_2, \cdots, x_m, u, \frac{\partial u}{\partial x_1}, \frac{\partial u}{\partial x_2}, \cdots, \frac{\partial u}{\partial x_m}\right) = 0$$

For examples, the equations ii), iv), v) and viii) are PDEs with a single dependent variable z and two independent variables (x, y) in equations ii), iv) and viii) and three independent variables (t, x, y) in equation v).

If numbers of dependent variables are more than one, then we have system of PDEs. For examples

$$u_t = -3uu_x + 3(v\varphi)_x + \frac{1}{2}u_{xxx}$$

$$v_t = 3uv_x - v_{xxx}$$

$$\varphi_t = 3u\varphi_x - \varphi_{xxx}$$

This system is famous Hirota–Satsuma coupled KdV system of PDEs with three dependent variables (u, v, φ) and two independent variables (t, x). This system is used to model physical phenomena of one-dimensional nonlinear waves in dispersion media without dissipation.

14.1.2 Order and Degree of Differential Equations

The order of a differential equation is the order of the highest order derivative term present in the differential equation. The order of differential equations i) and ii) is one; the order of equations iii)–vii) is two; while the last equation viii) is of order three.

The degree of a differential equation is the power of the highest order derivative term in differential equation (without any fractional power of dependent variable and its derivatives in the differential equation). The degree of the first equation is two (after eliminating the fraction of dependent variable, i.e., $y^{1/2}$). The degree of equation iv) is also two. All other equations are of degree one.

14.1.3 Homogeneous and Non-homogeneous Differential Equations

The first order differential equation $M(x, y)dx + N(x, y)dy = 0$ is said to be homogeneous if both the functions M and N are of the same degree in x and y. For examples $\left(x^3 + 3xy^2\right)dx + \left(yx^2\right)dy = 0$ is homogeneous first order differential equation, while $\left(x^3 + 3xy\right)dx + \left(yx^2\right)dy = 0$ is a nonhomogeneous differential equation.

In the case of higher order differential equation, if each term contains the dependent variable or its derivative then it is called homogeneous differential equation. Otherwise, it is a nonhomogeneous differential equation. For examples, following differential equations (i–iv) are homogeneous and (v–vii) are nonhomogeneous differential equations

i) $x\dfrac{d^3 y}{dx^3} + \left(\dfrac{dy}{dx}\right)^2 + xy = 0$

ii) $\dfrac{\partial^2 u}{\partial t^2} = x^{1/3}\dfrac{\partial^2 u}{\partial y^2} + \dfrac{\partial^2 u}{\partial x^2}$

iii) $\dfrac{d^2 y}{dx^2} + \sin(y) = 0$

iv) $\dfrac{d^2 y}{dx^2} + p(x)\dfrac{dy}{dx} + q(x)y = 0$

v) $y^2\left(\dfrac{\partial^2 z}{\partial x^2}\right)^2 + x\dfrac{\partial z}{\partial y} + z\left(\dfrac{\partial z}{\partial x}\right)^3 + \sin y = 0$

vi) $\dfrac{d^3 y}{dx^3} + \sin(y) + e^x = 0$

vii) $\dfrac{d^2 y}{dx^2} + p(x)\dfrac{dy}{dx} + q(x)y = r(x)$

14.1.4 Constant and Variable Coefficient Differential Equations

In constant coefficient differential equation, the coefficients of the terms containing the dependent variables and their derivatives are constants only. Otherwise, the differential equation is said to be variable coefficient differential equation. Consider the following differential equations

i) $3\dfrac{\partial^2 u}{\partial t^2} - \dfrac{\partial^2 u}{\partial y^2} - 2\dfrac{\partial^2 u}{\partial x^2} = u$ ii) $\dfrac{d^2 y}{dx^2} - 3\dfrac{dy}{dx} + 2y = r(x)$

iii) $\dfrac{d^2 y}{dx^2} + \sin(y) = 0$ iv) $y^2\left(\dfrac{\partial^2 z}{\partial x^2}\right)^2 + x\dfrac{\partial z}{\partial y} + z\left(\dfrac{\partial z}{\partial x}\right)^3 + \sin y = 0$

v) $x\dfrac{d^3 y}{dx^3} + \sin(y) + e^x = 0$ vi) $\dfrac{d^2 y}{dx^2} + p(x)\dfrac{dy}{dx} + q(x)y = 0$

The differential equations (i–iii) are with constant coefficients, while equations (iv–vi) are variable coefficient differential equations.

14.1.5 Linear and Nonlinear Differential Equations

In a differential equation, if the dependent variable and its derivatives are in linear form (degree of all these is one), then it is called as linear differential equation. Otherwise, it is nonlinear. Consider following differential equations.

i) $\dfrac{d^2 y}{dx^2} + x^2\dfrac{dy}{dx} + 3y = \sin(x)$ ii) $\dfrac{d^2 y}{dx^2} + \sin(y) = 0$

iii) $y^2\dfrac{\partial^2 z}{\partial x^2} + x\dfrac{\partial z}{\partial y} + z\dfrac{\partial z}{\partial x} + \sin y = 0$ iv) $\dfrac{dy}{dx} + y^2 = x$

The differential equation (i) is a linear differential equation as dependent variable y, and its derivative terms are of the linear form. While, the equations (ii) – (iv) are nonlinear differential equations. In equation (ii), we have sine function of dependent variable y; in the 3rd equation, there is a multiplicative term of dependent variable z and its derivative $\dfrac{\partial z}{\partial x}$; in last equation, the term y^2 is nonlinear.

The law of superposition is applicable to the homogeneous linear systems only, and hence these systems are easy to solve. Some methods have been introduced in the literature, to solve the linear systems analytically. But as far as nonlinear systems are concerned, only a few methods have been developed to solve some very particular types of nonlinear systems analytically. The mathematical models, arriving from the real world problems, are inherently nonlinear. Hence a very strong desire to have solutions of nonlinear systems leads us to numerical techniques. Numerical techniques are initially not easy to implement due to cumbersome computational work involved. But now with the advent of high-speed computers and software, these techniques are easy to implement and provide solutions for those problems which we are not able to solve analytically and in fact, sometimes provide better results than the analytical methods.

14.1.6 General, Particular and Singular Solutions

The general solution of an ODE of order n is a solution, which contains n arbitrary constants. For examples

i) $\dfrac{dy}{dx} = 3y$ $y = ce^{3x}$ is general solution

ii) $y'' + y = 0$ $y = c_1 \sin(x) + c_2 \cos(x)$ is a general solution

iii) $yy' + 4x = 0$ $4x^2 + y^2 = c^2$ is a general solution

Any solution obtained from the general solution by assigning particular values to arbitrary constants is called as a particular solution.

i) $\dfrac{dy}{dx} = 3y$ $y = 2e^{3x}$ is particular solution

ii) $y'' + y = 0$ $y = \sin(x) - 2\cos(x)$ is particular solution

iii) $yy' + 4x = 0$ $4x^2 + y^2 = 1$ is a particular solution

In the case of linear differential equations, each solution of a differential equation is deducible from the general solution by assigning some particular values to the arbitrary constants. But in the case of nonlinear differential equations, sometimes there may exist solutions which are not deducible from general solution. These solutions are known as singular solutions. For example, Clairaut equation $y = xy' + y'^2$ has a general solution $y = cx + c^2$, but it also has a singular solution $x^2 + 4y = 0$, which cannot be obtained from the general solution.

The solution of a differential equation is also called as the integral of the differential equations. The curves representing the solutions of the ordinary differential equation are the integral curves. The general integral of ordinary differential equation is also called as complete integral or complete primitive.

14.1.7 Initial Value Problem (IVP) and Boundary Value Problem (BVP)

To describe any physical phenomenon, we must define the initial or boundary conditions or both. For example, we take a very simple model of population growth. The rate of change of population is directly proportional to population itself. If $X(t)$ is population of any town at any time t, then the rate of change of population $\left(\dfrac{dX}{dt}\right)$ at any time t is given by the following first order ordinary differential equation

$$\frac{dX}{dt} = kX$$

where constant k depends on nature of the population.

Say, we want to compute the population of the town after five years i.e. $X(5)$, then we must have the present population. i.e., $X(0)$. Let $X(0) = 10000$. Now the model is complete for the computation of the population of the town at any time t.

$$\frac{dX}{dt} = kX; \qquad X(0) = 10000$$

The condition $X(0) = 10000$ is an initial condition. The conditions are known as initial conditions if these are defined at a single point. If the conditions are at more than one point then these conditions are known as boundary conditions. Differential equation with initial conditions is known as initial value problem (IVP), and with boundary conditions is known as boundary value problem (BVP). For examples, the differential equations (i) – (iii) are IVPs, and equations (iv)–(v) are BVPs

i) $\quad \dfrac{dy}{dx} + x^2 y = 3; \qquad\qquad\qquad y(1) = 2$

ii) $\quad \dfrac{d^2 y}{dx^2} + x\dfrac{dy}{dx} + y = 3; \qquad\qquad y(0) = 1, \ y'(0) = 2$

iii) $\quad \dfrac{d^3 y}{dx^3} + \sin x \dfrac{d^2 y}{dx^2} + xy = \cos x; \qquad y(0) = 1, \ y'(0) = 2, \ y''(0) = 2$

iv) $\quad x^2 \dfrac{d^2 y}{dx^2} + (x-1)\dfrac{dy}{dx} + y = 3; \qquad y(0) = 1, \ y(1) = 3$

v) $\quad \dfrac{d^3 y}{dx^3} + \sin x \dfrac{d^2 y}{dx^2} + xy = \cos x; \qquad y(0) = 1, \ y(1) = 2, \ y(3) = -4$

14.1.8 Existence and Uniqueness of Solutions

It is not necessary that each differential equation has a solution, and also if a solution exists it may not be unique. Consider the following first order ODEs

i) $\quad \left|\dfrac{dy}{dx}\right| + |y| = 0$ has a trivial solution $y = 0$ only

ii) $\quad \left|\dfrac{dy}{dx}\right| + |y| + c = 0, \ c > 0$; has no solution

iii) $\quad \dfrac{dy}{dx} + y = 0$ has infinitely many solutions $y = ce^{-x}$

Similarly, in the case of second order ODEs, the solution may not exist, and if it exists, it is not necessarily unique.

i) $\quad y'' + y = 0; \qquad y(0) = 0, \ y'(0) = 1 \qquad\qquad$ has unique solution

ii) $\quad y'' + y = 0; \qquad y(0) = 1, \ y(\pi) = 2 \qquad\qquad$ has no solution

iii) $\quad y'' + y = 0; \qquad y(0) = 1, \ y(\pi) = -1 \qquad\quad$ has infinitely many solutions

Note that few theorems on existence and uniqueness of the solutions of differential equations are helpful in this regard, but these are beyond the scope of this book. So, we will discuss numerical methods for only those IVPs/ BVPs, which have unique solutions.

14.1.9 Comparison of Analytical and Numerical Methods

Exact solutions play a vital role in the theories of different physical and engineering problems. Analytical methods are important methods for deriving exact solutions of differential equations. But, these methods can solve limited categories of the problems, generally for linearized models or those having low dimensionality and simple geometry. Since most of the real time problems are nonlinear and complex; hence study of numerical methods is inevitable for approximate solutions of differential equations governing important physical phenomena.

In pre-computer era, a significant amount of energy is used in the implementation of the numerical technique rather than applications of the technique, as numerical methods require repeated applications of arithmetic operations. In last few decades, the widespread availability and evolution of cheaper digital computers have led to a veritable explosion in the use and development of numerical methods. Consider following first order ODE with initial condition

$$\frac{dy}{dx} = f(x, y)$$
$$y(x_0) = y_0 \tag{14.1}$$

In this chapter, we will discuss following numerical methods for the solutions of IVP (14.1)

i)   Picard method of successive approximations
ii)   Taylor series method
iii) Euler and modified Euler (Heun) methods
iv)   Runge–Kutta methods
v)   Milne method
vi)   Adams method (Adams–Bashforth predictor and Adams–Moulton corrector methods)

14.2 Picard Method of Successive Approximations

In this method, we reduce the IVP (14.1)

$$\frac{dy}{dx} = f(x, y), \quad y(x_0) = y_0$$

into the integral equation by integrating the differential equation from initial point x_0 to any general point x.

$$\int_{x_0}^{x} \frac{dy}{dx} dx = \int_{x_0}^{x} f(x, y) dx$$

$$y(x) - y(x_0) = \int_{x_0}^{x} f(x, y) dx$$

$$y(x) = y(x_0) + \int_{x_0}^{x} f(x, y) dx$$

The right-hand side of equation involves yet to be determined value $y(x)$. Therefore to start the iterations, let initial approximation be $y^{(0)}(x) = y_0$.

The next approximation $y^{(1)}(x)$ can be computed by the following formula

$$y^{(1)}(x) = y(x_0) + \int_{x_0}^{x} f(x, y^{(0)}(x)) dx$$

Similarly, the following iterative formula produces the higher approximations

$$y^{(n+1)}(x) = y(x_0) + \int_{x_0}^{x} f(x, y^{(n)}(x)) dx \qquad\qquad n = 0, 1, 2, \cdots \qquad\qquad (14.2)$$

Example ── **14.1**

Use Picard method to solve the following IVP

$$\frac{dy}{dx} = -2xy, \quad y(0) = 1$$

Ans.

We have initial values $x_0 = 0$, $y(x_0) = y_0 = 1$, and the function, $f(x, y) = -2xy$.

The Picard formula (14.2) is given by

$$y^{(n+1)}(x) = y(x_0) + \int_{x_0}^{x} f(x, y^{(n)}(x)) dx \qquad\qquad n = 0, 1, 2, \cdots$$

Using $n = 0$ and $y^{(0)}(x) = y_0 = 1$, we get

$$y^{(1)}(x) = y(x_0) + \int_{x_0}^{x} f(x, y^{(0)}(x)) dx = y_0 + \int_{0}^{x} -2x \, dx$$

$$= 1 - x^2$$

Similarly, for $n = 1$, we have

$$y^{(2)} = 1 + \int_0^x -2x(1-x^2)dx$$

$$= 1 - x^2 + \frac{x^4}{2}$$

For $n = 2, 3, \ldots$, we obtain

$$y^{(3)} = 1 + \int_0^x -2x(1-x^2+\frac{x^4}{2})dx$$

$$= 1 - x^2 + \frac{x^4}{2} - \frac{x^6}{6}$$

$$y^{(4)} = 1 - x^2 + \frac{x^4}{2} - \frac{x^6}{6} + \frac{x^8}{24}$$

$$y^{(5)} = 1 - x^2 + \frac{x^4}{2} - \frac{x^6}{6} + \frac{x^8}{24} - \frac{x^{10}}{120}$$

$$\vdots$$

It is easy to see that these successive approximations will converge to the exact solution, $y = e^{-x^2}$.

Example ──────────────────────────────────── **14.2**

Perform four iterations of Picard method to solve the following IVP

$\dfrac{dy}{dx} = x + y$, $y(0) = 1$. Hence, compute the value of $y(0.1)$.

Ans.
We have $x_0 = 0$, $y(x_0) = y_0 = 1$, $f(x, y) = x + y$.

Using the Picard formula (14.2) for $n = 0, 1, 2, \ldots$, we get

$$y^{(n+1)}(x) = y(x_0) + \int_{x_0}^x f(x, y^{(n)}(x))\,dx \qquad\qquad n = 0, 1, 2, \cdots$$

$$y^{(1)} = 1 + \int_0^x (x+1)dx = 1 + \frac{x^2}{2} + x$$

$$y^{(2)} = 1 + \int_0^x x + \left(1 + x + \frac{x^2}{2}\right) dx$$

$$= 1 + x + x^2 + \frac{x^3}{6}$$

$$y^{(3)} = 1 + \int_0^x x + \left(1 + x + x^2 + \frac{x^3}{6}\right) dx$$

$$= 1 + x + x^2 + \frac{x^3}{3} + \frac{x^4}{24}$$

$$y^{(4)} = 1 + \int_0^x x + \left(1 + x + x^2 + \frac{x^3}{3} + \frac{x^4}{24}\right) dx$$

$$= 1 + x + x^2 + \frac{x^3}{3} + \frac{x^4}{12} + \frac{x^5}{120}$$

The value of $y(0.1)$ is given by

$$y(0.1) = 1 + (0.1) + (0.1)^2 + \frac{(0.1)^3}{3} + \frac{(0.1)^4}{12} + \frac{(0.1)^5}{120} = 0.9900498$$

Note that the method is converging towards the following exact solution

$$y = -x - 1 + 2e^{+x} = -x - 1 + 2\left(1 + x + \frac{x^2}{2!} + \frac{x^3}{3!} + \frac{x^4}{4!} + \cdots\right)$$

The Picard method involves integration, which is very difficult to solve in many problems. For example

$$\frac{dy}{dx} = ye^{-x^2}, \ y(0) = 1$$

The Picard method for this IVP requires $y^{(1)} = 1 + \int_0^1 e^{-x^2} dx$. The integration $\int_0^1 e^{-x^2} dx$ is not solvable analytically till now.

14.3 Taylor Series Method

Taylor series expansion of function $y(x)$ about the point $x = x_0$ is given by

$$y(x) = y(x_0) + (x - x_0)y'(x_0) + \frac{(x - x_0)^2}{2!} y''(x_0) + \ldots \frac{(x - x_0)^n}{n!} y^{(n)}(x_0) + \cdots \tag{14.3}$$

Now, we have to compute the terms in right-hand side of equation (14.3). The first term in series expansion is $y(x_0)$, which is given by initial condition in IVP (14.1). The second term can be computed by $y'(x_0) = f(x_0, y_0)$ using $\dfrac{dy}{dx} = f(x, y)$. We can differentiate the equation $\dfrac{dy}{dx} = f(x, y)$, to compute higher order derivatives at the point (x_0, y_0).

Example ——————————————————————————————— **14.3**

Use Taylor series method to solve the IVP $\dfrac{dy}{dx} = x + y, \ y(0) = 1$. Also, compute the value $y(0.1)$.

Ans.

From initial condition, we have $x_0 = 0$ and $y_0 = 1$.

By using the equation $\dfrac{dy}{dx} = x + y$, we have following derivative terms at initial point $(x_0, y_0) = (0, 1)$

$y' = x + y \Rightarrow \qquad\qquad y'(0) = f(x_0, y_0) = x_0 + y_0 = 0 + 1 = 1$

$y'' = 1 + y' \Rightarrow \qquad\qquad y''(0) = 1 + 1 = 2$

$y''' = y'' \qquad\qquad\qquad\quad y'''(0) = 2$

$\vdots \qquad\qquad\qquad\qquad\qquad y^{iv}(0) = 2$

$\vdots \qquad\qquad\qquad\qquad\qquad y^{v}(0) = 2...$

Taylor series formula (14.3) is given by

$$y(x) = y(x_0) + (x - x_0)y'(x_0) + \frac{(x - x_0)^2}{2!}y''(x_0) + \cdots$$

$$y(x) = 1 + x(1) + \frac{x^2}{2!}2 + \frac{x^3}{3!}2 + \cdots$$

$$= -1 - x + 2\left(1 + x + \frac{x^2}{2!} + \frac{x^3}{3!} + \frac{x^4}{4!} + \cdots\right)$$

$$= -1 - x + 2e^x$$

We can easily compute $y(0.1) = 1.110342$ from this solution.

Example ── **14.4** ────

Use Taylor series method to solve the IVP $\dfrac{dy}{dx} = -2xy$, $y(0) = 1$ and obtain the value, $y(0.1)$.

Ans.

We have following derivative terms at initial point $(0, 1)$

$$y' = -2xy \qquad\qquad y'(0) = 0$$

$$y'' = -2xy' - 2y \qquad\qquad y'''(0) = -2$$

$$y''' = -2xy'' - 4y' \qquad\qquad y'''(0) = 0$$

$$y^{iv} = -2xy''' - 6y'' \qquad\qquad y^{iv}(0) = 12$$

$$y^{v} = -2xy^{iv} - 6y''' \qquad\qquad y^{iv}(0) = 0$$

$$y^{vi} = -2xy^{v} - 8y^{iv} \qquad\qquad y^{vi}(0) = -96$$

$$\vdots$$

Substituting these values in Taylor series formula (14.3), we obtain

$$y(x) = y(x_0) + (x - x_0)y'(x_0) + \frac{(x - x_0)^2}{2!}y''(x_0) + \cdots$$

$$= 1 + x(0) + \frac{x^2}{2!}(-2) + \frac{x^3}{3!}(0) + \frac{x^4}{4!}(12) + \frac{x^5}{5!}(0) + \frac{x^6}{6!}(-96) + \cdots$$

$$= 1 - x^2 + \frac{x^4}{2} - \frac{2x^6}{15} + \cdots$$

$$y(0.1) = 1 - (0.1)^2 + \frac{(0.1)^4}{2} - \frac{2(0.1)^6}{15} = 0.99004986$$

The IVPs of higher order can also be solved with the help of Taylor series method. Note that the number of initial conditions is equal to order of ODE in IVP. Consider the following example for this purpose.

Example ── **14.5** ────

Compute the value of $y(0.2)$ with the help of Taylor series method for the following IVP

$$2yy' - xy'' + y''' = 1 \qquad\qquad y(0) = 0, \; y'(0) = 1, \; y''(0) = 1$$

Ans.

We have following initial conditions

$$y(0) = 0, \ y'(0) = 1, \ y''(0) = 1$$

To compute the higher derivative terms of order three and more, we will use the differential equation $2yy' - xy'' + y''' = 1$.

$$y''' = 1 - 2yy' + xy'' \qquad\qquad y'''(0) = 1 - 2(0)(1) + 0(1) = 1$$

$$y^{iv} = -2yy'' - 2(y')^2 + y'' + xy''' \quad y^{iv}(0) = 0 - 2 + 1 + 0 = -1$$

$$y^{v} = -2yy''' - 6y'y'' + 2y''' + xy^{iv} \quad y^{v}(0) = 0 - 2 - 4 + 1 + 1 + 0 = -4$$

$$\vdots$$

Using these values in Taylor series expansion (14.3) for $x = 0.2$, we have

$$y(0.2) = y(0) + (0.2)y'(0) + \frac{(0.2)^2}{2!} y''(0) + \cdots$$

$$y(0.2) = 0 + (0.2)(1) + \frac{(0.2)^2}{2!}(1) + \frac{(0.2)^3}{3!}(1) + \frac{(0.2)^4}{4!}(-1) + \frac{(0.2)^5}{5!}(-4) + \cdots$$

$$y(0.2) = 0 + 0.2 + 0.02 + 0.00133333 - 0.000066667 - 0.000010667$$

$$= 0.221256$$

Picard and Taylor series methods have implementation problems on the computer, as these methods require integration and differentiation analytically. So these methods are not basically numerical methods. Now, we will discuss the methods, which only require simple arithmetic operations and hence can be implemented easily on the computer. Many explicit, embedded and implicit numerical methods have been developed for the solutions of first order ODEs with initial conditions. Some methods are as follows.

1. Euler method
2. Modified (or) Improved Euler method (or) Heun method
3. Runge–Kutta methods
4. Milne method
5. Adams method
6. Bogacki–Shampine method
7. Fehlberg method
8. Cash–Karp method
9. Dormand–Prince method
10. Backward Euler method
11. Implicit midpoint method
12. Lobatto IIIA, IIIB, IIIC methods

$$\vdots$$

In this book, only two explicit methods (Euler and Runge–Kutta methods) and three implicit methods (modified Euler, Milne and Adams) have been discussed. Interested readers may consult other books (exclusively based on numerical techniques for differential equations) for a detailed and exhaustive study of numerical methods for differential equations.

14.4 Euler Method

Forward Euler method (or simply known as Euler method) is the simplest method for computation of numerical solution of IVP (14.1) $y' = f(x, y)$, $y(x_0) = y_0$.

Taylor series expansion (14.3) for $x_1 = x_0 + h$ is as follows

$$y_1 = y(x_1) = y(x_0 + h) = y(x_0) + h\, y'(x_0) + \frac{(h)^2}{2!} y''(x_0) + \cdots + \frac{(h)^{n-1}}{(n-1)!} y^{n-1}(x_0) + \cdots$$

On neglecting second and higher order terms of h, we get

$$y_1 = y(x_1) = y(x_0) + h\, y'(x_0)$$

Using Eq(14.1), we have

$$y_1 = y_0 + h\, f(x_0, y_0)$$

Similarly, the values at other points are given by

$$y_2 = y_1 + h\, f(x_1,\ y_1)$$
$$y_3 = y_2 + h\, f(x_2,\ y_2) \qquad \text{(or)} \qquad y_{i+1} = y_i + h\, f(x_i,\ y_i); \quad i = 0, 1, \cdots, n \qquad (14.4)$$
$$\vdots$$
$$y_n = y_{n-1} + h\, f(x_{n-1},\ y_{n-1})$$

Example ─── **14.6**

Calculate value of $y(1)$ for following IVP

$$\frac{dy}{dx} = x + y,\ \ y(0) = 1$$

Use Euler method with step size $h = 0.1$.

Ans.
We have initial values $x_0 = 0$ and $y(x_0) = y_0 = 1$, function $f(x, y) = x + y$ and step size $h = 0.1$.

Euler formula (14.4) is as follows

$$y_{i+1} = y_i + h f(x_i, y_i); \qquad i = 0, 1, \cdots, n$$

Using $i = 0, 1, \cdots, 9$ in this formula, we have

$$y(x_1) = y(0.1) = y_1 = y_0 + h f(x_0, y_0)$$
$$= y_0 + h(x_0 + y_0) = 1 + (0.1)(0 + 1) = 1.1$$

$$y_2 = y(0.2) = y_1 + h(x_1 + y_1) = 1.1 + (0.1)(0.1 + 1.1) = 1.22$$

$$y_3 = y(0.3) = y_2 + h(x_2 + y_2) = 1.22 + (0.1)(0.2 + 1.22) = 1.362$$

Similarly, other iterations are as follows

$$y_4 = y(0.4) = 1.5282000 \qquad\qquad y_5 = y(0.5) = 1.7210200$$

$$y_6 = y(0.6) = 1.9431220 \qquad\qquad y_7 = y(0.7) = 2.1974342$$

$$y_8 = y(0.8) = 2.4871776 \qquad\qquad y_9 = y(0.9) = 2.8158953$$

$$y_{10} = y(1) = 3.1874850$$

Example   **14.7**

Solve the IVP $\dfrac{dy}{dx} = x - y^2$, $y(1) = 2$ to obtain the value of $y(1.5)$ for following IVP with the help of Euler method. Use step size $h = 0.1$.

Ans.

Given that $x_0 = 1$, $y(x_0) = y_0 = 2$, $f(x, y) = x - y^2$ and $h = 0.1$.

By using Euler formula (14.4), we have

$$y(x_1) = y(1.1) = y_1 = y_0 + h f(x_0, y_0) = y_0 + h(x_0 - y_0^2) = 2 + (0.1)(1 - 2^2) = 1.7$$

$$y_2 = y(1.2) = y_1 + h(x_1 - y_1^2) = 1.7 + (0.1)(1.1 - 1.7^2) = 1.521$$

Similarly, we can compute following iterations

$$y_3 = y(1.3) = 1.4096559$$

$$y_4 = y(1.4) = 1.3409430$$

$$y_5 = y(1.5) = 1.3011302$$

Example ──────────────────────────── 14.8 ──

Use Euler method for the IVP $\dfrac{dy}{dx} = x^2 - \sin y$, $y(0) = .5$, to compute $y(1)$ with step size $h = 0.2$.

Ans.

We have $x_0 = 0$, $y_0 = 0.5$, $f(x, y) = x^2 - \sin y$ and $h = 0.2$.

By using Euler method, we obtain

$y(x_1) = y(0.2) = y_1 = y_0 + h(x_0^2 - \sin y_0) = 0.5 + (0.2)(0 - \sin(0.5)) = 0.4041149$

$y_2 = y(0.4) = y_1 + h(x_1^2 - \sin y_1) = 0.4041149 + (0.2)(0.2 - \sin(0.4041149)) = 0.33347389$

Successive values are given by

$y_3 = y(0.6) = 0.30000839$

$y_4 = y(0.8) = 0.31290275$

$y_5 = y(1) = 0.37933841$

Geometrical Interpretation of Forward Euler Method

Consider first order Taylor series expansion of any function $g(x)$ about a point $x = x_i$. It approximate the function $g(x)$ by tangent at the point $x = x_i$. In Euler method, we are using first order Taylor series. It means, we are approximating the solution curve $y(x)$ with the tangent at initial point $x = x_0$.

Given $\dfrac{dy}{dx} = f(x, y)$, $y(x_0) = y_0$, we have

$y(x_1) = y_1 = CE = CD + DE = y_0 + h \tan\theta$

$= y_0 + h\left(\dfrac{dy}{dx}\right)_{(x_0, y_0)} = y_0 + hf(x_0, y_0)$

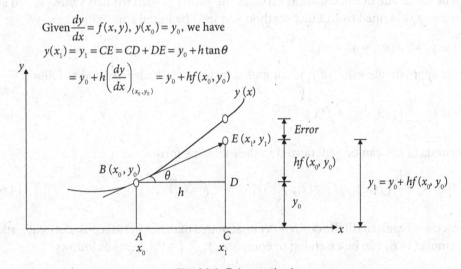

Fig. 14.1 Euler method

Note: In Euler method, we neglect the terms of h^2 and higher orders of Taylor series expansion. Therefore, a very small step size h is required to achieve even a moderate accuracy. For example, to achieve an accuracy of 5 decimal places, the step size is approximately $h = 0.001$ depending on the derivatives of the function, $f(x, y)$. Consider Example 14.6 with $h = 0.001$, we have to compute total $n = \dfrac{1-0}{0.001} = 1000$ iterations to find the value at the point $x = 1$. In these thousand iterations, a sufficiently large round-off error is generated to contaminate the final value to a large extent.

Exercise 14.1 Write a C-program with float variables (x, y) declaration and solve the Example 14.8 with different step sizes (10, 50, 1000, 10000, 50000) and interpret the results. We will see that increasing number of steps will increase the accuracy to a threshold point, and after that, there is a sharp decrease in accuracy.

It is worth mentioning here that the Euler method is conditionally stable as discussed in Section 14.11. We will discuss backward Euler method in Section 14.12, which is unconditionally stable, but implicit method.

14.5 Modified (or) Improved Euler Method (or) Heun Method

Euler method involves the slope at an initial point, (x_0, y_0). In modified Euler method, we use the average value of slopes at the initial point (x_0, y_0) and last point, (x_1, y_1). It improves the estimate of the slope for the interval (x_0, x_1)

$$y_1 = y_0 + \frac{h}{2}\left(f(x_0, y_0) + f(x_1, y_1)\right)$$

The right-hand side of the equation involves the yet-to-be-determined value, y_1. To start, we can use y_1 obtained from Euler method and let it be initial approximation, $y_1^{(0)}$.

$$y_1^{(0)} = y_0 + h f(x_0, y_0)$$

The next approximate value of y_1 is computed by modified Euler method as follows

$$y_1^{(1)} = y_0 + \frac{h}{2}\left(f(x_0, y_0) + f(x_1, y_1^{(0)})\right) \tag{14.5}$$

The formula (14.5) can be generalized in the following form

$$y_1^{(k+1)} = y_0 + \frac{h}{2}\left(f(x_0, y_0) + f(x_1, y_1^{(k)})\right), \qquad k = 0, 1, 2, \dots \tag{14.6}$$

The process is repeated till the desired decimal points matches in two consecutive iterations. The formula (14.6) can be extended to compute, y_{i+1}, $i = 0, 1, \cdots, n$ as follows

$$y_{i+1}^{(0)} = y_i + h f(x_i, y_i)$$

$$y_{i+1}^{(k+1)} = y_i + \frac{h}{2}\left(f(x_i, y_i) + f(x_{i+1}, y_{i+1}^{(k)})\right), \qquad k = 0, 1, 2, \ldots \qquad (14.7)$$

Example ――――――――――――――――――――――――――――――――――――――― 14.9 ――

Use modified Euler method to compute $y(1)$ for the following IVP

$$\frac{dy}{dx} = x + y, \quad y(0) = 1$$

Use step size, $h = 0.1$.

Ans.
We have

$$x_0 = 0, \ y_0 = 1, \ f(x, y) = x + y \text{ and } h = 0.1.$$

Value of $y(x_1) = y(0.1) = y_1$

Using Euler formula, we get following initial approximation $y_1^{(0)}$

$$y(x_1) = y(0.1) = y_1 = y_0 + h f(x_0, y_0) = y_0 + h(x_0 + y_0) = 1 + (0.1)(0+1) = 1.1 = y_1^{(0)}$$

Modified Euler method (14.6) can be used to improve the estimated value of $y(0.1)$ as follows

$$y_1^{(1)} = y_0 + \frac{h}{2}\left(f(x_0, y_0) + f(x_1, y_1^{(0)})\right) = y_0 + \frac{h}{2}\left((x_0 + y_0) + (x_1 + y_1^{(0)})\right)$$

$$= 1 + \frac{0.1}{2}\left((0+1) + (0.1+1.1)\right) = 1.11$$

$$y_1^{(2)} = y_0 + \frac{h}{2}\left(f(x_0, y_0) + f(x_1, y_1^{(1)})\right) = y_0 + \frac{h}{2}\left((x_0 + y_0) + (x_1 + y_1^{(1)})\right)$$

$$= 1 + \frac{0.1}{2}\left((0+1) + (0.1+1.11)\right) = 1.1105$$

$$y_1^{(3)} = y_0 + \frac{h}{2}\left(f(x_0, y_0) + f(x_1, y_1^{(2)})\right) = y_0 + \frac{h}{2}\left((x_0 + y_0) + (x_1 + y_1^{(2)})\right)$$

$$= 1 + \frac{0.1}{2}\left((0+1) + (0.1+1.1105)\right) = 1.110525$$

Value of $y(x_1) = y(0.1) = y_1 = 1.110525$

In these calculations, note that the superscripts are for the iterations of modified Euler method, while subscript denotes the variable.

Value of $y(x_2) = y(0.2) = y_2$

Using Euler formula, we have

$$y_2 = y(0.2) = y_1 + h(x_1 + y_1) = 1.110525 + (0.1)(0.1 + 1.110525) = 1.231578 = y_2^{(0)}$$

Modified Euler method (14.7) for $i = 1$ gives following iterations

$$y_2^{(1)} = y_1 + \frac{h}{2}\left(f(x_1, y_1) + f(x_2, y_2^{(0)})\right) = y_1 + \frac{h}{2}\left((x_1 + y_1) + (x_2 + y_2^{(0)})\right)$$

$$= 1.110525 + \frac{0.1}{2}\left((0.1 + 1.110525) + (0.2 + 1.231578)\right) = 1.242631$$

$$y_2^{(2)} = y_1 + \frac{h}{2}\left(f(x_1, y_1) + f(x_2, y_2^{(0)})\right) = y_1 + \frac{h}{2}\left((x_1 + y_1) + (x_2 + y_2^{(1)})\right)$$

$$= 1.110525 + \frac{0.1}{2}\left((0.1 + 1.110525) + (0.2 + 1.242631)\right) = 1.243184$$

$$y_2^{(3)} = y_1 + \frac{h}{2}\left(f(x_1, y_1) + f(x_2, y_2^{(0)})\right) = y_1 + \frac{h}{2}\left((x_1 + y_1) + (x_2 + y_2^{(2)})\right)$$

$$= 1.110525 + \frac{0.1}{2}\left((0.1 + 1.110525) + (0.2 + 1.243184)\right) = 1.243212$$

Value of $y(x_3) = y(0.3) = y_3$

The initial approximation for $y_3 = y(0.3)$ is given by

$$y_3^{(0)} = 1.387534$$

Using modified Euler formula, we get

$$y_3^{(1)} = 1.399750$$

$$y_3^{(2)} = 1.400361$$

$$y_3^{(3)} = 1.400392$$

Value of $y(x_4) = y(0.4) = y_4$

$$y_4^{(0)} = 1.570433$$

$$y_4^{(1)} = 1.583935$$

$y_4^{(2)} = 1.584610$

$y_4^{(3)} = 1.584643$

Similarly, we have the following values of y at $x = 0.5, 0.6, \ldots, 1$

$y_5^{(0)} = 1.783110 \quad y_5^{(1)} = 1.798033 \quad y_5^{(2)} = 1.798779 \quad y_5^{(3)} = 1.798816$

$y_6^{(0)} = 2.028700 \quad y_6^{(1)} = 2.045194 \quad y_6^{(2)} = 2.046019 \quad y_6^{(3)} = 2.046060$

$y_7^{(0)} = 2.310668 \quad y_7^{(1)} = 2.328898 \quad y_7^{(2)} = 2.329810 \quad y_7^{(3)} = 2.329856$

$y_8^{(0)} = 2.632844 \quad y_8^{(1)} = 2.652993 \quad y_8^{(2)} = 2.654000 \quad y_8^{(3)} = 2.654051$

$y_9^{(0)} = 2.999459 \quad y_9^{(1)} = 3.021729 \quad y_9^{(2)} = 3.022842 \quad y_9^{(3)} = 3.022898$

$y_{10}^{(0)} = 3.415191 \quad y_{10}^{(1)} = 3.439806 \quad y_{10}^{(2)} = 3.441036 \quad y_{10}^{(3)} = 3.441098$

It is worth mentioning here that all these iterations are obtained using C-Programs. It is very difficult and cumbersome to obtain all these manually or using a calculator. Hence it is advisable to solve these types of questions only for two or three iterations. For example, this question can be solved up to the value of y(0.3).

Example   **14.10**

Solve the IVP $\dfrac{dy}{dx} = x - y^2$, $y(1) = 2$ to compute $y(1.5)$. Use modified Euler method with step size $h = 0.1$.

Ans.
We have initial values $x_0 = 1$ and $y(x_0) = y_0 = 2$. Also, the function $f(x, y)$ is $x - y^2$, and the step size is $h = 0.1$.

Value of $y(x_1) = y(1.1) = y_1$

Use Euler formula to compute following initial approximation ($y_1^{(0)}$)

$$y(x_1) = y(1.1) = y_1 = y_0 + h f(x_0, y_0) = y_0 + h(x_0 - y_0^2) = 2 + (0.1)(1 - 2^2) = 1.7 = y_1^{(0)}$$

This value can be improved using modified Euler method (14.6) as follows

$$y_1^{(1)} = y_0 + \frac{h}{2}\left(f(x_0, y_0) + f(x_1, y_1^{(0)})\right) = y_0 + \frac{h}{2}\left((x_0 - y_0^2) + (x_1 - y_1^{(0)^2})\right)$$

$$= 2 + \frac{0.1}{2}\left((1 - 2^2) + (1.1 - 1.7^2)\right) = 1.760500$$

$$y_1^{(2)} = y_0 + \frac{h}{2}\left(f(x_0, y_0) + f(x_1, y_1^{(1)})\right) = y_0 + \frac{h}{2}\left((x_0 - y_0^2) + (x_1 - y_1^{(1)^2})\right)$$

$$= 2 + \frac{0.1}{2}\left((1-2^2) + (1.1 - 1.7605^2)\right) = 1.750032$$

$$y_1^{(3)} = y_0 + \frac{h}{2}\left(f(x_0, y_0) + f(x_1, y_1^{(2)})\right) = y_0 + \frac{h}{2}\left((x_0 - y_0^2) + (x_1 - y_1^{(2)^2})\right)$$

$$= 2 + \frac{0.1}{2}\left((1-2^2) + (1.1 - 1.750032^2)\right) = 1.751869$$

$$y_1^{(4)} = y_0 + \frac{h}{2}\left(f(x_0, y_0) + f(x_1, y_1^{(3)})\right) = y_0 + \frac{h}{2}\left((x_0 - y_0^2) + (x_1 - y_1^{(3)^2})\right)$$

$$= 2 + \frac{0.1}{2}\left((1-2^2) + (1.1 - 1.751869^2)\right) = 1.751548$$

$$y_1^{(5)} = y_0 + \frac{h}{2}\left(f(x_0, y_0) + f(x_1, y_1^{(4)})\right) = y_0 + \frac{h}{2}\left((x_0 - y_0^2) + (x_1 - y_1^{(4)^2})\right)$$

$$= 2 + \frac{0.1}{2}\left((1-2^2) + (1.1 - 1.751548^2)\right) = 1.751604$$

Proceeding in a similar manner as in the previous problem, we can obtain following values of y at different x

$$y_2^{(0)} = 1.554786 \quad y_2^{(1)} = 1.592322 \quad y_2^{(2)} = 1.586416$$
$$y_2^{(3)} = 1.587354 \quad y_2^{(4)} = 1.587205 \quad y_2^{(5)} = 1.587229$$

$$y_3^{(0)} = 1.455297 \quad y_3^{(1)} = 1.480367 \quad y_3^{(2)} = 1.476687 \quad y_3^{(3)} = 1.477231$$

$$y_4^{(0)} = 1.388953 \quad y_4^{(1)} = 1.406592 \quad y_4^{(2)} = 1.404127 \quad y_4^{(3)} = 1.404473$$

$$y_5^{(0)} = 1.347184 \quad y_5^{(1)} = 1.360059 \quad y_5^{(2)} = 1.358316 \quad y_5^{(3)} = 1.358553$$

Example 14.11

For the given IVP $\dfrac{dy}{dx} = x^2 - \sin y$, $y(0) = .5$, use modified Euler method to compute $y(1)$ with step size $h = 0.2$.

Ans.

Modified Euler formula (14.7) provides the following iterations

$$y_1^{(0)} = 0.404115 \quad y_1^{(1)} = 0.416737 \quad y_1^{(2)} = 0.415580 \quad y_1^{(3)} = 0.415685 = y(0.2)$$

$$y_2^{(0)} = 0.342914 \quad y_2^{(1)} = 0.361672 \quad y_2^{(2)} = 0.359911 \quad y_2^{(3)} = 0.360076 = y(0.4)$$

$$y_3^{(0)} = 0.321594 \quad y_3^{(1)} = 0.345219 \quad y_3^{(2)} = 0.342987 \quad y_3^{(3)} = 0.343197 = y(0.6)$$

$$y_4^{(0)} = 0.347881 \quad y_4^{(1)} = 0.375439 \quad y_4^{(2)} = 0.372861 \quad y_4^{(3)} = 0.373101 = y(0.8)$$

$$y_5^{(0)} = 0.428182 \quad y_5^{(1)} = 0.459109 \quad y_5^{(2)} = 0.456315 \quad y_5^{(3)} = 0.456566 = y(1)$$

14.6 Runge–Kutta (RK) Methods

The Euler method is the Taylor series method with terms containing only up to the first order of h. Similarly, we will see that the modified Euler method is also Taylor series method containing terms up to the order h^2. The inclusion of higher order terms is required to increase the accuracy. Higher order Taylor series requires the computation of partial derivatives of function, $f(x, y)$. In general, we cannot compute the partial derivatives of a function with the help of computer. Only some symbolic software like MAPLE, Mathematica, etc. can produce the partial derivatives symbolically, which cannot be used for computational purpose. So, Taylor series method cannot be implemented on a computer. Hence, Taylor series method is not suitable for practical applications.

In this section, we will derive certain higher order formulas known as Runge–Kutta methods, which do not involve the computations of derivative terms. Runge–Kutta methods (RK methods) are used to achieve the higher order accuracy of Taylor series without computing the higher order derivative terms. For this, we assume that the solution of the IVP

$$\frac{dy}{dx} = f(x, y), \ y(x_0) = y_0$$

is of the form

$$y_{i+1} = y_i + \lambda \tag{14.8}$$

where the general form of λ for an accuracy of $O(h^m)$ is given by the following expression

$$\lambda = w_1 k_1 + w_2 k_2 + w_3 k_3 + \cdots + w_m k_m \tag{14.9}$$

The aim is to determine the values of w_j's and k_j's in such a manner that we can achieve the desired accuracy. For this, let us assume k_j's of the forms

$$k_1 = h f(x_i, y_i)$$

$$k_2 = hf\left(x_i + a_1h,\ y_i + b_1k_1\right)$$

$$k_3 = hf\left(x_i + a_2h,\ y_i + b_2k_1 + b_3k_2\right)$$

$$k_4 = hf\left(x_i + a_3h,\ y_i + b_4k_1 + b_5k_2 + b_6k_3\right)$$

$$\vdots$$

$$(14.10)$$

where a_i's and b_i's are constants to determined.

First Order RK Method ($m = 1$)

Taylor series expansion is given by

$$y_{i+1} = y(x_i + h) = y(x_i) + h\,y'(x_i) + \frac{(h)^2}{2!}y''(x_i) + \cdots$$

By neglecting second and higher order terms, we get

$$y_{i+1} = y(x_i + h) = y(x_i) + h\,y'(x_i) = y(x_i) + h\,f(x_i, y_i)$$

On using $m = 1$ in Eqs. (14.8)–(14.10), we have

$$y_{i+1} = y_i + \lambda = y_i + \omega_1 k_1 = y_i + \omega_1 hf\left(x_i,\ y_i\right)$$

We get $w_1 = 1$ by comparing last two equations. So, RK method of order 1 is given by

$$y_{i+1} = y(x_i) + h\,f(x_i, y_i)$$

So, first order RK method is Euler method.

Second Order RK Method ($m = 2$)

Consider Eqs. (14.8–14.10) with $m = 2$, we have

$$y_{i+1} = y_i + \lambda,\ \text{with}\ \lambda = w_1 k_1 + w_2 k_2$$

where k_1 and k_2 are given by

$$k_1 = hf\left(x_i,\ y_i\right)$$

$$k_2 = hf\left(x_i + a_1h,\ y_i + b_1k_1\right)$$

Accordingly, we have

$$y_{i+1} = y_i + \lambda = y_i + w_1 k_1 + w_2 k_2$$

$$= y_i + w_1 hf\left(x_i,\ y_i\right) + w_2 hf\left(x_i + a_1h,\ y_i + b_1k_1\right) \qquad (14.11)$$

Expanding the term $f\left(x_i + a_1 h, \; y_i + b_1 k_1\right)$ by the Taylor series for the function of two variables

$$y_{i+1} = y_i + w_1 h f\left(x_i, \; y_i\right) + w_2 h \left(f\left(x_i, \; y_i\right) + a_1 h \frac{\partial f}{\partial x}\bigg|_{(x_i, y_i)} + b_1 k_1 \frac{\partial f}{\partial y}\bigg|_{(x_i, y_i)} + \cdots \right)$$

$$= y_i + w_1 h f\left(x_i, \; y_i\right) + w_2 h \left(f\left(x_i, \; y_i\right) + a_1 h \frac{\partial f}{\partial x}\bigg|_{(x_i, y_i)} + b_1 \left(h f\left(x_i, \; y_i\right)\right) \frac{\partial f}{\partial y}\bigg|_{(x_i, y_i)} + \cdots \right)$$

$$= y_i + \left(w_1 + w_2\right) h f\left(x_i, \; y_i\right) + w_2 h^2 \left(a_1 \frac{\partial f}{\partial x}\bigg|_{(x_i, y_i)} + b_1 \left(f\left(x_i, \; y_i\right)\right) \frac{\partial f}{\partial y}\bigg|_{(x_i, y_i)} + \cdots \right) \quad (14.12)$$

Since we have to achieve the accuracy up to $O(h^2)$, higher order terms can be avoided. Taylor series is given by

$$y_{i+1} = y(x_i + h) = y(x_i) + h\, y'(x_i) + \frac{(h)^2}{2!} y''(x_i) + \cdots \quad (14.13)$$

By using the given equation, $y' = f(x, y)$, we have

$$y'(x_i) = f(x_i, \; y_i)$$

$$y'' = \frac{\partial f}{\partial x} + \frac{\partial f}{\partial y} y'$$

$$\Rightarrow y''(x_i) = \frac{\partial f}{\partial x}\bigg|_{(x_i, y_i)} + \frac{\partial f}{\partial y}\bigg|_{(x_i, y_i)} y'(x_i) = \frac{\partial f}{\partial x}\bigg|_{(x_i, y_i)} + \frac{\partial f}{\partial y}\bigg|_{(x_i, y_i)} f(x_i, \; y_i) \quad (14.14)$$

Substituting the values of y' and y'' from Eqs. (14.14) in the Taylor series (14.13), we have

$$y_{i+1} = y(x_i + h) = y(x_i) + h\, f(x_i, \; y_i) + \frac{(h)^2}{2!} \left(\frac{\partial f}{\partial x}\bigg|_{(x_i, y_i)} + \frac{\partial f}{\partial y}\bigg|_{(x_i, y_i)} f(x_i, \; y_i) \right) + \cdots \quad (14.15)$$

Comparing the coefficients of $f(x_i, \; y_i)$, $\dfrac{\partial f}{\partial x}\bigg|_{(x_i, y_i)}$, $\dfrac{\partial f}{\partial y}\bigg|_{(x_i, y_i)} f(x_i, \; y_i)$ from Eqs. (14.12) and (14.15), we have

$$w_1 + w_2 = 1$$

$$w_2 a_1 = \frac{1}{2}$$

$$w_2 b_1 = \frac{1}{2} \quad (14.16)$$

The system (14.16) has three equations in four unknowns. One variable in system (14.16) can assume any value. Hence, infinite numbers of RK methods can be generated, here we are discussing only following two cases.

Case 1. $w_1 = \dfrac{1}{2}$ **(Modified Euler method)**

Let $w_1 = \dfrac{1}{2}$, then we have

$$w_2 = \frac{1}{2},\ a_1 = b_1 = 1.$$

Using values $w_1 = w_2 = \dfrac{1}{2},\ a_1 = b_1 = 1$, the formula (14.11) is given by

$$y_{i+1} = y_i + \lambda,\ \text{with}\ \lambda = \frac{1}{2}\big(k_1 + k_2\big)$$

where k_1 and k_2 are given by

$$k_1 = hf\big(x_i,\ y_i\big)$$

$$k_2 = hf\big(x_i + h,\ y_i + k_1\big) \qquad\qquad (14.17)$$

It is easy to see that it is modified Euler method.

Case 2 $w_1 = \dfrac{1}{3}$ **(Ralston and Rabinowitz Method)**

For second order RK method, Ralston and Rabinowitz obtained that if we select $w_1 = \dfrac{1}{3}$, then truncation error has a minimum bound. For this case, we have

$$w_2 = \frac{2}{3},\ a_1 = b_1 = \frac{3}{4}$$

On substituting the values $w_1 = \dfrac{1}{3},\ w_2 = \dfrac{2}{3},\ a_1 = b_1 = \dfrac{3}{4}$, the formula (14.11) produces following Ralston and Rabinowitz method for solution of IVP (14.1)

$$y_{i+1} = y_i + \lambda,\ \text{with}\ \lambda = \left(\frac{1}{3}k_1 + \frac{2}{3}k_2\right)$$

where k_1 and k_2 are given by

$$k_1 = hf\big(x_i,\ y_i\big)$$

$$k_2 = hf\left(x_i + \frac{3}{4}h,\ y_i + \frac{3}{4}k_1\right) \qquad\qquad (14.18)$$

Third Order RK Method ($m = 3$)

For $m = 3$, the formulas (14.8–14.10) are given by

$$y_{i+1} = y_i + \lambda, \text{ with } \lambda = w_1 k_1 + w_2 k_2 + w_3 k_3$$

where k_1, k_2 and k_3 are as follows

$$k_1 = h f(x_i,\, y_i)$$

$$k_2 = h f(x_i + a_1 h,\, y_i + b_1 k_1)$$

$$k_3 = h f(x_i + a_2 h,\, y_i + b_2 k_1 + b_3 k_2) \qquad (14.19)$$

Therefore, we have

$$y_{i+1} = y_i + \lambda = y_i + w_1 k_1 + w_2 k_2 + w_3 k_3$$

$$= y_i + w_1 h f(x_i,\, y_i) + w_2 h f(x_i + a_1 h,\, y_i + b_1 k_1) + w_3 h f(x_i + a_2 h,\, y_i + b_2 k_1 + b_3 k_2)$$

Expanding the term $f(x_i + a_1 h,\, y_i + b_1 k_1)$ and $f(x_i + a_2 h,\, y_i + b_2 k_1 + b_3 k_2)$ by the Taylor series of function of two variables

$$y_{i+1} = y_i + w_1 h f(x_i,\, y_i) + w_2 h \left(f(x_i,\, y_i) + a_1 h \left.\frac{\partial f}{\partial x}\right|_{(x_i, y_i)} + b_1 k_1 \left.\frac{\partial f}{\partial y}\right|_{(x_i, y_i)} + \cdots \right)$$

$$+ w_3 h \left(f(x_i,\, y_i) + a_2 h \left.\frac{\partial f}{\partial x}\right|_{(x_i, y_i)} + (b_2 k_1 + b_3 k_2) \left.\frac{\partial f}{\partial y}\right|_{(x_i, y_i)} + \cdots \right) \qquad (14.20)$$

Taylor series expansion is given by

$$y_{i+1} = y(x_i + h) = y(x_i) + h\, y'(x_i) + \frac{(h)^2}{2!} y''(x_i) + \frac{(h)^3}{3!} y'''(x_i) + \cdots$$

By using the given equation $y' = f(x,\, y)$ on a similar pattern as RK method of order 2, we have

$$y_{i+1} = y(x_i + h) = y(x_i) + h f(x_i,\, y_i) + \frac{(h)^2}{2!} \left(\left.\frac{\partial f}{\partial x}\right|_{(x_i, y_i)} + \left.\frac{\partial f}{\partial y}\right|_{(x_i, y_i)} f(x_i,\, y_i) \right) + \cdots \quad (14.21)$$

Comparing the different coefficients in Eqs. (14.20) and (14.21), we get the following six equations

$$w_1 + w_2 + w_3 = 1$$

$$b_1 - a_1 = 0$$

$$b_2 + b_3 - a_2 = 0$$

$$a_1 w_2 + a_2 w_3 = \frac{1}{2}$$

$$a_1^2 w_2 + a_2^2 w_3 = \frac{1}{3}$$

$$a_1 b_3 w_3 = \frac{1}{6} \tag{14.22}$$

The system (14.22) has six equations in eight unknowns, so any two variables can be set as free variables to obtain infinite numbers of solutions. One solution is given by

$$w_1 = \frac{1}{4}, \ w_2 = w_3 = \frac{3}{8}, \ a_1 = a_2 = b_1 = \frac{2}{3}, \ b_2 = 0, \ b_3 = \frac{2}{3} \tag{14.23}$$

So, the RK method of order three is given by

$$y_{i+1} = y_i + \lambda, \ \text{with} \ \lambda = \frac{1}{8}\left(2k_1 + 3k_2 + 3k_3\right)$$

where k_1, k_2 and k_3 are as follows

$$k_1 = h f\left(x_i, \ y_i\right)$$

$$k_2 = h f\left(x_i + \frac{2}{3}h, \ y_i + \frac{2}{3}k_1\right)$$

$$k_3 = h f\left(x_i + \frac{2}{3}h, \ y_i + \frac{2}{3}k_2\right) \tag{14.24}$$

Fourth Order Runge–Kutta Method
The solution is assumed to be of the following form

$$y_{i+1} = y_i + \lambda, \ \text{with} \ \lambda = w_1 k_1 + w_2 k_2 + w_3 k_3 + w_4 k_4$$

where k_1, k_2, k_3 and k_4 are given by

$$k_1 = h f\left(x_i, \ y_i\right)$$

$$k_2 = h f\left(x_i + a_1 h, \ y_i + b_1 k_1\right)$$

$$k_3 = h f\left(x_i + a_2 h, \ y_i + b_2 k_1 + b_3 k_2\right)$$

$$k_4 = h f\left(x_i + a_3 h, \ y_i + b_4 k_1 + b_5 k_2 + b_6 k_3\right)$$

Proceeding in a similar manner as in previous methods, following 11 equations in 13 unknowns are obtained

$$w_1 + w_2 + w_3 + w_4 = 1$$

$$b_1 - a_1 = 0$$

$$b_2 + b_3 - a_2 = 0$$

$$b_4 + b_5 + b_6 - a_3 = 0$$

$$a_1 w_2 + a_2 w_3 + a_3 w_4 = \frac{1}{2}$$

$$a_1^2 w_2 + a_2^2 w_3 + a_3^2 w_4 = \frac{1}{3}$$

$$a_1^3 w_2 + a_2^3 w_3 + a_3^3 w_4 = \frac{1}{4}$$

$$a_1 b_3 w_3 + a_1 b_5 w_4 + a_2 b_6 w_4 = \frac{1}{6}$$

$$a_1^2 b_3 w_3 + a_1^2 b_5 w_4 + a_2^2 b_6 w_4 = \frac{1}{12}$$

$$a_1 b_3 b_6 w_4 = \frac{1}{24}$$

$$a_1 a_2 b_3 w_3 + a_1 a_3 b_5 w_4 + a_2 a_3 b_6 w_4 = \frac{1}{8}$$

We can construct infinite numbers of 4th order RK method from solution of this system. But most commonly used method is classical RK method or simply known as RK fourth order method with the following values

$$w_1 = w_4 = \frac{1}{6}, \ w_2 = w_3 = \frac{1}{3}$$

$$a_1 = a_2 = \frac{1}{2}, \ a_3 = 1$$

$$b_1 = b_3 = \frac{1}{2}, \ b_2 = b_4 = b_5 = 0, \ b_6 = 1$$

RK fourth order method with these values is given by

$$y_{i+1} = y_i + \frac{1}{6}\left(k_1 + 2k_2 + 2k_3 + k_4\right)$$

where k_1, k_2, k_3 and k_4 are as follows

$$k_1 = h f\left(x_i, y_i\right)$$

$$k_2 = h f\left(x_i + \frac{1}{2}h, y_i + \frac{1}{2}k_1\right)$$

$$k_3 = h f\left(x_i + \frac{1}{2}h, y_i + \frac{1}{2}k_2\right)$$

$$k_4 = h f\left(x_i + h, y_i + k_3\right) \tag{14.25}$$

Example 14.12

Use Runge–Kutta second order method with minimum bound on truncation error (Ralston and Rabinowitz method) to solve the following IVP

$$\frac{dy}{dx} = x + y, \quad y(0) = 1$$

Compute $y(0.5)$ with step size $h = 0.1$.

Ans.
Given that $x_0 = 0$, $y(x_0) = y_0 = 1$, $f(x, y) = x + y$ and $h = 0.1$.

Ralston and Rabinowitz formula (14.18) is given by

$$y_{i+1} = y_i + \left(\frac{1}{3}k_1 + \frac{2}{3}k_2\right)$$

where k_1 and k_2 are as follows

$$k_1 = h f\left(x_i, y_i\right)$$

$$k_2 = h f\left(x_i + \frac{3}{4}h, y_i + \frac{3}{4}k_1\right)$$

Value of $y(0.1)$

$$k_1 = h f\left(x_0, y_0\right) = 0.1(x_0 + y_0) = 0.1(0 + 1) = 0.1$$

$$k_2 = h f\left(x_0 + \frac{3}{4}h, y_0 + \frac{3}{4}k_1\right) = 0.1(.075 + 1.075) = .115$$

$$y(0.1) = y_1 = y_0 + \left(\frac{1}{3}k_1 + \frac{2}{3}k_2\right) = 1 + \left(\frac{1}{3}(0.1) + \frac{2}{3}(0.115)\right) = 1.11$$

Value of $y(0.2)$

$$k_1 = h f\left(x_1, \ y_1\right) = 0.1(x_1 + y_1) = 0.1(0.1 + 1.11) = 0.121$$

$$k_2 = h f\left(x_1 + \frac{3}{4}h, \ y_1 + \frac{3}{4}k_1\right) = 0.1(.175 + 1.20075) = .137575$$

$$y(0.2) = y_2 = y_1 + \left(\frac{1}{3}k_1 + \frac{2}{3}k_2\right) = 1.11 + \left(\frac{1}{3}(0.121) + \frac{2}{3}(0.137575)\right) = 1.242050$$

Similarly, other iterations are as follows

$$k_1 = 0.144205 \qquad k_2 = 0.162520 \qquad y(0.3) = y_3 = 1.398465$$

$$k_1 = 0.169847 \qquad k_2 = 0.190085 \qquad y(0.4) = y_4 = 1.581804$$

$$k_1 = 0.198180 \qquad k_2 = 0.220544 \qquad y(0.5) = y_5 = 1.794894$$

Example ── **14.13**

Solve the IVP $\dfrac{dy}{dx} = x + y$, $y(0) = 1$, for the value $y(0.5)$. Use Runge–Kutta fourth order method with step size $h = 0.1$.

Ans.
RK fourth order method (14.25) is given by

$$y_{i+1} = y_i + \frac{1}{6}\left(k_1 + 2k_2 + 2k_3 + k_4\right)$$

where k_1, k_2, k_3 and k_4 are given by

$$k_1 = h f\left(x_i, \ y_i\right)$$

$$k_2 = h f\left(x_i + \frac{1}{2}h, \ y_i + \frac{1}{2}k_1\right)$$

$$k_3 = h f\left(x_i + \frac{1}{2}h, \ y_i + \frac{1}{2}k_2\right)$$

$$k_4 = h f\left(x_i + h, \ y_i + k_3\right)$$

Value of $y(0.1)$

$$k_1 = h f(x_0, y_0) = 0.1(x_0 + y_0) = 0.1(0+1) = 0.1$$

$$k_2 = h f\left(x_0 + \frac{1}{2}h, y_0 + \frac{1}{2}k_1\right) = 0.1(.05 + 1.05) = .11$$

$$k_3 = h f\left(x_0 + \frac{1}{2}h, y_0 + \frac{1}{2}k_2\right) = 0.1(.05 + 1.055) = .1105$$

$$k_4 = h f(x_0 + h, y_0 + k_3) = 0.1(0.1 + 1.1105) = 0.12105$$

$$y(0.1) = y_1 = y_0 + \frac{1}{6}(k_1 + 2k_2 + 2k_3 + k_4)$$

$$= 1 + \frac{1}{6}(0.1 + 2(0.11) + 2(0.1105) + (0.12105))$$

$$= 1.110342$$

Value of $y(0.2)$

$$k_1 = h f(x_1, y_1) = 0.1(x_1 + y_1) = 0.1(0.1 + 1.110342) = 0.121034$$

$$k_2 = h f\left(x_1 + \frac{1}{2}h, y_1 + \frac{1}{2}k_1\right) = 0.132086$$

$$k_3 = h f\left(x_1 + \frac{1}{2}h, y_1 + \frac{1}{2}k_2\right) = 0.132638$$

$$k_4 = h f(x_1 + h, y_1 + k_3) = 0.144298$$

$$y(0.2) = y_2 = y_1 + \frac{1}{6}(k_1 + 2k_2 + 2k_3 + k_4)$$

$$= 1.110342 + \frac{1}{6}(0.121034 + 2(0.132086) + 2(0.132638) + (0.144298))$$

$$= 1.242805$$

Similarly, the values at other iterations are as follows

$$k_1 = 0.144281 \qquad k_2 = 0.156495 \qquad k_3 = 0.157105 \qquad k_4 = 0.169991$$
$$y(0.3) = y_3 = 1.399717$$

$$k_1 = 0.169972 \qquad k_2 = 0.183470 \qquad k_3 = 0.184145 \qquad k_4 = 0.198386$$
$$y(0.4) = y_4 = 1.583648$$

$$k_1 = 0.198365 \qquad k_2 = 0.213283 \qquad k_3 = 0.214029 \qquad k_4 = 0.229768$$
$$y(0.5) = y_5 = 1.797441$$

Example ────────────────────────────────── **14.14**

Use Runge–Kutta fourth order method with step size $h = 0.1$ for the IVP $\dfrac{dy}{dx} = x - y^2$, $y(1) = 2$, to compute $y(1.2)$

Ans.

RK method of order 4 produces following iterations.

First Iteration

$$k_1 = h f(x_0, y_0) = 0.1(x_0 - y_0^2) = 0.1(1 - 2^2) = -0.3$$

$$k_2 = h f\left(x_0 + \frac{1}{2}h, \; y_0 + \frac{1}{2}k_1\right) = 0.1\left(1.05 - (1.85)^2\right) = -0.237250$$

$$k_3 = h f\left(x_0 + \frac{1}{2}h, \; y_0 + \frac{1}{2}k_2\right) = 0.1\left(1.05 - (1.881375)^2\right) = -0.248957$$

$$k_4 = h f(x_0 + h, \; y_0 + k_3) = 0.1\left(1.1 - (1.751043)^2\right) = -0.196615$$

$$y(1.1) = y_1 = y_0 + \frac{1}{6}\left(k_1 + 2k_2 + 2k_3 + k_4\right)$$

$$= 2 + \frac{1}{6}\left(-0.3 + 2(-0.23725) + 2(-0.248957) - 0.196615\right)$$

$$= 1.755162$$

Second Iteration

$$k_1 = h f(x_1, y_1) = 0.1\left(1.1 - (1.755162)^2\right) = -0.198059$$

$$k_2 = h f\left(x_1 + \frac{1}{2}h, \; y_1 + \frac{1}{2}k_1\right) = -0.159277$$

$$k_3 = h f\left(x_1 + \frac{1}{2}h, \; y_1 + \frac{1}{2}k_2\right) = -0.165738$$

$$k_4 = h f(x_1 + h, \; y_1 + k_3) = -0.132627$$

$$y(1.2) = y_2 = y_1 + \frac{1}{6}\left(k_1 + 2k_2 + 2k_3 + k_4\right) = 1.591709$$

Example ── 14.15 ──────

Given the IVP

$$\frac{dy}{dx} = x^2 - \sin y, \quad y(0) = .5$$

Estimate $y(0.4)$ using 4th order RK method with step size $h = 0.2$.

Ans.

Proceeding in a similar manner as in previous examples, one can easily obtain the following iterations

$k_1 = -0.095885 \qquad k_2 = -0.085363 \qquad k_3 = -0.086309 \qquad k_4 = -0.072398$

$y_1 = y(0.2) = 0.414729$

$k_1 = -0.072588 \qquad k_2 = -0.055893 \qquad k_3 = -0.057442 \qquad k_4 = -0.037947$

$y_2 = y(0.4) = 0.358528$

Predictor-Corrector Methods

In predictor-corrector methods, the initial value of y is estimated using a predictor formula; then corrector formula is used to improve this value. For example, modified Euler method

$$y_{i+1}^{(0)} = y_i + h f(x_i, y_i)$$

$$y_{i+1}^{(k+1)} = y_i + \frac{h}{2}\left(f(x_i, y_i) + f(x_{i+1}, y_{i+1}^{(k)})\right), \qquad k = 0, 1, 2, \ldots$$

The initial approximation $y_{i+1}^{(0)}$ is derived by Euler method (predictor formula), and this value is improved by using modified Euler method (corrector formula). In this section, we will derive two important predictor-corrector methods known as Milne and Adams methods.

The Euler, modified Euler, RK methods are single step methods as the values (x_k, y_k) at one previous point is used to compute the value y_{k+1} at next point. Once we have values at several points, then information at these points can be used to compute the values at successive points. The Milne and Adams methods require value at previous four points to compute values at sucessive point.

14.7 Milne Method (Milne Simpson Method)

This method requires three initial values, y_1, y_2, y_3, which are to be computed using any other method like Picard, Taylor series, Euler, Runge–Kutta method. These values of y are used to compute following function, $f(x, y)$ values

$$f_0 = f(x_0, y_0), f_1 = f(x_1, y_1), f_2 = f(x_2, y_2), f_3 = f(x_3, y_3)$$

Now, we will compute the value of y_4 using Milne method. Newton forward interpolation formula is given by

$$f(x, y) = f_0 + (s)\Delta f_0 + \frac{\Delta^2 f_0}{2!}(s)(s-1) + \frac{\Delta^3 f_0}{3!}(s)(s-1)(s-2) + \cdots \tag{14.26}$$

Integrating $\dfrac{dy}{dx} = f(x, y)$ from x_0 to x_4, we have

$$y\Big|_{x_0}^{x_4} = \int_{x_0}^{x_4} f(x, y)dx$$

$$y(x_4) - y(x_0) = \int_{x_0}^{x_4} f(x, y)\, dx$$

$$y_4 = y_0 + \int_{x_0}^{x_4} f(x, y)\, dx$$

Using Newton forward difference formula (14.26), we get

$$y_4 = y_0 + \int_{x_0}^{x_4} (f_0 + s\Delta f_0 + s\frac{(s-1)}{2}\Delta^2 f_0 + \ldots)\, dx$$

On differentiating $x = x_0 + sh$, we get $dx = h\, ds$.

$$\Rightarrow y_4 = y_0 + \int_0^4 (f_0 + s\Delta f_0 + s\frac{(s-1)}{2}\Delta^2 f_0 + \ldots)h\, ds$$

$$= y_0 + h(4f_0 + 8\Delta f_0 + \frac{20}{3}\Delta^2 f_0 + \frac{8}{3}\Delta^3 f_0 + \ldots) \tag{14.27}$$

The first, second, and third order forward differences have following expressions

$$\Delta f_0 = f_1 - f_0$$
$$\Delta^2 f_0 = f_2 - 2f_1 + f_0$$
$$\Delta^3 f_0 = f_3 - 3f_2 + 3f_1 - f_0 \tag{14.28}$$

On neglecting $\Delta^4 f_0$ and higher order differences and using (14.28) in (14.27), we get

$$y_4 = y_0 + \frac{4h}{3}(2f_1 - f_2 + 2f_3)$$

This is predictor formula for the value of y_4

$$y_4^{(p)} = y_0 + \frac{4h}{3}(2f_1 - f_2 + 2f_3) \tag{14.29}$$

The value of $f_4 = f(x_4, y_4)$ is computed using this value of $y_4^{(p)}$.

On integrating $\dfrac{dy}{dx} = f(x, y)$ from x_2 to x_4 and using Simpson 1/3 rule in the interval (x_2, x_4), we get

$$y(x_4) - y(x_2) = \int_{x_2}^{x_4} f(x, y)\, dx = \frac{h}{3}\left(f_2 + 4f_3 + f_4\right)$$

$$y(x_4) = y(x_2) + \frac{h}{3}\left(f_2 + 4f_3 + f_4\right)$$

$$y_4 = y_2 + \frac{h}{3}\left(f_2 + 4f_3 + f_4\right)$$

This expression is the corrector formula for the value of y_4.

$$y_4^{(c)} = y_2 + \frac{h}{3}(f_2 + 4f_3 + f_4)$$

We can use this formula iteratively as follows

$$y_4^{(n+1)} = y_2 + \frac{h}{3}(f_2 + 4f_3 + f_4^{(n)}); \quad n = 0, 1, 2, \dots \tag{14.30}$$

where $f_4^{(0)} = f\left(x_4, y_4^{(P)}\right)$

We can generalize the predictor formula (14.29) and corrector formula (14.30) to further compute the values of $y_5, y_6 \cdots$ as follows

$$y_{k+4}^{(P)} = y_k + \frac{4h}{3}(2f_{k+1} - f_{k+2} + 2f_{k+3})$$

$$y_{k+4}^{(n+1)} = y_{k+2} + \frac{h}{3}(f_{k+2} + 4f_{k+3} + f_{k+4}^{(n)}); \qquad n = 0, 1, 2, \dots \text{ and } k = 1, 2, 3, \dots$$

where $f_{k+4}^{(0)} = f\left(x_{k+4}, y_{k+4}^{(P)}\right)$

Stepwise Procedure

i)　To solve IVP $\dfrac{dy}{dx} = f(x, y);$ $\quad y(x_0) = y_0$; first compute the values of y_1, y_2, y_3

by using any method like Picard, Taylor series, Euler or Runge–Kutta method.

ii)　Use predictor formula (14.29) to compute the initial approximation $y_4^{(P)}$

$$y_4^{(P)} = y_0 + \frac{4h}{3}(2f_1 - f_2 + 2f_3)$$

The value $f_4^{(0)}$ can be computed at the point $y_4^{(P)}$. $f_4^{(0)} = f\left(x_4, y_4^{(P)}\right)$.

iii) Use corrector formula (14.30)

$$y_4^{(n+1)} = y_2 + \frac{h}{3}(f_2 + 4f_3 + f_4^{(n)}); \quad n = 0,1,2,...$$

to correct the y_4 till the desired accuracy achieved.

iv) Compute the values $y_5, y_6 \cdots$ by using following generalized formula

$$y_{k+4}^{(P)} = y_k + \frac{4h}{3}(2f_{k+1} - f_{k+2} + 2f_{k+3})$$

$$y_{k+4}^{(n+1)} = y_{k+2} + \frac{h}{3}(f_{k+2} + 4f_{k+3} + f_{k+4}^{(n)}); \qquad n = 0,1,2,... \ k = 1,2,3, ...$$

Where $f_{k+4}^{(0)} = f\left(x_{k+4}, y_{k+4}^{(P)}\right)$

Example 14.16

Compute the values of $y(0.1)$, $y(0.2)$ and $y(0.3)$ with Runge–Kutta fourth order method for the IVP $\frac{dy}{dx} = x + y$, $y(0) = 1$. Use Milne method to compute $y(0.5)$ with step size $h = 0.1$ till last two consecutive iterations are equal up to five decimal places.

Ans.
The following values of $y(0.1)$, $y(0.2)$ and $y(0.3)$ have already been computed with RK fourth order method in the Example 14.13

$y_1 = y(0.1) = 1.110342$

$y_2 = y(0.2) = 1.242805$

$y_3 = y(0.3) = 1.399717$

Now, we will compute $y_4 = y(0.4)$ and $y_5 = y(0.5)$ using Milne method.

Value of $y_4 = y(0.4)$

Value of $y_4 = y(0.4)$ using Milne predictor formula (14.29) is given by

$$y_4^{(P)} = y_0 + \frac{4h}{3}(2f_1 - f_2 + 2f_3)$$

$$= 1 + \frac{0.4}{3}\left(2(0.1 + 1.110342) - (0.2 + 1.242805) + 2(0.3 + 1.399717)\right)$$

$$= 1.583642$$

Using this value of $y_4^{(P)} = y_4^{(0)} = y(0.4) = 1.583642$ in Milne corrector formula (14.30), we obtain

$$y_4^{(1)} = y_2 + \frac{h}{3}\left(f_2 + 4f_3 + f_4^{(0)}\right)$$

$$= 1.242805 + \frac{0.1}{3}\left((0.2+1.242805) + 4(0.3+1.399717) + (0.4+1.583642)\right)$$

$$= 1.583649$$

Again using corrector formula with this new value of $y_4^{(1)} = y(0.4) = 1.583649$, we get

$$y_4^{(2)} = y_2 + \frac{h}{3}\left(f_2 + 4f_3 + f_4^{(1)}\right)$$

$$= 1.242805 + \frac{0.1}{3}\left((0.2+1.242805) + 4(0.3+1.399717) + (0.4+1.583649)\right)$$

$$= 1.583649$$

The last two approximations of $y_4 = y(0.4) = 1.583649$ are equal to five decimal places.

Value of $y_5 = y(0.5)$

Value of $y_5 = y(0.5)$ using Milne predictor formula is given by

$$y_5^{(P)} = y_1 + \frac{4h}{3}(2f_2 - f_3 + 2f_4)$$

$$= 1.110342 + \frac{0.4}{3}\left(2(0.2+1.242805) - (0.3+1.399717) + 2(0.4+1.583649)\right)$$

$$= 1.797434 = y_5^{(0)}$$

The value of $y_5^{(0)} = y(0.5)$ is used in Milne corrector formula to get

$$y_5^{(1)} = y_3 + \frac{h}{3}\left(f_3 + 4f_4 + f_5^{(0)}\right)$$

$$= 1.399717 + \frac{0.1}{3}\left((0.3+1.399717) + 4(0.4+1.583642) + (0.5+1.797434)\right)$$

$$= 1.797442$$

Using corrector formula with $y_5^{(1)}$, we get

$$y_5^{(2)} = y_3 + \frac{h}{3}\left(f_3 + 4f_4 + f_5^{(1)}\right)$$

$$= 1.399717 + \frac{0.1}{3}\left((0.3+1.399717) + 4(0.4+1.583642) + (0.5+1.797442)\right)$$

$$= 1.797442$$

$$\Rightarrow y_5 = y(0.5) = 1.797442$$

Note: We can also obtain following values if require

$y_6^{(0)} = 2.044228$ $y_6^{(1)} = 2.044237$ $y_6^{(2)} = 2.044238$

$y_7^{(0)} = 2.327494$ $y_7^{(1)} = 2.327505$ $y_7^{(2)} = 2.327505$

$y_8^{(0)} \doteq 2.651070$ $y_8^{(1)} = 2.651082$ $y_8^{(2)} = 2.651082$

$y_9^{(0)} = 3.019193$ $y_9^{(1)} = 3.019206$ $y_9^{(2)} = 3.019207$

$y_{10}^{(0)} = 3.436550$ $y_{10}^{(1)} = 3.436564$ $y_{10}^{(2)} = 3.436565$

Example 14.17

Use modified Euler method to solve the IVP $\dfrac{dy}{dx} = x - y^2$, $y(1) = 2$ for $x = 1.1, 1.2, 1.3$, and then compute $y(1.5)$ with step size $h = 0.1$ using Milne method.

Ans.
The values of $y(1.1)$, $y(1.2)$ and $y(1.3)$ have already been obtained in the Example 14.10 with modified Euler method

$y(1.1) = 1.751604$

$y(1.2) = 1.587229$

$y(1.3) = 1.477231$

Now, we will compute $y_4 = y(1.4)$ and $y_5 = y(1.5)$ using Milne method.

Value of $y_4 = y(1.4)$

Value of $y_4 = y(1.4)$ using Milne predictor formula (14.29) is given by

$$y_4^{(P)} = y_0 + \frac{4h}{3}(2f_1 - f_2 + 2f_3)$$

$$= 2 + \frac{0.4}{3}\left(2\left(1.1 - (1.751604)^2\right) - \left(1.2 - (1.587229)^2\right) + 2\left(1.3 - (1.477231)^2\right)\right)$$

$$= 1.415819 = y_4^{(0)}$$

The iterations of Milne corrector formula (14.30) are as follows

$$y_4^{(1)} = y_2 + \frac{h}{3}\left(f_2 + 4f_3 + f_4^{(0)}\right)$$

$$= 1.587229 + \frac{0.1}{3}\left(\left(1.2 - (1.587229)^2\right) + 4\left(1.3 - (1.477231)^2\right) + \left(1.4 - (1.415819)^2\right)\right)$$

$$= 1.405473$$

$$y_4^{(2)} = y_2 + \frac{h}{3}\left(f_2 + 4f_3 + f_4^{(1)}\right)$$

$$= 1.587229 + \frac{0.1}{3}\left(\left(1.2 - (1.587229)^2\right) + 4\left(1.3 - (1.477231)^2\right) + \left(1.4 - (1.405473)^2\right)\right)$$

$$= 1.406446$$

$$y_4^{(3)} = y_2 + \frac{h}{3}\left(f_2 + 4f_3 + f_4^{(2)}\right)$$

$$= 1.587229 + \frac{0.1}{3}\left(\left(1.2 - (1.587229)^2\right) + 4\left(1.3 - (1.477231)^2\right) + \left(1.4 - (1.406446)^2\right)\right)$$

$$= 1.406355$$

$$y_4^{(4)} = y_2 + \frac{h}{3}\left(f_2 + 4f_3 + f_4^{(3)}\right)$$

$$= 1.587229 + \frac{0.1}{3}\left(\left(1.2 - (1.587229)^2\right) + 4\left(1.3 - (1.477231)^2\right) + \left(1.4 - (1.406355)^2\right)\right)$$

$$= 1.406363$$

$$y_4^{(5)} = y_2 + \frac{h}{3}\left(f_2 + 4f_3 + f_4^{(4)}\right)$$

$$= 1.587229 + \frac{0.1}{3}\left(\left(1.2 - (1.587229)^2\right) + 4\left(1.3 - (1.477231)^2\right) + \left(1.4 - (1.406363)^2\right)\right)$$

$$= 1.406362$$

$$y_4^{(6)} = y_2 + \frac{h}{3}\left(f_2 + 4f_3 + f_4^{(5)}\right)$$

$$= 1.587229 + \frac{0.1}{3}\left(\left(1.2 - (1.587229)^2\right) + 4\left(1.3 - (1.477231)^2\right) + \left(1.4 - (1.406362)^2\right)\right).$$

$$= 1.406362$$

Value of $y_5 = y(1.5)$

Value of $y_5 = y(1.5)$ using Milne predictor formula is given by

$$y_5^{(P)} = y_1 + \frac{4h}{3}\left(2f_2 - f_3 + 2f_4\right)$$

$$= 1.751604 + \frac{0.4}{3}\left(2\left(1.2 - (1.587229)^2\right) - \left(1.3 - (1.477231)^2\right) + 2\left(1.4 - (1.406362)^2\right)\right)$$

$$= 1.363325 = y_5^{(0)}$$

On applying Milne corrector formula, we can easily obtain following values

$$y_5^{(1)} = y_3 + \frac{h}{3}\left(f_3 + 4f_4 + f_5^{(0)}\right)$$

$$= 1.477231 + \frac{0.1}{3}\left(\left(1.3 - (1.477231)^2\right) + 4\left(1.4 - (1.406362)^2\right) + \left(1.5 - (1.363325)^2\right)\right)$$

$$= 1.358821$$

$$y_5^{(2)} = y_3 + \frac{h}{3}\left(f_3 + 4f_4 + f_5^{(1)}\right)$$

$$= 1.477231 + \frac{0.1}{3}\left(\left(1.3 - (1.477231)^2\right) + 4\left(1.4 - (1.406362)^2\right) + \left(1.5 - (1.358821)^2\right)\right)$$

$$= 1.359230$$

$$y_5^{(3)} = y_3 + \frac{h}{3}\left(f_3 + 4f_4 + f_5^{(2)}\right)$$

$$= 1.477231 + \frac{0.1}{3}\left(\left(1.3 - (1.477231)^2\right) + 4\left(1.4 - (1.406362)^2\right) + \left(1.5 - (1.359230)^2\right)\right)$$

$$= 1.359193$$

$$y_5^{(4)} = y_3 + \frac{h}{3}\left(f_3 + 4f_4 + f_5^{(3)}\right)$$

$$= 1.477231 + \frac{0.1}{3}\left(\left(1.3 - (1.477231)^2\right) + 4\left(1.4 - (1.406362)^2\right) + \left(1.5 - (1.359193)^2\right)\right)$$

$$= 1.359196$$

$$y_5^{(5)} = y_3 + \frac{h}{3}\left(f_3 + 4f_4 + f_5^{(4)}\right)$$

$$= 1.477231 + \frac{0.1}{3}\left(\left(1.3 - (1.477231)^2\right) + 4\left(1.4 - (1.406362)^2\right) + \left(1.5 - (1.359196)^2\right)\right)$$

$$= 1.359196$$

Example 14.18

Use Milne method for the IVP $\dfrac{dy}{dx} = x^2 - \sin y$, $y(0) = .5$, to compute $y(1)$ with step size $h = 0.2$. Compute the prerequisite values with the aid of Runge–Kutta fourth order method.

Ans.
From Example 14.15, we have

$y_1 = y(0.2) = 0.414729$

$y_2 = y(0.4) = 0.358528$

$y_3 = y(0.6) = 0.341313$

Proceeding in a similar manner as in previous examples, we can easily obtain the following values using Milne predictor-corrector formula.

Value of $y_4 = y(0.8)$

$y_4^{(P)} = 0.370817 = y_4^{(0)}$

$y_4^{(1)} = 0.371050 \qquad y_4^{(2)} = 0.371035 \qquad y_4^{(3)} = 0.371036 \qquad y_4^{(4)} = 0.371036$

Value of $y_5 = y(1.0)$

$y_5^{(P)} = 0.454134 = y_5^{(0)}$

$y_5^{(1)} = 0.454397 \qquad y_5^{(2)} = 0.454382 \qquad y_5^{(3)} = 0.454383 \qquad y_5^{(4)} = 0.454383$

14.8 Adams Method (Adams–Bashforth Predictor and Adams– Moulton Corrector Formulas)

Newton forward difference formula is used to obtain Milne method. In Adams–Bashforth method, we will use Newton backward difference formula (10.12) with $n = 0$ as follows

$$f(x, y) = f_0 + (s)\nabla f_0 + \frac{\nabla^2 f_0}{2!}(s)(s+1) + \frac{\nabla^3 f_0}{3!}(s)(s+1)(s+2) + \cdots \tag{14.31}$$

Assume that IVP is of the following form

$$\frac{dy}{dx} = f(x, y), \quad y(x_{-3}) = y_{-3}$$

Solve this IVP to compute the values of y_{-2}, y_{-1} and y_0 by any method like Euler, etc.

On integrating $\dfrac{dy}{dx} = f(x, y)$ from x_0 to x_1, we have

$$y\Big|_{x_0}^{x_1} = \int_{x_0}^{x_1} f(x, y)dx$$

$$y_1 = y_0 + \int_{x_0}^{x_1} f(x, y)\, dx$$

By using Newton backward difference formula (14.31), we get

$$y_1 = y_0 + \int_{x_0}^{x_1}\left(f_0 + (s)\nabla f_0 + \frac{\nabla^2 f_0}{2!}(s)(s+1) + \frac{\nabla^3 f_0}{3!}(s)(s+1)(s+2) + \cdots\right)dx$$

Put $x = x_0 + sh$

$$\Rightarrow dx = h \, ds$$

$$y_1 = y_0 + \int_0^1 \left(f_0 + (s)\nabla f_0 + \frac{\nabla^2 f_0}{2!}(s)(s+1) + \frac{\nabla^3 f_0}{3!}(s)(s+1)(s+2) + \cdots \right) h \, ds$$

$$= y_0 + h\left(f_0 + \frac{1}{2}\nabla f_0 + \frac{5}{12}\nabla^2 f_0 + \frac{3}{8}\nabla^3 f_0 + \cdots \right) \tag{14.32}$$

The backward difference formulas are given by

$$\nabla f_0 = f_0 - f_{-1}$$
$$\nabla^2 f_0 = f_0 - 2f_{-1} + f_{-2}$$
$$\nabla^3 f_0 = f_0 - 3f_{-1} + 3f_{-2} - f_{-3} \tag{14.33}$$

Neglecting $\nabla^4 f_0$ and higher order differences and using the expressions (14.33) in equation (14.32), we obtain

$$y_1 = y_0 + \frac{h}{24}(55 f_0 - 59 f_{-1} + 37 f_{-2} - 9 f_{-3})$$

This expression is Adams–Bashforth predictor formula for the value of y_1

$$y_1^{(p)} = y_0 + \frac{h}{24}(55 f_0 - 59 f_{-1} + 37 f_{-2} - 9 f_{-3}) \tag{14.34}$$

At $x = x_1$, the Newton backward difference formula is given by

$$f(x, y) = f_1 + (s)\nabla f_1 + \frac{\nabla^2 f_1}{2!}(s)(s+1) + \frac{\nabla^3 f_1}{3!}(s)(s+1)(s+2) + \cdots \tag{14.35}$$

Integrating $\dfrac{dy}{dx} = f(x, y)$ from x_0 to x_1, we have

$$y(x_1) - y(x_0) = \int_{x_0}^{x_1} f(x, y)\, dx$$

Using Newton backward difference formula (14.35), we have

$$y(x_1) - y(x_0) = \int_{x_0}^{x_1} \left(f_1 + (s)\nabla f_1 + \frac{\nabla^2 f_1}{2!}(s)(s+1) + \frac{\nabla^3 f_1}{3!}(s)(s+1)(s+2) + \cdots \right) dx$$

Put $x = x_1 + sh$, $\Rightarrow dx = h\, ds$

$$y_1 = y_0 + \int_{-1}^{0}\left(f_1 + (s)\nabla f_1 + \frac{\nabla^2 f_1}{2!}(s)(s+1) + \frac{\nabla^3 f_1}{3!}(s)(s+1)(s+2) + \cdots \right) h\, ds$$

$$y_1 = y_0 + h\left(f_1 - \frac{1}{2}\nabla f_1 - \frac{1}{12}\nabla^2 f_1 - \frac{1}{24}\nabla^3 f_1 - \cdots \right)$$

Neglecting $\nabla^4 f_1$ and higher order differences, and using the similar expressions (14.33) for backward differences $\left(\nabla f_1, \nabla^2 f_1, \nabla^3 f_1\right)$, we get the following Adams–Moulton corrector formula

$$y_1 = y_0 + \frac{h}{24}\left(9 f_1 + 19 f_0 - 5 f_{-1} + f_{-2}\right)$$

$$y_1^{(c)} = y_0 + \frac{h}{24}\left(9 f_1 + 19 f_0 - 5 f_{-1} + f_{-2}\right)$$

(or) $\quad y_1^{(n+1)} = y_0 + \dfrac{h}{24}\left(9 f_1^{(n)} + 19 f_0 - 5 f_{-1} + f_{-2}\right); \qquad n = 0,1,2,\dots \qquad$ (14.36)

The function value $f_1^{(0)}$ can be computed at the point $y_1^{(P)}$.

The Adams–Bashforth predictor formula (14.34) and Adams–Moulton corrector formula (14.36) can be generalized to further compute the values of $y_2, y_3, \dots$ as follows

$$y_{k+1}^{(P)} = y_k + \frac{h}{24}\left(55 f_k - 59 f_{k-1} + 37 \cdot f_{k-2} - 9 f_{k-3}\right) = y_{k+1}^{(0)}$$

$$y_{k+1}^{(n+1)} = y_k + \frac{h}{24}\left(9 f_{k+1}^{(n)} + 19 f_k - 5 f_{k-1} + f_{k-2}\right); \qquad n = 0,1,2,\dots \qquad k = 0,1,2,\dots$$

where $f_{k+1}^{(0)} = f\left(x_{k+1}, y_{k+1}^{(0)}\right)$

Stepwise Procedure

i) Solve IVP $\dfrac{dy}{dx} = f(x,y)$; $\quad y(x_{-3}) = y_{-3}$; to compute the values of y_{-2}, y_{-1} and y_0

by using any other method like Taylor series, Euler or Runge–Kutta method.

ii) Use Adams–Bashforth predictor formula (14.34) to predict the initial approximation $y_1^{(P)}$

$$y_1^{(P)} = y_0 + \frac{h}{24}\left(55 f_0 - 59 f_{-1} + 37 f_{-2} - 9 f_{-3}\right)$$

The function value $f_1^{(0)}$ can be computed at the point $y_1^{(P)}$

$$f_1^{(0)} = f(x_1, y_1^{(P)})$$

iii) Use Adams–Moulton corrector formula (14.36)

$$y_1^{(n+1)} = y_0 + \frac{h}{24}\left(9f_1^{(n)} + 19f_0 - 5f_{-1} + f_{-2}\right); n = 0,1,2,...$$

to correct the y_1 till the desired accuracy achieved.

iv) The following generalized formula can be used to further compute the values of $y_2, y_3,...$

$$y_{k+1}^{(P)} = y_k + \frac{h}{24}\left(55f_k - 59f_{k-1} + 37f_{k-2} - 9f_{k-3}\right)$$

$$y_{k+1}^{(n+1)} = y_k + \frac{h}{24}\left(9f_{k+1}^{(n)} + 19f_k - 5f_{k-1} + f_{k-2}\right); \quad n = 0,1,2,... \quad k = 0,1,2,...$$

where $f_{k+1}^{(0)} = f\left(x_{k+1}, y_{k+1}^{(0)}\right)$

Example   **14.19**

Consider the following IVP

$$\frac{dy}{dx} = x + y, \quad y(0) = 1$$

Use Adams method to compute $y(0.5)$ with step size, $h = 0.1$. Compute the starting values $y(0.1)$, $y(0.2)$ and $y(0.3)$ with Runge–Kutta fourth order method.

Ans.

To apply Adams method, we will take initial condition $x_{-3} = 0$ and $y(x_{-3}) = y_{-3} = 1$. The function is $f(x, y) = x + y$, and the step size is $h = 0.1$.

We have already obtained the values of $y_{-2} = y(0.1)$, $y_{-1} = y(0.2)$, $y_0 = y(0.3)$ with RK fourth order method in Example 14.13. These values are given by

$$y_{-2} = y(0.1) = 1.110342$$
$$y_{-1} = y(0.2) = 1.242805$$
$$y_0 = y(0.3) = 1.399717$$

Now, we will compute $y_1 = y(0.4)$ and $y_2 = y(0.5)$ using the Adams method.

Value of $y_1 = y(0.4)$

Adams–Bashforth predictor formula (14.34) is given by

$$y_1^{(P)} = y_0 + \frac{h}{24}\left(55f_0 - 59f_{-1} + 37f_{-2} - 9f_{-3}\right)$$

$$= 1.399717 + \frac{0.1}{24}\big(55(0.3+1.399717) - 59(0.2+1.242805) + 37(0.1+1.110342) - 9(0+1)\big)$$

$$= 1.583640 = y_1^{(0)}$$

Now, we will use Adams–Moulton corrector formula (14.36) till the desired accuracy achieved

$$y_1^{(1)} = y_0 + \frac{h}{24}\big(9f_1^{(0)} + 19f_0 - 5f_{-1} + f_{-2}\big)$$

$$y_1^{(1)} = 1.399717 + \frac{0.1}{24}\left(\begin{array}{l}9(0.4+1.583640)+19(0.3+1.399717)\\ -5(0.2+1.242805)+(0.1+1.110342)\end{array}\right)$$

$$= 1.583649$$

$$y_1^{(2)} = 1.399717 + \frac{0.1}{24}\left(\begin{array}{l}9(0.4+1.583649)+19(0.3+1.399717)\\ -5(0.2+1.242805)+(0.1+1.110342)\end{array}\right)$$

$$= 1.583649$$

The last two approximations for y_1 are equal up to five decimal places, so

$$y_1 = y(0.4) = 1.583649$$

Value of $y_2 = y(0.5)$

Apply Adams–Bashforth predictor formula (14.34) for the value of y_2

$$y_2^{(P)} = y_1 + \frac{h}{24}(55f_1 - 59f_0 + 37f_{-1} - 9f_{-2})$$

$$= 1.583649 + \frac{0.1}{24}\left(\begin{array}{l}55(0.4+1.583649)-59(0.3+1.399717)\\ +37(0.2+1.242805)-9(0.1+1.110342)\end{array}\right)$$

$$= 1.797433 = y_2^{(0)}$$

Using Adams–Moulton corrector formula, we get

$$y_2^{(1)} = y_1 + \frac{h}{24}\big(9f_2^{(0)} + 19f_1 - 5f_0 + f_{-1}\big)$$

$$y_2^{(1)} = 1.583649 + \frac{0.1}{24}\left(\begin{array}{l}9(0.5+1.797433)+19(0.4+1.583649)\\ -5(0.3+1.399717)+(0.2+1.242805)\end{array}\right)$$

$$= 1.797443$$

$$y_2^{(2)} = 1.583649 + \frac{0.1}{24}\left(\begin{array}{l}9(0.5+1.797443)+19(0.4+1.583649)\\ -5(0.3+1.399717)+(0.2+1.242805)\end{array}\right)$$

$$= 1.797443$$

Since last two values are equal to six decimal places, hence

$$y_2 = y(0.5) = 1.797443$$

Note: On a similar pattern, we can also compute following values

$$y_3^{(0)} = y_3^P = y(0.6) = 2.044228 \qquad y_3^{(1)} = 2.044239 \qquad y_3^{(2)} = 2.044240$$

$$y_4^{(0)} = y_4^P = y(0.7) = 2.327496 \qquad y_4^{(1)} = 2.327508 \qquad y_4^{(2)} = 2.327508$$

$$y_5^{(0)} = y_5^P = y(0.8) = 2.651073 \qquad y_5^{(1)} = 2.651086 \qquad y_5^{(2)} = 2.651086$$

$$y_6^{(0)} = y_6^P = y(0.9) = 3.019197 \qquad y_6^{(1)} = 3.019212 \qquad y_6^{(2)} = 3.019212$$

$$y_7^{(0)} = y_7^P = y(1.0) = 3.436555 \qquad y_7^{(1)} = 3.436571 \qquad y_7^{(2)} = 3.436572$$

Example 14.20

Use modified Euler method to solve the IVP $\dfrac{dy}{dx} = x - y^2$, $y(1) = 2$ for $x = 1.1, 1.2, 1.3$, and then compute $y(1.5)$ with step size $h = 0.1$ using the Adams method.

Ans.
Consider initial condition $x_{-3} = 1$ and $y(x_{-3}) = y_{-3} = 2$. From Example 14.13, the values of $y(1.1)$, $y(1.2)$ and $y(1.3)$ with modified Euler method are given by

$$y_{-2} = y(1.1) = 1.751604$$
$$y_{-1} = y(1.2) = 1.587229$$
$$y_0 = y(1.3) = 1.477231$$

Now, we will compute $y_1 = y(1.4)$ and $y_2 = y(1.5)$ using the Adams method.

Value of $y_1 = y(1.4)$

Use Adams–Bashforth predictor formula (14.34) for initial value of y_1

$$y_1^{(P)} = y_0 + \frac{h}{24}(55f_0 - 59f_{-1} + 37f_{-2} - 9f_{-3})$$

$$y_1^{(P)} = 1.477231 + \frac{0.1}{24}\left(\begin{array}{l} 55\left((1.3 - (1.477231)^2\right) - 59\left((1.2 - (1.587229)^2\right) \\ +37\left(1.1 - (1.751604)^2\right) - 9(1 - 2^2) \end{array}\right)$$

$$y_1^{(P)} = 1.408466$$

Using Adams–Moulton corrector formula and proceeding in a similar manner as in the previous example, we have

$$y_1^{n+1} = y_0 + \frac{h}{24}\left(9f_1^{(n)} + 19f_0 - 5f_{-1} + f_{-2}\right)$$

$y_1^{(1)} = 1.404782$ $y_1^{(2)} = 1.405171$ $y_1^{(3)} = 1.405130$ $y_1^{(4)} = 1.405134$

Value of $y_2 = y(1.5)$

Similarly, predicted and corrected values of y_2 are given by

$y_2^P = 1.360790$

$y_2^{(1)} = 1.359353$ $y_2^{(2)} = 1.359499$ $y_2^{(3)} = 1.359484$ $y_2^{(4)} = 1.359486$

Example ── **14.21**

Use Adams method for the IVP $\dfrac{dy}{dx} = x^2 - \sin y$, $y(0) = .5$, to compute $y(1)$ with step size, $h = 0.2$. Compute the prerequisite values with the aid of Runge–Kutta fourth order method.

Ans.

Let $x_{-3} = 0$ and $y(x_{-3}) = y_{-3} = 0.5$. The values of $y(0.2)$, $y(0.4)$ and $y(0.6)$ with RK fourth order method obtained in Example 14.15 are as follows

$y_{-2} = y(0.2) = 0.414729$
$y_{-1} = y(0.4) = 0.358528$
$y_0 = y(0.6) = 0.341313$

The following results can easily be obtained with Adams method

Value of $y_1 = y(0.8)$

$y_1^P = 0.370805 = y_1^{(0)}$
$y_1^c = 0.371067$ $y_1^c = 0.371049$ $y_1^c = 0.371050$ $y_1^c = 0.371050$

Value of $y_2 = y(1)$

$y_2^P = 0.454128 = y_2^{(0)}$
$y_2^c = 0.454428$ $y_2^c = 0.454407$ $y_2^c = 0.454409$ $y_2^c = 0.454409$

14.9 Errors in Numerical Methods

Various numerical schemes have been discussed for the solutions of first order ordinary differential equations. Here, we will briefly discuss following five types of errors, which can occur during implementation of these numerical schemes for the solutions of first-order ordinary differential equations with initial conditions.

1. Errors in Initial Value Problem
2. Truncation Error
3. Round-off Error
4. Inherited Error
5. Propagated (or) Progressive Error

These errors and their effects on the obtained numerical solutions have much relevance and hence need to be discussed.

Errors in Initial Value Problem: During discussion of different numerical schemes for the IVP, we did not mention that whether a solution for IVP exists or not, and if it exists whether it is unique or not. An IVP may not have any solution or can have an infinite number of solutions. In these cases, our numerical schemes may oscillate or can produce diverging solutions, etc., thus needs more exploration. But these are not the parts of this book. Therefore, whenever we solve an IVP in this book, it is assumed that it has a unique solution. Also, the function $f(x, y)$ and initial values (x_0, y_0) may have errors (modeling and computational errors).

Truncation Error: We approximate the exact derivatives by finite differences in many numerical schemes for the solution of IVP. We truncate the Taylor series up to certain order, this error is known as truncation error. The method is said to be of order n if the error term of the method is of order $n + 1$.

For example, in Euler method, we approximate the solution from the Taylor series up to first order term.

$$y_1 = y(x_1) = y(x_0 + h) = y(x_0) + h\,y'(x_0) + \frac{(h)^2}{2!} y''(x_0) + \cdots + \frac{(h)^{n-1}}{(n-1)!} y^{n-1}(x_0) + \cdots$$

$$y_1 = y(x_0) + h\,y'(x_0) + O(h^2)$$

On neglecting second and higher order terms, we obtain Euler method

$$y_1 = y(x_1) = y(x_0) + h\,y'(x_0)$$
$$= y_0 + h\,f(x_0, y_0)$$

Hence, the scheme is accurate up to first order and has error term of order two. Truncation error depends on the step size h; it decreases as the step size decreases. The orders of different numerical schemes are listed in the Table 14.1.

Round-off Error: Round-off error has already been discussed in Chapter 2. We have to work with a finite number of digits in representing or storing a floating point number during the implementation of numerical algorithms with computing devices (mainly calculator and computer. It depends on the word length of the computer and hence machine dependent. There is a possibility of round-off error at any step. So, the rounding error increases as the number of computation increases. Therefore, before applying any numerical scheme for the solution of IVP, we must ensure that precision capability of the computer is such that it can maintain the desired number of significant digits in numerical calculations.

Inherited Errors: The mathematical model of any real-time problem always has coefficients that are imperfectly known. The reason is that the modeled problems often depend on some instruments whose measurements are up to only few precisions. Further, the model itself is not perfectly governing the situation itself. These kinds of errors are known as inherited error. Numerical methods cannot remove such type of errors, but we need to be more careful to avoid such uncertainty. To check the validity of such models in the real world problem, we may need to perform sensitivity analysis. Inherent errors can also be minimized using high precision computing machines and by taking better data.

Propagated or Progressive Errors: Propagated errors are the errors in the succeeding steps of a process due to an earlier error in the input. Previous iteration results are used to compute next iteration, so the error moves from one step to another step. All these errors (like round-off error, truncation error, and inherited error) propagate and tend to accumulate as computational numbers increase.

14.10 Order and Stability of Numerical Methods

We select a suitable step size, say h, in any numerical schemes like Euler, modified Euler, Runge–Kutta methods, Milne and Adams predictor-corrector methods. Then, we successively compute the values of y at different values of x with an interval of h. While computing successive iterations with the numerical schemes, we have to discuss following two important aspects for the numerical schemes

  1. Order
  2. Stability

Order of Numerical Scheme
A numerical scheme is said to be of order n if the error term of the scheme is of order $n+1$. For example, the Euler scheme is accurate up to first order and has error term of order two $O(h^2)$.

Stiff Equations and Stability of Numerical Scheme
A numerical scheme for solutions of initial value problem is said to be stable if the initial error introduced at some step does not increse indefinately at subsequent steps. The stiff equation is a differential equation, for which certain numerical method is stable only for very small step size. In fact, a precise definition of stiffness does not exist, but we have an idea that stiff equation has some terms that lead to variation in the solution. The concept of stiffness becomes clear from the following examples of first order IVPs.

Example ── **14.22**

Use Euler method for the IVP $\dfrac{dy}{dx} = -20y$, $y(0) = 1$, to compute $y(1)$ with step size $h = 0.1, 0.2$ and 0.02 and compare the solution with exact results.

Ans.

The exact solution of IVP is given by

$y = e^{-20x}$, which tends to zero as x tends to infinity.

We have initial condition $x_0 = 0$ and $y_0 = 1$, function $f(x, y) = -20y$. Let us discuss the Euler method with different spacing, $h = 0.1, 0.2$ and 0.02.

For step size $h = 0.1$, Euler formula produces following results

$$y(x_1) = y(0.1) = y_1 = y_0 + h\,f(x_0, y_0)$$
$$= y_0 + h(-20\,y_0) = 1 + (0.1)(-20) = -1$$

$$y_2 = y(0.2) = y_1 + h(-20\,y_1) = -1 + (0.1)(20) = 1$$

Similarly, $y_3 = -1$, $y_4 = 1$, $y_5 = -1$, $y_6 = 1$, $y_7 = -1$, and so on.

Solution oscillates between –1 and 1.

With step size $h = 0.2$, the iterations of Euler method are as follows

$y_1 = -3$	$y_2 = 9$	$y_3 = -27$
$y_4 = 81$	$y_5 = -243$	$y_6 = 729$
$y_7 = -2187$	$y_8 = 6561$	$y_9 = -19683$

$\vdots$

Solution oscillates wildly and increasing.

With step size $h = 0.02$, we have following iterations of Euler method

$y_1 = 0.6$	$y_2 = 0.36$	$y_3 = 0.216$
$y_4 = 0.1296$	$y_5 = 0.07776$	$y_6 = 0.046656$
$y_7 = 0.027994$	$y_8 = 0.016796$	$y_9 = 0.010078$

$\vdots$

The solution is decreasing monotonically as the exact solution.

It is clear from these computations that step size h plays a very significant role in stability of a numerical scheme. The small step size not only reduces the truncation error, but it is also crucial for convergent solution of the IVP. Note that very small step size (h) increases number of computations significantly, and hence round-off error also increases drastically. Therefore, we have to maintain a balance between round-off error, truncation error and stability of numerical scheme, while selecting step size h.

Example ── **14.23** ──

Solve the IVP in Example 14.22 with the help of Runge-Kutta 4th order method. Use step size $h = 0.1$ and 0.2

Ans.

Runge–Kutta method of order 4 with spacing $h = 0.1$ gives following iterations

$y_1 = 0.333333$ $y_2 = 0.111111$ $y_3 = 0.037037$

$y_4 = 0.012346$ $y_5 = 0.004115$ $y_6 = 0.001372$

$y_7 = 0.000457$ $y_8 = 0.000152$ $y_9 = 0.000051$

$\vdots$

The solution is decreasing monotonically as the exact solution and converging.

While with step size $h = 0.2$, we have following iterations of Runge–Kutta fourth order method

$y_1 = 5$ $y_2 = 25$ $y_3 = 125$

$y_4 = 625$ $y_5 = 3125$ $y_6 = 15625$

$\vdots$

The solution is increasing monotonously and diverging.

Note: *It is clear from examples 14.22 and 14.23 that for different IVPs and numerical schemes, we have different stability analysis. In next section, we will discuss stability analysis for a model IVP $y' = Ay$, $y(x_0) = y_0$, where constant A may be real or complex. Let $A = A_R + iA_I$, where A_R is real part and A_I is imaginary part. The exact solution of this IVP is $y = e^{At} y_0$. It is worth to mentioning that the real part (A_R) of constant A must be negative for a convergent solution. The solution $y \to 0$ as $t \to \infty$ for real part $A_R < 0$.*

14.11 Stability Analysis of IVP $y' = Ay$, $y(0) = y_0$

The IVP $y' = Ay$, $y(x_0) = y_0$ can always be converted into $y' = Ay$, $y(0) = y_0$ by a simple translation in x. So, in this section, we will discuss stability analysis of IVP $y' = Ay$, $y(0) = y_0$ for Euler method to have an understanding of the stability of numerical scheme.

Since the stability topic is very wide, it is not possible to cover it here; the reader further interested can consult any book which is exclusive on the topic of numerical solutions of differential equations.

Euler Method

The Euler method is given by

$$y_n = y_{n-1} + h f(x_{n-1}, y_{n-1}) \qquad\qquad n = 1, 2, 3, \dots .$$

For IVP, $y' = A y$, $y(0) = y_0$, we have

$$y_n = y_{n-1} + h(A y_{n-1}) = (1 + Ah) y_{n-1}$$

Using this expression recursively, we have

$$
\begin{aligned}
y_n &= y_{n-1} + h(A y_{n-1}) \\
&= (1 + Ah) y_{n-1} \\
&= (1 + Ah)^2 y_{n-2} \\
&\;\;\vdots \\
&= (1 + Ah)^n y_0
\end{aligned}
$$

The Euler scheme is stable and produces convergent solution, if $|1 + Ah| \le 1$. Otherwise, as n increases, solution will increase indefinitely. Consider the following different cases

a) A is real number; then Euler scheme is stable for $|1 + Ah| \le 1$ or $-2 \le Ah \le 0$

b) A is pure imaginary; let $A = ai$, then $|1 + ahi| = \sqrt{1 + a^2 h^2} \le 1$, which is not possible, hence for pure imaginary A, Euler scheme is unstable for each h.

c) A is complex number; $A = A_R + iA_I$, then

$$|1 + Ah| = |1 + (A_R + iA_I)h| = \sqrt{(1 + A_R h)^2 + (A_I h)^2} \le 1$$

This result implies that the region inside the unit circle is the stability region for Euler method. Thus the Euler scheme is conditionally stable. It is also clear from $\sqrt{(1 + A_r h)^2 + A_I h} \le 1$ that $A_r < 0$, other wise Euler scheme is unstable scheme

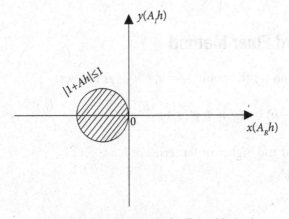

Fig. 14.2 Stability region for Euler Method

Example ── 14.24 ───

Discuss the stability of Euler method for the IVP in Example 14.22.

Ans.
The IVP is as follows

$$\frac{dy}{dx} = -20y, \ y(0) = 1, \text{ with step size } h = 0.1, 0.2 \text{ and } 0.02.$$

In this case, $A = -20$ is real number, so for stable Euler scheme, we have

$$-2 \leq A h \leq 0$$

$$-2 \leq (-20h) \leq 0, \text{ or } (0 \leq h \leq 0.1)$$

In case, step size $h = 0.02$, the Euler scheme is a stable scheme. But for step size $h = 0.2$, the scheme is unstable. For step size $h = 0.1$, the scheme is stable but oscillating, and It is not approaching to the solution.

Remark: It is worth mentioning here that stability of a scheme and accuracy are two different terms. Stability implies that numerical scheme gives a convergent result, but it is not always true that the result is accurate, the result may be inaccurate. If the total error (truncation error + round-off error) remains bounded as iterations tend to ∞, then the numerical scheme is stable, otherwise it is unstable. Stability implies that numerical scheme gives the results with a finite upper bound on error, but it is not necessary that the result is correct up to desired accuracy.

The explicit schemes are conditionally stable for stiff problems and require very small step sizes. But the implicit schemes are generally unconditionally stable schemes for stiff problems. Now, we will discuss backward Euler method as an example of implicit scheme. First, we will introduce the method and then discuss its stability. Note that backward Euler requires tedious computations.

14.12 Backward Euler Method

Taylor series expansion for the point $x_0 = x_1 - h$ is as follows

$$y_0 = y(x_0) = y(x_1 - h) = y(x_1) - h\, y'(x_1) + \frac{(h)^2}{2!} y''(x_1) + \cdots + \frac{(h)^{n-1}}{(n-1)!} y^{n-1}(x_1) + \cdots$$

On neglecting second and higher order terms of h, we get

$$y(x_0) = y(x_1) - h\, y'(x_1)$$

$$y(x_1) = y(x_0) + h\, y'(x_1)$$

By using $y' = f(x, y)$ from Eq. (14.1), we have

$$y_1 = y_0 + h f(x_1, y_1)$$

Similarly, the values at other points are given by

$$y_2 = y_1 + h f(x_2, y_2)$$
$$y_3 = y_2 + h f(x_3, y_3)$$
$$\vdots$$
$$y_n = y_{n-1} + h f(x_n, y_n)$$

Therefore, the backward Euler method is given by

$$y_{i+1} = y_i + h f(x_{i+1}, y_{i+1}); \qquad i = 0, 1, \cdots, n \tag{14.37}$$

The scheme (14.37) is an implicit scheme as it involves the unknown variable y_{i+1} on both sides of the equation. If the function $f(x, y)$ is nonlinear in variable y, then Eq. (14.37) produces nonlinear algebraic equation. We have to solve this nonlinear equation using the methods described in Chapter 3.

Example 14.25

Calculate value of $y(1)$ for following IVP

$$\frac{dy}{dx} = x + y, \quad y(0) = 1$$

Use backward Euler method with step size $h = 0.1$.

Ans.
From initial condition, we have initial values $x_0 = 0$ and $y(x_0) = y_0 = 1$. Also, the function $f(x, y) = x + y$ and the step size is $h = 0.1$.

Backward Euler formula (14.37) is as follows

$$y_{i+1} = y_i + h f(x_{i+1}, y_{i+1}); \quad i = 0, 1, \cdots, n$$

$$y_{i+1} = y_i + h(x_{i+1} + y_{i+1}); \quad i = 0, 1, \cdots, n$$

$$y_{i+1} = \frac{y_i + h x_{i+1}}{1 - h} = \frac{1}{0.9}(y_i + h x_{i+1}); \ i = 0, 1, \cdots, n$$

Using $i = 0, 1, \cdots, 9$ in this formula, we have

$$y(x_1) = y(0.1) = y_1 = \frac{y_0 + h x_1}{1 - h} = \frac{1 + 0.1(0.1)}{0.9} = 1.122222$$

$$y_2 = y(0.2) = \frac{y_1 + h x_2}{1 - h} = \frac{1.122222 + 0.1(0.2)}{0.9} = 1.269136$$

$$y_3 = y(0.3) = 1.443484$$
$$y_4 = y(0.4) = 1.648316$$
$$y_5 = y(0.5) = 1.887017$$
$$y_6 = y(0.6) = 2.163352$$
$$y_7 = y(0.7) = 2.481503$$
$$y_8 = y(0.8) = 2.846114$$
$$y_9 = y(0.9) = 3.262349$$
$$y_{10} = y(1) = 3.735943$$

Note: The function $f(x, y) = x + y$ is linear in y; therefore we have obtained the solution easily. The linear equations are easy to solve for analytical solutions, and we generally use the numerical methods for nonlinear equations. In case of nonlinear equations, it is very tedious to apply implicit schemes. So, explicit schemes are used much frequently as compared to implicit schemes. But, the explicit schemes are conditionally stable for some IVPs and require very small step sizes. Since the implicit schemes are generally unconditionally stable schemes, and hence preferred in such cases.

Example 14.26

Solve the IVP $\dfrac{dy}{dx} = x - y^2$, $y(1) = 2$ to obtain the value of $y(1.2)$ for following IVP with the help of backward Euler method. Use step size $h = 0.1$.

Ans.
Given that $x_0 = 1$, $y(x_0) = y_0 = 2$, $f(x, y) = x - y^2$ and $h = 0.1$.

By using backward Euler formula (14.37), we have

$$y_{i+1} = y_i + h f(x_{i+1}, y_{i+1}); \qquad i = 0, 1, \cdots, n$$
$$y_{i+1} = y_i + h(x_{i+1} - y_{i+1}^2); \qquad i = 0, 1, \cdots, n$$
$$y_{i+1}^2 + 10 y_{i+1} = 10 y_i + x_{i+1}; \qquad i = 0, 1, \cdots, n$$

This quadratic equation can be solved for $i = 0, 1$ to obtain the following results for $i = 0$

$$y_1^2 + 10 y_1 = 10 y_0 + x_1 = 21.1$$
$$y_1^2 + 10 y_1 - 21.1 = 0$$
$$y_1 = 1.789698$$

For $i = 1$

$$y_2^2 + 10y_2 = 10y_1 + x_2 = 19.09698$$

$$y_2^2 + 10y_2 - 19.09698 = 0$$

$$y_2 = 1.640556$$

Example

14.27

Use backward Euler method for the IVP $\dfrac{dy}{dx} = x^2 - \sin y$, $y(0) = .5$, to compute $y(0.4)$ with step size $h = 0.2$.

Ans.
We have $x_0 = 0$, $y_0 = 0.5$, $f(x, y) = x^2 - \sin y$ and $h = 0.2$.
By using backward Euler method, we obtain

$$y_{i+1} = y_i + h f(x_{i+1}, y_{i+1}); \qquad i = 0, 1, \cdots, n$$

$$y_{i+1} = y_i + h\left(x_{i+1}^2 - \sin(y_{i+1})\right); \qquad i = 0, 1, \cdots, n$$

$$5y_{i+1} + \sin(y_{i+1}) = 5y_i + x_{i+1}^2; \qquad i = 0, 1, \cdots, n$$

Using $i = 0$ in this equation, we get

$$5y_1 + \sin(y_1) = 5y_0 + x_1^2$$

$$5y_1 + \sin(y_1) = 2.54$$

We can use Newton–Raphson method to solve this nonlinear equation. The iterations of Newton–Raphson method with initial approximation $y_1^0 = 0.5$ are as follows

$$y_1^1 = 0.425237$$

$$y_1^2 = 0.425453$$

$$y_1^3 = 0.425453$$

So, value of $y_1 = y(0.2) = 0.425453$

Similarly, for $i = 1$, we get

$$5y_2 + \sin(y_2) = 5y_1 + x_2^2$$

$$5y_1 + \sin(y_1) = 2.287265$$

The Newton–Raphson method provides following iterations with initial approximation, $y_2^0 = 0.4$.

$y_2^1 = 0.382747$

$y_2^2 = 0.382757$

So, value of $y_2 = y(0.4) = 0.382757$

Stability Analysis of Backward Euler Method for IVP $y' = Ay$, $y(0) = y_0$

The backward Euler method (14.37) is given by

$$y_n = y_{n-1} + h f(x_n, y_n); \quad n = 1, 2, 3, \dots$$

For IVP $y' = A y$, $y(0) = y_0$, we have

$$y_n = y_{n-1} + h A y_n$$

$$y_n = \frac{1}{(1 - hA)} y_{n-1}$$

Using this expression recursively, we have

$$y_n = \frac{1}{(1 - hA)} y_{n-1}$$

$$= \frac{1}{(1 - hA)^2} y_{n-2}$$

$$\vdots$$

$$= \frac{1}{(1 - hA)^n} y_0$$

The backward Euler scheme is stable, if $\dfrac{1}{|1 - hA|} \leq 1$. Otherwise, as n increases, it will increase indefinitely. Consider following different cases

a) A is real number, then backward Euler scheme is stable for $Ah \leq 0$. The step size h is a positive constant. Also, the constant A must be negative for convergent solution. Therefore $Ah \leq 0$. It implies that backward Euler method is unconditionally stable for real $A \leq 0$.

b) A is complex number $A = A_R + iA_I$, then

$$\frac{1}{|1 - h A|} = \frac{1}{|1 - (A_R + iA_I)h|} = \frac{1}{\sqrt{(1 - A_R h)^2 + (A_I h)^2}}$$

The real part (A_R) of constant A must be negative for a convergent solution. Therefore, for a nonzero constant A

$$\frac{1}{\sqrt{(1-A_R h)^2 +(A_I h)^2}} < 1$$

Hence, backward Euler scheme is unconditionally stable for the IVP $y' = A\,y$, $y(0) = y_0$ with negative real part (A_R) of constant A.

Example ──────────────────────────────────── **14.28**

Use backward Euler method for the IVP $\dfrac{dy}{dx} = -20y$, $y(0) = 1$, to compute $y(1)$ with step size $h = 0.2$. Compare the result with forward Euler method in Example 14.22.

Ans.
The IVP is as follows

$\dfrac{dy}{dx} = -20y$, $y(0) = 1$, with step size $h = 0.2$

Backward Euler formula (14.37) is as follows

$$y_{i+1} = y_i + h\,f(x_{i+1}, y_{i+1}); \quad i = 0, 1, \cdots, n$$

$$y_{i+1} = y_i + (0.2)(-20y_{i+1}); \quad i = 0, 1, \cdots, n$$

$$y_{i+1} = \frac{1}{5}(y_i) = 0.2y_i; \quad i = 0, 1, \cdots, n$$

Using $i = 0(1)4$ in this formula, we have

$y(x_1) = y(0.2) = y_1 = 0.2y_0 = 0.2$

$y_2 = y(0.4) = 0.2y_1 = 0.04$

$y_3 = y(0.6) = 0.008$

$y_4 = y(0.8) = 0.0016$

$y_5 = y(1.0) = 0.00032$

It shows the stability of backward Euler scheme.

Table 14.1 Numerical Schemes for IVP

Sr. No.	Numerical Scheme for IVP $\dfrac{dy}{dx} = f(x, y), \; y(x_0) = y_0$	Order of Scheme	Order of Error Term	Programming	Value obtained in Iterations
1	Picard method $y^{(n+1)}(x) = y(x_0) + \displaystyle\int_{x_0}^{x} f(x, y^{(n)}(x))\,dx, \quad n = 0, 1, 2, \cdots$ where $y^{(0)}(x) = y(x_0)$	Approximation is not numerical	Approximation is not numerical	Not possible, as computation requires analytical integration	Accuracy depends on the number of iterations
2	Taylor series method of order n $y_1 = y(x_0) + hy'(x_0) + \dfrac{(h)^2}{2!} y''(x_0) + \cdots + \dfrac{(h)^n}{(n)!} y^n(x_0)$	n	$n+1$	Not possible, as computation requires derivative terms	1 iteration
3	Euler method $y_{i+1} = y(x_i) = y_i + hf(x_i, y_i), \; i = 0, 1, 2, \cdots$	1	2	Easy	1 iteration
4	Modified Euler method $y_{i+1}^{(0)} = y_i + hf(x_i, y_i)$ $y_{i+1}^{(k+1)} = y_i + \dfrac{h}{2}\Big(f(x_i, y_i) + f\big(x_{i+1}, y_{i+1}^{(k)}\big)\Big)$ $i = 0, 1, 2, \ldots; \; k = 0, 1, 2, \ldots$	2	3	Moderate	Accuracy depends on the number of iterations
5	Ralston and Rabinowitz method $y_{i+1} = y_i + \left(\dfrac{1}{3}k_1 + \dfrac{2}{3}k_2\right),$ where k_1 and k_2 are given by $k_1 = hf(x_i, y_i)$ $k_2 = hf\left(x_i + \dfrac{3}{4}h, \; y_i + \dfrac{3}{4}k_1\right)$	2	3	Easy	1 iteration

#	Method				
6	**3rd Order Runge–Kutta method** $$y_{i+1} = y_i + \frac{1}{8}(2k_1 + 3k_2 + 3k_3),$$ where k_1, k_2 and k_3 are given by $$k_1 = hf(x_i, y_i)$$ $$k_2 = hf\left(x_i + \frac{2}{3}h,\, y_i + \frac{2}{3}k_1\right)$$ $$k_3 = hf\left(x_i + \frac{2}{3}h,\, y_i + \frac{2}{3}k_2\right)$$	3	Easy	4	1 iteration
7	**Classical Runge–Kutta method (or) Runge–Kutta method order 4** $$y_{i+1} = y_i + \frac{1}{6}(k_1 + 2k_2 + 2k_3 + k_4)$$ where k_1, k_2, k_3 and k_4 are given by $$k_1 = hf(x_i, y_i)$$ $$k_2 = hf\left(x_i + \frac{1}{2}h,\, y_i + \frac{1}{2}k_1\right)$$ $$k_3 = hf\left(x_i + \frac{1}{2}h,\, y_i + \frac{1}{2}k_2\right)$$ $$k_4 = hf(x_i + h,\, y_i + k_3)$$	4	Easy	5	1 iteration
8	**Milne method (Milne–Simpson method)** $$y_4^{(0)} = y_0 + \frac{4h}{3}(2f_1 - f_2 + 2f_3)$$ $$y_4^{(n+1)} = y_2 + \frac{h}{2}(f_2 + 4f_3 + f_4^{(n)}); \quad n = 0,1,2,\dots$$	3	Moderate	4	Accuracy depends on the number of iterations
9	**Adams formula (Adams–Bashforth predictor and Adams–Moulton corrector).** $$y_1^{(0)} = y_0 + \frac{h}{24}(55f_0 - 59f_{-1} + 37f_{-2} - 9f_{-3})$$ $$y_1^{(n+1)} = y_0 + \frac{h}{24}(9f_1^{(n)} + 19f_0 - 5f_{-1} + f_{-2}), \quad n = 0, 1, 2\dots$$	3	Moderate	4	Accuracy depends on the number of iterations

Exercise 14

1. Classify the following differential equations according to sections (14.1.1–5)

i) $\dfrac{\partial^2 u}{\partial t^2} - \sin(u)\dfrac{\partial^2 u}{\partial y^2} - \left(\dfrac{\partial^2 u}{\partial x^2}\right)^2 = u$ ii) $\dfrac{d^2 y}{dx^2} - x\dfrac{dy}{dx} + y = \cos(x)$

iii) $\dfrac{d^2 y}{dx^2} + \sin(y) = 0$ iv) $y^2\dfrac{\partial^2 z}{\partial x^2} + x\dfrac{\partial z}{\partial y} + z\left(\dfrac{\partial z}{\partial x}\right)^3 + \sin y = 0$

v) $x\dfrac{d^3 y}{dx^3} + e^x y = 0$

Ans.
i) PDE, Order = 2, Degree = 2, Homogeneous, Constant Coefficient, Nonlinear
ii) ODE, Order = 2, Degree = 1, Nonhomogeneous, Variable Coefficient, Linear
iii) ODE, Order = 2, Degree = 1, Homogeneous, Constant Coefficient, Nonlinear
iv) PDE, Order = 2, Degree = 1, Nonhomogeneous, Variable Coefficient, Nonlinear
v) ODE, Order = 3, Degree = 1, Homogeneous, Variable Coefficient, Linear

2. Find the solution of first order IVP

$$\dfrac{dy}{dx} = x^2 + y; \ y(0) = 0 \text{, for } x = 0.1$$

with the help of Picard method. Compute only first four nonzero terms.

Ans. $y = \dfrac{x^3}{3} + \dfrac{x^4}{12} + \dfrac{x^5}{60} + \dfrac{x^6}{360}$, $y(0.1) = .000341837$

3. Obtain first four successive approximations of Picard method to the following IVP

$$\dfrac{dx}{dt} = t - x; \ x(0) = 1.$$

Use it to compute $x(0.2)$ and compare the result with the exact solution.

Ans. $1 - t + \dfrac{2t^2}{2!} - \dfrac{2t^3}{3!} + \dfrac{2t^4}{4!} - \dfrac{t^5}{5!}$, $x(0.2) = 0.837464$

Exact Solution, $x = -1 + t + 2e^{-t}$, $x(0.2) = 0.837462$

4. Use Taylor series of order four to estimate $y(0.1)$ for the following IVP
$y' = x^2 y - 1, \ y(0) = 1.$
Ans. $y(0.1) = 0.900308$

5. Compute the values of $y(0.1)$ and $y(0.2)$ using Taylor series method of order five for the following IVP

$$\dfrac{dy}{dx} = x + y; \ y(0) = 2$$

Use six decimal places arithmetic and compare the result with exact solutions.

Ans. $2 + 2x + 3\dfrac{x^2}{2!} + 3\dfrac{x^3}{3!} + 3\dfrac{x^4}{4!} + 3\dfrac{x^5}{5!}$, $y(0.1) = 2.215513$, $y(0.2) = 2.464208$

Exact Solution, $y = -1 - x + 3e^x$, $y(0.1) = 2.215512754$, $y(0.2) = 2.464208274$

6. Evaluate the first five terms of Taylor series for the second order IVP

    $y'' + 2xy' - y^2 = x$; $y(0) = 0$, $y'(0) = 0$.

    Also, compute $y(0.1)$.

    Ans. $\dfrac{x^3}{6} - \dfrac{x^5}{20}$, 0.000166167

7. Find the first six terms of Taylor series to solve the following IVP
    $y''' + yy'' - xy^2 = 0$; $y(0) = 1$, $y'(0) = 1$, $y''(0) = 1$ and hence find $y(0.1)$.

    Ans. $1 + x + \dfrac{x^2}{2!} - \dfrac{x^3}{3!} + \dfrac{x^4}{4!} - 3\dfrac{x^6}{6!}$, $y(0.1) = 1.1048375$

8. Solve the IVP $\dfrac{dy}{dx} = -2(x + y^2)$, $y(1) = 0$; in the range $1 \le x \le 2$ using Euler method. Use step size $h = 0.2$.

    Ans. $x = 1.2$, $y = -0.400000$
    $x = 1.4$, $y = -0.944000$
    $x = 1.6$, $y = -1.860454$
    $x = 1.8$, $y = -3.884971$
    $x = 2.0$, $y = -10.642171$

9. Calculate the solution of IVP $\dfrac{dy}{dx} = e^x - \sin(y)$, $y(0) = 0$, at $x = 0.2$ with the aid of Euler method. Use step size $h = 0.05$.

    Ans. $x = 0.05$, $y = 0.05$
    $x = 0.10$, $y = 0.100065$
    $x = 0.15$, $y = 0.150328$
    $x = 0.20$, $y = 0.200932$

10. Given $\dfrac{dx}{dt} = \dfrac{t - x^2}{t + x^2}$, $x(0) = 1$, compute the values of $x(0.05)$, $x(0.1)$, $x(0.15)$ with the aid of Euler method.

    Ans. $t = 0.05$, $x = 0.95$
    $t = 0.10$, $x = 0.905249$
    $t = 0.15$, $x = 0.866125$

11. Use Euler method to estimate the value of $y(2)$ for the IVP $y' = e^x$, $y(1) = 1$ with the step sizes $h = 0.25$, 0.5. Compare the solutions with the exact solution and discuss the performance of Euler method.

    Ans. For $h = 0.25$, we have
    $x = 1.25$, $y = 1.679570$
    $x = 1.50$, $y = 2.552156$
    $x = 1.75$, $y = 3.672579$
    $x = 2.00$, $y = 5.111229$
    For $h = 0.5$, we have
    $x = 1.5$, $y = 2.359141$
    $x = 2.0$, $y = 4.599986$
    Exact solution is $y = e^x$, hence $y(2) = e^2 = 7.389056$

12. One of the first order nonlinear differential equation of the form $\dfrac{dy}{dx} + P(x)y = Q(x)y^n$ is known

as Bernoulli differential equation. For example, $\dfrac{dy}{dx} + y = xy^3$, solve this equation to obtain $y(1)$

for given initial condition $y(0) = 1$. Use Euler method with step size $h = 0.2$.

Ans.

$x = 0.2, y = 0.800000$
$x = 0.4, y = 0.660480$
$x = 0.6, y = 0.551434$
$x = 0.8, y = 0.461269$
$x = 1.0, y = 0.384718$

13. The population of a given city is governed by the law, $\dfrac{dx}{dt} = (0.02)x - (.0001)x^2$, where time (t) and

city population (x) are measured in years and thousands, respectively. Estimate the population of the city in year 2012, while it was given that population of the city in the year 2011 is 28,750. Apply Euler formula with step size $h = 3$ months.

Ans. Given $t_0 = 2011$, $x_0 = 28.750$, $h = 0.25$, we have

$t = 2011.250, x = 28.873085$
$t = 2011.500, x = 28.996609$
$t = 2011.750, x = 29.120571$
$t = 2012.000, x = 29.244974$

14. Solve the IVP $\dfrac{dy}{dx} = x - y^2$, $y(1) = 0$ in the range $1 \le x \le 1.2$ using Euler and modified Euler

methods. Take step size $h = 0.1$.

Ans. Euler method

$x = 1.1,$ $y = 0.1$
$x = 1.2,$ $y = 0.209$

Modified Euler method

$x = 1.1,\ y_1^{(0)} = 0.1,\ y_1^{(1)} = 0.104500,\ y_1^{(2)} = 0.104454,\ y_1^{(3)} = 0.104454$

$x = 1.2,\ y_2^{(0)} = 0.213363,\ y_2^{(1)} = 0.216633,\ y_2^{(2)} = 0.216562,\ y_2^{(3)} = 0.216564$

15. Use modified Euler method to find an approximate value of variable u when $x = 1.2$ for the IVP

$\dfrac{du}{dx} = x^2 + u;\ u(1) = 2$.

Compute only four iterations of modified Euler method. Take $h = 0.1$.

Ans.

$u_1^{(0)} = 2.300000,\ u_1^{(1)} = 2.325500,\ u_1^{(2)} = 2.326775,\ u_1^{(3)} = 2.326839,\ u_1^{(4)} = 2.326842$

$u_2^{(0)} = 2.680526,\ u_2^{(1)} = 2.709710,\ u_2^{(2)} = 2.711169,\ u_2^{(3)} = 2.711242,\ u_2^{(4)} = 2.711246$

16. Calculate the solution of $\dfrac{dy}{dx} = e^x + x\sin(y)$, $y(0) = 0$, at $x = 0.2$ with the aid of modified Euler

method. Use step size $h = 0.05$. Compute the iterations till the difference between last two consecutive iterations is less than 0.00005.

Ans.

$$y_1^{(0)} = 0.05,\ y_1^{(1)} = 0.051344,\ y_1^{(2)} = 0.051346$$

$$y_2^{(0)} = 0.104038,\ y_2^{(1)} = 0.105581,\ y_2^{(2)} = 0.105585$$

$$y_3^{(0)} = 0.161370,\ y_3^{(1)} = 0.163126,\ y_3^{(2)} = 0.163132$$

$$y_4^{(0)} = 0.222442,\ y_4^{(1)} = 0.224425,\ y_4^{(2)} = 0.224435$$

17. Find the values of $y(0.1)$ and (0.2) for Riccati equation $y' = y^2 + xy - 1;\ y(0) = 0$ with the help of modified Euler method.

Ans.

$$y_1^0 = -0.1,\ y_1^1 = -0.1$$

$$y_2^0 = -0.2,\ y_2^1 = -0.2$$

18. Use Runge–Kutta second order method with minimum bound on truncation error (Ralston and Rabinowitz method) to solve the IVP $\dfrac{dy}{dx} = x - y^2,\ y(0) = 1.25$. Compute $y(0.2)$ with step size $h = 0.1$.

Ans.

$k_1 = -0.156250$ $k_2 = -0.120826$ $y(0.1) = 1.117366$

$k_1 = -0.114851$ $k_2 = -0.088843$ $y(0.2) = 1.019853$

19. Find the value of $y(1.3)$ for Bernoulli equation $y' = x^2 y - y^3$ with initial condition, $y(1) = 2$. Use Ralston and Rabinowitz method with step size 0.1.

Ans.

$k_1 = -0.600000$ $k_2 = -0.193266$ $y(1.1) = 1.671156$

$k_1 = -0.264504$ $k_2 = -0.116121$ $y(1.2) = 1.505574$

$k_1 = -0.124474$ $k_2 = -0.052074$ $y(1.3) = 1.429367$

20. Solve the initial value problem $\dfrac{dy}{dx} = y - \dfrac{1}{x} y^2,\ y(1) = 1$ to find $y(1.2)$ by Runge–Kutta method of order four with step size 0.1.

Ans.

$k_1 = 0$ $k_2 = 0.004762$ $k_3 = 0.004546$ $k_4 = 0.008717$

$y(1.1) = 1.004555$

$k_1 = 0.008716$ $k_2 = 0.012378$ $k_3 = 0.012239$ $k_4 = 0.015524$

$y(1.2) = 1.016801$

21. Solve the initial value problem $\frac{dy}{dx} = x^2 - \sin(y)$, $y(0) = 1$ to compute $y(0.3)$ by Runge–Kutta fourth order method with step size 0.1.

Ans.

$k_1 = -0.084147 \quad k_2 = -0.081550 \quad k_3 = -0.081625 \quad k_4 = -0.078462$

$y(0.1) = 0.918507$

$k_1 = -0.078470 \quad k_2 = -0.074777 \quad k_3 = -0.074895 \quad k_4 = -0.070705$

$y(0.2) = 0.843754$

$k_1 = -0.070714 \quad k_2 = -0.066068 \quad k_3 = -0.066228 \quad k_4 = -0.061152$

$y(0.3) = 0.777677$

22. A body of weight 10 kg falls from rest toward the earth with a velocity v. Air resistance on the body that is dependent on the velocity of a body is approximately $2v$. Find the velocity of the body after time $t = 2$ sec. Use fourth order Runge–Kutta method with step size 1 sec.

(Hint: Newton's second law $F = ma$; where $a = \dfrac{dv}{dt}$ and $m = 10/9.8 = 1.02$.

Two forces acting on the body are given by
i) Gravitational force ($F_1 = mg = 10$)
ii) Air resistance ($F_2 = -2v$, negative sign as it opposes the motion)
Since body falls from rest i.e. $v(0) = 0$.
Finally, we have the following IVP

$$m\frac{dv}{dt} = F_1 + F_2 \text{ (or) } 1.02\frac{dv}{dt} = 10 - 2v$$

$v(0) = 0$.

On solving this IVP with Runge–Kutta fourth order method, we have

$k_1 = 9.803922 \quad k_2 = 0.192234 \quad k_3 = 9.615458 \quad k_4 = -9.049916$

$v(1) = 3.394898$

$k_1 = 3.147259 \quad k_2 = 0.061711 \quad k_3 = 3.086758 \quad k_4 = -2.905207$

$v(2) = 4.484730$

23. Solve the initial value problem $\frac{dx}{dt} = \frac{t - 2x}{2t - x}$, $x(0) = 1$ on the interval $[0, 0.4]$ with the aid of fourth order Runge–Kutta method. Take $h = 0.2$.

Ans.

$k_1 = 0.400000 \quad k_2 = 0.460000 \quad k_3 = 0.458252 \quad k_4 = 0.513395$

$x(0.2) = 1.458317$

$k_1 = 0.513388 \quad k_2 = 0.561433 \quad k_3 = 0.558029 \quad k_4 = 0.597312$

$x(0.4) = 2.016587$

24. Solve the initial value problem $\dfrac{dr}{d\theta} + r^2 = \sin 2\theta$; $r(0) = 0$ by using 4th order Runge–Kutta method to compute the values of $r(0.2)$ and $r(0.4)$.

Ans.

$k_1 = 0.000000$ $k_2 = 0.039734$ $k_3 = 0.039655$ $k_4 = 0.077569$

$r(0.2) = 0.039391$

$k_1 = 0.077573$ $k_2 = 0.111706$ $k_3 = 0.111114$ $k_4 = 0.138941$

$r(0.4) = 0.149750$

25. Use Milne predictor-corrector method to compute $y(0.4)$ from differential equation $\dfrac{dy}{dx} = y^2 - x^2$ and following values

x:	0	0.1	0.2	0.3
y:	1	1.11	1.25	1.42

Ans.

$y_4^{(P)} = 1.636600 = y_4^{(0)}$

$y_4^{(1)} = 1.641552$, $y_4^{(2)} = 1.642093$, $y_4^{(3)} = 1.642152$, $y_4^{(4)} = 1.642159$, $y_4^{(5)} = 1.642159$

26. The values of $y(1.1)$, $y(1.2)$ and $y(1.3)$ are obtained in exercise 19 for Bernoulli equation $y' = x^2 y - y^3$ with initial condition $y(1) = 2$. Use these values in Milne predictor-corrector method to obtain the value of $y(1.5)$ with $h = 0.1$.

Ans.

$y(1.1) = 1.671156$ $y(1.2) = 1.505574$ $y(1.3) = 1.429367$

$y_4^{(P)} = 1.326035 = y_4^{(0)}$

$y_4^{(1)} = 1.405702$, $y_4^{(2)} = 1.396040$, $y_4^{(3)} = 1.397305$, $y_4^{(4)} = 1.397141$, $y_4^{(5)} = 1.397162$, $y_4^{(6)} = 1.397160$

$y_5^{P} = 1.409478 = y_5^{(0)}$

$y_5^{(1)} = 1.426398$, $y_5^{(2)} = 1.424265$, $y_5^{(3)} = 1.424538$, $y_5^{(4)} = 1.424503$, $y_5^{(5)} = 1.424508$

27. Use Adams predictor-corrector method in exercise 25.

Ans.

$y_1^{(P)} = 1.638092 = y_1^{(0)}$

$y_1^{(1)} = 1.640505$, $y_1^{(2)} = 1.640802$, $y_1^{(3)} = 1.640839$, $y_1^{(4)} = 1.640843$

28. Solve the initial value problem, $\dfrac{dy}{dx} = x^2 - \sin(y)$, $y(0) = 1$ to find $y(0.1)$, $y(0.2)$ and $y(0.3)$ by Runge–Kutta fourth order method. Further, obtain the values of $y(0.4)$ and $y(0.5)$ using the Adams method.

Ans.

$y(0.1) = 0.918507$ $y(0.2) = 0.843754$ $y(0.3) = 0.777677$

$y_1^{(P)} = 0.721933$

$y_1^{(1)} = 0.721938$, $y_1^{(2)} = 0.721938$

$y_2^{(P)} = 0.677928$

$y_2^{(1)} = 0.677939$, $y_2^{(2)} = 0.677939$

Systems of First Order ODEs and Higher Order ODEs: Initial and Boundary Value Problems

Mathematics compares the most diverse phenomena and discovers the secret analogies that unite them.

Jean-Baptiste Joseph Fourier
(March 21, 1768–May 16, 1830)
He was a great mathematician, who discovered the Fourier series and the greenhouse effect.

In Chapter 14, we have discussed the numerical methods for the solutions of the first order ordinary differential equation with initial condition (14.1)

$$\frac{dy}{dx} = f(x, y); \quad y(x_0) = y_0$$

In this chapter, we will discuss the numerical solutions of systems of first order ordinary differential equations (ODEs) and the higher order ODEs with initial and boundary conditions. Some examples of these ODEs are as follows

System of first order ODEs with initial conditions	Second and higher order ODEs with initial conditions Initial value problems (IVPs)	
$\dfrac{dy}{dx} = x\,y - \sin(z)$ $\dfrac{dz}{dx} = x^2 \cos(y) e^z$ $y(1) = 2, z(1) = -1$	i) $\dfrac{d^2 y}{dx^2} + x\dfrac{dy}{dx} + y = 3;$ ii) $\dfrac{d^3 y}{dx^3} + \sin x \dfrac{d^2 y}{dx^2} + xy = \cos x;$	$y(0) = 1,\ y'(0) = 2$ $y(0) = 1,\ y'(0) = 2,\ y''(0) = 2$

Second and higher order ODEs with boundary conditions Boundary value problems (BVPs)
i) $x^2 \dfrac{d^2 y}{dx^2} + (x-1)\dfrac{dy}{dx} + y = 3;$ $y(0) = 1,\ \ y(1) = 3$
ii) $\dfrac{d^3 y}{dx^3} + \sin x \dfrac{d^2 y}{dx^2} + xy = \cos x;$ $y(0) = 1,\ \ y(1) = 2,\ \ y(3) = -4$

These differential equations arise in the mathematical formulations of various scientific and engineering problems (for examples, refer to exercises 15.7 and 15.8). In this chapter, following three types of problems are discussed for numerical solutions.

System of first order ODEs with initial conditions
The numerical methods discussed in Chapter 14 can be generalized easily to the following systems of first order ODEs with initial conditions

$$\frac{dy}{dx} = f(x, y, z)$$

$$\frac{dz}{dx} = g(x, y, z)$$

$$y(x_0) = y_0, z(x_0) = z_0 \tag{15.1}$$

This chapter contains following four methods for the solution of system (15.1)

   i)   Picard method
   ii)   Taylor series method
   iii)   Euler method
   iv)   Classical Runge–Kutta method

Note that we have discussed these methods for solutions of a system of two first order ODES only. But, these methods can be generalized for a general system of any finite number of first order ODEs.

Second and higher order ODEs with initial conditions
Initial value problems (IVPs)
The higher order ODE with initial conditions can be converted into the system of first order ODEs of the types (15.1). For example

$$y'' + xy' + y^2 = \sin(x); \qquad y(0) = 1,\ y'(0) = 1$$

Let $y' = z$, using this in above equation, we have

$$z' + xz + y^2 = \sin(x) \text{ (or) } z' = \sin(x) - xz - y^2$$

Equivalently, the system of first order ODEs with initial conditions is as follows

$$\frac{dy}{dx} = y' = z$$
$$\frac{dz}{dx} = z' = \sin(x) - xz - y^2$$; $\quad y(0) = 1,\ z(0) = 1$

Therefore, solutions for second and higher order ODEs with initial conditions can be obtained by converting them into the system of first order ODEs like system 15.1.

Second and higher order ODEs with boundary conditions
Boundary value problems (BVPs)
General linear second order ODE with boundary conditions is given by

$$\frac{d^2 y}{dx^2} + p(x)\frac{dy}{dx} + q(x)y = r(x)$$
$$c_1 y(a) + c_2 y'(a) = c_3,\ c_4 y(b) + c_5 y'(b) = c_6$$

The following methods are discussed for the numerical solutions of the linear boundary value problems (BVPs).

i)   Shooting method
ii) Finite difference method

In Shooting method, the BVP is converted into an equivalent system of IVPs, then the solution of BVP can be obtained from the solution of these IVPs. In the finite difference method, the various derivative terms in the BVP are approximated with the help of finite differences. The BVP is converted into a system of linear algebraic equations, which can be solved easily to give the solution of BVP.

15.1 Picard Method

The Picard method has been discussed in Section 14.2 for a single first order ODE with the initial condition. The method can be generalized to find the numerical solutions of the system of first order ODEs (15.1).

On integrating the system (15.1) from initial point x_0 to any general point x, we get

$$y(x) = y(x_0) + \int_{x_0}^{x} f(x, y, z)\, dx$$

$$z(x) = z(x_0) + \int_{x_0}^{x} g(x, y, z)\, dx$$

Let y_0 and z_0 have initial approximations $y^{(0)}(x)$ and $z^{(0)}(x)$ respectively. The next approximations $y^{(1)}(x)$ and $z^{(1)}(x)$ are given by

$$y^{(1)}(x) = y(x_0) + \int_{x_0}^{x} f(x, y^{(0)}(x), z^{(0)}(x)) dx$$

$$z^{(1)}(x) = z(x_0) + \int_{x_0}^{x} g(x, y^{(0)}(x), z^{(0)}(x)) dx$$

Similarly, the higher approximations are as follows

$$y^{(n+1)}(x) = y(x_0) + \int_{x_0}^{x} f(x, y^{(n)}(x), z^{(n)}(x)) dx$$

$$z^{(n+1)}(x) = z(x_0) + \int_{x_0}^{x} g(x, y^{(n)}(x), z^{(n)}(x)) dx \qquad n = 0, 1, 2,... \qquad (15.2)$$

Example _____ **15.1**

Solve the following IVP with the help of Picard method. Obtain only first three approximations and find the values of $y(0.1)$ and $z(0.1)$.

$$\frac{dy}{dx} = z + x$$

$$\frac{dz}{dx} = y^2 + x + z$$

$$y(0) = 0, z(0) = 0$$

Ans.

Comparing the given system with the system (15.1), we get

$$\frac{dy}{dx} = z + x = f(x, y, z)$$

$$\frac{dz}{dx} = y^2 + x + z = g(x, y, z)$$

$$x_0 = y_0 = z_0 = 0$$

The initial approximation is $y^{(0)}(x) = y_0 = 0$ and $z^{(0)}(x) = z_0 = 0$. Using Picard method (15.2), the iterations are given by

First approximation

$$y^{(1)}(x) = y(x_0) + \int_{x_0}^{x} f(x, y^{(0)}(x), z^{(0)}(x)) dx$$

$$z^{(1)}(x) = z(x_0) + \int_{x_0}^{x} g(x, y^{(0)}(x), z^{(0)}(x)) dx$$

$$y^{(1)} = 0 + \int x\,dx = \frac{x^2}{2}$$

$$z^{(1)} = 0 + \int (0^2 + x + 0) = \frac{x^2}{2}$$

Second approximation

$$y^{(2)}(x) = y(x_0) + \int_{x_0}^{x} f(x, y^{(1)}(x), z^{(1)}(x))\,dx$$

$$z^{(2)}(x) = z(x_0) + \int_{x_0}^{x} g(x, y^{(1)}(x), z^{(1)}(x))\,dx$$

$$y^{(2)} = 0 + \int \left(\frac{x^2}{2} + x \right) dx = \frac{x^3}{6} + \frac{x^2}{2}$$

$$z^{(2)} = 0 + \int \left(\left(\frac{x^2}{2} \right)^2 + x + \frac{x^2}{2} \right) dx = \frac{x^5}{20} + \frac{x^2}{2} + \frac{x^5}{6}$$

Third approximation

$$y^{(3)}(x) = y(x_0) + \int_{x_0}^{x} f(x, y^{(2)}(x), z^{(2)}(x))\,dx$$

$$z^{(3)}(x) = z(x_0) + \int_{x_0}^{x} g(x, y^{(2)}(x), z^{(2)}(x))\,dx$$

$$y^{(3)} = 0 + \int \left(\left(\frac{x^5}{20} + \frac{x^2}{2} + \frac{x^3}{6} \right) + x \right) dx = \frac{x^6}{120} + \frac{x^3}{6} + \frac{x^4}{24} + \frac{x^2}{2}$$

$$z^{(3)} = \int \left(\left(\frac{x^3}{6} + \frac{x^2}{2} \right)^2 + x + \frac{x^5}{20} + \frac{x^2}{2} + \frac{x^3}{6} \right) dx = \frac{x^7}{252} + \frac{x^6}{45} + \frac{x^5}{20} + \frac{x^4}{24} + \frac{x^3}{6} + \frac{x^2}{2}$$

Values of $y(0.1)$ and $z(0.1)$

$$y(0.1) = \frac{(0.1)^6}{120} + \frac{(0.1)^3}{6} + \frac{(0.1)^4}{24} + \frac{(0.1)^2}{2} = 0.00517084$$

$$z(0.1) = \frac{(0.1)^7}{252} + \frac{(0.1)^6}{45} + \frac{(0.1)^5}{20} + \frac{(0.1)^4}{24} + \frac{(0.1)^3}{6} + \frac{(0.1)^2}{2} = 0.00517091$$

Note that we can use latest value $y^{(n+1)}(x)$ in 2nd equation of formula (15.2) to get faster convergence. Therefore, we have

$$z^{(n+1)}(x) = z(x_0) + \int_{x_0}^{x} g\left(x, y^{(n+1)}(x), z^{(n)}(x)\right) dx$$

15.2 Taylor Series Method

The Taylor series method discussed in Section 14.3 can be extended easily for the solution of system (15.1). The Taylor series expansions for $y(x)$ and $z(x)$ are as follows

$$y(x) = y(x_0) + (x - x_0)y'(x_0) + \frac{(x - x_0)^2}{2!}y''(x_0) + \dots \frac{(x - x_0)^n}{n!}y^{(n)}(x_0) + \dots$$

$$z(x) = z(x_0) + (x - x_0)z'(x_0) + \frac{(x - x_0)^2}{2!}z''(x_0) + \dots \frac{(x - x_0)^n}{n!}z^{(n)}(x_0) + \dots \qquad (15.3)$$

We can obtain the following equivalent form by using $x = x_0 + h$ in the expressions (15.3)

$$y(x_0 + h) = y(x_0) + h\,y'(x_0) + \frac{(h)^2}{2!}y''(x_0) + \dots + \frac{(h)^n}{(n)!}y^n(x_0) + \dots$$

$$z(x_0 + h) = z(x_0) + h\,z'(x_0) + \frac{(h)^2}{2!}z''(x_0) + \dots + \frac{(h)^n}{(n)!}z^n(x_0) + \dots$$

The higher order derivative terms at $\left(x_0, y_0, z_0\right)$ can be computed by differentiating the equations $\dfrac{dy}{dx} = f(x, y, z)$ and $\dfrac{dz}{dx} = g(x, y, z)$ from the system (15.1).

Example ─── **15.2** ─

Use Taylor series method to compute the values of $y(0.1)$ and $z(0.1)$ for the following system

$$\frac{dy}{dx} = z + x$$

$$\frac{dz}{dx} = y^2 + x + z$$

$$y(0) = 0,\ z(0) = 0$$

Compute only first five terms of Taylor series.

Ans.

The initial conditions are given by

$$x_0 = 0,\ y(x_0) = y_0 = 0,\ z(x_0) = z_0 = 0$$

Computing various derivative terms for the given system, we have

$$\begin{aligned} y' &= z + x & y_0' &= 0 \\ z' &= y^2 + x + z & z_0' &= 0 \\[6pt] y'' &= z' + 1 & y_0'' &= 1 \\ z'' &= 2yy' + 1 + z & z_0'' &= 1 \end{aligned}$$

$$y''' = z$$
$$z''' = 2yy'' + 2y'^2 + z''$$

$$y_0''' = 1$$
$$z_0''' = 1$$

$$y^{iv} = z'''$$
$$z^{iv} = 6y'y'' + 2yy''' + z'''$$

$$y^{iv}_0 = 1$$
$$z^{iv}_0 = 1$$

$$y^v = z^{iv}$$
$$z^v = 6y''^2 + 8y'y''' + 2yy^{iv} + z^{iv}$$

$$y^v_0 = z^{iv}_0 = 1$$
$$z^v_0 = 7$$

By using these derivative terms in the Taylor series (15.3), we get

$$y(x) = 0 + (x-0)(0) + \frac{(x-0)^2}{2!}(1) + \frac{(x-0)^3}{3!}(1) + \frac{(x-0)^4}{4!}(1) + \frac{(x-0)^5}{5!}(1) + \cdots$$

$$z(x) = 0 + (x-0)(0) + \frac{(x-0)^2}{2!}(1) + \frac{(x-0)^3}{3!}(1) + \frac{(x-0)^4}{4!}(1) + \frac{(x-0)^5}{5!}(7) + \cdots$$

The final expressions for Taylor series expansions up to first five terms are as follows

$$y(x) = \frac{x^2}{2} + \frac{x^3}{6} + \frac{x^4}{24} + \frac{x^5}{120}$$

$$z(x) = \frac{x^2}{2} + \frac{x^3}{6} + \frac{x^4}{24} + \frac{7x^5}{120}$$

Values of $y(0.1)$ and $z(0.1)$ are given by

$$y(0.1) = \frac{(0.1)^2}{2} + \frac{(0.1)^3}{6} + \frac{(0.1)^4}{24} + \frac{(0.1)^5}{120} = 0.005170916$$

$$z(0.1) = \frac{(0.1)^2}{2} + \frac{(0.1)^3}{6} + \frac{(0.1)^4}{24} + \frac{7(0.1)^5}{120} = 0.005171416$$

15.3 Euler Method

Euler method is the simplest method for the computation of numerical solution of system (15.1). The Taylor series expansions for $y(x)$ and $z(x)$ are as follows

$$y(x_0 + h) = y(x_0) + h\, y'(x_0) + \frac{(h)^2}{2!} y''(x_0) + \cdots + \frac{(h)^n}{(n)!} y^n(x_0) + \cdots$$

$$z(x_0 + h) = z(x_0) + h z'(x_0) + \frac{(h)^2}{2!} z''(x_0) + \cdots + \frac{(h)^n}{(n)!} z^n(x_0) + \cdots$$

Let $x_1 = x_0 + h$. Neglecting second and higher order terms in Taylor series expansions and using system (15.1), we get

$$y_1 = y(x_1) = y(x_0) + h\,y'(x_0)$$
$$= y_0 + h\,f(x_0, y_0, z_0)$$
$$z_1 = z(x_1) = z(x_0) + h\,z'(x_0)$$
$$= z_0 + h\,g(x_0, y_0, z_0)$$

Similarly, the values at other points are given by

$$y_2 = y_1 + h\,f(x_1, y_1, z_1)$$
$$z_2 = z_1 + h\,g(x_1, y_1, z_1)$$

$$y_3 = y_2 + h\,f(x_2, y_2, z_2)$$
$$z_3 = z_2 + h\,g(x_2, y_2, z_2)$$
$$\vdots$$

$$y_n = y_{n-1} + h\,f(x_{n-1}, y_{n-1}, z_{n-1})$$
$$z_n = z_{n-1} + h\,g(x_{n-1}, y_{n-1}, z_{n-1})$$

$$y_{i+1} = y_i + h\,f(x_i, y_i, z_i)$$
$$\text{(or)} \quad z_{i+1} = z_i + h\,g(x_i, y_i, z_i); \qquad i = 0, 1, \cdots, n \tag{15.4}$$

Example ——————————————————————————————————— **15.3**

Find the values of $y(0.75)$ and $z(0.75)$ for the following system of first order ODEs. Use Euler method with step size $h = 0.25$.

$$\frac{dy}{dx} = z + x$$

$$\frac{dz}{dx} = y^2 + x + z$$

$$y(0) = 0,\, z(0) = 0$$

Ans.

For given system, the initial conditions are as follows

$$x_0 = 0,\ y(x_0) = y_0 = 0,\ z(x_0) = z_0 = 0$$

The functions are $f(x, y, z) = x + z$ and $g(x, y, z) = y^2 + x + z$. The step size is $h = 0.25$.

Using Euler formula (15.4), the approximations are given by

$$y_1 = y(x_1) = y(0.25) = y_0 + h\,f(x_0, y_0, z_0)$$
$$= y_0 + h(x_0 + z_0) = 0 + 0.25(0 + 0) = 0$$

$$z_1 = z(x_1) = z(0.25) = z_0 + h\,g(x_0, y_0, z_0)$$
$$= z_0 + h(y_0^2 + x_0 + z_0) = 0 + 0.25(0) = 0$$

$$y_2 = y(x_2) = y(0.5) = y_1 + h\,f(x_1, y_1, z_1)$$
$$= y_1 + h(x_1 + z_1) = 0 + 0.25(0.25 + 0) = 0.0625$$
$$z_2 = z(x_2) = z(0.5) = z_1 + h\,g(x_1, y_1, z_1)$$
$$= z_1 + h(y_1^2 + x_1 + z_1) = 0 + 0.25(0 + .25 + 0) = 0.0625$$

$$y_3 = y(x_3) = y(0.75) = y_2 + h\,f(x_2, y_2, z_2)$$
$$= y_2 + h(x_2 + z_2) = 0.0625 + 0.25(0.5 + 0.0625) = 0.203125$$
$$z_3 = z(x_3) = z(0.75) = z_2 + h\,g(x_2, y_2, z_2)$$
$$= z_2 + h(y_2^2 + x_2 + z_2) = 0.0625 + 0.25((0.0625)^2 + 0.5 + 0.0625)$$
$$= 0.2041015625$$

Example ——————————————————— **15.4**

Use Euler method to compute the values of $y(1.5)$ and $z(1.5)$ for the following system of first order ODEs with step size $h = 0.1$.

$$\frac{dy}{dx} = x + yz^2 = f(x, y, z)$$

$$\frac{dz}{dx} = x^2 - yz = g(x, y, z) \quad y(1) = -1, \ z(1) = 2$$

Ans.

Given initial conditions, $x_0 = 1$, $y(x_0) = y_0 = -1$, $z(x_0) = z_0 = 2$,

functions $f(x, y, z) = x + yz^2$ and $g(x, y, z) = x^2 - yz$

step size $h = 0.1$.

By using Euler formula (15.4), we have

$$y_1 = y(x_1) = y(1.1) = y_0 + h\,f(x_0, y_0, z_0)$$
$$= y_0 + h(x_0 + y_0 z_0^2) = -1 + 0.1(1 + (-1)2^2) = -1.3$$
$$z_1 = z(x_1) = z(1.1) = z_0 + h\,g(x_0, y_0, z_0)$$
$$= z_0 + h(x_0^2 - y_0 z_0) = 2 + 0.1(1^2 - (-1)(2)) = 2.3$$

$$y_2 = y(x_2) = y(1.2) = y_1 + h f(x_1, y_1, z_1)$$
$$= y_1 + h(x_1 + y_1 z_1^2) = -1.3 + 0.1(1.1 + (-1.3)(2.3)^2) = -1.8777$$
$$z_2 = z(x_2) = z(1.2) = z_1 + h g(x_1, y_1, z_1)$$
$$= z_1 + h(x_1^2 - y_1 z_1) = 2.3 + 0.1((1.1)^2 - (-1.3)(2.3)) = 2.72$$

$$y_3 = y(x_3) = y(1.3) = y_2 + h f(x_2, y_2, z_2)$$
$$= y_2 + h(x_2 + y_2 z_2^2) = -1.8777 + 0.1(1.2 + (-1.8777)(2.72)^2) = -3.146897$$
$$z_3 = z(x_3) = z(1.3) = z_2 + h g(x_2, y_2, z_2)$$
$$= z_2 + h(x_2^2 - y_2 z_2) = 2.72 + 0.1((1.2)^2 - (-1.8777)(2.72)) = 3.374734$$

Similarly, other values can be computed as follows
$$y_4 = -6.600846 \qquad z_4 = 4.605729$$
$$y_5 = -20.463047 \qquad z_5 = 7.841899$$

Example — 15.5

Given the second order ODE $y'' + xy' + y^2 = \sin(x)$ with initial conditions $y(0) = 1$, $y'(0) = 1$. Use Euler method to find the values of y and y' at $x = 0.5$. Let the step size be 0.1.

Ans.

On substituting $y' = z$ in the given second-order ODE, the following system of first order ODEs is obtained

$$\frac{dy}{dx} = y' = z$$

$$\frac{dz}{dx} = z' = \sin(x) - xz - y^2$$

$$y(0) = 1, \ z(0) = 1$$

For this system, we have

initial conditions $x_0 = 0$, $y(x_0) = y_0 = 1$, $z(x_0) = z_0 = 1$,
functions $f(x, y, z) = z$ and $g(x, y, z) = \sin x - xz - y^2$
step size $h = 0.1$.

On applying Euler formula as in previous examples, we get

$$y_1 = y(x_1) = y(0.1) = y_0 + h f(x_0, y_0, z_0)$$
$$= y_0 + h(z_0) = 1 + 0.1(1) = 1.1$$

$$z_1 = z(x_1) = z(0.1) = z_0 + h\,g(x_0, y_0, z_0)$$
$$= z_0 + h(\sin x_0 - x_0 z_0 - y_0^2) = 1 + 0.1(0 - 0 - 1) = 0.9$$

Similarly, other iterations are as follows

$$y_2 = 1.190000 \qquad z_2 = 0.779983$$

$$y_3 = 1.267998 \qquad z_3 = 0.642641$$

$$y_4 = 1.332262 \qquad z_4 = 0.492131$$

$$y_5 = 1.381476 \qquad z_5 = 0.333896$$

Note that $y(0.5) = y_5 = 1.381476$ and $y'(0.5) = z(0.5) = z_5 = 0.333896$.

15.4 Runge–Kutta Fourth Order Method

Runge–Kutta method of different orders can be used to solve the system (15.1), but here we are presenting only classical Runge–Kutta method (also known as Runge–Kutta method of order 4). Keeping in mind the method for a single equation, classical Runge–Kutta method for the solution of the system (15.1) is as follows

$$y_{i+1} = y_i + \frac{1}{6}\left(k_1 + 2k_2 + 2k_3 + k_4\right)$$

$$z_{i+1} = z_i + \frac{1}{6}\left(l_1 + 2l_2 + 2l_3 + l_4\right) \qquad i = 0,1,2,\ldots,n \tag{15.4}$$

where k_1, k_2, k_3, k_4 and l_1, l_2, l_3, l_4 are given by the following expression

$$k_1 = h\,f(x_i,\, y_i,\, z_i) \qquad\qquad\qquad l_1 = h\,g(x_i,\, y_i,\, z_i)$$

$$k_2 = h\,f\left(x_i + \frac{h}{2},\, y_i + \frac{k_1}{2},\, z_i + \frac{l_1}{2}\right) \qquad l_2 = h\,g\left(x_i + \frac{h}{2},\, y_i + \frac{k_1}{2},\, z_i + \frac{l_1}{2}\right)$$

$$k_3 = h\,f\left(x_i + \frac{h}{2},\, y_i + \frac{k_2}{2},\, z_i + \frac{l_2}{2}\right) \qquad l_3 = h\,g\left(x_i + \frac{h}{2},\, y_i + \frac{k_2}{2},\, z_i + \frac{l_2}{2}\right)$$

$$k_4 = h\,f\left(x_i + h,\, y_i + k_3,\, z_i + l_3\right) \qquad l_4 = h\,g\left(x_i + h,\, y_i + k_3,\, z_i + l_3\right) \tag{15.5}$$

Example
15.6

Use classical Runge–Kutta method to find the values of $y(0.2)$ and $z(0.2)$ for the following system of first order ODEs. Use step size $h = 0.1$.

$$\frac{dy}{dx} = z + x$$

$$\frac{dz}{dx} = y^2 + x + z$$

$$y(0) = 0, z(0) = 0$$

Ans.

By using the following values

$$x_0 = 0, \ y(x_0) = y_0 = 0, \ z(x_0) = z_0 = 0$$

$$f(x, y, z) = x + z, \ g(x, y, z) = y^2 + x + z$$

$$h = 0.1$$

in the equations (15.4) and (15.5), the following results are obtained.

Values of $y(0.1)$ and $z(0.1)$

On computing the values of k_1, k_2, k_3, k_4 and l_1, l_2, l_3, l_4 from equations (15.5), we have

$$k_1 = h f(x_0, y_0, z_0) = h(x_0 + z_0) = 0.1(0 + 0) = 0$$

$$l_1 = h g(x_0, y_0, z_0) = h(y_0^2 + x_0 + z_0) = 0.1(0) = 0$$

$$k_2 = h f\left(x_0 + \frac{h}{2}, y_0 + \frac{k_1}{2}, z_0 + \frac{l_1}{2}\right) = h\left(x_0 + \frac{h}{2} + z_0 + \frac{l_1}{2}\right) = 0.1\left(0 + \frac{0.1}{2} + 0 + 0\right) = .005$$

$$l_2 = h g\left(x_0 + \frac{h}{2}, y_0 + \frac{k_1}{2}, z_0 + \frac{l_1}{2}\right) = h\left(\left(y_0 + \frac{k_1}{2}\right)^2 + x_0 + \frac{h}{2} + z_0 + \frac{l_1}{2}\right)$$

$$= 0.1\left(0 + 0 + \frac{0.1}{2} + 0 + 0\right) = 0.005$$

$$k_3 = h f\left(x_0 + \frac{h}{2}, y_0 + \frac{k_2}{2}, z_0 + \frac{l_2}{2}\right) = h\left(x_0 + \frac{h}{2} + z_0 + \frac{l_2}{2}\right) = 0.1\left(0 + \frac{0.1}{2} + 0 + \frac{0.005}{2}\right) = 0.005250$$

$$l_3 = h g\left(x_0 + \frac{h}{2}, y_0 + \frac{k_2}{2}, z_0 + \frac{l_2}{2}\right) = h\left(\left(y_0 + \frac{k_2}{2}\right)^2 + x_0 + \frac{h}{2} + z_0 + \frac{l_2}{2}\right)$$

$$= 0.1\left(\left(\frac{0.005}{2}\right)^2 + 0 + \frac{0.1}{2} + 0 + \frac{0.005}{2}\right) = 0.005250625 \approx 0.005251$$

$$k_4 = h f \left(x_0 + h, \; y_0 + k_3, \; z_0 + l_3 \right) = h \left(x_0 + h + z_0 + l_3 \right)$$

$$= 0.1 \left(0 + 0.1 + 0 + .005250625 \right) = 0.0105250625 \approx 0.010525$$

$$l_4 = h g \left(x_0 + h, \; y_0 + k_3, \; z_0 + l_3 \right) = h \left(\left(y_0 + k_3 \right)^2 + x_0 + h + z_0 + l_3 \right)$$

$$= 0.1 \left(\left(0.00525 \right)^2 + 0 + 0.1 + 0 + .005250625 \right) = 0.01052775875 \approx 0.010528$$

The following values of $y(0.1)$ and $z(0.1)$ are computed by using above obtained values in Eqs. (15.4).

$$y_1 = y(0.1) = y_0 + \frac{1}{6} \left(k_1 + 2k_2 + 2k_3 + k_4 \right)$$

$$= 0 + \frac{1}{6} \left(0 + 2(0.005) + 2(0.00525) + 0.0105250625 \right)$$

$$= 0.00517084375 \approx 0.005171$$

$$z_1 = z(0.1) = z_0 + \frac{1}{6} \left(l_1 + 2l_2 + 2l_3 + l_4 \right)$$

$$= 0 + \frac{1}{6} \left(0 + 2(0.005) + 2(0.005250625) + 0.01052775875 \right)$$

$$= 0.005171501458333 \approx 0.005172$$

Values of $y(0.2)$ and $z(0.2)$

Using the values of $y(0.1) = 0.005171$ and $z(0.1) = 0.005172$ in Eqs. (15.5), we have following results

$$k_1 = h f(x_1, \; y_1, \; z_1) = h(x_1 + z_1) = 0.1(0.1 + 0.005172) \approx 0.010517$$

$$l_1 = h g(x_1, \; y_1, \; z_1) = h(y_1^2 + x_1 + z_1) = 0.1 \left((0.005171)^2 + 0.1 + 0.005172 \right) \approx 0.010520$$

$$k_2 = h f \left(x_1 + \frac{h}{2}, \; y_1 + \frac{k_1}{2}, \; z_1 + \frac{l_1}{2} \right) = h \left(x_1 + \frac{h}{2} + z_1 + \frac{l_1}{2} \right) = 0.016043$$

$$l_2 = h g \left(x_1 + \frac{h}{2}, \; y_1 + \frac{k_1}{2}, \; z_1 + \frac{l_1}{2} \right) = h \left(\left(y_1 + \frac{k_1}{2} \right)^2 + x_1 + \frac{h}{2} + z_1 + \frac{l_1}{2} \right) = 0.016054$$

$$k_3 = h f \left(x_1 + \frac{h}{2}, \; y_1 + \frac{k_2}{2}, \; z_1 + \frac{l_2}{2} \right) = h \left(x_1 + \frac{h}{2} + z_1 + \frac{l_2}{2} \right) = 0.016320$$

$$l_3 = h g \left(x_1 + \frac{h}{2}, \; y_1 + \frac{k_2}{2}, \; z_1 + \frac{l_2}{2} \right) = h \left(\left(y_1 + \frac{k_2}{2} \right)^2 + x_1 + \frac{h}{2} + z_1 + \frac{l_2}{2} \right) = 0.016337$$

$$k_4 = h f\left(x_1 + h,\ y_1 + k_3,\ z_1 + l_3\right) = h\left(x_1 + h + z_1 + l_3\right) = 0.022151$$

$$l_4 = h g\left(x_0 + h,\ y_0 + k_3,\ z_0 + l_3\right) = h\left(\left(y_0 + k_3\right)^2 + x_0 + h + z_0 + l_3\right) = 0.022197$$

We can compute the values of $y(0.2)$ and $z(0.2)$ from equation (15.4) as follows

$$y_2 = y(0.2) = y_1 + \frac{1}{6}\left(k_1 + 2k_2 + 2k_3 + k_4\right)$$

$$= 0.005171 + \frac{1}{6}\left(0.010517 + 2(0.016043) + 2(0.016320) + 0.022151\right) = 0.021403$$

$$z_2 = z(0.2) = z_1 + \frac{1}{6}\left(l_1 + 2l_2 + 2l_3 + l_4\right)$$

$$= 0.005172 + \frac{1}{6}\left(0.010520 + 2(0.016054) + 2(0.016337) + 0.022197\right) = 0.021421$$

Example ──────────────────────────────────── **15.7**

Use classical Runge–Kutta method to compute the values of $y(1.2)$ and $z(1.2)$ for the following system of first order ODEs with step size $h = 0.1$.

$$\frac{dy}{dx} = x + yz^2 = f(x, y, z)$$

$$\frac{dz}{dx} = x^2 - yz = g(x, y, z) \quad y(1) = -1,\ z(1) = 2$$

Ans.

Values of $y(1.1)$ and $z(1.1)$

Proceeding like the previous example, we get the following values of k_1, k_2, k_3, k_4 and l_1, l_2, l_3, l_4

$$k_1 = h f(x_0,\ y_0,\ z_0) = h(x_0 + y_0 z_0^2) = 0.1(1 + (-1)(2)^2) = -0.3$$

$$l_1 = h g(x_0,\ y_0,\ z_0) = h(x_0^2 - y_0 z_0) = 0.1(1 - (-1)(2)) = 0.3$$

$$k_2 = h f\left(x_0 + \frac{h}{2},\ y_0 + \frac{k_1}{2},\ z_0 + \frac{l_1}{2}\right) = -0.426588$$

$$l_2 = h g\left(x_0 + \frac{h}{2},\ y_0 + \frac{k_1}{2},\ z_0 + \frac{l_1}{2}\right) = 0.357500$$

$$k_3 = h f\left(x_0 + \frac{h}{2}, \ y_0 + \frac{k_2}{2}, \ z_0 + \frac{l_2}{2}\right) = -0.470945$$

$$l_3 = h g\left(x_0 + \frac{h}{2}, \ y_0 + \frac{k_2}{2}, \ z_0 + \frac{l_2}{2}\right) = 0.374596$$

$$k_4 = h f\left(x_0 + h, \ y_0 + k_3, \ z_0 + l_3\right) = -0.719423$$

$$l_4 = h g\left(x_0 + h, \ y_0 + k_3, \ z_0 + l_3\right) = 0.470290$$

The values of $y(1.1)$ and $z(1.1)$ are given by

$$y_1 = y(1.1) = y_0 + \frac{1}{6}\left(k_1 + 2k_2 + 2k_3 + k_4\right) = -1.469081$$

$$z_1 = z(1.1) = z_0 + \frac{1}{6}\left(l_1 + 2l_2 + 2l_3 + l_4\right) = 2.372414$$

Values of $y(1.2)$ and $z(1.2)$

By using the values of $y(1.1) = -1.469081$ and $z(1.1) = 2.372414$ in Eq. (15.5), we have

$$k_1 = h f(x_1, \ y_1, \ z_1) = -0.716850$$

$$l_1 = h g(x_1, \ y_1, \ z_1) = 0.469527$$

$$k_2 = h f\left(x_1 + \frac{h}{2}, \ y_1 + \frac{k_1}{2}, \ z_1 + \frac{l_1}{2}\right) = -1.127224$$

$$l_2 = h g\left(x_1 + \frac{h}{2}, \ y_1 + \frac{k_1}{2}, \ z_1 + \frac{l_1}{2}\right) = 0.608713$$

$$k_3 = h f\left(x_1 + \frac{h}{2}, \ y_1 + \frac{k_2}{2}, \ z_1 + \frac{l_2}{2}\right) = -1.341445$$

$$l_3 = h g\left(x_1 + \frac{h}{2}, \ y_1 + \frac{k_2}{2}, \ z_1 + \frac{l_2}{2}\right) = 0.676355$$

$$k_4 = h f\left(x_1 + h, \ y_1 + k_3, \ z_1 + l_3\right) = -2.492383$$

$$l_4 = h g\left(x_0 + h, \ y_0 + k_3, \ z_0 + l_3\right) = 1.000865$$

The values of $y(1.2)$ and $z(1.2)$ are as follows

$$y_2 = y(1.2) = y_1 + \frac{1}{6}\left(k_1 + 2k_2 + 2k_3 + k_4\right) = -2.826843$$

$$z_2 = z(1.2) = z_1 + \frac{1}{6}\left(l_1 + 2l_2 + 2l_3 + l_4\right) = 3.045835$$

Example 15.8

Solve the following IVP

$$y'' + xy' + y^2 = \sin(x); \quad y(0) = 1, \; y'(0) = 1$$

Use Runge–Kutta fourth order method to find the value of y at $x = 0.5$ with step size 0.25.

Ans.

From Example 15.5, the given IVP is equivalent to the following system

$$\frac{dy}{dx} = y' = z$$

$$\frac{dz}{dx} = z' = \sin(x) - xz - y^2; \qquad y(0) = 1, \; z(0) = 1$$

Continuing in the same way as in the previous examples, we can easily compute the following iterations

Values of $y(0.25)$ and $z(0.25)$

The values of k_1, k_2, k_3, k_4 and l_1, l_2, l_3, l_4 are given by

$$k_1 = 0.250000 \qquad k_2 = 0.218750 \qquad k_3 = 0.210927 \qquad k_4 = 0.174281$$
$$l_1 = -0.250000 \qquad l_2 = -0.312581 \qquad l_3 = -0.302875 \qquad l_4 = -0.348306$$

The values of $y(0.25)$ and $z(0.25)$ are as follows

$$y_1 = y(0.25) = y_0 + \frac{1}{6}(k_1 + 2k_2 + 2k_3 + k_4) = 1.213939$$

$$z_1 = z(0.25) = z_0 + \frac{1}{6}(l_1 + 2l_2 + 2l_3 + l_4) = 0.695130$$

Values of $y(0.5)$ and $z(0.5)$

The values of k_1, k_2, k_3, k_4 and l_1, l_2, l_3, l_4 are given by

$$k_1 = 0.173783 \qquad k_2 = 0.130032 \qquad k_3 = 0.126253 \qquad k_4 = 0.082605$$
$$l_1 = -0.350007 \qquad l_2 = -0.380234 \qquad l_3 = -0.364708 \qquad l_4 = -0.370475$$

The values of $y(0.5)$ and $z(0.5)$ are as follows

$$y_2 = y(0.5) = y_1 + \frac{1}{6}(k_1 + 2k_2 + 2k_3 + k_4) = 1.342099$$

$$z_2 = z(0.5) = z_1 + \frac{1}{6}(l_1 + 2l_2 + 2l_3 + l_4) = 0.326736$$

Table 15.1 Formulations for solutions of IVPs

Sr. No.	Numerical Scheme for the solution of IVP $\dfrac{dy}{dx} = f(x,y,z)$ $\dfrac{dz}{dx} = g(x,y,z)$ $y(x_0) = y_0, z(x_0) = z_0$
1	Picard method $$y^{(n+1)}(x) = y(x_0) + \int_{x_0}^{x} f(x, y^{(n)}(x), z^{(n)}(x))\, dx$$ $$z^{(n+1)}(x) = z(x_0) + \int_{x_0}^{x} g(x, y^{(n)}(x), z^{(n)}(x))\, dx \qquad n = 0, 1, 2, \cdots$$ where y_0 and z_0 are initial approximations $y^{(0)}(x)$ and $z^{(0)}(x)$ respectively
2	Taylor series method of order n $$y(x_0 + h) = y(x_0) + h\, y'(x_0) + \frac{(h)^2}{2!} y''(x_0) + \cdots + \frac{(h)^n}{(n)!} y^n(x_0) + \cdots$$ $$z(x_0 + h) = z(x_0) + h\, z'(x_0) + \frac{(h)^2}{2!} z''(x_0) + \cdots + \frac{(h)^n}{(n)!} z^n(x_0) + \cdots$$
3	Euler method $$y_{i+1} = y_i + h f(x_i, y_i, z_i)$$ $$z_{i+1} = z_i + h g(x_i, y_i, z_i); \qquad i = 0, 1, \cdots, n$$
4	Classical Runge–Kutta method (or) Runge–Kutta 4th order method $$y_{i+1} = y_i + \frac{1}{6}(k_1 + 2k_2 + 2k_3 + k_4)$$ $$z_{i+1} = z_i + \frac{1}{6}(l_1 + 2l_2 + 2l_3 + l_4) \qquad i = 0,1,2,\cdots,n$$ where k_1, k_2, k_3, k_4 and l_1, l_2, l_3, l_4 are given by the following expression $$k_1 = h f(x_i, y_i, z_i) \qquad\qquad l_1 = h g(x_i, y_i, z_i)$$ $$k_2 = h f\left(x_i + \frac{h}{2}, y_i + \frac{k_1}{2}, z_i + \frac{l_1}{2}\right) \qquad l_2 = h g\left(x_i + \frac{h}{2}, y_i + \frac{k_1}{2}, z_i + \frac{l_1}{2}\right)$$ $$k_3 = h f\left(x_i + \frac{h}{2}, y_i + \frac{k_2}{2}, z_i + \frac{l_2}{2}\right) \qquad l_3 = h g\left(x_i + \frac{h}{2}, y_i + \frac{k_2}{2}, z_i + \frac{l_2}{2}\right)$$ $$k_4 = h f\left(x_i + h, y_i + k_3, z_i + l_3\right) \qquad l_4 = h g\left(x_i + h, y_i + k_3, z_i + l_3\right)$$

15.5 Boundary Value Problem: Shooting Method

This section deals with the shooting method for the solutions of BVPs. In shooting method, the given BVP is converted into a set of IVPs. Then, we construct the solution of given BVP by using the solutions of these IVPs.

Only linear BVPs are considered in this chapter. It is worth mentioning that nonlinear BVPs can also be solved using Shooting method. But, nonlinear BVPs are not discussed due to the complexity of solution procedure.

Consider the following linear BVP

$$y'' + p(x)y' + q(x)y = r(x), \qquad a \le x \le b$$
$$y(a) = \alpha \text{ and } y(b) = \beta \tag{15.6}$$

where $p(x)$, $q(x)$ and $r(x)$ are continuous functions of x in the interval $[a, b]$.

Let $u(x)$ be the solution of the following nonhomogeneous IVP

$$u'' + p(x)u' + q(x)u = r(x); \quad u(a) = \alpha \text{ and } u'(a) = 0 \tag{15.7}$$

and $v(x)$ be the solution of the following homogeneous IVP

$$v'' + p(x)v' + q(x)v = 0; \quad v(a) = 0 \text{ and } v'(a) = \beta \tag{15.8}$$

Let the linear combination $y(x) = c_1 u(x) + c_2 v(x)$ be a solution of BVP (15.6) for some constants c_1 and c_2. Using $y(x) = c_1 u(x) + c_2 v(x)$ in Eq. (15.6) and then using Eqs. (15.7) and (15.8), we get

$$c_1 = 1 \tag{15.9}$$

Now, we require that the solution $y(x) = u(x) + c_2 v(x)$ also satisfies boundary conditions, i.e.,

$$y(a) = u(a) + c_2 v(a) = \alpha$$
$$y(b) = u(b) + c_2 v(b) = \beta$$

Using conditions from IVPs (15.7) and (15.8), it is easy to show that first equation is identically satisfied, and second equation gives the following result

$$c_2 = \frac{\beta - u(b)}{v(b)} \tag{15.10}$$

On substituting the values from Eqs. (15.9) and (15.10) in $y(x) = c_1 u(x) + c_2 v(x)$, the solution of BVP (15.6) is given by

$$y(x) = u(x) + \frac{\beta - u(b)}{v(b)} v(x) \tag{15.11}$$

Stepwise Procedure: In this method, we have following three steps

i)   Convert the given BVP (15.6) in to two IVPs (15.7) and (15.8).

ii)   Solve these two IVPs from any methods like Taylor series method, Runge–Kutta fourth order method, etc.

iii)   Use these two solutions $u(x)$ and $v(x)$ in Eq. (15.11) to obtain the solution of given BVP.

Example ———————————————————————————————— **15.9**

Solve the differential equation with the help of shooting method,

$$y'' - y = x; \qquad 0 \le x \le 0.2$$

subject to boundary conditions

$$y(0) = 0, \quad y(0.2) = -0.2$$

Solve the corresponding IVPs by Runge–Kutta fourth order method with step size $h = 0.1$.

Ans.

Given BVP is as follows

$$y'' - y = x; \qquad y(0) = 0, \quad y(0.2) = -0.2 \tag{15.12}$$

Step 1

First, we have to create the following two IVPs corresponding to the given BVP (15.12)

$$u'' - u = x; \qquad u(0) = 0, \ u'(0) = 0 \tag{15.13}$$

$$v'' - v = 0; \qquad v(0) = 0, \ v'(0) = -0.2 \tag{15.14}$$

Step 2

On solving the IVPs (15.13) and (15.14) with the help of Runge–Kutta fourth order method with step size $h = 0.1$, we obtain following results.

Solution of IVP (15.13)

The second order ODE (15.13) is equivalent to the following system

$$\frac{du}{dx} = u' = z$$

$$\frac{dz}{dx} = z' = x + u; \qquad u(0) = 0, \ z(0) = 0$$

Proceeding in a similar manner as in the examples (15.6–15.8), the values of k_1, k_2, k_3, k_4, l_1, l_2, l_3, l_4 and $u(0.1), u(0.2)$ are given by

$k_1 = 0$	$k_2 = 0$	$k_3 = 0.00025$	$k_4 = 0.0005$
$l_1 = 0$	$l_2 = 0.005$	$l_3 = 0.005$	$l_4 = 0.010025$

$u(0.1) = 0.000167$

$k_1 = 0.0005$	$k_2 = 0.001001$	$k_3 = 0.001253$	$k_4 = 0.002007$
$l_1 = 0.010017$	$l_2 = 0.015042$	$l_3 = 0.015067$	$l_4 = 0.020142$

$u(0.2) = 0.001336$

Solution of IVP (15.14)

The IVP (15.14) is equivalent to the following system

$$\frac{dv}{dx} = v' = z$$

$$\frac{dz}{dx} = z' = v; \quad v(0) = 0, \ v'(0) = -0.2$$

On solving this system with Runge–Kutta fourth order method, we get

$k_1 = -0.02$	$k_2 = -0.02$	$k_3 = -0.02005$	$k_4 = -0.0201$
$l_1 = 0$	$l_2 = -0.001$	$l_3 = -0.001$	$l_4 = -0.002005$

$v(0.1) = -0.020033$

$k_1 = -0.0201$	$k_2 = -0.0202$	$k_3 = -0.020251$	$k_4 = -0.020401$
$l_1 = -0.002003$	$l_2 = -0.003008$	$l_3 = -0.003013$	$l_4 = -0.004028$

$v(0.2) = -0.040267$

Step 3

By using the values $\beta = -0.2$, $u(b) = u(0.2) = 0.001336$ and $v(b) = v(0.2) = -0.040267$ in Eq. (15.11), we have

$$y(x) = u(x) + 5.000025 v(x)$$

$$y(0.1) = u(0.1) + 5.000025 v(0.1)$$

$$= 0.000167 + 5.000025(-0.020033)$$

$$= -0.099998$$

Note that the exact solution of BVP is $y(x) = -x$. So the solution obtained by shooting method $y(0.1) = -0.099998$ is in good approximation with the exact result.

15.6 Finite Difference Approximations for Derivatives

The finite difference method is a simple and most commonly used numerical method to solve differential equations with boundary conditions. In this method, various derivative terms in the differential equation, and the derivative boundary conditions are replaced by their finite difference approximations. Consequently, the differential equation is converted into a set of algebraic equations. If the differential equation and boundary conditions are linear, then the system of algebraic equations is also linear. This system of linear equations can be solved by any direct or iterative procedure discussed in Chapter 5. Then, the solution of the differential equation is the solution to this system of linear equations. An important advantage of this method is that most of the calculations can be carried out on the computer and hence the solution is easy to obtain. In this section, we will discuss the finite difference approximations of the derivatives using Taylor series.

Consider the following Taylor series expansions

$$y(x+h) = y(x) + hy'(x) + \frac{h^2}{2!}y''(x) + \cdots \qquad (15.15)$$

$$y(x-h) = y(x) - hy'(x) + \frac{h^2}{2!}y''(x) - \cdots \qquad (15.16)$$

$$y(x+2h) = y(x) + 2hy'(x) + \frac{(2h)^2}{2!}y''(x) + \cdots \qquad (15.17)$$

$$y(x-2h) = y(x) - 2hy'(x) + \frac{(2h)^2}{2!}y''(x) - \cdots \qquad (15.18)$$

On using these Taylor expansions, Eqs. (15.15–15.18), we can easily obtain following approximations for first and second orders derivatives in subsections 15.6.1 and 15.6.2, respectively.

15.6.1 First Order Derivatives

Forward Difference: On using the expression (15.15), we get

$$y'(x) = \frac{y(x+h) - y(x)}{h} - \frac{h}{2}y''(x) - \cdots$$

On using Big O notation, we can rewrite it as follows

$$y'(x) = \frac{y(x+h) - y(x)}{h} + O(h)$$

On neglecting the terms contained in notation, $O(h)$, the forward difference approximation for first order derivative is given by

$$y'(x) = \frac{y(x+h) - y(x)}{h}$$

At any point $x = x_i$, we have

$$y'(x_i) = \frac{y(x_i + h) - y(x_i)}{h} + O(h)$$

$$= \frac{y(x_{i+1}) - y(x_i)}{h} + O(h) \qquad (15.19)$$

$$= \frac{y_{i+1} - y_i}{h} + O(h)$$

Backward Difference: From Taylor expansion (15.16), the backward difference approximation for first order derivative is as follows

$$y'(x) = \frac{y(x) - y(x-h)}{h} + \tilde{O}(h)$$

At $x = x_i$, we get

$$y'(x_i) = \frac{y(x_i) - y(x_i - h)}{h} + O(h)$$

$$= \frac{y(x_i) - y(x_{i-1})}{h} + O(h) \tag{15.20}$$

$$= \frac{y_i - y_{i-1}}{h} + O(h)$$

Central Difference: Taylor expansions (15.15) and (15.16) provide following central difference approximation for first order derivative

$$y'(x) = \frac{y(x+h) - y(x-h)}{2h} + O(h^2)$$

At $x = x_i$, we have

$$y'(x_i) = \frac{y(x_i + h) - y(x_i - h)}{2h} + O(h^2)$$

$$= \frac{y(x_{i+1}) - y(x_{i-1})}{2h} + O(h^2) \tag{15.21}$$

$$= \frac{y_{i+1} - y_{i-1}}{2h} + O(h^2)$$

15.6.2 Second Order Derivatives

Forward Difference: On using the expressions (15.15) and (15.17), the forward difference approximation for second order derivative is given by

$$y''(x) = \frac{y(x+2h) - 2y(x+h) + y(x)}{h^2} + O(h)$$

At $x = x_i$, we have

$$y''(x_i) = \frac{y(x_i + 2h) - 2y(x_i + h) + y(x_i)}{h^2} + O(h)$$

$$= \frac{y(x_{i+2}) - 2y(x_{i+1}) + y(x_i)}{h^2} + O(h) \tag{15.22}$$

$$= \frac{y_{i+2} - 2y_{i+1} + y_i}{h^2} + O(h)$$

Backward Difference: From Taylor expansions (15.16) and (15.18), the backward difference approximation is as follows

$$y''(x) = \frac{y(x) - 2y(x-h) + y(x-2h)}{h^2} + O(h)$$

At $x = x_i$, we have

$$y''(x_i) = \frac{y(x_i) - 2y(x_i - h) + y(x_i - 2h)}{h^2} + O(h)$$

$$= \frac{y(x_i) - 2y(x_{i-1}) + y(x_{i-2})}{h^2} + O(h) \tag{15.23}$$

$$= \frac{y_i - 2y_{i-1} + y_{i-2}}{h^2} + O(h)$$

Central Difference: Taylor expansions (15.15) and (15.16) provide following central difference approximation for second order derivative

$$y''(x) = \frac{y(x+h) - 2y(x) + y(x-h)}{h^2} + O(h^2)$$

At $x = x_i$, we have

$$y''(x_i) = \frac{y(x_i + h) - 2y(x_i) + y(x_i - h)}{h^2} + O(h^2)$$

$$= \frac{y(x_{i+1}) - 2y(x_i) + y(x_{i-1})}{h^2} + O(h^2) \tag{15.24}$$

$$= \frac{y_{i+1} - 2y_i + y_{i-1}}{h^2} + O(h^2)$$

Note: The neglecting terms $[O(h^2)]$ are of order h^2 in central differences as compared to the order of h in forward and backward differences. So, the central differences for derivatives are better approximations compared to other differences. Therefore, central difference formulas are preferred over forward and backward difference formulas.

15.7 Boundary Value Problem: Finite Difference Method

Consider the following second-order linear differential equation

$$\frac{d^2 y}{dx^2} + p(x)\frac{dy}{dx} + q(x)y = r(x), \qquad a \le x \le b$$

(or) $\qquad y'' + p(x)y' + q(x)y = r(x), \qquad a \le x \le b \tag{15.25}$

where the functions $p(x)$, $q(x)$ and $r(x)$ are continuous functions of x. Let the boundary conditions at the boundary points ($x = a$ and $x = b$) be given by

$$l_1 y(a) + m_1 y'(a) = n_1$$
$$l_2 y(b) + m_2 y'(b) = n_2 \tag{15.26}$$

where $l_i, m_i, n_i;$ $i = 1, 2$ are constants. We have following two particular cases for the boundary conditions

1. Let $m_i = 0$, then the boundary condition is said to be of Dirichlet type. In this case, only function value is prescribed on the boundary.
2. Let $l_i = 0$, then the condition is said to be of Neumann type. The boundary condition contains only the value of y' (derivative value).

Equation (15.25), along with the boundary conditions (15.26) are known as boundary value problem. For the solution of BVP, the first step is to divide the interval $[a, b]$ into n numbers of sub-domains;

$$(x_i, x_{i+1}), \quad 0 \le i \le n-1, \text{ where } a = x_0 \text{ and } b = x_n$$

Each sub-domains are of equal length, $h = \dfrac{(b-a)}{n}$. Also, the points $x_i = x_0 + ih$ are known as mesh/nodal/pivotal points. The aim is to obtain the values of $y_i = y(x_i)$ at the internal node points $x_i, 1 \le i \le n-1$ with the help of finite difference approximations. Various derivatives in the differential equation are replaced by their corresponding finite differences to obtain a linear system of simultaneous algebraic equations in the variables $y_i, 0 \le i \le n$. This linear system of equations can be easily solved to get the solution of BVP.

Example   **15.10**

Solve the differential equation

$$(1-x)y'' + xy' - y = (1-x)^2; \qquad 0 \le x \le 1$$

subject to boundary conditions

$$y(0) = 1, \quad y(1) = 3$$

by taking four equal subintervals (or) step size $h = 0.25$. Use central difference approximations.

Ans.

The nodal points for the interval $(0, 1)$ with step size $h = 0.25$ are as follows

$$x_0 = 0, \ x_1 = 0.25, \ x_2 = 0.5, \ x_3 = 0.75, \ x_4 = 1$$

The given function values are $y_0 = y(0) = 1$, $y_4 = y(1) = 3$. Now, we have to compute the values of $y(0.25)$, $y(0.5)$, $y(0.75)$ (or y_1, y_2, y_3 respectively) for the following BVP

$$(1-x)y'' + xy' - y = (1-x)^2$$

At any point $x = x_i$, we have

$$(1-x_i)y''(x_i) + x_i y'(x_i) - y(x_i) = (1-x_i)^2$$

On using the central difference formulas (15.21) and (15.24) for derivative terms, we get

$$(1-x_i)\frac{y_{i+1}-2y_i+y_{i-1}}{h^2}+x_i\frac{y_{i+1}-y_{i-1}}{2h}-y_i=(1-x_i)^2$$

$$2(1-x_i)[y_{i+1}-2y_i+y_{i-1}]+hx_i[y_{i+1}-y_{i-1}]-2h^2y_i=2h^2(1-x_i)^2$$

$$[2(1-x_i)+hx_i]y_{i+1}-2[2(1-x_i)+h^2]y_i+[2(1-x_i)-hx_i]y_{i-1}=2h^2(1-x_i)^2 \qquad (15.25)$$

Using $i = 1, 2, 3$ in this equation, the following set of three algebraic equations is obtained

$$[2(1-x_1)+hx_1]y_2-2[2(1-x_1)+h^2]y_1+[2(1-x_1)-hx_1]y_0=2h^2(1-x_1)^2$$

$$[2(1-x_2)+hx_2]y_3-2[2(1-x_2)+h^2]y_2+[2(1-x_2)-hx_2]y_1=2h^2(1-x_2)^2$$

$$[2(1-x_3)+hx_3]y_4-2[2(1-x_3)+h^2]y_3+[2(1-x_3)-hx_3]y_2=2h^2(1-x_3)^2$$

By using $h = 0.25$; $x_0 = 0$, $x_1 = 0.25$, $x_2 = 0.5$, $x_3 = 0.75$, $x_4 = 1$ and $y_0 = 1$, $y_4 = 3$, in the above three equations, the following three simplified equations are obtained

$$1.5625y_2 - 3.125y_1 = -1.3671875$$

$$1.125y_3 - 2.125y_2 + 0.875y_1 = 0.03125$$

$$-1.125y_3 + 0.3125y_2 = -2.0546875$$

This system of three equations can be solved easily by any method. Here, Gauss–Seidel method is used for the solution, and the result obtained after 14 iterations is as follows

$$y(0.25) = y_1 = 1.312497$$

$$y(0.5) = y_2 = 1.749998$$

$$y(0.75) = y_3 = 2.312500$$

Note. It is worth mentioning here that the exact solution of the above BVP is $y(x) = 1 + x + x^2$; therefore the exact result is

$$y(0.25) = y_1 = 1.3125$$
$$y(0.5) = y_2 = 1.75$$
$$y(0.75) = y_3 = 2.3125$$

. Also, note that the central differences

are used for approximations of derivatives as they provide a higher order of accuracy as compared to forward and backward differences.

Example ——————————————————————————— **15.11**

Solve the differential equation

$$y'' - xy' + 2y = x + 4; \qquad 0 \le x \le 0.6$$

subject to boundary conditions

$y'(0) = 1$, $y(0.6) = 1.96$, by taking step size $h = 0.2$. Use central difference approximations for the differential equation and derivative boundary condition.

Ans.

The nodal points are $x_0 = 0$, $x_1 = 0.2$, $x_2 = 0.4$, $x_3 = 0.6$. Discretization of BVP at $x = x_i$ gives

$$\frac{y_{i+1} - 2y_i + y_{i-1}}{h^2} - x_i \frac{y_{i+1} - y_{i-1}}{2h} + 2y_i = x_i + 4$$

$$2(y_{i+1} - 2y_i + y_{i-1}) - x_i h(y_{i+1} - y_{i-1}) + 4h^2 y_i = 2h^2 (x_i + 4)$$

$$(2 - x_i h) y_{i+1} + (2 + x_i h) y_{i-1} - 4(1 - h^2) y_i = 2h^2 (x_i + 4) \qquad (15.26)$$

The central difference approximation of derivative boundary condition $y'(0) = 1$ provides the following result

$$\frac{y_1 - y_{-1}}{2h} = 1, \text{ or } y_{-1} = y_1 - 0.4 \qquad (15.27)$$

Now, we will evaluate Eq. (15.26) at different values of x.

At $x = x_0 = 0$; Eq. (15.26) is given by

$$(2 - x_0 h) y_1 + (2 + x_0 h) y_{-1} - 4(1 - h^2) y_0 = 2h^2 (x_0 + 4)$$

$$2y_1 + 2y_{-1} - 3.84 y_0 = .32$$

Replacing $y_{-1} = y_1 - 0.4$ from Eq. (15.27), we get

$$4y_1 - 3.84 y_0 = 1.12 \qquad (15.28)$$

At $x = x_1 = 0.2$; the expression (15.26) provides

$$(2 - x_1 h) y_2 + (2 + x_1 h) y_0 - 4(1 - h^2) y_1 = 2h^2 (x_1 + 4)$$

$$1.96 y_2 + 2.04 y_0 - 3.84 y_1 = .3216 \qquad (15.29)$$

Similarly at $x = x_2 = 0.4$, we have

$$(2 - x_2 h) y_3 + (2 + x_2 h) y_1 - 4(1 - h^2) y_2 = 2h^2 (x_2 + 4)$$

$$1.92 y_3 + 2.08 y_1 - 3.84 y_2 = .3232$$

Using value of $y_3 = y(0.6) = 1.96$, from given boundary condition, we get

$$2.08 y_1 - 3.84 y_2 = -3.44 \qquad (15.30)$$

On solving the system of equations (15.28–15.30), the values of y at different nodal points are as follows

$$y_0 = y(0) = 1.046396684$$

$$y_1 = y(0.2) = 1.284540816$$

$$y_2 = y(0.4) = 1.591626276$$

Note that the exact analytical solution is $x^2 + x + 1$. We can easily compare the solution obtained from finite difference method and exact solution.

15.8 Finite Difference Approximations for Unequal Intervals

So far, we have discussed the BVPs with equally spaced node points, which is not always practically suitable. In many scientific and engineering problems, the changes near the boundaries are much more as compare to the other points in the domains. Therefore, the spacing between nodes near the boundaries has to be much less in comparison with other nodes in the domain. So, computations with unequally spaced nodes produce more accurate results as compared to equally spaced nodes.

Consider the values of function $y(x)$ are given at unequal intervals, e.g., $x - h_2$, $x, x + h_1$. Then, the various difference formulas for first and second derivatives are as follows

Forward Difference

Let the values of function be on the points with unequal length, $x, x + h_1, x + h_1 + h_2$, then using the Taylor expansions, we have

$$y(x + h_1) = y(x) + h_1 y'(x) + \frac{h_1^2}{2!} y''(x) + \cdots \tag{15.31}$$

$$y(x + h_1 + h_2) = y(x) + (h_1 + h_2) y'(x) + \frac{(h_1 + h_2)^2}{2!} y''(x) + \cdots \tag{15.32}$$

The finite difference approximation for the first order derivative term from Eq. (15.31) is as follows

$$y'(x) = \frac{y(x + h_1) - y(x)}{h_1} + O(h_1) \tag{15.33}$$

This formula has order one as the error term is of the order $O(h_1)$. A more accurate approximation can be obtained from Eqs. (15.31) and (15.32), where the order is two.

$$y'(x) = \frac{-h_1^2 y(x + h_1 + h_2) + (h_1 + h_2)^2 y(x + h_1) - \{(h_1 + h_2)^2 - h_1^2\} y(x)}{h_1 h_2 (h_1 + h_2)} + O(h_1^2, h_1 h_2) \tag{15.34}$$

Similarly, the forward difference approximation for the second order derivative is given by

$$y''(x) = \frac{2\{h_1 y(x + h_1 + h_2) - (h_1 + h_2) y(x + h_1) + h_2 y(x)\}}{h_1 h_2 (h_1 + h_2)} + O(h_1, h_2) \tag{15.35}$$

Backward Difference

Let the values of function be on the points, $x - (h_1 + h_2)$, $x - h_1$, x, then using the Taylor expansions, we have

$$y(x - h_1) = y(x) - h_1 y'(x) + \frac{h_1^2}{2!} y''(x) - \cdots \tag{15.36}$$

$$y\{x - (h_1 + h_2)\} = y(x) - (h_1 + h_2) y'(x) + \frac{(h_1 + h_2)^2}{2!} y''(x) - \cdots \tag{15.37}$$

On using the Taylor expansion (15.36), we get following backward difference formula for first order derivative

$$y'(x) = \frac{y(x) - y(x - h_1)}{h_1} + O(h_1) \tag{15.38}$$

This approximation has order one, we can compute the backward difference approximation of order two by using both the Taylor expansions (15.36) and (15.37)

$$y'(x) = \frac{\{(h_1 + h_2)^2 - h_1^2\} y(x) - (h_1 + h_2)^2 y(x - h_1) + h_1^2 y(x - h_1 - h_2)}{h_1 h_2 (h_1 + h_2)} + O(h_1^2, h_1 h_2) \tag{15.39}$$

Similarly, the second order derivative term can be approximated by the following formula

$$y''(x) = \frac{2\{h_2 y(x) - (h_1 + h_2) y(x - h_1) + h_1 y(x - h_1 - h_2)\}}{h_1 h_2 (h_1 + h_2)} + O(h_1, h_2) \tag{15.40}$$

Central Difference

On using Taylor expansions at the points $x - h_2$ and $x + h_1$, we have

$$y(x + h_1) = y(x) + h_1 y'(x) + \frac{h_1^2}{2!} y''(x) + \cdots$$

$$y(x - h_2) = y(x) - h_2 y'(x) + \frac{h_2^2}{2!} y''(x) - \cdots \tag{15.41}$$

We can easily compute the following central difference approximation for the first order derivative term

$$y'(x) = \frac{h_2^2 y(x + h_1) + (h_1^2 - h_2^2) y(x) - h_1^2 y(x - h_2)}{h_1 h_2 (h_1 + h_2)} + O(h_1 h_2) \tag{15.42}$$

Second order derivative can be approximated by following central difference formula

$$y''(x) = \frac{2\{h_2 y(x + h_1) - (h_1 + h_2) y(x) + h_1 y(x - h_2)\}}{h_1 h_2 (h_1 + h_2)} - \frac{1}{6}(h_1 - h_2) y'''(x) + O\{h^2 (h_1, h_2)\} \tag{15.43}$$

Example ——————————————————————————— 15.12

Solve the differential equation

$$(1 - x) y'' + x y' - y = (1 - x)^2; \qquad 0 \le x \le 1$$

subject to boundary conditions

$$y(0) = 1, \qquad y(1) = 3$$

by taking nodal points 0.2, 0.5 and 0.8.

Ans.

The nodal points are given by

$$x_0 = 0, \ x_1 = 0.2, \ x_2 = 0.5, \ x_3 = 0.8, \ x_4 = 1$$
$$y_0 = y(0), \ y_1 = y(0.2), \ y_2 = y(0.5), \ y_3 = y(0.8), \ y_4 = y(1)$$

The given differential equation is as follows

$$(1-x)y'' + xy' - y = (1-x)^2$$

On using the finite differences (16.28) and (16.29), at $x = x_i$, we get the following expression

$$(1-x_i)\left(\frac{2\{h_2 y(x_i + h_1) - (h_1 + h_2)y(x_i) + h_1 y(x_i - h_2)\}}{h_1 h_2 (h_1 + h_2)}\right)$$
$$+ x_i \left(\frac{h_2^2 y(x_i + h_1) + (h_1^2 - h_2^2)y(x_i) - h_1^2 y(x_i - h_2)}{h_1 h_2 (h_1 + h_2)}\right) - y(x_i) = (1-x_i)^2$$

$$2(1-x_i)\{h_2 y(x_i + h_1) - (h_1 + h_2)y(x_i) + h_1 y(x_i - h_2)\}$$
$$+ x_i \left(h_2^2 y(x_i + h_1) + (h_1^2 - h_2^2)y(x_i) - h_1^2 y(x_i - h_2)\right)$$
$$- y(x_i)h_1 h_2 (h_1 + h_2) = (1-x_i)^2 h_1 h_2 (h_1 + h_2)$$

$$\left(2(1-x_i)h_2 + x_i h_2^2\right)y(x_i + h_1)$$
$$+ \left(-2(1-x_i)(h_1 + h_2) + x_i(h_1^2 - h_2^2) - h_1 h_2 (h_1 + h_2)\right)y(x_i)$$
$$+ \left(2(1-x_i)h_1 - x_i h_1^2\right)y(x_i - h_2) = (1-x_i)^2 h_1 h_2 (h_1 + h_2) \tag{15.44}$$

Using the values $i = 1, 2, 3$ in Eq. (15.44), the following set of three algebraic equations is obtained

$i = 1, \ h_2 = 0.2, \ h_1 = 0.3, \ y_0 = 1$

$$\left(2(1-0.2)0.2 + (0.2)(0.2)^2\right)y_2$$
$$+ \left(-2(1-0.2)(0.3+0.2) + 0.2((0.3)^2 - (0.2)^2) - (0.3)(0.2)(0.3+0.2)\right)y_1 \tag{15.45}$$
$$+ \left(2(1-0.2)0.3 - (0.2)(0.3)^2\right)(1) = (1-0.2)^2 (0.3)(0.2)(0.3+0.2)$$

$0.328 y_2 - 0.829 y_1 = -0.4428$

$i = 2, \ h_2 = 0.3, \ h_1 = 0.3$

$$\left(2(1-0.5)0.3 + 0.5(0.3)^2\right)y_3$$
$$+ \left(-2(1-0.5)(0.3+0.3) + 0.5(0.3^2 - 0.3^2) - 0.3(0.3)(0.3+0.3)\right)y_2$$
$$+ \left(2(1-0.5)0.3 - 0.5(0.3)^2\right)y_1 = (1-0.5)^2 (0.3)(0.3)(0.3+0.3) \tag{15.46}$$

$0.345 y_3 - 0.654 y_2 + 0.255 y_1 = 0.0135$

$i = 3, h_2 = 0.3, h_1 = 0.2, y_4 = 3$

$\left(2(1-0.8)(0.3) + 0.8(0.3)^2\right)(3)$

$$+\left(-2(1-0.8)(0.2+0.3) + 0.8((0.2)^2 - (0.3)^2) - (0.2)(0.3)(0.2+0.3)\right)y_3$$

(15.47)

$$+\left(2(1-0.8)0.2 - 0.8(0.2)^2\right)y_2 = (1-0.8)^2(0.2)(0.3)(0.2+0.3)$$

$-0.27y_3 + 0.048y_2 = -0.5748$

Solving equations (15.45–15.47), we get

$$y(0.2) = y_1 = 1.223776$$
$$y(0.5) = y_2 = 1.743020$$
$$y(0.8) = y_3 = 2.438759$$

15.9 Discussion

Differential equations arise in the modeling of many physical phenomena. So, the solutions of these differential equations are of great importance in the understanding of these physical phenomena. In practical problems, the solutions of differential equations are not an easy task. The existence and uniqueness of the solution of boundary value problem also needs to be addressed. For example, exercise 15.14 has two such simple examples of boundary value problems without unique solutions.

During the implementation of the numerical schemes to the boundary value problems, we face many problems. Here, we point out some major aspects of implementing the finite difference method to boundary value problems.

1. We have neglected the terms containing powers of step size (h) in finite difference approximations of derivative terms. In central difference approximations of derivative terms in Example 15.10, we have neglected the terms containing $h^2 = (0.25)^2 = 0.0625$ and higher power of h. This error is very large in scientific and engineering computations. We have to reduce the step size h to decrease this error, for example, step size $h = 0.001$. In this case, we have a linear system of 999 equations in 999 variables. The problems in the solution of this large system are discussed below.

2. Programming is the only way to solve such a large system. The solution procedure involves two major steps
    i.   Construction of linear system from the boundary value problem
    ii. To apply the numerical method for the solution of this linear system
    These two steps require very large numbers of computations. To handle these computations and to reduce round-off errors, we require high-end computers and latest software.

3. We can solve this large system using direct methods like Gauss elimination, and then a large round-off error is produced during computation. We can use different strategies like pivoting to reduce this error, but still, a large error remains in computations.

4. If we solve the large system using an iterative procedure like Gauss–Seidel, then following two problems need to addressed properly:

   i. Convergence of iterative procedure to the exact solution, see discussion section of Chapter 5 for more details.

   ii. The number of computations increases significantly as compared to the direct method, which increases the run-time of the algorithm as well as required high configuration computer for computations.

5. It is worth mentioning that the generated linear system is tridiagonal. So, instead of using algorithms for general systems, we can use some efficient algorithms specifically designed to solve tridiagonal systems. For example, Thomas algorithm in Example 5.16 is used to solve the linear system of Example 15.10.

Exercise 15

1. Obtain first two approximations of Picard method for the following system of ODEs and hence find the values of $y(0.2)$ and $z(0.2)$.

$$\frac{dy}{dx} = x + z$$

$$\frac{dz}{dx} = x - y^2; \quad y(0) = 2, z(0) = 1$$

Ans.

$$y^{(1)}(x) = 2 + x + \frac{x^2}{2} \qquad z^{(1)}(x) = 1 - 4x + \frac{x^2}{2}$$

$$y^{(2)}(x) = 2 + x - \frac{3x^2}{2} + \frac{x^3}{6} \qquad z^{(2)}(x) = 1 - 4x - \frac{3x^2}{2} - x^3 - \frac{x^4}{4} - \frac{x^5}{20}$$

$$y(0.2) = 2.141333 \qquad z(0.2) = 0.131584$$

2. Find the third approximation of Picard method for the following IVP and hence find the value of $y(0.1)$.

$$\frac{d^2 y}{dx^2} - xy = 1$$

$$y(0) = 1, \ y'(0) = 0$$

Ans. $y_3 = 1 + \dfrac{x^2}{2} + \dfrac{x^3}{6}; \quad y(0.1) = 1.005167$

3. Solve the following system of first order ODEs with the aid of Taylor series method. Obtain first five terms of Taylor series and hence find the values of $y(0.1)$ and $z(0.1)$.

$$\frac{dy}{dx} = 2z + y - x^2$$

$$\frac{dz}{dx} = y + \cos(x)$$

$$y(0) = 0, z(0) = 0$$

Ans.

$$y(x) = x^2 + \frac{x^4}{12} + \frac{x^5}{60} \qquad z(x) = x + \frac{x^3}{6} + \frac{x^5}{40}$$

$$y(0.1) = 0.0100085 \qquad z(0.1) = 0.1001669$$

4. Find the values of $y(0.5)$ and $z(0.5)$ for the following system of first order ordinary differential equations with the help of Euler method. Use step size $h=0.1$

$$\frac{dy}{dx} = x + \sin(y) + z^2$$

$$\frac{dz}{dx} = \cos(x) - z + y^2; \quad y(0) = 1, z(0) = 1$$

Ans.
$y(0.1) = 1.184147$ $z(0.1) = 1.100000$
$y(0.2) = 1.407765$ $z(0.2) = 1.229721$
$y(0.3) = 1.677660$ $z(0.3) = 1.402936$
$y(0.4) = 2.003913$ $z(0.4) = 1.639630$
$y(0.5) = 2.403517$ $z(0.5) = 1.969340$

5. Find the values of $y(0.4)$ and $z(0.4)$ for the following system of first order ODEs with the help of Runge–Kutta method of order 4. Use step size $h = 0.2$.

$$\frac{dy}{dx} = x + 2y + z$$

$$\frac{dz}{dx} = y + z^2$$

$$y(0) = 0, z(0) = 0$$

Ans.

$k_1 = 0$ $k_2 = 0.02$ $k_3 = 0.024$ $k_4 = 0.05$
$l_1 = 0$ $l_2 = 0$ $l_3 = 0.002$ $l_4 = 0.004801$
$y(0.2) = 0.023$ $z(0.2) = 0.001467$

$k_1 = 0.049493$ $k_2 = 0.079852$ $k_3 = 0.086419$ $k_4 = 0.126580$
$l_1 = 0.004600$ $l_2 = 0.009552$ $l_3 = 0.012593$ $l_4 = 0.021923$
$y(0.4) = 0.107769$ $z(0.4) = 0.013269$

6. Apply the Runge–Kutta fourth order method to the following IVP

$$x^2 y'' - 3xy' + 4y = 8, \quad y(1) = 3, \quad y'(1) = 2$$

to estimate $y(1.4)$ with $h = 0.2$ and compare the results with the exact solution.

Ans.

The equivalent system of first order ODEs is as follows

$$\frac{dy}{dx} = z$$

$$\frac{dz}{dx} = \frac{8 + 3xz - 4y}{x^2}$$

$$y(1) = 3, \quad z(1) = 2$$

On solving the given system with the help of Runge–Kutta fourth order method, we get

$k_1 = 0.400000$ $k_2 = 0.440000$ $k_3 = 0.440661$ $k_4 = 0.479038$
$l_1 = 0.400000$ $l_2 = 0.406612$ $l_3 = 0.395192$ $l_4 = 0.397229$

$y(1.2) = 3.440060 \quad z(1.2) = 2.400139$

$k_1 = 0.480028$ $k_2 = 0.520031$ $k_3 = 0.520505$ $k_4 = 0.559307$
$l_1 = 0.400036$ $l_2 = 0.404771$ $l_3 = 0.396396$ $l_4 = 0.398284$

$y(1.4) = 3.960128 \quad z(1.4) = 2.800248$

Exact Solution: $y(x) = x^2 + 2$

7. Consider a series circuit with resistor R, inductor L, and capacitor C, along with electromotive force (voltage supply) E.

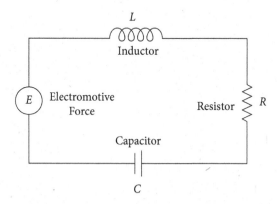

The resistance R, inductance L and capacitance C are measured in ohms (Ω), henrys (H) and farads (F) respectively. Electromotive force E is measured in volts (V) and it may be a function of time.

Charge (q) and current $\left(I = \dfrac{dq}{dt} \right)$ in the circuit are measured in coulombs and amperes respectively. Kirchhoff's second law can be used to study current flow in the circuit. It states that the electromotive force E around a closed circuit loop is equal to the sum of voltage drops across every component in the circuit.

The inductor L, resistor R and capacitor C produce the voltage drops of $L\dfrac{d^2q}{dt^2}$, $R\dfrac{dq}{dt}$ and $\dfrac{1}{C}q$ respectively. So, Kirchhoff's second law gives the following equation

$$L\frac{d^2q}{dt^2}+R\frac{dq}{dt}+\frac{1}{C}q=E$$

Now, we will discuss the application of this model to investigate the current flow in a closed series circuit with the help of the following example.

A series circuit has an inductor of 0.2 H, a resistor of 2Ω and a capacitor of $\dfrac{1}{20}$ F as shown in the following figure. An electromotive force $E=10\sin(2t)$ is applied to the

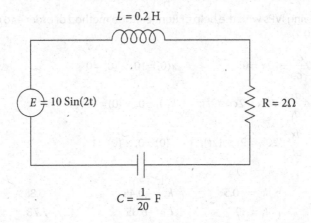

$$L = 0.2\ \text{H}$$

$$E = 10\ \text{Sin}(2t) \qquad R = 2\Omega$$

$$C = \frac{1}{20}\ \text{F}$$

If the initial charge on the capacitor and initial current are zero, then the charge (q) on the capacitor at any time $t > 0$ is given by the following differential equation

$$\frac{1}{5}\frac{d^2q}{dt^2}+2\frac{dq}{dt}+20q=10\sin(2t); \qquad q(0)=q'(0)=0$$

Solve the above differential equation with the aid of Runge–Kutta method of order 4 to compute the charges on the capacitor at time $t = 0.1$ and 0.2 [i.e., find the values of $q(0.1)$ and $q(0.2)$].

Ans.

$k_1 = 0$ $k_2 = 0$ $k_3 = 0.024958$ $k_4 = 0.024958$

$l_1 = 0$ $l_2 = 0.499167$ $l_3 = 0.249584$ $l_4 = 0.494180$

$q(0.1) = 0.012479$

$k_1 = 0.033195$ $k_2 = 0.060025$ $k_3 = 0.062524$ $k_4 = 0.075939$

$l_1 = 0.536608$ $l_2 = 0.586585$ $l_3 = 0.427445$ $l_4 = 0.437669$

$q(0.2) = 0.071518$

8. A coil spring, with spring constant k, is suspended from a fixed support; weight (mg) is attached to the lower end of the coil. The spring stretched l_1 metres from its equilibrium position due to attached weight. The weight further pulled down l_2 metres below equilibrium position and release from an initial velocity v_0 at time $t = 0$. The resistance of the medium is given by $a\dfrac{dx}{dt}$, where a is a constant depending on medium viscosity and $\dfrac{dx}{dt}$ is the velocity of the spring. If an external force $F(t)$ is applied, then the displacement x of the weight at any time t is given by following initial value problem

$$m\frac{d^2x}{dt^2}+a\frac{dx}{dt}+kx=F(t); \quad x(0)=l_2, \left.\frac{dx}{dt}\right|_{t=0}=v(0)=v_0$$

The following initial value problems represent a few of such kind of models.

Solve the following IVPs with the help of Runge–Kutta method of order 4 to compute the values of $x(0.1)$ and $x(0.2)$.

a) $\dfrac{d^2x}{dt^2}+2\dfrac{dx}{dt}+10x=0;$ $x(0)=10,\ x'(0)=0$

b) $\dfrac{d^2x}{dt^2}+4\dfrac{dx}{dt}+20x=2\cos(2t);$ $x(0)=0,\ x'(0)=0$

c) $\dfrac{d^2x}{dt^2}+4\dfrac{dx}{dt}+20x=2\cos(2t);$ $x(0)=0,\ x'(0)=1$

Ans.

a) $k_1=0$ $k_2=-0.5$ $k_3=-0.45$ $k_4=-0.885$
   $l_1=-10$ $l_2=-9$ $l_3=-8.85$ $l_4=-7.78$
   $x(0.1)=9.535833$

   $k_1=-0.891333$ $k_2=-1.278992$ $k_3=-1.217943$ $k_4=-1.537379$
   $l_1=-7.753166$ $l_2=-6.532184$ $l_3=-6.460453$ $l_4=-5.243134$
   $x(0.2)=8.298737$

b) $k_1=0$ $k_2=0.01$ $k_3=0.007950$ $k_4=0.015720$
   $l_1=0.2$ $l_2=0.159001$ $l_3=0.157201$ $l_4=0.117233$
   $x(0.1)=0.008603$

   $k_1=0.015827$ $k_2=0.021602$ $k_3=0.019409$ $k_4=0.023290$
   $l_1=0.115498$ $l_2=0.071625$ $l_3=0.074624$ $l_4=0.035030$
   $x(0.2)=0.028793$

c) $k_1=0.1$ $k_2=0.09$ $k_3=0.086950$ $k_4=0.076120$
   $l_1=-0.2$ $l_2=-0.260999$ $l_3=-0.238799$ $l_4=-0.282367$
   $x(0.1)=0.088337$

   $k_1=0.075301$ $k_2=0.061207$ $k_3=0.060014$ $k_4=0.046614$
   $l_1=-0.281862$ $l_2=-0.305737$ $l_3=-0.286869$ $l_4=-0.298944$
   $x(0.2)=0.149063$

9. Compute the approximate solution of following differential equation using finite difference method,

$$xy'' - y' + 2xy = x; \quad 1 \le x \le 2$$

subject to boundary conditions

$$y(1) = 1, \quad y(2) = 3$$

Subdivide the interval [1, 2] into four equal parts.

Ans.

$y(1.25) = 1.5093, y(1.5) = 2.0355, y(1.75) = 2.5422$

10. Use finite difference approximation to find the values of $y(0.25)$, $y(0.5)$, and y (0.75) for the following BVP

$$y'' - 2xy' + y = x^2; \quad y(0) = 0, \quad y(1) = 0$$

Ans.

$y(0.25) = -0.02191, y(0.5) = -0.04045, y(0.75) = -0.04296$

11. Use finite difference approximation to find the values of $y(0.25)$, $y(0.5)$, and y (0.75) for the following BVP

$$y'' - y' + y = 0; \qquad y(0) = 0, \quad y(1) = 1$$

Ans.

$y(0.25) = 0.1938, y(0.5) = 0.4290, y(0.75) = 0.7007$

12. Solve the following differential equation with boundary conditions with the aid of finite difference method

$$y'' + xy' - y = 1; \quad 0 \le x \le 1$$
$$y(0) = -1, \qquad y(1) = 1$$

Subdivide the interval [0, 1] into five equal parts.

Ans.

$y(0.2) = -0.6, \, y(0.4) = -0.2, \, y(0.6) = 0.2, \, y(0.8) = 0.6$

13. Solve the following BVP with the aid of finite difference method

$$y'' - 2xy' + 4y = 2x - 2; \quad 0 \le x \le 0.6$$
$$y'(0) = 1, \qquad\qquad y(0.6) = -0.04$$

Replace derivative boundary condition by central difference and divide the interval into three equal subintervals (or $h = 0.2$).

Ans.

$y(0) = -1, \, y(0.2) = -0.76, \, y(0.4) = -0.44$

14. Try to solve the following boundary value problems with the help of finite difference method and interpret the results.

   i) $y'' + y = 0;$ $y(0) = 1, \, y(\pi) = 2$ (BVP has no solution)

   ii) $y'' + y = 0;$ $y(0) = 1, \, y(\pi) = 1$ (BVP has infinitely many solutions)

15. Find the approximate solution of the following differential equation with the help of shooting method,

$$xy'' - y' + 2xy = x; \quad 1 \le x \le 2$$

subject to boundary conditions

$$y(1) = 1, \quad y(2) = 3$$

Subdivide the interval [1, 2] into two equal parts and use Runge–Kutta 4th order method to solve the IVPs.

Ans.

$u(1.5) = 0.861875 \quad u(2) = 0.467040$

$v(1.5) = 1.721875 \quad v(2) = 3.163212$

On using these values in the following expression, we get

$$y(x) = u(x) + \frac{\beta - u(b)}{v(b)} v(x)$$

16. Use shooting method to find the values of $y(0.2)$ and $y(0.4)$ for the following BVP

$$y'' - 2xy' + 4y = 2x + 6; \qquad y(0) = 1, \ y(0.6) = 1.96$$

Use Runge–Kutta 4th order method to solve the corresponding IVPs. Compare the result with the exact solutions $y(x) = 1 + x + x^2$.

Ans.

$u(0.2) = 1.042699 \quad u(0.4) = 1.181706 \quad u(0.6) = 1.434736$

$v(0.2) = 0.386721 \quad v(0.4) = 0.741434 \quad v(0.6) = 1.029416$

By using these values, we have

$$y(0.2) = 1.240025 \quad y(0.4) = 1.560026$$

The exact solution is given by

$$y(0.2) = 1.24 \qquad y(0.4) = 1.56$$

Partial Differential Equations: Finite Difference Methods

> *There's no sense in being precise when you don't even know what you're talking about.*
>
> **John von Neumann**
> (December 28, 1903–February 8, 1957)
> He was a great mathematician, physicist and computer scientist who had many research investigations to his credit.
>
> *Mathematics is the science which draws necessary conclusions.*
>
> **Benjamin Peirce**
> (April 4, 1809–October 6, 1880)
> He was a mathematician who had contributed to algebra, number theory, statistics and mechanics.

Parabolic Equation (Heat Conduction or Diffusion Equation)
Elliptic Equation (Laplace and Poisson Equations)
Hyperbolic Equation (Wave Equation)

Most of the problems posed by nature, and which are of interest to physicists and mathematicians are usually governed by a single or a system of differential equations. In general, a physical system involves more than one independent variable; in that case, our mathematical model contains partial differential equations (PDEs). PDEs play a vital role in the study of many branches of applied sciences and engineering; for example, fluid dynamics, heat transfer, elasticity, electromagnetic theory, optics, plasma physics, quantum mechanics, etc. In fact, the theories of modern physics, generally involve a mathematical model, as far as possible it is a set of PDEs. We first solve the mathematical model for solutions and then come to mathematical and physical interpretations of these solutions. So it is necessary to solve the mathematical model to study the physical system. Often, it is very difficult to solve these sets of PDEs explicitly for exact solutions. Consequently, numerical methods are applied to obtain approximate solutions of these equations. In fact, there is much current interest in obtaining numerical solutions of the PDEs.

The finite difference method is a simple and most commonly used method to solve PDEs. In this method, various derivatives in the partial differential equation are replaced by their finite difference approximations, and the PDE is converted to a set of linear algebraic equations. This system of linear equations can be solved by any iterative procedure discussed in Chapter 5. Then, the solution of PDE is the solution of this system of linear equations. An important advantage of this method is that the most of the calculations can be carried out on the computer, and hence the solution is easy to obtain.

16.1 Classification of Second-Order Quasi-Linear PDEs

Many physical phenomena like heat flow in a metal rod, waves in the string can be well described by a second order PDE of type

$$R(x, y)\frac{\partial^2 u}{\partial x^2} + S(x, y)\frac{\partial^2 u}{\partial x \partial y} + T(x, y)\frac{\partial^2 u}{\partial y^2} + L\left(x, y, u, \frac{\partial u}{\partial x}, \frac{\partial u}{\partial y}\right) = 0 \qquad (16.1)$$

where the functions R, S, T are continuous functions of the variables x and y only, while L is a continuous function of $x, y, u, \dfrac{\partial u}{\partial x}, \dfrac{\partial u}{\partial y}$. The equation is linear in highest derivatives (second derivative), therefore it is quasi-linear. The PDE (16.1) is classified according to the value of $S^2 - 4RT$ as follows

a) Elliptic;   if $S^2 - 4RT < 0$
b) Parabolic;   if $S^2 - 4RT = 0$
c) Hyperbolic; if $S^2 - 4RT > 0$

It is similar to the general second-degree equation $ax^2 + bxy + cy^2 + fx + gy + h = 0$ in coordinate geometry, where it represents an ellipse, parabola and hyperbola if the value of $b^2 - 4ac$ is negative, zero and positive respectively. Similar to an ellipse with no asymptote, the elliptic equation has no characteristic; like parabola has one asymptote, the parabolic equation has one characteristic; and the hyperbolic equation has two characteristics similar to the hyperbola, which has two asymptotes.

The functions R, S, T are continuous functions of the variables x and y, hence $S^2 - 4RT$ is also a function of the variables x and y. Therefore, the domain of PDE (16.1) is important in describing the classification of PDE, consider the following example of a PDE

$$x\frac{\partial^2 u}{\partial x^2} + 2x\frac{\partial^2 u}{\partial x \partial y} + (x + y)\frac{\partial^2 u}{\partial y^2} + xy\frac{\partial u}{\partial y} + u = 0$$

For this PDE, we have

$$S^2 - 4RT = (2x)^2 - 4(x)(x + y) = -4xy$$

This PDE is parabolic on x and y-axes, elliptic in 1st and 3rd quadrants and hyperbolic in 2nd and 4th quadrants.

Standard P.D.E.:

Equation (16.1) can be classified as discussed above, but certain particular equations govern some of the important physical systems. These standard PDEs are as follows

a) Parabolic Equation: Heat conduction or diffusion equation is an important example of a parabolic equation, which in one dimension can be read as follows

$$\frac{\partial u}{\partial t} = c \frac{\partial^2 u}{\partial x^2}$$

where the constant $c = \dfrac{K}{\rho\omega}$ is known as diffusivity constant or coefficient of heat conduction, and it depends on the nature of the medium. The constants K, ρ, and ω are the conductivity, density and specific heat of the medium, respectively. This equation represents heat flow in a homogeneous medium along the x-axis. The variable $u(x, t)$ denotes the temperature at any point x and at any time t. This equation describes many other physical phenomena like fluid flow, diffusion, etc.

The equation represents the change with respect to temporal (time variable t) and spatial (space variable x) variables. Therefore it involves both initial as well as boundary conditions. It contains the first derivative of the time variable, so only one initial condition is present. Two boundary conditions are required due to presence of second derivative of space variable.

Similarly, two-dimensional heat conduction equation is as follows

$$\frac{\partial u}{\partial t} = c\left(\frac{\partial^2 u}{\partial x^2} + \frac{\partial^2 u}{\partial y^2}\right) = c\nabla^2 u$$

where $\nabla^2 \equiv \nabla.\nabla \equiv \dfrac{\partial^2}{\partial x^2} + \dfrac{\partial^2}{\partial y^2}$ is a two dimensional Laplace operator.

Similarly three-dimensional heat conduction equation is as follows

$$\frac{\partial u}{\partial t} = c\nabla^2 u$$

where $\nabla^2 \equiv \dfrac{\partial^2}{\partial x^2} + \dfrac{\partial^2}{\partial y^2} + \dfrac{\partial^2}{\partial y^2}$ is Laplace operator in three dimensions.

b) Hyperbolic Equation: Vibrations of a tightly stretched string between two points are well described by the hyperbolic equation known as wave equation in one dimension

$$\frac{\partial^2 u}{\partial t^2} = c \frac{\partial^2 u}{\partial x^2}$$

Here $c = \dfrac{T}{m}$ (T is tension in the string and m is the mass per unit length) is a positive constant, and it depends on the nature of string. The string is homogeneous, i.e., uniform and elastic. The dependent variable $u(x, t)$ is the displacement of the string at any point x and at any time t from equilibrium position.

Similar expressions for two and three dimensional wave equations are as follows

$$\frac{\partial^2 u}{\partial t^2} = c\left(\frac{\partial^2 u}{\partial x^2} + \frac{\partial^2 u}{\partial y^2}\right) = c\nabla^2 u$$

$$\frac{\partial^2 u}{\partial t^2} = c\left(\frac{\partial^2 u}{\partial x^2} + \frac{\partial^2 u}{\partial y^2} + \frac{\partial^2 u}{\partial z^2}\right)$$

c) Elliptic Equation: All time-dependent problems are known as "transient" problems. As time increases, transient problem tends to steady state i.e. independent of time (mathematically, $\frac{\partial u}{\partial t} = 0$). In many cases, the process attains its final stage very soon. In such cases, we are interested in the final stage of the problem, i.e., the steady state. Two important cases of elliptic equations are Laplace and Poisson equations. These equations also represent heat and wave equations in steady states.

$$\nabla^2 u \equiv \frac{\partial^2 u}{\partial x^2} + \frac{\partial^2 u}{\partial y^2} = 0 \qquad \text{(Laplace equation)}$$

$$\nabla^2 u \equiv \frac{\partial^2 u}{\partial x^2} + \frac{\partial^2 u}{\partial y^2} = f(x, y) \qquad \text{(Poisson equation)}$$

These equations also describe other physically important phenomena in different branches of science, like electromagnetic theory and torsion problems etc. These equations are also known as potential equations as the variable u represents the gravitational potential, velocity potential, and electromagnetic potential in various relevant fields of science.

16.2 Initial and Boundary Conditions

The mathematical model for any physical phenomenon contains the adequate number of constraints, such that we can determine a unique solution for the problem. It is worth mentioning here that if the mathematical model has derivative term with respect to time variable, then the restrictions are initial conditions. Therefore, in case of one-dimensional heat conduction equation only one initial condition [$u(x, 0) = f(x)$] is required. It is quite obvious as initially ($t = 0$), the composition of system is well known. Similarly, two initial conditions ($u(x, 0) = f(x)$ and $u_t(x, 0) = g(x)$) are given in case of wave equation as the time derivative is up to second order.

If the differential equation contains the derivative terms on spatial variables, then the boundary conditions are prescribed. Since all three types of equations (Heat, Wave, and Laplace and Poisson), have second derivatives w.r.t. space variable, so two boundary conditions must be given for each spatial variable. Also, the boundary condition can contain term up to the first derivative. Boundary conditions at the boundary points a and b must be of term form

$$l_1 u(a,t) + m_1 \frac{\partial u}{\partial n}(a,t) = k_1$$

$$l_2 u(b,t) + m_2 \frac{\partial u}{\partial n}(b,t) = k_2$$

where $k_i, l_i, m_i, \ i = 1, 2$; are the functions of t only and n is the direction of the outward drawn normal to the surface (in two and three dimensions). In the case of one dimension, the direction of the outward drawn normal is the direction of increasing x. This type of boundary conditions is known as mixed type boundary conditions, as it involves both values of the function $u(x, t)$ and its normal derivative. Particular types of the boundary condition are classified as follows

i)   Dirichlet conditions: If only function value $u(x, t)$ is prescribed on the boundary.
ii) Neumann conditions: If only normal derivative (flux) of $u(x, t)$ is prescribed on the boundary.

16.3 Finite Difference Approximations for Partial Derivatives

The finite difference method is simple and most commonly used numerical method to solve differential equations with boundary conditions. In this method, various derivative terms in the differential equation are replaced by their finite difference approximations, and the differential equation is converted into a set of algebraic equations. Then, the solution of the system of linear equations provides the solution of the differential equation. This system of linear equations can be solved by any direct or iterative procedure discussed in Chapter 5. We have already discussed finite difference method for solutions of boundary value problems of ODEs in Chapter 15. An important advantage of this method is that it can be implemented on the computer to obtain the solution easily. In this section, we will discuss the finite difference approximations of partial derivatives using Taylor series.

The Taylor series expansions for the function $u = u(x, y)$ are as follows

$$u(x+h, y) = u(x,y) + h\frac{\partial u}{\partial x} + \frac{h^2}{2!}\frac{\partial^2 u}{\partial x^2} + \frac{h^3}{3!}\frac{\partial^3 u}{\partial x^3} + \cdots$$

$$u(x-h, y) = u(x,y) - h\frac{\partial u}{\partial x} + \frac{h^2}{2!}\frac{\partial^2 u}{\partial x^2} - \frac{h^3}{3!}\frac{\partial^3 u}{\partial x^3} + \cdots$$

$$u(x+2h, y) = u(x,y) + 2h\frac{\partial u}{\partial x} + \frac{(2h)^2}{2!}\frac{\partial^2 u}{\partial x^2} + \frac{(2h)^3}{3!}\frac{\partial^3 u}{\partial x^3} + \cdots$$

$$u(x-2h, y) = u(x,y) - 2h\frac{\partial u}{\partial x} + \frac{(2h)^2}{2!}\frac{\partial^2 u}{\partial x^2} - \frac{(2h)^3}{3!}\frac{\partial^3 u}{\partial x^3} + \cdots$$

$$u(x, y+k) = u(x,y) + k\frac{\partial u}{\partial y} + \frac{k^2}{2!}\frac{\partial^2 u}{\partial y^2} + \frac{k^3}{3!}\frac{\partial^3 u}{\partial y^3} + \cdots$$

$$u(x, y-k) = u(x,y) - k\frac{\partial u}{\partial y} + \frac{k^2}{2!}\frac{\partial^2 u}{\partial y^2} - \frac{k^3}{3!}\frac{\partial^3 u}{\partial y^3} + \cdots$$

$$u(x, y+2k) = u(x, y)+2k\frac{\partial u}{\partial y}+\frac{(2k)^2}{2!}\frac{\partial^2 u}{\partial y^2}+\frac{(2k)^3}{3!}\frac{\partial^3 u}{\partial y^3}+\cdots$$

$$u(x, y-2k) = u(x, y)-2k\frac{\partial u}{\partial y}+\frac{(2k)^2}{2!}\frac{\partial^2 u}{\partial y^2}-\frac{(2k)^3}{3!}\frac{\partial^3 u}{\partial y^3}+\cdots$$

$$u(x+h, y+k) = u(x, y)+\left(h\frac{\partial}{\partial x}+k\frac{\partial}{\partial y}\right)u+\frac{1}{2!}\left(h\frac{\partial}{\partial x}+k\frac{\partial}{\partial y}\right)^2 u+\cdots$$

We can easily obtain following approximations for first and second orders partial derivatives by using these Taylor expansions.

First Order Partial Derivatives:

$$\frac{\partial u(x, y)}{\partial x} = \frac{u(x+h, y)-u(x, y)}{h}+O(h)$$

$$\frac{\partial u(x, y)}{\partial y} = \frac{u(x, y+k)-u(x, y)}{k}+O(k)$$

Forward Differences

$$\frac{\partial u(x, y)}{\partial x} = \frac{u(x, y)-u(x-h, y)}{h}+O(h)$$

$$\frac{\partial u(x, y)}{\partial y} = \frac{u(x, y)-u(x, y-k)}{k}+O(k)$$

Backward Differences

$$\frac{\partial u(x, y)}{\partial x} = \frac{u(x+h, y)-u(x-h, y)}{2h}+O(h^2)$$

$$\frac{\partial u(x, y)}{\partial y} = \frac{u(x, y+k)-u(x, y-k)}{2k}+O(k^2)$$

Central Differences

Second Order Partial Derivatives:
Forward Differences:

$$\frac{\partial^2 u(x, y)}{\partial x^2} = \frac{u(x+2h, y)-2u(x+h, y)+u(x, y)}{h^2}+O(h)$$

$$\frac{\partial^2 u(x, y)}{\partial y^2} = \frac{u(x, y+2k)-2u(x, y+k)+u(x, y)}{k^2}+O(k)$$

$$\frac{\partial^2 u(x, y)}{\partial x \partial y} = \frac{u(x+h, y+k)-u(x, y+k)-u(x+h, y)+u(x, y)}{hk}+O(h, k)$$

Last mixed derivative is obtained from finite difference approximations of first order derivatives. Similarly, backward and central differences for second order derivatives are as follows

Backward Differences:

$$\frac{\partial^2 u(x,y)}{\partial x^2} = \frac{u(x,y) - 2u(x-h,y) + u(x-2h,y)}{h^2} + O(h)$$

$$\frac{\partial^2 u(x,y)}{\partial y^2} = \frac{u(x,y) - 2u(x,y-k) + u(x,y-2k)}{k^2} + O(k)$$

$$\frac{\partial^2 u(x,y)}{\partial x \partial y} = \frac{u(x,y) - u(x,y-k) - u(x-h,y) + u(x-h,y-k)}{hk} + O(h,k)$$

Central Differences:

$$\frac{\partial^2 u(x,y)}{\partial x^2} = \frac{u(x+h,y) - 2u(x,y) + u(x-h,y)}{h^2} + O(h^2)$$

$$\frac{\partial^2 u(x,y)}{\partial y^2} = \frac{u(x,y+k) - 2u(x,y) + u(x,y-k)}{k^2} + O(k^2)$$

$$\frac{\partial^2 u(x,y)}{\partial x \partial y} = \frac{u(x+h,y+k) - u(x+h,y-k) - u(x-h,y+k) + u(x-h,y-k)}{4hk} + O(hk)$$

To replace various derivative terms by their finite difference approximations, we discretize the derivative terms at any point (x_i, y_j). Let the variables x and y be equidistant with spacing h and k respectively. Let us assume that the node points are $(x_i, y_j); i = 1(1)n, j = 1(1)m$. Consider the function $u(x_i, y_j) = u_{i,j}$. So, we can obtain following approximations for first and second order derivative terms at a point (x_i, y_j) by using the above expressions.

First Order Partial Derivatives:

Forward Differences:

$$\frac{\partial u(x_i, y_j)}{\partial x} = \frac{u(x_i + h, y_j) - u(x_i, y_j)}{h} + O(h)$$

$$= \frac{u(x_{i+1}, y_j) - u(x_i, y_j)}{h} + O(h)$$

$$= \frac{u_{i+1,j} - u_{i,j}}{h} + O(h)$$

$$\frac{\partial u(x_i, y_j)}{\partial y} = \frac{u(x_i, y_j + k) - u(x_i, y_j)}{k} + O(k)$$

$$= \frac{u(x_i, y_{j+1}) - u(x_i, y_j)}{k} + O(k)$$

$$= \frac{u_{i,j+1} - u_{i,j}}{k} + O(k)$$

Backward Differences:

$$\frac{\partial u(x_i, y_j)}{\partial x} = \frac{u(x_i, y_j) - u(x_i - h, y_j)}{h} + O(h)$$

$$= \frac{u(x_i, y_j) - u(x_{i-1}, y_j)}{h} + O(h)$$

$$= \frac{u_{i,j} - u_{i-1,j}}{h} + O(h)$$

$$\frac{\partial u(x_i, y_j)}{\partial y} = \frac{u(x_i, y_j) - u(x_i, y_j - k)}{k} + O(k)$$

$$= \frac{u(x_i, y_j) - u(x_i, y_{j-1})}{k} + O(k)$$

$$= \frac{u_{i,j} - u_{i,j-1}}{k} + O(k)$$

Central Differences:

$$\frac{\partial u(x_i, y_j)}{\partial x} = \frac{u(x_i + h, y_j) - u(x_i - h, y_j)}{2h} + O(h^2)$$

$$= \frac{u(x_{i+1}, y_j) - u(x_{i-1}, y_j)}{2h} + O(h^2)$$

$$= \frac{u_{i+1,j} - u_{i-1,j}}{2h} + O(h^2)$$

$$\frac{\partial u(x_i, y_j)}{\partial y} = \frac{u(x_i, y_j + k) - u(x_i, y_j - k)}{2k} + O(k^2)$$

$$= \frac{u(x_i, y_{j+1}) - u(x_i, y_{j-1})}{2k} + O(k^2)$$

$$= \frac{u_{i,j+1} - u_{i,j-1}}{2k} + O(k^2)$$

Second Order Partial Derivatives:
Forward Differences:

$$\frac{\partial^2 u(x_i, y_j)}{\partial x^2} = \frac{u(x_i + 2h, y_j) - 2u(x_i + h, y_j) + u(x_i, y_j)}{h^2} + O(h)$$

$$= \frac{u(x_{i+2}, y_j) - 2u(x_{i+1}, y_j) + u(x_i, y_j)}{h^2} + O(h)$$

$$= \frac{u_{i+2,j} - 2u_{i+1,j} + u_{i,j}}{h^2} + O(h)$$

$$\frac{\partial^2 u(x_i, y_j)}{\partial y^2} = \frac{u(x_i, y_j + 2k) - 2u(x_i, y_j + k) + u(x_i, y_j)}{k^2} + O(k)$$

$$= \frac{u(x_i, y_{j+2}) - 2u(x_i, y_{j+1}) + u(x_i, y_j)}{k^2} + O(k)$$

$$= \frac{u_{i,j+2} - 2u_{i,j+1} + u_{i,j}}{k^2} + O(k)$$

$$\frac{\partial^2 u(x_i, y_j)}{\partial x \partial y} = \frac{u(x_i + h, y_j + k) - u(x_i + h, y_j) - u(x_i, y_j + k) + u(x_i, y_j)}{hk} + O(h, k)$$

$$= \frac{u(x_{i+1}, y_{j+1}) - u(x_{i+1}, y_j) - u(x_i, y_{j+1}) + u(x_i, y_j)}{hk} + O(h, k)$$

$$= \frac{u_{i+1,j+1} - u_{i+1,j} - u_{i,j+1} + u_{i,j}}{hk} + O(h, k)$$

Backward Differences:

$$\frac{\partial^2 u(x_i, y_j)}{\partial x^2} = \frac{u(x_i, y_j) - 2u(x_i - h, y_j) + u(x_i - 2h, y_j)}{h^2} + O(h)$$

$$= \frac{u(x_i, y_j) - 2u(x_{i-1}, y_j) + u(x_{i-2}, y_j)}{h^2} + O(h)$$

$$= \frac{u_{i,j} - 2u_{i-1,j} + u_{i-2,j}}{h^2} + O(h)$$

$$\frac{\partial^2 u(x_i, y_j)}{\partial y^2} = \frac{u(x_i, y_j) - 2u(x_i, y_j - k) + u(x_i, y_j - 2k)}{k^2} + O(k)$$

$$= \frac{u(x_i, y_j) - 2u(x_i, y_{j-1}) + u(x_i, y_{j-2})}{k^2} + O(k)$$

$$= \frac{u_{i,j} - 2u_{i,j-1} + u_{i,j-2}}{k^2} + O(k)$$

$$\frac{\partial^2 u(x_i, y_j)}{\partial x \partial y} = \frac{u(x_i, y_j) - u(x_i - h, y_j) - u(x_i, y_j - k) + u(x_i - h, y_j - k)}{hk} + O(h, k)$$

$$= \frac{u(x_i, y_j) - u(x_{i-1}, y_j) - u(x_i, y_{j-1}) + u(x_{i-1}, y_{j-1})}{hk} + O(h, k)$$

$$= \frac{u_{i,j} - u_{i-1,j} - u_{i,j-1} + u_{i-1,j-1}}{hk} + O(h, k)$$

Central Differences:

$$\frac{\partial^2 u(x_i, y_j)}{\partial x^2} = \frac{u(x_i + h, y_j) - 2u(x_i, y_j) + u(x_i - h, y_j)}{h^2} + O(h^2)$$

$$= \frac{u(x_{i+1}, y_j) - 2u(x_i, y_j) + u(x_{i-1}, y_j)}{h^2} + O(h^2)$$

$$= \frac{u_{i+1,j} - 2u_{i,j} + u_{i-1,j}}{h^2} + O(h^2)$$

$$\frac{\partial^2 u(x_i, y_j)}{\partial y^2} = \frac{u(x_i, y_j + k) - 2u(x_i, y_j) + u(x_i, y_j - k)}{k^2} + O(k^2)$$

$$= \frac{u(x_i, y_{j+1}) - 2u(x_i, y_j) + u(x_i, y_{j-1})}{k^2} + O(k^2)$$

$$= \frac{u_{i,j+1} - 2u_{i,j} + u_{i,j-1}}{k^2} + O(k^2)$$

$$\frac{\partial^2 u(x_i, y_j)}{\partial x \partial y} = \frac{u(x_i + h, y_j + k) - u(x_i - h, y_j + k) - u(x_i + h, y_j - k) + u(x_i - h, y_j - k)}{4hk} + O(hk)$$

$$= \frac{u(x_{i+1}, y_{j+1}) - u(x_{i-1}, y_{j+1}) - u(x_{i+1}, y_{j-1}) + u(x_{i-1}, y_{j-1})}{4hk} + O(hk)$$

$$= \frac{u_{i+1,j+1} - u_{i+1,j-1} - u_{i-1,j+1} + u_{i-1,j-1}}{4hk} + O(hk)$$

Now, we will discuss various finite difference methods (FDM) for the solutions of some standard PDEs representing important physical systems. These PDEs include heat conduction (diffusion) equation, Laplace and Poisson equations, and wave equation.

Note: The central differences are better approximations compared to other differences, as the neglecting terms are of order h^2 in central differences compared to the order of h in forward and backward differences. Hence, central difference formulas are preferred over forward and backward differences.

16.4 Parabolic Equation (1-dimensional Heat Conduction Equation)

The heat conduction (or diffusion) equation in 1-dimension is a parabolic equation of following form

$$\frac{\partial u}{\partial t} = c \frac{\partial^2 u}{\partial x^2} \tag{16.2}$$

The following five finite difference schemes for 1-dimensional heat conduction Eq. (16.2) will be discussed.

i)   Bender–Schmidt Explicit Scheme
ii)   Crank–Nicolson (CN) Scheme
iii) General Implicit Scheme
iv)   Richardson Scheme
v)   Du-Fort and Frankel Scheme

16.4.1 Bender–Schmidt Explicit Scheme

Let us assume that the value of $u(x, t)$ at point (x_i, t_j) is $u_{i,j}$ i.e. $u(x_i, t_j) = u_{i,j}$. Also assume that step size for the variable t is $\Delta t = k$, and step size for the variable x is $\Delta x = h$.

Discretizing Eq. (16.2) at point (x_i, t_j), we get

$$\left.\frac{\partial u}{\partial t}\right|_{(x_i,t_j)} = c \left.\frac{\partial^2 u}{\partial x^2}\right|_{(x_i,t_j)}$$

The forward difference formula of first order derivative term for time variable and central difference formula of second order derivative term for space variable (from Section 16.3) are given by

$$\frac{\partial u(x_i,t_j)}{\partial t} = \frac{u_{i,j+1} - u_{i,j}}{k} + O(k)$$

$$\frac{\partial^2 u(x_i,t_j)}{\partial x^2} = \frac{u_{i+1,j} - 2u_{i,j} + u_{i-1,j}}{h^2} + O(h^2)$$

Using these formulas in heat equation, and neglecting the error terms ($O(k)$ and $O(h^2)$), we have

$$\frac{u_{i,j+1} - u_{i,j}}{k} = c \frac{u_{i+1,j} - 2u_{i,j} + u_{i-1,j}}{h^2}$$

(or) $u_{i,j+1} - u_{i,j} = \dfrac{ck}{h^2}\left(u_{i+1,j} - 2u_{i,j} + u_{i-1,j}\right)$

Let $r = \dfrac{ck}{h^2}$, then the Bender–Schmidt explicit scheme for the solution of Eq. (16.2) is as follows

$$u_{i,j+1} = r u_{i-1,j} + (1-2r)u_{i,j} + r u_{i+1,j} \tag{16.3}$$

This scheme (16.3) is known as explicit scheme.

Here, forward difference formula is used to approximate the term $\dfrac{\partial u}{\partial t}$. Therefore, the error term in time derivative approximation is of linear order $O(k)$. In next section, we will show

that the explicit scheme is stable for the value $r = \dfrac{ck}{h^2} \le \dfrac{1}{2}$. *Now we will move towards more*
accurate and unconditionally stable scheme, i.e., Crank–Nicolson scheme, in which, both the
derivative terms are replaced by central difference formulas.

16.4.2 Crank–Nicolson (CN) Scheme

Let us discretize Eq. (16.2) at point $(x_i, t_{j+1/2})$

$$\left. \frac{\partial u}{\partial t} \right|_{(x_i, t_{j+1/2})} = c \left. \frac{\partial^2 u}{\partial x^2} \right|_{(x_i, t_{j+1/2})} \tag{16.4}$$

Assume the node points in time variable t with step size $k/2$ are $t_j - k/2, t_j, t_j + k/2$ (or
$t_{j-1/2}, t_j, t_{j+1/2}$). Using central difference approximation for $\dfrac{\partial u}{\partial t}$ with these points, we have

$$\frac{\partial u(x_i, t_{j+1/2})}{\partial t} = \frac{u_{i,j+1} - u_{i,j}}{2(k/2)} = \frac{u_{i,j+1} - u_{i,j}}{k} \tag{16.5a}$$

As the values at the point $(x_i, t_{j+1/2})$ are not known therefore the central difference

formula $\dfrac{\partial^2 u(x_i, t_{j+1/2})}{\partial x^2} = \dfrac{u_{i+1,j+1/2} - 2u_{i,j+1/2} + u_{i-1,j+1/2}}{h^2}$ cannot be used for the derivative term

$\left. \dfrac{\partial^2 u}{\partial x^2} \right|_{(x_i, t_{j+1/2})}$ in Eq. (16.4). So, we will approximate the term $\left. \dfrac{\partial^2 u}{\partial x^2} \right|_{(x_i, t_{j+1/2})}$ with average values

of $\dfrac{\partial^2 u}{\partial x^2}$ at (x_i, t_j) and (x_i, t_{j+1}). i.e.

$$\frac{\partial^2 u(x_i, t_{j+1/2})}{\partial x^2} = \frac{1}{2} \left(\frac{\partial^2 u(x_i, t_j)}{\partial x^2} + \frac{\partial^2 u(x_i, t_{j+1})}{\partial x^2} \right)$$

$$= \frac{1}{2} \left(\frac{u_{i+1,j} - 2u_{i,j} + u_{i-1,j}}{h^2} + \frac{u_{i+1,j+1} - 2u_{i,j+1} + u_{i-1,j+1}}{h^2} \right) \tag{16.5b}$$

Using Eqs. (16.5a) and (16.5b) in Eq. (16.4), we get

$$\frac{u_{i,j+1} - u_{i,j}}{k} = c \frac{1}{2} \left(\frac{u_{i+1,j} - 2u_{i,j} + u_{i-1,j}}{h^2} + \frac{u_{i+1,j+1} - 2u_{i,j+1} + u_{i-1,j+1}}{h^2} \right)$$

Let $r = \dfrac{ck}{h^2}$, then the Crank–Nicolson (CN) implicit scheme for the solution of Eq. (16.2) is
as follows

$$-r\,u_{i-1,j+1} + 2(1+r)u_{i,j+1} - r\,u_{i+1,j+1} = r\,u_{i-1,j} + 2(1-r)u_{i,j} + r\,u_{i+1,j} \qquad (16.6)$$

It is easy to see that this scheme is implicit scheme, as we can't obtain the solution $u_{i,\,j+1}$ directly from the scheme. First, a set of equations is obtained, then we solve this set for values of $u_{i,\,j+1}$ in a row.

16.4.3 General Implicit Scheme

In CN-scheme, equal weightage is given to both j and $j+1$ levels, consider Eq. (16.5b)

$$\frac{\partial^2 u(x_i, t_{j+1/2})}{\partial x^2} = \frac{1}{2}\left(\frac{\partial^2 u(x_i, t_j)}{\partial x^2} + \frac{\partial^2 u(x_i, t_{j+1})}{\partial x^2} \right)$$

Let us construct an implicit scheme by giving different weight to different levels, say θ and $(1-\theta)$ to levels j and $j+1$, respectively, then

$$\frac{\partial^2 u(x_i, t_{j+1/2})}{\partial x^2} = \theta \frac{\partial^2 u(x_i, t_j)}{\partial x^2} + (1-\theta)\frac{\partial^2 u(x_i, t_{j+1})}{\partial x^2}$$

It is worth to note down that CN method can be obtained for $\theta = \dfrac{1}{2}$. Using central difference formulas, we have

$$\frac{u_{i,j+1} - u_{i,j}}{k} = c\left(\theta \frac{u_{i+1,j} - 2u_{i,j} + u_{i-1,j}}{h^2} + (1-\theta)\frac{u_{i+1,j+1} - 2u_{i,j+1} + u_{i-1,j+1}}{h^2} \right)$$

Assume, $r = \dfrac{ck}{h^2}$, the general implicit scheme for the solution of Eq. (16.2) is as follows

$$u_{i,j+1} - u_{i,j} = r\left(\theta\left(u_{i+1,j} - 2u_{i,j} + u_{i-1,j}\right) + (1-\theta)\left(u_{i+1,j+1} - 2u_{i,j+1} + u_{i-1,j+1}\right) \right)$$
$$\text{(or)}\ -r(1-\theta)u_{i-1,j+1} + (1+2r(1-\theta))u_{i,j+1} - r(1-\theta)u_{i+1,j+1} = r\,\theta u_{i-1,j} + (1-2r\theta)u_{i,j} + r\theta u_{i+1,j}$$
$$(16.7)$$

This scheme (16.7) is known as general implicit scheme.

So far, we have discussed Bender–Schmidt explicit scheme and CN and general implicit schemes for the solutions of 1-dimensional heat conduction Eq. (16.2). It is worth to mentioning here that these schemes are two-level schemes as these involve only two levels (j and $j+1$) of time variable (clear from formulas 16.3, 16.6 and 16.7 of these schemes).

Now, we will discuss three level schemes for the solution of Eq. (16.2) known as Richardson scheme and Du-Fort & Frankel scheme.

16.4.4 Richardson Scheme

Consider Eq. (16.2) at point (x_i, t_j)

$$\left.\frac{\partial u}{\partial t}\right|_{(x_i, t_j)} = c\left.\frac{\partial^2 u}{\partial x^2}\right|_{(x_i, t_j)}$$

Using central difference formulas for $\dfrac{\partial u(x_i, t_j)}{\partial t} = \dfrac{u_{i,j+1} - u_{i,j-1}}{2k} + O(k^2)$ and

$\dfrac{\partial^2 u(x_i, t_j)}{\partial x^2} = \dfrac{u_{i+1,j} - 2u_{i,j} + u_{i-1,j}}{h^2} + O(h^2)$, we have

$$\frac{u_{i,j+1} - u_{i,j-1}}{2k} = c\frac{u_{i+1,j} - 2u_{i,j} + u_{i-1,j}}{h^2}$$

$$\text{(or)} \quad u_{i,j+1} = u_{i,j-1} + 2\frac{ck}{h^2}\left(u_{i+1,j} - 2u_{i,j} + u_{i-1,j}\right)$$

$$u_{i,j+1} = u_{i,j-1} + 2r\left(u_{i+1,j} - 2u_{i,j} + u_{i-1,j}\right) \qquad (16.8)$$

This expression is known as Richardson scheme.

Note that this scheme is unstable for all values of $r = \dfrac{ck}{h^2}$. *So, we will construct an unconditionally stable Du-Fort and Frankel scheme, which is stable for all values of* r.

16.4.5 Du-Fort and Frankel Scheme

Richardson scheme (16.8) for the solution of Eq. (16.2) is as follows

$$u_{i,j+1} = u_{i,j-1} + 2r\left(u_{i+1,j} - 2u_{i,j} + u_{i-1,j}\right)$$

Replace the term $u_{i,j} = \dfrac{1}{2}\left(u_{i,j-1} + u_{i,j+1}\right)$ with average value at $j-1$ and $j+1$ levels as follows

$$u_{i,j+1} = u_{i,j-1} + 2r\left(u_{i+1,j} - (u_{i,j-1} + u_{i,j+1}) + u_{i-1,j}\right)$$

Rearranging the terms, Du-Fort and Frankel scheme is given by

$$u_{i,j+1} = \frac{1-2r}{1+2r}u_{i,j-1} + \frac{2r}{1+2r}\left(u_{i+1,j} + u_{i-1,j}\right) \qquad (16.9)$$

So far, we have discussed five finite difference schemes for the solution of 1-dimensional heat conduction Eq. (16.2).

We will use all these schemes to compute the values of $u_{i,j} = u(x_i, t_j)$ at various node points. We will construct the following table to collect all the values at one place.

Table for values of $u_{i,j} = u(x_i, t_j)$

j \ i	0	1	2	...	n
0	u_{00}	u_{10}	u_{20}	...	u_{n0}
1	u_{01}	u_{11}	u_{21}	...	u_{n1}
2	u_{02}	u_{12}	u_{22}	...	u_{n2}
.	.	.	.	.	.
.	.	.	.	.	.
.	.	.	.		

Note that the initial condition provides the first row of the table, and boundary conditions give first and last columns of the table.

Example ———————————————————————————————— **16.1**

Solve the 1-dimensional heat conduction equation

$$\frac{\partial u}{\partial t} = 2\frac{\partial^2 u}{\partial x^2}; \qquad 0 \le x \le 1 \text{ with}$$

initial condition $u(x,0) = x(2-x)$, and
boundary conditions $u(0,t) = 0$ and $u(1,t) = 1$.

Use Explicit scheme to find the value $u(x,t)$ up to $t = 0.02$, with $\Delta x = 0.2$ and $\Delta t = 0.005$.

Ans.
Since $\Delta x = 0.2$ for $0 \le x \le 1$, hence our node points are as follows

$$x_0 = 0, x_1 = 0.2, x_2 = 0.4, x_3 = 0.6, x_4 = 0.8, x_5 = 1$$

Let $u_{i,j} = u(x_i, t_j)$. The initial condition $u(x,0) = x(2-x)$ at $t_0 = 0$ provides

$u_{0,0} = u(x_0, t_0) = u(0,0) = 0(2-0) = 0$

$u_{1,0} = u(x_1, t_0) = u(0.2,0) = 0.2(2-0.2) = 0.36$

$u_{2,0} = u(x_2, t_0) = u(0.4,0) = 0.4(2-0.4) = 0.64$

$u_{3,0} = u(x_3, t_0) = u(0.6,0) = 0.6(2-0.6) = 0.84$

$u_{4,0} = u(x_4, t_0) = u(0.8,0) = 0.8(2-0.8) = 0.96$

$u_{5,0} = u(x_5, t_0) = u(1,0) = 1(2-1) = 1$ \hfill (16.10)

For t = 0

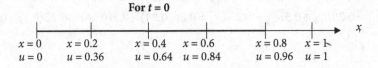

$x = 0 \qquad x = 0.2 \qquad x = 0.4 \quad x = 0.6 \qquad x = 0.8 \quad x = 1$
$u = 0 \qquad u = 0.36 \qquad u = 0.64 \quad u = 0.84 \qquad u = 0.96 \quad u = 1$

Similarly, boundary conditions are $u(0,t) = 0$ and $u(1,t) = 1$, so

$$u_{0,j} = u(x_0, t_j) = u(0, t) = 0$$
$$u_{5,j} = u(x_5, t_j) = u(1, t) = 1; \quad \text{for } \forall\, j = 0,1,2,3,\cdots \tag{16.11}$$

In the table form, initial values (16.10) and boundary values (16.11) are as follows

$i\,(x)$ $j\,(t)$	0(0)	1(0.2)	2(0.4)	3(0.6)	4(0.8)	5(1)
0(0)	0	0.36	0.64	0.84	0.96	1
1(0.005)	0					1
2(0.01)	0					1
3(0.015)	0					1
4(0.02)	0					1

After utilizing the initial and boundary conditions, we will now use the explicit method to find the values of $u(x, t)$ at node points. With $\Delta x = 0.2$, $\Delta t = 0.005$ and $c = 2$, our r is given by

$$r = \frac{c\Delta t}{\Delta x^2} = \frac{ck}{h^2} = \frac{1(0.005)}{(0.2)^2} = 0.25$$

Bender–Schmidt Explicit Scheme (16.3) is given by

$$u_{i,j+1} = r\,u_{i-1,j} + (1-2r)u_{i,j} + r\,u_{i+1,j}$$

For $r = 0.25$, we have

$$u_{i,j+1} = 0.25\,u_{i-1,j} + 0.5\,u_{i,j} + 0.25\,u_{i+1,j}$$

Using $j = 0$ in the above formula, we have

$$u_{i,1} = 0.25\,u_{i-1,0} + 0.5\,u_{i,0} + 0.25\,u_{i+1,0}$$

Computing the values for $i = 1, 2, 3$ and 4, we get

$$u_{1,1} = 0.25\,u_{0,0} + 0.5\,u_{1,0} + 0.25\,u_{2,0} = 0.25(0) + 0.5(0.36) + 0.25(0.64) = 0.34$$
$$u_{2,1} = 0.25\,u_{1,0} + 0.5\,u_{2,0} + 0.25\,u_{3,0} = 0.25(0.36) + 0.5(0.64) + 0.25(0.84) = 0.62$$
$$u_{3,1} = 0.25\,u_{2,0} + 0.5\,u_{3,0} + 0.25\,u_{4,0} = 0.25(0.64) + 0.5(0.84) + 0.25(0.96) = 0.82$$
$$u_{4,1} = 0.25\,u_{3,0} + 0.5\,u_{4,0} + 0.25\,u_{5,0} = 0.25(0.84) + 0.5(0.96) + 0.25(0.1) = 0.94$$

These values give the second row of the table.

i (x) j (t)	0(0)	1(0.2)	2(0.4)	3(0.6)	4(0.8)	5(1)
0(0)	0	0.36	0.64	0.84	0.96	1
1(0.005)	0	0.34	0.62	0.82	0.94	1
2(0.01)	0					1
3(0.015)	0					1
4(0.02)	0					1

Proceeding in a similar manner, for $j = 1, 2, 3, 4$ and 5, we will get different rows of the table as follows (up to six decimal digits):

i (x) j (t)	0(0)	1(0.2)	2(0.4)	3(0.6)	4(0.8)	5(1)
0(0)	0	0.36	0.64	0.84	0.96	1
1(0.005)	0	0.34	0.62	0.82	0.94	1
2(0.01)	0	0.325	0.60	0.80	0.925	1
3(0.015)	0	0.3125	0.58125	0.78125	0.9125	1
4(0.02)	0	0.301563	0.564063	0.764063	0.901563	1

Example 16.2

Use Explicit scheme to compute temperature distribution in a uniform insulated rod of length $1m$ with diffusivity constant of the material of the rod is given 1. Both ends of the rod are kept at zero temperature, and initial temperature distribution in the rod is given by the function $u(x,0) = \sin(\pi x)$. Take $\Delta x = 1/4$, $\Delta t = 1/16$; solve up to $t = 1/8$.

Ans.

The diffusivity constant of the material of the rod is 1, i.e. $c = 1$. So, the temperature distribution is given by following heat conduction equation

$$\frac{\partial u}{\partial t} = \frac{\partial^2 u}{\partial x^2}; \qquad 0 \le x \le 1$$

Both ends of the rod are kept at zero temperature; therefore boundary conditions are given by

$$u(0, t) = u(1, t) = 0$$

Also, initial temperature distribution gives following initial condition

$$u(x, 0) = \sin(\pi x)$$

The mathematical model for the given physical problem is complete.

Now, we will compute the temperature distribution at various nodes using explicit scheme

With $\Delta x = 1/4$, $t = 0$, the initial condition $u(x, 0) = \sin(\pi x)$ gives

$$u_{1,0} = u(x_1, t_0) = u(1/4, 0) = \sin(\pi/4) = 1/\sqrt{2} = 0.707107$$
$$u_{2,0} = u(x_2, t_0) = u(1/2, 0) = \sin(\pi/2) = 1$$
$$u_{3,0} = u(x_3, t_0) = u(3/4, 0) = \sin(3\pi/4) = 1/\sqrt{2} = 0.707107 \qquad (16.12)$$

For t = 0

$x = 0$ $x = 1/4$ $x = 1/2$ $x = 3/4$ $x = 1$
$u = 0$ $u = 0.707107$ $u = 1$ $u = 0.707107$ $u = 0$

The boundary conditions $u(0, t) = u(1, t) = 0$ provide

$$u_{0,j} = u(x_0, t_j) = u(0, t) = 0$$
$$u_{4,j} = u(x_4, t_j) = u(1, t) = 0; \qquad \text{for } \forall j = 0, 1, 2, 3, \cdots \qquad (16.13)$$

In the table form, (16.12) and (16.13) are as follows

i (x) \ j (t)	0(0)	1(1/4)	2(1/2)	3(3/4)	4(1)
0(0)	0	0.707107	1	0.707107	0
1(1/16)	0				0
2(1/8)	0				0

With $\Delta x = 1/4$, $\Delta t = 1/16$ and $c = 1$, the value of constant r is given by

$$r = \frac{c\Delta t}{\Delta x^2} = \frac{1(1/16)}{(1/4)^2} = 1$$

Explicit scheme (16.3) for $r = 1$ is as follows

$$u_{i,j+1} = u_{i-1,j} - u_{i,j} + u_{i+1,j}$$

Using $j = 0$ in the above formula, we have

$$u_{i,1} = u_{i-1,0} - u_{i,0} + u_{i+1,0}$$

Computing the different values of $u_{i,1}$ for $i = 1, 2$ and 3, we get second row of following table

$i(x)$ $j(t)$	0(0)	1(1/4)	2(1/2)	3(3/4)	4(1)
0(0)	0	0.707107	1	0.707107	0
1(1/16)	0	0.292803	0.414214	0.292803	0
2(1/8)	0				0

By using, $j = 1$ in the explicit formula, we get

$$u_{i,2} = u_{i-1,1} - u_{i,1} + u_{i+1,1}$$

For $i = 1, 2, 3$, we can find values at time $t = 1/8$. These values are given in the third row of the following table.

$i(x)$ $j(t)$	0(0)	1(1/4)	2(1/2)	3(3/4)	4(1)
0(0)	0	0.707107	1	0.707107	0
1(1/16)	0	0.292803	0.414214	0.292803	0
2(1/8)	0	0.121411	0.171392	0.121411	0

Example 16.3

Solve the PDE of Example 16.2 using Crank–Nicolson scheme.

Ans.

Values of $u_{i,j}$ for $t = 1/16$ (or $j = 1$):

Crank–Nicolson formula (16.6) is as follows

$$-ru_{i+1,j+1} - ru_{i-1,j+1} + 2(1+r)u_{i,j+1} = ru_{i+1,j} + ru_{i-1,j} + 2(1-r)u_{i,j}$$

With $r = 1, j = 0$, the CN-formula gives following equations for $i = 1, 2, 3$

$$-u_{2,1} - u_{0,1} + 4u_{1,1} = u_{2,0} + u_{0,0}$$

$$-u_{3,1} - u_{1,1} + 4u_{2,1} = u_{3,0} + u_{1,0}$$

$$-u_{4,1} - u_{2,1} + 4u_{3,1} = u_{4,0} + u_{2,0} \qquad (16.14)$$

Using values from Eqs. (16.12) and (16.13) in the system (16.14), we get

$$-u_{2,1} + 4u_{1,1} = 1 + 0 = 1$$

$$-u_{3,1} - u_{1,1} + 4u_{2,1} = 1/\sqrt{2} + 1/\sqrt{2} = \sqrt{2}$$

$$-u_{2,1} + 4u_{3,1} = 0 + 1 = 1 \qquad (16.15)$$

On solving the system (16.15), we have

$$u_{11} = u_{31} = 0.386729, \quad u_{21} = 0.546916 \qquad (16.16)$$

$j(t)$ \ $i(x)$	0(0)	1(1/4)	2(1/2)	3(3/4)	4(1)
0(0)	0	0.707107	1	0.707107	0
1(1/16)	0	0.386729	0.546916	0.386729	0
2(1/8)	0				0

Values of $u_{i,j}$ for $t = 1/8$ (or $j = 2$):

With $r = 1, j = 1, i = 1, 2, 3$, the CN-formula (16.6) provides the following linear system

$$-u_{2,2} - u_{0,2} + 4u_{1,2} = u_{2,1} + u_{0,1}$$

$$-u_{3,2} - u_{1,2} + 4u_{2,2} = u_{3,1} + u_{1,1}$$

$$-u_{4,2} - u_{2,2} + 4u_{3,2} = u_{4,1} + u_{2,1} \qquad (16.17)$$

Using the known values in the above system, and solving the resulting system of equations, we get

$$u_{12} = u_{32} = 0.211509, \quad u_{21} = 0.29912$$

$j(t)$ \ $i(x)$	0(0)	1(1/4)	2(1/2)	3(3/4)	4(1)
0(0)	0	0.707107	1	0.707107	0
1(1/16)	0	0.386729	0.546916	0.386729	0
2(1/8)	0	0.211509	0.29912	0.211509	0

Note: The systems 16.15 and 16.17 are tridiagonal systems of linear equations. These small systems can be solved using any method discussed in Chapter 5. We have neglected the terms containing $(\Delta x)^2 = (0.25)^2 = 0.0625$ and $(\Delta t)^2 = (0.0625)^2 = 0.00390625$. These errors are very large in scientific and engineering computations. We have to reduce the step sizes Δx and Δt to decrease these errors, for example, let $\Delta x = \Delta t = 0.001$. Then, we have to solve 124 linear systems of 999 equations each to compute the temperature distribution up to time $t = 0.125$. Programming is the only way to solve so many large systems. The solution procedure requires very large numbers of computations. To handle these computations and to reduce round-off errors, we require high-end computers and latest software. Also, it is worth to mention here that these systems are tridiagonal systems, so some computationally efficient algorithms for tridiagonal systems will be helpful to reduce round-off error to large extent. Since, these systems are diagonally dominant, so iterative procedures like Gauss–Seidel method can be used.

Example 16.4

Solve the PDE in Example 16.2 with the help of Richardson scheme and Du-Fort & Frankel scheme.

Ans.
The Richardson scheme (16.8)

$$u_{i,j+1} = u_{i,j-1} + 2r\left(u_{i+1,j} - 2u_{i,j} + u_{i-1,j}\right)$$

is a three level scheme. To start with Richardson scheme, we must have values for at least $j = 0$ and $j = 1$. Let us use values for $t = 1/16$ (or $j = 1$) from CN-scheme of Example 16.3.

$i\,(x)$ \ $j\,(t)$	0(0)	1(1/4)	2(1/2)	3(3/4)	4(1)
0(0)	0	0.707107	1	0.707107	0
1(1/16)	0	0.386729	0.546916	0.386729	0
2(1/8)	0				0

The values of $u_{i,j}$ for $t = 1/8$ (or $j = 2$) with the aid of Richardson scheme (16.8) for $i = 1, 2, 3$ are given by

$i\,(x)$ \ $j\,(t)$	0(0)	1(1/4)	2(1/2)	3(3/4)	4(1)
0(0)	0	0.707107	1	0.707107	0
1(1/16)	0	0.386729	0.546916	0.386729	0
2(1/8)	0	0.254025	0.359246	0.254025	0

Similarly, we can compute values for $j = 3, 4$.

$i (x)$ \ $j (t)$	0(0)	1(1/4)	2(1/2)	3(3/4)	4(1)
0(0)	0	0.707107	1	0.707107	0
1(1/16)	0	0.386729	0.546916	0.386729	0
2(1/8)	0	0.254025	0.359246	0.254025	0
3(3/16)	0	0.149613	0.211586	0.149613	0
4(1/4)	0	0.089121	0.126036	0.089121	0

Du-Fort and Frankel Scheme

This scheme is also three level scheme, again using values for $t = 1/16$ (or $j = 1$) from CN-scheme

$i (x)$ \ $j (t)$	0(0)	1(1/4)	2(1/2)	3(3/4)	4(1)
0(0)	0	0.707107	1	0.707107	0
1(1/16)	0	0.386729	0.546916	0.386729	0
2(1/8)	0				0

On applying Du-Fort and Frankel scheme (16.9) to calculate the values of $u_{i,j}$ for $t = 1/8$ (or $j = 2$), we have

$$u_{i,j+1} = \frac{1-2r}{1+2r} u_{i,j-1} + \frac{2r}{1+2r} \left(u_{i+1,j} + u_{i-1,j} \right)$$

$$u_{i,2} = \frac{1-2r}{1+2r} u_{i,0} + \frac{2r}{1+2r} \left(u_{i+1,1} + u_{i-1,1} \right)$$

The values for $i = 1, 2, 3$ are given by third row of the following table

$i (x)$ \ $j (t)$	0(0)	1(1/4)	2(1/2)	3(3/4)	4(1)
0(0)	0	0.707107	1	0.707107	0
1(1/16)	0	0.386729	0.546916	0.386729	0
2(1/8)	0	0.128910	0.182306	0.128910	0

Similarly, values for $j = 3$ and 4 are as follows

$i\,(x)$ ⟍ $j\,(t)$	0(0)	1(1/4)	2(1/2)	3(3/4)	4(1)
0(0)	0	0.707107	1	0.707107	0
1(1/16)	0	0.386729	0.546916	0.386729	0
2(1/8)	0	0.128910	0.182306	0.128910	0
3(3/16)	0	− 0.007372	− 0.010426	− 0.007372	0
4(1/4)	0	− 0.049921	− 0.070599	− 0.049921	0

16.5 Consistency, Convergence and Stability of Explicit and Crank–Nicolson Schemes

In this section, following three important features of finite difference method have been discussed, which are useful to select finite difference scheme for solutions of PDEs

1. Consistency
2. Convergence and Order
3. Stability

Consider any linear PDE of the following form

$$Lu = 0 \tag{16.18}$$

For example, heat conduction equation

$$\frac{\partial u}{\partial t} - c\frac{\partial^2 u}{\partial x^2} = 0$$

$$\left(\frac{\partial}{\partial t} - c\frac{\partial^2}{\partial x^2}\right)u = Lu = 0$$

where linear operator $L \equiv \dfrac{\partial}{\partial t} - c\dfrac{\partial^2}{\partial x^2}$

In any finite difference scheme, we approximate derivative terms of the differential equation with finite differences to obtain discretized equation. The discretized equation

$$L_D(u) = 0 \tag{16.19}$$

is finite difference equation containing parameters Δt and Δx. During computation of solution from this finite difference equation, various other errors like rounding error, also occur. Let u be the exact solution of Eq. (16.18) and u_D be the exact solution of the discretized Eq. (16.19). Let us assume that u_C is the final solution obtained from discretized Eq. (16.19). The solution u_C contains following two types of errors.

a) Truncation Error: In any finite difference scheme, we approximate partial derivatives of given PDE with finite difference approximations (like forward, central differences, etc.). These finite differences are Taylor series approximations of partial derivatives, therefore these finite differences have truncation error. So, the finite difference scheme has truncation error also known as discretization error. The difference $u-u_D$ is called as "discretization error." A finite difference scheme is consistent with a PDE if the discretization error vanishes as the grid spacing goes to zero independently.

b) Stability Error: Once a finite difference scheme is created, it is used to obtain the solution of given linear PDE. We have already discussed the stability of Euler method in Section 14.11 that the spacing h must be small enough to produce bounded solutions. Similarly, a finite difference scheme is stable if it gives the bounded solution for a stable PDE. The difference u_C-u_D is called as "stability error."

Since u is the exact solution of differential Eq. (16.18) and u_C is the final solution obtained from the scheme (16.19), so the total error is given by

$$\text{Total error} = u - u_C = (u-u_D)+(u_D-u_C)$$

The total error is the sum of discretization error and stability error.

16.5.1 Consistency

A finite difference scheme is consistent with a PDE if the difference $(u-u_D)$ between the solutions of both the equations (i.e., truncation error) vanishes as the sizes of the grid spacing go to zero independently. If T is truncation error, then we can write Eq. (16.18) as follows

$$Lu = L_D(u)+T$$

If the finite difference approximation $L_D(u)$ tends to Lu (truncation error vanishes) as the grid spacing between different independent variables tend to zero; then the finite difference scheme is said to be compatible or consistent with the original differential equation. In any finite difference scheme, we approximate partial derivatives of given PDE with finite difference approximations. These finite differences are nothing, but Taylor series approximations of partial derivatives. Hence, these finite differences have truncation error. For example, consider following Taylor series expansion

$$u(x,t+\Delta t) = u(x,t)+\Delta t \frac{\partial u}{\partial t}+\frac{\Delta t^2}{2!}\frac{\partial^2 u}{\partial t^2}+\cdots$$

On rearranging the terms, we get

$$\frac{\partial u}{\partial t} = \frac{u(x,t+\Delta t)-u(x,t)}{\Delta t}-\frac{\Delta t}{2!}\frac{\partial^2 u}{\partial t^2}-\cdots$$

On neglecting the terms containing grid spacing Δt and its higher powers, we get following forward difference approximation of the partial derivative $\dfrac{\partial u}{\partial t}$

$$\frac{\partial u}{\partial t} = \frac{u(x, t + \Delta t) - u(x, t)}{\Delta t} + O(\Delta t)$$

The truncation of Taylor series produces the truncation error. A finite difference method is compatible with a PDE if the truncation error vanishes if the grid spacing between different independent variables tends to zero. For example, consider Bender–Schmidt explicit scheme (16.3) for 1-dimensional heat conduction equation $\dfrac{\partial u}{\partial t} = c \dfrac{\partial^2 u}{\partial x^2}$

$$u_{i,j+1} = r u_{i-1,j} + (1 - 2r)u_{i,j} + r u_{i+1,j}$$

The truncation error (T) of finite difference approximation tends to zero with $\Delta t, \Delta x \to 0$. Therefore, the scheme is consistent with heat conduction equation in one space variable.

16.5.2 Consistency of Explicit Scheme

Let us assume that the grid spacings for variables t and x are $k = \Delta t$ and $h = \Delta x$, respectively. Consider following Taylor series expansions

$$u(x, t + k) = u(x, t) + k\frac{\partial u}{\partial t} + \frac{k^2}{2!}\frac{\partial^2 u}{\partial t^2} + \cdots \tag{16.20}$$

$$u(x + h, t) = u(x, t) + h\frac{\partial u}{\partial x} + \frac{h^2}{2!}\frac{\partial^2 u}{\partial x^2} + \frac{h^3}{3!}\frac{\partial^3 u}{\partial x^3} + \cdots \tag{16.21}$$

$$u(x - h, t) = u(x, t) - h\frac{\partial u}{\partial x} + \frac{h^2}{2!}\frac{\partial^2 u}{\partial x^2} - \frac{h^3}{3!}\frac{\partial^3 u}{\partial x^3} + \cdots \tag{16.22}$$

Using Eq. (16.20), we have

$$\frac{u(x, t + k) - u(x, t)}{k} = \frac{\partial u}{\partial t} + \frac{k}{2}\frac{\partial^2 u}{\partial t^2} + \frac{k^2}{6}\frac{\partial^3 u}{\partial t^3} + \cdots \tag{16.23}$$

Addition of Eqs. (16.21) and (16.22) gives the following expression for $\dfrac{\partial^2 u}{\partial x^2}$

$$\frac{u(x + h, t) - 2u(x, t) + u(x - h, t)}{h^2} = \frac{\partial^2 u}{\partial x^2} + \frac{h^2}{12}\frac{\partial^4 u}{\partial x^4} + \frac{h^4}{360}\frac{\partial^6 u}{\partial x^6} + \cdots \tag{16.24}$$

From Eqs. (16.23) and (16.24), we obtain

$$\frac{u(x, t + k) - u(x, t)}{k} - c\frac{u(x + h, t) - 2u(x, t) + u(x - h, t)}{h^2} + T = \frac{\partial u}{\partial t} - c\frac{\partial^2 u}{\partial x^2} \tag{16.25}$$

where T is truncation error. It is given by

$$T = c\left(\frac{h^2}{12}\frac{\partial^4 u}{\partial x^4} + \frac{h^4}{360}\frac{\partial^6 u}{\partial x^6} + \cdots\right) - \left(\frac{k}{2}\frac{\partial^2 u}{\partial t^2} + \frac{k^2}{6}\frac{\partial^3 u}{\partial t^3} + \cdots\right)$$

The scheme (16.25) is Bender–Schmidt Explicit Scheme. It is easy to see that truncation error $T \to 0$ as $k, h \to 0$. So the explicit scheme is compatible with heat conduction equation.

We can minimize the truncation error by selecting the suitable value of r. Using the expression $\dfrac{\partial}{\partial t} = c\dfrac{\partial^2}{\partial x^2}$ from heat conduction equation $\dfrac{\partial u}{\partial t} = c\dfrac{\partial^2 u}{\partial x^2}$, we have

$$\frac{\partial^2 u}{\partial t^2} = c^2\frac{\partial^4 u}{\partial x^4}, \quad \frac{\partial^3 u}{\partial t^3} = c^3\frac{\partial^6 u}{\partial x^6}\cdots$$

Consider the truncation error T in the following form

$$T = \left(c\frac{h^2}{12}\frac{\partial^4 u}{\partial x^4} - \frac{k}{2}\frac{\partial^2 u}{\partial t^2}\right) + \left(c\frac{h^4}{360}\frac{\partial^6 u}{\partial x^6} - \frac{k^2}{6}\frac{\partial^3 u}{\partial t^3}\right) + \cdots$$

$$= \left(c\frac{h^2}{12}\frac{\partial^4 u}{\partial x^4} - \frac{k}{2}c^2\frac{\partial^4 u}{\partial x^4}\right) + \left(c\frac{h^4}{360}\frac{\partial^6 u}{\partial x^6} - \frac{k^2}{6}c^3\frac{\partial^6 u}{\partial x^6}\right) + \cdots$$

Let $r = \dfrac{ck}{h^2} = \dfrac{1}{6}$, then the first term vanishes, and truncation error reduces to the following expression

$$T = \frac{-k^2 c^3}{15}\frac{\partial^6 u}{\partial x^6} + \cdots$$

So, the truncation error is of the highest order for the value $r = \dfrac{ck}{h^2} = \dfrac{1}{6}$

16.5.3 Convergence and Order

The finite difference scheme is said to be convergent to the exact solution if the discretization error $(u - u_D)$ tends to zero as step sizes of various independent variables approach zero. For example, the explicit scheme of one-dimensional heat conduction equation is convergent to the exact solution as $u_D \to u$ for $\Delta t, \Delta x \to 0$.

The order is the rate at which the finite difference scheme tends to the exact solutions as the grid sizes go to zero. In other words, the minimum degree of error terms present in the finite difference approximation of the derivative terms of the differential equation is the order. For example, in Bender–Schmidt Explicit Scheme (16.3), we approximate the time derivative term with forward difference

$$\frac{\partial u(x_i, t_j)}{\partial t} = \frac{u_{i,j+1} - u_{i,j}}{k} + O(k)$$

While the derivative term containing spatial variable is approximated with central difference formula

$$\frac{\partial^2 u(x_i, t_j)}{\partial x^2} = \frac{u_{i+1,j} - 2u_{i,j} + u_{i-1,j}}{h^2} + O(h^2)$$

Hence, the explicit scheme has truncation error of order $O(k) + O(h^2)$.

16.5.4 Stability

If the stability error $u_C - u_D$ tends to zero with $\Delta t, \Delta x \to 0$, then the scheme is said to be a stable scheme. The exact solutions of a given PDE must be examined to discuss the stability of any finite difference scheme. If the PDE is itself unstable, then the solution obtained by finite difference scheme is also unstable. But the numerical solution must be bounded for a stable PDE. Stability analysis is performed only for linear PDEs. So, nonlinear PDEs must be linearized locally for stability analysis. There are several methods for stability analysis of a finite difference scheme. The following three methods are most commonly used methods for stability analysis of any finite difference scheme.

1. Matrix method
2. Von Neumann method
3. Discrete perturbation method

In this book, only matrix method and Neumann method have been discussed for the stability of two finite difference schemes (only for explicit and CN schemes).

16.5.5 Matrix Method for Stability of Explicit Scheme

The explicit scheme is given by

$$u_{i,j+1} = r u_{i-1,j} + (1 - 2r) u_{i,j} + r u_{i+1,j} \qquad i = 1, 2, \dots n-1 \qquad (16.26)$$

The values $u_{i,j}$ computed by this method have certain error, let these approximate values be $u^*_{i,j}$

$$u^*_{i,j+1} = r u^*_{i-1,j} + (1 - 2r) u^*_{i,j} + r u^*_{i+1,j} \qquad i = 1, 2, \dots n-1 \qquad (16.27)$$

Consider the error $e_{i,j} = u_{i,j} - u^*_{i,j}$. On subtracting Eq. (16.27) from Eq. (16.26), we get

$$e_{i,j+1} = r e_{i-1,j} + (1 - 2r) e_{i,j} + r e_{i+1,j} \qquad i = 1, 2, \dots n-1 \qquad (16.28)$$

Note that for linear finite difference scheme; the error equation is same as that of the scheme itself. Let the boundary conditions be prescribed at end points, so the error in these values is zero, i.e.

$$e_{0,j} = e_{n,j} = 0 \qquad \forall j = 1, 2, \dots, n$$

The equations (16.28) can be written in matrix form as follows

$$
\begin{bmatrix}
e_{1,j+1} \\
e_{2,j+1} \\
e_{3,j+1} \\
\vdots \\
e_{n-1,j+1}
\end{bmatrix}
=
\begin{bmatrix}
1-2r & r & 0 & 0 & \cdots & 0 & 0 \\
r & 1-2r & r & 0 & \cdots & 0 & 0 \\
0 & r & 1-2r & r & \cdots & 0 & 0 \\
\vdots & & & & & & \\
0 & 0 & 0 & 0 & \cdots & r & 1-2r
\end{bmatrix}
\begin{bmatrix}
e_{1,j} \\
e_{2,j} \\
e_{3,j} \\
\vdots \\
e_{n-1,j}
\end{bmatrix}
$$

In compact form, let

$$E_{j+1} = AE_j$$

where

$$
E_{j+1} =
\begin{bmatrix}
e_{1,j+1} \\
e_{2,j+1} \\
e_{3,j+1} \\
\vdots \\
e_{n-1,j+1}
\end{bmatrix},
A =
\begin{bmatrix}
1-2r & r & 0 & 0 & \cdots & 0 & 0 \\
r & 1-2r & r & 0 & \cdots & 0 & 0 \\
0 & r & 1-2r & r & \cdots & 0 & 0 \\
\vdots & & & & & & \\
0 & 0 & 0 & 0 & \cdots & r & 1-2r
\end{bmatrix},
E_j =
\begin{bmatrix}
e_{1,j} \\
e_{2,j} \\
e_{3,j} \\
\vdots \\
e_{n-1,j}
\end{bmatrix}
$$

Let error vector at initial time $t = t_0$ be E_0, so $E_1 = AE_0$. Similarly $E_2 = AE_1$. After m numbers of steps, we have

$$E_m = A^m E_0 \tag{16.29}$$

The initial error vector E_0 is finite quantity. The scheme is stable, if $A^m \to 0$ as $m \to \infty$. Let us study the behavior of matrix A containing the elements dependent on the value $r = \dfrac{ck}{h^2}$.

Let the eigenvalues $\lambda_1, \lambda_2, \cdots, \lambda_{n-1}$ of matrix A be distinct with corresponding eigenvectors $X_1, X_2, \cdots, X_{n-1}$.

$$AX_i = \lambda_i X_i \quad i = 1, 2, \ldots n-1$$

The error vector E_0 can be written as linear combination of eigenvectors as follows

$$E_0 = c_1 X_1 + c_2 X_2 + \cdots + c_{n-1} X_{n-1}$$

Pre-multiplying with matrix A, we get

$$
\begin{aligned}
A E_0 &= A\left(c_1 X_1 + c_2 X_2 + \cdots + c_{n-1} X_{n-1}\right) \\
&= c_1 A X_1 + c_2 A X_2 + \cdots + c_{n-1} A X_{n-1} && \left(c_i; 1 \le i \le n-1\right) \\
&= c_1 \lambda_1 X_1 + c_2 \lambda_2 X_2 + \cdots + c_n \lambda_{n-1} X_{n-1} && \left(AX_i = \lambda X_i; 1 \le i \le n-1\right)
\end{aligned}
$$

Again, pre-multiplying with A, we have

$$A^2 E_0 = c_1 \lambda_1^2 X_1 + c_2 \lambda_2^2 X_2 + \ldots + c_{n-1} \lambda_{n-1}^2 X_{n-1} \qquad (c_i, \lambda_i; 1 \le i \le n \text{ are scalars})$$

Repeating the multiplication m-times successively, we obtain

$$A^m E_0 = c_1 \lambda_1^m X_1 + c_2 \lambda_2^m X_2 + \dots + c_{n-1} \lambda_{n-1}^m X_{n-1}$$

The scheme is stable if the error term $E_m = A^m E_0$ is finite as limit $m \to \infty$. So, we must have

$$|\lambda_i| \le 1; \ \forall 1 \le i \le n \ \text{(or)} \ |\lambda_{max}| \le 1$$

The maximum absolute eigenvalue must be less than or equal to unity. The quantity $r = \dfrac{ck}{h^2}$ is always positive. Using Brauer theorem 6.2 for matrix A, we have

$$|\lambda - (1 - 2r)| \le r \ \text{and} \ |\lambda - (1 - 2r)| \le 2r$$

On simplifying, we get

$$1 - 3r \le \lambda \le 1 - r \ \text{and} \ 1 - 4r \le \lambda \le 1$$

For $|\lambda| \le 1$, we must have

$$-1 \le 1 - 3r \le 1, \ -1 \le 1 - r \le 1, \ \text{and} \ -1 \le 1 - 4r \le 1$$

$$0 \le r \le \frac{2}{3}, \ 0 \le r \le 2 \ \text{and} \ 0 \le r \le \frac{1}{2}$$

All these conditions satisfy for $0 \le r \le \dfrac{1}{2}$. So, the explicit scheme is a stable scheme for

$$0 \le r \le \frac{1}{2}.$$

16.5.6 Matrix Method for Stability of CN Scheme

The CN scheme is given by

$$-r u_{i-1,j+1} + 2(1+r) u_{i,j+1} - r u_{i+1,j+1} = r u_{i-1,j} + 2(1-r) u_{i,j} + r u_{i+1,j}$$

Proceeding in a similar manner as in explicit scheme, we have

$$A E_{j+1} = B E_j \tag{16.30}$$

where

$$A = \begin{bmatrix} 2(1+r) & -r & 0 & 0 & \dots & 0 & 0 \\ -r & 2(1+r) & -r & 0 & \dots & 0 & 0 \\ 0 & -r & 2(1+r) & -r & \dots & 0 & 0 \\ \vdots & & & & & & \\ 0 & 0 & 0 & 0 & \dots & -r & 2(1+r) \end{bmatrix}, \ E_{j+1} = \begin{bmatrix} e_{1,j+1} \\ e_{2,j+1} \\ e_{3,j+1} \\ \vdots \\ e_{n-1,j+1} \end{bmatrix}$$

$$B = \begin{bmatrix} 2(1-r) & r & 0 & 0 & \cdots & 0 & 0 \\ r & 2(1-r) & r & 0 & \cdots & 0 & 0 \\ 0 & r & 2(1-r) & r & \cdots & 0 & 0 \\ \vdots & & & & & & \\ 0 & 0 & 0 & 0 & \cdots & r & 2(1-r) \end{bmatrix}, \quad E_j = \begin{bmatrix} e_{1,j} \\ e_{2,j} \\ e_{3,j} \\ \vdots \\ e_{n-1,j} \end{bmatrix}$$

From Eq. (16.30), we have

$$E_{j+1} = A^{-1} B E_j$$

The CN scheme is stable if the modulus of eigenvalues of $A^{-1}B$ is less than or equal to unity. For simplification, consider the following matrix

$$C = \begin{bmatrix} 2 & -1 & 0 & \cdots & 0 & 0 \\ -1 & 2 & -1 & \cdots & 0 & 0 \\ \vdots & & & & & \\ 0 & 0 & 0 & \cdots & -1 & 2 \end{bmatrix}$$

Let us assume that the eigenvalue of matrix C is μ. Using Brauer theorem 6.2 for matrix C, we have

$$-2 \le |\mu - 2| \le 2 \implies 0 \le \mu \le 4$$

The matrix $A^{-1}B = (2I + rC)^{-1}(2I - rC)$. Let the eigenvalue of matrix $A^{-1}B$ be λ, then

$$\lambda = \frac{2 - r\mu}{2 + r\mu}$$

Since r and μ both are non-negative quantities, so we have $|\lambda| \le 1$. Hence, the CN scheme is stable for each r, i.e., unconditionally stable scheme.

16.5.7 Neumann Method for Stability of Explicit Scheme

The exact solution of heat conduction equation by the method of separation of variables is as follows

$$u(x,t) = A \exp(\alpha t) \exp(I \beta x) \tag{16.31}$$

where A, α and β are constants and $I = \sqrt{-1}$. The error Eq. (16.28) for explicit scheme is given by

$$e_{i,j+1} = r e_{i-1,j} + (1 - 2r) e_{i,j} + r e_{i+1,j} \qquad i = 1, 2, \ldots n-1 \tag{16.32}$$

Since the heat conduction equation is linear, therefore explicit scheme and error equation have the same expressions. From Eq. (16.31), we have

$$e_{i,j} = A \exp(\alpha j \Delta t) \exp(I \beta i \Delta x) \tag{16.33}$$

The error term must be bounded as $j \to \infty$, so $|\exp(\alpha \Delta t)| \le 1$. Using Eq. (16.33) in (16.32) and canceling constant A from both sides of the equation, we have

$$\exp(\alpha(j+1)\Delta t)\exp(I\beta i\Delta x)$$
$$= r\exp(\alpha j\Delta t)\exp(I\beta (i-1)\Delta x)+(1-2r)\exp(\alpha j\Delta t)\exp(I\beta i\Delta x)+r\exp(\alpha j\Delta t)\exp(I\beta(i+1)\Delta x)$$

On simplifying, we obtain

$$\exp(\alpha\Delta t)= r\exp(-I\beta\Delta x)+(1-2r)+r\exp(I\beta\Delta x)$$
$$=2r\cos(\beta\Delta x)+1-2r$$
$$=1-4r\sin^2\left(\frac{\beta\Delta x}{2}\right)$$

For stability, we must have

$$\left|\exp(\alpha\Delta t)\right|\le 1$$

or $$\left|1-4r\sin^2\left(\frac{\beta\Delta x}{2}\right)\right|\le 1$$

or $$-1\le 1-4r\sin^2\left(\frac{\beta\Delta x}{2}\right)\le 1$$

Using $\sin^2\left(\dfrac{\beta\Delta x}{2}\right)\le 1$, we get

$$0\le r\le\frac{1}{2}$$

This is required condition for the stability of the explicit method.

16.5.8 Neumann Method for Stability of CN Scheme

The CN scheme is given by

$$-r u_{i-1,j+1}+2(1+r)u_{i,j+1}-r u_{i+1,j+1}=r u_{i-1,j}+2(1-r)u_{i,j}+r u_{i+1,j}$$

The error equation for CM scheme is as follows

$$-re_{i-1,j+1}+2(1+r)e_{i,j+1}-re_{i+1,j+1}=re_{i-1,j}+2(1-r)e_{i,j}+re_{i+1,j}$$

Using Eq. (16.33) and simplifying the expressions, we have

$$\exp(\alpha\Delta t)\left(2+2r-2r\cos(\beta\Delta x)\right)=2-2r+2r\cos(\beta\Delta x)$$

Rearranging the terms, we have

$$\exp(\alpha\Delta t)=\frac{2-2r+2r\cos(\beta\Delta x)}{2+2r-2r\cos(\beta\Delta x)}$$

$$=\frac{1-2r\sin^2\left(\dfrac{\beta\Delta x}{2}\right)}{1+2r\sin^2\left(\dfrac{\beta\Delta x}{2}\right)}$$

For stability, we must have $\left|\exp(\alpha\Delta t)\right|\le 1$. It is true for all values of the variable $r\ge 0$. So the CN scheme is unconditionally stable.

Table 16.1 Summary Table of Finite Difference Methods for 1-Dimensional Heat Conduction Equation

Method	Formulation	Level	Stability	Difficulty	Truncation Error
Explicit	$u_{i,j+1} = r u_{i-1,j} + (1-2r)u_{i,j} + r u_{i+1,j}$	2	Stable for $r \le 1/2$	Easy (u_{ij} can be obtained directly)	$O(\Delta t) + O(\Delta x^2)$
Crank–Nicolson	$-r u_{i-1,j+1} + 2(1+r)u_{i,j+1} - r u_{i+1,j+1}$ $= r u_{i-1,j} + 2(1-r)u_{i,j} + r u_{i+1,j}$	2	Unconditionally stable	Difficult (system of equations in u_{ij} has to solve)	$O(\Delta t^2) + O(\Delta x^2)$
General Implicit	$-r(1-\theta)u_{i-1,j+1} + (1 + 2r(1-\theta))u_{i,j+1} - r(1-\theta)u_{i+1,j+1}$ $= r\theta u_{i-1,j} + (1-2r\theta)u_{i,j} + r\theta u_{i+1,j}$	2	For $1/2 \le \theta \le 1$ Stable if $r \le \frac{1}{2}(2\theta - 1)$ For $0 \le \theta \le 1/2$ Unconditionally stable	Difficult (system of equations in u_{ij} has to solve)	$O(\Delta t^2) + O(\Delta x^2)$
Richardson	$u_{i,j+1} = u_{i,j-1} + 2r\left(u_{i+1,j} - 2u_{i,j} + u_{i-1,j}\right)$	3	Unstable	Easy (u_{ij} can be obtained directly)	$O(\Delta t^2) + O(\Delta x^2)$
Du-Fort and Frankel	$u_{i,j+1} = \dfrac{1-2r}{1+2r}u_{i,j-1} + \dfrac{2r}{1+2r}\left(u_{i+1,j} + u_{i-1,j}\right)$	3	Unconditionally stable	Easy (u_{ij} can be obtained directly)	$O(\Delta t^2) + O(\Delta x^2)$

16.6 2-Dimensional Heat Conduction Equation

The 2–dimensional heat conduction equation is as follows

$$\frac{\partial u}{\partial t} = c\left(\frac{\partial^2 u}{\partial x^2} + \frac{\partial^2 u}{\partial y^2}\right) \tag{16.34}$$

The following schemes will be discussed for the numerical solutions of Eq. (16.34)

- i) Explicit Scheme
- ii) Crank–Nicolson Scheme
- iii) Alternating Direction Implicit (ADI) Scheme

16.6.1 Explicit Scheme

Let the value of $u(x, y, t)$ at point (x_i, y_j, t_k) be $u_{i,j,k}$ i.e. $u(x_i, y_j, t_k) = u_{i,j,k}$. Also assume that step sizes for the variable t, x and y are Δt, Δx, and Δy respectively.

Discretize Eq. (16.34) at point (x_i, y_j, t_k), i.e.

$$\left.\frac{\partial u}{\partial t}\right|_{(x_i, y_j, t_k)} = c\left(\frac{\partial^2 u}{\partial x^2} + \frac{\partial^2 u}{\partial y^2}\right)\bigg|_{(x_i, y_j, t_k)}$$

Using forward difference formula for $\dfrac{\partial u(x_i, y_j, t_k)}{\partial t} = \dfrac{u_{i,j,k+1} - u_{i,j,k}}{\Delta t} + O(\Delta t)$ and

central difference formulas for $\dfrac{\partial^2 u(x_i, y_j, t_k)}{\partial x^2} = \dfrac{u_{i+1,j,k} - 2u_{i,j,k} + u_{i-1,j,k}}{\Delta x^2} + O(\Delta x^2)$ and

$\dfrac{\partial^2 u(x_i, y_j, t_k)}{\partial y^2} = \dfrac{u_{i,j+1,k} - 2u_{i,j,k} + u_{i,j-1,k}}{\Delta y^2} + O(\Delta y^2)$ in this equation and neglecting the error

terms in discretization, we have

$$\frac{u_{i,j,k+1} - u_{i,j,k}}{\Delta t} = c\left(\frac{u_{i+1,j,k} - 2u_{i,j,k} + u_{i-1,j,k}}{\Delta x^2} + \frac{u_{i,j+1,k} - 2u_{i,j,k} + u_{i,j-1,k}}{\Delta y^2}\right)$$

$$\text{(or)}\quad u_{i,j,k+1} - u_{i,j,k} = \frac{c\Delta t}{\Delta x^2}\left(u_{i+1,j,k} - 2u_{i,j,k} + u_{i-1,j,k}\right) + \frac{c\Delta t}{\Delta y^2}\left(u_{i,j+1,k} - 2u_{i,j,k} + u_{i,j-1,k}\right)$$

Let $r_1 = \dfrac{c\Delta t}{\Delta x^2}$ and $r_2 = \dfrac{c\Delta t}{\Delta y^2}$, then the explicit scheme for the solution of Eq. (16.34) is as follows

$$u_{i,j,k+1} - u_{i,j,k} = r_1\left(u_{i+1,j,k} - 2u_{i,j,k} + u_{i-1,j,k}\right) + r_2\left(u_{i,j+1,k} - 2u_{i,j,k} + u_{i,j-1,k}\right)$$

$$u_{i,j,k+1} = r_1\left(u_{i+1,j,k} + u_{i-1,j,k}\right) + r_2\left(u_{i,j+1,k} + u_{i,j-1,k}\right) + (1 - 2r_1 - 2r_2)u_{i,j,k} \qquad (16.35)$$

The scheme (16.35) is known as explicit scheme.

If the spacing in x and y are equal, i.e., $\Delta x = \Delta y$, then

$$r = \frac{c\Delta t}{\Delta x^2} = \frac{c\Delta t}{\Delta y^2}$$

The explicit scheme (16.35) is simplified in the following form

$$u_{i,j,k+1} = r\left(u_{i+1,j,k} + u_{i-1,j,k} + u_{i,j+1,k} + u_{i,j-1,k}\right) + (1 - 4r)u_{i,j,k} \qquad (16.36)$$

Forward difference formula is used for the term $\dfrac{\partial u}{\partial t}$, *so the error term in time t is of order*

$O(\Delta t)$. *Also, the explicit scheme is stable if* $c\Delta t\left(\dfrac{1}{\Delta x^2} + \dfrac{1}{\Delta y^2}\right) \leq \dfrac{1}{2}$. *Let us discuss more accurate*

and unconditionally stable scheme, i.e., Crank–Nicolson scheme, in which, we replace all the derivative terms by central difference formulas.

16.6.2 Crank–Nicolson (CN) Scheme

Discretize Eq. (16.34) at point $\left(x_i, y_j, t_{k+\frac{1}{2}}\right)$, i.e.,

$$\left.\frac{\partial u}{\partial t}\right|_{\left(x_i, y_j, t_{k+\frac{1}{2}}\right)} = c\left(\frac{\partial^2 u}{\partial x^2} + \frac{\partial^2 u}{\partial y^2}\right)\Bigg|_{\left(x_i, y_j, t_{k+\frac{1}{2}}\right)}$$

Let the points be $t_k - \Delta t/2$, t_k, $t_k + \Delta t/2$ (or $t_{k-1/2}$, t_k, $t_{k+1/2}$) with step size $\Delta t/2$. The central

difference formula for $\dfrac{\partial u}{\partial t}$ is as follows

$$\frac{\partial u(x_i, y_j, t_{k+1/2})}{\partial t} = \frac{u_{i,j,k+1} - u_{i,j,k}}{2(\Delta t/2)} + O\left(\Delta t^2\right) \qquad (16.37)$$

We will approximate value of $\dfrac{\partial^2 u}{\partial x^2}$ at $(x_i, y_j, t_{k+1/2})$ with average values of $\dfrac{\partial^2 u}{\partial x^2}$ at (x_i, y_j, t_k) and (x_i, y_j, t_{k+1}), i.e.,

$$\frac{\partial^2 u(x_i, y_j, t_{k+1/2})}{\partial x^2} = \frac{1}{2}\left(\frac{u_{i+1, j, k} - 2u_{i, j, k} + u_{i-1, j,k}}{\Delta x^2} + \frac{u_{i+1, j, k+1} - 2u_{i, j, k+1} + u_{i-1, j,k+1}}{\Delta x^2} \right) + O(\Delta x^2)$$

(16.38)

Similarly, the value of $\dfrac{\partial^2 u}{\partial y^2}$ at $(x_i, y_j, t_{k+1/2})$ will be approximated by the average values of

$\dfrac{\partial^2 u}{\partial y^2}$ at (x_i, y_j, t_k) and (x_i, y_j, t_{k+1})

$$\frac{\partial^2 u(x_i, y_j, t_{k+1/2})}{\partial y^2} = \frac{1}{2}\left(\frac{u_{i, j+1, k} - 2u_{i, j, k} + u_{i, j-1,k}}{\Delta y^2} + \frac{u_{i, j+1, k+1} - 2u_{i, j, k+1} + u_{i, j-1,k+1}}{\Delta y^2} \right) + O(\Delta y^2)$$

(16.39)

Using Eqs. (16.37–16.39) in Eq. (16.34) and neglecting the error terms, we have

$$\frac{u_{i, j, k+1} - u_{i,j, k}}{\Delta t} = c \left(\begin{array}{l} \dfrac{1}{2}\left(\dfrac{u_{i+1, j, k} - 2u_{i, j, k} + u_{i-1, j,k}}{\Delta x^2} + \dfrac{u_{i+1, j, k+1} - 2u_{i, j, k+1} + u_{i-1, j,k+1}}{\Delta x^2} \right) \\[3mm] + \dfrac{1}{2}\left(\dfrac{u_{i, j+1, k} - 2u_{i, j, k} + u_{i, j-1,k}}{\Delta y^2} + \dfrac{u_{i, j+1, k+1} - 2u_{i, j, k+1} + u_{i, j-1,k+1}}{\Delta y^2} \right) \end{array} \right)$$

Let $r_1 = \dfrac{c\Delta t}{\Delta x^2}$ and $r_2 = \dfrac{c\Delta t}{\Delta y^2}$, then we get

$$u_{i, j, k+1} - u_{i,j, k} = \left(\begin{array}{l} \dfrac{1}{2}r_1\left(u_{i+1, j, k} - 2u_{i, j, k} + u_{i-1, j,k} + u_{i+1, j, k+1} - 2u_{i, j, k+1} + u_{i-1, j,k+1} \right) \\[3mm] + \dfrac{1}{2}r_2\left(u_{i, j+1, k} - 2u_{i, j, k} + u_{i, j-1,k} + u_{i, j+1, k+1} - 2u_{i, j, k+1} + u_{i, j-1,k+1} \right) \end{array} \right)$$

On collecting the terms at $(k+1)^{\text{th}}$ level in L.H.S. and terms at k^{th} level in R.H.S., the Crank–Nicolson (CN) Scheme is as follows

$$-r_1\left(u_{i+1, j, k+1} + u_{i-1, j,k+1} \right) - r_2\left(u_{i, j+1, k+1} + u_{i, j-1,k+1} \right) + 2\left(1 + r_1 + r_2 \right)u_{i, j, k+1}$$

$$= r_1\left(u_{i+1, j, k} + u_{i-1, j,k} \right) + r_2\left(u_{i, j+1, k} + u_{i, j-1,k} \right) + 2\left(1 - r_1 - r_2 \right)u_{i,j, k}$$

(16.40)

Let the spacing in x and y be equal, i.e. $\Delta x = \Delta y$. Assume, $r = \dfrac{c\Delta t}{\Delta x^2} = \dfrac{c\Delta t}{\Delta y^2}$, then CN scheme (16.40) is simplified as follows

$$-r\left(u_{i+1,\,j,\,k+1} + u_{i-1,\,j,k+1} + u_{i,\,j+1,\,k+1} + u_{i,\,j-1,k+1}\right) + 2\left(1+2r\right)u_{i,\,j,\,k+1}$$

$$= r\left(u_{i+1,\,j,\,k} + u_{i-1,\,j,k} + u_{i,\,j+1,\,k} + u_{i,\,j-1,k}\right) + 2\left(1-2r\right)u_{i,j,\,k} \tag{16.41}$$

It is easy to see that this scheme is implicit scheme, as we can't obtain the solution at $(k + 1)^{\text{th}}$ level directly from the values at k^{th} level. First, a set of equations is obtained then we solve this set for final values $u_{i,\,j,\,k+1}$.

The major problem in CN-Scheme is that a set of a large number of equations is obtained. For example, if we have 20 nodes for both the variables x and y, then total 400 simultaneous equations are obtained in 400 variables. It is very difficult to solve, and round off error is very high for such a large system. But, this scheme is an unconditionally stable scheme.

In 1955, Peaceman and Rachford suggested a scheme known as alternating direction implicit (ADI) scheme, which is unconditionally stable and computational work is much less than CN-scheme.

In the next section, we will discuss alternating direction implicit method. This scheme involves two phases: in the first phase of computation, we move from k^{th} level to $(k+1)^{\text{th}}$ level, in which one space derivative $\left(\text{say } \dfrac{\partial^2 u}{\partial x^2}\right)$ is approximated at $(k+1)^{\text{th}}$ level, and another space derivative $\left(\dfrac{\partial^2 u}{\partial y^2}\right)$ is approximated at k^{th} level. In next phase, we move from $(k+1)^{\text{th}}$ level to $(k+2)^{\text{th}}$ level, in which alternating space derivative $\left(\dfrac{\partial^2 u}{\partial y^2}\right)$ is approximated at $(k+2)^{\text{th}}$ level, and another space derivative $\left(\dfrac{\partial^2 u}{\partial x^2}\right)$ is approximated at $(k+1)^{\text{th}}$ level.

Hence, in ADI scheme, say we have 20 nodes for the variables x and y, then instead of 400 simultaneous equations as in CN-scheme, we have 20 sets of 20 simultaneous equations, which reduces the computational efforts to large extent and hence round off error.

16.6.3 Alternating Direction Implicit (ADI) Scheme

This method contains two phases: In phase-1 (from k^{th} level to $(k+1)^{\text{th}}$ level), one space derivative term $\left(\text{say } \dfrac{\partial^2 u}{\partial x^2}\right)$ is approximated at a lower level (k^{th} level), while another derivative term $\left(\dfrac{\partial^2 u}{\partial y^2}\right)$ at a higher level ($(k+1)^{\text{th}}$ level). In phase-2 (from $(k+1)^{\text{th}}$ level to

$(k+2)^{th}$ level), the approximation levels are interchanged. Therefore the derivative term $\left(\dfrac{\partial^2 u}{\partial y^2}\right)$ is approximated at a higher level $[(k+2)^{th}$ level], while another derivative term $\left(\dfrac{\partial^2 u}{\partial x^2}\right)$ is approximated at a lower level $[(k+1)^{th}$ level]. These two phases combined to form one iteration of ADI method.

Phase-1 (k^{th} level to $(k+1)^{th}$ level)

On discretizing Eq. (16.34) at point $\left(x_i, y_j, t_{k+\frac{1}{2}}\right)$, we get

$$\left.\frac{\partial u}{\partial t}\right|_{x_i, y_j, t_{k+\frac{1}{2}}} = c\left(\frac{\partial^2 u}{\partial x^2} + \frac{\partial^2 u}{\partial y^2}\right)\Bigg|_{x_i, y_j, t_{k+\frac{1}{2}}} \qquad (16.42)$$

The central difference formula for $\dfrac{\partial u}{\partial t}$ with step size $\Delta t/2$ (as in CN-scheme) is as follows

$$\frac{\partial u(x_i, y_j, t_{k+1/2})}{\partial t} = \frac{u_{i,j,k+1} - u_{i,j,k}}{2(\Delta t/2)} + O\left(\Delta t^2\right) \qquad (16.43)$$

In CN-scheme, we approximated both the space derivatives $\left(\dfrac{\partial^2 u}{\partial x^2} \text{ and } \dfrac{\partial^2 u}{\partial y^2}\right)$ with an average value at points (x_i, y_j, t_k) and (x_i, y_j, t_{k+1}). But in ADI scheme, we approximate one space derivative $\left(\text{say } \dfrac{\partial^2 u}{\partial x^2}\right)$ at $(k+1)^{th}$ level and another derivative $\left(\dfrac{\partial^2 u}{\partial y^2}\right)$ at k^{th} level.

$$\frac{\partial^2 u(x_i, y_j, t_{k+1/2})}{\partial x^2} = \frac{u_{i+1,j,k+1} - 2u_{i,j,k+1} + u_{i-1,j,k+1}}{\Delta x^2} + O(\Delta x^2)$$

$$\frac{\partial^2 u(x_i, y_j, t_{k+1/2})}{\partial y^2} = \frac{u_{i,j+1,k} - 2u_{i,j,k} + u_{i,j-1,k}}{\Delta y^2} + O(\Delta y^2) \qquad (16.44)$$

Using equations (16.43) and (16.44) in Eq. (16.42), and neglecting the error terms, we get

$$\frac{u_{i,j,k+1} - u_{i,j,k}}{\Delta t} = c\left(\left(\frac{u_{i+1,j,k+1} - 2u_{i,j,k+1} + u_{i-1,j,k+1}}{\Delta x^2}\right) + \left(\frac{u_{i,j+1,k} - 2u_{i,j,k} + u_{i,j-1,k}}{\Delta y^2}\right)\right)$$

Let $r_1 = \dfrac{c\Delta t}{\Delta x^2}$ and $r_2 = \dfrac{c\Delta t}{\Delta y^2}$, then the ADI scheme for the solution of Eq. (16.34) is as follows

$$u_{i,j,k+1} - u_{i,j,k} = r_1\left(u_{i+1,j,k+1} - 2u_{i,j,k+1} + u_{i-1,j,k+1}\right) + r_2\left(u_{i,j+1,k} - 2u_{i,j,k} + u_{i,j-1,k}\right)$$

On collecting the terms at $(k+1)^{th}$ level in L.H.S. and k^{th} level terms in R.H.S., we obtain

$$-r_1 u_{i+1,j,k+1} + (1+2r_1)u_{i,j,k+1} - r_1 u_{i-1,j,k+1} = r_2 u_{i,j+1,k} + (1-2r_2)u_{i,j,k} + r_2 u_{i,j-1,k} \qquad (16.45)$$

Phase-2 $[(k+1)^{th}$ level to $(k+2)^{th}$ level$]$

Discretize Eq. (16.34) at point, $\left(x_i,\ y_j,\ t_{k+\frac{3}{2}} \right)$, i.e.,

$$\left. \frac{\partial u}{\partial t} \right|_{x_i, y_j, t_{k+\frac{3}{2}}} = c\left(\frac{\partial^2 u}{\partial x^2} + \frac{\partial^2 u}{\partial y^2} \right)_{x_i, y_j, t_{k+\frac{3}{2}}} \qquad (16.46)$$

For the points $t_{k+1} - \Delta t/2,\ t_{k+1},\ t_{k+1} + \Delta t/2$, the central difference formula for $\dfrac{\partial u}{\partial t}$ is as follows

$$\frac{\partial u(x_i, y_j, t_{k+3/2})}{\partial t} = \frac{u_{i,j,k+2} - u_{i,j,k+1}}{2(\Delta t/2)} + O\left(\Delta t^2 \right) \qquad (16.47)$$

On approximating $\dfrac{\partial^2 u}{\partial x^2}$ at $(k+1)^{th}$ level and $\dfrac{\partial^2 u}{\partial y^2}$ at $(k+2)^{th}$ level, we get

$$\frac{\partial^2 u(x_i, y_j, t_{k+3/2})}{\partial x^2} = \frac{u_{i+1,j,k+1} - 2u_{i,j,k+1} + u_{i-1,j,k+1}}{\Delta x^2} + O(\Delta x^2)$$

$$\frac{\partial^2 u(x_i, y_j, t_{k+3/2})}{\partial y^2} = \frac{u_{i,j+1,k+2} - 2u_{i,j,k+2} + u_{i,j-1,k+2}}{\Delta y^2} + O(\Delta y^2) \qquad (16.48)$$

Using equations (16.47) and (16.48) in Eq. (16.46), we have

$$\frac{u_{i,j,k+2} - u_{i,j,k+1}}{\Delta t} = c\left(\left(\frac{u_{i+1,j,k+1} - 2u_{i,j,k+1} + u_{i-1,j,k+1}}{\Delta x^2} \right) + \left(\frac{u_{i,j+1,k+2} - 2u_{i,j,k+2} + u_{i,j-1,k+2}}{\Delta y^2} \right) \right)$$

Let $r_1 = \dfrac{c\Delta t}{\Delta x^2}$ and $r_2 = \dfrac{c\Delta t}{\Delta y^2}$, then

$$u_{i,j,k+2} - u_{i,j,k+1} = r_1\left(u_{i+1,j,k+1} - 2u_{i,j,k+1} + u_{i-1,j,k+1} \right) + r_2\left(u_{i,j+1,k+2} - 2u_{i,j,k+2} + u_{i,j-1,k+2} \right)$$

On collecting the terms, we have

$$-r_2 u_{i,j+1,k+2} - r_2 u_{i,j-1,k+2}\left(1+2r_2\right)u_{i,j,k+2} = r_1 u_{i+1,j,k+1} + (1-2r_1)u_{i,j,k+1} + r_1 u_{i-1,j,k+1} \qquad (16.49)$$

Equations (16.45) and (16.49) are collectively form one iteration of ADI scheme for the solution of 2-dimensional heat conduction Eq. (16.34)

Table 16.2 Summary Table of Finite Difference Methods for 2-Dimensional Heat Conduction Equation

Method	Formulation	Stability	Difficulty	Truncation Error
Explicit	$u_{i,j,k+1} = r_1\left(u_{i+1,j,k} + u_{i-1,j,k}\right) + r_2\left(u_{i,j+1,k} + u_{i,j-1,k}\right) + (1 - 2r_1 - 2r_2)u_{i,j,k}$	Stable for $r_1 + r_2 \le \dfrac{1}{2}$	Easy ($u_{i,j,k}$ can be obtained directly)	$O(\Delta t) + O(\Delta x^2)$
Crank-Nicolson	$-r_1\left(u_{i+1,j,k+1} + u_{i-1,j,k+1}\right) - r_2\left(u_{i,j+1,k+1} + u_{i,j-1,k+1}\right) + 2\left(1 + r_1 + r_2\right)u_{i,j,k+1}$ $= r_1\left(u_{i+1,j,k} + u_{i-1,j,k}\right) + r_2\left(u_{i,j+1,k} + u_{i,j-1,k}\right) + 2\left(1 - r_1 - r_2\right)u_{i,j,k}$	Unconditionally stable	Difficult (system of large number of equations in $u_{i,j,k}$ has to solve)	$O(\Delta t^2) + O(\Delta x^2)$
ADI Scheme	$-r_1 u_{i+1,j,k+1} + (1 + 2r_1)u_{i,j,k+1} - r_1 u_{i-1,j,k+1} = r_2 u_{i,j+1,k} + (1 - 2r_2)u_{i,j,k} + r_2 u_{i,j-1,k}$ $-r_2 u_{i,j+1,k+2} + (1 + 2r_2)u_{i,j,k+2} - r_2 u_{i,j-1,k+2} = r_1 u_{i+1,j,k+1} + (1 - 2r_1)u_{i,j,k+1} + r_1 u_{i-1,j,k+1}$	Unconditionally stable	Moderate (systems of small number of equations in $u_{i,j,k}$ have to solve)	$O(\Delta t^2) + O(\Delta x^2)$

Example ———————————————————————————— 16.5

Solve the following initial-boundary value problem

$$\frac{\partial u}{\partial t} = \frac{\partial^2 u}{\partial x^2} + \frac{\partial^2 u}{\partial y^2}$$

$$u(x, y, 0) = \cos\left(\frac{\pi}{2}x\right)\cos\left(\frac{\pi}{2}y\right) \quad -1 \le x, y \le 1$$

$$u = 0, \quad x = \pm 1, \ y = \pm 1, \ t \ge 0$$

using the Explicit method with step size $\Delta x = \Delta y = 2/3$. Take $r_1 = r_2 = \dfrac{1}{6}$. Integrate for two steps only.

Ans.

We have $\Delta x = \Delta y = 2/3$, $c = 1$, and $r_1 = r_2 = \dfrac{1}{6}$

As $r_1 = r_2 = \dfrac{c\Delta t}{\Delta x^2} \quad \Rightarrow \Delta t = \dfrac{1}{6} \times \dfrac{4}{9} = \dfrac{2}{27}$

The initial condition

$$u(x, y, 0) = \cos\left(\frac{\pi}{2}x\right)\cos\left(\frac{\pi}{2}y\right), \quad -1 \le x, y \le 1$$

gives the values of $u(x_i, \ y_j, \ 0)$ These values are presented in the following table at $t = 0$ ($k = 0$).

i (x) \ j (y)	0 (−1)	1 (−1/3)	2 (1/3)	3 (1)
0 (−1)	0	0	0	0
1 (−1/3)	0	0.75	0.75	0
2 (1/3)	0	0.75	0.75	0
3 (1)	0	0	0	0

From the given boundary conditions

$$u = 0, \ x = \pm 1, \ y = \pm 1, \ t \ge 0, \text{ we have following values at } t = \frac{2}{27} \ (k = 1).$$

i (x) \ j (y)	0 (−1)	1 (−1/3)	2 (1/3)	3 (1)
0 (−1)	0	0	0	0
1 (−1/3)	0			0
2 (1/3)	0			0
3 (1)	0	0	0	0

At time $t = \dfrac{2}{27}$ $(k = 1)$, the explicit scheme to compute remaining values is as follows:

First Iteration:
The explicit scheme (16.35) is given by

$$u_{i,j,k+1} = r_1 \left(u_{i+1,j,k} + u_{i-1,j,k} \right) + r_2 \left(u_{i,j+1,k} + u_{i,j-1,k} \right) + (1 - 2r_1 - 2r_2)u_{i,j,k}$$

Using $r_1 = r_2 = \dfrac{1}{6}$ and $k = 0$, we have

$$u_{i,j,1} = \frac{1}{6}\left(u_{i+1,j,0} + u_{i-1,j,0} + u_{i,j+1,0} + u_{i,j-1,0} \right) + \frac{1}{3}u_{i,j,0}$$

We can compute the values of u at nodal points by putting the values i, $j = 1, 2$ in the above equation.

$$u_{1,1,1} = \frac{1}{6}\left(u_{2,1,0} + u_{0,1,0} + u_{1,2,0} + u_{1,0,0} \right) + \frac{1}{3}u_{1,1,0}$$
$$= \frac{1}{6}(0.75 + 0 + 0.75 + 0) + \frac{1}{3}0.75 = 0.5 \qquad (i,\ j = 1)$$

$$u_{2,1,1} = \frac{1}{6}\left(u_{3,1,0} + u_{1,1,0} + u_{2,2,0} + u_{2,0,0} \right) + \frac{1}{3}u_{2,1,0}$$
$$= \frac{1}{6}(0 + 0.75 + 0.75 + 0) + \frac{1}{3}0.75 = 0.5 \qquad (i = 2,\ j = 1)$$

$$u_{1,2,1} = \frac{1}{6}\left(u_{2,2,0} + u_{0,2,0} + u_{1,3,0} + u_{1,1,0} \right) + \frac{1}{3}u_{1,2,0}$$
$$= \frac{1}{6}(0.75 + 0 + 0 + 0.75) + \frac{1}{3}0.75 = 0.5 \qquad (i = 1,\ j = 2)$$

$$u_{2,2,1} = \frac{1}{6}\left(u_{3,2,0} + u_{1,2,0} + u_{2,3,0} + u_{2,1,0} \right) + \frac{1}{3}u_{2,2,0}$$
$$= \frac{1}{6}(0 + 0.75 + 0 + 0.75) + \frac{1}{3}0.75 = 0.5 \qquad (i = 2,\ j = 2)$$

On collecting all these values in a table, we have following values for $t = \dfrac{2}{27}$ $(k = 1)$

$j\ (y)$ \ $i\ (x)$	0 (−1)	1 (−1/3)	2 (1/3)	3 (1)
0 (−1)	0	0	0	0
1 (−1/3)	0	0.5	0.5	0
2 (1/3)	0	0.5	0.5	0
3 (1)	0	0	0	0

Second Iteration: The boundary condition $u = 0$, $x = \pm 1$, $y = \pm 1$, $t \geq 0$ provides the values of $u(x_i, y_j, t_2)$ for $t = \dfrac{4}{27}$ $(k = 2)$.

i (x) j (y)	0 (−1)	1 (−1/3)	2 (1/3)	3 (1)
0 (−1)	0	0	0	0
1 (−1/3)	0			0
2 (1/3)	0			0
3 (1)	0	0	0	0

The explicit scheme (16.35) for $r_1 = r_2 = \dfrac{1}{6}$ and $k = 1$, is given by

$$u_{i,j,2} = \frac{1}{6}\left(u_{i+1,j,1} + u_{i-1,j,1} + u_{i,j+1,1} + u_{i,j-1,1}\right) + \frac{1}{3}u_{i,j,1}$$

Proceeding in a similar manner as in iteration 1, we can easily obtain following values for $t = \dfrac{4}{27}$ $(k = 2)$ by using $i, j = 1, 2$ in the above equation.

i (x) j (y)	0 (−1)	1 (−1/3)	2 (1/3)	3 (1)
0 (−1)	0	0	0	0
1 (−1/3)	0	0.333333	0.333333	0
2 (1/3)	0	0.333333	0.333333	0
3 (1)	0	0	0	0

Note: We can use the symmetry over x and y – axes. It will reduce our computational efforts. We will elaborate the symmetric problem in case elliptic equations.

Example ———————————————————— 16.6

Solve the problem in Example 16.5 with CN (Crank–Nicolson) method.

Ans.
CN scheme (16.40) is as follows:

$$-r_1\left(u_{i+1,j,k+1} + u_{i-1,j,k+1}\right) - r_2\left(u_{i,j+1,k+1} + u_{i,j-1,k+1}\right) + 2\left(1 + r_1 + r_2\right)u_{i,j,k+1}$$

$$= r_1\left(u_{i+1,j,k} + u_{i-1,j,k}\right) + r_2\left(u_{i,j+1,k} + u_{i,j-1,k}\right) + 2\left(1 - r_1 - r_2\right)u_{i,j,k}$$

First Iteration:

Using $r_1 = r_2 = \dfrac{1}{6}$ and $k = 0$, we have

$$-\left(u_{i+1,\,j,\,1} + u_{i-1,\,j,1} + u_{i,\,j+1,\,1} + u_{i,\,j-1,1}\right) + 16u_{i,\,j,\,1}$$
$$= \left(u_{i+1,\,j,\,0} + u_{i-1,\,j,0} + u_{i,\,j+1,\,0} + u_{i,\,j-1,0}\right) + 8u_{i,j,\,0}$$

By using the initial values of Example 16.5, at $t = 0$ and $i, j = 1, 2$, we get following set of linear equations

$$-\left(u_{2,1,\,1} + u_{0,1,1} + u_{1,\,2,1} + u_{1,\,0,1}\right) + 16u_{1,1,\,1}$$
$$= \left(u_{2,1,\,0} + u_{0,1,0} + u_{1,\,2,\,0} + u_{1,\,0,0}\right) + 8u_{1,1,\,0} \qquad (i,\, j = 1)$$

$$-\left(u_{2,1,\,1} + 0 + u_{1,\,2,1} + 0\right) + 16u_{1,1,\,1} = \left(0.75 + 0 + 0.75 + 0\right) + 8(0.75)$$

$$-u_{2,1,1} - u_{1,\,2,1} + 16u_{1,1,\,1} = 7.5 \qquad\qquad (16.50)$$

$$-\left(u_{3,1,\,1} + u_{1,\,1,1} + u_{2,\,2,1} + u_{2,\,0,1}\right) + 16u_{2,1,\,1}$$
$$= \left(u_{3,1,\,0} + u_{1,\,1,0} + u_{2,\,2,\,0} + u_{2,\,0,0}\right) + 8u_{2,1,\,0} \qquad (i = 2,\, j = 1)$$

$$-\left(0 + u_{1,1,1} + u_{2,\,2,1} + 0\right) + 16u_{2,\,1,1} = \left(0 + 0.75 + 0.75 + 0\right) + 8(0.75)$$

$$-u_{1,\,1,1} - u_{2,\,2,1} + 16u_{2,\,1,1} = 7.5 \qquad\qquad (16.51)$$

$$-\left(u_{2,\,2,1} + u_{0,\,2,1} + u_{1,\,3,\,1} + u_{1,\,1,1}\right) + 16u_{1,\,2,1}$$
$$= \left(u_{2,\,2,\,0} + u_{0,\,2,0} + u_{1,\,3,\,0} + u_{1,\,1,0}\right) + 8u_{1,2,\,0} \qquad (i = 1,\, j = 2)$$

$$-\left(u_{2,\,2,1} + 0 + 0 + u_{1,\,1,1}\right) + 16u_{1,\,2,1} = \left(0.75 + 0 + 0 + 0.75\right) + 8(0.75)$$

$$-u_{2,\,2,1} - u_{1,\,1,1} + 16u_{1,\,2,1} = 7.5 \qquad\qquad (16.52)$$

$$-\left(u_{3,\,2,1} + u_{1,\,2,1} + u_{2,\,3,\,1} + u_{2,\,1,1}\right) + 16u_{2,\,2,1}$$
$$= \left(u_{3,\,2,\,0} + u_{1,\,2,0} + u_{2,\,3,\,0} + u_{2,\,1,0}\right) + 8u_{2,2,\,0} \qquad (i = 2,\, j = 2)$$

$$-\left(0 + u_{1,\,2,1} + 0 + u_{2,\,1,1}\right) + 16u_{2,\,2,1} = \left(0 + 0.75 + 0 + 0.75\right) + 8(0.75)$$

$$-u_{1,\,2,1} - u_{2,\,1,1} + 16u_{2,\,2,1} = 7.5 \qquad\qquad (16.53)$$

The solution of the Eqs. (16.50–16.53) is as follows

$$u_{1,\,1,1} = u_{1,\,2,1} = u_{2,\,1,1} = u_{2,\,2,\,1} = 0.535714$$

So, for $t = \dfrac{2}{27}(k = 1)$, we have following values

i (x) \\ j (y)	0 (−1)	1 (−1/3)	2 (1/3)	3 (1)
0 (−1)	0	0	0	0
1 (−1/3)	0	0.535714	0.535714	0
2 (1/3)	0	0.535714	0.535714	0
3 (1)	0	0	0	0

Second Iteration: The CN scheme (16.40) for $r_1 = r_2 = \dfrac{1}{6}$ and $k = 1$ is given by

$$-\left(u_{i+1,\,j,\,2} + u_{i-1,\,j,2} + u_{i,\,j+1,\,2} + u_{i,\,j-1,2}\right) + 16u_{i,\,j,\,2}$$
$$= \left(u_{i+1,\,j,\,1} + u_{i-1,\,j,1} + u_{i,\,j+1,\,1} + u_{i,\,j-1,1}\right) + 8u_{i,j,1}$$

Similar to First iteration, we can obtain following four equations

$$-u_{2,1,2} - u_{1,2,2} + 16u_{1,1,2} = 5.35714 \tag{16.54}$$

$$-u_{1,1,2} - u_{2,2,2} + 16u_{2,1,2} = 5.35714 \tag{16.55}$$

$$-u_{2,2,2} - u_{1,1,2} + 16u_{1,2,2} = 5.35714 \tag{16.56}$$

$$-u_{1,2,2} - u_{2,1,2} + 16u_{2,2,2} = 5.35714 \tag{16.57}$$

The solution of these four equations is given by

$$u_{1,1,2} = u_{1,2,2} = u_{2,1,2} = u_{2,2,2} = 0.382653$$

For $t = \dfrac{4}{27}$ $(k = 2)$, we have

i (x) \\ j (y)	0 (−1)	1 (−1/3)	2 (1/3)	3 (1)
0 (−1)	0	0	0	0
1 (−1/3)	0	0.382653	0.382653	0
2 (1/3)	0	0.382653	0.382653	0
3 (1)	0	0	0	0

Note: The linear systems are diagonally dominant systems, therefore the solution can also be obtained using iterative methods.

Example 16.7

Solve the problem in Example 16.5 with Alternating Direction Implicit (ADI) method.

Ans.
ADI scheme has two phases, phase-1 is from level k to level $k + 1$, and phase-2 is for level $k + 1$ to level $k + 2$.

Phase-1
ADI scheme (16.45) for phase-1 is as follows

$$-r_1 u_{i+1, j, k+1} +(1+2r_1)u_{i, j, k+1} - r_1 u_{i-1, j, k+1} = r_2 u_{i, j+1, k} +(1-2r_2)u_{i, j, k} +r_2 u_{i, j-1, k}$$

On putting $r_1 = r_2 = \dfrac{1}{6}$ and $k = 0$ in the above equation, we get

$$-u_{i+1, j, 1} +8u_{i, j, 1} - u_{i-1, j, 1} = u_{i, j+1, 0} + 4u_{i, j, 0} + u_{i, j-1, 0} \tag{16.58}$$

For $i = 1, 2$ and $j = 1$, we get following two linear equations

$$-u_{2, 1, 1} +8u_{1, 1, 1} - u_{0, 1, 1} = u_{1, 2, 0} + 4u_{1, 1, 0} + u_{1, 0, 0} \qquad (i = 1)$$

$$-u_{2, 1, 1} +8u_{1, 1, 1} - 0 = 0.75 + 4(0.75) + 0 \tag{16.59}$$

$$-u_{3, 1, 1} +8u_{2, 1, 1} - u_{1, 1, 1} = u_{2, 2, 0} + 4u_{2, 1, 0} + u_{2, 0, 0} \qquad (i = 2)$$

$$0 + 8u_{2, 1, 1} - u_{1, 1, 1} = 0.75 + 4(0.75) + 0 \tag{16.60}$$

On solving equations (16.59) and (16.60), we get

$$u_{2, 1, 1} = u_{1, 1, 1} = 0.535714 \tag{16.61}$$

Similarly, using $i = 1, 2$ and $j = 2$ in Eq. (16.58), we have

$$-u_{2, 2, 1} +8u_{1, 2, 1} - u_{0, 2, 1} = u_{1, 3, 0} + 4u_{1, 2, 0} + u_{1, 1, 0} \qquad (i = 1)$$

$$-u_{2, 2, 1} +8u_{1, 2, 1} - 0 = 0 + 4(0.75) + 0.75 \tag{16.62}$$

$$-u_{3, 2, 1} +8u_{2, 2, 1} - u_{1, 2, 1} = u_{2, 3, 0} + 4u_{2, 2, 0} + u_{2, 1, 0} \qquad (i = 2)$$

$$-0 + 8u_{2, 2, 1} - u_{1, 2, 1} = 0 + 4(0.75) + 0.75 \tag{16.63}$$

The solution of equations (16.62) and (16.63) is given by

$$u_{2, 2, 1} = u_{1, 2, 1} = 0.535714 \tag{16.64}$$

The following table contains the results from (16.61) and (16.64) for $t = \dfrac{2}{27}$ ($k = 1$).

i (x) ╲ j (y)	0 (−1)	1 (−1/3)	2 (1/3)	3 (1)
0 (−1)	0	0	0	0
1 (−1/3)	0	0.535714	0.535714	0
2 (1/3)	0	0.535714	0.535714	0
3 (1)	0	0	0	0

Phase – 2

The ADI scheme (16.49) is given by

$$-r_2 u_{i,\,j+1,\,k+2} + \left(1 + 2r_2\right) u_{i,\,j,\,k+2} - r_2 u_{i,\,j-1,\,k+2} = r_1 u_{i+1,\,j,\,k+1} + \left(1 - 2r_1\right) u_{i,\,j,\,k+1} + r_1 u_{i-1,\,j,k+1}$$

For $r_1 = r_2 = \dfrac{1}{6}$ and $k = 2$, we have

$$-u_{i,\,j+1,\,2} + 8u_{i,\,j,\,2} - u_{i,\,j-1,\,2} = u_{i+1,\,j,\,k+1} + 4u_{i,\,j,\,k+1} + u_{i-1,\,j,k+1} \tag{16.65}$$

Using $j = 1,2$ and $i = 1$, we have

$$-u_{1,\,2,\,2} + 8u_{1,\,1,\,2} - u_{1,\,0,\,2} = u_{2,\,1,\,1} + 4u_{1,\,1,\,1} + u_{0,\,1,1} \qquad (j = 1)$$

$$-u_{1,\,2,\,2} + 8u_{1,\,1,\,2} - 0 = 0.535714 + 4(0.535714) \tag{16.66}$$

$$-u_{1,\,3,\,2} + 8u_{1,\,2,\,2} - u_{1,\,1,\,2} = u_{2,\,2,\,1} + 4u_{1,\,2,\,1} + u_{0,\,2,1} \qquad (j = 2)$$

$$0 + 8u_{1,\,2,\,2} - u_{1,\,1,\,2} = 0.535714 + 4(0.535714) + 0 \tag{16.67}$$

The solution of Eqs. (16.66–16.67) gives

$$u_{1,\,2,\,2} = u_{1,\,1,\,2} = 0.382653 \tag{16.68}$$

Using $j = 1,2$ and $i = 2$ in Eq. (16.65), we have

$$-u_{2,\,2,\,2} + 8u_{2,\,1,\,2} - u_{2,\,0,\,2} = u_{3,\,1,\,1} + 4u_{2,\,1,\,1} + u_{1,\,1,1} \qquad (j = 1)$$

$$-u_{2,\,2,\,2} + 8u_{2,\,1,\,2} - 0 = 0 + 4(0.535714) + 0.535714 \tag{16.69}$$

$$-u_{2,\,3,\,2} + 8u_{2,\,2,\,2} - u_{2,\,1,\,2} = u_{3,\,2,\,1} + 4u_{2,\,2,\,1} + u_{1,\,2,1} \qquad (j = 2)$$

$$-0 + 8u_{2,\,2,\,2} - u_{2,\,1,\,2} = 0 + 4(0.535714) + 0.535714 \tag{16.70}$$

On solving Eqs. (16.69–16.70), we get

$$u_{2,\,2,\,2} = u_{2,\,1,\,2} = 0.382653 \tag{16.71}$$

The solutions (16.68), (16.71) and boundary conditions are given in the following table.

For $t = \dfrac{4}{27}$ $(k = 2)$

$i\,(x)$ / $j\,(y)$	0 (−1)	1 (−1/3)	2 (1/3)	3 (1)
0 (−1)	0	0	0	0
1 (−1/3)	0	0.382653	0.382653	0
2 (1/3)	0	0.382653	0.382653	0
3 (1)	0	0	0	0

16.7 Elliptic Equations (Laplace and Poisson Equations)

All time-dependent problems are known as "transient" problems. As time increases, all transient problems tend to steady state, i.e., the problems are independent of time (mathematically, $\dfrac{\partial u}{\partial t} = 0$). Some physical processes come to the steady state in a very short span of time. Therefore, we are interested in final stage for these processes. Parabolic and hyperbolic equations in two and three dimensions tend to elliptic equations in their steady state conditions. For example, heat conduction and wave equations reduce to the following Laplace and Poisson equations.

$$\nabla^2 u \equiv \frac{\partial^2 u}{\partial x^2} + \frac{\partial^2 u}{\partial y^2} = 0 \qquad \text{(Laplace equation)}$$

$$\nabla^2 u \equiv \frac{\partial^2 u}{\partial x^2} + \frac{\partial^2 u}{\partial y^2} = f(x, y) \qquad \text{(Poisson equation)}$$

These equations are also known as potential equations as the variable u represents the gravitational potential, velocity potential, and electromagnetic potential in various relevant fields of science. Finite difference approximation will be used for the solutions of various types of these elliptic equations.

In this section, Laplace and Poisson equations for the following cases are discussed with the help of finite difference approximations to obtain solutions at the pivotal points.

i)   **Dirichlet conditions only**
ii) **Symmetric problem**
iii) **Mixed type boundary conditions (Dirichlet and Neumann both)**
iv) **Non-rectangular domain**

16.7.1 Laplace Equation

Consider the Laplace equation

$$\frac{\partial^2 u}{\partial x^2} + \frac{\partial^2 u}{\partial y^2} = 0 \qquad\qquad (16.72)$$

Let x_i and y_j be equally spaced points with spacing h and k, respectively, and $u(x_i, y_j) = u_{i,j}$. The central differences for derivative terms are given by

$$\frac{\partial^2 u(x_i, y_j)}{\partial x^2} = \frac{u(x_i + h, y_j) - 2u(x_i, y_j) + u(x_i - h, y_j)}{h^2} + O(h^2)$$

$$= \frac{u(x_{i+1}, y_j) - 2u(x_i, y_j) + u(x_{i-1}, y_j)}{h^2} + O(h^2)$$

$$= \frac{u_{i+1,j} - 2u_{i,j} + u_{i-1,j}}{h^2} + O(h^2)$$

$$\frac{\partial^2 u(x_i, y_j)}{\partial y^2} = \frac{u(x_i, y_j + k) - 2u(x_i, y_j) + u(x_i, y_j - k)}{k^2} + O(k^2)$$

$$= \frac{u(x_i, y_{j+1}) - 2u(x_i, y_j) + u(x_i, y_{j-1})}{k^2} + O(k^2)$$

$$= \frac{u_{i,j+1} - 2u_{i,j} + u_{i,j-1}}{k^2} + O(k^2)$$

Note that both h and k are spacing in the values of spatial variables, therefore we can consider the square meshes for our calculations. Let $h = k$. Using central difference approximations in Laplace equation at point, (x_i, y_j), we get

$$\frac{\partial^2 u}{\partial x^2} + \frac{\partial^2 u}{\partial y^2} = \frac{u_{i+1,j} - 2u_{i,j} + u_{i-1,j}}{h^2} + \frac{u_{i,j+1} - 2u_{i,j} + u_{i,j-1}}{h^2} + O(h^2) = 0$$

On neglecting the terms in $O(h^2)$, we have

$$\frac{u_{i+1,j} - 2u_{i,j} + u_{i-1,j}}{h^2} + \frac{u_{i,j+1} - 2u_{i,j} + u_{i,j-1}}{h^2} = 0$$

$$u_{i+1,j} + u_{i-1,j} + u_{i,j+1} + u_{i,j-1} - 4u_{i,j} = 0$$

$$u_{i,j} = \frac{1}{4}\left(u_{i+1,j} + u_{i-1,j} + u_{i,j+1} + u_{i,j-1}\right) \qquad\qquad (16.73)$$

This formula is known as standard 5-points formula. It becomes clearer from the following figure that the value at any point is the average of four other points, situated at the lower, upper, left and right sides.

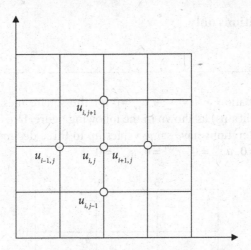

Fig. 16.1 Standard five-points formula

Similarly, we can obtain following diagonal 5-points formula by considering diagonal points for finite differences

$$u_{i,j} = \frac{1}{4}\left(u_{i+1,j+1} + u_{i-1,j-1} + u_{i-1,j+1} + u_{i+1,j-1}\right) \tag{16.74}$$

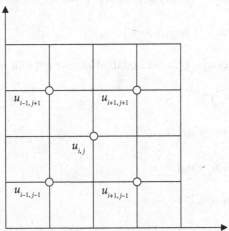

Fig. 16.2 Diagonal five-points formula

It is easy to see from the Fig. 16.2, that the value of the function at any point is an average of the values at the diagonal points of the square meshes. But the value obtained from diagonal 5-points formula is less accurate, in general, as the points under discussion are at more distance in this case as compared to the standard 5-points formula. So, we will use the diagonal formula only when the standard formula is not applicable to the grid, or to ease the computation.

Case i) Dirichlet conditions only

| Example | 16.8 |

Solve the Laplace equation $\nabla^2 u = u_{xx} + u_{yy} = 0$ for the square mesh with the boundary values (Dirichlet conditions) as shown in the following figure. Use Gauss–Seidel method till two consecutive iterations have same values up to three decimal points. Take initial approximation, $u_1^{(0)} = 0$, $u_2^{(0)} = 0$, $u_3^{(0)} = 0$, $u_4^{(0)} = 0$.

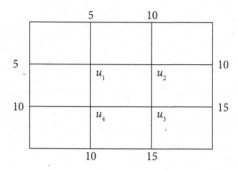

Ans.
Dirichlet conditions (function value, $u(x, t)$) are given at both the boundaries for the variables x and y. It is easy to apply standard 5-points formula (16.73)

$$u_{i,j} = \frac{1}{4}\left(u_{i+1,j} + u_{i-1,j} + u_{i,j+1} + u_{i,j-1}\right)$$

Using the function values $u(x, t)$ at the nodal points, we get following four equations

$$u_1 = \frac{1}{4}\left(5 + 5 + u_2 + u_4\right)$$

$$u_2 = \frac{1}{4}\left(10 + 10 + u_1 + u_3\right)$$

$$u_3 = \frac{1}{4}\left(15 + 15 + u_2 + u_4\right)$$

$$u_4 = \frac{1}{4}\left(10 + 10 + u_1 + u_3\right)$$

On solving this system of simultaneous linear equations by Gauss–Seidel method with initial approximation $[0, 0, 0, 0]^T$, the following iterations $\begin{bmatrix} u_1 & u_2 & u_3 & u_4 \end{bmatrix}$ are obtained.

Iteration 1

| 2.500000 | 5.625000 | 8.906250 | 7.851562 |

Iteration 2

| 5.869141 | 8.693848 | 11.636353 | 9.376373 |

Iteration 3

| 7.017555 | 9.663477 | 12.259962 | 9.819380 |

Iteration 4

| 7.370714 | 9.907669 | 12.431763 | 9.950619 |

Iteration 5

| 7.464572 | 9.974084 | 12.481175 | 9.986437 |

Iteration 6

| 7.490130 | 9.992826 | 12.494816 | 9.996237 |

Iteration 7

| 7.497266 | 9.998020 | 12.498564 | 9.998958 |

Iteration 8

| 7.499245 | 9.999453 | 12.499602 | 9.999712 |

Iteration 9

| 7.499791 | 9.999848 | 12.499890 | 9.999920 |

Iteration 10

| 7.499942 | 9.999958 | 12.499969 | 9.999978 |

The last two iterations are equal up to three decimal digits. So the solution is given by values at 10th iteration. Note that exact answer is given by

$$u_1 = 7.5 \quad u_2 = 10 \quad u_3 = 12.5 \quad u_4 = 10$$

Note: The result can be justified by the following figure. The standard 5-points formula is exactly true for each nodal point.

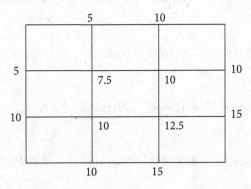

Example ———————————————— 16.9

Find the steady state temperature distribution in a thin rectangular plate, whose edges $x = 0$, $x = 0.8$, $y = 0$, $y = 0.6$ are kept at the temperature shown in the following figure. Find the values of temperature at the nodal points of the rectangular region with mess length 0.2. Use the Gauss–Seidel method to compute the temperature at nodal points until the difference between successive iterations is less than 0.005.

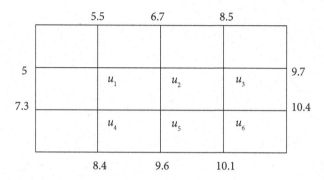

Ans.

On applying standard 5-points formula at each nodal point, we get

$$u_1 = \frac{1}{4}\left(5 + 5.5 + u_2 + u_4\right) \qquad u_2 = \frac{1}{4}\left(6.7 + u_1 + u_3 + u_5\right)$$

$$u_3 = \frac{1}{4}\left(8.5 + 9.7 + u_2 + u_6\right) \qquad u_4 = \frac{1}{4}\left(7.3 + 8.4 + u_1 + u_5\right)$$

$$u_5 = \frac{1}{4}\left(9.6 + u_2 + u_4 + u_6\right) \qquad u_6 = \frac{1}{4}\left(10.1 + 10.4 + u_3 + u_5\right)$$

On solving the above system by Gauss–Seidel method with initial approximation $[0, 0, 0, 0, 0, 0]^T$, the following iterations $\begin{bmatrix} u_1 & u_2 & u_3 & u_4 & u_5 & u_6 \end{bmatrix}$ are obtained

Iteration 1

| 2.625000 | 2.331250 | 5.132812 | 4.581250 | 4.128125 | 7.440234 |

Iteration 2

| 4.353125 | 5.078516 | 7.679688 | 6.045313 | 7.041016 | 8.805176 |

Iteration 3

| 5.405957 | 6.706665 | 8.427960 | 7.036743 | 8.037146 | 9.241277 |

Iteration 4

| 6.060852 | 7.306490 | 8.686942 | 7.449499 | 8.399317 | 9.396564 |

Iteration 5

6.313997	7.525064	8.780407	7.603329	8.531240	9.452911

Iteration 6

6.407098	7.604687	8.814400	7.659584	8.579296	9.473424

Iteration 7

6.441068	7.633691	8.826778	7.680091	8.596802	9.480895

Iteration 8

6.453445	7.644257	8.831288	7.687562	8.603178	9.483617

Iteration 9

6.457954	7.648106	8.832931	7.690283	8.605501	9.484608

Iteration 10

6.459597	7.649508	8.833529	7.691275	8.606348	9.484969

Final solution is as follows

$$u_1 = 6.459597 \quad u_2 = 7.649508 \quad u_3 = 8.833529$$
$$u_4 = 7.691275 \quad u_5 = 8.606348 \quad u_6 = 9.484969$$

Case ii) Symmetric Problem

Example

16.10

Find the steady state temperature distribution in a thin rectangular plate, whose edges $x = 0$, $x = 2$ are kept at 0°C (in ice), and edges $y = 0$, $y = 2$ are kept at temperature 100°C (in boiling water). Find the values of temperature at the nodal points of the rectangular region with mess length 0.5. Compute ten iterations of Gauss–Seidel method for temperature distribution on the grid.

Ans.
The temperature at the edges $x = 0$, $x = 2$ is 0°C, and the edges $y = 0$, $y = 2$ are kept at temperature 100°C. The grid with spacing $h = 0.5$ is shown in the following figure.

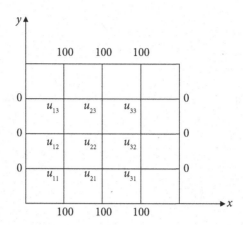

The steady state temperature distribution in a thin rectangular plate is defined by the Laplace equation. The boundary conditions and Laplace equation ($u_{xx} + u_{yy} = 0$) both are symmetrical around the lines $x = 1$ and $y = 1$. Therefore, the values about these lines are equal, i.e.

$$u_{11} = u_{31}, \; u_{12} = u_{32}, \; u_{13} = u_{33} \qquad \text{(Symmetry about } x = 1)$$

$$u_{11} = u_{13}, \; u_{21} = u_{23}, \; u_{31} = u_{33} \qquad \text{(Symmetry about } y = 1) \qquad (16.75)$$

So, the values u_{11}, u_{21}, u_{12}, u_{22} need to be computed only. Using standard 5-points formula and symmetries [from Eq. (16.75)], we have following simplified equations for $i, j = 1, 2$.

At (1, 1) $4u_{11} - u_{21} - u_{12} = 100$

At (2, 1) $4u_{21} - 2u_{11} - u_{22} = 100$

At (1, 2) $4u_{12} - u_{22} - 2u_{11} = 0$

At (2, 2) $2u_{22} - u_{12} - u_{21} = 0$

The ten iterations of Gauss–Seidel method for u_{11}, u_{21}, u_{12}, u_{22} are as follows. The zero vector is used as initial approximation.

Iteration 1

| 25.000000 | 37.500000 | 12.500000 | 25.000000 |

Iteration 2

| 37.500000 | 50.000000 | 25.000000 | 37.500000 |

Iteration 3

| 43.750000 | 56.250000 | 31.250000 | 43.750000 |

Iteration 4

| 46.875000 | 59.375000 | 34.375000 | 46.875000 |

Iteration 5

| 48.437500 | 60.937500 | 35.937500 | 48.437500 |

Iteration 6

| 49.218750 | 61.718750 | 36.718750 | 49.218750 |

Iteration 7

| 49.609375 | 62.109375 | 37.109375 | 49.609375 |

Iteration 8

| 49.804688 | 62.304688 | 37.304688 | 49.804688 |

Iteration 9

| 49.902344 | 62.402344 | 37.402344 | 49.902344 |

Iteration 10

| 49.951172 | 62.451172 | 37.451172 | 49.951172 |

The final iteration is approximately given by

$$u_{11} = 50, \ u_{21} = 62.5, \ u_{12} = 37.5, \ u_{22} = 50$$

The solution is as follows

$$u_{11} = u_{13} = u_{31} = u_{33} = 50$$
$$u_{21} = u_{23} = 62.5$$
$$u_{12} = u_{32} = 37.5$$
$$u_{22} = 50$$

Note: There are discontinuities at corner points, as the temperature at the corner points are 0°C from the boundary condition at edges, $x = 0$, $x = 2$, and 100°C from the boundary conditions at the edges, $y = 0$, $y = 2$. To avoid discontinuity, the values of temperature at the corner points are not used. Also, the correctness of the result is easily verified by the following figure, where the standard 5-points formula holds for these values.

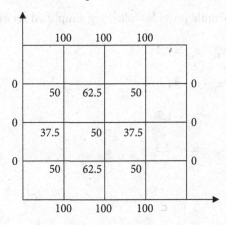

Example ──────────────────────────────────── 16.11 ──

Solve the Dirichlet problem

$$\nabla^2 u = 0 \text{ in } \Re, \text{ where } \Re \text{ is the square } 0 \le x, \; y \le 0.75.$$

Given that $u = x + y$ on the boundary of the square $\Re$. Take $h = 0.25$.

Ans.

The node points are as follows

$$x_0 = 0, \; x_1 = 0.25, \; x_2 = 0.5, \; x_3 = 0.75$$
$$y_0 = 0, \; y_1 = 0.25, \; y_2 = 0.5, \; y_3 = 0.75$$

Let $u_{ij} = u(x_i, \; y_j)$

The boundary condition $u = x + y$ provides the following results

$u_{10} = u_{01} = 0.25 \qquad u_{02} = u_{20} = 0.5$

$u_{13} = u_{31} = 1.0 \qquad u_{32} = u_{23} = 1.25$

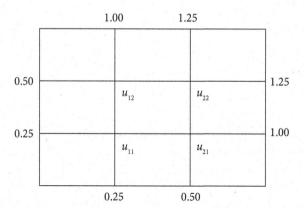

The standard 5-points formula provides following simplified set of linear equations

At $(1, 1)$ $\qquad 4u_{11} - u_{21} - u_{12} = 0.5$

At $(2, 1)$ $\qquad 4u_{21} - u_{11} - u_{22} = 1.5$

At $(1, 2)$ $\qquad 4u_{12} - u_{22} - u_{11} = 1.5$

At $(2, 2)$ $\qquad 4u_{22} - u_{12} - u_{21} = 2.5$

The following table contains 6 iterations $\begin{bmatrix} u_{11} & u_{12} & u_{21} & u_{22} \end{bmatrix}$ of Gauss–Seidel method with initial approximation $[0\ 0\ 0\ 0]^T$ for the above diagonally dominant system.

Iteration 1	Iteration 2	Iteration 3
0.125000	0.328125	0.457031
0.406250	0.664062	0.728516
0.406250	0.664062	0.728516
0.828125	0.957031	0.989258
Iteration 4	Iteration 5	Iteration 6
0.489258	0.497314	0.499329
0.744629	0.748657	0.749664
0.744629	0.748657	0.749664
0.997314	0.999329	0.999832

The exact answer is as follows
$u_{11} = 0.5$, $u_{21} = 0.75$, $u_{12} = 0.75$, $u_{22} = 1.00$

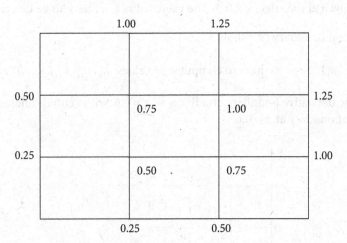

Case iii) Mixed type boundary conditions (Dirichlet and Neumann both)

Example ───────────────────────── **16.12** ─────

The Laplace equation $\nabla^2 u = u_{xx} + u_{yy} = 0$ is defined over the square region $\{0 \le x \le 0.6; \ 0 \le y \le 0.6\}$. The boundary conditions are defined by

   i) $u = 0$ over the edges $x = 0$, $y = 0$, $y = 0.6$ (Dirichlet condition)

   ii) $\dfrac{\partial u}{\partial x} = 1$ at the edge $x = 0.6$. (Neumann condition)

Find the values of $u(x, y)$ at the nodal points of the square region with mess length 0.2 with the aid of Gauss–Seidel method. Replace the derivative boundary condition with their central difference approximation.

Ans.

The nodal points of the square region $\{0 \le x \le 0.6; \ 0 \le y \le 0.6\}$ with mess length 0.2 are given by

$$x_0 = 0, \ x_1 = 0.2, \ x_2 = 0.4, \ x_3 = 0.6$$
$$y_0 = 0, \ y_1 = 0.2, \ y_2 = 0.4, \ y_3 = 0.6$$

The Dirichlet condition ($u = 0$) is given at $x = 0$, $y = 0$, $y = 0.6$, and Neumann condition $\left(\dfrac{\partial u}{\partial x} = 1\right)$ is given at the edge, $x = 0.6$. The values of $u(x, y)$ have to be determined at the mess points and boundary ($x = 0.6$).

Let $u_{ij} = u(x_i, y_j)$. Now, we have to compute the values $u_{11}, u_{12}, u_{21}, u_{22}, u_{31}, u_{32}$.

To replace the derivative boundary condition at $x = 0.6$ with central difference, we have to extend the boundary at $x = 0.6$.

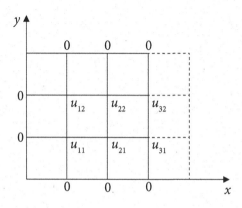

On replacing the boundary condition $\dfrac{\partial u}{\partial x} = 1$ at any point (x_i, y_j) with central difference, we get

$$\left. \frac{\partial u}{\partial x} \right|_{(x_i, y_j)} = \frac{u_{i+1,j} - u_{i-1,j}}{2h} = 1$$

At the edge $x = 0.6\,(i = 3)$ and for $j = 1, 2$, we have

$$\frac{u_{4,1} - u_{2,1}}{2(0.2)} = 1 \qquad\qquad \Rightarrow u_{4,1} = 0.4 + u_{2,1} \qquad\qquad (16.76)$$

$$\frac{u_{4,2} - u_{2,2}}{2(0.2)} = 1 \qquad\qquad \Rightarrow u_{4,2} = 0.4 + u_{2,2} \qquad\qquad (16.77)$$

On applying standard 5-points formula for calculations at each nodal point, we have

At $(1, 1)$ $\qquad u_{21} + u_{12} - 4u_{11} = 0$

At $(2, 1)$ $\qquad u_{31} + u_{11} + u_{22} - 4u_{21} = 0$

At $(3, 1)$ $\qquad u_{21} + u_{32} + u_{41} - 4u_{31} = 0$

$\qquad\qquad\qquad u_{21} + u_{32} + (u_{21} + 0.4) - 4u_{31} = 0 \qquad$ From Eq. (16.76)

At $(1, 2)$ $\qquad u_{22} + u_{11} - 4u_{12} = 0$

At $(2, 2)$ $\qquad u_{12} + u_{21} + u_{32} - 4u_{22} = 0$

At $(3, 2)$ $\qquad u_{31} + u_{22} + u_{42} - 4u_{32} = 0$

$\qquad\qquad\qquad u_{31} + u_{22} + (0.4 + u_{22}) - 4u_{32} = 0 \qquad$ From Eq. (16.77)

Equations (16.76) and (16.77) are used to replace values of u_{41} and u_{42} from equations at nodes $(3, 1)$ and $(3, 2)$. Now, we are left with following six equations.

$$u_{11} = \frac{1}{4}\left(u_{21} + u_{12}\right)$$

$$u_{21} = \frac{1}{4}\left(u_{31} + u_{11} + u_{22}\right)$$

$$u_{31} = \frac{1}{4}\left(u_{21} + u_{32} + u_{21} + 0.4\right)$$

$$u_{12} = \frac{1}{4}\left(u_{22} + u_{11}\right)$$

$$u_{22} = \frac{1}{4}\left(u_{12} + u_{21} + u_{32}\right)$$

$$u_{32} = \frac{1}{4}\left(u_{31} + u_{22} + 0.4 + u_{22}\right)$$

These equations are diagonally dominant, so we can apply Gauss–Seidel method. On solving this system by Gauss–Seidel with initial approximation $[0, 0, 0, 0, 0, 0]^T$, we get following iterations

Iteration 1

| 0.000000 | 0.000000 | 0.100000 | 0.000000 | 0.000000 | 0.125000 |

Iteration 2

| 0.000000 | 0.025000 | 0.143750 | 0.000000 | 0.037500 | 0.154687 |

Iteration 3

| 0.006250 | 0.046875 | 0.162109 | 0.010938 | 0.053125 | 0.167090 |

Iteration 4

| 0.014453 | 0.057422 | 0.170483 | 0.016895 | 0.060352 | 0.172797 |

Iteration 5

| 0.018579 | 0.062354 | 0.174376 | 0.019733 | 0.063721 | 0.175454 |

Iteration 6

| 0.020522 | 0.064655 | 0.176191 | 0.021061 | 0.065292 | 0.176694 |

Iteration 7

| 0.021429 | 0.065728 | 0.177037 | 0.021680 | 0.066026 | 0.177272 |

Iteration 8

| 0.021852 | 0.066229 | 0.177432 | 0.021969 | 0.066368 | 0.177542 |

The final solution of the Laplace equation is given by

$$u_{11} = 0.021852 \quad u_{21} = 0.066229 \quad u_{31} = 0.177432$$
$$u_{12} = 0.021969 \quad u_{22} = 0.066368 \quad u_{32} = 0.177542$$

Case iv) Non-rectangular domain

Example 16.13

The Laplace equation $\nabla^2 u = u_{xx} + u_{yy} = 0$ is defined over the following triangular region, $\{x = 0;\ y = 0;\ x + y = 5\}$. The following boundary conditions are prescribed

    i) $u = 0$ over the edges $x = 0,\ y = 0$

    ii) $u = 25 - x^2 - y^2$ at the edge $x + y = 5$

Find the values of $u(x, y)$ at the nodal points of the square region with mess length 1, with the aid of Gauss–Seidel method till the last two iterations have same values up to three decimal points.

Ans.

Since we have to compute the values of $u(x, y)$ at the nodal points of the triangular region $\{x = 0;\ y = 0;\ x + y = 5\}$ with mess length 1. Therefore, mesh points are given by

$$x_0 = 0,\ x_1 = 1,\ x_2 = 2,\ x_3 = 3,\ x_4 = 4,\ x_5 = 5$$
$$y_0 = 0,\ y_1 = 1,\ y_2 = 2,\ y_3 = 3,\ y_4 = 4,\ y_5 = 5$$

Let $u_{ij} = u(x_i, y_j)$. We have to compute the values of $u_{11},\ u_{21},\ u_{31},\ u_{12},\ u_{22},\ u_{13}$.

The following boundary conditions are defined

  i) $u = 0$ over the edges $x = 0,\ y = 0$
  ii) $u = 25 - x^2 - y^2$ at the edge $x + y = 5$

So, we have

$$u_{00} = u_{10} = u_{20} = u_{30} = u_{40} = u_{50} = 0$$
$$u_{01} = u_{02} = u_{03} = u_{04} = u_{05} = 0$$
$$u_{14} = u_{41} = 8$$
$$u_{23} = u_{32} = 12$$

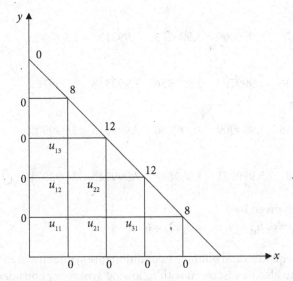

It is easy to apply standard 5-points formula for calculations at each nodal point to get following equations

At $(1, 1)$ $u_{21} + u_{12} - 4u_{11} = 0$

At $(2, 1)$ $u_{31} + u_{11} + u_{22} - 4u_{21} = 0$

At $(3, 1)$ $u_{21} - 4u_{31} = -20$

At $(1, 2)$ $u_{22} + u_{11} + u_{13} - 4u_{12} = 0$

At $(2, 2)$ $u_{12} + u_{21} - 4u_{22} = -24$

At $(1, 3)$ $u_{12} - 4u_{13} = -20$

We get following iterations of Gauss–Seidel method for $u_{11}, u_{21}, u_{31}, u_{12}, u_{22}, u_{13}$

Iteration 1

0.000000   0.000000   5.000000   0.000000   6.000000   5.000000

Iteration 2

0.000000   2.750000   5.687500   2.750000   7.375000   5.687500

Iteration 3

1.375000   3.609375   5.902344   3.609375   7.804688   5.902344

Iteration 4

1.804688   3.877930   5.969482   3.877930   7.938965   5.969482

Iteration 5

1.938965   3.961853   5.990463   3.961853   7.980927   5.990463

Iteration 6

1.980927   3.988079   5.997020   3.988079   7.994040   5.997020

Iteration 7

1.994040   3.996275   5.999069   3.996275   7.998137   5.999069

Iteration 8

1.998137   3.998836   5.999709   3.998836   7.999418   5.999709

Iteration 9

1.999418   3.999636   5.999909   3.999636   7.999818   5.999909

Iteration 10

1.999818   3.999887   5.999971   3.999887   7.999943   5.999971

The final solution is given by

$u_{11} = 2, u_{21} = 4, u_{31} = 6, u_{12} = 4, u_{22} = 8, u_{13} = 6$

Note: The Laplace equation and boundary conditions are symmetrical about the line $y = x$. So, the problem can also be discussed with following symmetry consideration

$$u_{21} = u_{12} \text{ and } u_{31} = u_{13}$$

16.7.2 Poisson Equation

Let us consider the Poisson equation

$$\frac{\partial^2 u}{\partial x^2} + \frac{\partial^2 u}{\partial y^2} = f(x, y)$$

where $f(x, y)$ is any function of the variables x and y. Assuming that the variables x and y are equally spaced points with spacing h and k, respectively, the central differences for the derivative terms are as follows

$$\frac{\partial^2 u(x_i, y_j)}{\partial x^2} = \frac{u_{i+1,j} - 2u_{i,j} + u_{i-1,j}}{h^2} + O(h^2)$$

$$\frac{\partial^2 u(x_i, y_j)}{\partial y^2} = \frac{u_{i,j+1} - 2u_{i,j} + u_{i,j-1}}{k^2} + O(k^2)$$

The Poisson equation at any point (x_i, y_j) is as follows

$$\left.\frac{\partial^2 u}{\partial x^2}\right|_{(x_i, y_j)} + \left.\frac{\partial^2 u}{\partial y^2}\right|_{(x_i, y_j)} = \frac{u_{i+1,j} - 2u_{i,j} + u_{i-1,j}}{h^2} + \frac{u_{i,j+1} - 2u_{i,j} + u_{i,j-1}}{k^2} = f(x_i, y_j)$$

Let $h = k$, then

$$u_{i+1,j} + u_{i-1,j} + u_{i,j+1} + u_{i,j-1} - 4u_{i,j} = h^2 f(x_i, y_j) \tag{16.78}$$

Now, we will discuss few examples of Poisson equation.

Case i) Dirichlet conditions only

Example   **16.14**

Solve the Poisson equation $\nabla^2 u = u_{xx} + u_{yy} = x + y$ for the square mesh, whose edges, $x = 0$, $x = 0.8$, $y = 0$, $y = 0.6$ are kept at the temperature shown in the following figure. Find the values of $u(x, y)$ at the nodal points of the rectangular region with mess length 0.2. Use Gauss–Seidel iterative method to compute values at nodal points until the difference between successive values at each point is less than 0.005.

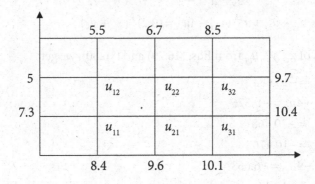

Ans.

The edges of the rectangular region are $x = 0$, $x = 0.8$, $y = 0$, $y = 0.6$ with the mess length 0.2. So, values of x and y are as follows

x	y
$x_0 = 0$	$y_0 = 0$
$x_1 = 0.2$	$y_1 = 0.2$
$x_2 = 0.4$	$y_2 = 0.4$
$x_3 = 0.6$	$y_3 = 0.6$
$x_4 = 0.8$	

$$(16.79)$$

Let the value of $u(x_i, y_j) = u_{ij}$, we have

$$u_{10} = 8.4, \ u_{20} = 9.6, \ u_{30} = 10.1, \ u_{01} = 7.3, \ u_{02} = 5$$
$$u_{13} = 5.5, \ u_{23} = 6.7, \ u_{33} = 8.5, \ u_{41} = 10.4, \ u_{42} = 9.7 \tag{16.80}$$

On replacing the derivative terms with central differences in the Poisson equation $u_{xx} + u_{yy} = x + y$ at the point, (x_i, y_j), we have

$$\frac{u_{i+1,j} - 2u_{i,j} + u_{i-1,j}}{h^2} + \frac{u_{i,j+1} - 2u_{i,j} + u_{i,j-1}}{h^2} = x_i + y_j$$

(or) $\quad u_{i+1,j} + u_{i-1,j} + u_{i,j+1} + u_{i,j-1} - 4u_{i,j} = h^2 \left(x_i + y_j \right)$

For $i = 1, 2, 3$ and $j = 1, 2$, we obtain

At (1, 1) $\qquad u_{21} + u_{01} + u_{12} + u_{10} - 4u_{11} = (0.2)^2 \left(x_1 + y_1 \right)$

At (2, 1) $\qquad u_{31} + u_{11} + u_{22} + u_{20} - 4u_{21} = (0.2)^2 \left(x_2 + y_1 \right)$

At (3, 1) $\qquad u_{41} + u_{21} + u_{32} + u_{30} - 4u_{31} = (0.2)^2 \left(x_3 + y_1 \right)$

At (1, 2) $\qquad u_{22} + u_{02} + u_{13} + u_{11} - 4u_{12} = (0.2)^2 \left(x_1 + y_2 \right)$

At (2, 2) $\qquad u_{32} + u_{12} + u_{23} + u_{21} - 4u_{22} = (0.2)^2 \left(x_2 + y_2 \right)$

At (3, 2) $\qquad u_{42} + u_{22} + u_{33} + u_{31} - 4u_{32} = (0.2)^2 \left(x_3 + y_2 \right)$

Using the values of x_i, y_j, u_{ij} from Eqs. (16.79) and (16.80), we get

$$u_{21} + u_{12} - 4u_{11} = -15.684$$
$$u_{31} + u_{11} + u_{22} - 4u_{21} = -9.576$$
$$u_{21} + u_{32} - 4u_{31} = -20.468$$
$$u_{22} + u_{11} - 4u_{12} = -10.476$$
$$u_{32} + u_{12} + u_{21} - 4u_{22} = -6.668$$
$$u_{22} + u_{31} - 4u_{32} = -18.16$$

On solving the above system of simultaneous linear equations by Gauss–Seidel with initial approximation $[0, 0, 0, 0, 0, 0]^T$, the following iterations $[u_{11}, u_{21}, u_{31}, u_{12}, u_{22}, u_{13}]$ are obtained

Iteration 1

3.921000 3.374250 5.960563 3.599250 3.410375 6.882734

Iteration 2

5.664375 6.152828 8.375891 4.887688 6.147813 8.170926

Iteration 3

6.681129 7.695209 9.083534 5.826235 7.090093 8.583406

Iteration 4

7.301361 8.262747 9.328539 6.216864 7.432755 8.730324

Iteration 5

7.540903 8.469549 9.416968 6.362414 7.557571 8.783635

Iteration 6

7.628991 8.544883 9.449129 6.415641 7.603040 8.803042

Iteration 7

7.661131 8.572325 9.460842 6.435042 7.619602 8.810111

Iteration 8

7.672842 8.582321 9.465109 6.442111 7.625636 8.812686

Iteration 9

7.677108 8.585963 9.466662 6.444686 7.627833 8.813623

Iteration 10

7.678662 8.587290 9.467228 6.445624 7.628634 8.813966

The final solution is as follows

$$u_{11} = 7.678662 \qquad u_{21} = 8.587290 \qquad u_{31} = 9.467228$$
$$u_{12} = 6.445624 \qquad u_{22} = 7.628634 \qquad u_{32} = 8.813966$$

Case ii) Symmetric problem

Example 16.15

Solve the Poisson equation $u_{xx} + u_{yy} = x^2 + y^2$ for a thin rectangular plate, whose edges $x = 0$, $x = 2$ are kept at $0°C$ (in ice) and edges $y = 0$, $y = 2$ are kept at temperature $100°C$ (in boiling water). Find the values of $u(x, y)$ at the nodal points of the rectangular region with mess length 0.5. Use Gauss–Seidel iterative method to compute values at nodal points until the difference between successive values at each point is less than 0.005. Use symmetry.

Ans.

The edges of the square region are $x = 0$, $x = 2$, $y = 0$, $y = 2$ with the mess length 0.5. So, the values of x and y are as follows

x	y
$x_0 = 0$	$y_0 = 0$
$x_1 = 0.5$	$y_1 = 0.5$
$x_2 = 1$	$y_2 = 1$
$x_3 = 1.5$	$y_3 = 1.5$
$x_4 = 2$	$y_4 = 2$

$$(16.81)$$

Let the value of $u(x_i, y_j) = u_{ij}$, we have

$$
\begin{aligned}
&u_{01} = 0, \quad u_{02} = 0, \ u_{03} = 0 \quad &(x = 0)\\
&u_{41} = 0, \quad u_{42} = 0, \ u_{43} = 0 \quad &(x = 2)\\
&u_{10} = 100, \quad u_{20} = 100, \ u_{30} = 100 \quad &(y = 0)\\
&u_{14} = 100, \quad u_{24} = 100, \ u_{34} = 100 \quad &(y = 2)
\end{aligned}
$$

$$(16.82)$$

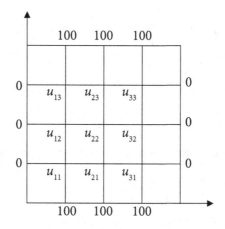

Let us solve the Poisson equation $u_{xx} + u_{yy} = x^2 + y^2$ with symmetry consideration. Since the boundary conditions and the Poisson equation are symmetrical around the lines $x = 1$ and $y = 1$; so we can assume the values about these lines are equal; i.e.

$$
\begin{aligned}
&u_{11} = u_{31}, \ u_{12} = u_{32}, \ u_{13} = u_{33} \quad &\text{(Symmetry about } x = 1)\\
&u_{11} = u_{13}, \ u_{21} = u_{23}, \ u_{31} = u_{33} \quad &\text{(Symmetry about } y = 1)
\end{aligned}
$$

$$(16.83)$$

So, we need to find values of u_{11}, u_{21}, u_{12}, u_{22} only.

The Poisson equation $\nabla^2 u = u_{xx} + u_{yy} = x^2 + y^2$ at point (x_i, y_j) is given by

$$\frac{u_{i+1,j} - 2u_{i,j} + u_{i-1,j}}{h^2} + \frac{u_{i,j+1} - 2u_{i,j} + u_{i,j-1}}{h^2} = x_i^2 + y_j^2$$

(or) $\quad u_{i+1,j} + u_{i-1,j} + u_{i,j+1} + u_{i,j-1} - 4u_{i,j} = h^2 \left(x_i^2 + y_j^2 \right)$

We need to find values of u_{11}, u_{21}, u_{12}, u_{22}, so

At $(1, 1)$ $\qquad u_{21} + u_{01} + u_{12} + u_{10} - 4u_{11} = (0.5)^2 \left(x_1^2 + y_1^2 \right)$

At $(2, 1)$ $\qquad u_{31} + u_{11} + u_{22} + u_{20} - 4u_{21} = (0.5)^2 \left(x_2^2 + y_1^2 \right)$

At $(1, 2)$ $\qquad u_{22} + u_{02} + u_{13} + u_{11} - 4u_{12} = (0.5)^2 \left(x_1^2 + y_2^2 \right)$

At $(2, 2)$ $\qquad u_{32} + u_{12} + u_{23} + u_{21} - 4u_{22} = (0.5)^2 \left(x_2^2 + y_2^2 \right)$ $\qquad\qquad\qquad$ (16.84)

Using symmetries from Eq. (16.81), the values of x_i, y_j, u_{ij} from Eqs. (16.82) and (16.83), the set of Eqs. (16.84) becomes

$$u_{21} + u_{12} - 4u_{11} = -100 + (0.5)^2 (0.5) = -99.875$$

$$2u_{11} + u_{22} - 4u_{21} = -100 + (0.5)^2 (1.25) = -99.6875$$

$$u_{22} + 2u_{11} - 4u_{12} = (0.5)^2 (1.25) = 0.3125$$

$$2u_{12} + 2u_{21} - 4u_{22} = (0.5)^2 (2) = 0.5$$

On solving this system of linear equations with the help of Gauss–Seidel method for u_{11}, u_{21}, u_{12}, u_{22}, we have

Iteration 1

24.968750   37.406250   12.406250   24.781250

Iteration 2

37.421875   49.828125   24.828125   37.203125

Iteration 3

43.632812   56.039062   31.039062   43.414062

Iteration 4

46.738281   59.144531   34.144531   46.519531

Iteration 5

48.291016   60.697266   35.697266   48.072266

Iteration 6

49.067383   61.473633   36.473633   48.848633

Iteration 7

49.455566   61.861816   36.861816   49.236816

Iteration 8

49.649658   62.055908   37.055908   49.430908

Iteration 9

49.746704   62.152954   37.152954   49.527954

Iteration 10

49.795227   62.201477   37.201477   49.576477

Iteration 11

49.819489   62.225739   37.225739   49.600739

Iteration 12

49.831619   62.237869   37.237869   49.612869

Iteration 13

49.837685 62.243935 37.243935 49.618935

Iteration 14

49.840717 62.246967 37.246967 49.621967

Iteration 15

49.842232 62.248482 37.248482 49.623482

Iteration 16

49.842991 62.249241 37.249241 49.624241

The Final solution is given by

$$u_{11} = u_{13} = u_{31} = u_{33} = 49.842991$$
$$u_{21} = u_{23} = 62.249241$$
$$u_{12} = u_{32} = 37.249241$$
$$u_{22} = 49.624241$$

Example 16.16

Solve the following Poisson problem

$$\nabla^2 u = e^{x+y} \text{ in } \Re, \text{ where } \Re \text{ is the square } 0 \le x, \ y \le 0.75.$$

Given that $u = x^2 + y^2$ on the boundary of the square $\Re$. Take $h = 0.25$.

Ans.

The node points are as follows

$$x_0 = 0, \ x_1 = 0.25, \ x_2 = 0.5, \ x_3 = 0.75$$
$$y_0 = 0, \ y_1 = 0.25, \ y_2 = 0.5, \ y_3 = 0.75$$

The boundary condition $u = x^2 + y^2$ provides the following values

$$u_{10} = u_{01} = 0.0625 \qquad u_{02} = u_{20} = 0.25$$
$$u_{13} = u_{31} = 0.625 \qquad u_{32} = u_{23} = 0.8125$$

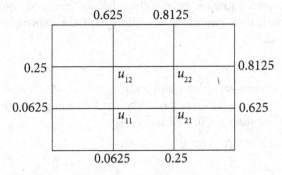

The system of linear equations is obtained by applying standard 5-points formula. Then, Gauss–Seidel method provides following 6 iterations

Iteration 1

0.005489   0.187044   0.187044   0.457299

Iteration 2

0.099011   0.324749   0.324749   0.526151

Iteration 3

0.167863   0.359176   0.359176   0.543365

Iteration 4

0.185077   0.367782   0.367782   0.547668

Iteration 5

0.189380   0.369934   0.369934   0.548744

Iteration 6

0.190456   0.370472   0.370472   0.549013

After 6 iterations, the solution is as follows

$$u_{11} = 0.190456 \qquad u_{12} = 0.370472 \qquad u_{21} = 0.370472 \qquad u_{22} = 0.549013$$

Case iii) Mixed type boundary conditions (Dirichlet and Neumann both)

Example ——————————————————————————————————— **16.17**

A Poisson equation $u_{xx} + u_{yy} = 4(x + y)$ is defined over a domain $0 \leq x, y \leq 0.75$ with the following boundary conditions,

$\qquad u = 0$ on the sides $x = 0, 0.75$ and $y = 0$ \qquad (Dirichlet conditions)

$\qquad \dfrac{\partial u}{\partial y} = u$ on $y = 0.75$ \qquad\qquad\qquad (Neumann condition)

Solve the given Poisson equation by dividing the domain into squares of side 0.25. Approximate the derivative boundary condition with the central difference, and use Gauss–Seidel method.

Ans.

Nodes in the directions of variables x and y are $x_0 = 0$, $x_1 = 0.25$, $x_2 = 0.5$, $x_3 = 0.75$ and $y_0 = 0$, $y_1 = 0.25$, $y_2 = 0.5$, $y_3 = 0.75$ respectively. The following figure shows the boundary conditions $u = 0$ on the sides $x = 0, 0.75$ and $y = 0$.

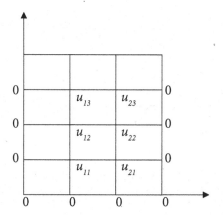

On replacing the derivative boundary condition at the boundary $y = 0.75$ ($y = y_3$) with the central difference, we have

$$\frac{\partial u(x_i, y_j)}{\partial y} = \frac{u_{i, j+1} - u_{i, j-1}}{2h} = u_{ij}$$

$$\frac{\partial u(x_i, y_3)}{\partial y} = \frac{u_{i,4} - u_{i,2}}{0.5} = u_{i3}$$

$$u_{i,4} = u_{i,2} + 0.5\, u_{i3} \tag{16.85}$$

This formula can be used for the hypothetical values at $j = 4$. In another way, we are extending the boundaries to compute the required mesh values at $j = 3$. Using the central differences for derivative terms in the Poisson equation, we have

$$u_{i+1,j} + u_{i-1,j} + u_{i,j+1} + u_{i,j-1} - 4u_{i,j} = h^2 f(x_i, y_j)$$
$$= (0.25)^2 \, 4(x_i + y_j)$$
$$= (0.25)(x_i + y_j)$$

Using different values of i, j at the mesh points, we have

At (1, 1) $\qquad u_{21} + u_{01} + u_{12} + u_{10} - 4u_{11} = .125$

At (2, 1) $\qquad u_{31} + u_{11} + u_{22} + u_{20} - 4u_{21} = .1875$

At (1, 2) $\qquad u_{22} + u_{02} + u_{13} + u_{11} - 4u_{12} = .1875$

At (2, 2) $\qquad u_{32} + u_{12} + u_{23} + u_{21} - 4u_{22} = .25$

At (1, 3) $\qquad u_{23} + u_{03} + u_{14} + u_{12} - 4u_{13} = .25$

At (2, 3) $\qquad u_{33} + u_{13} + u_{24} + u_{22} - 4u_{23} = .3125$

Equation (16.85) is used to replace the values of u_{14} and u_{24} from last two equations. Therefore, we are left with following six equations with six unknowns.

$$u_{21} + u_{12} - 4u_{11} = .125$$
$$u_{11} + u_{22} - 4u_{21} = .1875$$
$$u_{22} + u_{13} + u_{11} - 4u_{12} = .1875$$
$$u_{12} + u_{23} + u_{21} - 4u_{22} = .25$$
$$u_{23} + 2u_{12} - 3.5u_{13} = .25$$
$$u_{13} + 2u_{22} - 3.5u_{23} = .3125$$

Applying Gauss–Seidel method, we have following iterations

Iteration 1

-0.031250 -0.054688 -0.054688 -0.089844 -0.087054 -0.139828

Iteration 2

-0.058594 -0.083984 -0.105748 -0.144890 -0.141593 -0.171138

Iteration 3

-0.078683 -0.102768 -0.138167 -0.165518 -0.159801 -0.182234

Iteration 4

-0.091484 -0.111125 -0.151076 -0.173609 -0.166660 -0.186505

Iteration 5

-0.096800 -0.114477 -0.156142 -0.176781 -0.169328 -0.188174

Iteration 6

-0.098905 -0.115797 -0.158129 -0.178025 -0.170372 -0.188828

Iteration 7

| -0.099731 | -0.116314 | -0.158907 | -0.178512 | -0.170781 | -0.189084 |

Iteration 8

| -0.100055 | -0.116517 | -0.159212 | -0.178703 | -0.170942 | -0.189184 |

After 8 iterations, we have

$$u_{11} = -0.100055 \quad u_{21} = -0.116517 \quad u_{12} = -0.159212$$
$$u_{22} = -0.178703 \quad u_{13} = -0.170942 \quad u_{23} = -0.189184$$

Discussion:

1. We have discussed the elliptic equations with rectangular and triangular boundaries with equal grid spacing in both the directions. These equations can also be discussed with irregular boundaries with different grid spacing like circular etc. We have already discussed such case for ordinary differential equation in Section 15.8. Due to complexity of such problems, here we have avoided it.

2. Laplace and Poisson equations in 3-dimensions can also be solved on similar pattern. Here, we have avoided 3-dimensional cases due to cumbersome computation.

3. There is truncation error in finite difference approximations of derivative terms. We have to reduce the grid sizes to decrease this error. The number of equations becomes very large with very small grid sizes. The problems increase manifold during solution of such a large system. Few such problems have already been discussed in Section 15.9.

4. Here, we have discussed only Gauss–Seidel method for solutions of systems of linear equations. The relaxation method can also be used to solve these systems, and in some cases it can reduce computational efforts.

16.8 Hyperbolic Equation (Wave Equation)

Vibrations in a tightly stretched string between two points are well described by the following hyperbolic equation known as 1-dimensional wave equation

$$\frac{\partial^2 u}{\partial t^2} = c \frac{\partial^2 u}{\partial x^2} \tag{16.86}$$

where $c = \dfrac{T}{m}$ (T is tension in the string and m is the mass per unit length) is a positive constant and it depends on the nature of string. Second order derivative terms in the variables x and t are present in Eq. 16.86. Therefore, two boundary conditions (corresponding to x) and two initial conditions (corresponding to t) are required for a well-posed problem.

Similar expressions for two and three-dimensional wave equations are as follows

$$\frac{\partial^2 u}{\partial t^2} = c \left(\frac{\partial^2 u}{\partial x^2} + \frac{\partial^2 u}{\partial y^2} \right) = c \nabla^2 u$$

$$\frac{\partial^2 u}{\partial t^2} = c \left(\frac{\partial^2 u}{\partial x^2} + \frac{\partial^2 u}{\partial y^2} + \frac{\partial^2 u}{\partial z^2} \right) = c \nabla^2 u$$

We will discuss explicit and implicit schemes for the solution of wave Eq. (16.86). Higher dimensional equations are not discussed in this book.

16.8.1 Explicit Scheme

Let the value of $u(x, t)$ at point (x_i, t_j) be $u_{i,j}$ i.e. $u(x_i, t_j) = u_{i,j}$. Also assume that step sizes for the variable x and t are $\Delta x = h$ and $\Delta t = k$, respectively.

Let us discretize Eq. (16.86) at point (x_i, t_j)

$$\left. \frac{\partial^2 u}{\partial t^2} \right|_{(x_i, t_j)} = c \left. \frac{\partial^2 u}{\partial x^2} \right|_{(x_i, t_j)} \tag{16.87}$$

Using central difference formulas in this equation, and neglecting the error terms in discretization, we have

$$\frac{u_{i,j+1} - 2u_{i,j} + u_{i,j-1}}{k^2} = c \frac{u_{i+1,j} - 2u_{i,j} + u_{i-1,j}}{h^2}$$

(or) $\quad u_{i,j+1} = \dfrac{ck^2}{h^2} \left(u_{i+1,j} - 2u_{i,j} + u_{i-1,j} \right) + 2u_{i,j} - u_{i,j-1}$

Let $r = \dfrac{ck^2}{h^2}$, then we have

$$u_{i,j+1} = r \left(u_{i+1,j} + u_{i-1,j} \right) + 2(1-r)u_{i,j} - u_{i,j-1} \tag{16.88}$$

The explicit scheme (16.88) is stable for $r \leq 1$; Let us discuss the unconditionally stable implicit scheme.

16.8.2 Implicit Scheme

In this scheme, the central difference formula for the time derivative and average of central differences at $j - 1$ and $j + 1$ levels for space derivative are used in Eq. (16.87). After neglecting the error terms in discretization, we have

$$\frac{u_{i,j+1}-2u_{i,j}+u_{i,j-1}}{k^2}=c\frac{1}{2}\left(\frac{u_{i+1,j-1}-2u_{i,j-1}+u_{i-1,j-1}}{h^2}+\frac{u_{i+1,j+1}-2u_{i,j+1}+u_{i-1,j+1}}{h^2}\right) \tag{16.89}$$

Let $r=\dfrac{ck^2}{h^2}$, then we have

$$-r\,u_{i-1,j+1}+2(1+r)u_{i,j+1}-r\,u_{i+1,j+1}=r\,u_{i-1,j-1}-2(1+r)u_{i,j-1}+r\,u_{i+1,j-1}+4u_{i,j} \tag{16.90}$$

The scheme (16.90) is known as implicit scheme.

Note: In the implicit scheme (16.90), equal weightage is given to $(j{-}1)^{\text{th}}$ and $(j{+}1)^{\text{th}}$ levels and zero weightage to $(j)^{\text{th}}$ level. Different implicit schemes can be obtained depending on the weightage given to different levels. For example, θ_1 to $(j{-}1)^{\text{th}}$ level, θ_2 to $(j)^{\text{th}}$ level and $1-\theta_1-\theta_2$ to $(j{+}1)^{\text{th}}$ level, then Eq. (16.87) is given by

$$\frac{u_{i,j+1}-2u_{i,j}+u_{i,j-1}}{k^2}=c\left(\begin{array}{c}\theta_1\left(\dfrac{u_{i+1,j-1}-2u_{i,j-1}+u_{i-1,j-1}}{h^2}\right)+\theta_2\left(\dfrac{u_{i+1,j}-2u_{i,j}+u_{i-1,j}}{h^2}\right)\\[2mm]+(1-\theta_1-\theta_2)\dfrac{u_{i+1,j+1}-2u_{i,j+1}+u_{i-1,j+1}}{h^2}\end{array}\right)$$

Note that $0\le\theta_1,\theta_2\le1$. On rearranging the terms and using $r=\dfrac{ck^2}{h^2}$, we can easily obtain following the general implicit scheme.

$$u_{i,j+1}-2u_{i,j}+u_{i,j-1}=r\left(\begin{array}{c}\theta_1\left(u_{i+1,j-1}-2u_{i,j-1}+u_{i-1,j-1}\right)+\theta_2\left(u_{i+1,j}-2u_{i,j}+u_{i-1,j}\right)\\[2mm]+(1-\theta_1-\theta_2)\left(u_{i+1,j+1}-2u_{i,j+1}+u_{i-1,j+1}\right)\end{array}\right)$$

$$-(1-\theta_1-\theta_2)r\,u_{i-1,j+1}+\left(1+2r(1-\theta_1-\theta_2)\right)u_{i,j+1}-(1-\theta_1-\theta_2)r\,u_{i+1,j+1}$$
$$=r\theta_1\left(u_{i+1,j-1}+u_{i-1,j-1}\right)-(1+2r\theta_1)u_{i,j-1}+r\theta_2\left(u_{i+1,j}+u_{i-1,j}\right)-(1-2r\theta_2)u_{i,j}$$

Table for $u_{i,j}=u(x_i,\,t_j)$

$j\,(t)$ \ $i\,(x)$	0	1	2	...	n
0	u_{00}	u_{10}	u_{20}	...	u_{n0}
1	u_{01}	u_{11}	u_{21}	...	u_{n1}
2	u_{02}	u_{12}	u_{22}	...	u_{n2}
.	.	.	.	.	.
.	.	.	.	.	.
.	.	.	.	.	.

Example 16.18

Solve the wave equation $\dfrac{\partial^2 u}{\partial t^2} = 4\dfrac{\partial^2 u}{\partial x^2}$ with

initial conditions $u(x,0) = \begin{cases} 0.1(x) & 0 \le x \le 1/2 \\ 0.1(1-x) & 1/2 \le x \le 1 \end{cases}$ and $\dfrac{\partial u}{\partial t}\Big|_{t=0} = 0$

boundary conditions $\quad u(0,\,t) = u(1,\,t) = 0 \qquad t \ge 0.$

Take the step size for t is 0.1, and step size for x is 0.25. Use explicit scheme to compute the solution up to time $t = 0.3$. Use central difference formula for the derivative term in the initial condition.

Ans.

It is given that step size for t is 0.1 ($k = 0.1$) and step size for x is 0.25 ($h = 0.25$). So, we have $x = 0,\ 0.25,\ 0.5,\ 0.75,\ 1$ and $t = 0,\ 0.1,\ 0.2,\ 0.3$.

Using the initial condition

$$u(x,0) = \begin{cases} 0.1(x) & 0 \le x \le 1/2 \\ 0.1(1-x) & 1/2 \le x \le 1 \end{cases}$$

and boundary conditions

$$u(0,\,t) = u(1,\,t) = 0 \qquad t \ge 0,$$

we have following table for values of $u_{i,j} = u(x_i, t_j)$

$j\,(t)$ \ $i\,(x)$	0(0)	1(0.25)	2(0.5)	3(0.75)	4(1)
0 (0)	0	0.025	0.05	0.025	0
1 (0.1)	0	.	.	.	0
2 (0.2)	0	.	.	.	0
3(0.3)	0	.	.	.	0
4(0.4)	0				0

The explicit scheme (16.88) for the solution of wave equation is as follows

$$u_{i,j+1} = r\left(u_{i+1,j} + u_{i-1,j}\right) + 2(1-r)u_{i,j} - u_{i,j-1}$$

Using $r = \dfrac{ck^2}{h^2} = 4\dfrac{(0.1)^2}{(0.25)^2} = 0.64$, we have

$$u_{i,j+1} = 0.64\left(u_{i+1,j} + u_{i-1,j}\right) + 0.72u_{i,j} - u_{i,j-1} \tag{16.91}$$

Equation (16.91) can be used to compute various nodal values for different j.

$j = 1$.

At time $t = 0$, the initial condition is $\left.\dfrac{\partial u}{\partial t}\right|_{t=0} = 0$. The central difference formula provides the following equation

$$\left.\frac{\partial u}{\partial t}\right|_{t=0} = \left.\frac{u_{i,j+1} - u_{i,j-1}}{2k}\right|_{j=0} = \left.\frac{u_{i,1} - u_{i,-1}}{2k}\right|_{j=0} = 0$$

$$\Rightarrow u_{i,1} = u_{i,-1} \tag{16.92}$$

The scheme (16.91) for $j = 0$ is given by

$$u_{i,1} = 0.64\left(u_{i+1,0} + u_{i-1,0}\right) + 0.72u_{i,0} - u_{i,-1}$$

Using Eq. (16.92), we have

$$u_{i,1} = 0.32\left(u_{i+1,0} + u_{i-1,0}\right) + 0.36u_{i,0}$$

For $i = 1, 2, 3$, we can easily obtain the following values

$$u_{1,1} = 0.32\left(u_{2,0} + u_{0,0}\right) + 0.36u_{1,0} = 0.32(0.05 + 0) + 0.36(0.025) = 0.025$$

$$u_{2,1} = 0.32\left(u_{3,0} + u_{1,0}\right) + 0.36u_{2,0} = 0.32(0.025 + 0.025) + 0.36(0.05) = 0.034$$

$$u_{3,1} = 0.32\left(u_{4,0} + u_{2,0}\right) + 0.36u_{3,0} = 0.32(0 + 0.05) + 0.36(0.025) = 0.025$$

The following table shows these values in the second row.

$i\,(x)$ / $j\,(t)$	0(0)	1(0.25)	2(0.5)	3(0.75)	4(1)
0 (0)	0	0.025	0.05	0.025	0
1 (0.1)	0	0.025	0.034	0.025	0
2 (0.2)	0				0
3(0.3)	0				0

$j = 2,$

The explicit scheme (16.91) for $j = 1$ gives following equation

$$u_{i,2} = 0.64\left(u_{i+1,1} + u_{i-1,1}\right) + 0.72u_{i,1} - u_{i,0}$$

For $i = 1, 2, 3$, we can easily obtain the following values

$$u_{1,2} = 0.64\left(u_{2,1} + u_{0,1}\right) + 0.72u_{1,1} - u_{1,0} = 0.64\left(0.034 + 0\right) + 0.72(0.025) - 0.025 = 0.01476$$

$$u_{2,2} = 0.64\left(u_{3,1} + u_{1,1}\right) + 0.72u_{2,1} - u_{2,0} = 0.64\left(0.025 + 0.025\right) + 0.72(0.034) - 0.05 = 0.00648$$

$$u_{3,2} = 0.64\left(u_{4,1} + u_{2,1}\right) + 0.72u_{3,1} - u_{3,0} = 0.64\left(0 + 0.034\right) + 0.72(0.025) - 0.025 = 0.01476$$

$i\,(x)$ $j\,(t)$	0(0)	1(0.25)	2(0.5)	3(0.75)	4(1)
0 (0)	0	0.025	0.05	0.025	0
1 (0.1)	0	0.025	0.034	0.025	0
2 (0.2)	0	0.01476	0.00648	0.01476	0
3(0.3)	0				0

$j = 3,$

For $j = 2$, the explicit scheme (16.91) becomes

$$u_{i,3} = 0.64\left(u_{i+1,2} + u_{i-1,2}\right) + 0.72u_{i,2} - u_{i,1}$$

For $i = 1, 2, 3$, the following values are obtained

$$u_{1,3} = 0.64\left(u_{2,2} + u_{0,2}\right) + 0.72u_{1,2} - u_{1,1} = 0.64\left(0.00648 + 0\right) + 0.72(0.01476) - 0.025$$
$$= -0.0102256$$

$$u_{2,3} = 0.64\left(u_{3,2} + u_{1,2}\right) + 0.72u_{2,2} - u_{2,1} = 0.64\left(0.01476 + 0.01476\right) + 0.72(0.00648) - 0.034$$
$$= -0.0104416$$

$$u_{3,3} = 0.64\left(u_{4,2} + u_{2,2}\right) + 0.72u_{3,2} - u_{3,1} = 0.64\left(0 + 0.00648\right) + 0.72(0.01476) - 0.025$$
$$= -0.0102256$$

i (x) / j (t)	0(0)	1(0.25)	2(0.5)	3(0.75)	4(1)
0 (0)	0	0.025	0.05	0.025	0
1 (0.1)	0	0.025	0.034	0.025	0
2 (0.2)	0	0.01476	0.00648	0.01476	0
3(0.3)	0	-0.0102256	-0.0104416	-0.0102256	0

Example 16.19

A tightly stretched flexible string has its ends fixed at $x = 0$ and $x = 1$. The string is plucked at middle point by an initial displacement 0.05 and then released from this position. Find the transverse displacement of a point at a distance x from one end and at any time t of the vibrating string. The displacement at any time t and at a distance x satisfies the wave equation $\dfrac{\partial^2 u}{\partial t^2} = 4\dfrac{\partial^2 u}{\partial x^2}$.

Take the step size for t is 0.1 and step size for x is 0.25. Use explicit scheme to compute the solution up to time $t = 0.3$. Use central difference formula for the derivative term in the initial condition.

Ans.

The mathematical model for this problem is exactly same as in Example 16.18. So, the solution to this problem is the solution of Example 16.18.

Example 16.20

Solve the wave equation, $\dfrac{\partial^2 u}{\partial t^2} = \dfrac{\partial^2 u}{\partial x^2}$; $0 \le x \le 1, t \ge 0$

with

initial conditions $u(x,0) = \sin(\pi x)$ and $\dfrac{\partial u}{\partial t}\bigg|_{t=0} = 0$

boundary conditions $u(0, t) = u(1, t) = 0$ $t \ge 0$

Take the step size for time t is 1/6, and step size for x is 1/3. Use implicit scheme to compute the solution up to time $t = 1/3$. Use central difference formula for the derivative term in the initial condition.

Ans.

The spacing $h = 1/3$ and $k = 1/6$ for variables x and t provide the following grid points

$x = 0, 1/3, 2/3, 1$ and $t = 0, 1/6, 1/3$.

The initial and boundary conditions are summarize in the following table

$i\,(x)$ $j\,(t)$	0(0)	1(1/3)	2(2/3)	3(1)
0 (0)	0	0.8660	0.8660	0
1 (1/6)	0			0
2 (1/3)	0			0

The implicit scheme for the solution of wave equation is as follows

$$-r u_{i-1,j+1} + 2(1+r)u_{i,j+1} - r u_{i+1,j+1} = r u_{i-1,j-1} - 2(1+r)u_{i,j-1} + r u_{i+1,j-1} + 4u_{i,j}$$

Using $r = \dfrac{ck^2}{h^2} = \dfrac{(1/6)^2}{(1/3)^2} = 1/4$, we get

$$-u_{i-1,j+1} + 10 u_{i,j+1} - u_{i+1,j+1} = u_{i-1,j-1} - 10 u_{i,j-1} + u_{i+1,j-1} + 16 u_{i,j} \qquad (16.93)$$

This implicit scheme will be used to compute solution for different j.

$j = 1$.

The initial condition is $\left.\dfrac{\partial u}{\partial t}\right|_{t=0} = 0$. Using central difference formula, we get

$$\left.\frac{\partial u}{\partial t}\right|_{t=0} = \left.\frac{u_{i,j+1} - u_{i,j-1}}{2k}\right|_{j=0} = \left.\frac{u_{i,1} - u_{i,-1}}{2k}\right|_{j=0} = 0$$

$$\Rightarrow u_{i,1} = u_{i,-1} \qquad (16.94)$$

Implicit scheme (16.93) for $j = 0$ is given by

$$-u_{i-1,1} + 10 u_{i,1} - u_{i+1,1} = u_{i-1,-1} - 10 u_{i,-1} + u_{i+1,-1} + 16 u_{i,0}$$

Using Eq. (16.94), we have

$$-u_{i-1,1} + 10 u_{i,1} - u_{i+1,1} = 8 u_{i,0}$$

For $i = 1, 2$, we get following two equations respectively

$$-u_{0,1} + 10u_{1,1} - u_{2,1} = 8u_{1,0}$$

$$-u_{1,1} + 10u_{2,1} - u_{3,1} = 8u_{2,0}$$

Using the values $u_{0,1} = u_{3,1} = 0$ and $u_{1,0} = u_{2,0} = 0.8660$, and on solving resulting two equations, we get

$$u_{1,1} = u_{2,1} = 0.7698$$

So, we have

i (x) \ j (t)	0(0)	1(1/3)	2(2/3)	3(1)
0 (0)	0	0.8660	0.8660	0
1 (1/6)	0	0.7698	0.7698	0
2 (1/3)	0			0

j = 2.

Implicit scheme (16.93) for $j = 1$ is given by

$$-u_{i-1,2} + 10u_{i,2} - u_{i+1,2} = u_{i-1,0} - 10u_{i,0} + u_{i+1,0} + 16u_{i,1}$$

The following equations are obtained for $i = 1, 2$

$$-u_{0,2} + 10u_{1,2} - u_{2,2} = u_{0,0} - 10u_{1,0} + u_{2,0} + 16u_{1,1}$$

$$-u_{1,2} + 10u_{2,2} - u_{3,2} = u_{1,0} - 10u_{2,0} + u_{3,0} + 16u_{2,1}$$

Use $u_{0,0} = u_{0,2} = u_{3,2} = u_{3,0} = 0$, $u_{1,1} = u_{2,1} = 0.7698$ and $u_{1,0} = u_{2,0} = 0.8660$ in these two equations, then solution of the equations are given by

$$u_{1,2} = u_{2,2} = 0.5025$$

So, we have

i (x) \ j (t)	0(0)	1(1/3)	2(2/3)	3(1)
0 (0)	0	0.8660	0.8660	0
1 (1/6)	0	0.7698	0.7698	0
2 (1/3)	0	0.5025	0.5025	0

16.9 Creating Own Scheme for a Problem

An example will be discussed in this section to understand that how we can create schemes for linear PDEs. For this, we simply replace derivative terms with their finite difference approximations in the given linear PDE and then use initial and boundary conditions. Here, a very simple example has been discussed from the explanation point of view.

Example 16.21

Obtain an explicit finite difference scheme for the solution of following variable coefficient problem

$$\frac{\partial u}{\partial t} = \frac{\partial^2 u}{\partial x^2} + x\frac{\partial u}{\partial x} \qquad 0 \le x \le 1,\ t \ge 0$$

with following initial and boundary conditions

$$u(x,\ 0) = x(2-x)$$

$$u(0,t) = 0, \quad u(1,t) = 1$$

Replace temporal derivative term with forward difference and spatial derivative terms by central differences. Use this scheme for the solution over a rectangular grid defined by spacing $\Delta x = 0.2$ and $\Delta t = 0.005$. Solve up to $t = 0.02$ only.

Ans.

The spacing is $\Delta x = 0.2$ for $0 \le x \le 1$; so, our node points are given by

$$x_0 = 0, x_1 = 0.2, x_2 = 0.4, x_3 = 0.6, x_4 = 0.8, x_5 = 1$$

Let $u_{i,j} = u(x_i, t_j)$. The initial condition is $u(x,0) = x(2-x)$, so we have

$$u_{0,0} = u(x_0,t_0) = u(0,0) = 0(2-0) = 0$$

$$u_{1,0} = u(x_1,t_0) = u(0.2,0) = 0.2(2-0.2) = 0.36$$

$$u_{2,0} = u(x_2,t_0) = u(0.4,0) = 0.4(2-0.4) = 0.64$$

$$u_{3,0} = u(x_3,t_0) = u(0.6,0) = 0.6(2-0.6) = 0.84$$

$$u_{4,0} = u(x_4,t_0) = u(0.8,0) = 0.8(2-0.8) = 0.96$$

$$u_{5,0} = u(x_5,t_0) = u(1,0) = 1(2-1) = 1 \qquad (16.95)$$

For t = 0

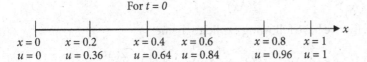

$x = 0$	$x = 0.2$	$x = 0.4$	$x = 0.6$	$x = 0.8$	$x = 1$
$u = 0$	$u = 0.36$	$u = 0.64$	$u = 0.84$	$u = 0.96$	$u = 1$

Similarly, boundary conditions are $u(0,t)=0$ and $u(1,t)=1$. This implies

$$u_{0,j} = u(x_0, t_j) = u(0,\ t) = 0$$

$$u_{5,j} = u(x_5, t_j) = u(1,\ t) = 1; \quad \text{for } \forall\ j = 0,1,2,3,\cdots \tag{16.96}$$

The following table shows the initial (16.95) and boundary (16.96) conditions.

i (x) / j (t)	0(0)	1(0.2)	2(0.4)	3(0.6)	4(0.8)	5(1)
0(0)	0	0.36	0.64	0.84	0.96	1
1(0.005)	0					1
2(0.01)	0					1
3(0.015)	0					1
4(0.02)	0					1

Let the value of $u(x, t)$ at point (x_i, t_j) be $u_{i,j}$ i.e. $u(x_i, t_j) = u_{i,j}$. Discretizing the given variable coefficient problem at point (x_i, t_j), we get

$$\left.\frac{\partial u}{\partial t}\right|_{(x_i,t_j)} = \left.\frac{\partial^2 u}{\partial x^2}\right|_{(x_i,t_j)} + x\left.\frac{\partial u}{\partial x}\right|_{(x_i,t_j)} \tag{16.97}$$

The following forward difference (for time derivative term) and central difference (for space derivative terms) formulas can be used to generate the explicit scheme

$$\frac{\partial u(x_i,t_j)}{\partial t} = \frac{u_{i,j+1} - u_{i,j}}{\Delta t} + O(\Delta t)$$

$$\frac{\partial u(x_i,t_j)}{\partial x} = \frac{u_{i+1,j} - u_{i-1,j}}{2(\Delta x)} + O(\Delta x^2)$$

$$\frac{\partial^2 u(x_i,t_j)}{\partial x^2} = \frac{u_{i+1,j} - 2u_{i,j} + u_{i-1,j}}{(\Delta x)^2} + O(\Delta x^2) \tag{16.98}$$

Putting the finite differences (16.98) in Eq. (16.97) and neglecting the error terms in discretization, we have

$$\frac{u_{i,j+1} - u_{i,j}}{\Delta t} = \frac{u_{i+1,j} - 2u_{i,j} + u_{i-1,j}}{(\Delta x)^2} + x_i\frac{u_{i+1,j} - u_{i-1,j}}{2(\Delta x)}$$

On rearranging the terms, we have following explicit scheme

$$u_{i,j+1} = \frac{\Delta t}{2(\Delta x)^2}\left((2 + x_i(\Delta x))u_{i+1,j} + (2 - x_i(\Delta x))u_{i-1,j}\right) + \left(1 - \frac{2(\Delta t)}{(\Delta x)^2}\right)u_{i,j}$$

Using $\Delta x = 0.2$ and $\Delta t = 0.005$, we have

$$u_{i,j+1} = 0.0625\big((2+0.2x_i)u_{i+1,j} + (2-0.2x_i)u_{i-1,j}\big) + 0.75u_{i,j} \tag{16.99}$$

For $j = 0$, we get

$$u_{i,1} = 0.0625\big((2+0.2x_i)u_{i+1,0} + (2-0.2x_i)u_{i-1,0}\big) + 0.75u_{i,0}$$

Computing the values for $i = 1, 2, 3$ and 4, we get

$$u_{1,1} = 0.0625\big((2+0.2x_1)u_{2,0} + (2-0.2x_1)u_{0,0}\big) + 0.75u_{1,0} = 0.351600$$

$$u_{2,1} = 0.0625\big((2+0.2x_2)u_{3,0} + (2-0.2x_2)u_{1,0}\big) + 0.75u_{2,0} = 0.632400$$

$$u_{3,1} = 0.0625\big((2+0.2x_3)u_{4,0} + (2-0.2x_3)u_{2,0}\big) + 0.75u_{3,0} = 0.832400$$

$$u_{4,1} = 0.0625\big((2+0.2x_4)u_{5,0} + (2-0.2x_4)u_{3,0}\big) + 0.75u_{4,0} = 0.951600$$

These values give the second row of the table.
Proceeding in a similar manner, for $j = 1, 2$, and 3, we will get different rows of the table.
The results are presented in the following table (only six decimal digits)

i (x) $\diagdown$ j (t)	0(0)	1(0.2)	2(0.4)	3(0.6)	4(0.8)	5(1)
0(0)	0	0.36	0.64	0.84	0.96	1
1(0.005)	0	0.351600	0.632400	0.832400	0.951600	1
2(0.01)	0	0.344331	0.624704	0.824694	0.944426	1
3(0.015)	0	0.337898	0.617058	0.817060	0.938159	1
4(0.02)	0	0.332098	0.609559	0.809605	0.932581	1

Exercise 16.1

Parabolic Equation (Heat Conduction or Diffusion Equation)

1. Classify the following PDEs

a) $\dfrac{\partial^2 u}{\partial x^2} + \dfrac{\partial^2 u}{\partial y^2} + u = 0$

b) $\dfrac{\partial^2 u}{\partial x^2} + y\dfrac{\partial u}{\partial y} + u = 0$

c) $2\dfrac{\partial^2 u}{\partial x^2} + 3\dfrac{\partial^2 u}{\partial x \partial y} + 4\dfrac{\partial^2 u}{\partial y^2} + u = 0$

d) $\dfrac{\partial^2 u}{\partial x^2} + (x+y)\dfrac{\partial^2 u}{\partial y^2} + xu = 0$

e) $2x\dfrac{\partial^2 u}{\partial x \partial y} + (x+y)\dfrac{\partial^2 u}{\partial y^2} + \dfrac{\partial u}{\partial x} + e^x u = \sin x$

f) $(y+1)\dfrac{\partial^2 u}{\partial x^2} + 2x\dfrac{\partial^2 u}{\partial x \partial y} + (1-y)\dfrac{\partial^2 u}{\partial y^2}$

Ans.
a) Elliptic
b) Parabolic
c) Hyperbolic
d) Elliptic in the region $x+y>0$, Parabolic on the line $x+y=0$, Hyperbolic in the region $x+y<0$
e) Hyperbolic
f) Elliptic inside the circle $x^2+y^2<1$, Parabolic on the circle $x^2+y^2=1$, and Hyperbolic outside the circle $x^2+y^2>1$

2. Solve the 1-dimensional heat conduction equation

$$\frac{\partial u}{\partial t}=2\frac{\partial^2 u}{\partial x^2};\qquad 0\le x\le 1 \text{ with}$$

initial condition $u(x,0)=x^2(2-x)$, and

boundary conditions $u(0,t)=0$ and $u(1,t)=1$.

Use Explicit scheme to find the values of $u(x, t)$ up to $t=0.02$, with $\Delta x=0.25$ and $\Delta t=0.005$.

Ans. $r=0.160$

i (x) j (t)	0(0)	1(0.25)	2(0.5)	3(0.75)	4(1)
0(0)	0	0.109375	0.375000	0.703125	1
1(0.005)	0	0.134375	0.385000	0.538125	1
2(0.01)	0	0.152975	0.369400	0.427525	1
3(0.015)	0	0.163127	0.344072	0.349821	1
4(0.02)	0	0.165978	0.316041	0.292930	1

3. Given the following initial-boundary value problem

$$\frac{\partial u}{\partial t}=\frac{\partial^2 u}{\partial x^2};\qquad 0\le x\le 0.5$$

$$u(x,0)=\cos(\pi x)$$

$u(0,t)=1$ and $u(0.5,t)=0$.

Compute the values of $u(x, t)$ up to $t=0.02$, with $\Delta x=0.1$ and $\Delta t=0.005$. Use Explicit scheme.

Ans. $r=0.5$

i (x) j (t)	0(0)	1(0.1)	2(0.2)	3(0.3)	4(0.4)	5(0.5)
0(0)	1	0.951057	0.809017	0.587785	0.309017	0
1(0.005)	1	0.904509	0.769421	0.559017	0.293893	0
2(0.01)	1	0.884710	0.731763	0.531657	0.279509	0
3(0.015)	1	0.865881	0.708184	0.505636	0.265828	0
4(0.02)	1	0.854092	0.685759	0.487006	0.252818	0

4. Use Bender–Schmidt Explicit Scheme to compute temperature distribution in a uniform insulated rod of length $1m$ with diffusivity constant of the material of the rod is 1. Both ends of the rod are kept at zero temperature, and initial temperature distribution in the rod is $u(x,0) = x(1-x)$. Take $h = 1/5$ and $k = 0.005$. Compute till $t = 0.02$.

Ans. The temperature distribution in the rod is given by

$$\frac{\partial u}{\partial t} = \frac{\partial^2 u}{\partial x^2}; \qquad 0 \le x \le 1.$$

Both ends of the rod are kept at zero temperature, so we have following boundary conditions
$$u(0, t) = u(1, t) = 0$$
Also, initial temperature distribution gives following initial condition
$$u(x,0) = x(1-x)$$
$r = 0.125$

$i\,(x)$ / $j\,(t)$	0(0)	1(0.2)	2(0.4)	3(0.6)	4(0.8)	5(1)
0(0)	0	0.16	0.24	0.24	0.16	0
1(0.005)	0	0.15	0.23	0.23	0.15	0
2(0.01)	0	0.14125	0.22	0.22	0.14125	0
3(0.015)	0	0.133437	0.210156	0.210156	0.133437	0
4(0.02)	0	0.126348	0.200566	0.200566	0.126348	0

5. Use Crank–Nicolson scheme to compute numerical solution of parabolic equation

$$\frac{\partial u}{\partial t} = 2\frac{\partial^2 u}{\partial x^2}; \qquad 0 \le x \le 1 \text{ with}$$

initial condition $u(x,0) = x^2(2-x)$, and

boundary conditions $u(0,t) = 0$ and $u(1,t) = 1$.
Find the values of $u(x, t)$ up to $t = 0.01$, with $\Delta x = 0.25$ and $\Delta t = 0.005$.

Ans.

$i\,(x)$ / $j\,(t)$	0(0)	1(0.25)	2(0.5)	3(0.75)	4(1)
0(0)	0	0.109375	0.375000	0.703125	1
1(0.005)	0	0.131610	0.384904	0.699498	1
2(0.01)	0	0.149054	0.394474	0.698214	1

6. Consider following initial-boundary value problem

$$\frac{\partial u}{\partial t} = \frac{\partial^2 u}{\partial x^2}; \qquad 0 \le x \le 0.5 \text{ with}$$

initial condition $u(x,0) = \cos(\pi x)$, and
boundary conditions $u(0,t) = 1$ and $u(0.5,t) = 0$.

Use Crank–Nicolson scheme to find the values of $u(x, t)$ up to $t = 0.01$, with $\Delta x = 0.125$ and $\Delta t = 0.005$.

Ans. $r = 0.5$

$j\,(t)$ \ $i\,(x)$	0(0)	1(0.125)	2(0.250)	3(0.375)	4(0.5)
0(0)	1	0.923880	0.707107	0.382683	0
1(0.005)	1	0.885793	0.674198	0.364571	0
2(0.01)	1	0.858657	0.645101	0.347724	0

7. Solve the following heat conduction equation

$$16\frac{\partial u}{\partial t} = \frac{\partial^2 u}{\partial x^2}; \qquad 0 \le x \le 1$$

$u(x,0) = 2\sin(2\pi x)$
$u(0,t) = u(1,t) = 0.$

Use Crank–Nicolson scheme to find the values of $u(x, t)$ up to $t = 0.5$. The step sizes are $\Delta x = \Delta t = 0.25$.

Ans. $r = 0.25$

$j\,(t)$ \ $i\,(x)$	0(0)	1(0.25)	2(0.5)	3(0.75)	4(1)
0(0)	0	2	0	-2	0
1(0.25)	0	1.2	0	-1.2	0
2(0.5)	0	0.72	0	-0.72	0

8. Solve the 1-dimensional heat conduction equation

$$\frac{\partial u}{\partial t} = 2\frac{\partial^2 u}{\partial x^2}; \qquad 0 \le x \le 1$$

$u(x,0) = x^2(2-x),$
$u(0, t) = 0, u(1, t) = 1$

Use Crank–Nicolson scheme to find the values of $u(x, t)$ at time $t = 0.005$. Further use Richardson scheme to compute the values up to time $t = 0.02$. Take $\Delta x = 0.25$ and $\Delta t = 0.005$.

Ans.

$j\,(t)$ \ $i\,(x)$	0(0)	1(0.25)	2(0.5)	3(0.75)	4(1)
0(0)	0	0.109375	0.375000	0.703125	1
1(0.005)	0	0.131610	0.384904	0.699498	1
2(0.01)	0	0.148314	0.394616	0.698616	1
3(0.015)	0	0.162966	0.403367	0.698661	1
4(0.02)	0	0.173093	0.412181	0.700550	1

9. Consider following initial-boundary value problem

$$\frac{\partial u}{\partial t} = \frac{\partial^2 u}{\partial x^2}; \qquad 0 \le x \le 0.5 \text{ with}$$

initial condition $u(x,0) = \cos(\pi x)$, and
boundary conditions $u(0,t) = 1$ and $u(0.5,t) = 0$.
Use Crank–Nicolson scheme to find the values of $u(x, t)$ at time $t = 0.005$. Then, use Du-Fort and Frankel scheme up to $t = 0.02$, with $\Delta x = 0.125$ and $\Delta t = 0.005$.

Ans. $r = 0.5$

i (x) / j (t)	0(0)	1(0.125)	2(0.250)	3(0.375)	4(0.5)
0(0)	1	0.923880	0.707107	0.382683	0
1(0.005)	1	0.885793	0.674198	0.364571	0
2(0.01)	1	0.856148	0.643165	0.347105	0
3(0.015)	1	0.835678	0.617557	0.331019	0
4(0.02)	1	0.819177	0.596479	0.317192	0

10. Consider following heat conduction equation

$$16\frac{\partial u}{\partial t} = \frac{\partial^2 u}{\partial x^2}; \qquad 0 \le x \le 1 \text{ with}$$

initial condition $u(x,0) = 2\sin(2\pi x)$, and
boundary conditions $u(0,t) = u(1,t) = 0$.
Use Crank–Nicolson method to find the values of $u(x, t)$ at time $t = 0.25$. Further use Du-Fort and Frankel scheme to compute other values of $u(x, t)$ up to $t = 1.0$. Take $\Delta x = \Delta t = 0.25$.

Ans. $r = 0.25$

i (x) / j (t)	0(0)	1(0.25)	2(0.5)	3(0.75)	4(1)
0(0)	0	2	0	−2	0
1(0.25)	0	1.2	0	−1.2	0
2(0.5)	0	0.666667	0	−0.666667	0
3(0.75)	0	0.4	0	−0.04	0
4(1.0)	0	0.222222	0	−0.222222	0

11. Check the stability conditions of Bender–Schmidt Explicit Scheme for heat conduction equation of order two in one space variable.

12. Prove that the Crank–Nicolson scheme is always convergent for a heat conduction equation of order two in one space variable.

13. The parabolic equation $\dfrac{\partial u}{\partial t} = \dfrac{\partial^2 u}{\partial x^2}$ is approximated by finite difference scheme

$$u_{i,j+1} - u_{i,j} = r\left(\theta(u_{i-1,j+1} - 2u_{i,j+1} + u_{i+1,j+1}) + (1-\theta)(u_{i-1,j} - 2u_{i,j} + u_{i+1,j})\right)$$

where $r = \dfrac{\Delta t}{\Delta x^2}$, and θ is a parameter $0 \le \theta \le 1$.

Prove that the scheme is stable for $r \le \dfrac{1}{2}(1-2\theta)$ if $0 \le \theta \le \dfrac{1}{2}$ and unconditionally stable when $\dfrac{1}{2} \le \theta \le 1$. It may be assumed that Dirichlet conditions are prescribed at the end points.

14. Use explicit scheme to solve the 2-dimensional equation

$$\frac{\partial u}{\partial t} = \frac{\partial^2 u}{\partial x^2} + \frac{\partial^2 u}{\partial y^2}$$

with initial condition
$$u(x,y,0) = \sin(2\pi x)\sin(2\pi y) \quad 0 \le x, y \le 1$$

and boundary conditions
$$u = 0 \text{ for } x, y = 0;\ x, y = 1;\ t > 0$$

Take step size $\Delta x = \Delta y = \dfrac{1}{3}$ and $r_1 = r_2 = r = \dfrac{1}{8}$. Integrate for one-time step only.

Ans.

We have $\Delta x = \Delta y = 1/3$, $c = 1$, and $r_1 = r_2 = \dfrac{1}{8}$.

As $r_1 = r_2 = \dfrac{c\Delta t}{\Delta x^2}$ $\Rightarrow \Delta t = \dfrac{1}{8} \times \dfrac{1}{9} = \dfrac{1}{72}$.

The initial condition
$$u(x,y,0) = \sin(2\pi x)\sin(2\pi y) \quad\quad 0 \le x, y \le 1;\ t \ge 0$$
gives the following values of $u(x_i, y_j, 0)$ in the table at $t = 0$ ($k = 0$).

j (y) \ i (x)	0 (0)	1 (1/3)	2 (2/3)	3 (1)
0 (0)	0	0	0	0
1 (1/3)	0	0.75	−0.75	0
2 (2/3)	0	−0.75	0.75	0
3 (1)	0	0	0	0

Using the given boundary conditions and explicit scheme, we have following values for $t = \dfrac{1}{72}$ ($k = 1$).

j (y) \ i (x)	0 (1)	1 (1/3)	2 (2/3)	3 (1)
0 (1)	0	0	0	0
1 (1/3)	0	0.1875	−0.1875	0
2 (2/3)	0	−0.1875	0.1875	0
3 (1)	0	0	0	0

15. Consider the following two-dimensional heat conduction equation,

$$\frac{\partial u}{\partial t} = \frac{\partial^2 u}{\partial x^2} + \frac{\partial^2 u}{\partial y^2}$$

with following initial and boundary conditions

$$u(x,y,0) = 3(x(0.75 - x) + y(0.75 - y)), \qquad 0 \le x, y \le 0.75; \quad t = 0$$

$$u = 0 \text{ on } x = 0, x = 0.75, y = 0 \text{ and } y = 0.75.$$

Use Crank–Nicolson (CN) method with step size $\Delta x = \Delta y = 0.25$ to evaluate various mesh values. Take $r_1 = r_2 = \dfrac{1}{6}$. Integrate for one-time step.

Ans. We have $\Delta x = \Delta y = 0.25$, $c = 1$, and $r_1 = r_2 = \dfrac{1}{6}$.

As $r_1 = r_2 = \dfrac{c\Delta t}{\Delta x^2}$ $\qquad \Rightarrow \Delta t = \dfrac{1}{6} \times \dfrac{1}{16} = \dfrac{1}{96}$.

The initial condition

$$u(x,y,0) = 3(x(0.75 - x) + y(0.75 - y)), \qquad 0 \le x, y \le 0.75; \quad t = 0$$

gives the following values of $u(x_i, y_j, 0)$ in the table at $t = 0$ ($k = 0$).

i (x) / j (y)	0 (0)	1 (0.25)	2 (0.5)	3 (0.75)
0 (0)	0	0	0	0
1 (0.25)	0	0.75	0.75	0
2 (0.5)	0	0.75	0.75	0
3 (0.75)	0	0	0	0

By using the given boundary conditions and CN scheme, we have following values for $t = \dfrac{1}{96}$ ($k = 1$).

i (x) / j (y)	0 (0)	1 (0.25)	2 (0.5)	3 (0.75)
0 (0)	0	0	0	0
1 (0.25)	0	0.535714	0.535714	0
2 (0.5)	0	0.535714	0.535714	0
3 (0.75)	0	0	0	0

16. A heat flow equation in two dimensions is given by

$$\frac{\partial u}{\partial t} = \frac{\partial^2 u}{\partial x^2} + \frac{\partial^2 u}{\partial y^2}, \qquad 0 \le x, y \le 1; \ t \ge 0$$

with initial condition $u(x,y,0) = \sin(\pi x)\sin(\pi y)$ and boundary conditions $u(x,y,t) = 0$ on $x, y = 0, 1$. Use ADI (Alternating Direction Implicit) method with the step sizes $\Delta x = \Delta y = \dfrac{1}{3}$ to evaluate various mesh values. Take $r_1 = r_2 = \dfrac{1}{6}$. Integrate for two-time steps.

Ans. We have $\Delta x = \Delta y = 1/3$, $c = 1$, and $r_1 = r_2 = \dfrac{1}{6}$.

As $r_1 = r_2 = \dfrac{c\Delta t}{\Delta x^2}$ $\qquad \Rightarrow \Delta t = \dfrac{1}{6} \times \dfrac{1}{9} = \dfrac{1}{54}$.

The initial condition

$$u(x,y,0) = \sin(\pi x)\sin(\pi y) \qquad\qquad 0 \le x, y \le 1$$

gives the following values of $u(x_i, y_j, 0)$ in the table at $t = 0$ $(k = 0)$.

i (x) / j (y)	0 (0)	1 (1/3)	2 (2/3)	3 (1)
0 (0)	0	0	0	0
1 (1/3)	0	0.75	0.75	0
2 (2/3)	0	0.75	0.75	0
3 (1)	0	0	0	0

By using the given boundary conditions and ADI scheme, we obtain following values for $t = \dfrac{1}{54}$ $(k = 1)$.

i (x) / j (y)	0 (0)	1 (1/3)	2 (2/3)	3 (1)
0 (0)	0	0	0	0
1 (1/3)	0	0.535714	0.535714	0
2 (2/3)	0	0.535714	0.535714	0
3 (1)	0	0	0	0

ADI scheme gives following solution for $t = \dfrac{1}{27}$ $(k = 2)$.

i (x) / j (y)	0 (0)	1 (1/3)	2 (2/3)	3 (1)
0 (0)	0	0	0	0
1 (1/3)	0	0.382653	0.382653	0
2 (2/3)	0	0.382653	0.382653	0
3 (1)	0	0	0	0

17. Use ADI method to evaluate various mesh values for the following initial-boundary value problem,

$$\frac{\partial u}{\partial t} = \frac{\partial^2 u}{\partial x^2} + \frac{\partial^2 u}{\partial y^2},$$

$$u(x,y,0) = \cos\left(\frac{\pi}{2}x\right)\cos\left(\frac{\pi}{2}y\right) \qquad -1 \le x, y \le 1$$

$$u = 0,\ x = \pm 1,\ y = \pm 1,\ t > 0$$

Take $r_1 = r_2 = \dfrac{1}{6}$ and step size $\Delta x = 1/2$. Integrate for one-time step only. Use symmetry over x and y – axes.

Ans.

The symmetry over x and y – axes gives

$$u_{11} = u_{31} = u_{33} = u_{13}$$

$$u_{12} = u_{21} = u_{23} = u_{32}$$

So, we have to compute only u_{11}, u_{12} and u_{22}.

Using the given initial and boundary conditions, we have following values at node points u_{ij} at initial time $t = 0$ ($k = 0$)

$i(x)$ $j(y)$	0 (−1)	1 (−1/2)	2 (0)	3 (1/2)	4 (1)
0 (−1)	0	0	0	0	0
1 (−1/2)	0	0.5	0.7071	0.5	0
2 (0)	0	0.7071	1	0.7071	0
3 (1/2)	0	0.5	0.7071	0.5	0
4 (1)	0	0	0	0	0

Using the given boundary conditions and ADI scheme, we have following equations for the values at node points

$$8u_{11} - u_{21} = 2.7071$$

$$-u_{11} + u_{21} = 1.9142$$

$$8u_{22} - 2u_{21} = 5.4142$$

On solving these equations, we get following values of u_{ij} at $t = \dfrac{1}{24}$ ($k = 1$).

$i(x)$ $j(y)$	0 (−1)	1 (−1/2)	2 (0)	3 (1/2)	4 (1)
0 (−1)	0	0	0	0	0
1 (−1/2)	0	0.4110	0.5813	0.4110	0
2 (0)	0	0.5813	0.8210	0.5813	0
3 (1/2)	0	0.4110	0.5813	0.4110	0
4 (1)	0	0	0	0	0

Exercise 16.2

Elliptic Equation (Laplace and Poisson Equations)

18. Solve the Laplace equation $\nabla^2 u = u_{xx} + u_{yy} = 0$ for the square mesh with the boundary values (Dirichlet conditions) as shown in the following figure. Use Gauss elimination method.

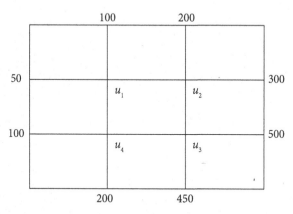

Ans. $u_1 = 150$ $u_2 = 250$ $u_3 = 350$ $u_4 = 200$

19. Find the steady state temperature distribution in a thin rectangular plate, whose edges $x = 0$, $x = 0.8$, $y = 0$, $y = 0.6$ are kept at the temperature shown in the following figure. Find the values of temperature at the nodal points of the rectangular region with mess length 0.2. Use Gauss–Seidel iterative method to compute values at nodal points until the difference between successive values at each point is less than 0.005.

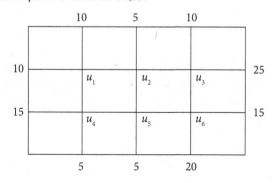

Ans. Exact solution $u_1 = 10$ $u_2 = 10$ $u_3 = 15$ $u_4 = 10$ $u_5 = 10$ $u_6 = 15$.

20. Find the steady state temperature distribution in a thin rectangular plate, whose edges $x = 0$, $x = 2$ are kept at 20°C (cold water), and edges $y = 0$, $y = 2$ are kept at temperature 100°C (in boiling water). Find the values of temperature at the nodal points of the rectangular region with mess length 0.5. Use Gauss–Seidel iterative method to compute values at nodal points until the difference between successive values at each point is less than 0.005. Use symmetry about $x = 1$ and $y = 1$ lines.

Ans.

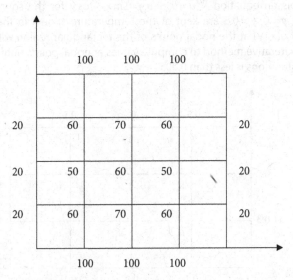

21. Solve the Laplace equation $\nabla^2 u = 0$ in $0 \le x, y \le 0.6$, over a grid of mesh length 0.2. Given $u = e^{-x} \sin(y)$ on the boundary of the square region $0 \le x, y \le 0.6$. Use Gauss–Seidel method with zero vector as an initial solution.

Ans.
The node points are as follows

$$x_0 = 0, \ x_1 = 0.2, \ x_2 = 0.4, \ x_3 = 0.6$$

$$y_0 = 0, \ y_1 = 0.2, \ y_2 = 0.4, \ y_3 = 0.6$$

Let $u_{ij} = u(x_i, y_j)$.

The boundary condition $u = e^{-x} \sin(y)$ provides following values

$u_{10} = 0$ $u_{20} = 0$ $u_{01} = 0.198669$

$u_{02} = 0.389418$ $u_{13} = 0.462290$ $u_{31} = 0.109032$

$u_{32} = 0.213717$ $u_{23} = 0.378491$

Using Gauss–Seidel method, we get following iterations for $\left[u_{11}, u_{12}, u_{21}, u_{22}\right]$

Iteration 1	Iteration 2	Iteration 3
0.049667	0.115922	0.150992
0.225344	0.295484	0.313019
0.039675	0.109815	0.127350
0.214307	0.249377	0.258145
Iteration 4	Iteration 5	Iteration 6
0.159760	0.161952	0.162500
0.317403	0.318499	0.318773
0.131734	0.132830	0.133104
0.260336	0.260884	0.261021

22. Solve the Poisson equation $\nabla^2 u = u_{xx} + u_{yy} = \sin x + \cos y$ for the square mesh, whose edges $x = 0$, $x = 0.8$, $y = 0$, $y = 0.6$ are kept at the temperature shown in the following figure. Find the values of $u(x, y)$ at the nodal points of the rectangular region with mess length 0.2. Use Gauss–Seidel iterative method to compute values at nodal points until the difference between consecutive iterations is less than 0.005.

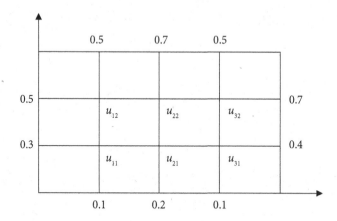

Ans. The result is obtained after seven iterations of the Gauss–Seidel method with zero initial approximation. The values at 7th iteration are given by

$u_{11} = 0.265608$ $u_{12} = 0.419638$ $u_{21} = 0.292323$

$u_{22} = 0.458575$ $u_{31} = 0.301323$ $u_{32} = 0.475118$

23. Solve the Poisson equation $u_{xx} + u_{yy} = \sin(x + y)$ for a thin rectangular plate, whose edges, $x = 0$, $x = 0.6$ are kept at 0°C, and edges $y = 0$, $y = 0.6$ are kept at temperature 100°C. Find the values of $u(x, y)$ at the nodal points of the square region with mess length 0.2. Use Gauss–Seidel iterative method to compute values at nodal points until the difference between successive values at each point is less than 0.005. Use $[0\ 0\ 0\ 0]^T$ as initial approximation.

Ans. The result is obtained after nine iterations of the Gauss–Seidel method. The ninth iteration is given by

$u_{11} = 49.968647$ $u_{12} = 49.966934$ $u_{21} = 49.966228$ $u_{22} = 49.964371$

24. Solve the Poisson equation $\nabla^2 u = x^2 + y^2$ in $0 \le x$, $y \le 0.6$, over a grid of mesh length 0.2. Given $u = e^{-y} \sin(x + y)$ on the boundary of the square region $0 \le x$, $y \le 0.6$. Use Gauss–Seidel method with zero vector as an initial solution.

Ans.

The node points are as follows

$x_0 = 0$, $x_1 = 0.2$, $x_2 = 0.4$, $x_3 = 0.6$

$y_0 = 0$, $y_1 = 0.2$, $y_2 = 0.4$, $y_3 = 0.6$

The boundary condition $u = e^{-y} \sin(x + y)$ provides the following results

$u_{10} = 0.198669$ $u_{20} = 0.389418$ $u_{01} = 0.162657$

$u_{02} = 0.261035$ $u_{13} = 0.393693$ $u_{31} = 0.587322$

$u_{32} = 0.564055$ $u_{23} = 0.461809$

The Gauss–Seidel iterations for $[u_{11}, u_{12}, u_{21}, u_{22}]$ are given by

Iteration 1	Iteration 2	Iteration 3
0.089532	0.201690	0.261387
0.184065	0.303461	0.333309
0.264568	0.383963	0.413812
0.365424	0.425122	0.440046
Iteration 4	Iteration 5	Iteration 6
0.276312	0.280043	0.280976
0.340772	0.342637	0.343104
0.421275	0.423140	0.423607
0.443778	0.444710	0.444944

Exercise 16.3

Hyperbolic Equation (Wave Equation)

25. Solve the wave equation $\dfrac{\partial^2 u}{\partial t^2} = 4\dfrac{\partial^2 u}{\partial x^2}$; $0 \le x \le 1, t \ge 0$ with following initial and boundary conditions.

$u(x,0) = \sin(\pi x)$ and $\dfrac{\partial u}{\partial t}\Big|_{t=0} = 0$

$u(0, t) = u(1, t) = 0$ $\quad t \ge 0$.

Use explicit scheme to compute the solution up to time $t = 0.375$ with the step sizes for variables t and x are 0.125 and 0.25 respectively. Approximate the derivative term of initial condition with central difference formula.

Ans. $r = 1$

$i\,(x)$ \ $j\,(t)$	0(0)	1(0.25)	2(0.5)	3(0.75)	4(1)
0 (0)	0	0.707107	1.000000	0.707107	0
1 (0.125)	0	0.588388	0.832107	0.588388	0
2 (0.25)	0	0.125000	0.176777	0.125000	0
3(0.375)	0	−0.411612	−0.582107	−0.411612	0

26. Solve the wave equation $\dfrac{\partial^2 u}{\partial t^2} = \dfrac{\partial^2 u}{\partial x^2}$; $\quad 0 \le x \le 1, t \ge 0$

with

initial conditions $u(x,0) = \cos(\pi x)$ and $\dfrac{\partial u}{\partial t}\Big|_{t=0} = 0$

boundary conditions $\quad u(0, t) = 1 \quad u(1, t) = -1 \quad t \ge 0$.

Use explicit scheme to compute the solution up to time $t = 0.5$. Take the step size for t is 0.125 and step size for x is 0.25. Use central difference formula for derivative term in the initial condition.

Ans.

$r = 0.250000$

i (x) j (t)	0(0)	1(0.25)	2(0.5)	3(0.75)	4(1)
0 (0)	1.000000	0.707107	0.000000	−0.707107	−1.000000
1 (0.125)	1.000000	0.743718	0.000000	−0.743718	−1.000000
2 (0.25)	1.000000	0.658471	0.000000	−0.658471	−1.000000
3(0.375)	1.000000	0.493988	0.000000	−0.493988	−1.000000
4(0.5)	1.000000	0.332511	0.000000	−0.332511	−1.000000

27. A tightly stretched flexible string has its ends fixed at $x = 0$ and $x = 1$. The string is plucked at middle point by an initial displacement ½ and then released from this position. The transverse displacement of a point at a distance x from one end and at any time t of the vibrating string satisfies the wave equation $\dfrac{\partial^2 u}{\partial t^2} = 2\dfrac{\partial^2 u}{\partial x^2}$. Evaluate the pivotal values for the interval $0 \le t \le 0.4$. Take $\Delta x = 0.2$, $\Delta t = 0.1$. Use explicit method.

Ans. The mathematical formulation is given by following wave equation

$$\frac{\partial^2 u}{\partial t^2} = 2\frac{\partial^2 u}{\partial x^2}$$

with following initial and boundary conditions

$$u(x,0) = \begin{cases} x & 0 \le x \le 1/2 \\ 1-x & 1/2 \le x \le 1 \end{cases}$$

$$\frac{\partial u}{\partial t}(x,0) = 0$$

$$u(0, t) = u(1, t) = 0 \ .$$

The solution with spacing $\Delta x = 0.2$, $\Delta t = 0.1$ by using the explicit method is as follows
$r = 0.500000$

i (x) j (t)	0(0)	1(0.2)	2(0.4)	3(0.6)	4(0.8)	5(1)
0 (0)	0.000000	0.200000	0.400000	0.400000	0.200000	0.000000
1 (0.1)	0.000000	0.220000	0.390000	0.390000	0.220000	0.000000
2 (0.2)	0.000000	0.215000	0.295000	0.295000	0.215000	0.000000
3(0.3)	0.000000	0.142500	0.160000	0.160000	0.142500	0.000000
4(0.4)	0.000000	0.007500	0.016250	0.016250	0.007500	0.000000

28. Solve the initial boundary value problem in cylindrical coordinates

$$\frac{\partial u}{\partial t} = \sin(r)\frac{\partial^2 u}{\partial r^2} + r^2 \frac{\partial u}{\partial r}$$

$$u(r,0) = r(1-r), \qquad 0 \le r \le 1$$

$$u(0, t) = 0, \qquad u(1, t) = 0 \qquad t > 0$$

Create own explicit scheme with $\Delta r = 0.2$ and $\Delta t = 0.02$. Solve up to $t = 0.08$.

Ans. The spacing is $\Delta r = 0.2$ for the interval $0 \le r \le 1$. The initial condition and boundary conditions give following tabulated values

i (x) j (t)	0(0)	1(0.2)	2(0.4)	3(0.6)	4(0.8)	5(1)
0(0)	0	0.16	0.24	0.24	0.16	0
1(0.02)	0					0
2(0.04)	0					0
3(0.06)	0					0
4(0.08)	0					0

Let the value of $u(r, t)$ at point (r_i, t_j) is $u_{i,j}$ i.e.

$$u(r_i, t_j) = u_{i,j}$$

On discretizing the given variable coefficient problem at point (r_i, t_j), we get

$$\left.\frac{\partial u}{\partial t}\right|_{(r_i, t_j)} = \sin(r_i)\left.\frac{\partial^2 u}{\partial r^2}\right|_{(r_i, t_j)} + r_i^2 \left.\frac{\partial u}{\partial r}\right|_{(r_i, t_j)}$$

Using forward difference for time derivative term and central difference for space derivative terms, we have following explicit scheme

$$\frac{u_{i,j+1} - u_{i,j}}{\Delta t} = \sin(r_i)\frac{u_{i+1,j} - 2u_{i,j} + u_{i-1,j}}{(\Delta r)^2} + r_i^2 \frac{u_{i+1,j} - u_{i-1,j}}{2(\Delta r)}$$

Using $\Delta r = 0.2$ and $\Delta t = 0.02$, and rearranging the terms

$$u_{i,j+1} = u_{i,j} + (0.5)\sin(r_i)\left(u_{i+1,j} - 2u_{i,j} + u_{i-1,j}\right) + (0.05)r_i^2\left(u_{i+1,j} - u_{i-1,j}\right)$$

$$u_{i,j+1} = \left((0.5)\sin(r_i) + (0.05)r_i^2\right)u_{i+1,j} + \left((0.5)\sin(r_i) - (0.05)r_i^2\right)u_{i-1,j} + (1 - \sin(r_i))u_{i,j}$$

For $i = 1, 2, 3, 4$ and $j = 0, 1, 2$, we get

i (x) j (t)	0(0)	1(0.2)	2(0.4)	3(0.6)	4(0.8)	5(1)
0(0)	0	0.16	0.24	0.24	0.16	0
1(0.02)	0	0.152533	0.225063	0.215974	0.123626	0
2(0.04)	0	0.145036	0.209679	0.190642	0.105496	0
3(0.06)	0	0.137470	0.193751	0.170103	0.092097	0
4(0.08)	0	0.129792	0.178449	0.152927	0.081599	0

29. Consider the Poisson problem

$$\frac{\partial^2 u}{\partial x^2} + (1 - x^2 - y^2)\frac{\partial^2 u}{\partial y^2} = 1 \qquad |x|, |y| \le 0.2;$$

with following boundary conditions

$$u = 0 \text{, for } |x| = 0.2, |y| = 0.2$$

Obtain a suitable finite difference scheme and discuss symmetry for the nodal points shown in the following figure.

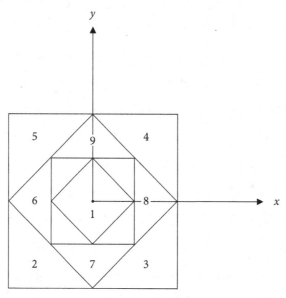

Ans.

The node points are as follows

$$x_0 = -0.2, \ x_1 = -0.1, \ x_2 = 0, \ x_3 = 0.1, \ x_4 = 0.2$$
$$y_0 = -0.2, \ y_1 = -0.1, \ y_2 = 0, \ y_3 = 0.1, \ y_4 = 0.2$$

Let $u_{ij} = u(x_i, y_j)$. The boundary condition is $u = 0$, on the boundary $|x| = 0.2, |y| = 0.2$. The differential equation and boundary conditions are symmetrical about x and y-axes, therefore

$$u_7 = u_9, \ u_6 = u_8, \ u_2 = u_3 = u_4 = u_5$$

So, we have to compute only u_1, u_2, u_6 and u_7. On replacing the derivative terms with central difference in the given PDE and neglecting error terms, we have

$$\frac{u_{i+1,j} - 2u_{i,j} + u_{i-1,j}}{h^2} + (1 - x_i^2 - y_j^2)\frac{u_{i,j+1} - 2u_{i,j} + u_{i,j-1}}{h^2} = 1$$

$$u_{i+1,j} - 2u_{i,j} + u_{i-1,j} + (1 - x_i^2 - y_j^2)\left(u_{i,j+1} - 2u_{i,j} + u_{i,j-1}\right) = h^2$$

$$u_{i+1,j} + u_{i-1,j} + (1 - x_i^2 - y_j^2)\left(u_{i,j+1} + u_{i,j-1}\right) + 2(x_i^2 + y_j^2 - 2)u_{i,j} = h^2$$

This scheme is required finite difference scheme. For computation purpose, note that

$$u_1 = u_{22}, \ u_2 = u_{11}, \ u_3 = u_{31}, \ u_4 = u_{33}, \ u_5 = u_{13}$$
$$u_6 = u_{12}, \ u_7 = u_{21}, \ u_8 = u_{32}, \ u_9 = u_{23}$$

30. Obtain a suitable finite difference scheme for following boundary value problem

$$u_{xx} - 4u_{xy} + u_{yy} = 0; \quad 0 \le x, y \le 1$$

$$u(0, y) = y^2, \quad u(1, y) = 1 + y^2 + y$$
$$u(x, 0) = x^2, \quad u(x, 1) = 1 + x^2 + x$$

Use symmetry about the line $y = x$ and take the step sizes $h = k = 0.25$. Note that the exact solution is $u(x, y) = x^2 + y^2 + xy$.

Ans. The boundary conditions provide following values at boundaries of the square region $0 \le x, y \le 1$

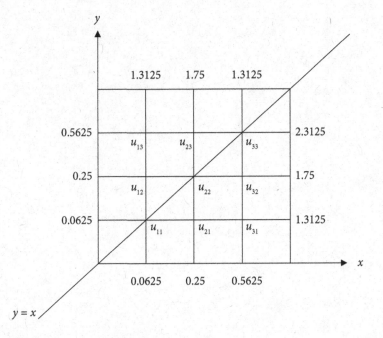

Using symmetry about line, $y = x$, we get

$$u_{12} = u_{21}, \quad u_{13} = u_{31}, \quad u_{23} = u_{32}$$

So, we have to compute only $u_{11}, u_{12}, u_{13}, u_{22}, u_{23}$ and u_{33}. On replacing the derivative terms with central difference in the given PDE and neglecting error terms, we have

$$\frac{u_{i+1,j} - 2u_{i,j} + u_{i-1,j}}{h^2} - 4\frac{u_{i+1,j+1} - u_{i+1,j-1} - u_{i-1,j+1} + u_{i-1,j-1}}{4hk} + \frac{u_{i,j+1} - 2u_{i,j} + u_{i,j-1}}{h^2} = 0$$

$$u_{i+1,j} + u_{i-1,j} - \frac{h}{k}\left(u_{i+1,j+1} - u_{i+1,j-1} - u_{i-1,j+1} + u_{i-1,j-1}\right) + u_{i,j+1} - 4u_{i,j} + u_{i,j-1} = 0$$

The scheme is required finite difference scheme.

Comparison of Analytical and Numerical Techniques

In this section, we discuss some advantages and disadvantages of numerical, analytical and graphical techniques.

a) Analytical methods, which provide exact solutions, are important methods so far for deriving solutions of different problems. These exact solutions play a vital role in the theory of different physical and engineering problems and provide a deep insight into various systems. But, these methods have their applicability to limited categories of the problems, generally for linearized models or those having low dimensionality and simple geometry. Since most of the real time problems are nonlinear and complex; hence these methods have limited applications in the present scenario.

b) Graphical methods require plotting the problem for solutions. These solutions can be obtained even for complex problems, but the results obtained are not up to the desired accuracy. These methods are limited to the problems with three or fewer dimensions, and are very difficult to implement. However, these graphical solutions provide information about the behavior of the physical phenomena.

c) Numerical techniques require applications of an unambiguous set of arithmetic operations to solve mathematical models numerically. Computations involved in implementation of numerical methods are tedious and cumbersome, and these methods provide approximate solutions to the problems. In pre-computer era, the mathematical problems are linear and simple, which are easily solvable with the help of analytical methods. In last few decades, the nature of problems are becoming nonlinear and complex, and we are unable to solve these problems analytically. Numerical methods are now frequently used to solve such problems. Also, due to the widespread availability of digital computers, these numerical techniques are easy to implement. In fact, the widespread availability and evolution of cheaper digital computers have led to a veritable explosion in the use and development of the numerical methods.

Comparison of Numerical and Analytical Techniques

i) In many practical problems, we have set of data points instead of the mathematical model for the problem. Analytical methods do not work for these cases, but we can apply various numerical techniques for approximate solutions.

ii)   Integrals like $\int\limits_{0}^{\pi}\sqrt{1+\cos^2 x}\,dx$, $\int\limits_{1}^{1}e^{-x^2}dx$, and $\int\limits_{2}^{3}\dfrac{\sin x}{x}dx$ have no available exact solutions, but numerical techniques provide numerical answers to these problems.

iii)   In the case of transcendental equations, e.g., $xe^x + \cos x = 0$, $x - e^{-x} = 0$, etc., the roots are not possible to obtain using direct analytical methods. The numerical methods are helpful in obtaining the approximate roots of these equations.

iv)   It is comparatively easy to obtain exact solutions of linear differential equations. But in the case of nonlinear differential equations, it is almost impossible to examine these differential equations for exact solutions except some particular cases. Numerical methods provide an easy and comprehensive way to get approximate solutions of these nonlinear differential equations.

v)   The analytical methods are not easy to implement on a computer, but almost all the numerical methods are algorithmic and require simple arithmetic operations. Hence, computers and numerical techniques make a perfect combination.

vi)   It is worth mentioning here that numerical techniques provide approximate solutions, while analytical methods provide an exact answer. Hence we prefer analytical solutions over numerical solutions if available.

vii)   Analytical methods usually give results containing mathematical functions; these can be evaluated for specific instances. Thus, there is an advantage to the analytic result that the behavior and properties of the function are often apparent; this is not the case for purely numerical results. However, numerical results can be plotted to show some properties of the solution.

viii)   Number of arithmetic operations (+, -, *, /) is an important factor in analyzing the efficiency of a numerical method. The round-off error and the execution time of the algorithm, both are directly related to the number of arithmetic operations involved in the numerical computation.

ix)   In direct analytical methods, we apply the method without any repetition. If any error occurs at any step, there is no remedy to recover it. But in the case of numerical techniques, we can increase the iterations and results can be made accurate up to some desired tolerance. The level of accuracy achievable in numerical methods is limited, because of the way that computers do arithmetic. We will discuss these limitations in Chapter 2.

x)   In some cases, especially when cumbersome calculations are involved, errors in obtaining exact solutions from the direct method are higher than numerical methods. In numerical methods, the desired accuracy can be achieved by increasing number of iterations.

xi)   Convergence and stability are two major aspects of any numerical method. Before implementing numerical schemes to a mathematical model, we need to consider these two properties.

To summarize, the computer-oriented numerical techniques have methods to solve the scientific and engineering problems; have algorithms to implement on the computer; and have solutions to those problems which we are not able to solve analytically.

Numerical methods require tedious, cumbersome and repetitive arithmetic operations for large problems. It is almost impossible to do these cumbersome arithmetic operations manually. The development of information technology enhances the potential of these numerical techniques, and various software can handle algebra involved in these techniques in a very simple and sophisticated manner. Since most of the numerical techniques are algorithmic in nature and require repetitive cumbersome iterations, so it is practical to apply these algorithms to a computer. Of course, a computer must be given detailed and complete instructions for each step. During the formulation of any algorithm, we must keep in mind the following main features of the computer.

i) The computer is capable of performing only the basic arithmetic operations. Hence, each problem must be reduced to problems of these arithmetic operations. Numerical techniques provide these algorithms for a wide range of problems.

ii) The memory of the computer stores the algorithms and results of computations, and this enables the repetitive execution and results can be retrieved as per requirement.

iii) Computer's memory facilitates the alteration in the execution of instructions depending on results obtained during the execution.

A systematic stepwise set of instructions utilizing above features of the computer enables us to solve complicated and cumbersome problems. Our efforts aim at the search for such algorithms. We will see that for a specific type of problems, there are several algorithms (For example, to solve nonlinear equations, we can choose among various algorithms like Bisection, Regula–Falsi, Newton–Raphson, etc.). We have a lot of choices depending on our requirement for speed, accuracy, and convergence, etc. A combination of two algorithms can also be used. For example, we can find a close approximation to the root of an equation by Bisection method keeping in mind the convergence; and then continue with the Newton–Raphson method from this approximation onwards keeping in mind the speed of the method. For modest size problem, we can easily implement any algorithm with high configuration computer. But in the case of large-scale problems, slow algorithms need to be rejected.

So far, many algorithms are developed for different kinds of problems. As we discussed above, there are so many reasons to select an algorithm over others. Many algorithms developed in the past are now obsolete, just because we get better new algorithms compared to these old algorithms. For example, Gauss–Seidel method replaces the Jacobi method for the solution of system of linear equations, as the rate of convergence of Gauss–Seidel method is double than the Jacobi method.

The knowledge of programming is must for the implementation of numerical techniques. This book contains solutions for various complicated problems using numerical techniques, to illustrate the strength of the numerical methods over analytical methods. Discussion on mathematical software is not a part of this book, but we must know about the software with user-friendly interfaces. The software provides not only graphical and numerical outputs but also capable of generating algebraic, closed form analytical solutions to a wide range of mathematical problems and development in this field continues at a rapid rate. Derive, Maple, Mathematica, Matlab, and Reduce are some important software.

Many numerical methods are based on Taylor series expansions of functions. If the function $f(x)$ has infinite continuous derivatives on any interval I and a is some point in I, then for all $x \in I$, Taylor series expansion for function $f(x)$ at center a is given by

$$f(x) = f(a) + (x-a)f'(a) + \frac{(x-a)^2}{2!} f''(a) + \dots + \frac{(x-a)^n}{(n)!} f^{(n)}(a) + \cdots$$

If we use first (n) terms of this series expansion, then Taylor series expansion of the function $f(x)$ is of order (n), and truncation error is said to be of order $(n + 1)$.

Let the function $f(x)$ have $(n + 1)$ continuous derivatives on any interval I and a is some point in the interval I, then for all $x \in I$

$$f(x) = f(a) + (x-a)f'(a) + \frac{(x-a)^2}{2!} f''(a) + \cdots + \frac{(x-a)^n}{(n)!} f^{(n)}(a) + R_n(\xi) \tag{1}$$

where $R_n(\xi) = \dfrac{(x-a)^{n+1}}{(n+1)!} f^{(n+1)}(\xi)$; for some $\xi \in (a, x)$. $R_n(\xi)$ is known as remainder term.

For a small change in the value of x (say δx), the function value at the point $x + \delta x$ can be represented by the following form of Taylor series

$$f(x + \delta x) = f(x) + \delta x \, f'(x) + \frac{\delta x^2}{2!} f''(x) + \cdots$$

The above Taylor series up to n-terms is as follows

$$f(x + \delta x) = f(x) + \delta x \, f'(x) + \frac{\delta x^2}{2!} f''(x) + \cdots + \frac{\delta x^n}{(n)!} f^{(n)}(x) + R_n(\xi) \tag{2}$$

where $R_n(\xi) = \dfrac{(\delta x)^{n+1}}{(n+1)!} f^{(n+1)}(\xi)$; $\quad x < \xi < x + \delta x$

If we replace $x = a + h$ in Taylor expansion (1), then another form of Taylor series is given by

$$f(a + h) = f(a) + (h)f'(a) + \frac{(h)^2}{2!} f''(a) + \cdots + \frac{(h)^n}{(n)!} f^{(n)}(a) + R_n(\xi) \tag{3}$$

where $R_n(\xi) = \dfrac{(h)^{n+1}}{(n+1)!} f^{(n+1)}(\xi); \quad a < \xi < a+h$

For a convergent series, $R_n(\xi) \to 0$ as $n \to \infty$. Expressions (1–3) are different forms of Taylor series expansion. Since it is not possible to compute an infinite number of terms, we approximate the function $f(x)$ by first $n+1$-terms by neglecting higher order terms. It is known as approximation by Taylor polynomial of degree n. The remainder term $R_n(\xi)$ gives the error term in this approximation. It is not possible to compute the exact value of ξ, hence the value of ξ is such that the error term is maximum.

The Taylor series expressions of some basic functions about the point $x = 0$ (Maclaurin series) are as follows.

$$\sin x = \sum_{n=0}^{\infty} (-1)^n \frac{x^{2n+1}}{(2n+1)!} = x - \frac{x^3}{3!} + \frac{x^5}{5!} - \frac{x^7}{7!} + \cdots \qquad \text{for all } x$$

$$\cos x = \sum_{n=0}^{\infty} (-1)^n \frac{x^{2n}}{(2n)!} = 1 - \frac{x^2}{2!} + \frac{x^4}{4!} - \frac{x^6}{6!} + \cdots \qquad \text{for all } x$$

$$\sinh x = \sum_{n=0}^{\infty} \frac{x^{2n+1}}{(2n+1)!} = x + \frac{x^3}{3!} + \frac{x^5}{5!} + \frac{x^7}{7!} + \cdots \qquad \text{for all } x$$

$$\cosh x = \sum_{n=0}^{\infty} \frac{x^{2n}}{(2n)!} = 1 + \frac{x^2}{2!} + \frac{x^4}{4!} + \frac{x^6}{6!} + \cdots \qquad \text{for all } x$$

$$\sin^{-1} x = \sum_{n=0}^{\infty} \frac{1}{4^n} \binom{2n}{n} \frac{x^{2n+1}}{2n+1} = x + \frac{1}{2}\left(\frac{x^3}{3}\right) + \frac{1.3}{2.4}\left(\frac{x^5}{5}\right) + \frac{1.3.5}{2.4.6}\left(\frac{x^7}{7}\right) + \cdots \qquad \text{for } |x| < 1$$

$$\cos^{-1} x = \frac{\pi}{2} - \sin^{-1} x \qquad \text{for } |x| < 1$$

$$\tan^{-1} x = \sum_{n=0}^{\infty} (-1)^n \left(\frac{x^{2n+1}}{2n+1}\right) = x - \frac{x^3}{3} + \frac{x^5}{5} - \frac{x^7}{7} + \cdots \qquad \text{for } |x| < 1$$

$$\cot^{-1} x = \frac{\pi}{2} - \tan^{-1} x \qquad \text{for } |x| < 1$$

$$\sec^{-1} x = \cos^{-1}\left(\frac{1}{x}\right) \qquad \text{for } |x| \geq 1$$

$$\csc^{-1}(x) = \sin^{-1}\left(\frac{1}{x}\right) \qquad \text{for } |x| \geq 1$$

$$e^x = \sum_{n=0}^{\infty} \frac{x^n}{n!} = 1 + x + \frac{x^2}{2!} + \frac{x^3}{3!} + \frac{x^4}{4!} + \cdots \qquad \text{for all } x$$

$$\ln(1+x) = \sum_{n=1}^{\infty} (-1)^{n+1} \frac{x^n}{n} = x - \frac{x^2}{2} + \frac{x^3}{3} - \frac{x^4}{4} + \cdots \qquad \text{for } |x| < 1$$

$$\frac{1}{1-x} = \sum_{n=0}^{\infty} x^n = 1 + x + x^2 + x^3 + \cdots \qquad\qquad \text{for } |x| < 1$$

$$(1+x)^n = 1 + nx + \frac{n(n-1)}{2!}x^2 + \frac{n(n-1)(n-2)}{3!}x^3 + \cdots \qquad\qquad \text{for } |x| < 1$$

Note: Since only arithmetic operations are possible on computer, hence the series expansions are very useful. For example, we want to compute value of sin(0.23), then our computer will put $x = 0.23$ in series expansion of sin(x), and compute specified number of terms (depending up on the accuracy required) of series and give back the result.

Taylor Series for the Functions of More than One Variable

The Taylor series expansions for the functions of one independent variable can be generalized to the functions of more than one independent variable. Let $y = f(x_1, x_2, ..., x_n)$ be a function of n-independent variables $x_1, x_2, ..., x_n$. Let $\delta x_1, \delta x_2, ..., \delta x_n$ be the errors in calculating the $x_1, x_2, ..., x_n$ respectively. Then, Taylor series expansion is given by

$$y + \delta y = f(x_1 + \delta x_1, x_2 + \delta x_2, ..., x_n + \delta x_n)$$

$$y + \delta y = f(x_1, x_2, ..., x_n) + \left(\delta x_1 \frac{\partial}{\partial x_1} + \frac{\partial}{\partial x_2} \delta x_2 + ... + \frac{\partial}{\partial x_n} \delta x_n \right) f$$

$$+ \frac{1}{2!} \left(\delta x_1 \frac{\partial}{\partial x_1} + \frac{\partial}{\partial x_2} \delta x_2 + ... + \frac{\partial}{\partial x_n} \delta x_n \right)^2 f$$

$$+ \text{ terms involving third and higher powers of } \delta x_1, \delta x_2, ..., \delta x_n$$

Lagrange Mean Value (LMV) Theorem

Taylor series formula up to first order term provides Lagrange mean value (LMV) theorem. LMV theorem states as follows

If the function $f(x)$ is continuous on the finite interval $[a, b]$ and differentiable on (a, b), then there exists a point $\xi \in (a, b)$ such that

$$\frac{f(b) - f(a)}{b - a} = f'(\xi)$$

Rolle Theorem

A further particular case of LMV-theorem is Rolle theorem, which states as follows

Let the function $f(x)$ be continuous on the finite interval $[a, b]$ and differentiable on (a, b), if $f(a) = f(b) = 0$; then there exists a point $\xi \in (a, b)$ such that

$$f'(\xi) = 0$$

Any function of one variable which does not graph as a straight line in two dimensions or any function of two variables which does not graph as a plane in three dimensions is said to be nonlinear.

The function $y = f(x)$ is a linear function if the dependent variable y changes in direct proportion to the change in an independent variable x. For example, $y = 2x + 10$ is a linear function as the variable x appears only with degree one. But the functions $y = x^2 + 3$ and $y = x - e^{-x}$ are nonlinear functions, as the terms x^2 and e^{-x} are nonlinear. In the case of two variables, $f(x, y) = x + 3y - 8$ is a linear function, while $f(x, y) = x^2 + y, f(x, y) = xy + 3x + 2y + 2$ and $f(x, y) = x - e^{-x} + \sin(y)$ are nonlinear functions.

In the case of single equations, $2x + 3 = 0$ is a linear equation and $x^2 - 4x + 5 = 0$ is a nonlinear equation. Similarly, the system of equations $\begin{matrix} x + y = 3 \\ 2x - 3y = 5 \end{matrix}$ is a linear system of equations and $\begin{matrix} x^2 + y = 3 \\ 2x - 3y = 5 \end{matrix}$ is a system of nonlinear equations since the first equation is nonlinear. In broader sense, if the unknown variables involved in the system have only linear terms (i.e. with degree one only), then the system is a linear system.

A differential equation is linear if the dependent variable and its derivatives are in linear form, otherwise, it is nonlinear. For example, the differential equation $\dfrac{d^2 y}{dx^2} + x^2 \dfrac{dy}{dx} + 3y = \sin(x)$ is a linear differential equation as dependent variable y, and its derivative terms are of the linear form (degree of all these is one). The differential equations, $\dfrac{d^2 y}{dx^2} + \sin(y) = 0$, $y\dfrac{d^2 y}{dx^2} + x\dfrac{dy}{dx} = 3$ and $\dfrac{dy}{dx} + y^2 = x$ are nonlinear differential equations. In the first equation, we have sine function of y; in the second equation, there is a multiplicative term of dependent variable y and its derivative; in third equation, the term y^2 is nonlinear.

Many analytical methods have been introduced in the literature to solve the linear systems of different kinds. Also, the superposition principle implies that if X, Y (functions, vectors, vector fields or any other mathematical quantities) are two solutions of a homogeneous linear system, then $aX + bY$ is also a solution for scalars a, b. Therefore, the analytical methods and superposition principle form a great combination for the solution of linear systems. But, very few analytical methods have been developed to solve some very particular types of nonlinear systems. Therefore, a number of complex nonlinear systems remain unsolved by analytical methods.

Since all the mathematical models governing real world problems are inherently nonlinear, hence a very strong desire to have solutions to nonlinear systems leads us to numerical techniques. A few decades ago, numerical techniques were not easy to implement due to the cumbersome computational work involved. Now days, with the advent of high-speed computers and software, these techniques are easy to implement. Numerical techniques provide solutions for those problems which we are not able to solve analytically and in fact, sometimes provide better results than the analytical methods.

Graphs of Standard Functions

Algebraic Functions

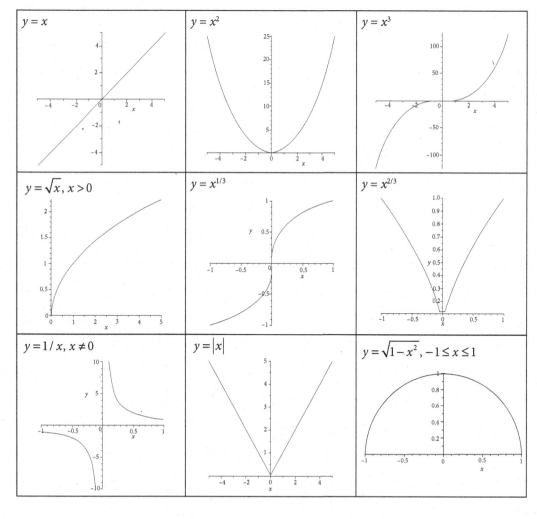

Transcendental Functions

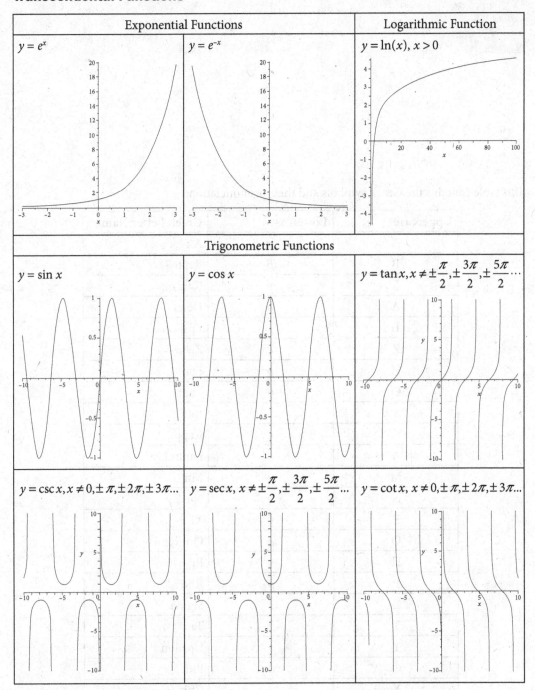

| Exponential Functions | | Logarithmic Function |

$y = e^x$ $y = e^{-x}$ $y = \ln(x), x > 0$

Trigonometric Functions

$y = \sin x$

$y = \cos x$

$y = \tan x, x \neq \pm\dfrac{\pi}{2}, \pm\dfrac{3\pi}{2}, \pm\dfrac{5\pi}{2}\cdots$

$y = \csc x, x \neq 0, \pm\pi, \pm 2\pi, \pm 3\pi\cdots$

$y = \sec x, x \neq \pm\dfrac{\pi}{2}, \pm\dfrac{3\pi}{2}, \pm\dfrac{5\pi}{2}\cdots$

$y = \cot x, x \neq 0, \pm\pi, \pm 2\pi, \pm 3\pi\cdots$

Greek Letters

This table contains the Greek symbols and their pronunciation.

Uppercase	Lowercase	Greek Letter Name
A	α	Alpha
B	β	Beta
Γ	γ	Gamma
Δ	δ	Delta
E	ε	Epsilon
Z	ζ	Zeta
H	η	Eta
Θ	θ	Theta
I	ι	Iota
K	κ	Kappa
Λ	λ	Lambda
M	μ	Mu
N	ν	Nu
Ξ	ξ	Xi
O	o	Omicron
Π	π	Pi
P	ρ	Rho
Σ	σ	Sigma
T	τ	Tau
Υ	υ	Upsilon
Φ	ϕ	Phi
X	χ	Chi
Ψ	ψ	Psi
Ω	ω	Omega

Index

Printed in the United States
by Baker & Taylor Publisher Services